# NIOSH POCKET GUIDE TO CHEMICAL HAZARDS

DEPARTMENT OF HEALTH AND HUMAN SERVICES
Centers for Disease Control and Prevention
National Institute for Occupational Safety and Health

**September 2007**

**DHHS (NIOSH) Publication No. 2005-149**

*First Printing* – September 2005
*Second Printing* – August 2006, with minor technical changes
*Third Printing* – September 2007, with minor technical changes

## DISCLAIMER

Mention of the name of any company or product does not constitute endorsement by the National Institute for Occupational Safety and Health (NIOSH). In addition, citations to Web sites external to NIOSH do not constitute NIOSH endorsement of the sponsoring organizations or their programs or products. Furthermore, NIOSH is not responsible for the content of these Web sites.

**Published by Books Express Publishing**

**ISBN 978-1-78039-852-5**

**Books Express publications are available from all good retail and online booksellers. For publishing proposals and direct ordering please contact us at: info@books-express.com**

# PREFACE

The *NIOSH Pocket Guide to Chemical Hazards* presents information taken from the *NIOSH/OSHA Occupational Health Guidelines for Chemical Hazards*, from National Institute for Occupational Safety and Health (NIOSH) criteria documents and Current Intelligence Bulletins, and from recognized references in the fields of industrial hygiene, occupational medicine, toxicology, and analytical chemistry. The information is presented in tabular form to provide a quick, convenient source of information on general industrial hygiene practices. The information in the *Pocket Guide* includes chemical structures or formulas, identification codes, synonyms, exposure limits, chemical and physical properties, incompatibilities and reactivities, measurement methods, respirator selections, signs and symptoms of exposure, and procedures for emergency treatment.

The information assembled in the original 1978 printing of the *Pocket Guide* was the result of the Standards Completion Program, a joint effort by NIOSH and the Department of Labor to develop supplemental requirements for the approximately 380 workplace environmental exposure standards adopted by the Occupational Safety and Health Administration (OSHA) in 1971.

Listed below are changes that were made for this edition (2005-149) of the *Pocket Guide*:

- New layout for the Chemical Listing section.
- Recommendations for particulate respirators have been revised to incorporate "Part 84" terminology. See "Recommendations for Respirator Selection" on page xiv for a more thorough explanation of these changes.
- The Synonym and Trade Name Index has been expanded. This index is now called the Chemical, Synonym, and Trade Name Index (page 383).
- Some ID and Guide Numbers were changed to reflect changes made in the *2004 Emergency Response Guidebook* (http://hazmat.dot.gov/pubs/erg/gydebook.htm).
- Appendix E (page 351) has been revised. It now contains OSHA respirator requirements for 28 chemicals or hazardous substances that were identified in the preamble to the OSHA Respiratory Protection Standard (29 CFR 1910.134).
- Other minor technical changes have also been made since the February 2004 edition. (For the most current information and updates, consult the electronic version on the NIOSH Web site: http://www.cdc.gov/niosh/npg/npg.html.)

Listed below are changes made for the 3rd printing of this edition of the *Pocket Guide*:

- Changes were made to reflect the new OSHA PEL for hexavalent chromium.
- The NIOSH REL for coal mine dust was added to the coal dust entry.
- A few other minor technical changes have been made.

## CONTENTS

## ACKNOWLEDGMENTS

The Education and Information Division (EID), National Institute for Occupational Safety and Health (NIOSH), has primary responsibility for the development of the *Pocket Guide*. There have been many people who have contributed to the preparation and development of this document since it was first published in 1978. I would like to express my appreciation to the following people within EID for their efforts: Vern Anderson for general guidance; Guss Hasbani (Constella Group, Inc.) for computer programming and database development that has been vital to the production of this new edition; Heinz Ahlers, Barb Dames, Charles Geraci, Richard Niemeier, David Votaw, Alan Weinrich, and Ralph Zumwalde for policy review; David Case, Laura Delaney, and Rolland Rogers for reformatting and computerization; Vanessa Becks, Anne Hamilton, and Rodger Tatken for editorial review; Clayton Doak, Eileen Kuempel, Leela Murthy, Henryka Nagy, John Palassis, Faye Rice, and David Votaw for assistance in updating and adding information; Lawrence Foster, Vicki Reuss, Lucy Schoolfield, and Ronald Schuler for data acquisition; Kent Hatfield for consultation on toxicology issues; Charlene Maloney for publication dissemination and general guidance; and Oliver F. Cobb and Associates (Carla Brooks, George Brown, Sherri Diana, and Jesse Romans) for answering requests and mailing thousands of copies of the *Pocket Guide*.

The following people, who constitute the *Pocket Guide* Editorial Board, have contributed greatly by providing guidance and review of the content and style of this new edition: Steven Ahrenholz (Division of Surveillance, Hazard Evaluations and Field Studies, DSHEFS), Roland BerryAnn (National Personal Protective Technology Laboratory, NPPTL), Joseph Bowman (Division of Applied Research and Technology, DART), Pamela Drake (Spokane Research Laboratory, SRL), Gerald Joy (Pittsburgh Research Laboratory, PRL), Alan Lunsford (DART), Nancy Nilsen (DSHEFS), Paula Fey O'Connor (DART), Carl Ornot (Office of Administrative and Management Services, OAMS), Jay Snyder (NPPTL), Sidney Soderholm (Health Effects Laboratory Division, HELD), David Sylvain (DSHEFS), Ainsley Weston (HELD), and Anthony Zimmer (DART).

In addition, the following people also have contributed greatly to the *Pocket Guide*: Mary Ellen Cassinelli (DART), Donald Dollberg (DART), and Paula Fey O'Connor (DART) for the development of the measurement methods section; Roland BerryAnn (NPPTL), Nancy Bollinger (HELD), Christopher Coffey (Division of Respiratory Disease Studies, DRDS) for the development of respirator recommendations; Laurence Reed (DART) and John Whalen (DART) for policy review; Crystal Ellison (Office of Compensation Analysis and Support, OCAS) for assistance in updating and adding information; and Henry Chan and Howard Ludwig (former *Pocket Guide* Technical Editors) for general guidance.

Also, thanks are due to all of the people who have reviewed and commented on the *Pocket Guide* during its initial development and subsequent revisions.

Michael E. Barsan
(Technical Editor)

## INTRODUCTION

The *NIOSH Pocket Guide to Chemical Hazards* provides a concise source of general industrial hygiene information for workers, employers, and occupational health professionals. The *Pocket Guide* presents key information and data in abbreviated tabular form for 677 chemicals or substance groupings commonly found in the work environment (e.g., manganese compounds, tellurium compounds, inorganic tin compounds, etc.). The industrial hygiene information found in the *Pocket Guide* assists users to recognize and control occupational chemical hazards. The chemicals or substances contained in this revision include all substances for which the National Institute for Occupational Safety and Health (NIOSH) has recommended exposure limits (RELs) and those with permissible exposure limits (PELs) as found in the Occupational Safety and Health Administration (OSHA) Occupational Safety and Health Standards (29 CFR 1910.1000 – 1052).

### Background

In 1974, NIOSH (which is responsible for recommending health and safety standards) joined OSHA (whose jurisdictions include promulgation and enforcement activities) in developing a series of occupational health standards for substances with existing PELs. This joint effort was labeled the Standards Completion Program and involved the cooperative efforts of several contractors and personnel from various divisions within NIOSH and OSHA. The Standards Completion Program developed 380 substance-specific draft standards with supporting documentation that contained technical information and recommendations needed for the promulgation of new occupational health regulations. The *Pocket Guide* was developed to make the technical information in those draft standards more conveniently available to workers, employers, and occupational health professionals. The *Pocket Guide* is updated periodically to reflect new data regarding the toxicity of various substances and any changes in exposure standards or recommendations. (For the most current information and updates, consult the electronic version on the NIOSH Web site: http://www.cdc.gov/niosh/npg/npg.html.)

### Data Collection and Application

The data collected for this revision were derived from a variety of sources, including NIOSH policy documents such as Criteria Documents and Current Intelligence Bulletins (CIBs), and recognized references in the fields of industrial hygiene, occupational medicine, toxicology, and analytical chemistry.

## NIOSH RECOMMENDATIONS

Acting under the authority of the Occupational Safety and Health Act of 1970 (29 USC Chapter 15) and the Federal Mine Safety and Health Act of 1977 (30 USC Chapter 22), NIOSH develops and periodically revises recommended exposure limits (RELs) for hazardous substances or conditions in the workplace. NIOSH also recommends appropriate preventive measures to reduce or eliminate the adverse health and safety effects of these hazards. To formulate these recommendations, NIOSH evaluates all known and available medical, biological, engineering, chemical, trade, and other information relevant to the hazard. These recommendations are then published and transmitted to OSHA and the Mine Safety and Health Administration (MSHA) for use in promulgating legal standards.

NIOSH recommendations are published in a variety of documents. Criteria documents recommend workplace exposure limits and appropriate preventive measures to reduce or eliminate adverse health effects and accidental injuries.

Current Intelligence Bulletins (CIBs) are issued to disseminate new scientific information about occupational hazards. A CIB may draw attention to a formerly unrecognized hazard, report new data on a known hazard, or present information on hazard control.

Alerts, Special Hazard Reviews, Occupational Hazard Assessments, and Technical Guidelines support and complement the other standard development activities of the Institute. Their purpose is to assess the safety and health problems associated with a given agent or hazard (e.g., the potential for injury or for carcinogenic, mutagenic, or teratogenic effects) and to recommend appropriate control and surveillance methods. Although these documents are not intended to supplant the more comprehensive criteria documents, they are prepared in order to assist OSHA and MSHA in the formulation of regulations.

In addition to these publications, NIOSH periodically presents testimony before various Congressional committees and at OSHA and MSHA rulemaking hearings.

Recommendations made through 1992 are available in a single compendium entitled *NIOSH Recommendations for Occupational Safety and Health: Compendium of Policy Documents and Statements* [DHHS (NIOSH) Publication No. 92-100] (http://www.cdc.gov/niosh/92-100.html). More recent recommendations are available on the NIOSH Web site (http://www.cdc.gov/niosh). Copies of the *Compendium* may be ordered from the NIOSH Publications office (800-356-4674).

## HOW TO USE THIS POCKET GUIDE

The *Pocket Guide* has been designed to provide chemical-specific data to supplement general industrial hygiene knowledge. Individual tables for each chemical present this data in the Chemical Listing section (page 1). To maximize the amount of data provided in the limited space in these tables, abbreviations and codes have been used extensively. These abbreviations and codes, which have been designed to permit rapid comprehension by the regular user, are discussed for each field in these chemical tables in the following subsections.

### Chemical Name

The chemical name found in the OSHA General Industry Air Contaminants Standard (29 CFR 1910.1000) is listed in the blue box in the top left portion of each chemical table. This name is referred to as the "primary name" in the Chemical, Synonym, and Trade Name Index (page 383).

### Structure/Formula

The chemical structure or formula is listed in the field to the right of the chemical name in each chemical table. Carbon-carbon double bonds (-C=C-) and carbon-carbon triple bonds (-C≡C-) have been indicated where applicable.

### CAS Number

This section lists the Chemical Abstracts Service (CAS) registry number. The CAS number, in the format xxx-xx-x, is unique for each chemical and allows efficient searching on computerized data bases. A page index for all CAS registry numbers listed is included at the back of the *Pocket Guide* (page 374) to help the user locate a specific substance.

### RTECS Number

This section lists the NIOSH Registry of Toxic Effects of Chemical Substances (RTECS®) number, in the format ABxxxxxxx. RTECS® may be useful for obtaining additional toxicologic information on a specific substance.

RTECS® is a compendium of data extracted from the open scientific literature. On December 18, 2001, CDC's Technology Transfer Office, on behalf of NIOSH, successfully completed negotiating a "PHS Trademark Licensing Agreement" for RTECS®. This non-exclusive licensing agreement provides for the transfer and continued development of the "RTECS® database and its trademark" to MDL Information Systems, Inc. (MDL), a wholly owned subsidiary of Elsevier Science, Inc. Under this agreement, MDL will be responsible for updating, licensing, marketing, and distributing RTECS®. For more information visit the MDL Web site (http://www.mdli.com).

The RTECS® entries for chemicals listed in the *Pocket Guide* can be viewed on the NIOSH Web site (http://www.cdc.gov/niosh/npg/npg.html) or on the CD-ROM version of the *Pocket Guide* (see page iii for ordering information).

### IDLH

This section lists the immediately dangerous to life or health concentrations (IDLHs). For the June 1994 Edition of the *Pocket Guide*, NIOSH reviewed and in many cases revised the IDLH values. The criteria utilized to determine the adequacy of the original IDLH values were a combination of those used during the Standards Completion Program and a newer methodology developed by NIOSH. These "interim" criteria formed a tiered approach, preferentially using acute human toxicity data, followed by acute animal inhalation toxicity data, and then by acute animal oral toxicity data to determine a preliminary updated IDLH value. When relevant acute toxicity data were insufficient or unavailable, NIOSH also considered using chronic toxicity data or an analogy to a chemically similar substance. NIOSH then compared these preliminary values with the following criteria to determine the updated IDLH value: 10% of lower explosive limit (LEL); acute animal respiratory irritation data ($RD_{50}$); other short-term exposure guidelines; and the *NIOSH Respirator Selection Logic* (DHHS [NIOSH] Publication No. 2005-100;

http://www.cdc.gov/niosh/docs/2005-100). The *Documentation for Immediately Dangerous to Life or Health Concentrations* (NTIS Publication Number PB-94-195047) further describes these criteria and provides information sources for both the original and revised IDLH values (http://www.cdc.gov/niosh/idlh/idlh-1.html). NIOSH currently is assessing the various uses of IDLHs, whether the criteria used to derive the IDLH values are valid, and if other information or criteria should be utilized.

The purpose for establishing an IDLH value in the Standards Completion Program was to determine the airborne concentration from which a worker could escape without injury or irreversible health effects from an IDLH exposure in the event of the failure of respiratory protection equipment. The IDLH was considered a maximum concentration above which only a highly reliable breathing apparatus providing maximum worker protection should be permitted. In determining IDLH values, NIOSH considered the ability of a worker to escape without loss of life or irreversible health effects along with certain transient effects, such as severe eye or respiratory irritation, disorientation, and incoordination, which could prevent escape. As a safety margin, IDLH values are based on effects that might occur as a consequence of a 30-minute exposure. However, the 30-minute period was NOT meant to imply that workers should stay in the work environment any longer than necessary; in fact, EVERY EFFORT SHOULD BE MADE TO EXIT IMMEDIATELY!

The *NIOSH Respirator Selection Logic* defines IDLH exposure conditions as "conditions that pose an immediate threat to life or health, or conditions that pose an immediate threat of severe exposure to contaminants, such as radioactive materials, which are likely to have adverse cumulative or delayed effects on health." The purpose of establishing an IDLH exposure concentration is to ensure that the worker can escape from a given contaminated environment in the event of failure of the respiratory protection equipment. The *Respirator Selection Logic* uses IDLH values as one of several respirator selection criteria. Under the *Respirator Selection Logic*, the most protective respirators (e.g., a self-contained breathing apparatus equipped with a full facepiece and operated in a pressure-demand or other positive-pressure mode) would be selected for firefighting, exposure to carcinogens, entry into oxygen-deficient atmospheres, in emergency situations, during entry into an atmosphere that contains a substance at a concentration greater than 2,000 times the NIOSH REL or OSHA PEL, and for entry into IDLH atmospheres. IDLH values are listed in the *Pocket Guide* for over 380 substances.

The notation "**Ca**" appears in the IDLH field for all substances that NIOSH considers potential occupational carcinogens. However, IDLH values that were originally determined in the Standards Completion Program or were subsequently revised are shown in brackets following the "**Ca**" designations. "**10%LEL**" indicates that the IDLH was based on 10% of the lower explosive limit for safety considerations even though the relevant toxicological data indicated that irreversible health effects or impairment of escape existed only at higher concentrations. "**N.D.**" indicates that an IDLH value has not been determined for that substance. Appendix F (page 361) contains an explanation of the "Effective" IDLHs used for four chloronaphthalene compounds.

## Conversion Factors

This section lists factors for the conversion of ppm (parts of vapor or gas per million parts of contaminated air by volume) to $mg/m^3$ (milligrams of vapor or gas per cubic meter of contaminated air) at 25°C and 1 atmosphere for chemicals with exposure limits expressed in ppm.

## DOT ID and Guide Number

This section lists the U.S. Department of Transportation (DOT) Identification numbers and the corresponding Guide numbers. Their format is xxxx yyy. The Identification (ID) number (xxxx) indicates that the chemical is regulated by DOT. The Guide number (yyy) refers to actions to be taken to stabilize an emergency situation; this information can be found in the *2004 Emergency Response Guidebook* (Office of Hazardous Materials Initiatives and Training [DHM-50], Research and Special Programs Administration, U.S. Department of Transportation, 400 7th Street, S.W., Washington, D.C. 20590-0001; for sale by the U.S. Government Printing Office, Superintendent of Documents, P.O. Box 371954, Pittsburgh, PA 15250-7954). This information is also available on the CD-ROM and Web site versions of the *Pocket Guide* (http://www.cdc.gov/niosh/npg/npg.html). A page index for all DOT ID numbers listed is provided on page 379 to help the user locate a specific substance; please note however, that many DOT numbers are **not** unique for a specific substance.

## Synonyms and Trade Names

This section contains an alphabetical list of common synonyms and trade names for each chemical. A page index for all chemical names, synonyms, and trade names listed in the *Pocket Guide* is included on page 383.

## Exposure Limits

The NIOSH recommended exposure limits (**REL**s) are listed first in this section. For NIOSH RELs, "**TWA**" indicates a time-weighted average concentration for up to a 10-hour workday during a 40-hour workweek. A short-term exposure limit (STEL) is designated by "**ST**" preceding the value; unless noted otherwise, the STEL is a 15-minute TWA exposure that should not be exceeded at any time during a workday. A ceiling REL is designated by "**C**" preceding the value; unless noted otherwise, the ceiling value should not be exceeded at any time. Any substance that NIOSH considers to be a potential occupational carcinogen is designated by the notation "**Ca**" (see Appendix A [page 342], which contains a brief discussion of potential occupational carcinogens).

The OSHA permissible exposure limits (**PEL**s), as found in Tables Z-1, Z-2, and Z-3 of the OSHA General Industry Air Contaminants Standard (29 CFR 1910.1000), that were effective on July 1, 1993* and which are currently enforced by OSHA are listed next.

> *In July 1992, the 11th Circuit Court of Appeals in its decision in AFL-CIO v. OSHA, 965 F.2d 962 (11th Cir., 1992) vacated more protective PELs set by OSHA in 1989 for 212 substances, moving them back to PELs

established in 1971. The appeals court also vacated new PELs for 164 substances that were not previously regulated. Enforcement of the court decision began on June 30, 1993. Although OSHA is currently enforcing exposure limits in Tables Z-1, Z-2, and Z-3 of 29 CFR 1910.1000 which were in effect before 1989, violations of the "general duty clause" as contained in Section 5(a) (1) of the Occupational Safety and Health Act may be considered when worker exposures exceed the 1989 PELs for the 164 substances that were not previously regulated. The substances for which OSHA PELs were vacated on June 30, 1993 are indicated by the symbol "**†**" following OSHA in this section and previous values (the PELs that were vacated) are listed in Appendix G (page 362).

TWA concentrations for OSHA **PEL**s must not be exceeded during any 8-hour workshift of a 40-hour workweek. A STEL is designated by "**ST**" preceding the value and is measured over a 15-minute period unless noted otherwise. OSHA ceiling concentrations (designated by "**C**" preceding the value) must not be exceeded during any part of the workday; if instantaneous monitoring is not feasible, the ceiling must be assessed as a 15-minute TWA exposure. In addition, there are a number of substances from Table Z-2 (e.g., beryllium, ethylene dibromide) that have PEL ceiling values that must not be exceeded except for specified excursions. For example, a "5-minute maximum peak in any 2 hours" means that a 5-minute exposure above the ceiling value, but never above the maximum peak, is allowed in any 2 hours during an 8-hour workday. Appendix B (page 344) contains a brief discussion of substances regulated as carcinogens by OSHA.

Concentrations are given in ppm, $mg/m^3$, mppcf (millions of particles per cubic foot of air as determined from counting an impinger sample), or $fibers/cm^3$ (fibers per cubic centimeter). The "**[skin]**" designation indicates the potential for dermal absorption; skin exposure should be prevented as necessary through the use of good work practices, gloves, coveralls, goggles, and other appropriate equipment. The "**(total)**" designation indicates that the REL or PEL listed is for "total particulate" versus the "**(resp)**" designation which refers to the "respirable fraction" of the airborne particulate.

Appendix C (page 345) contains more detailed discussions of the specific exposure limits for certain low-molecular-weight aldehydes, asbestos, various dyes (benzidine-, o-tolidine-, and o-dianisidine-based), carbon black, chloroethanes, the various chromium compounds (chromic acid and chromates, chromium(II) and chromium(III) compounds, and chromium metal), coal tar pitch volatiles, coke oven emissions, cotton dust, lead, mineral dusts, NIAX® Catalyst ESN, trichloroethylene, and tungsten carbide (cemented). Appendix D (page 350) contains a brief discussion of substances included in the *Pocket Guide* with no established RELs at this time. Appendix F (page 361) contains miscellaneous notes regarding the OSHA PEL for benzene, and Appendix G (page 362) lists the OSHA PELS that were vacated on June 30, 1993.

## Measurement Methods

The section provides a source (NIOSH or OSHA) and the corresponding method number for measurement methods which can be used to determine the exposure for the chemical or substance. Unless otherwise noted, the NIOSH methods are from the 4th edition of the *NIOSH Manual of Analytical Methods* (DHHS [NIOSH] Publication No. 94-113 [http://www.cdc.gov/niosh/nmam]) and supplements. If a different edition of the *NIOSH Manual of Analytical Methods* is cited, the appropriate edition and, where applicable, the volume number are noted [e.g., II-4 (2nd edition, volume 4)]. The OSHA methods are from the OSHA Web site (http://www.osha-slc.gov/dts/sltc/methods). "None available" means that no method is available from NIOSH or OSHA. Table 1 (page xvii) lists the editions, volumes, and supplements of the *NIOSH Manual of Analytical Methods*.

Each method listed is the recommended method for the analysis of the compound of interest. However, the method may not have been fully optimized to meet the specific sampling situation. Note that some methods are only partially evaluated and have been used in very limited sampling situations. Review the details of the method and consult with the laboratory performing the analysis regarding the applicability of the method and the need for further modifications to the method in order to adjust for the particular conditions.

## Physical Description

A brief description of the appearance and odor of each substance is provided in the physical description section. Notations are made as to whether a substance can be shipped as a liquefied compressed gas or whether it has major use as a pesticide.

## Chemical and Physical Properties

The following abbreviations are used for the chemical and physical properties given for each substance. "NA" indicates that a property is not applicable, and a question mark (?) indicates that it is unknown.

MW................ Molecular weight

BP...................Boiling point at 1 atmosphere, °F

Sol.................. Solubility in water at 68°F*, % by weight (i.e., g/100 ml)

Fl.P.................Flash point (i.e., the temperature at which the liquid phase gives off enough vapor to flash when exposed to an external ignition source), closed cup (unless annotated "(oc)" for open cup), °F

IP.....................Ionization potential, eV (electron volts) (Ionization potentials are given as a guideline for the selection of photoionization detector lamps used in some direct-reading instruments.)

Sp.Gr.............. Specific gravity at 68°F* referenced to water at 39.2°F (4°C)

RGasD...........Relative density of gases referenced to air = 1 (indicates how many times a gas is heavier than air at the same temperature)

VP.................. Vapor pressure at 68°F*, mm Hg; "approx" indicates approximately

FRZ................ Freezing point for liquids and gases, °F

MLT............... Melting point for solids, °F

UEL................Upper explosive (flammable) limit in air, % by volume (at room temperature*)

LEL................ Lower explosive (flammable) limit in air, % by volume (at room temperature*)

MEC...............Minimum explosive concentration, g/m$^3$ (when available)

*If noted after a specific entry, these properties may be reported at other temperatures.

When available, the flammability/combustibility of a substance is listed at the bottom of the chemical and physical properties section. The following OSHA criteria (29 CFR 1910.106) were used to classify flammable or combustible liquids:

Class IA flammable liquid..................... Fl.P below 73°F and BP below 100°F.
Class IB flammable liquid..................... Fl.P below 73°F and BP at or above 100°F.
Class IC flammable liquid..................... Fl.P at or above 73°F and below 100°F.
Class II combustible liquid.................... Fl.P at or above 100°F and below 140°F.
Class IIIA combustible liquid................ Fl.P at or above 140°F and below 200°F.
Class IIIB combustible liquid................ Fl.P at or above 200°F.

### Personal Protection and Sanitation

This section presents a summary of recommended practices for each substance. These recommendations supplement general work practices (e.g., no eating, drinking, or smoking where chemicals are used) and should be followed if additional controls are needed after using all feasible process, equipment, and task controls. Table 2 (page xviii) explains the codes used. Each category is described as follows:

SKIN................ Recommends the need for personal protective clothing.
EYES................ Recommends the need for eye protection.
WASH SKIN...... Recommends when workers should wash the spilled chemical from the body in addition to normal washing (e.g., before eating).
REMOVE.............Advises workers when to remove clothing that has accidentally become wet or significantly contaminated.
CHANGE.............Recommends whether routine changing of clothing is needed.
PROVIDE............ Recommends the need for eyewash fountains and/or quick drench facilities.

### Recommendations for Respirator Selection

This section provides a condensed table of allowable respirators to be used for those substances for which IDLH values have been determined, or for which NIOSH has previously provided respirator recommendations (e.g., in criteria documents or Current Intelligence Bulletins) for certain chemicals. There are, however, 186 chemicals listed in the *Pocket Guide* for which IDLH values have yet to be determined. Since the IDLH is a critical component for completing the Respirator Selection Logic for a given chemical, the *Pocket Guide* does not provide respiratory recommendations for those 186 chemicals without IDLH values. As new or revised IDLH values are developed for those and other chemicals, NIOSH will provide appropriate respirator recommendations. (Updated information on the *Pocket Guide* can be found on the NIOSH Web site (http://www.cdc.gov/niosh/npg/npg.html) and incorporated into subsequent editions of the *Pocket Guide.* [Appendix F (page 361) contains an explanation of the "Effective" IDLHs used for four chloronaphthalene compounds.]

In 1995, NIOSH developed a new set of regulations in 42 CFR 84 (also referred to as "Part 84") for testing and certifying non-powered, air-purifying, particulate-filter respirators. The new Part 84 respirators have passed a more demanding certification test than the old respirators (e.g., dust; dust and mist; dust, mist, and fume; spray paint; pesticide) certified under 30 CFR 11 (also referred to as "Part 11"). Recommendations for non-powered, air-purifying particulate respirators have been updated from previous editions of the *Pocket Guide* to incorporate Part 84 respirators; Part 11 terminology has been removed. See Table 4 (page xxv) for information concerning the selection of N-, R-, or P-series (Part 84) particulate respirators.

In January 1998, OSHA revised its respiratory protection standard (29 CFR 1910.134). Among the provisions in the revised standard is the requirement for an end-of-service-life indicator (ESLI) or a change schedule when air-purifying respirators with chemical cartridges or canisters are used for protection against gases and vapors [29 CFR 1910.134(d)(3)(iii)]. *(Note: All respirator codes containing "**Ccr**" or "**Ov**" are covered by this requirement.)* In the *Pocket Guide*, air-purifying respirators (without ESLIs) for protection against gases and vapors are recommended only for chemicals with adequate warning properties, but now these respirators may be selected regardless of the warning properties. Respirator recommendations in the *Pocket Guide* have not been revised in this edition to reflect the OSHA requirements for ESLIs or change schedules.

Appendix A (page 342) lists the NIOSH carcinogen policy. Respirator recommendations for carcinogens in the *Pocket Guide* have not been revised to reflect this policy; these recommendations will be revised in future editions.

The first line in the entry indicates whether the "NIOSH" or the "OSHA" exposure limit is used on which to base the respirator recommendations. The more protective limit between the NIOSH REL or the OSHA PEL is always used. "NIOSH/OSHA" indicates that the limits are equivalent.

Each subsequent line lists a maximum use concentration (MUC) followed by the classes of respirators that are acceptable for use up to the MUC. Codes for the various categories of respirators, and Assigned Protection Factors (APFs) for these respirators, are listed in Table 3 (page xx). Individual respirator classes are separated by diagonal lines (/). More protective respirators may be worn. The symbol "**§**" is followed by the classes of respirators that are acceptable for emergency or planned entry into unknown concentrations or entry into IDLH conditions. "**Escape**" indicates that the respirators are to be used only for escape purposes. For each MUC or condition, this entry lists only those respirators with the required APF and other use restrictions based on the *NIOSH Respirator Selection Logic*.

All respirators selected must be approved by NIOSH under the provisions of 42 CFR 84. The current listing of NIOSH/MSHA certified respirators can be found in the *NIOSH Certified Equipment List*, which is available on the NIOSH Web site (http://www.cdc.gov/niosh/npptl/topics/respirators/cel).

A complete respiratory protection program must be implemented and must fulfill all requirements of 29 CFR 1910.134. A respiratory protection program must include a written standard operating procedure covering regular training, fit-testing, fit-checking, periodic environmental monitoring, maintenance, medical monitoring, inspection, cleaning, storage and periodic program evaluation. Selection of a specific respirator within a given class of recommended respirators depends on the particular situation; this choice should be made only by a knowledgeable person. REMEMBER: Air-purifying respirators will not protect users against oxygen-deficient atmospheres, and they are not to be used in IDLH conditions. The only respirators recommended for fire fighting are self-contained breathing apparatuses that have full facepieces and are operated in a pressure-demand or other positive-pressure mode. Additional information on the selection and use of respirators can be found in the *NIOSH Respirator Selection Logic* (DHHS [NIOSH] Publication No. 2005-100) and the *NIOSH Guide to Industrial Respiratory Protection* (DHHS [NIOSH] Publication No. 87-116), which are available on the Respirator Topic Page on the NIOSH Web site (http://www.cdc.gov/niosh/npptl/topics/respirators).

## Incompatibilities and Reactivities

This section lists important hazardous incompatibilities or reactivities for each substance.

## Exposure Routes, Symptoms, and Target Organs

The first row for each substance in this section lists the toxicologically important entry routes (**ER**) and whether contact with the skin or eyes is potentially hazardous. The second row lists the potential symptoms of exposure (**SY**) and whether NIOSH considers the substance a potential occupational carcinogen (**[carc]**). The third row lists target organs (**TO**) affected by exposure to the substance (for carcinogens, the types of cancer are listed in brackets). Information in this section reflects human data unless otherwise noted. Abbreviations are defined in Table 5 (page xxvi).

## First Aid

This section lists emergency procedures for eye and skin contact, inhalation, and ingestion of the toxic substance. Abbreviations are defined in Table 6 (page xxviii).

**Table 1**
**NIOSH Manual of Analytical Methods**

| Edition | Volume | Supplement | Publication No. |
|---|---|---|---|
| 2 | 1 | | 77-157-A |
| 2 | 2 | | 77-157-B |
| 2 | 3 | | 77-157-C |
| 2 | 4 | | 78-175 |
| 2 | 5 | | 79-141 |
| 2 | 6 | | 80-125 |
| 2 | 7 | | 82-100 |
| 3 | | | 84-100 |
| 3 | | 1 | 85-117 |
| 3 | | 2 | 87-117 |
| 3 | | 3 | 89-127 |
| 3 | | 4 | 90-121 |
| 4 | | | 94-113 |
| 4 | | 1 | 96-135 |
| 4 | | 2 | 98-119 |
| 4 | | 3 | 2003-154 |

See **Measurement Methods** section on page xii for more information. The *NIOSH Manual of Analytical Methods* is available on the NIOSH Web site (http://www.cdc.gov/niosh/nmam).

## Table 2
## Personal Protection and Sanitation Codes

| Code | | Definition |
|---|---|---|
| **Skin:** | Prevent skin contact | Wear appropriate personal protective clothing to prevent skin contact. |
| | Frostbite | Compressed gases may create low temperatures when they expand rapidly. Leaks and uses that allow rapid expansion may cause a frostbite hazard. Wear appropriate personal protective clothing to prevent the skin from becoming frozen. |
| | N.R | No recommendation is made specifying the need for personal protective equipment for the body. |
| **Eyes:** | Prevent eye contact | Wear appropriate eye protection to prevent eye contact. |
| | Frostbite | Wear appropriate eye protection to prevent eye contact with the liquid that could result in burns or tissue damage from frostbite. |
| | N.R. | No recommendation is made specifying the need for eye protection. |
| **Wash skin:** | When contam | The worker should immediately wash the skin when it becomes contaminated. |
| | Daily | The worker should wash daily at the end of each work shift, and prior to eating, drinking, smoking, etc. |
| | N.R. | No recommendation is made specifying the need for washing the substance from the skin (either immediately or at the end of the work shift). |
| **Remove:** | When wet or contam | Work clothing that becomes wet or significantly contaminated should be removed and replaced. |
| | When wet (flamm) | Work clothing that becomes wet should be immediately removed due to its flammability hazard (i.e., for liquids with a flash point <100°F). |
| | N.R. | No recommendation is made specifying the need for removing clothing that becomes wet or contaminated. |

**Table 2 (Continued)**
**Personal Protection and Sanitation Codes**

| Code | | Definition |
|---|---|---|
| **Change:** | Daily | Workers whose clothing may have become contaminated should change into uncontaminated clothing before leaving the work premises. |
| | N.R. | No recommendation is made specifying the need for the worker to change clothing after the workshift. |
| **Provide:** | Eyewash | Eyewash fountains should be provided in areas where there is any possibility that workers could be exposed to the substances; this is irrespective of the recommendation involving the wearing of eye protection. |
| | Quick drench | Facilities for quickly drenching the body should be provided within the immediate work area for emergency use where there is a possibility of exposure. [**Note:** It is intended that these facilities provide a sufficient quantity or flow of water to quickly remove the substance from any body areas likely to be exposed. The actual determination of what constitutes an adequate quick drench facility depends on the specific circumstances. In certain instances, a deluge shower should be readily available, whereas in others, the availability of water from a sink or hose could be considered adequate.] |
| | Frostbite wash | Quick drench facilities and/or eyewash fountains should be provided within the immediate work area for emergency use where there is any possibility of exposure to liquids that are extremely cold or rapidly evaporating. |
| **Other codes:** | Liq | Liquid |
| | Molt | Molten |
| | Sol | Solid |
| | Soln | Solution containing the contaminant |
| | Vap | Vapor |

## Table 3
## Symbols, Code Components, and Codes Used for Respirator Selection

| Symbol | Description |
|---|---|
| ¥ | At concentrations above the NIOSH REL, or where there is no REL, at any detectable concentration |
| § | Emergency or planned entry into unknown concentrations or IDLH conditions |
| * | Substance reported to cause eye irritation or damage; may require eye protection |
| £ | Substance causes eye irritation or damage; eye protection needed |
| ¿ | Only nonoxidizable sorbents allowed (not charcoal) |
| † | End of service life indicator (ESLI) required |
| APF | Assigned protection factor |

| Code Component | Description |
|---|---|
| 95 | Particulate respirator or filter that is 95% efficient. See **Table 4** (page xxv) to select N95, R95, or P95. |
| 99 | Particulate respirator or filter that is 99% efficient. See **Table 4** (page xxv) to select N99, R99, or P99. |
| 100 | Particulate respirator or filter that is 99.97% efficient. See **Table 4** (page xxv) to select N100, R100, or P100. |
| Ccr | Chemical cartridge respirator |
| F | Full facepiece |
| GmF | Air-purifying, full-facepiece respirator (gas mask) with a chin-style, front- or back-mounted canister |
| Papr | Powered, air-purifying respirator |
| Sa | Supplied-air respirator |
| Scba | Self-contained breathing apparatus |
| Ag | Acid gas cartridge or canister |
| Cf | Continuous flow mode |
| Hie | High-efficiency particulate filter |
| Ov | Organic vapor cartridge or canister |
| Pd,Pp | Pressure-demand or other positive-pressure mode |
| Qm | Quarter-mask respirator |
| S | Chemical cartridge or canister providing protection against the compound of concern |
| T | Tight-fitting facepiece |
| XQ | Except quarter-mask respirator |

## Table 3 (Continued)
## Symbols, Code Components, and Codes Used for Respirator Selection

| Code | APF | Description |
|---|---|---|
| 95F | 10 | Any air-purifying full-facepiece respirator equipped with an N95, R95, or P95 filter. The following filters may also be used: N99, R99, P99, N100, R100, P100. See **Table 4** (page xxv) for information on selection of N, R, or P filters. |
| 95XQ | 10 | Any particulate respirator equipped with an N95, R95, or P95 filter (including N95, R95, and P95 filtering facepieces) except quarter-mask respirators. The following filters may also be used: N99, R99, P99, N100, R100, P100. See **Table 4** (page xxv) for information on selection of N, R, or P filters. |
| 100F | 50 | Any air-purifying, full-facepiece respirator with an N100, R100, or P100 filter. See **Table 4** (page xxv) for information on selection of N, R, or P filters. |
| 100XQ | 10 | Any air-purifying respirator with an N100, R100, or P100 filter (including N100, R100, and P100 filtering facepieces) except quarter-mask respirators. See **Table 4** (page xxv) for information on selection of N, R, or P filters. |
| CcrFAg100 | 50 | Any chemical cartridge respirator with a full facepiece and acid gas cartridge(s) in combination with an N100, R100, or P100 filter. See **Table 4** (page xxv) for information on selection of N, R, or P filters. |
| CcrFOv | 50 | Any air-purifying full-facepiece respirator equipped with organic vapor cartridge(s). |
| CcrFOv95 | 10 | Any full-facepiece respirator with organic vapor cartridge(s) in combination with an N95, R95, or P95 filter. The following filters may also be used: N99, R99, P99, N100, R100, P100. See **Table 4** (page xxv) for information on selection of N, R, or P filters. |
| CcrFOv100 | 50 | Any air-purifying full-facepiece respirator equipped with organic vapor cartridge(s) in combination with an N100, R100, or P100 filter. See **Table 4** (page xxv) for information on selection of N, R, or P filters. |
| CcrFS | 50 | Any air-purifying full-facepiece respirator equipped with cartridge(s) providing protection against the compound of concern. |
| CcrFS100 | 50 | Any air-purifying full-facepiece respirator equipped with cartridge(s) providing protection against the compound of concern in combination with an N100, R100, or P100 filter. See **Table 4** (page xxv) for information on selection of N, R, or P filters. |
| CcrOv | 10 | Any air-purifying half-mask respirator equipped with organic vapor cartridge(s). |

## Table 3 (Continued)
## Symbols, Code Components, and Codes
## Used for Respirator Selection

| Code | APF | Description |
|---|---|---|
| CcrOv95 | 10 | Any air-purifying half-mask respirator with organic vapor cartridge(s) in combination with an N95, R95, or P95 filter. The following filters may also be used: N99, R99, P99, N100, R100, P100. See **Table 4** (page xxv) for information on selection of N, R, or P filters. |
| CcrOv100 | 10 | Any air-purifying half-mask respirator with organic vapor cartridge(s) in combination with an N100, R100, or P100 filter. See **Table 4** (page xxv) for information on selection of N, R, or P filters. |
| CcrOvAg | 10 | Any air-purifying half-mask respirator equipped with organic vapor and acid gas cartridge(s). |
| CcrS | 10 | Any air-purifying half-mask respirator equipped with cartridge(s) providing protection against the compound of concern. |
| GmFAg | 50 | Any air-purifying, full-facepiece respirator (gas mask) with a chin-style, front- or back-mounted acid gas canister. |
| GmFAg100 | 50 | Any air-purifying, full-facepiece respirator (gas mask) with a chin-style, front- or back-mounted acid gas canister having an N100, R100, or P100 filter. See **Table 4** (page xxv) for information on selection of N, R, or P filters. |
| GmFOv | 50 | Any air-purifying, full-facepiece respirator (gas mask) with a chin-style, front- or back-mounted organic vapor canister. |
| GmFOv95 | 10 | Any air-purifying, full-facepiece respirator (gas mask) with a chin-style, front- or back-mounted organic vapor canister in combination with an N95, R95, or P95 filter. The following filters may also be used: N99, R99, P99, N100, R100, P100. See **Table 4** (page xxv) for information on selection of N, R, or P filters. |
| GmFOv100 | 50 | Any air-purifying, full-facepiece respirator (gas mask) with a chin-style, front- or back-mounted organic vapor canister having an N100, R100, or P100 filter. See **Table 4** (page xxv) for information on selection of N, R, or P filters. |
| GmFOvAg | 50 | Any air-purifying, full facepiece respirator (gas mask) with a chin-style, front- or back-mounted organic vapor and acid gas canister. |
| GmFOvAg100 | 50 | Any air-purifying, full facepiece respirator (gas mask) with a chin-style, front- or back-mounted organic vapor and acid gas canister having an N100, R100, or P100 filter. See **Table 4** (page xxv) for information on selection of N, R, or P filters. |
| GmFS | 50 | Any air-purifying, full-facepiece respirator (gas mask) with a chin-style, front- or back-mounted canister providing protection against the compound of concern. |

## Table 3 (Continued)
## Symbols, Code Components, and Codes Used for Respirator Selection

| Code | APF | Description |
| --- | --- | --- |
| GmFS100 | 50 | Any air-purifying, full-facepiece respirator (gas mask) with a chin-style, front- or back-mounted canister providing protection against the compound of concern and having an N100, R100, or P100 filter. See **Table 4** (page xxv) for information on selection of N, R, or P filters. |
| PaprAg | 25 | Any powered air-purifying respirator with acid gas cartridge(s). |
| PaprAgHie | 25 | Any powered air-purifying respirator with acid gas cartridge(s) in combination with a high-efficiency particulate filter. |
| PaprHie | 25 | Any powered air-purifying respirator with a high-efficiency particulate filter. |
| PaprOv | 25 | Any powered air-purifying respirator with organic vapor cartridge(s). |
| PaprOvAg | 25 | Any powered air-purifying respirator with organic vapor and acid gas cartridge(s). |
| PaprOvHie | 25 | Any powered air-purifying respirator with an organic vapor cartridge in combination with a high-efficiency particulate filter. |
| PaprS | 25 | Any powered air-purifying respirator with cartridge(s) providing protection against the compound of concern. |
| PaprTHie | 50 | Any powered air-purifying respirator with a tight-fitting facepiece and a high-efficiency particulate filter. |
| PaprTOv | 50 | Any powered air-purifying respirator with a tight-fitting facepiece and organic vapor cartridge(s). |
| PaprTOvHie | 50 | Any powered air-purifying respirator with a tight-fitting facepiece and organic vapor cartridge(s) in combination with a high-efficiency particulate filter. |
| PaprTS | 50 | Any powered air-purifying respirator with a tight-fitting facepiece and cartridge(s) providing protection against the compound of concern. |
| Qm | 5 | Any quarter-mask respirator. See **Table 4** (page xxv) for information on selection of N, R, or P particulate filters. |
| Sa | 10 | Any supplied-air respirator. |
| Sa:Cf | 25 | Any supplied-air respirator operated in a continuous-flow mode. |
| Sa:Pd,Pp | 1,000 | Any supplied-air respirator operated in a pressure-demand or other positive-pressure mode. |

## Table 3 (Continued)
## Symbols, Code Components, and Codes Used for Respirator Selection

| Code | APF | Description |
|---|---|---|
| SaF | 50 | Any supplied-air respirator with a full facepiece. |
| SaF:Pd,Pp | 2,000 | Any supplied-air respirator that has a full facepiece and is operated in a pressure-demand or other positive pressure mode. |
| SaF:Pd,Pp:AScba | 10,000 | Any supplied-air respirator that has a full-facepiece and is operated in a pressure-demand or other positive-pressure mode in combination with an auxiliary self-contained breathing apparatus operated in pressure-demand or other positive-pressure mode. |
| SaT:Cf | 50 | Any supplied-air respirator that has a tight-fitting facepiece and is operated in a continuous-flow mode. |
| ScbaE | | Any appropriate escape-type, self-contained breathing apparatus. |
| ScbaF | 50 | Any self-contained breathing apparatus with a full facepiece. |
| ScbaF:Pd,Pp | 10,000 | Any self-contained breathing apparatus that has a full facepiece and is operated in a pressure-demand or other positive-pressure mode. |

## Table 4
## Selection of N-, R-, or P-Series Particulate Respirators

1. The selection of N-, R-, and P-series filters depends on the presence of oil particles as follows:

   - If no oil particles are present in the work environment, use a filter of any series (i.e., N-, R-, or P-series).
   - If oil particles (e.g., lubricants, cutting fluids, glycerine) are present, use an R- or P-series filter. **Note: N-series filters cannot be used if oil particles are present.**
   - If oil particles are present and the filter is to be used for more than one work shift, use only a P-series filter.

> **Note:** To help you remember the filter series, use the following guide:
>
> **N** for **N**ot resistant to oil,
> **R** for **R**esistant to oil,
> **P** for oil **P**roof.

2. Selection of filter efficiency (i.e., 95%, 99%, or 99.97%) depends on how much filter leakage can be accepted. Higher filter efficiency means lower filter leakage.

3. The choice of facepiece depends on the level of protection needed – that is, the assigned protection factor (APF) needed. See **Table 3** (page xx) for APFs of respirator classes, and see **Recommendations for Respirator Selection** (page xiv) for more information.

**Table 5**
**Abbreviations for Exposure Routes, Symptoms, and Target Organs**

| Code | Definition | Code | Definition |
|---|---|---|---|
| abdom | Abdominal | dizz | Dizziness |
| abnor | Abnormal/Abnormalities | drow | Drowsiness |
| abs | Skin absorption | dysp | Dyspnea (breathing difficulty) |
| album | Albuminuria | emphy | Emphysema |
| anes | Anesthesia | eosin | Eosinophilia |
| anor | Anorexia | epilep | Epileptiform |
| anos | Anosmia (loss of the sense of smell) | epis | Epistaxis (nosebleed) |
| anxi | Anxiety | equi | Equilibrium |
| arrhy | Arrhythmias | eryt | Erythema (skin redness) |
| aspir | Aspiration | euph | Euphoria |
| asphy | Asphyxia | fail | Failure |
| BP | Blood pressure | fasc | Fasciculation |
| breath | Breath/breathing | FEV | Forced expiratory volume |
| bron | Bronchitis | fib | Fibrosis |
| BUN | Blood urea nitrogen | ftg | Fatigue |
| [carc] | Potential occupational carcinogen | func | Function |
| card | Cardiac | GI | Gastrointestinal |
| chol | Cholinesterase | halu | Hallucinations |
| cirr | Cirrhosis | head | Headache |
| CNS | Central nervous system | hema | Hematuria (blood in the urine) |
| conc | Concentration | hemato | Hematopoietic |
| con | Skin and/or eye contact | hemorr | Hemorrhage |
| conf | Confusion | hyperpig | Hyperpigmentation |
| conj | Conjunctivitis | hypox | Hypoxemia (reduced $O_2$ in the blood) |
| constip | Constipation | inco | Incoordination |
| convuls | Convulsions | incr | Increased |
| corn | Corneal | inebri | Inebriation |
| CVS | Cardiovascular system | inflamm | Inflammation |
| cyan | Cyanosis | ing | Ingestion |
| decr | Decreased | inh | Inhalation |
| depres | Depressed/Depression | inj | Injury |
| derm | Dermatitis | insom | Insomnia |
| diarr | Diarrhea | irreg | Irregular/Irregularities |
| dist | Disturbance | irrit | Irritation |

**Table 5 (Continued)**
**Abbreviations for Exposure Routes, Symptoms, and Target Organs**

| Code | Definition | Code | Definition |
|---|---|---|---|
| irrity | Irritability | prot | Proteinuria |
| jaun | Jaundice | pulm | Pulmonary |
| kera | Keratitis (inflammation of the cornea) | RBC | Red blood cell |
| lac | Lacrimation (discharge of tears) | repro | Reproductive |
| lar | Laryngeal | resp | Respiratory/respiration |
| lass | Lassitude (weakness, exhaustion) | restless | Restlessness |
| leucyt | Leukocytosis (increased blood leukocytes) | retster | Retrosternal (occurring behind the sternum) |
| leupen | Leukopenia (reduced blood leukocytes) | rhin | Rhinorrhea (discharge of thin nasal mucus) |
| liq | Liquid | | |
| local | Localized | salv | Salivation |
| low-wgt | Weight loss | sens | Sensitization |
| mal | Malaise (vague feeling of discomfort) | short | Shortness |
| malnut | Malnutrition | sneez | Sneezing |
| methemo | Methemoglobinemia | sol | Solid |
| muc memb | Mucous membrane | soln | Solution |
| musc | Muscle | subs | Substernal (occurring beneath the sternum) |
| narco | Narcosis | | |
| nau | Nausea | sweat | Sweating |
| nec | Necrosis | swell | Swelling |
| neph | Nephritis | sys | System |
| numb | Numb/numbness | tacar | Tachycardia |
| opac | Opacity | tend | Tenderness |
| palp | Palpitations | terato | Teratogenic |
| para | Paralysis | throb | Throbbing |
| pares | Paresthesia | tight | Tightness |
| perf | Perforation | twitch | Twitching |
| peri neur | Peripheral neuropathy | uncon | Unconsciousness |
| periorb | Periorbital (situated around the eye) | vap | Vapor |
| phar | Pharyngeal | vesic | Vesiculation |
| photo | Photophobia (abnormal visual intolerance to light) | vis | Visual |
| | | vomit | Vomiting |
| pneu | Pneumonitis | weak | Weak/weakness |
| PNS | Peripheral nervous system | wheez | Wheezing |
| polyneur | Polyneuropathy | | |

**Table 6**
**Codes for First Aid Data**

| Code | Definition |
|---|---|
| **Eye:** | |
| Irr immed | If this chemical contacts the eyes, immediately wash (irrigate) the eyes with large amounts of water, occasionally lifting the lower and upper lids. Get medical attention immediately. |
| Irr prompt | If this chemical contacts the eyes, promptly wash (irrigate) the eyes with large amounts of water, occasionally lifting the lower and upper lids. Get medical attention if any discomfort continues. |
| Frostbite | If eye tissue is frozen, seek medical attention immediately; if tissue is not frozen, immediately and thoroughly flush the eyes with large amounts of water for at least 15 minutes, occasionally lifting the lower and upper eyelids. If irritation, pain, swelling, lacrimation, or photophobia persist, get medical attention as soon as possible. |
| Medical attention | Get medical attention. |
| **Skin:** | |
| Blot/brush away | If irritation occurs, gently blot or brush away excess. |
| Dust off solid; water flush | If this solid chemical contacts the skin, dust it off immediately and then flush the contaminated skin with water. If this chemical or liquids containing this chemical penetrate the clothing, promptly remove the clothing and flush the skin with water. Get medical attention immediately. |
| Frostbite | If frostbite has occurred, seek medical attention immediately; do NOT rub the affected areas or flush them with water. In order to prevent further tissue damage, do NOT attempt to remove frozen clothing from frostbitten areas. If frostbite has NOT occurred, immediately and thoroughly wash contaminated skin with soap and water. |
| Molten flush immed/ sol-liq soap wash prompt | If this molten chemical contacts the skin, immediately flush the skin with large amounts of water. Get medical attention immediately. If this chemical (or liquids containing this chemical) contacts the skin, promptly wash the contaminated skin with soap and water. If this chemical or liquids containing this chemical penetrate the clothing, immediately remove the clothing and wash the skin with soap and water. If irritation persists after washing, get medical attention. |
| Soap flush immed | If this chemical contacts the skin, immediately flush the contaminated skin with soap and water. If this chemical penetrates the clothing, immediately remove the clothing and flush the skin with water. If irritation persists after washing, get medical attention. |
| Soap flush prompt | If this chemical contacts the skin, promptly flush the contaminated skin with soap and water. If this chemical penetrates the clothing, promptly remove the clothing and flush the skin with water. If irritation persists after washing, get medical attention. |

## Table 6 (Continued)
## Codes for First Aid Data

| Code | Definition |
|---|---|
| **Skin (continued):** | |
| Soap prompt/molten flush immed | If this solid chemical or a liquid containing this chemical contacts the skin, promptly wash the contaminated skin with soap and water. If irritation persists after washing, get medical attention. If this molten chemical contacts the skin or nonimpervious clothing, immediately flush the affected area with large amounts of water to remove heat. Get medical attention immediately. |
| Soap wash | If this chemical contacts the skin, wash the contaminated skin with soap and water. |
| Soap wash immed | If this chemical contacts the skin, immediately wash the contaminated skin with soap and water. If this chemical penetrates the clothing, immediately remove the clothing, wash the skin with soap and water, and get medical attention promptly. |
| Soap wash prompt | If this chemical contacts the skin, promptly wash the contaminated skin with soap and water. If this chemical penetrates the clothing, promptly remove the clothing and wash the skin with soap and water. Get medical attention promptly. |
| Water flush | If this chemical contacts the skin, flush the contaminated skin with water. Where there is evidence of skin irritation, get medical attention. |
| Water flush immed | If this chemical contacts the skin, immediately flush the contaminated skin with water. If this chemical penetrates the clothing, immediately remove the clothing and flush the skin with water. Get medical attention promptly. |
| Water flush prompt | If this chemical contacts the skin, flush the contaminated skin with water promptly. If this chemical penetrates the clothing, immediately remove the clothing and flush the skin with water promptly. If irritation persists after washing, get medical attention. |
| Water wash | If this chemical contacts the skin, wash the contaminated skin with water. |
| Water wash immed | If this chemical contacts this skin, immediately wash the contaminated skin with water. If this chemical penetrates the clothing, immediately remove the clothing and wash the skin with water. If symptoms occur after washing, get medical attention immediately. |
| Water wash prompt | If this chemical contacts the skin, promptly wash the contaminated skin with water. If this chemical penetrates the clothing, promptly remove the clothing and wash the skin with water. If irritation persists after washing, get medical attention. |

## Table 6 (Continued)
## Codes for First Aid Data

| Code | Definition |
|---|---|
| **Breath:** | |
| Resp support | If a person breathes large amounts of this chemical, move the exposed person to fresh air at once. If breathing has stopped, perform artificial respiration. Keep the affected person warm and at rest. Get medical attention as soon as possible. |
| Fresh air | If a person breathes large amounts of this chemical, move the exposed person to fresh air at once. Other measures are usually unnecessary. |
| Fresh air; 100% $O_2$ | If a person breathes large amounts of this chemical, move the exposed person to fresh air at once. If breathing has stopped, perform artificial respiration. When breathing is difficult, properly trained personnel may assist the affected person by administering 100% oxygen. Keep the affected person warm and at rest. Get medical attention as soon as possible. |
| **Swallow:** | |
| Medical attention immed | If this chemical has been swallowed, get medical attention immediately. |

# CHEMICAL LISTING

## Acetaldehyde

| Acetaldehyde | Formula: $CH_3CHO$ | CAS#: 75-07-0 | RTECS#: AB1925000 | IDLH: Ca [2000 ppm] |
|---|---|---|---|---|
| **Conversion:** 1 ppm = 1.80 mg/$m^3$ | **DOT:** 1089 129 | | | |

**Synonyms/Trade Names:** Acetic aldehyde, Ethanal, Ethyl aldehyde

**Exposure Limits:**
**NIOSH REL:** Ca
See Appendix A
See Appendix C (Aldehydes)
**OSHA PEL†:** TWA 200 ppm (360 mg/$m^3$)

**Measurement Methods (see Table 1):**
**NIOSH** 2018, 2538, 3507
**OSHA** 68

**Physical Description:** Colorless liquid or gas (above 69°F) with a pungent, fruity odor.

**Chemical & Physical Properties:**
**MW:** 44.1
**BP:** 69°F
**Sol:** Miscible
**Fl.P:** -36°F
**IP:** 10.22 eV
**Sp.Gr:** 0.79
**VP:** 740 mmHg
**FRZ:** -190°F
**UEL:** 60%
**LEL:** 4.0%
Class IA Flammable Liquid

**Personal Protection/Sanitation (see Table 2):**
**Skin:** Prevent skin contact
**Eyes:** Prevent eye contact
**Wash skin:** When contam
**Remove:** When wet (flamm)
**Change:** N.R.
**Provide:** Eyewash
Quick drench

**Respirator Recommendations (see Tables 3 and 4):**
**NIOSH**
**¥:** ScbaF:Pd,Pp/SaF:Pd,Pp:AScba
**Escape:** GmFOv/ScbaE

**Incompatibilities and Reactivities:** Strong oxidizers, acids, bases, alcohols, ammonia & amines, phenols, ketones, HCN, $H_2S$ [**Note:** Prolonged contact with air may cause formation of peroxides that may explode and burst containers; easily undergoes polymerization.]

**Exposure Routes, Symptoms, Target Organs (see Table 5):**
**ER:** Inh, Ing, Con
**SY:** Irrit eyes, nose, throat; eye, skin burns; derm; conj; cough; CNS depres; delayed pulm edema; in animals: kidney, repro, terato effects; [carc]
**TO:** Eyes, skin, resp sys, kidneys, CNS, repro sys [in animals: nasal cancer]

**First Aid (see Table 6):**
**Eye:** Irr immed
**Skin:** Water flush prompt
**Breath:** Resp support
**Swallow:** Medical attention immed

## Acetic acid

| Acetic acid | Formula: $CH_3COOH$ | CAS#: 64-19-7 | RTECS#: AF1225000 | IDLH: 50 ppm |
|---|---|---|---|---|
| **Conversion:** 1 ppm = 2.46 mg/$m^3$ | **DOT:** 2790 153 (10-80% acid); 2789 132 (>80% acid) | | | |

**Synonyms/Trade Names:** Acetic acid (aqueous), Ethanoic acid, Glacial acetic acid (pure compound), Methanecarboxylic acid [**Note:** Can be found in concentrations of 5-8% in vinegar.]

**Exposure Limits:**
**NIOSH REL:** TWA 10 ppm (25 mg/$m^3$)
ST 15 ppm (37 mg/$m^3$)
**OSHA PEL:** TWA 10 ppm (25 mg/$m^3$)

**Measurement Methods (see Table 1):**
**NIOSH** 1603
**OSHA** ID186SG

**Physical Description:** Colorless liquid or crystals with a sour, vinegar-like odor. [**Note:** Pure compound is a solid below 62°F. Often used in an aqueous solution.]

**Chemical & Physical Properties:**
**MW:** 60.1
**BP:** 244°F
**Sol:** Miscible
**Fl.P:** 103°F
**IP:** 10.66 eV
**Sp.Gr:** 1.05
**VP:** 11 mmHg
**FRZ:** 62°F
**UEL(200°F):** 19.9%
**LEL:** 4.0%
Class II Combustible Liquid

**Personal Protection/Sanitation (see Table 2):**
**Skin:** Prevent skin contact (>10%)
**Eyes:** Prevent eye contact
**Wash skin:** When contam (>10%)
**Remove:** When wet or contam (>10%)
**Change:** N.R.
**Provide:** Eyewash (>5%)
Quick drench (>50%)

**Respirator Recommendations (see Tables 3 and 4):**
**NIOSH/OSHA**
**50 ppm:** Sa:Cf£/PaprOv£/CcrFOv/GmFOv/ScbaF/SaF
**§:** ScbaF:Pd,Pp/SaF:Pd,Pp:AScba
**Escape:** GmFOv/ScbaE

**Incompatibilities and Reactivities:** Strong oxidizers (especially chromic acid, sodium peroxide & nitric acid), strong caustics [**Note:** Corrosive to metals.]

**Exposure Routes, Symptoms, Target Organs (see Table 5):**
**ER:** Inh, Con
**SY:** Irrit eyes, skin, nose, throat; eye, skin burns; skin sens; dental erosion; black skin, hyperkeratosis; conj, lac; phar edema, chronic bron
**TO:** Eyes, skin, resp sys, teeth

**First Aid (see Table 6):**
**Eye:** Irr immed
**Skin:** Water flush immed
**Breath:** Resp support
**Swallow:** Medical attention immed

## Acetic anhydride

| Acetic anhydride | Formula: $(CH_3CO)_2O$ | CAS#: 108-24-7 | RTECS#: AK1925000 | IDLH: 200 ppm |
|---|---|---|---|---|
| **Conversion:** 1 ppm = 4.18 mg/$m^3$ | **DOT:** 1715 137 | | | |

**Synonyms/Trade Names:** Acetic acid anhydride, Acetic oxide, Acetyl oxide, Ethanoic anhydride

**Exposure Limits:**
**NIOSH REL:** C 5 ppm (20 mg/$m^3$)
**OSHA PEL†:** TWA 5 ppm (20 mg/$m^3$)

**Measurement Methods (see Table 1):**
**NIOSH** 3506
**OSHA** 82, 102

**Physical Description:** Colorless liquid with a strong, pungent, vinegar-like odor.

| Chemical & Physical Properties: | Personal Protection/Sanitation (see Table 2): | Respirator Recommendations (see Tables 3 and 4): |
|---|---|---|
| **MW:** 102.1<br>**BP:** 282°F<br>**Sol:** 12%<br>**Fl.P:** 120°F<br>**IP:** 10.00 eV<br>**Sp.Gr:** 1.08<br>**VP:** 4 mmHg<br>**FRZ:** -99°F<br>**UEL:** 10.3%<br>**LEL:** 2.7%<br>Class II Combustible Liquid | **Skin:** Prevent skin contact<br>**Eyes:** Prevent eye contact<br>**Wash skin:** When contam<br>**Remove:** When wet or contam<br>**Change:** N.R.<br>**Provide:** Eyewash<br>Quick drench | **NIOSH/OSHA**<br>**125 ppm:** Sa:Cf£/PaprOv£<br>**200 ppm:** CcrFOv/GmFOv/PaprTOv£/ScbaF/SaF<br>**§:** ScbaF:Pd,Pp/SaF:Pd,Pp:AScba<br>**Escape:** GmFOv/ScbaE |

**Incompatibilities and Reactivities:** Water, alcohols, strong oxidizers (especially chromic acid), amines, strong caustics [**Note:** Corrosive to iron, steel & other metals. Reacts with water to form acetic acid.]

| Exposure Routes, Symptoms, Target Organs (see Table 5): | First Aid (see Table 6): |
|---|---|
| **ER:** Inh, Ing, Con<br>**SY:** Conj, lac, corn edema, opac, photo; nasal, phar irrit; cough, dysp, bron; skin burns, vesic, sens derm<br>**TO:** Eyes, skin, resp sys | **Eye:** Irr immed<br>**Skin:** Water flush immed<br>**Breath:** Resp support<br>**Swallow:** Medical attention immed |

## Acetone

| Acetone | Formula: $(CH_3)_2CO$ | CAS#: 67-64-1 | RTECS#: AL3150000 | IDLH: 2500 ppm [10%LEL] |
|---|---|---|---|---|
| **Conversion:** 1 ppm = 2.38 mg/$m^3$ | **DOT:** 1090 127 | | | |

**Synonyms/Trade Names:** Dimethyl ketone, Ketone propane, 2-Propanone

**Exposure Limits:**
**NIOSH REL:** TWA 250 ppm (590 mg/$m^3$)
**OSHA PEL†:** TWA 1000 ppm (2400 mg/$m^3$)

**Measurement Methods (see Table 1):**
**NIOSH** 1300, 2555, 3800
**OSHA** 69

**Physical Description:** Colorless liquid with a fragrant, mint-like odor.

| Chemical & Physical Properties: | Personal Protection/Sanitation (see Table 2): | Respirator Recommendations (see Tables 3 and 4): |
|---|---|---|
| **MW:** 58.1<br>**BP:** 133°F<br>**Sol:** Miscible<br>**Fl.P:** 0°F<br>**IP:** 9.69 eV<br>**Sp.Gr:** 0.79<br>**VP:** 180 mmHg<br>**FRZ:** -140°F<br>**UEL:** 12.8%<br>**LEL:** 2.5%<br>Class IB Flammable Liquid | **Skin:** Prevent skin contact<br>**Eyes:** Prevent eye contact<br>**Wash skin:** When contam<br>**Remove:** When wet (flamm)<br>**Change:** N.R. | **NIOSH**<br>**2500 ppm:** CcrOv*/PaprOv*/GmFOv/Sa*/ScbaF<br>**§:** ScbaF:Pd,Pp/SaF:Pd,Pp:AScba<br>**Escape:** GmFOv/ScbaE |

**Incompatibilities and Reactivities:** Oxidizers, acids

| Exposure Routes, Symptoms, Target Organs (see Table 5): | First Aid (see Table 6): |
|---|---|
| **ER:** Inh, Ing, Con<br>**SY:** Irrit eyes, nose, throat; head, dizz, CNS depres; derm<br>**TO:** Eyes, skin, resp sys, CNS | **Eye:** Irr immed<br>**Skin:** Soap wash immed<br>**Breath:** Resp support<br>**Swallow:** Medical attention immed |

A

## Acetone cyanohydrin

| | |
|---|---|
| **Formula:** | $CH_3C(OH)CNCH_3$ |
| **CAS#:** | 75-86-5 |
| **RTECS#:** | OD9275000 |
| **IDLH:** | N.D. |
| **Conversion:** | 1 ppm = 3.48 mg/m$^3$ |
| **DOT:** | 1541 155 (stabilized) |

**Synonyms/Trade Names:** Cyanohydrin-2-propanone, 2-Cyano-2-propanol, α-Hydroxyisobutyronitrile, 2-Hydroxy-2-methyl-propionitrile, 2-Methyllactonitrile

**Exposure Limits:**
**NIOSH REL:** C 1 ppm (4 mg/m$^3$) [15-minute]
**OSHA PEL:** none

**Measurement Methods (see Table 1):**
**NIOSH** 2506

**Physical Description:** Colorless liquid with a faint odor of bitter almond.
[**Note:** Forms cyanide in the body.]

**Chemical & Physical Properties:**
**MW:** 85.1
**BP:** 203°F
**Sol:** Miscible
**Fl.P:** 165°F
**IP:** ?
**Sp.Gr(77°F):** 0.93
**VP:** 0.8 mmHg
**FRZ:** -4°F
**UEL:** 12.0%
**LEL:** 2.2%
Class IIIA Combustible Liquid

**Personal Protection/Sanitation (see Table 2):**
**Skin:** Prevent skin contact
**Eyes:** Prevent eye contact
**Wash skin:** When contam
**Remove:** When wet or contam
**Change:** N.R.
**Provide:** Eyewash
Quick drench

**Respirator Recommendations (see Tables 3 and 4):**
**NIOSH**
**10 ppm:** Sa
**25 ppm:** Sa:Cf
**50 ppm:** ScbaF/SaF
**250 ppm:** SaF:Pd,Pp
**§:** ScbaF:Pd,Pp/SaF:Pd,Pp:AScba
**Escape:** GmFOv/ScbaE

**Incompatibilities and Reactivities:** Sulfuric acid, caustics [**Note:** Slowly decomposes to acetone & HCN at room temperatures; rate is accelerated by an increase in pH, water content, or temperature.]

**Exposure Routes, Symptoms, Target Organs (see Table 5):**
**ER:** Inh, Abs, Ing, Con
**SY:** Irrit eyes, skin, resp sys; dizz, lass, head, conf, convuls; liver, kidney inj; pulm edema, asphy
**TO:** Eyes, skin, resp sys, CNS, CVS, liver, kidneys, GI tract

**First Aid (see Table 6):**
**Eye:** Irr immed
**Skin:** Water flush immed
**Breath:** Resp support
**Swallow:** Medical attention immed

## Acetonitrile

| | |
|---|---|
| **Formula:** | $CH_3CN$ |
| **CAS#:** | 75-05-8 |
| **RTECS#:** | AL7700000 |
| **IDLH:** | 500 ppm |
| **Conversion:** | 1 ppm = 1.68 mg/m$^3$ |
| **DOT:** | 1648 127 |

**Synonyms/Trade Names:** Cyanomethane, Ethyl nitrile, Methyl cyanide [**Note:** Forms cyanide in the body.]

**Exposure Limits:**
**NIOSH REL:** TWA 20 ppm (34 mg/m$^3$)
**OSHA PEL†:** TWA 40 ppm (70 mg/m$^3$)

**Measurement Methods (see Table 1):**
**NIOSH** 1606

**Physical Description:** Colorless liquid with an aromatic odor.

**Chemical & Physical Properties:**
**MW:** 41.1
**BP:** 179°F
**Sol:** Miscible
**Fl.P(oc):** 42°F
**IP:** 12.20 eV
**Sp.Gr:** 0.78
**VP:** 73 mmHg
**FRZ:** -49°F
**UEL:** 16.0%
**LEL:** 3.0%
Class IB Flammable Liquid

**Personal Protection/Sanitation (see Table 2):**
**Skin:** Prevent skin contact
**Eyes:** Prevent eye contact
**Wash skin:** When contam
**Remove:** When wet (flamm)
**Change:** N.R.
**Provide:** Quick drench

**Respirator Recommendations (see Tables 3 and 4):**
**NIOSH**
**200 ppm:** CcrOv/Sa
**500 ppm:** Sa:Cf/PaprOv/CcrFOv/GmFOv/ScbaF/SaF
**§:** ScbaF:Pd,Pp/SaF:Pd,Pp:AScba
**Escape:** GmFOv/ScbaE

**Incompatibilities and Reactivities:** Strong oxidizers

**Exposure Routes, Symptoms, Target Organs (see Table 5):**
**ER:** Inh, Abs, Ing, Con
**SY:** Irrit nose, throat; asphy; nau, vomit; chest pain; lass; stupor, convuls; in animals: liver, kidney damage
**TO:** Resp sys, CVS, CNS, liver, kidneys

**First Aid (see Table 6):**
**Eye:** Irr immed
**Skin:** Water flush immed
**Breath:** Resp support
**Swallow:** Medical attention immed

## 2-Acetylaminofluorene

**Formula:** $C_{15}H_{13}NO$ | **CAS#:** 53-96-3 | **RTECS#:** AB9450000 | **IDLH:** Ca [N.D.]

**Conversion:** | **DOT:**

**Synonyms/Trade Names:** AAF, 2-AAF, 2-Acetaminofluorene, N-Acetyl-2-aminofluorene, FAA, 2-FAA, 2-Fluorenylacetamide

**Exposure Limits:**
**NIOSH REL:** Ca
See Appendix A
**OSHA PEL:** [1910.1014] See Appendix B

**Measurement Methods (see Table 1):**
None available

**Physical Description:** Tan, crystalline powder.

**Chemical & Physical Properties:**
**MW:** 223.3
**BP:** ?
**Sol:** Insoluble
**Fl.P:** ?
**IP:**?
**Sp.Gr:** ?
**VP:** ?
**MLT:** 381°F
**UEL:** ?
**LEL:** ?
Combustible Solid

**Personal Protection/Sanitation (see Table 2):**
**Skin:** Prevent skin contact
**Eyes:** Prevent eye contact
**Wash skin:** When contam/Daily
**Remove:** When wet or contam
**Change:** Daily
**Provide:** Eyewash
Quick drench

**Respirator Recommendations (see Tables 3 and 4):**
**NIOSH**
**¥:** ScbaF:Pd,Pp/SaF:Pd,Pp:AScba
**Escape:** 100F/ScbaE

**See Appendix E** (page 351)

**Incompatibilities and Reactivities:** None reported

**Exposure Routes, Symptoms, Target Organs (see Table 5):**
**ER:** Inh, Abs, Ing, Con
**SY:** Reduced function of liver, kidneys, bladder, pancreas; [carc]
**TO:** Liver, bladder, kidneys, pancreas, skin [in animals: tumors of the liver, bladder, lungs, skin & pancreas]

**First Aid (see Table 6):**
**Eye:** Irr immed
**Skin:** Soap wash immed
**Breath:** Resp support
**Swallow:** Medical attention immed

## Acetylene

**Formula:** HC≡CH | **CAS#:** 74-86-2 | **RTECS#:** AO9600000 | **IDLH:** N.D.

**Conversion:** 1 ppm = 1.06 mg/m$^3$ | **DOT:** 1001 116

**Synonyms/Trade Names:** Ethine, Ethyne [**Note:** A compressed gas used in the welding & cutting of metals.]

**Exposure Limits:**
**NIOSH REL:** C 2500 ppm (2662 mg/m$^3$)
**OSHA PEL:** none

**Measurement Methods (see Table 1):**
**NIOSH**
Acetylene Criteria Document

**Physical Description:** Colorless gas with a faint, ethereal odor. [**Note:** Commercial grade has a garlic-like odor. Shipped under pressure dissolved in acetone.]

**Chemical & Physical Properties:**
**MW:** 26.0
**BP:** Sublimes
**Sol:** 2%
**Fl.P:** NA (Gas)
**IP:** 11.40 eV
**RGasD:** 0.91
**VP:** 44.2 atm
**FRZ:** -119°F (Sublimes)
**UEL:** 100%
**LEL:** 2.5%
Flammable Gas

**Personal Protection/Sanitation (see Table 2):**
**Skin:** Frostbite
**Eyes:** Frostbite
**Wash skin:** N.R.
**Remove:** When wet (flamm)
**Change:** N.R.
**Provide:** Frostbite wash

**Respirator Recommendations (see Tables 3 and 4):**
Not available.

**Incompatibilities and Reactivities:** Zinc; oxygen & other oxidizing agents such as halogens [**Note:** Forms explosive acetylide compounds with copper, mercury, silver & brasses (containing more than 66% copper).]

**Exposure Routes, Symptoms, Target Organs (see Table 5):**
**ER:** Inh, Con (liquid)
**SY:** Head, dizz; asphy; liquid: frostbite
**TO:** CNS, resp sys

**First Aid (see Table 6):**
**Eye:** Frostbite
**Skin:** Frostbite
**Breath:** Fresh air

A

## Acetylene tetrabromide

| | | | | |
|---|---|---|---|---|
| **Acetylene tetrabromide** | **Formula:** $CHBr_2CHBr_2$ | **CAS#:** 79-27-6 | **RTECS#:** KI8225000 | **IDLH:** 8 ppm |
| **Conversion:** 1 ppm = 14.14 mg/m$^3$ | **DOT:** 2504 159 | | | |

**Synonyms/Trade Names:** Symmetrical tetrabromoethane, TBE, Tetrabromoacetylene, Tetrabromoethane, 1,1,2,2-Tetrabromoethane

**Exposure Limits:**
**NIOSH REL:** See Appendix D
**OSHA PEL:** TWA 1 ppm (14 mg/m$^3$)

**Measurement Methods (see Table 1):**
NIOSH 2003

**Physical Description:** Pale-yellow liquid with a pungent odor similar to camphor or iodoform. [**Note:** A solid below 32°F.]

**Chemical & Physical Properties:**
**MW:** 345.7
**BP:** 474°F (Decomposes)
**Sol:** 0.07%
**Fl.P:** NA
**IP:** ?
**Sp.Gr:** 2.97
**VP:** 0.02 mmHg
**FRZ:** 32°F
**UEL:** NA
**LEL:** NA
Noncombustible Liquid

**Personal Protection/Sanitation (see Table 2):**
**Skin:** Prevent skin contact
**Eyes:** Prevent eye contact
**Wash skin:** When contam
**Remove:** When wet or contam
**Change:** N.R.

**Respirator Recommendations (see Tables 3 and 4):**
**OSHA**
**8 ppm:** Sa/ScbaF
**§:** ScbaF:Pd,Pp/SaF:Pd,Pp:AScba
**Escape:** GmFOv/ScbaE

**Incompatibilities and Reactivities:** Strong caustics; hot iron; reducing metals such as aluminum, magnesium, and zinc

**Exposure Routes, Symptoms, Target Organs (see Table 5):**
**ER:** Inh, Ing, Con
**SY:** Irrit eyes, nose; anor, nau; head; abdom pain; jaun; leucyt; CNS depres
**TO:** Eyes, resp sys, liver, CNS

**First Aid (see Table 6):**
**Eye:** Irr immed
**Skin:** Water flush prompt
**Breath:** Resp support
**Swallow:** Medical attention immed

## Acetylsalicylic acid

| | | | | |
|---|---|---|---|---|
| **Acetylsalicylic acid** | **Formula:** $CH_3COOC_6H_4COOH$ | **CAS#:** 50-78-2 | **RTECS#:** VO0700000 | **IDLH:** N.D. |
| **Conversion:** | **DOT:** | | | |

**Synonyms/Trade Names:** o-Acetoxybenzoic acid, 2-Acetoxybenzoic acid, Aspirin

**Exposure Limits:**
**NIOSH REL:** TWA 5 mg/m$^3$
**OSHA PEL†:** none

**Measurement Methods (see Table 1):**
NIOSH 0500

**Physical Description:** Odorless, colorless to white, crystal-line powder. [aspirin]
[**Note:** Develops the vinegar-like odor of acetic acid on contact with moisture.]

**Chemical & Physical Properties:**
**MW:** 180.2
**BP:** 284°F (Decomposes)
**Sol(77°F):** 0.3%
**Fl.P:** NA
**IP:** NA
**Sp.Gr:** 1.35
**VP:** 0 mmHg (approx)
**MLT:** 275°F
**UEL:** NA
**LEL:** NA
**MEC:** 40 g/m$^3$
Combustible Powder; explosion hazard if dispersed in air.

**Personal Protection/Sanitation (see Table 2):**
**Skin:** Prevent skin contact
**Eyes:** Prevent eye contact
**Wash skin:** When contam
**Remove:** N.R.
**Change:** Daily
**Provide:** Eyewash
Quick drench

**Respirator Recommendations (see Tables 3 and 4):**
Not available.

**Incompatibilities and Reactivities:** Solutions of alkali hydroxides or carbonates, strong oxidizers, moisture
[**Note:** Slowly hydrolyzes in moist air to salicyclic & acetic acids.]

**Exposure Routes, Symptoms, Target Organs (see Table 5):**
**ER:** Inh, Ing, Con
**SY:** Irrit eyes, skin, upper resp sys; incr blood clotting time; nau, vomit; liver, kidney inj
**TO:** Eyes, skin, resp sys, blood, liver, kidneys

**First Aid (see Table 6):**
**Eye:** Irr immed
**Skin:** Soap wash
**Breath:** Resp support
**Swallow:** Medical attention immed

## Acrolein

| | |
|---|---|
| **Formula:** $CH_2{=}CHCHO$ | **CAS#:** 107-02-8 |
| **RTECS#:** AS1050000 | **IDLH:** 2 ppm |
| **Conversion:** 1 ppm = 2.29 mg/m³ | **DOT:** 1092 131P (inhibited) |

**Synonyms/Trade Names:** Acraldehyde, Acrylaldehyde, Acrylic aldehyde, Allyl aldehyde, Propenal, 2-Propenal

**Exposure Limits:**
**NIOSH REL:** TWA 0.1 ppm (0.25 $mg/m^3$)
ST 0.3 ppm (0.8 $mg/m^3$)
See Appendix C (Aldehydes)
**OSHA PEL†:** TWA 0.1 ppm (0.25 $mg/m^3$)

**Measurement Methods (see Table 1):**
**NIOSH** 2501
**OSHA** 52

**Physical Description:** Colorless or yellow liquid with a piercing, disagreeable odor.

**Chemical & Physical Properties:**
**MW:** 56.1
**BP:** 127°F
**Sol:** 40%
**Fl.P:** -15°F
**IP:** 10.13 eV
**Sp.Gr:** 0.84
**VP:** 210 mmHg
**FRZ:** -126°F
**UEL:** 31%
**LEL:** 2.8%
Class IB Flammable Liquid

**Personal Protection/Sanitation (see Table 2):**
**Skin:** Prevent skin contact
**Eyes:** Prevent eye contact
**Wash skin:** When contam
**Remove:** When wet (flamm)
**Change:** N.R.
**Provide:** Eyewash
Quick drench

**Respirator Recommendations (see Tables 3 and 4):**
**NIOSH/OSHA**
**2 ppm:** Sa:Cf*/PaprOv*/CcrFOv/GmFOv/ScbaF/SaF
**§:** ScbaF:Pd,Pp/SaF:Pd,Pp:AScba
**Escape:** GmFOv/ScbaE

**Incompatibilities and Reactivities:** Oxidizers, acids, alkalis, ammonia, amines [**Note:** Polymerizes readily unless inhibited--usually with hydroquinone. May form shock-sensitive peroxides over time.]

**Exposure Routes, Symptoms, Target Organs (see Table 5):**
**ER:** Inh, Ing, Con
**SY:** Irrit eyes, skin, muc memb; decr pulm func; delayed pulm edema; chronic resp disease
**TO:** Eyes, skin, resp sys, heart

**First Aid (see Table 6):**
**Eye:** Irr immed
**Skin:** Water flush immed
**Breath:** Resp support
**Swallow:** Medical attention immed

## Acrylamide

| | |
|---|---|
| **Formula:** $CH_2{=}CHCONH_2$ | **CAS#:** 79-06-1 |
| **RTECS#:** AS3325000 | **IDLH:** Ca [60 mg/m³] |
| **Conversion:** | **DOT:** 2074 153P |

**Synonyms/Trade Names:** Acrylamide monomer, Acrylic amide, Propenamide, 2-Propenamide

**Exposure Limits:**
**NIOSH REL:** Ca
TWA 0.03 $mg/m^3$ [skin]
See Appendix A
**OSHA PEL†:** TWA 0.3 $mg/m^3$ [skin]

**Measurement Methods (see Table 1):**
**OSHA** 21, PV2004

**Physical Description:** White crystalline, odorless solid.

**Chemical & Physical Properties:**
**MW:** 71.1
**BP:** 347-572°F (Decomposes)
**Sol(86°F):** 216%
**Fl.P:** 280°F
**IP:** 9.50 eV
**Sp.Gr:** 1.12
**VP:** 0.007 mmHg
**MLT:** 184°F
**UEL:** ?
**LEL:** ?
Combustible Solid (may also be dissolved in flammable liquids).

**Personal Protection/Sanitation (see Table 2):**
**Skin:** Prevent skin contact
**Eyes:** Prevent eye contact
**Wash skin:** When contam/Daily
**Remove:** When wet or contam
**Change:** Daily
**Provide:** Eyewash
Quick drench

**Respirator Recommendations (see Tables 3 and 4):**
**NIOSH**
**¥:** ScbaF:Pd,Pp/SaF:Pd,Pp:AScba
**Escape:** GmFOv/ScbaE

**Incompatibilities and Reactivities:** Strong oxidizers [**Note:** May polymerize violently upon melting.]

**Exposure Routes, Symptoms, Target Organs (see Table 5):**
**ER:** Inh, Abs, Ing, Con
**SY:** Irrit eyes, skin; ataxia, numb limbs, pares; musc weak; absent deep tendon reflex; hand sweat; lass, drow; repro effects; [carc]
**TO:** Eyes, skin, CNS, PNS, repro sys [in animals: tumors of the lungs, testes, thyroid & adrenal glands]

**First Aid (see Table 6):**
**Eye:** Irr immed
**Skin:** Water flush immed
**Breath:** Resp support
**Swallow:** Medical attention immed

A

## Acrylic acid

| | |
|---|---|
| **Formula:** $CH_2$=CHCOOH | **CAS#:** 79-10-7 |
| **RTECS#:** AS4375000 | **IDLH:** N.D. |
| **Conversion:** 1 ppm = 2.95 mg/$m^3$ | **DOT:** 2218 132P (inhibited) |

**Synonyms/Trade Names:** Acroleic acid, Aqueous acrylic acid (technical grade is 94%), Ethylenecarboxylic acid, Glacial acrylic acid (98% in aqueous solution), 2-Propenoic acid

**Exposure Limits:**
**NIOSH REL:** TWA 2 ppm (6 mg/$m^3$) [skin]
**OSHA PEL†:** none

**Measurement Methods (see Table 1):**
**OSHA** 28, PV2005

**Physical Description:** Colorless liquid or solid (below 55°F) with a distinctive, acrid odor.
[**Note:** Shipped with an inhibitor (e.g., hydroquinone) since it readily polymerizes.]

**Chemical & Physical Properties:**
**MW:** 72.1
**BP:** 286°F
**Sol:** Miscible
**Fl.P:** 121°F
**IP:** ?
**Sp.Gr:** 1.05
**VP:** 3 mmHg
**FRZ:** 55°F
**UEL:** 8.02%
**LEL:** 2.4%
Class II Combustible Liquid

**Personal Protection/Sanitation (see Table 2):**
**Skin:** Prevent skin contact
**Eyes:** Prevent eye contact
**Wash skin:** When contam
**Remove:** When wet or contam
**Change:** N.R.
**Provide:** Eyewash
Quick drench

**Respirator Recommendations (see Tables 3 and 4):**
Not available.

**Incompatibilities and Reactivities:** Oxidizers, amines, alkalis, ammonium hydroxide, chloro-sulfonic acid, oleum, ethylene diamine, ethyleneimine, 2-aminoethanol [**Note:** Corrosive to many metals.]

**Exposure Routes, Symptoms, Target Organs (see Table 5):**
**ER:** Inh, Abs, Ing, Con
**SY:** Irrit eyes, skin, resp sys; eye, skin burns; skin sens; in animals: lung, liver, kidney inj
**TO:** Eyes, skin, resp sys

**First Aid (see Table 6):**
**Eye:** Irr immed
**Skin:** Water flush immed
**Breath:** Resp support
**Swallow:** Medical attention immed

## Acrylonitrile

| | |
|---|---|
| **Formula:** $CH_2$=CHCN | **CAS#:** 107-13-1 |
| **RTECS#:** AT5250000 | **IDLH:** Ca [85 ppm] |
| **Conversion:** 1 ppm = 2.17 mg/$m^3$ | **DOT:** 1093 131P (inhibited) |

**Synonyms/Trade Names:** Acrylonitrile monomer, AN, Cyanoethylene, Propenenitrile, 2-Propenenitrile, VCN, Vinyl cyanide

**Exposure Limits:**
**NIOSH REL:** Ca
TWA 1 ppm
C 10 ppm [15-minute] [skin]
See Appendix A
**OSHA PEL:** [1910.1045] TWA 2 ppm
C 10 ppm [15-minute] [skin]

**Measurement Methods (see Table 1):**
**NIOSH** 1604
**OSHA** 37

**Physical Description:** Colorless to pale-yellow liquid with an unpleasant odor. [**Note:** Odor can only be detected above the PEL.]

**Chemical & Physical Properties:**
**MW:** 53.1
**BP:** 171°F
**Sol:** 7%
**Fl.P:** 30°F
**IP:** 10.91 eV
**Sp.Gr:** 0.81
**VP:** 83 mmHg
**FRZ:** -116°F
**UEL:** 17%
**LEL:** 3.0%
Class IB Flammable Liquid

**Personal Protection/Sanitation (see Table 2):**
**Skin:** Prevent skin contact
**Eyes:** Prevent eye contact
**Wash skin:** When contam
**Remove:** When wet (flamm)
**Change:** N.R.
**Provide:** Eyewash
Quick drench

**Respirator Recommendations (see Tables 3 and 4):**
**NIOSH**
**¥:** ScbaF:Pd,Pp/SaF:Pd,Pp:AScba
**Escape:** GmFOv/ScbaE

**See Appendix E** (page 351)

**Incompatibilities and Reactivities:** Strong oxidizers, acids & alkalis; bromine; amines [**Note:** Unless inhibited (usually with methylhydroquinone), may polymerize spontaneously or when heated or in presence of strong alkali. Attacks copper.]

**Exposure Routes, Symptoms, Target Organs (see Table 5):**
**ER:** Inh, Abs, Ing, Con
**SY:** Irrit eyes, skin; asphy; head; sneez; nau, vomit; lass, dizz; skin vesic; scaling derm; [carc]
**TO:** Eyes, skin, CVS, liver, kidneys, CNS [brain tumors, lung & bowel cancer]

**First Aid (see Table 6):**
**Eye:** Irr immed
**Skin:** Water wash immed
**Breath:** Resp support
**Swallow:** Medical attention immed

## Adiponitrile

| Formula: | CAS#: | RTECS#: | IDLH: |
|---|---|---|---|
| $NC(CH_2)_4CN$ | 111-69-3 | AV2625000 | N.D. |

**Conversion:** 1 ppm = 4.43 mg/$m^3$ | **DOT:** 2205 153

**Synonyms/Trade Names:** 1,4-Dicyanobutane, Hexanedinitrile, Tetramethylene cyanide

**Exposure Limits:**
**NIOSH REL:** TWA 4 ppm (18 mg/$m^3$) **OSHA PEL:** none

**Measurement Methods (see Table 1):**
**NIOSH**
Nitriles Criteria Document

**Physical Description:** Water-white, practically odorless, oily liquid.
[**Note:** A solid below 34°F. Forms cyanide in the body.]

**Chemical & Physical Properties:**
**MW:** 108.2
**BP:** 563°F
**Sol:** 4.5%
**Fl.P(oc):** 199°F
**IP:** ?
**Sp.Gr:** 0.97
**VP:** 0.002 mmHg
**FRZ:** 34°F
**UEL:** 5.0%
**LEL:** 1.7%
Class IIIA Combustible Liquid

**Personal Protection/Sanitation (see Table 2):**
**Skin:** Prevent skin contact
**Eyes:** Prevent eye contact
**Wash skin:** When contam
**Remove:** When wet or contam
**Change:** Daily

**Respirator Recommendations (see Tables 3 and 4):**
**NIOSH**
**40 ppm:** Sa
**100 ppm:** Sa:Cf
**200 ppm:** ScbaF/SaF
**250 ppm:** SaF:Pd,Pp
**§:** ScbaF:Pd,Pp/SaF:Pd,Pp:AScba
**Escape:** GmFOv/ScbaE

**Incompatibilities and Reactivities:** Oxidizers (e.g., perchlorates, nitrates), strong acids (e.g., sulfuric acid)
[**Note:** Decomposes above 194°F, forming hydrogen cyanide.]

**Exposure Routes, Symptoms, Target Organs (see Table 5):**
**ER:** Inh, Abs, Ing, Con
**SY:** Irrit eyes, skin, resp sys; head, dizz, lass, conf, convuls; blurred vision; dysp; abdom pain, nau, vomit
**TO:** Eyes, skin, resp sys, CNS, CVS

**First Aid (see Table 6):**
**Eye:** Irr immed
**Skin:** Soap wash immed
**Breath:** Resp support
**Swallow:** Medical attention immed

## Aldrin

| Formula: | CAS#: | RTECS#: | IDLH: |
|---|---|---|---|
| $C_{12}H_8Cl_6$ | 309-00-2 | IO2100000 | Ca [25 mg/$m^3$] |

**Conversion:** | **DOT:** 2761 151

**Synonyms/Trade Names:** HHDN, Octalene,
1,2,3,4,10,10-Hexachloro-1,4,4a,5,8,8a-hexahydro-endo-1,4-exo-5,8-dimethanonaphthalene

**Exposure Limits:**
**NIOSH REL:** Ca
TWA 0.25 mg/$m^3$ [skin]
See Appendix A
**OSHA PEL:** TWA 0.25 mg/$m^3$ [skin]

**Measurement Methods (see Table 1):**
**NIOSH** 5502

**Physical Description:** Colorless to dark-brown crystalline solid with a mild chemical odor. [**Note:** Formerly used as an insecticide.]

**Chemical & Physical Properties:**
**MW:** 364.9
**BP:** Decomposes
**Sol:** 0.003%
**Fl.P:** NA
**IP:** ?
**Sp.Gr:** 1.60
**VP:** 0.00008 mmHg
**MLT:** 219°F
**UEL:** NA
**LEL:** NA
Noncombustible Solid, but may be dissolved in flammable liquids.

**Personal Protection/Sanitation (see Table 2):**
**Skin:** Prevent skin contact
**Eyes:** Prevent eye contact
**Wash skin:** When contam/Daily
**Remove:** When wet or contam
**Change:** Daily
**Provide:** Eyewash
Quick drench

**Respirator Recommendations (see Tables 3 and 4):**
**NIOSH**
**¥:** ScbaF:Pd,Pp/SaF:Pd,Pp:AScba
**Escape:** GmFOv100/ScbaE

**Incompatibilities and Reactivities:** Concentrated mineral acids, active metals, acid catalysts, acid oxidizing agents, phenol

**Exposure Routes, Symptoms, Target Organs (see Table 5):**
**ER:** Inh, Abs, Ing, Con
**SY:** Head, dizz; nau, vomit, mal; myoclonic jerks of limbs; clonic, tonic convuls; coma; hema, azotemia; [carc]
**TO:** CNS, liver, kidneys, skin [in animals: tumors of the lungs, liver, thyroid & adrenal glands]

**First Aid (see Table 6):**
**Eye:** Irr immed
**Skin:** Soap wash immed
**Breath:** Resp support
**Swallow:** Medical attention immed

| Allyl alcohol | Formula: $CH_2{=}CHCH_2OH$ | CAS#: 107-18-6 | RTECS#: BA5075000 | IDLH: 20 ppm |
|---|---|---|---|---|
| **Conversion:** 1 ppm = 2.38 mg/$m^3$ | **DOT:** 1098 131 | | | |
| **Synonyms/Trade Names:** AA, Allylic alcohol, Propenol, 1-Propen-3-ol, 2-Propenol, Vinyl carbinol | | | | |
| **Exposure Limits:**<br>**NIOSH REL:** TWA 2 ppm (5 mg/$m^3$)<br>ST 4 ppm (10 mg/$m^3$) [skin]<br>**OSHA PEL†:** TWA 2 ppm (5 mg/$m^3$) [skin] | **Measurement Methods (see Table 1):**<br>**NIOSH** 1402, 1405 | | | |
| **Physical Description:** Colorless liquid with a pungent, mustard-like odor. | | | | |
| **Chemical & Physical Properties:**<br>**MW:** 58.1<br>**BP:** 205°F<br>**Sol:** Miscible<br>**Fl.P:** 70°F<br>**IP:** 9.63 eV<br>**Sp.Gr:** 0.85<br>**VP:** 17 mmHg<br>**FRZ:** -200°F<br>**UEL:** 18.0%<br>**LEL:** 2.5%<br>Class IB Flammable Liquid | **Personal Protection/Sanitation (see Table 2):**<br>**Skin:** Prevent skin contact<br>**Eyes:** Prevent eye contact<br>**Wash skin:** When contam<br>**Remove:** When wet (flamm)<br>**Change:** N.R.<br>**Provide:** Quick drench | **Respirator Recommendations (see Tables 3 and 4):**<br>**NIOSH/OSHA**<br>**20 ppm:** Sa:Cf*/PaprOv*/CcrFOv/ GmFOv/ScbaF/SaF<br>**§:** ScbaF:Pd,Pp/SaF:Pd,Pp:AScba<br>**Escape:** GmFOv/ScbaE | | |
| **Incompatibilities and Reactivities:** Strong oxidizers, acids, carbon tetrachloride<br>[**Note:** Polymerization may be caused by elevated temperatures, oxidizers, or peroxides.] | | | | |
| **Exposure Routes, Symptoms, Target Organs (see Table 5):**<br>**ER:** Inh, Abs, Ing, Con<br>**SY:** Eye irrit, tissue damage; irrit upper resp sys, skin; pulm edema<br>**TO:** Eyes, skin, resp sys | **First Aid (see Table 6):**<br>**Eye:** Irr immed<br>**Skin:** Water flush immed<br>**Breath:** Resp support<br>**Swallow:** Medical attention immed | | | |

| Allyl chloride | Formula: $CH_2{=}CHCH_2Cl$ | CAS#: 107-05-1 | RTECS#: UC7350000 | IDLH: 250 ppm |
|---|---|---|---|---|
| **Conversion:** 1 ppm = 3.13 mg/$m^3$ | **DOT:** 1100 131 | | | |
| **Synonyms/Trade Names:** 3-Chloropropene, 1-Chloro-2-propene, 3-Chloropropylene | | | | |
| **Exposure Limits:**<br>**NIOSH REL:** TWA 1 ppm (3 mg/$m^3$)<br>ST 2 ppm (6 mg/$m^3$)<br>**OSHA PEL†:** TWA 1 ppm (3 mg/$m^3$) | **Measurement Methods (see Table 1):**<br>**NIOSH** 1000<br>**OSHA** 7 | | | |
| **Physical Description:** Colorless, brown, yellow, or purple liquid with a pungent, unpleasant odor. | | | | |
| **Chemical & Physical Properties:**<br>**MW:** 76.5<br>**BP:** 113°F<br>**Sol:** 0.4%<br>**Fl.P:** -25°F<br>**IP:** 10.05 eV<br>**Sp.Gr:** 0.94<br>**VP:** 295 mmHg<br>**MLT:** -210°F<br>**UEL:** 11.1%<br>**LEL:** 2.9%<br>Class IB Flammable Liquid | **Personal Protection/Sanitation (see Table 2):**<br>**Skin:** Prevent skin contact<br>**Eyes:** Prevent eye contact<br>**Wash skin:** When contam<br>**Remove:** When wet (flamm)<br>**Change:** N.R.<br>**Provide:** Quick drench | **Respirator Recommendations (see Tables 3 and 4):**<br>**NIOSH/OSHA**<br>**25 ppm:** Sa:Cf*<br>**50 ppm:** ScbaF/SaF<br>**250 ppm:** SaF:Pd,Pp<br>**§:** ScbaF:Pd,Pp/SaF:Pd,Pp:AScba<br>**Escape:** GmFOv/ScbaE | | |
| **Incompatibilities and Reactivities:** Strong oxidizers, acids, amines, iron & aluminum chlorides, magnesium, zinc | | | | |
| **Exposure Routes, Symptoms, Target Organs (see Table 5):**<br>**ER:** Inh, Abs, Ing, Con<br>**SY:** Irrit eyes, skin, nose, muc memb; pulm edema; in animals: liver, kidney inj<br>**TO:** Eyes, skin, resp sys, liver, kidneys | **First Aid (see Table 6):**<br>**Eye:** Irr immed<br>**Skin:** Soap wash immed<br>**Breath:** Resp support<br>**Swallow:** Medical attention immed | | | |

A

## Allyl glycidyl ether

| Formula: | CAS#: | RTECS#: | IDLH: |
|---|---|---|---|
| $C_6H_{10}O_2$ | 106-92-3 | RR0875000 | 50 ppm |

**Conversion:** 1 ppm = 4.67 mg/m$^3$ | **DOT:** 2219 129

**Synonyms/Trade Names:** AGE, 1-Allyloxy-2,3-epoxypropane, Glycidyl allyl ether, [(2-Propenyloxy)methyl] oxirane

**Exposure Limits:**
**NIOSH REL:** TWA 5 ppm (22 mg/m$^3$) [skin]
ST 10 ppm (44 mg/m$^3$)
**OSHA PEL†:** C 10 ppm (45 mg/m$^3$)

**Measurement Methods (see Table 1):**
**NIOSH** 2545

**Physical Description:** Colorless liquid with a pleasant odor.

**Chemical & Physical Properties:**
**MW:** 114.2
**BP:** 309°F
**Sol:** 14%
**Fl.P:** 135°F
**IP:** ?
**Sp.Gr:** 0.97
**VP:** 2 mmHg
**FRZ:** -148°F [forms glass]
**UEL:** ?
**LEL:** ?
Class II Combustible Liquid

**Personal Protection/Sanitation (see Table 2):**
**Skin:** Prevent skin contact
**Eyes:** Prevent eye contact
**Wash skin:** When contam
**Remove:** When wet or contam
**Change:** N.R.
**Provide:** Eyewash

**Respirator Recommendations (see Tables 3 and 4):**
**NIOSH**
**50 ppm:** CcrOv/PaprOv/GmFOv/Sa/ScbaF
**§:** ScbaF:Pd,Pp/SaF:Pd,Pp:AScba
**Escape:** GmFOv/ScbaE

**Incompatibilities and Reactivities:** Strong oxidizers

**Exposure Routes, Symptoms, Target Organs (see Table 5):**
**ER:** Inh, Abs, Ing, Con
**SY:** Irrit eyes, skin, nose, resp sys; derm; pulm edema; narco; possible hemato, repro effects
**TO:** Eyes, skin, resp sys, blood, repro sys

**First Aid (see Table 6):**
**Eye:** Irr immed
**Skin:** Water flush prompt
**Breath:** Resp support
**Swallow:** Medical attention immed

## Allyl propyl disulfide

| Formula: | CAS#: | RTECS#: | IDLH: |
|---|---|---|---|
| $H_2C=CHCH_2S_2CH_2CH_2CH_3$ | 2179-59-1 | JO0350000 | N.D. |

**Conversion:** 1 ppm = 6.07 mg/m$^3$ | **DOT:**

**Synonyms/Trade Names:** 4,5-Dithia-1-octene, Onion oil, 2-Propenyl propyl disulfide, Propyl allyl disulfide

**Exposure Limits:**
**NIOSH REL:** TWA 2 ppm (12 mg/m$^3$)
ST 3 ppm (18 mg/m$^3$)
**OSHA PEL†:** TWA 2 ppm (12 mg/m$^3$)

**Measurement Methods (see Table 1):**
**OSHA** PV2086

**Physical Description:** Pale-yellow liquid with a strong & irritating onion-like odor.
[**Note:** The chief volatile component of onion oil.]

**Chemical & Physical Properties:**
**MW:** 148.3
**BP:** ?
**Sol:** Insoluble
**Fl.P:** ?
**IP:** ?
**Sp.Gr(59°F):** 0.93
**VP:** ?
**FRZ:** 5°F
**UEL:** ?
**LEL:** ?
Combustible Liquid

**Personal Protection/Sanitation (see Table 2):**
**Skin:** N.R.
**Eyes:** Prevent eye contact
**Wash skin:** N.R.
**Remove:** When wet or contam
**Change:** N.R.

**Respirator Recommendations (see Tables 3 and 4):**
Not available.

**Incompatibilities and Reactivities:** Oxidizers

**Exposure Routes, Symptoms, Target Organs (see Table 5):**
**ER:** Inh, Ing, Con
**SY:** Irrit eyes, nose, resp sys; lac
**TO:** Eyes, resp sys

**First Aid (see Table 6):**
**Eye:** Irr immed
**Skin:** Soap wash immed
**Breath:** Resp support
**Swallow:** Medical attention immed

## α-Alumina

**Formula:** $Al_2O_3$ | **CAS#:** 1344-28-1 | **RTECS#:** BD1200000 | **IDLH:** N.D.

**Conversion:**

**DOT:**

**Synonyms/Trade Names:** Alumina, Aluminum oxide, Aluminum trioxide [**Note:** α-Alumina is the main component of technical grade alumina. Corundum is natural $Al_2O_3$. Emery is an impure crystalline variety of $Al_2O_3$.]

**Exposure Limits:**
**NIOSH REL:** See Appendix D
**OSHA PEL†:** TWA 15 mg/m$^3$ (total)
TWA 5 mg/m$^3$ (resp)

**Measurement Methods (see Table 1):**
**NIOSH** 0500, 0600
**OSHA** ID109SG, ID198SG

**Physical Description:** White, odorless, crystalline powder.

**Chemical & Physical Properties:**
**MW:** 101.9
**BP:** 5396°F
**Sol:** Insoluble
**Fl.P:** NA
**IP:** NA
**Sp.Gr:** 4.0
**VP:** 0 mmHg (approx)
**MLT:** 3632°F
**UEL:** NA
**LEL:** NA
Noncombustible solid, but dusts may form explosive mixtures in air.

**Personal Protection/Sanitation (see Table 2):**
**Skin:** N.R.
**Eyes:** N.R.
**Wash skin:** N.R.
**Remove:** N.R.
**Change:** N.R.

**Respirator Recommendations (see Tables 3 and 4):**
Not available.

**Incompatibilities and Reactivities:** Chlorine trifluoride, hot chlorinated rubber, acids, oxidizers [**Note:** Hydrogen gas may be formed when finely divided iron contacts moisture during crushing & milling operations.]

**Exposure Routes, Symptoms, Target Organs (see Table 5):**
**ER:** Inh, Ing, Con
**SY:** Irrit eyes, skin, resp sys
**TO:** Eyes, skin, resp sys

**First Aid (see Table 6):**
**Eye:** Irr immed
**Skin:** Blot/brush away
**Breath:** Fresh air
**Swallow:** Medical attention immed

## Aluminum

**Formula:** Al | **CAS#:** 7429-90-5 | **RTECS#:** BD0330000 | **IDLH:** N.D.

**Conversion:**

**DOT:** 1309 170 (powder, coated); 1396 138 (powder, uncoated); 9260 169 (molten)

**Synonyms/Trade Names:** Aluminium, Aluminum metal, Aluminum powder, Elemental aluminum

**Exposure Limits:**
**NIOSH REL:** TWA 10 mg/m$^3$ (total)
TWA 5 mg/m$^3$ (resp)
**OSHA PEL:** TWA 15 mg/m$^3$ (total)
TWA 5 mg/m$^3$ (resp)

**Measurement Methods (see Table 1):**
**NIOSH** 7013, 7300, 7301, 7303
**OSHA** ID121

**Physical Description:** Silvery-white, malleable, ductile, odorless metal.

**Chemical & Physical Properties:**
**MW:** 27.0
**BP:** 4221°F
**Sol:** Insoluble
**Fl.P:** NA
**IP:** NA
**Sp.Gr:** 2.70
**VP:** 0 mmHg (approx)
**MLT:** 1220°F
**UEL:** NA
**LEL:** NA
Combustible Solid, finely divided dust is easily ignited; may cause explosions.

**Personal Protection/Sanitation (see Table 2):**
**Skin:** N.R.
**Eyes:** N.R.
**Wash skin:** N.R.
**Remove:** N.R.
**Change:** N.R.

**Respirator Recommendations (see Tables 3 and 4):**
Not available.

**Incompatibilities and Reactivities:** Strong oxidizers & acids, halogenated hydrocarbons [**Note:** Corrodes in contact with acids & other metals. Ignition may occur if powders are mixed with halogens, carbon disulfide, or methyl chloride.]

**Exposure Routes, Symptoms, Target Organs (see Table 5):**
**ER:** Inh, Con
**SY:** Irrit eyes, skin, resp sys
**TO:** Eyes, skin, resp sys

**First Aid (see Table 6):**
**Eye:** Irr immed
**Breath:** Fresh air

A

| Aluminum (pyro powders and welding fumes, as Al) | Formula: | CAS#: | RTECS#: | IDLH: N.D. |
|---|---|---|---|---|
| **Conversion:** | **DOT:** 1383 135 (powder, pyrophoric) | | | |

**Synonyms/Trade Names:** Synonyms vary depending upon the specific aluminum compound.

**Exposure Limits:**
**NIOSH REL:** TWA 5 $mg/m^3$
**OSHA PEL†:** none

**Measurement Methods (see Table 1):**
**NIOSH** 7300, 7301, 7303

**Physical Description:** Appearance and odor vary depending upon the specific aluminum compound.

**Chemical & Physical Properties:**
Properties vary depending upon the specific aluminum compound.

**Personal Protection/Sanitation (see Table 2):**
**Skin:** N.R.
**Eyes:** N.R.
**Wash skin:** N.R.
**Remove:** N.R.
**Change:** N.R.

**Respirator Recommendations (see Tables 3 and 4):**
Not available.

**Incompatibilities and Reactivities:** Varies

**Exposure Routes, Symptoms, Target Organs (see Table 5):**
**ER:** Inh, Ing, Con
**SY:** Irrit skin, resp sys; pulm fib
**TO:** Skin, resp sys

**First Aid (see Table 6):**
**Eye:** Irr immed
**Skin:** Water flush immed
**Breath:** Resp support
**Swallow:** Medical attention immed

| Aluminum (soluble salts and alkyls, as Al) | Formula: | CAS#: | RTECS#: | IDLH: N.D. |
|---|---|---|---|---|
| **Conversion:** | **DOT:** 3051 135 (Aluminum alkyls) | | | |

**Synonyms/Trade Names:** Synonyms vary depending upon the specific aluminum compound.

**Exposure Limits:**
**NIOSH REL:** TWA 2 $mg/m^3$
**OSHA PEL†:** none

**Measurement Methods (see Table 1):**
**NIOSH** 7013, 7300, 7301, 7303
**OSHA** ID121

**Physical Description:** Appearance and odor vary depending upon the specific aluminum compound.

**Chemical & Physical Properties:**
Properties vary depending upon the specific aluminum compound.

**Personal Protection/Sanitation (see Table 2):**
**Skin:** Prevent skin contact
**Eyes:** Prevent eye contact
**Wash skin:** When contam
**Remove:** When wet or contam
**Change:** Daily

**Respirator Recommendations (see Tables 3 and 4):**
Not available.

**Incompatibilities and Reactivities:** Varies

**Exposure Routes, Symptoms, Target Organs (see Table 5):**
**ER:** Inh, Ing, Con
**SY:** Irrit skin, resp sys; skin burns
**TO:** Skin, resp sys

**First Aid (see Table 6):**
**Eye:** Irr immed
**Skin:** Water flush immed
**Breath:** Resp support
**Swallow:** Medical attention immed

## 4-Aminodiphenyl

| | | | |
|---|---|---|---|
| **Formula:** $C_6H_5C_6H_4NH_2$ | **CAS#:** 92-67-1 | **RTECS#:** DU8925000 | **IDLH:** Ca [N.D.] |

**Conversion:**

**DOT:**

**Synonyms/Trade Names:** 4-Aminobiphenyl, p-Aminobiphenyl, p-Aminodiphenyl, 4-Phenylaniline

**Exposure Limits:**
**NIOSH REL:** Ca
See Appendix A
**OSHA PEL:** [1910.1011] See Appendix B

**Measurement Methods (see Table 1):**
**NIOSH** P&CAM269 (II-4)
**OSHA** 93

**Physical Description:** Colorless crystals with a floral odor.
[**Note:** Turns purple on contact with air.]

**Chemical & Physical Properties:**
**MW:** 169.2
**BP:** 576°F
**Sol:** Slight
**Fl.P:** ?
**IP:** ?
**Sp.Gr:** 1.16
**VP(227°F):** 1 mmHg
**MLT:** 127°F
**UEL:** ?
**LEL:** ?
Combustible Solid, but must be preheated before ignition possible.

**Personal Protection/Sanitation (see Table 2):**
**Skin:** Prevent skin contact
**Eyes:** Prevent eye contact
**Wash skin:** When contam/Daily
**Remove:** When wet or contam
**Change:** Daily
**Provide:** Eyewash
Quick drench

**Respirator Recommendations (see Tables 3 and 4):**
**NIOSH**
**¥:** ScbaF:Pd,Pp/SaF:Pd,Pp:AScba
**Escape:** 100F/ScbaE

**See Appendix E** (page 351)

**Incompatibilities and Reactivities:** Oxidized by air

**Exposure Routes, Symptoms, Target Organs (see Table 5):**
**ER:** Inh, Abs, Ing, Con
**SY:** Head, dizz; drow, dysp; ataxia, lass; methemo; urinary burning; acute hemorrhagic cystitis; [carc]
**TO:** Bladder, skin [bladder cancer]

**First Aid (see Table 6):**
**Eye:** Irr immed
**Skin:** Soap wash immed
**Breath:** Resp support
**Swallow:** Medical attention immed

## 2-Aminopyridine

| | | | |
|---|---|---|---|
| **Formula:** $NH_2C_5H_4N$ | **CAS#:** 504-29-0 | **RTECS#:** US1575000 | **IDLH:** 5 ppm |

**Conversion:** 1 ppm = 3.85 mg/m$^3$

**DOT:** 2671 153

**Synonyms/Trade Names:** α-Aminopyridine, α-Pyridylamine

**Exposure Limits:**
**NIOSH REL:** TWA 0.5 ppm (2 mg/m$^3$)
**OSHA PEL:** TWA 0.5 ppm (2 mg/m$^3$)

**Measurement Methods (see Table 1):**
**NIOSH** S158 (II-4)

**Physical Description:** White powder, leaflets, or crystals with a characteristic odor.

**Chemical & Physical Properties:**
**MW:** 94.1
**BP:** 411°F
**Sol:** >100%
**Fl.P:** 154°F
**IP:** 8.00 eV
**Sp.Gr:** ?
**VP(77°F):** 0.8 mmHg
**MLT:** 137°F
**UEL:** ?
**LEL:** ?
Combustible Solid

**Personal Protection/Sanitation (see Table 2):**
**Skin:** Prevent skin contact
**Eyes:** Prevent eye contact
**Wash skin:** When contam
**Remove:** When wet or contam
**Change:** Daily
**Provide:** Quick drench

**Respirator Recommendations (see Tables 3 and 4):**
**NIOSH/OSHA**
**5 ppm:** Sa*/ScbaF
**§:** ScbaF:Pd,Pp/SaF:Pd,Pp:AScba
**Escape:** GmFOv100/ScbaE

**Incompatibilities and Reactivities:** Strong oxidizers

**Exposure Routes, Symptoms, Target Organs (see Table 5):**
**ER:** Inh, Abs, Ing, Con
**SY:** Irrit eyes, nose, throat; head, dizz; excitement; nau; high BP; resp distress; lass; convuls; stupor
**TO:** CNS, resp sys

**First Aid (see Table 6):**
**Eye:** Irr immed
**Skin:** Water flush immed
**Breath:** Resp support
**Swallow:** Medical attention immed

## Amitrole

**Formula:** $C_2H_4N_4$ | **CAS#:** 61-82-5 | **RTECS#:** XZ3850000 | **IDLH:** Ca [N.D.]

**Conversion:**

**DOT:**

**Synonyms/Trade Names:** Aminotriazole; 3-Aminotriazole; 2-Amino-1,3,4-triazole; 3-Amino-1,2,4-triazole

**Exposure Limits:**
**NIOSH REL:** Ca
TWA 0.2 mg/m$^3$
See Appendix A

**OSHA PEL†:** none

**Measurement Methods (see Table 1):**
**NIOSH** 0500
**OSHA** PV2006

**Physical Description:** Colorless to white, crystalline powder. [herbicide]
[**Note:** Odorless when pure.]

**Chemical & Physical Properties:**
**MW:** 84.1
**BP:** ?
**Sol(77°F):** 28%
**Fl.P:** NA
**IP:** ?
**Sp.Gr:** 1.14
**VP:** <0.000008 mmHg
**MLT:** 318°F
**UEL:** NA
**LEL:** NA
Noncombustible Solid, but may be dissolved in flammable liquids.

**Personal Protection/Sanitation (see Table 2):**
**Skin:** Prevent skin contact
**Eyes:** Prevent eye contact
**Wash skin:** Daily
**Remove:** When wet or contam
**Change:** Daily
**Provide:** Eyewash
Quick drench

**Respirator Recommendations (see Tables 3 and 4):**
**NIOSH**
**¥:** ScbaF:Pd,Pp/SaF:Pd,Pp:AScba
**Escape:** GmFOv100/ScbaE

**Incompatibilities and Reactivities:** Light (decomposes), strong oxidizers
[**Note:** Corrosive to iron, aluminum & copper.]

**Exposure Routes, Symptoms, Target Organs (see Table 5):**
**ER:** Inh, Ing, Con
**SY:** Irrit eyes, skin; dysp, musc spasms, ataxia, anor, salv, incr body temperature; lass, skin dryness, depres (thyroid func suppression)
**TO:** Eyes, skin, thyroid [in animals: liver, thyroid & pituitary gland tumors]

**First Aid (see Table 6):**
**Eye:** Irr immed
**Skin:** Water wash immed
**Breath:** Resp support
**Swallow:** Medical attention immed

## Ammonia

**Formula:** $NH_3$ | **CAS#:** 7664-41-7 | **RTECS#:** BO0875000 | **IDLH:** 300 ppm

**Conversion:** 1 ppm = 0.70 mg/m$^3$

**DOT:** 1005 125 (anhydrous); 2672 154 (10-35% solution); 2073 125 (>35-50% solution); 1005 125 (>50% solution)

**Synonyms/Trade Names:** Anhydrous ammonia, Aqua ammonia, Aqueous ammonia
[**Note:** Often used in an aqueous solution.]

**Exposure Limits:**
**NIOSH REL:** TWA 25 ppm (18 mg/m$^3$)
ST 35 ppm (27 mg/m$^3$)

**OSHA PEL†:** TWA 50 ppm (35 mg/m$^3$)

**Measurement Methods (see Table 1):**
**NIOSH** 3800, 6015, 6016
**OSHA** ID188

**Physical Description:** Colorless gas with a pungent, suffocating odor.
[**Note:** Shipped as a liquefied compressed gas. Easily liquefied under pressure.]

**Chemical & Physical Properties:**
**MW:** 17.0
**BP:** -28°F
**Sol:** 34%
**Fl.P:** NA (Gas)
**IP:** 10.18 eV
**RGasD:** 0.60
**VP:** 8.5 atm
**FRZ:** -108°F
**UEL:** 28%
**LEL:** 15%
[**Note:** Although $NH_3$ does not meet the DOT definition of a Flammable Gas (for labeling purposes), it should be treated as one.]

**Personal Protection/Sanitation (see Table 2):**
**Skin:** Prevent skin contact
**Eyes:** Prevent eye contact
**Wash skin:** When contam (solution)
**Remove:** When wet or contam (solution)
**Change:** N.R.
**Provide:** Eyewash (>10%)
Quick drench (>10%)

**Respirator Recommendations (see Tables 3 and 4):**
**NIOSH**
**250 ppm:** CcrS*/Sa*
**300 ppm:** Sa:Cf*/PaprS*/CcrFS/GmFS/ScbaF/SaF
**§:** ScbaF:Pd,Pp/SaF:Pd,Pp:AScba
**Escape:** GmFS/ScbaE

**Incompatibilities and Reactivities:** Strong oxidizers, acids, halogens, salts of silver & zinc
[**Note:** Corrosive to copper & galvanized surfaces.]

**Exposure Routes, Symptoms, Target Organs (see Table 5):**
**ER:** Inh, Ing (solution), Con (solution/liquid)
**SY:** Irrit eyes, nose, throat; dysp, wheez, chest pain; pulm edema; pink frothy sputum; skin burns, vesic; liquid: frostbite
**TO:** Eyes, skin, resp sys

**First Aid (see Table 6):**
**Eye:** Irr immed (solution/liquid)
**Skin:** Water flush immed (solution/liquid)
**Breath:** Resp support
**Swallow:** Medical attention immed (solution)

## Ammonium chloride fume

| Formula: | CAS#: | RTECS#: | IDLH: |
|---|---|---|---|
| $NH_4Cl$ | 12125-02-9 | BP4550000 | N.D. |

**Conversion:**

**DOT:**

**Synonyms/Trade Names:** Ammonium chloride, Ammonium muriate fume, Sal ammoniac fume

**Exposure Limits:**
**NIOSH REL:** TWA 10 mg/m$^3$
ST 20 mg/m$^3$
**OSHA PEL†:** none

**Measurement Methods (see Table 1):**
**OSHA** ID188

**Physical Description:** Finely divided, odorless, white particulate dispersed in air.

**Chemical & Physical Properties:**
**MW:** 53.5
**BP:** Sublimes
**Sol:** 37%
**Fl.P:** NA
**IP:** NA
**Sp.Gr:** 1.53
**VP(321°F):** 1 mmHg
**MLT:** 662°F (Sublimes)
**UEL:** NA
**LEL:** NA
Noncombustible Solid

**Personal Protection/Sanitation (see Table 2):**
**Skin:** Prevent skin contact
**Eyes:** Prevent eye contact
**Wash skin:** When contam
**Remove:** When wet or contam
**Change:** Daily
**Provide:** Eyewash
Quick drench

**Respirator Recommendations (see Tables 3 and 4):**
Not available.

**Incompatibilities and Reactivities:** Alkalis & their carbonates, lead & silver salts, strong oxidizers, ammonium nitrate, potassium chlorate, bromine trifluoride [**Note:** Corrodes most metals at high (i.e., fire) temperatures.]

**Exposure Routes, Symptoms, Target Organs (see Table 5):**
**ER:** Inh, Con
**SY:** Irrit eyes, skin, resp sys; cough, dysp, pulm sens
**TO:** Eyes, skin, resp sys

**First Aid (see Table 6):**
**Eye:** Irr immed
**Skin:** Soap wash immed
**Breath:** Resp support

## Ammonium sulfamate

| Formula: | CAS#: | RTECS#: | IDLH: |
|---|---|---|---|
| $NH_4OSO_2NH_2$ | 7773-06-0 | WO6125000 | 1500 mg/m$^3$ |

**Conversion:**

**DOT:**

**Synonyms/Trade Names:** Ammate herbicide, Ammonium amidosulfonate, AMS, Monoammonium salt of sulfamic acid, Sulfamate

**Exposure Limits:**
**NIOSH REL:** TWA 10 mg/m$^3$ (total)
TWA 5 mg/m$^3$ (resp)
**OSHA PEL†:** TWA 15 mg/m$^3$ (total)
TWA 5 mg/m$^3$ (resp)

**Measurement Methods (see Table 1):**
**NIOSH** S348 (II-5)

**Physical Description:** Colorless to white crystalline, odorless solid. [herbicide]

**Chemical & Physical Properties:**
**MW:** 114.1
**BP:** 320°F (Decomposes)
**Sol:** 200%
**Fl.P:** NA
**IP:** ?
**Sp.Gr:** 1.77
**VP:** 0 mmHg (approx)
**MLT:** 268°F
**UEL:** NA
**LEL:** NA
Noncombustible Solid

**Personal Protection/Sanitation (see Table 2):**
**Skin:** N.R.
**Eyes:** N.R.
**Wash skin:** N.R.
**Remove:** N.R.
**Change:** N.R.

**Respirator Recommendations (see Tables 3 and 4):**
**NIOSH**
**50 mg/m$^3$:** Qm
**100 mg/m$^3$:** 95XQ/Sa
**250 mg/m$^3$:** Sa:Cf/PaprHie
**500 mg/m$^3$:** SaT:Cf/PaprTHie/100F/ScbaF/SaF
**1500 mg/m$^3$:** Sa:Pd,Pp
**§:** ScbaF:Pd,Pp/SaF:Pd,Pp:AScba
**Escape:** 100F/ScbaE

**Incompatibilities and Reactivities:** Acids, hot water
[**Note:** Elevated temperatures cause a highly exothermic reaction with water.]

**Exposure Routes, Symptoms, Target Organs (see Table 5):**
**ER:** Inh, Con
**SY:** Irrit eyes, nose, throat; cough, dysp
**TO:** Eyes, resp sys

**First Aid (see Table 6):**
**Eye:** Irr immed
**Skin:** Soap wash prompt
**Breath:** Resp support
**Swallow:** Medical attention immed

## n-Amyl acetate

| n-Amyl acetate | Formula: $CH_3COO[CH_2]_4CH_3$ | CAS#: 628-63-7 | RTECS#: AJ9250000 | IDLH: 1000 ppm |
|---|---|---|---|---|
| **Conversion:** 1 ppm = 5.33 mg/m$^3$ | **DOT:** 1104 129 | | | |

**Synonyms/Trade Names:** Amyl acetic ester, Amyl acetic ether, 1-Pentanol acetate, Pentyl ester of acetic acid, Primary amyl acetate

**Exposure Limits:**
**NIOSH REL:** TWA 100 ppm (525 mg/m$^3$)
**OSHA PEL:** TWA 100 ppm (525 mg/m$^3$)

**Measurement Methods (see Table 1):**
**NIOSH** 1450, 2549
**OSHA** 7

**Physical Description:** Colorless liquid with a persistent banana-like odor.

| Chemical & Physical Properties: | Personal Protection/Sanitation (see Table 2): | Respirator Recommendations (see Tables 3 and 4): |
|---|---|---|
| **MW:** 130.2<br>**BP:** 301°F<br>**Sol:** 0.2%<br>**Fl.P:** 77°F<br>**IP:** ?<br>**Sp.Gr:** 0.88<br>**VP:** 4 mmHg<br>**FRZ:** -95°F<br>**UEL:** 7.5%<br>**LEL:** 1.1%<br>Class IC Flammable Liquid | **Skin:** Prevent skin contact<br>**Eyes:** Prevent eye contact<br>**Wash skin:** When contam<br>**Remove:** When wet (flamm)<br>**Change:** N.R. | **NIOSH/OSHA**<br>**1000 ppm:** CcrOv*/GmFOv/PaprOv*/ Sa*/ScbaF<br>**§:** ScbaF:Pd,Pp/SaF:Pd,Pp:AScba<br>**Escape:** GmFOv/ScbaE |

**Incompatibilities and Reactivities:** Nitrates; strong oxidizers, alkalis & acids

| Exposure Routes, Symptoms, Target Organs (see Table 5): | First Aid (see Table 6): |
|---|---|
| **ER:** Inh, Ing, Con<br>**SY:** Irrit eyes, nose; derm; possible CNS depres, narco<br>**TO:** Eyes, skin, resp sys, CNS | **Eye:** Irr immed<br>**Skin:** Water flush prompt<br>**Breath:** Resp support<br>**Swallow:** Medical attention immed |

## sec-Amyl acetate

| sec-Amyl acetate | Formula: $CH_3COOCH(CH_3)C_3H_7$ | CAS#: 626-38-0 | RTECS#: AJ2100000 | IDLH: 1000 ppm |
|---|---|---|---|---|
| **Conversion:** 1 ppm = 5.33 mg/m$^3$ | **DOT:** 1104 129 | | | |

**Synonyms/Trade Names:** 1-Methylbutyl acetate, 2-Pentanol acetate, 2-Pentyl ester of acetic acid

**Exposure Limits:**
**NIOSH REL:** TWA 125 ppm (650 mg/m$^3$)
**OSHA PEL:** TWA 125 ppm (650 mg/m$^3$)

**Measurement Methods (see Table 1):**
**NIOSH** 1450, 2549
**OSHA** 7

**Physical Description:** Colorless liquid with a mild odor.

| Chemical & Physical Properties: | Personal Protection/Sanitation (see Table 2): | Respirator Recommendations (see Tables 3 and 4): |
|---|---|---|
| **MW:** 130.2<br>**BP:** 249°F<br>**Sol:** Slight<br>**Fl.P:** 89°F<br>**IP:** ?<br>**Sp.Gr:** 0.87<br>**VP:** 7 mmHg<br>**FRZ:** -109°F<br>**UEL:** 7.5%<br>**LEL:** 1%<br>Class IC Flammable Liquid | **Skin:** Prevent skin contact<br>**Eyes:** Prevent eye contact<br>**Wash skin:** When contam<br>**Remove:** When wet (flamm)<br>**Change:** N.R. | **NIOSH/OSHA**<br>**1000 ppm:** CcrOv*/GmFOv/PaprOv*/ Sa*/ScbaF<br>**§:** ScbaF:Pd,Pp/SaF:Pd,Pp:AScba<br>**Escape:** GmFOv/ScbaE |

**Incompatibilities and Reactivities:** Nitrates; strong oxidizers, alkalis & acids

| Exposure Routes, Symptoms, Target Organs (see Table 5): | First Aid (see Table 6): |
|---|---|
| **ER:** Inh, Ing, Con<br>**SY:** Irrit eyes, skin, nose; narco; derm; possible kidney, liver inj; possible CNS depres<br>**TO:** Eyes, skin, resp sys, kidneys, liver, CNS | **Eye:** Irr immed<br>**Skin:** Water flush prompt<br>**Breath:** Resp support<br>**Swallow:** Medical attention immed |

A

## Aniline (and homologs)

| Aniline (and homologs) | Formula: $C_6H_5NH_2$ | CAS#: 62-53-3 | RTECS#: BW6650000 | IDLH: Ca [100 ppm] |
|---|---|---|---|---|
| **Conversion:** 1 ppm = 3.81 mg/m$^3$ | **DOT:** 1547 153 | | | |

**Synonyms/Trade Names:** Aminobenzene, Aniline oil, Benzenamine, Phenylamine

**Exposure Limits:**
**NIOSH REL:** Ca
See Appendix A
**OSHA PEL†:** TWA 5 ppm (19 mg/m$^3$) [skin]

**Measurement Methods (see Table 1):**
**NIOSH** 2002, 2017, 8317
**OSHA** PV2079

**Physical Description:** Colorless to brown, oily liquid with an aromatic amine-like odor. [**Note:** A solid below 21°F.]

**Chemical & Physical Properties:**
**MW:** 93.1
**BP:** 363°F
**Sol:** 4%
**Fl.P:** 158°F
**IP:** 7.70 eV
**Sp.Gr:** 1.02
**VP:** 0.6 mmHg
**FRZ:** 21°F
**UEL:** 11%
**LEL:** 1.3%
Class IIIA Combustible Liquid

**Personal Protection/Sanitation (see Table 2):**
**Skin:** Prevent skin contact
**Eyes:** Prevent eye contact
**Wash skin:** When contam
**Remove:** When wet or contam
**Change:** N.R.
**Provide:** Quick drench

**Respirator Recommendations (see Tables 3 and 4):**
**NIOSH**
**¥:** ScbaF:Pd,Pp/SaF:Pd,Pp:AScba
**Escape:** GmFOv/ScbaE

**Incompatibilities and Reactivities:** Strong oxidizers, strong acids, toluene diisocyanate, alkalis

**Exposure Routes, Symptoms, Target Organs (see Table 5):**
**ER:** Inh, Abs, Ing, Con
**SY:** Head, lass, dizz; cyan; ataxia; dysp on effort; tacar; irrit eyes; methemo; cirr; [carc]
**TO:** Blood, CVS, eyes, liver, kidneys, resp sys [bladder cancer]

**First Aid (see Table 6):**
**Eye:** Irr immed
**Skin:** Soap wash prompt
**Breath:** Resp support
**Swallow:** Medical attention immed

## o-Anisidine

| o-Anisidine | Formula: $NH_2C_6H_4OCH_3$ | CAS#: 90-04-0 | RTECS#: BZ5410000 | IDLH: Ca [50 mg/m$^3$] |
|---|---|---|---|---|
| **Conversion:** | **DOT:** 2431 153 | | | |

**Synonyms/Trade Names:** ortho-Aminoanisole, 2-Anisidine, o-Methoxyaniline
[**Note:** o-Anisidine has been used as a basis for many dyes.]

**Exposure Limits:**
**NIOSH REL:** Ca
0.5 mg/m$^3$ [skin]
See Appendix A
**OSHA PEL:** TWA 0.5 mg/m$^3$ [skin]

**Measurement Methods (see Table 1):**
**NIOSH** 2514

**Physical Description:** Red or yellow, oily liquid with an amine-like odor. [**Note:** A solid below 41°F.]

**Chemical & Physical Properties:**
**MW:** 123.2
**BP:** 437°F
**Sol(77°F):** 1%
**Fl.P(oc):** 244°F
**IP:** 7.44 eV
**Sp.Gr:** 1.10
**VP:** <0.1 mmHg
**FRZ:** 41°F
**UEL:** ?
**LEL:** ?
Class IIIB Combustible Liquid

**Personal Protection/Sanitation (see Table 2):**
**Skin:** Prevent skin contact
**Eyes:** Prevent eye contact
**Wash skin:** When contam
**Remove:** When wet or contam
**Change:** Daily
**Provide:** Eyewash
Quick drench

**Respirator Recommendations (see Tables 3 and 4):**
**NIOSH**
**¥:** ScbaF:Pd,Pp/SaF:Pd,Pp:AScba
**Escape:** GmFOv/ScbaE

**Incompatibilities and Reactivities:** Strong oxidizers

**Exposure Routes, Symptoms, Target Organs (see Table 5):**
**ER:** Inh, Abs, Ing, Con
**SY:** Head, dizz; cyan; RBC Heinz bodies; [carc]
**TO:** Blood, kidneys, liver, CVS, CNS [in animals: tumors of the thyroid gland, bladder & kidneys]

**First Aid (see Table 6):**
**Eye:** Irr immed
**Skin:** Soap wash immed
**Breath:** Resp support
**Swallow:** Medical attention immed

## p-Anisidine

| p-Anisidine | Formula: $NH_2C_6H_4OCH_3$ | CAS#: 104-94-9 | RTECS#: BZ5450000 | IDLH: 50 mg/m³ |
|---|---|---|---|---|
| Conversion: | DOT: 2431 153 | | | |

**Synonyms/Trade Names:** para-Aminoanisole, 4-Anisidine, p-Methoxyaniline

**Exposure Limits:**
**NIOSH REL:** TWA 0.5 mg/m³ [skin]
**OSHA PEL:** TWA 0.5 mg/m³ [skin]

**Measurement Methods (see Table 1):**
**NIOSH** 2514

**Physical Description:** Yellow to brown, crystalline solid with an amine-like odor.

| Chemical & Physical Properties: | Personal Protection/Sanitation (see Table 2): | Respirator Recommendations (see Tables 3 and 4): |
|---|---|---|
| **MW:** 123.2<br>**BP:** 475°F<br>**Sol:** Moderate<br>**Fl.P:** ?<br>**IP:** 7.44 eV<br>**Sp.Gr:** 1.07<br>**VP(77°F):** 0.006 mmHg<br>**MLT:** 135°F<br>**UEL:** ?<br>**LEL:** ?<br>Combustible Solid | **Skin:** Prevent skin contact<br>**Eyes:** Prevent eye contact<br>**Wash skin:** When contam<br>**Remove:** When wet or contam<br>**Change:** Daily<br>**Provide:** Quick drench | **NIOSH/OSHA**<br>**5 mg/m³:** 95XQ/Sa<br>**12.5 mg/m³:** Sa:Cf/PaprHie<br>**25 mg/m³:** 100F/PaprTHie*/ScbaF/SaF<br>**50 mg/m³:** Sa:Pd,Pp*<br>**§:** ScbaF:Pd,Pp/SaF:Pd,Pp:AScba<br>**Escape:** 100F/ScbaE |

**Incompatibilities and Reactivities:** Strong oxidizers

| Exposure Routes, Symptoms, Target Organs (see Table 5): | First Aid (see Table 6): |
|---|---|
| **ER:** Inh, Abs, Ing, Con<br>**SY:** Head, dizz; cyan; RBC Heinz bodies<br>**TO:** Blood, kidneys, liver, CVS, CNS | **Eye:** Irr immed<br>**Skin:** Soap wash immed<br>**Breath:** Resp support<br>**Swallow:** Medical attention immed |

## Antimony

| Antimony | Formula: Sb | CAS#: 7440-36-0 | RTECS#: CC4025000 | IDLH: 50 mg/m³ (as Sb) |
|---|---|---|---|---|
| Conversion: | DOT: 1549 157 (inorganic compounds, n.o.s.); 2871 170 (powder); 3141 157 (inorganic liquid compounds, n.o.s.) | | | |

**Synonyms/Trade Names:** Antimony metal, Antimony powder, Stibium

**Exposure Limits:**
**NIOSH REL*:** TWA 0.5 mg/m³
**OSHA PEL*:** TWA 0.5 mg/m³
[***Note:** The REL and PEL also apply to other antimony compounds (as Sb).]

**Measurement Methods (see Table 1):**
**NIOSH** 7301, 7303, P&CAM 261 (II-4)
**OSHA** ID121, ID125G, ID206

**Physical Description:** Silver-white, lustrous, hard, brittle solid; scale-like crystals; or a dark-gray, lustrous powder.

| Chemical & Physical Properties: | Personal Protection/Sanitation (see Table 2): | Respirator Recommendations (see Tables 3 and 4): |
|---|---|---|
| **MW:** 121.8<br>**BP:** 2975°F<br>**Sol:** Insoluble<br>**Fl.P:** NA<br>**IP:** NA<br>**Sp.Gr:** 6.69<br>**VP:** 0 mmHg (approx)<br>**MLT:** 1166°F<br>**UEL:** NA<br>**LEL:** NA | **Skin:** Prevent skin contact<br>**Eyes:** Prevent eye contact<br>**Wash skin:** When contam<br>**Remove:** When wet or contam<br>**Change:** Daily | **NIOSH/OSHA**<br>**5 mg/m³:** 95XQ/Sa<br>**12.5 mg/m³:** Sa:Cf/PaprHie<br>**25 mg/m³:** 100F/SaT:Cf/PaprTHie/ScbaF/SaF<br>**50 mg/m³:** Sa:Pd,Pp<br>**§:** ScbaF:Pd,Pp/SaF:Pd,Pp:AScba<br>**Escape:** 100F/ScbaE |

Noncombustible Solid in bulk form, but a moderate explosion hazard in the form of dust when exposed to flame.

**Incompatibilities and Reactivities:** Strong oxidizers, acids, halogenated acids
[**Note:** Stibine is formed when antimony is exposed to nascent (freshly formed) hydrogen.]

| Exposure Routes, Symptoms, Target Organs (see Table 5): | First Aid (see Table 6): |
|---|---|
| **ER:** Inh, Ing, Con<br>**SY:** Irrit eyes, skin, nose, throat, mouth; cough; dizz; head; nau, vomit, diarr; stomach cramps; insom; anor; unable to smell properly<br>**TO:** Eyes, skin, resp sys, CVS | **Eye:** Irr immed<br>**Skin:** Soap wash immed<br>**Breath:** Resp support<br>**Swallow:** Medical attention immed |

## ANTU

| ANTU | Formula: $C_{10}H_7NHC(NH_2)S$ | CAS#: 86-88-4 | RTECS#: YT9275000 | IDLH: 100 mg/m$^3$ |
|---|---|---|---|---|
| **Conversion:** | **DOT:** 1651 153 | | | |

**Synonyms/Trade Names:** α-Naphthyl thiocarbamide, 1-Naphthyl thiourea, α-Naphthyl thiourea

**Exposure Limits:**
**NIOSH REL:** TWA 0.3 mg/m$^3$
**OSHA PEL:** TWA 0.3 mg/m$^3$

**Measurement Methods (see Table 1):**
**NIOSH** S276 (II-5)

**Physical Description:** White crystalline or gray, odorless powder. [rodenticide]

**Chemical & Physical Properties:**
**MW:** 202.3
**BP:** Decomposes
**Sol:** 0.06%
**Fl.P:** NA
**IP:** ?
**Sp.Gr:** ?
**VP:** Low
**MLT:** 388°F
**UEL:** NA
**LEL:** NA
Noncombustible Solid

**Personal Protection/Sanitation (see Table 2):**
**Skin:** N.R.
**Eyes:** N.R.
**Wash skin:** N.R.
**Remove:** N.R.
**Change:** Daily

**Respirator Recommendations (see Tables 3 and 4):**
**NIOSH/OSHA**
**3 mg/m$^3$:** CcrOv95/Sa
**7.5 mg/m$^3$:** Sa:Cf/PaprOvHie
**15 mg/m$^3$:** CcrFOv100/GmFOv100/PaprTOvHie/SaT:Cf/ScbaF/SaF
**100 mg/m$^3$:** Sa:Pd,Pp
**§:** ScbaF:Pd,Pp/SaF:Pd,Pp:AScba
**Escape:** GmFOv100/ScbaE

**Incompatibilities and Reactivities:** Strong oxidizers, silver nitrate

**Exposure Routes, Symptoms, Target Organs (see Table 5):**
**ER:** Inh, Ing
**SY:** After ingestion of large doses: vomit, dysp, cyan, coarse pulm rales; liver damage
**TO:** Resp sys, blood, liver

**First Aid (see Table 6):**
**Eye:** Irr immed
**Skin:** Soap wash prompt
**Breath:** Resp support
**Swallow:** Medical attention immed

## Arsenic (inorganic compounds, as As)

| Arsenic (inorganic compounds, as As) | Formula: As (metal) | CAS#: 7440-38-2 (metal) | RTECS#: CG0525000 (metal) | IDLH: Ca [5 mg/m$^3$ (as As)] |
|---|---|---|---|---|
| **Conversion:** | **DOT:** 1558 152 (metal); 1562 152 (dust) | | | |

**Synonyms/Trade Names:** **Arsenic metal:** Arsenia
Other synonyms vary depending upon the specific As compound. [**Note:** OSHA considers "Inorganic Arsenic" to mean copper acetoarsenite & all inorganic compounds containing arsenic except ARSINE.]

**Exposure Limits:**
**NIOSH REL:** Ca
C 0.002 mg/m$^3$ [15-minute]
See Appendix A
**OSHA PEL:** [1910.1018] TWA 0.010 mg/m$^3$

**Measurement Methods (see Table 1):**
**NIOSH** 7300, 7301, 7303, 9102, 7900
**OSHA** ID105

**Physical Description:** Metal: Silver-gray or tin-white, brittle, odorless solid.

**Chemical & Physical Properties:**
**MW:** 74.9
**BP:** Sublimes
**Sol:** Insoluble
**Fl.P:** NA
**IP:** NA
**Sp.Gr:** 5.73 (metal)
**VP:** 0 mmHg (approx)
**MLT:** 1135°F (Sublimes)
**UEL:** NA
**LEL:** NA

**Personal Protection/Sanitation (see Table 2):**
**Skin:** Prevent skin contact
**Eyes:** Prevent eye contact
**Wash skin:** When contam/Daily
**Remove:** When wet or contam
**Change:** Daily
**Provide:** Eyewash
Quick drench

**Respirator Recommendations (see Tables 3 and 4):**
**NIOSH**
**¥:** ScbaF:Pd,Pp/SaF:Pd,Pp:AScba
**Escape:** GmFAg100/ScbaE

**See Appendix E** (page 351)

Metal: Noncombustible Solid in bulk form, but a slight explosion hazard in the form of dust when exposed to flame.

**Incompatibilities and Reactivities:** Strong oxidizers, bromine azide
[**Note:** Hydrogen gas can react with inorganic arsenic to form the highly toxic gas arsine.]

**Exposure Routes, Symptoms, Target Organs (see Table 5):**
**ER:** Inh, Abs, Con, Ing
**SY:** Ulceration of nasal septum, derm, GI disturbances, peri neur, resp irrit, hyperpig of skin, [carc]
**TO:** Liver, kidneys, skin, lungs, lymphatic sys [lung & lymphatic cancer]

**First Aid (see Table 6):**
**Eye:** Irr immed
**Skin:** Soap wash immed
**Breath:** Resp support
**Swallow:** Medical attention immed

| Arsenic (organic compounds, as As) | Formula: | CAS#: | RTECS#: | IDLH: N.D. |
|---|---|---|---|---|
| **Conversion:** | **DOT:** | | | |

**Synonyms/Trade Names:** Synonyms vary depending upon the specific organic arsenic compound.

**Exposure Limits:**
**NIOSH REL:** none
**OSHA PEL:** TWA 0.5 mg/m$^3$

**Measurement Methods (see Table 1):**
**NIOSH** 5022

**Physical Description:** Appearance and odor vary depending upon the specific organic arsenic compound.

**Chemical & Physical Properties:**
Properties vary depending upon the specific organic arsenic compound.

**Personal Protection/Sanitation (see Table 2):**
Recommendations regarding personal protective clothing vary depending upon the specific compound.

**Respirator Recommendations (see Tables 3 and 4):**
Not available.

**Incompatibilities and Reactivities:** Varies

**Exposure Routes, Symptoms, Target Organs (see Table 5):**
**ER:** Inh, Ing, Con
**SY:** In animals: irrit skin, possible derm; resp distress; diarr; kidney damage; musc tremor, convuls; possible GI tract, repro effects; possible liver damage
**TO:** Skin, resp sys, kidneys, CNS, liver, GI tract, repro sys

**First Aid (see Table 6):**
**Eye:** Irr immed
**Skin:** Soap wash immed
**Breath:** Resp support
**Swallow:** Medical attention immed

| Arsine | Formula: $AsH_3$ | CAS#: 7784-42-1 | RTECS#: CG6475000 | IDLH: Ca [3 ppm] |
|---|---|---|---|---|
| **Conversion:** 1 ppm = 3.19 mg/m$^3$ | **DOT:** 2188 119 | | | |

**Synonyms/Trade Names:** Arsenic hydride, Arsenic trihydride, Arseniuretted hydrogen, Arsenous hydride, Hydrogen arsenide

**Exposure Limits:**
**NIOSH REL:** Ca
C 0.002 mg/m$^3$ [15-minute]
See Appendix A
**OSHA PEL:** TWA 0.05 ppm (0.2 mg/m$^3$)

**Measurement Methods (see Table 1):**
**NIOSH** 6001
**OSHA** ID105

**Physical Description:** Colorless gas with a mild, garlic-like odor.
[**Note:** Shipped as a liquefied compressed gas.]

**Chemical & Physical Properties:**
**MW:** 78.0
**BP:** -81°F
**Sol:** 20%
**Fl.P:** NA (Gas)
**IP:** 9.89 eV
**RGasD:** 2.69
**VP(70°F):** 14.9 atm
**FRZ:** -179°F
**UEL:** 78%
**LEL:** 5.1%
Flammable Gas

**Personal Protection/Sanitation (see Table 2):**
**Skin:** Frostbite
**Eyes:** Frostbite
**Wash skin:** N.R.
**Remove:** When wet (flamm)
**Change:** N.R.
**Provide:** Frostbite wash

**Respirator Recommendations (see Tables 3 and 4):**
**NIOSH**
**¥:** ScbaF:Pd,Pp/SaF:Pd,Pp:AScba
**Escape:** GmFS/ScbaE

**Incompatibilities and Reactivities:** Strong oxidizers, chlorine, nitric acid
[**Note:** Decomposes above 446°F. There is a high potential for the generation of arsine gas when inorganic arsenic is exposed to nascent (freshly formed) hydrogen.]

**Exposure Routes, Symptoms, Target Organs (see Table 5):**
**ER:** Inh, Con (liquid)
**SY:** Head, mal, lass, dizz; dysp; abdom, back pain; nau, vomit; bronze skin; hema; jaun; peri neur; liquid: frostbite; [carc]
**TO:** Blood, kidneys, liver [lung & lymphatic cancer]

**First Aid (see Table 6):**
**Eye:** Frostbite
**Skin:** Frostbite
**Breath:** Resp support

## Asbestos

| Asbestos | Formula: Hydrated mineral silicates | CAS#: 1332-21-4 | RTECS#: CI6475000 | IDLH: Ca [N.D.] |
|---|---|---|---|---|
| **Conversion:** | **DOT:** 2212 171 (blue, brown); 2590 171 (white) | | | |

**Synonyms/Trade Names:** Actinolite, Actinolite asbestos, Amosite (cummingtonite-grunerite), Anthophyllite, Anthophyllite asbestos, Chrysotile, Crocidolite (Riebeckite), Tremolite, Tremolite asbestos

**Exposure Limits:**
**NIOSH REL:** Ca
See Appendix A
See Appendix C
**OSHA PEL:** [1910.1001] [1926.1101] See Appendix C

**Measurement Methods (see Table 1):**
**NIOSH** 7400, 7402
**OSHA** ID160, ID191

**Physical Description:** White or greenish (chrysotile), blue (crocidolite), or gray-green (amosite) fibrous, odorless solids.

**Chemical & Physical Properties:**
**MW:** Varies
**BP:** Decomposes
**Sol:** Insoluble
**Fl.P:** NA
**IP:** NA
**Sp.Gr:** ?
**VP:** 0 mmHg (approx)
**MLT:** 1112°F (Decomposes)
**UEL:** NA
**LEL:** NA
Noncombustible Solids

**Personal Protection/Sanitation (see Table 2):**
**Skin:** Prevent skin contact
**Eyes:** Prevent eye contact
**Wash skin:** Daily
**Remove:** N.R.
**Change:** Daily

**Respirator Recommendations (see Tables 3 and 4):**
**NIOSH**
**¥:** ScbaF:Pd,Pp/SaF:Pd,Pp:AScba
**Escape:** 100F/ScbaE

**See Appendix E** (page 351)

**Incompatibilities and Reactivities:** None reported

**Exposure Routes, Symptoms, Target Organs (see Table 5):**
**ER:** Inh, Ing, Con
**SY:** Asbestosis (chronic exposure): dysp, interstitial fib, restricted pulm function, finger clubbing; irrit eyes; [carc]
**TO:** Resp sys, eyes [lung cancer]

**First Aid (see Table 6):**
**Eye:** Irr immed
**Breath:** Fresh air

## Asphalt fumes

| Asphalt fumes | Formula: | CAS#: 8052-42-4 | RTECS#: CI9900000 | IDLH: Ca [N.D.] |
|---|---|---|---|---|
| **Conversion:** | **DOT:** 1999 130 (asphalt) | | | |

**Synonyms/Trade Names: Asphalt:** Asphaltum, Bitumen (European term), Petroleum asphalt, Petroleum bitumen, Road asphalt, Roofing asphalt

**Exposure Limits:**
**NIOSH REL:** Ca
C 5 mg/m$^3$ [15-minute]
See Appendix A
See Appendix C
**OSHA PEL:** none

**Measurement Methods (see Table 1):**
**NIOSH** 5042

**Physical Description:** Fumes generated during the production or application of asphalt (a dark-brown to black cement-like substance manufactured by the vacuum distillation of crude petroleum oil).

**Chemical & Physical Properties:**
Properties vary depending upon the specific asphalt formulation or mixture.

Asphalt: Combustible Solid

**Personal Protection/Sanitation (see Table 2):**
**Skin:** Prevent skin contact
**Eyes:** Prevent eye contact
**Wash skin:** Daily
**Remove:** N.R.
**Change:** Daily

**Respirator Recommendations (see Tables 3 and 4):**
**NIOSH**
**¥:** ScbaF:Pd,Pp/SaF:Pd,Pp:AScba
**Escape:** GmFOv100/ScbaE

**Incompatibilities and Reactivities:** None reported [**Note:** Asphalt becomes molten at about 200°F.]

**Exposure Routes, Symptoms, Target Organs (see Table 5):**
**ER:** Inh, Abs, Con
**SY:** Irrit eyes, resp sys; [carc]
**TO:** Eyes, resp sys [in animals: skin tumors]

**First Aid (see Table 6):**
**Eye:** Irr immed
**Breath:** Resp support

## Atrazine

| Formula: | CAS#: | RTECS#: | IDLH: |
|---|---|---|---|
| $C_8H_{14}ClN_5$ | 1912-24-9 | XY5600000 | N.D. |

**Conversion:**

**DOT:** 2763 151 (triazine pesticide)

**Synonyms/Trade Names:** 2-Chloro-4-ethylamino-6-isopropylamino-s-triazine; 6-Chloro-N-ethyl-N'-(1-methylethyl)-1,3,5-triazine-2,4-diamine

**Exposure Limits:**
**NIOSH REL:** TWA 5 mg/m$^3$
**OSHA PEL†:** none

**Measurement Methods (see Table 1):**
**NIOSH** 5602, 8315

**Physical Description:** Colorless or white, odorless, crystalline powder. [herbicide]

**Chemical & Physical Properties:**
**MW:** 215.7
**BP:** Decomposes
**Sol:** 0.003%
**Fl.P:** NA
**IP:** NA
**Sp.Gr:** 1.19
**VP:** 0.0000003 mmHg
**MLT:** 340°F
**UEL:** NA
**LEL:** NA
Noncombustible Solid, but may be mixed with flammable liquids.

**Personal Protection/Sanitation (see Table 2):**
**Skin:** Prevent skin contact
**Eyes:** Prevent eye contact
**Wash skin:** When contam
**Remove:** When wet or contam
**Change:** Daily
**Provide:** Eyewash
Quick drench

**Respirator Recommendations (see Tables 3 and 4):**
Not available.

**Incompatibilities and Reactivities:** Strong acids, strong bases

**Exposure Routes, Symptoms, Target Organs (see Table 5):**
**ER:** Inh, Ing, Con
**SY:** Irrit eyes, skin; derm, sens skin; dysp, lass, inco, salv; hypothermia; liver inj
**TO:** Eyes, skin, resp sys, CNS, liver

**First Aid (see Table 6):**
**Eye:** Irr immed
**Skin:** Soap wash immed
**Breath:** Resp support
**Swallow:** Medical attention immed

## Azinphos-methyl

| Formula: | CAS#: | RTECS#: | IDLH: |
|---|---|---|---|
| $C_{10}H_{12}O_3PS_2N_3$ [$(CH_3O)_2P(S)SCH_2(N_3C_7H_4O)$] | 86-50-0 | TE1925000 | 10 mg/m$^3$ |

**Conversion:**

**DOT:** 2783 152 (organophosphorus pesticide, solid, toxic)

**Synonyms/Trade Names:** O,O-Dimethyl-S-4-oxo-1,2,3-benzotriazin-3(4H)-ylmethyl phosphorodithioate; Guthion®; Methyl azinphos

**Exposure Limits:**
**NIOSH REL:** TWA 0.2 mg/m$^3$ [skin]
**OSHA PEL:** TWA 0.2 mg/m$^3$ [skin]

**Measurement Methods (see Table 1):**
**NIOSH** 5600
**OSHA** PV2087

**Physical Description:** Colorless crystals or a brown, waxy solid. [insecticide]

**Chemical & Physical Properties:**
**MW:** 317.3
**BP:** Decomposes
**Sol:** 0.003%
**Fl.P:** NA
**IP:** ?
**Sp.Gr:** 1.44
**VP:** 8 x 10$^{-9}$ mmHg
**MLT:** 163°F
**UEL:** NA
**LEL:** NA
Noncombustible Solid

**Personal Protection/Sanitation (see Table 2):**
**Skin:** Prevent skin contact
**Eyes:** Prevent eye contact
**Wash skin:** When contam
**Remove:** When wet or contam
**Change:** Daily
**Provide:** Quick drench

**Respirator Recommendations (see Tables 3 and 4):**
**NIOSH/OSHA**
**2 mg/m$^3$:** CcrOv95/Sa
**5 mg/m$^3$:** Sa:Cf/PaprOvHie
**10 mg/m$^3$:** CcrFOv100/GmFOv100/PaprTOvHie/SaT:Cf/ScbaF/SaF
**§:** ScbaF:Pd,Pp/SaF:Pd,Pp:AScba
**Escape:** GmFOv100/ScbaE

**Incompatibilities and Reactivities:** Strong oxidizers, acids

**Exposure Routes, Symptoms, Target Organs (see Table 5):**
**ER:** Inh, Abs, Ing, Con
**SY:** Miosis; ache eyes; blurred vision, lac, rhin; head; chest tight, wheez, lar spasm; salv; cyan; anor; nau, vomit, diarr; sweat; twitch, para, convuls; low BP, card irreg
**TO:** Resp sys, CNS, CVS, blood chol

**First Aid (see Table 6):**
**Eye:** Irr immed
**Skin:** Soap wash immed
**Breath:** Resp support
**Swallow:** Medical attention immed

## Barium chloride (as Ba)

| Barium chloride (as Ba) | Formula: $BaCl_2$ | CAS#: 10361-37-2 | RTECS#: CQ8750000 | IDLH: 50 mg/m$^3$ (as Ba) |
|---|---|---|---|---|
| **Conversion:** | **DOT:** 1564 154 (barium compound, n.o.s.) | | | |

**Synonyms/Trade Names:** Barium dichloride

**Exposure Limits:**
**NIOSH REL*:** TWA 0.5 mg/m$^3$
**OSHA PEL*:** TWA 0.5 mg/m$^3$
[***Note:** The REL and PEL also apply to other soluble barium compounds (as Ba) except Barium sulfate.]

**Measurement Methods (see Table 1):**
**NIOSH** 7056, 7303
**OSHA** ID121

**Physical Description:** White, odorless solid.

**Chemical & Physical Properties:**
**MW:** 208.2
**BP:** 2840°F
**Sol:** 38%
**Fl.P:** NA
**IP:** ?
**Sp.Gr:** 3.86
**VP:** Low
**MLT:** 1765°F
**UEL:** NA
**LEL:** NA
Noncombustible Solid

**Personal Protection/Sanitation (see Table 2):**
**Skin:** Prevent skin contact
**Eyes:** Prevent eye contact
**Wash skin:** When contam
**Remove:** When wet or contam
**Change:** Daily

**Respirator Recommendations (see Tables 3 and 4):**
**NIOSH/OSHA**
**5 mg/m$^3$:** 95XQ/Sa
**12.5 mg/m$^3$:** Sa:Cf/PaprHie
**25 mg/m$^3$:** 100F/SaT:Cf/PaprTHie/ScbaF/SaF
**50 mg/m$^3$:** SaF:Pd,Pp
**§:** ScbaF:Pd,Pp/SaF:Pd,Pp:AScba
**Escape:** 100F/ScbaE

**Incompatibilities and Reactivities:** Acids, oxidizers

**Exposure Routes, Symptoms, Target Organs (see Table 5):**
**ER:** Inh, Ing, Con
**SY:** Irrit eyes, skin, upper resp sys; skin burns; gastroenteritis; musc spasm; slow pulse, extrasystoles; hypokalemia
**TO:** Eyes, skin, resp sys, heart, CNS

**First Aid (see Table 6):**
**Eye:** Irr immed
**Skin:** Water flush immed
**Breath:** Resp support
**Swallow:** Medical attention immed

## Barium nitrate (as Ba)

| Barium nitrate (as Ba) | Formula: $Ba(NO_3)_2$ | CAS#: 10022-31-8 | RTECS#: CQ9625000 | IDLH: 50 mg/m$^3$ (as Ba) |
|---|---|---|---|---|
| **Conversion:** | **DOT:** 1446 141 | | | |

**Synonyms/Trade Names:** Barium dinitrate, Barium(II) nitrate (1:2), Barium salt of nitric acid

**Exposure Limits:**
**NIOSH REL*:** TWA 0.5 mg/m$^3$
**OSHA PEL*:** TWA 0.5 mg/m$^3$
[***Note:** The REL and PEL also apply to other soluble barium compounds (as Ba) except Barium sulfate.]

**Measurement Methods (see Table 1):**
**NIOSH** 7056
**OSHA** ID121

**Physical Description:** White, odorless solid.

**Chemical & Physical Properties:**
**MW:** 261.4
**BP:** Decomposes
**Sol:** 9%
**Fl.P:** NA
**IP:** ?
**Sp.Gr:** 3.24
**VP:** Low
**MLT:** 1094°F
**UEL:** NA
**LEL:** NA
Noncombustible Solid, but will accelerate the burning of combustible materials.

**Personal Protection/Sanitation (see Table 2):**
**Skin:** Prevent skin contact
**Eyes:** Prevent eye contact
**Wash skin:** When contam
**Remove:** When wet or contam
**Change:** Daily

**Respirator Recommendations (see Tables 3 and 4):**
**NIOSH/OSHA**
**5 mg/m$^3$:** 95XQ/Sa
**12.5 mg/m$^3$:** Sa:Cf/PaprHie
**25 mg/m$^3$:** 100F/SaT:Cf/PaprTHie/ScbaF/SaF
**50 mg/m$^3$:** SaF:Pd,Pp
**§:** ScbaF:Pd,Pp/SaF:Pd,Pp:AScba
**Escape:** 100F/ScbaE

**Incompatibilities and Reactivities:** Acids, oxidizers, aluminum-magnesium alloys, (barium dioxide + zinc)
[**Note:** Contact with combustible material may cause fire.]

**Exposure Routes, Symptoms, Target Organs (see Table 5):**
**ER:** Inh, Ing, Con
**SY:** Irrit eyes, skin, upper resp sys; skin burns; gastroenteritis; musc spasm; slow pulse, extrasystoles; hypokalemia
**TO:** Eyes, skin, resp sys, heart, CNS

**First Aid (see Table 6):**
**Eye:** Irr immed
**Skin:** Water flush immed
**Breath:** Resp support
**Swallow:** Medical attention immed

## Barium sulfate

| Barium sulfate | Formula: $BaSO_4$ | CAS#: 7727-43-7 | RTECS#: CR0600000 | IDLH: N.D. |
|---|---|---|---|---|
| **Conversion:** | **DOT:** 1564 154 (barium compound, n.o.s.) | | | |

B

**Synonyms/Trade Names:** Artificial barite, Barite, Barium salt of sulfuric acid, Barytes (natural)

**Exposure Limits:**
**NIOSH REL:** TWA 10 mg/m³ (total)
TWA 5 mg/m³ (resp)
**OSHA PEL†:** TWA 15 mg/m³ (total)
TWA 5 mg/m³ (resp)

**Measurement Methods (see Table 1):**
**NIOSH** 0500, 0600

**Physical Description:** White or yellowish, odorless powder.

**Chemical & Physical Properties:**
**MW:** 233.4
**BP:** 2912°F (Decomposes)
**Sol(64°F):** 0.0002%
**Fl.P:** NA
**IP:** NA
**Sp.Gr:** 4.25-4.5
**VP:** 0 mmHg (approx)
**MLT:** 2876°F
**UEL:** NA
**LEL:** NA
Noncombustible Solid

**Personal Protection/Sanitation (see Table 2):**
**Skin:** Prevent skin contact
**Eyes:** Prevent eye contact
**Wash skin:** Daily
**Remove:** N.R.
**Change:** N.R.
**Provide:** Eyewash
Quick drench

**Respirator Recommendations (see Tables 3 and 4):**
Not available.

**Incompatibilities and Reactivities:** Phosphorus, aluminum
[**Note:** Aluminum in the presence of heat can cause an explosion.]

**Exposure Routes, Symptoms, Target Organs (see Table 5):**
**ER:** Inh, Con
**SY:** Irrit eyes, nose, upper resp sys; benign pneumoconiosis (baritosis)
**TO:** Eyes, resp sys

**First Aid (see Table 6):**
**Eye:** Irr immed
**Skin:** Soap wash
**Breath:** Resp support
**Swallow:** Medical attention immed

## Benomyl

| Benomyl | Formula: $C_{14}H_{18}N_4O_3$ | CAS#: 17804-35-2 | RTECS#: DD6475000 | IDLH: N.D. |
|---|---|---|---|---|
| **Conversion:** | **DOT:** 2757 151 (carbamate pesticide, solid) | | | |

**Synonyms/Trade Names:** Methyl 1-(butylcarbamoyl)-2-benzimidazolecarbamate

**Exposure Limits:**
**NIOSH REL:** See Appendix D
**OSHA PEL†:** TWA 15 mg/m³ (total)
TWA 5 mg/m³ (resp)

**Measurement Methods (see Table 1):**
**NIOSH** 0500, 0600
**OSHA** PV2107

**Physical Description:** White crystalline solid with a faint, acrid odor. [fungicide]
[**Note:** Decomposes without melting above 572°F.]

**Chemical & Physical Properties:**
**MW:** 290.4
**BP:** Decomposes
**Sol:** 0.0004%
**Fl.P:** NA
**IP:** NA
**Sp.Gr:** ?
**VP:** <0.00001 mmHg
**MLT:** >572°F (Decomposes)
**UEL:** NA
**LEL:** NA
Noncombustible Solid

**Personal Protection/Sanitation (see Table 2):**
**Skin:** Prevent skin contact
**Eyes:** Prevent eye contact
**Wash skin:** When contam
**Remove:** When wet or contam
**Change:** Daily
**Provide:** Eyewash
Quick drench

**Respirator Recommendations (see Tables 3 and 4):**
Not available.

**Incompatibilities and Reactivities:** Heat, strong acids, strong alkalis

**Exposure Routes, Symptoms, Target Organs (see Table 5):**
**ER:** Inh, Ing, Con
**SY:** Irrit eyes, skin, upper resp sys; skin sens; possible repro, terato effects
**TO:** Eyes, skin, resp sys, repro sys

**First Aid (see Table 6):**
**Eye:** Irr immed
**Skin:** Soap wash immed
**Breath:** Resp support
**Swallow:** Medical attention immed

B

| Benzene | Formula: $C_6H_6$ | CAS#: 71-43-2 | RTECS#: CY1400000 | IDLH: Ca [500 ppm] |
|---|---|---|---|---|
| **Conversion:** 1 ppm = 3.19 mg/m$^3$ | **DOT:** 1114 130 | | | |

**Synonyms/Trade Names:** Benzol, Phenyl hydride

**Exposure Limits:**
**NIOSH REL:** Ca
TWA 0.1 ppm
ST 1 ppm
See Appendix A

**OSHA PEL:** [1910.1028]
TWA 1 ppm
ST 5 ppm
See Appendix F

**Measurement Methods (see Table 1):**
**NIOSH** 1500, 1501, 3700, 3800
**OSHA** 12, 1005

**Physical Description:** Colorless to light-yellow liquid with an aromatic odor. [**Note:** A solid below 42°F.]

**Chemical & Physical Properties:**
**MW:** 78.1
**BP:** 176°F
**Sol:** 0.07%
**Fl.P:** 12°F
**IP:** 9.24 eV
**Sp.Gr:** 0.88
**VP:** 75 mmHg
**FRZ:** 42°F
**UEL:** 7.8%
**LEL:** 1.2%
Class IB Flammable Liquid

**Personal Protection/Sanitation (see Table 2):**
**Skin:** Prevent skin contact
**Eyes:** Prevent eye contact
**Wash skin:** When contam
**Remove:** When wet (flamm)
**Change:** N.R.
**Provide:** Eyewash
Quick drench

**Respirator Recommendations (see Tables 3 and 4):**
**NIOSH**
**¥:** ScbaF:Pd,Pp/SaF:Pd,Pp:AScba
**Escape:** GmFOv/ScbaE

**See Appendix E** (page 351)

**Incompatibilities and Reactivities:** Strong oxidizers, many fluorides & perchlorates, nitric acid

**Exposure Routes, Symptoms, Target Organs (see Table 5):**
**ER:** Inh, Abs, Ing, Con
**SY:** Irrit eyes, skin, nose, resp sys; dizz; head, nau, staggered gait; anor, lass; derm; bone marrow depres; [carc]
**TO:** Eyes, skin, resp sys, blood, CNS, bone marrow [leukemia]

**First Aid (see Table 6):**
**Eye:** Irr immed
**Skin:** Soap wash immed
**Breath:** Resp support
**Swallow:** Medical attention immed

| Benzenethiol | Formula: $C_6H_5SH$ | CAS#: 108-98-5 | RTECS#: DC0525000 | IDLH: N.D. |
|---|---|---|---|---|
| **Conversion:** 1 ppm = 4.51 mg/m$^3$ | **DOT:** 2337 131 | | | |

**Synonyms/Trade Names:** Mercaptobenzene, Phenyl mercaptan, Thiophenol

**Exposure Limits:**
**NIOSH REL:** C 0.1 ppm (0.5 mg/m$^3$) [15-minute]
**OSHA PEL†:** none

**Measurement Methods (see Table 1):**
**OSHA** PV2075

**Physical Description:** Water-white liquid with an offensive, garlic-like odor.
[**Note:** A solid below 5°F.]

**Chemical & Physical Properties:**
**MW:** 110.2
**BP:** 336°F
**Sol(77°F):** 0.08%
**Fl.P:** 132°F
**IP:** 8.33 eV
**Sp.Gr:** 1.08
**VP(65°F):** 1 mmHg
**FRZ:** 5°F
**UEL:** ?
**LEL:** ?
Class II Combustible Liquid

**Personal Protection/Sanitation (see Table 2):**
**Skin:** Prevent skin contact
**Eyes:** Prevent eye contact
**Wash skin:** When contam
**Remove:** When wet or contam
**Change:** N.R.
**Provide:** Eyewash
Quick drench

**Respirator Recommendations (see Tables 3 and 4):**
**NIOSH**
**1 ppm:** CcrOv/Sa
**2.5 ppm:** Sa:Cf/PaprOv
**5 ppm:** CcrFOv/GmFOv/PaprTOv/ScbaF/SaF
**§:** ScbaF:Pd,Pp/SaF:Pd,Pp:AScba
**Escape:** GmFOv/ScbaE

**Incompatibilities and Reactivities:** Strong acids & bases, calcium hypochlorite, alkali metals
[**Note:** Oxidizes on exposure to air.]

**Exposure Routes, Symptoms, Target Organs (see Table 5):**
**ER:** Inh, Abs, Ing, Con
**SY:** Irrit eyes, skin, resp sys; derm; cyan; cough, wheez, dysp, pulm edema, pneu; head, dizz, CNS depres; nau, vomit; kidney, liver, spleen damage
**TO:** Eyes, skin, resp sys, CNS, kidneys, liver, spleen

**First Aid (see Table 6):**
**Eye:** Irr immed
**Skin:** Soap wash immed
**Breath:** Resp support
**Swallow:** Medical attention immed

B

## Benzidine

| Formula: | CAS#: | RTECS#: | IDLH: |
|---|---|---|---|
| $NH_2C_6H_4C_6H_4NH_2$ | 92-87-5 | DC9625000 | Ca [N.D.] |

**Conversion:**

**DOT:** 1885 153

**Synonyms/Trade Names:** Benzidine-based dyes; 4,4'-Bianiline; 4,4'-Biphenyldiamine; 1,1'-Biphenyl-4,4'-diamine; 4,4'-Diaminobiphenyl; p-Diaminodiphenyl [**Note:** Benzidine has been used as a basis for many dyes.]

**Exposure Limits:**
**NIOSH REL:** Ca
See Appendix A
See Appendix C
**OSHA PEL:** [1910.1010]
See Appendix B
See Appendix C

**Measurement Methods (see Table 1):**
**NIOSH** 5509
**OSHA** 65

**Physical Description:** Grayish-yellow, reddish-gray, or white crystalline powder.
[**Note:** Darkens on exposure to air and light.]

**Chemical & Physical Properties:**
**MW:** 184.3
**BP:** 752°F
**Sol(54°F):** 0.04%
**Fl.P:** ?
**IP:** ?
**Sp.Gr:** 1.25
**VP:** Low
**MLT:** 239°F
**UEL:** ?
**LEL:** ?
Combustible Solid, but difficult to burn.

**Personal Protection/Sanitation (see Table 2):**
**Skin:** Prevent skin contact
**Eyes:** Prevent eye contact
**Wash skin:** When contam/Daily
**Remove:** When wet or contam
**Change:** Daily
**Provide:** Eyewash
Quick drench

**Respirator Recommendations (see Tables 3 and 4):**
**NIOSH**
**¥:** ScbaF:Pd,Pp/SaF:Pd,Pp:AScba
**Escape:** 100F/ScbaE

**See Appendix E** (page 351)

**Incompatibilities and Reactivities:** Red fuming nitric acid

**Exposure Routes, Symptoms, Target Organs (see Table 5):**
**ER:** Inh, Abs, Ing, Con
**SY:** Hema; secondary anemia from hemolysis; acute cystitis; acute liver disorders; derm; painful, irreg urination; [carc]
**TO:** Bladder, skin, kidneys, liver, blood [liver, kidney & bladder cancer]

**First Aid (see Table 6):**
**Eye:** Irr immed
**Skin:** Soap wash immed
**Breath:** Resp support
**Swallow:** Medical attention immed

## Benzoyl peroxide

| Formula: | CAS#: | RTECS#: | IDLH: |
|---|---|---|---|
| $(C_6H_5CO)_2O_2$ | 94-36-0 | DM8575000 | 1500 mg/m$^3$ |

**Conversion:**

**DOT:**

**Synonyms/Trade Names:** Benzoperoxide, Dibenzoyl peroxide

**Exposure Limits:**
**NIOSH REL:** TWA 5 mg/m$^3$
**OSHA PEL:** TWA 5 mg/m$^3$

**Measurement Methods (see Table 1):**
**NIOSH** 5009

**Physical Description:** Colorless to white crystals or a granular powder with a faint, benzaldehyde-like odor.

**Chemical & Physical Properties:**
**MW:** 242.2
**BP:** Decomposes explosively
**Sol:** <1%
**Fl.P:** 176°F
**IP:** ?
**Sp.Gr:** 1.33
**VP:** <1 mmHg
**MLT:** 217°F
**UEL:** ?
**LEL:** ?
Combustible Solid (easily ignited and burns very rapidly).

**Personal Protection/Sanitation (see Table 2):**
**Skin:** Prevent skin contact
**Eyes:** Prevent eye contact
**Wash skin:** When contam
**Remove:** When wet or contam
**Change:** Daily

**Respirator Recommendations (see Tables 3 and 4):**
**NIOSH/OSHA**
**50 mg/m$^3$:** 95XQ*/Sa*
**125 mg/m$^3$:** Sa:Cf*/PaprHie*
**250 mg/m$^3$:** 100F/PaprTHie*/ScbaF/SaF
**1500 mg/m$^3$:** SaF:Pd,Pp
**§:** ScbaF:Pd,Pp/SaF:Pd,Pp:AScba
**Escape:** 100F/ScbaE

**Incompatibilities and Reactivities:** Combustible substances (wood, paper, etc.), acids, alkalis, alcohols, amines, ethers [**Note:** Containers may explode when heated. Extremely explosion-sensitive to shock, heat & friction.]

**Exposure Routes, Symptoms, Target Organs (see Table 5):**
**ER:** Inh, Ing, Con
**SY:** Irrit eyes, skin, muc memb; sens derm
**TO:** Eyes, skin, resp sys

**First Aid (see Table 6):**
**Eye:** Irr immed
**Skin:** Soap wash prompt
**Breath:** Resp support
**Swallow:** Medical attention immed

B

## Benzyl chloride

| Formula: | CAS#: | RTECS#: | IDLH: |
|---|---|---|---|
| $C_6H_5CH_2Cl$ | 100-44-7 | XS8925000 | 10 ppm |

**Conversion:** 1 ppm = 5.18 mg/m$^3$ | **DOT:** 1738 156

**Synonyms/Trade Names:** Chloromethylbenzene, α-Chlorotoluene

**Exposure Limits:**
**NIOSH REL:** C 1 ppm (5 mg/m$^3$) [15-minute]
**OSHA PEL:** TWA 1 ppm (5 mg/m$^3$)

**Measurement Methods (see Table 1):**
**NIOSH** 1003
**OSHA** 7

**Physical Description:** Colorless to slightly yellow liquid with a pungent, aromatic odor.

**Chemical & Physical Properties:**
**MW:** 126.6
**BP:** 354°F
**Sol:** 0.05%
**Fl.P:** 153°F
**IP:** ?
**Sp.Gr:** 1.10
**VP:** 1 mmHg
**FRZ:** -38°F
**UEL:** ?
**LEL:** 1.1%
Class IIIA Combustible Liquid

**Personal Protection/Sanitation (see Table 2):**
**Skin:** Prevent skin contact
**Eyes:** Prevent eye contact
**Wash skin:** When contam
**Remove:** When wet or contam
**Change:** N.R.
**Provide:** Eyewash
Quick drench

**Respirator Recommendations (see Tables 3 and 4):**
**NIOSH/OSHA**
**10 ppm:** CcrOvAg*/GmFOvAg/
PaprOvAg*/Sa*/ScbaF
**§:** ScbaF:Pd,Pp/SaF:Pd,Pp:AScba
**Escape:** GmFOvAg/ScbaE

**Incompatibilities and Reactivities:** Oxidizers, acids, copper, aluminum, magnesium, iron, zinc, tin [**Note:** Can polymerize when in contact with all common metals except nickel & lead. Hydrolyzes in $H_2O$ to benzyl alcohol.]

**Exposure Routes, Symptoms, Target Organs (see Table 5):**
**ER:** Inh, Ing, Con
**SY:** Irrit eyes, skin, nose; lass; irrity; head; skin eruption; pulm edema
**TO:** Eyes, skin, resp sys, CNS

**First Aid (see Table 6):**
**Eye:** Irr immed
**Skin:** Soap wash immed
**Breath:** Resp support
**Swallow:** Medical attention immed

## Beryllium & beryllium compounds (as Be)

| Formula: | CAS#: | RTECS#: | IDLH: |
|---|---|---|---|
| Be (metal) | 7440-41-7 (metal) | DS1750000 (metal) | Ca [4 mg/m$^3$ (as Be)] |

**Conversion:** | **DOT:** 1566 154 (compounds); 1567 134 (powder)

**Synonyms/Trade Names: Beryllium metal:** Beryllium
Other synonyms vary depending upon the specific beryllium compound.

**Exposure Limits:**
**NIOSH REL:** Ca
Not to exceed 0.0005 mg/m$^3$
See Appendix A
**OSHA PEL:** TWA 0.002 mg/m$^3$
C 0.005 mg/m$^3$ 0.025 mg/m$^3$ [30-minute maximum peak]

**Measurement Methods (see Table 1):**
**NIOSH** 7102, 7300, 7301, 7303, 9102
**OSHA** ID125G, ID206

**Physical Description:** Metal: A hard, brittle, gray-white solid.

**Chemical & Physical Properties:**
**MW:** 9.0
**BP:** 4532°F
**Sol:** Insoluble
**Fl.P:** NA
**IP:** NA
**Sp.Gr:** 1.85 (metal)
**VP:** 0 mmHg (approx)
**MLT:** 2349°F
**UEL:** NA
**LEL:** NA

**Personal Protection/Sanitation (see Table 2):**
**Skin:** Prevent skin contact
**Eyes:** Prevent eye contact
**Wash skin:** Daily
**Remove:** When wet or contam
**Change:** Daily
**Provide:** Eyewash

**Respirator Recommendations (see Tables 3 and 4):**
**NIOSH**
**¥:** ScbaF:Pd,Pp/SaF:Pd,Pp:AScba
**Escape:** 100F/ScbaE

Metal: Noncombustible Solid in bulk form, but a slight explosion hazard in the form of a powder or dust.

**Incompatibilities and Reactivities:** Acids, caustics, chlorinated hydrocarbons, oxidizers, molten lithium

**Exposure Routes, Symptoms, Target Organs (see Table 5):**
**ER:** Inh, Con
**SY:** Berylliosis (chronic exposure): anor, low-wgt, lass, chest pain, cough, clubbing of fingers, cyan, pulm insufficiency; irrit eyes; derm; [carc]
**TO:** Eyes, skin, resp sys [lung cancer]

**First Aid (see Table 6):**
**Eye:** Irr immed
**Breath:** Fresh air

## Bismuth telluride, doped with Selenium sulfide (as $Bi_2Te_3$)

| Formula: | CAS#: | RTECS#: | IDLH: |
|---|---|---|---|
| | | | N.D. |

**Conversion:**

**DOT:**

**Synonyms/Trade Names:** Doped bismuth sesquitelluride, Doped bismuth telluride, Doped bismuth tritelluride, Doped tellurobismuthite [**Note:** Doped with selenium sulfide. Commercial mix may contain 80% $Bi_2Te_3$, 20% stannous telluride, plus some tellurium.]

**Exposure Limits:**
**NIOSH REL:** TWA 5 mg/m$^3$
**OSHA PEL†:** none

**Measurement Methods (see Table 1):**
**NIOSH** 0500
**OSHA** ID121

**Physical Description:** Gray, crystalline solid that has been enhanced (doped) with a small amount of selenium sulfide (SeS).
[**Note:** Doping alters the conductivity of a semiconductor.]

**Chemical & Physical Properties:**
Properties are unavailable but should be similar to Bismuth telluride, undoped.

**Sp.Gr:** ?

Noncombustible Solid

**Personal Protection/Sanitation (see Table 2):**
**Skin:** Prevent skin contact
**Eyes:** Prevent eye contact
**Wash skin:** When contam
**Remove:** When wet or contam
**Change:** N.R.
**Provide:** Eyewash
Quick drench

**Respirator Recommendations (see Tables 3 and 4):**
Not available.

**Incompatibilities and Reactivities:** Strong oxidizers, moisture

**Exposure Routes, Symptoms, Target Organs (see Table 5):**
**ER:** Inh, Con
**SY:** Irrit eyes, skin, upper resp sys; garlic breath; in animals: pulm lesions (nonfibrotic)
**TO:** Eyes, skin, resp sys

**First Aid (see Table 6):**
**Eye:** Irr immed
**Skin:** Soap wash immed
**Breath:** Resp support
**Swallow:** Medical attention immed

## Bismuth telluride, undoped

| Formula: | CAS#: | RTECS#: | IDLH: |
|---|---|---|---|
| $Bi_2Te_3$ | 1304-82-1 | EB3110000 | N.D. |

**Conversion:**

**DOT:**

**Synonyms/Trade Names:** Bismuth sesquitelluride, Bismuth telluride, Bismuth tritelluride, Tellurobismuthite

**Exposure Limits:**
**NIOSH REL:** TWA 10 mg/m$^3$ (total)
TWA 5 mg/m$^3$ (resp)
**OSHA PEL:** TWA 15 mg/m$^3$ (total)
TWA 5 mg/m$^3$ (resp)

**Measurement Methods (see Table 1):**
**NIOSH** 0500, 0600
**OSHA** ID121

**Physical Description:** Gray, crystalline solid.

**Chemical & Physical Properties:**
**MW:** 800.8
**BP:** ?
**Sol:** Insoluble
**Fl.P:** NA
**IP:** NA
**Sp.Gr:** 7.7
**VP:** 0 mmHg (approx)
**MLT:** 1063°F
**UEL:** NA
**LEL:** NA
Noncombustible Solid

**Personal Protection/Sanitation (see Table 2):**
**Skin:** Prevent skin contact
**Eyes:** Prevent eye contact
**Wash skin:** When contam
**Remove:** When wet or contam
**Change:** N.R.
**Provide:** Eyewash
Quick drench

**Respirator Recommendations (see Tables 3 and 4):**
Not available.

**Incompatibilities and Reactivities:** Strong oxidizers (e.g., bromine, chlorine, or fluorine), moisture, nitric acid (decomposes)

**Exposure Routes, Symptoms, Target Organs (see Table 5):**
**ER:** Inh, Con
**SY:** Irrit eyes, skin, upper resp sys; garlic breath
**TO:** Eyes, skin, resp sys

**First Aid (see Table 6):**
**Eye:** Irr immed
**Skin:** Soap wash immed
**Breath:** Resp support
**Swallow:** Medical attention immed

B

| Borates, tetra, sodium salts (Anhydrous) | Formula: $Na_2B_4O_7$ | CAS#: 1330-43-4 | RTECS#: ED4588000 | IDLH: N.D. |
|---|---|---|---|---|
| Conversion: | DOT: | | | |

**Synonyms/Trade Names:** Anhydrous borax, Borax dehydrated, Disodium salt of boric acid, Disodium tetrabromate, Fused borax, Sodium borate (anhydrous) Sodium tetraborate

**Exposure Limits:**
**NIOSH REL:** TWA 1 mg/m$^3$
**OSHA PEL†:** none

**Measurement Methods (see Table 1):**
**NIOSH** 0500
**OSHA** ID125G

**Physical Description:** White to gray, odorless powder. [herbicide]
[**Note:** Becomes opaque on exposure to air.]

| Chemical & Physical Properties: | Personal Protection/Sanitation (see Table 2): | Respirator Recommendations (see Tables 3 and 4): |
|---|---|---|
| **MW:** 201.2<br>**BP:** 2867°F (Decomposes)<br>**Sol:** 4%<br>**Fl.P:** NA<br>**IP:** NA<br>**Sp.Gr:** 2.37<br>**VP:** 0 mmHg (approx)<br>**MLT:** 1366°F<br>**UEL:** NA<br>**LEL:** NA<br>Noncombustible Solid | **Skin:** N.R.<br>**Eyes:** N.R.<br>**Wash skin:** Daily<br>**Remove:** N.R.<br>**Change:** Daily | Not available. |

**Incompatibilities and Reactivities:** Moisture [**Note:** Forms partial hydrate in moist air.]

| Exposure Routes, Symptoms, Target Organs (see Table 5): | First Aid (see Table 6): |
|---|---|
| **ER:** Inh, Ing, Con<br>**SY:** Irrit eyes, skin, upper resp sys; derm; epis; cough, dysp<br>**TO:** Eyes, skin, resp sys | **Eye:** Irr immed<br>**Skin:** Soap wash<br>**Breath:** Resp support<br>**Swallow:** Medical attention immed |

| Borates, tetra, sodium salts (Decahydrate) | Formula: $Na_2B_4O_7 \times 10H_2O$ | CAS#: 1303-96-4 | RTECS#: VZ2275000 | IDLH: N.D. |
|---|---|---|---|---|
| Conversion: | DOT: | | | |

**Synonyms/Trade Names:** Borax, Borax decahydrate, Sodium borate decahydrate, Sodium tetraborate decahydrate

**Exposure Limits:**
**NIOSH REL:** TWA 5 mg/m$^3$
**OSHA PEL†:** none

**Measurement Methods (see Table 1):**
**NIOSH** 0500
**OSHA** ID125G

**Physical Description:** White, odorless, crystalline solid. [herbicide]
[**Note:** Becomes anhydrous at 608°F.]

| Chemical & Physical Properties: | Personal Protection/Sanitation (see Table 2): | Respirator Recommendations (see Tables 3 and 4): |
|---|---|---|
| **MW:** 381.4<br>**BP:** 608°F<br>**Sol:** 6%<br>**Fl.P:** NA<br>**IP:** NA<br>**Sp.Gr:** 1.73<br>**VP:** 0 mmHg (approx)<br>**MLT:** 167°F<br>**UEL:** NA<br>**LEL:** NA<br>Noncombustible Solid (an inherent fire retardant). | **Skin:** N.R.<br>**Eyes:** N.R.<br>**Wash skin:** Daily<br>**Remove:** N.R.<br>**Change:** Daily | Not available. |

**Incompatibilities and Reactivities:** Zirconium, strong acids, metallic salts

| Exposure Routes, Symptoms, Target Organs (see Table 5): | First Aid (see Table 6): |
|---|---|
| **ER:** Inh, Ing, Con<br>**SY:** Irrit eyes, skin, upper resp sys; derm; epis; cough, dysp<br>**TO:** Eyes, skin, resp sys | **Eye:** Irr immed<br>**Skin:** Soap wash<br>**Breath:** Resp support<br>**Swallow:** Medical attention immed |

B

| Borates, tetra, sodium salts (Pentahydrate) | Formula: $Na_2B_4O_7 \times 5H_2O$ | CAS#: 12179-04-3 | RTECS#: VZ2540000 | IDLH: N.D. |
|---|---|---|---|---|
| **Conversion:** | **DOT:** | | | |

**Synonyms/Trade Names:** Borax pentahydrate, Sodium borate pentahydrate, Sodium tetraborate pentahydrate

**Exposure Limits:**
**NIOSH REL:** TWA 1 mg/m$^3$
**OSHA PEL†:** none

**Measurement Methods (see Table 1):**
**NIOSH** 0500
**OSHA** ID125G

**Physical Description:** Colorless or white, odorless crystals or free-flowing powder. [herbicide] [**Note:** Begins to lose water of hydration at 252°F.]

| Chemical & Physical Properties: | Personal Protection/Sanitation (see Table 2): | Respirator Recommendations (see Tables 3 and 4): |
|---|---|---|
| **MW:** 291.4<br>**BP:** ?<br>**Sol:** 3%<br>**Fl.P:** NA<br>**IP:** NA<br>**Sp.Gr:** 1.82<br>**VP:** 0 mmHg (approx)<br>**MLT:** 392°F<br>**UEL:** NA<br>**LEL:** NA<br>Noncombustible Solid | **Skin:** N.R.<br>**Eyes:** N.R.<br>**Wash skin:** Daily<br>**Remove:** N.R.<br>**Change:** Daily | Not available. |

**Incompatibilities and Reactivities:** None reported
[**Note:** See the reactivities & incompatibilities reported for the related substance Borax decahydrate above.]

| Exposure Routes, Symptoms, Target Organs (see Table 5): | First Aid (see Table 6): |
|---|---|
| **ER:** Inh, Ing, Con<br>**SY:** Irrit eyes, skin, upper resp sys; derm; epis; cough, dysp<br>**TO:** Eyes, skin, resp sys | **Eye:** Irr immed<br>**Skin:** Soap wash<br>**Breath:** Resp support<br>**Swallow:** Medical attention immed |

| Boron oxide | Formula: $B_2O_3$ | CAS#: 1303-86-2 | RTECS#: ED7900000 | IDLH: 2000 mg/m$^3$ |
|---|---|---|---|---|
| **Conversion:** | **DOT:** | | | |

**Synonyms/Trade Names:** Boric anhydride, Boric oxide, Boron trioxide

**Exposure Limits:**
**NIOSH REL:** TWA 10 mg/m$^3$
**OSHA PEL†:** TWA 15 mg/m$^3$

**Measurement Methods (see Table 1):**
**NIOSH** 0500

**Physical Description:** Colorless, semitransparent lumps or hard, white, odorless crystals.

| Chemical & Physical Properties: | Personal Protection/Sanitation (see Table 2): | Respirator Recommendations (see Tables 3 and 4): |
|---|---|---|
| **MW:** 69.6<br>**BP:** 3380°F<br>**Sol:** 3%<br>**Fl.P:** NA<br>**IP:** 13.50 eV<br>**Sp.Gr:** 2.46<br>**VP:** 0 mmHg (approx)<br>**MLT:** 842°F<br>**UEL:** NA<br>**LEL:** NA<br>Noncombustible Solid | **Skin:** Prevent skin contact<br>**Eyes:** Prevent eye contact<br>**Wash skin:** When contam<br>**Remove:** When wet or contam<br>**Change:** N.R. | **NIOSH**<br>**50 mg/m$^3$:** Qm*<br>**100 mg/m$^3$:** 95XQ*/Sa*<br>**250 mg/m$^3$:** Sa:Cf*/PaprHie*<br>**500 mg/m$^3$:** 100F/PaprTHie*/ScbaF/SaF<br>**2000 mg/m$^3$:** SaF:Pd,Pp<br>**§:** ScbaF:Pd,Pp/SaF:Pd,Pp:AScba<br>**Escape:** 100F/ScbaE |

**Incompatibilities and Reactivities:** Water [**Note:** Reacts slowly with water to form boric acid.]

| Exposure Routes, Symptoms, Target Organs (see Table 5): | First Aid (see Table 6): |
|---|---|
| **ER:** Inh, Ing, Con<br>**SY:** Irrit eyes, skin, resp sys; cough; conj; skin eryt<br>**TO:** Eyes, skin, resp sys | **Eye:** Irr immed<br>**Skin:** Water flush prompt<br>**Breath:** Fresh air<br>**Swallow:** Medical attention immed |

B

## Boron tribromide

**Formula:** $BBr_3$ | **CAS#:** 10294-33-4 | **RTECS#:** ED7400000 | **IDLH:** N.D.

**Conversion:** 1 ppm = 10.25 mg/m$^3$ | **DOT:** 2692 157

**Synonyms/Trade Names:** Boron bromide, Tribromoborane

**Exposure Limits:**
**NIOSH REL:** C 1 ppm (10 mg/m$^3$)
**OSHA PEL†:** none

**Measurement Methods (see Table 1):**
None available

**Physical Description:** Colorless, fuming liquid with a sharp, irritating odor.

**Chemical & Physical Properties:**
**MW:** 250.5
**BP:** 194°F
**Sol:** Decomposes
**Fl.P:** NA
**IP:** 9.70 eV
**Sp.Gr(65°F):** 2.64
**VP(57°F):** 40 mmHg
**FRZ:** -51°F
**UEL:** NA
**LEL:** NA
Noncombustible Liquid

**Personal Protection/Sanitation (see Table 2):**
**Skin:** Prevent skin contact
**Eyes:** Prevent eye contact
**Wash skin:** When contam
**Remove:** When wet or contam
**Change:** N.R.
**Provide:** Eyewash
Quick drench

**Respirator Recommendations (see Tables 3 and 4):**
Not available.

**Incompatibilities and Reactivities:** Moisture, water, heat, potassium, sodium, alcohols
[**Note:** Attacks metals, wood & rubber. Reacts with water to form boric acid and hydrogen bromide.]

**Exposure Routes, Symptoms, Target Organs (see Table 5):**
**ER:** Inh, Ing, Con
**SY:** Irrit eyes, skin, resp sys; eye, skin burns; dysp, pulm edema
**TO:** Eyes, skin, resp sys

**First Aid (see Table 6):**
**Eye:** Irr immed
**Skin:** Water flush immed
**Breath:** Resp support
**Swallow:** Medical attention immed

## Boron trifluoride

**Formula:** $BF_3$ | **CAS#:** 7637-07-2 | **RTECS#:** ED2275000 | **IDLH:** 25 ppm

**Conversion:** 1 ppm = 2.77 mg/m$^3$ | **DOT:** 1008 125

**Synonyms/Trade Names:** Boron fluoride, Trifluoroborane

**Exposure Limits:**
**NIOSH REL:** C 1 ppm (3 mg/m$^3$)
**OSHA PEL:** C 1 ppm (3 mg/m$^3$)

**Measurement Methods (see Table 1):**
None available

**Physical Description:** Colorless gas with a pungent, suffocating odor.
[**Note:** Forms dense white fumes in moist air. Shipped as a nonliquefied compressed gas.]

**Chemical & Physical Properties:**
**MW:** 67.8
**BP:** -148°F
**Sol:** 106% (in cold $H_2O$)
**Fl.P:** NA
**IP:** 15.50 eV
**RGasD:** 2.38
**VP:** >50 atm
**FRZ:** -196°F
**UEL:** NA
**LEL:** NA
Nonflammable Gas

**Personal Protection/Sanitation (see Table 2):**
**Skin:** N.R.
**Eyes:** N.R.
**Wash skin:** N.R.
**Remove:** N.R.
**Change:** N.R.

**Respirator Recommendations (see Tables 3 and 4):**
**NIOSH/OSHA**
**10 ppm:** Sa*
**25 ppm:** Sa:Cf*/ScbaF/SaF
**§:** ScbaF:Pd,Pp/SaF:Pd,Pp:AScba
**Escape:** GmFS/ScbaE

**Incompatibilities and Reactivities:** Alkali metals, calcium oxide
[**Note:** Hydrolyzes in moist air or hot water to form boric acid, hydrogen fluoride & fluoboric acid.]

**Exposure Routes, Symptoms, Target Organs (see Table 5):**
**ER:** Inh, Con
**SY:** Irrit eyes, skin, nose, resp sys; epis; eye, skin burns; in animals: pneu; kidney damage
**TO:** Eyes, skin, resp sys, kidneys

**First Aid (see Table 6):**
**Eye:** Irr immed
**Skin:** Water flush immed
**Breath:** Resp support

## Bromacil

| Bromacil | Formula: $C_9H_{13}BrN_2O_2$ | CAS#: 314-40-9 | RTECS#: YQ9100000 | IDLH: N.D. |
|---|---|---|---|---|
| **Conversion:** 1 ppm = 10.68 mg/m$^3$ | **DOT:** | | | |

**Synonyms/Trade Names:** 5-Bromo-3-sec-butyl-6-methyluracil, 5-Bromo-6-methyl-3-(1-methylpropyl)uracil

**Exposure Limits:**
**NIOSH REL:** TWA 1 ppm (10 mg/m$^3$)
**OSHA PEL†:** none

**Measurement Methods (see Table 1):**
**NIOSH** 0500

**Physical Description:** Odorless, colorless to white, crystalline solid. [herbicide]
[**Note:** Commercially available as a wettable powder or in liquid formulations.]

**Chemical & Physical Properties:**
**MW:** 261.2
**BP:** Sublimes
**Sol(77°F):** 0.08%
**Fl.P:** NA
**IP:** ?
**Sp.Gr:** 1.55
**VP(212°F):** 0.0008 mmHg
**MLT:** 317°F (Sublimes)
**UEL:** NA
**LEL:** NA
Noncombustible Solid, but may be dissolved in flammable liquids.

**Personal Protection/Sanitation (see Table 2):**
**Skin:** Prevent skin contact
**Eyes:** Prevent eye contact
**Wash skin:** When contam
**Remove:** When wet or contam
**Change:** Daily
**Provide:** Eyewash
Quick drench

**Respirator Recommendations (see Tables 3 and 4):**
Not available.

**Incompatibilities and Reactivities:** Strong acids (decomposes slowly), oxidizers, heat, sparks, open flames

**Exposure Routes, Symptoms, Target Organs (see Table 5):**
**ER:** Inh, Ing, Con
**SY:** Irrit eyes, skin, upper resp sys; in animals: thyroid inj
**TO:** Eyes, skin, resp sys, thyroid

**First Aid (see Table 6):**
**Eye:** Irr immed
**Skin:** Soap wash immed
**Breath:** Resp support
**Swallow:** Medical attention immed

## Bromine

| Bromine | Formula: $Br_2$ | CAS#: 7726-95-6 | RTECS#: EF9100000 | IDLH: 3 ppm |
|---|---|---|---|---|
| **Conversion:** 1 ppm = 6.54 mg/m$^3$ | **DOT:** 1744 154 | | | |

**Synonyms/Trade Names:** Molecular bromine

**Exposure Limits:**
**NIOSH REL:** TWA 0.1 ppm (0.7 mg/m$^3$)
ST 0.3 ppm (2 mg/m$^3$)
**OSHA PEL†:** TWA 0.1 ppm (0.7 mg/m$^3$)

**Measurement Methods (see Table 1):**
**NIOSH** 6011
**OSHA** ID108

**Physical Description:** Dark reddish-brown, fuming liquid with suffocating, irritating fumes.

**Chemical & Physical Properties:**
**MW:** 159.8
**BP:** 139°F
**Sol:** 4%
**Fl.P:** NA
**IP:** 10.55 eV
**Sp.Gr:** 3.12
**VP:** 172 mmHg
**FRZ:** 19°F
**UEL:** NA
**LEL:** NA
Noncombustible Liquid, but accelerates the burning of combustibles.

**Personal Protection/Sanitation (see Table 2):**
**Skin:** Prevent skin contact
**Eyes:** Prevent eye contact
**Wash skin:** When contam
**Remove:** When wet or contam
**Change:** N.R.
**Provide:** Eyewash
Quick drench

**Respirator Recommendations (see Tables 3 and 4):**
**NIOSH/OSHA**
**2.5 ppm:** Sa:Cf£/PaprS¿£
**3 ppm:** CcrFS¿/GmFS¿/PaprTS¿£/ScbaF/SaF
**§:** ScbaF:Pd,Pp/SaF:Pd,Pp:AScba
**Escape:** GmFS¿/ScbaE

**Incompatibilities and Reactivities:** Combustible organics (sawdust, wood, cotton, straw, etc.), aluminum, readily oxidizable materials, ammonia, hydrogen, acetylene, phosphorus, potassium, sodium
[**Note:** Corrodes iron, steel, stainless steel & copper.]

**Exposure Routes, Symptoms, Target Organs (see Table 5):**
**ER:** Inh, Ing, Con
**SY:** Dizz, head; lac, epis; cough, feeling of oppression, pulm edema, pneu; abdom pain, diarr; measle-like eruptions; eye, skin burns
**TO:** Resp sys, eyes, CNS, skin

**First Aid (see Table 6):**
**Eye:** Irr immed
**Skin:** Soap wash immed
**Breath:** Resp support
**Swallow:** Medical attention immed

B

## Bromine pentafluoride

| Formula: | CAS#: | RTECS#: | IDLH: |
|---|---|---|---|
| $BrF_5$ | 7789-30-2 | EF9350000 | N.D. |

**Conversion:** 1 ppm = 7.15 mg/m$^3$ | **DOT:** 1745 144

**Synonyms/Trade Names:** Bromine fluoride

**Exposure Limits:**
**NIOSH REL:** TWA 0.1 ppm (0.7 mg/m$^3$)
**OSHA PEL†:** none

**Measurement Methods (see Table 1):**
None available

**Physical Description:** Colorless to pale-yellow, fuming liquid with a pungent odor.
[**Note:** A colorless gas above 105°F. Shipped as a compressed gas.]

**Chemical & Physical Properties:**
**MW:** 174.9
**BP:** 105°F
**Sol:** Reacts violently
**Fl.P:** NA
**IP:** ?
**Sp.Gr:** 2.48
**VP:** 328 mmHg
**FRZ:** -77°F
**UEL:** NA
**LEL:** NA
Noncombustible Liquid, but a very powerful oxidizer.

**Personal Protection/Sanitation (see Table 2):**
**Skin:** Prevent skin contact
**Eyes:** Prevent eye contact
**Wash skin:** When contam
**Remove:** When wet or contam
**Change:** N.R.
**Provide:** Eyewash
Quick drench

**Respirator Recommendations (see Tables 3 and 4):**
Not available.

**Incompatibilities and Reactivities:** Acids, halogens, arsenic, selenium, sulfur, glass, organic materials, water
[**Note:** Reacts with all elements except inert gases, nitrogen & oxygen.]

**Exposure Routes, Symptoms, Target Organs (see Table 5):**
**ER:** Inh, Ing, Con
**SY:** Irrit eyes, skin, resp sys; corn nec; skin burns; cough, dysp, pulm edema; liver, kidney inj
**TO:** Eyes, skin, resp sys, liver, kidneys

**First Aid (see Table 6):**
**Eye:** Irr immed
**Skin:** Water flush immed
**Breath:** Resp support
**Swallow:** Medical attention immed

## Bromoform

| Formula: | CAS#: | RTECS#: | IDLH: |
|---|---|---|---|
| $CHBr_3$ | 75-25-2 | PB5600000 | 850 ppm |

**Conversion:** 1 ppm = 10.34 mg/m$^3$ | **DOT:** 2515 159

**Synonyms/Trade Names:** Methyl tribromide, Tribromomethane

**Exposure Limits:**
**NIOSH REL:** TWA 0.5 ppm (5 mg/m$^3$) [skin]
**OSHA PEL:** TWA 0.5 ppm (5 mg/m$^3$) [skin]

**Measurement Methods (see Table 1):**
**NIOSH** 1003
**OSHA** 7

**Physical Description:** Colorless to yellow liquid with a chloroform-like odor.
[**Note:** A solid below 47°F.]

**Chemical & Physical Properties:**
**MW:** 252.8
**BP:** 301°F
**Sol:** 0.1%
**Fl.P:** NA
**IP:** 10.48 eV
**Sp.Gr:** 2.89
**VP:** 5 mmHg
**FRZ:** 47°F
**UEL:** NA
**LEL:** NA
Noncombustible Liquid

**Personal Protection/Sanitation (see Table 2):**
**Skin:** Prevent skin contact
**Eyes:** Prevent eye contact
**Wash skin:** When contam
**Remove:** When wet or contam
**Change:** N.R.

**Respirator Recommendations (see Tables 3 and 4):**
**NIOSH/OSHA**
**12.5 ppm:** Sa:Cf£/PaprOv£
**25 ppm:** CcrFOv/GmFOv/PaprTOv£/ScbaF/SaF
**850 ppm:** SaF:Pd,Pp
**§:** ScbaF:Pd,Pp/SaF:Pd,Pp:AScba
**Escape:** GmFOv/ScbaE

**Incompatibilities and Reactivities:** Lithium, sodium, potassium, calcium, aluminum, zinc, magnesium, strong caustics, acetone [**Note:** Gradually decomposes, acquiring yellow color; air & light accelerate decomposition.]

**Exposure Routes, Symptoms, Target Organs (see Table 5):**
**ER:** Inh, Abs, Ing, Con
**SY:** Irrit eyes, skin, resp sys; CNS depres; liver, kidney damage
**TO:** Eyes, skin, resp sys, CNS, liver, kidneys

**First Aid (see Table 6):**
**Eye:** Irr immed
**Skin:** Soap wash prompt
**Breath:** Resp support
**Swallow:** Medical attention immed

## 1,3-Butadiene

| 1,3-Butadiene | Formula: $CH_2$=CHCH=$CH_2$ | CAS#: 106-99-0 | RTECS#: EI9275000 | IDLH: Ca [2000 ppm] [10%LEL] |
|---|---|---|---|---|

**Conversion:** 1 ppm = 2.21 mg/$m^3$ **DOT:** 1010 116P (inhibited)

**Synonyms/Trade Names:** Biethylene, Bivinyl, Butadiene, Divinyl, Erythrene, Vinylethylene

**Exposure Limits:**
**NIOSH REL:** Ca
See Appendix A
**OSHA PEL:** [1910.1051] TWA 1 ppm
ST 5 ppm

**Measurement Methods (see Table 1):**
**NIOSH** 1024
**OSHA** 56

**Physical Description:** Colorless gas with a mild aromatic or gasoline-like odor.
[**Note:** A liquid below 24°F. Shipped as a liquefied compressed gas.]

**Chemical & Physical Properties:**
**MW:** 54.1
**BP:** 24°F
**Sol:** Insoluble
**Fl.P:** NA (Gas)
-105°F (Liquid)
**IP:** 9.07 eV
**RGasD:** 1.88
**Sp.Gr:** 0.65 (Liquid at 24°F)
**VP:** 2.4 atm
**FRZ:** -164°F
**UEL:** 12.0%
**LEL:** 2.0%
Flammable Gas

**Personal Protection/Sanitation (see Table 2):**
**Skin:** Frostbite
**Eyes:** Frostbite
**Wash skin:** N.R.
**Remove:** When wet (flamm)
**Change:** N.R.
**Provide:** Frostbite wash

**Respirator Recommendations (see Tables 3 and 4):**
**NIOSH**
**¥:** ScbaF:Pd,Pp/SaF:Pd,Pp:AScba
**Escape:** GmFS/ScbaE

**See Appendix E** (page 351)

**Incompatibilities and Reactivities:** Phenol, chlorine dioxide, copper, crotonaldehyde [**Note:** May contain inhibitors (e.g., tributylcatechol) to prevent self-polymerization. May form explosive peroxides upon exposure to air.]

**Exposure Routes, Symptoms, Target Organs (see Table 5):**
**ER:** Inh, Con (liquid)
**SY:** Irrit eyes, nose, throat; drow, dizz; liquid: frostbite; terato, repro effects; [carc]
**TO:** Eyes, resp sys, CNS, repro sys [hemato cancer]

**First Aid (see Table 6):**
**Eye:** Frostbite
**Skin:** Frostbite
**Breath:** Resp support

## n-Butane

| n-Butane | Formula: $CH_3CH_2CH_2CH_3$ | CAS#: 106-97-8 | RTECS#: EJ4200000 | IDLH: N.D. |
|---|---|---|---|---|

**Conversion:** 1 ppm = 2.38 mg/$m^3$ **DOT:** 1011 115; 1075 115

**Synonyms/Trade Names:** normal-Butane, Butyl hydride, Diethyl, Methylethylmethane
[**Note:** Also see specific listing for Isobutane.]

**Exposure Limits:**
**NIOSH REL:** TWA 800 ppm (1900 mg/$m^3$)
**OSHA PEL†:** none

**Measurement Methods (see Table 1):**
**OSHA** 56

**Physical Description:** Colorless gas with a gasoline-like or natural gas odor.
[**Note:** Shipped as a liquefied compressed gas. A liquid below 31°F.]

**Chemical & Physical Properties:**
**MW:** 58.1
**BP:** 31°F
**Sol:** Slight
**Fl.P:** NA (Gas)
**IP:** 10.63 eV
**RGasD:** 2.11
**Sp.Gr:** 0.6 (Liquid at 31°F)
**VP:** 2.05 atm
**FRZ:** -217°F
**UEL:** 8.4%
**LEL:** 1.6%
Flammable Gas

**Personal Protection/Sanitation (see Table 2):**
**Skin:** Frostbite
**Eyes:** Frostbite
**Wash skin:** N.R.
**Remove:** When wet (flamm)
**Change:** N.R.
**Provide:** Frostbite wash

**Respirator Recommendations (see Tables 3 and 4):**
Not available.

**Incompatibilities and Reactivities:** Strong oxidizers (e.g., nitrates and perchlorates), chlorine, fluorine, (nickel carbonyl + oxygen)

**Exposure Routes, Symptoms, Target Organs (see Table 5):**
**ER:** Inh, Con (liquid)
**SY:** Drow, narco, asphy; liquid: frostbite
**TO:** CNS

**First Aid (see Table 6):**
**Eye:** Frostbite
**Skin:** Frostbite
**Breath:** Resp support

B

| 2-Butanone | Formula: $CH_3COCH_2CH_3$ | CAS#: 78-93-3 | RTECS#: EL6475000 | IDLH: 3000 ppm |
|---|---|---|---|---|
| **Conversion:** 1 ppm = 2.95 mg/m$^3$ | **DOT:** 1193 127 | | | |

**Synonyms/Trade Names:** Ethyl methyl ketone, MEK, Methyl acetone, Methyl ethyl ketone

**Exposure Limits:**
**NIOSH REL:** TWA 200 ppm (590 mg/m$^3$)
ST 300 ppm (885 mg/m$^3$)
**OSHA PEL†:** TWA 200 ppm (590 mg/m$^3$)

**Measurement Methods (see Table 1):**
**NIOSH** 2500, 2555, 3800
**OSHA** 16, 84, 1004

**Physical Description:** Colorless liquid with a moderately sharp, fragrant, mint- or acetone-like odor.

| Chemical & Physical Properties: | Personal Protection/Sanitation (see Table 2): | Respirator Recommendations (see Tables 3 and 4): |
|---|---|---|
| **MW:** 72.1<br>**BP:** 175°F<br>**Sol:** 28%<br>**Fl.P:** 16°F<br>**IP:** 9.54 eV<br>**Sp.Gr:** 0.81<br>**VP:** 78 mmHg<br>**FRZ:** -123°F<br>**UEL(200°F):** 11.4%<br>**LEL(200°F):** 1.4%<br>Class IB Flammable Liquid | **Skin:** Prevent skin contact<br>**Eyes:** Prevent eye contact<br>**Wash skin:** When contam<br>**Remove:** When wet (flamm)<br>**Change:** N.R.<br>**Provide:** Eyewash | **NIOSH/OSHA**<br>**3000 ppm:** Sa:Cf£/PaprOv£/CcrFOv/GmFOv/ScbaF/SaF<br>**§:** ScbaF:Pd,Pp/SaF:Pd,Pp:AScba<br>**Escape:** GmFOv/ScbaE |

**Incompatibilities and Reactivities:** Strong oxidizers, amines, ammonia, inorganic acids, caustics, isocyanates, pyridines

| Exposure Routes, Symptoms, Target Organs (see Table 5): | First Aid (see Table 6): |
|---|---|
| **ER:** Inh, Ing, Con<br>**SY:** Irrit eyes, skin, nose; head; dizz; vomit; derm<br>**TO:** Eyes, skin, resp sys, CNS | **Eye:** Irr immed<br>**Skin:** Water wash immed<br>**Breath:** Fresh air<br>**Swallow:** Medical attention immed |

| 2-Butoxyethanol | Formula: $C_4H_9OCH_2CH_2OH$ | CAS#: 111-76-2 | RTECS#: KJ8575000 | IDLH: 700 ppm |
|---|---|---|---|---|
| **Conversion:** 1 ppm = 4.83 mg/m$^3$ | **DOT:** 2369 152 | | | |

**Synonyms/Trade Names:** Butyl Cellosolve®, Butyl oxitol, Dowanol® EB, EGBE, Ektasolve EB®, Ethylene glycol monobutyl ether, Jeffersol EB

**Exposure Limits:**
**NIOSH REL:** TWA 5 ppm (24 mg/m$^3$) [skin]
**OSHA PEL†:** TWA 50 ppm (240 mg/m$^3$) [skin]

**Measurement Methods (see Table 1):**
**NIOSH** 1403
**OSHA** 83

**Physical Description:** Colorless liquid with a mild, ether-like odor.

| Chemical & Physical Properties: | Personal Protection/Sanitation (see Table 2): | Respirator Recommendations (see Tables 3 and 4): |
|---|---|---|
| **MW:** 118.2<br>**BP:** 339°F<br>**Sol:** Miscible<br>**Fl.P:** 143°F<br>**IP:** 10.00 eV<br>**Sp.Gr:** 0.90<br>**VP:** 0.8 mmHg<br>**FRZ:** -107°F<br>**UEL(275°F):** 12.7%<br>**LEL(200°F):** 1.1%<br>Class IIIA Combustible Liquid | **Skin:** Prevent skin contact<br>**Eyes:** Prevent eye contact<br>**Wash skin:** When contam<br>**Remove:** When wet or contam<br>**Change:** N.R.<br>**Provide:** Quick drench | **NIOSH**<br>**50 ppm:** CcrOv*/Sa*<br>**125 ppm:** Sa:Cf*/PaprOv*<br>**250 ppm:** CcrFOv/GmFOv/PaprTOv*/ScbaF/SaF<br>**700 ppm:** SaF:Pd,Pp<br>**§:** ScbaF:Pd,Pp/SaF:Pd,Pp:AScba<br>**Escape:** GmFOv/ScbaE |

**Incompatibilities and Reactivities:** Strong oxidizers, strong caustics

| Exposure Routes, Symptoms, Target Organs (see Table 5): | First Aid (see Table 6): |
|---|---|
| **ER:** Inh, Abs, Ing, Con<br>**SY:** Irrit eyes, skin, nose, throat; hemolysis, hema; CNS depres, head; vomit<br>**TO:** Eyes, skin, resp sys, CNS, hemato sys, blood, kidneys, liver, lymphoid sys | **Eye:** Irr immed<br>**Skin:** Soap wash prompt<br>**Breath:** Resp support<br>**Swallow:** Medical attention immed |

B

| 2-Butoxyethanol acetate | Formula: $C_4H_9O(CH_2)_2OCOCH_3$ | CAS#: 112-07-2 | RTECS#: KJ8925000 | IDLH: N.D. |
|---|---|---|---|---|
| **Conversion:** 1 ppm = 6.55 mg/m$^3$ | **DOT:** | | | |

**Synonyms/Trade Names:** 2-Butoxyethyl acetate, Butyl Cellosolve® acetate, Butyl glycol acetate, EGBEA, Ektasolve EB® acetate, Ethylene glycol monobutyl ether acetate

| Exposure Limits | Measurement Methods (see Table 1) |
|---|---|
| **NIOSH REL:** TWA 5 ppm (33 mg/m$^3$)<br>**OSHA PEL:** none | **OSHA** 83 |
| **Physical Description:** Colorless liquid with a pleasant, sweet, fruity odor. | |

| Chemical & Physical Properties | Personal Protection/Sanitation (see Table 2) | Respirator Recommendations (see Tables 3 and 4) |
|---|---|---|
| **MW:** 160.2<br>**BP:** 378°F<br>**Sol:** 1.5%<br>**Fl.P:** 160°F<br>**IP:** ?<br>**Sp.Gr:** 0.94<br>**VP:** 0.3 mmHg<br>**FRZ:** -82°F<br>**UEL(275°F):** 8.54%<br>**LEL(200°F):** 0.88%<br>Class IIIA Combustible Liquid | **Skin:** Prevent skin contact<br>**Eyes:** Prevent eye contact<br>**Wash skin:** When contam<br>**Remove:** When wet (flamm)<br>**Change:** N.R. | **NIOSH**<br>**50 ppm:** CcrOv*/Sa*<br>**125 ppm:** Sa:Cf*/PaprOv*<br>**250 ppm:** CcrFOv/GmFOv/PaprTOv*/ ScbaF/SaF<br>**700 ppm:** SaF:Pd,Pp<br>**§:** ScbaF:Pd,Pp/SaF:Pd,Pp:AScba<br>**Escape:** GmFOv/ScbaE |

**Incompatibilities and Reactivities:** Oxidizers

| Exposure Routes, Symptoms, Target Organs (see Table 5) | First Aid (see Table 6) |
|---|---|
| **ER:** Inh, Abs, Ing, Con<br>**SY:** Irrit eyes, skin, nose, throat; hemolysis, hema; CNS depres, head; vomit<br>**TO:** Eyes, skin, resp sys, CNS, hemato sys, blood, kidneys, liver, lymphoid sys | **Eye:** Irr immed<br>**Skin:** Soap wash prompt<br>**Breath:** Resp support<br>**Swallow:** Medical attention immed |

| n-Butyl acetate | Formula: $CH_3COO[CH_2]_3CH_3$ | CAS#: 123-86-4 | RTECS#: AF7350000 | IDLH: 1700 ppm [10%LEL] |
|---|---|---|---|---|
| **Conversion:** 1 ppm = 4.75 mg/m$^3$ | **DOT:** 1123 129 | | | |

**Synonyms/Trade Names:** Butyl acetate, n-Butyl ester of acetic acid, Butyl ethanoate

| Exposure Limits | Measurement Methods (see Table 1) |
|---|---|
| **NIOSH REL:** TWA 150 ppm (710 mg/m$^3$)<br>ST 200 ppm (950 mg/m$^3$)<br>**OSHA PEL†:** TWA 150 ppm (710 mg/m$^3$) | **NIOSH** 1450<br>**OSHA** 7 |
| **Physical Description:** Colorless liquid with a fruity odor. | |

| Chemical & Physical Properties | Personal Protection/Sanitation (see Table 2) | Respirator Recommendations (see Tables 3 and 4) |
|---|---|---|
| **MW:** 116.2<br>**BP:** 258°F<br>**Sol:** 1%<br>**Fl.P:** 72°F<br>**IP:** 10.00 eV<br>**Sp.Gr:** 0.88<br>**VP:** 10 mmHg<br>**FRZ:** -107°F<br>**UEL:** 7.6%<br>**LEL:** 1.7%<br>Class IB Flammable Liquid | **Skin:** Prevent skin contact<br>**Eyes:** Prevent eye contact<br>**Wash skin:** When contam<br>**Remove:** When wet (flamm)<br>**Change:** N.R. | **NIOSH/OSHA**<br>**1500 ppm:** CcrOv*/Sa*<br>**1700 ppm:** Sa:Cf*/PaprOv*/CcrFOv/ GmFOv/ScbaF/SaF<br>**§:** ScbaF:Pd,Pp/SaF:Pd,Pp:AScba<br>**Escape:** GmFOv/ScbaE |

**Incompatibilities and Reactivities:** Nitrates; strong oxidizers, alkalis & acids

| Exposure Routes, Symptoms, Target Organs (see Table 5) | First Aid (see Table 6) |
|---|---|
| **ER:** Inh, Ing, Con<br>**SY:** Irrit eyes, skin, upper resp sys; head, drow, narco<br>**TO:** Eyes, skin, resp sys, CNS | **Eye:** Irr immed<br>**Skin:** Water flush prompt<br>**Breath:** Resp support<br>**Swallow:** Medical attention immed |

B

## sec-Butyl acetate

| | | | | |
|---|---|---|---|---|
| **sec-Butyl acetate** | **Formula:** $CH_3COOCH(CH_3)CH_2CH_3$ | **CAS#:** 105-46-4 | **RTECS#:** AF7380000 | **IDLH:** 1700 ppm [10%LEL] |
| **Conversion:** 1 ppm = 4.75 mg/m$^3$ | **DOT:** 1123 129 | | | |

**Synonyms/Trade Names:** sec-Butyl ester of acetic acid, 1-Methylpropyl acetate

**Exposure Limits:**
**NIOSH REL:** TWA 200 ppm (950 mg/m$^3$)
**OSHA PEL:** TWA 200 ppm (950 mg/m$^3$)

**Measurement Methods (see Table 1):**
**NIOSH** 1450
**OSHA** 7

**Physical Description:** Colorless liquid with a pleasant, fruity odor.

| **Chemical & Physical Properties:** | **Personal Protection/Sanitation (see Table 2):** | **Respirator Recommendations (see Tables 3 and 4):** |
|---|---|---|
| **MW:** 116.2<br>**BP:** 234°F<br>**Sol:** 0.8%<br>**Fl.P:** 62°F<br>**IP:** 9.91 eV<br>**Sp.Gr:** 0.86<br>**VP:** 10 mmHg<br>**FRZ:** -100°F<br>**UEL:** 9.8%<br>**LEL:** 1.7%<br>Class IB Flammable Liquid | **Skin:** Prevent skin contact<br>**Eyes:** Prevent eye contact<br>**Wash skin:** When contam<br>**Remove:** When wet (flamm)<br>**Change:** N.R. | **NIOSH/OSHA**<br>**1700 ppm:** Sa:Cf£/PaprOv£/CcrFOv/GmFOv/ScbaF/SaF<br>**§:** ScbaF:Pd,Pp/SaF:Pd,Pp:AScba<br>**Escape:** GmFOv/ScbaE |

**Incompatibilities and Reactivities:** Nitrates; strong oxidizers, alkalis & acids

| **Exposure Routes, Symptoms, Target Organs (see Table 5):** | **First Aid (see Table 6):** |
|---|---|
| **ER:** Inh, Ing, Con<br>**SY:** Irrit eyes; head; drow; dryness upper resp sys, skin; narco<br>**TO:** Eyes, skin, resp sys, CNS | **Eye:** Irr immed<br>**Skin:** Water flush prompt<br>**Breath:** Resp support<br>**Swallow:** Medical attention immed |

## tert-Butyl acetate

| | | | | |
|---|---|---|---|---|
| **tert-Butyl acetate** | **Formula:** $CH_3COOC(CH_3)_3$ | **CAS#:** 540-88-5 | **RTECS#:** AF7400000 | **IDLH:** 1500 ppm [10%LEL] |
| **Conversion:** 1 ppm = 4.75 mg/m$^3$ | **DOT:** 1123 129 | | | |

**Synonyms/Trade Names:** tert-Butyl ester of acetic acid

**Exposure Limits:**
**NIOSH REL:** TWA 200 ppm (950 mg/m$^3$)
**OSHA PEL:** TWA 200 ppm (950 mg/m$^3$)

**Measurement Methods (see Table 1):**
**NIOSH** 1450
**OSHA** 7

**Physical Description:** Colorless liquid with a fruity odor.

| **Chemical & Physical Properties:** | **Personal Protection/Sanitation (see Table 2):** | **Respirator Recommendations (see Tables 3 and 4):** |
|---|---|---|
| **MW:** 116.2<br>**BP:** 208°F<br>**Sol:** Insoluble<br>**Fl.P:** 72°F<br>**IP:** ?<br>**Sp.Gr:** 0.87<br>**VP:** ?<br>**FRZ:** ?<br>**UEL:** ?<br>**LEL:** 1.5%<br>Class IB Flammable Liquid | **Skin:** Prevent skin contact<br>**Eyes:** Prevent eye contact<br>**Wash skin:** When contam<br>**Remove:** When wet (flamm)<br>**Change:** N.R. | **NIOSH/OSHA**<br>**1500 ppm:** Sa:Cf£/PaprOv£/CcrFOv/GmFOv/ScbaF/SaF<br>**§:** ScbaF:Pd,Pp/SaF:Pd,Pp:AScba<br>**Escape:** GmFOv/ScbaE |

**Incompatibilities and Reactivities:** Nitrates; strong oxidizers, alkalis & acids

| **Exposure Routes, Symptoms, Target Organs (see Table 5):** | **First Aid (see Table 6):** |
|---|---|
| **ER:** Inh, Ing, Con<br>**SY:** Itch, inflamm eyes; irrit upper resp tract; head; narco; derm<br>**TO:** Resp sys, eyes, skin, CNS | **Eye:** Irr immed<br>**Skin:** Water flush prompt<br>**Breath:** Resp support<br>**Swallow:** Medical attention immed |

B

## Butyl acrylate

| Butyl acrylate | Formula: $CH_2=CHCOOC_4H_9$ | CAS#: 141-32-2 | RTECS#: UD3150000 | IDLH: N.D. |
|---|---|---|---|---|
| **Conversion:** 1 ppm = 5.24 mg/m$^3$ | **DOT:** 2348 130P | | | |

**Synonyms/Trade Names:** n-Butyl acrylate, Butyl ester of acrylic acid, Butyl-2-propenoate

**Exposure Limits:**
**NIOSH REL:** TWA 10 ppm (55 mg/m$^3$)
**OSHA PEL†:** none

**Measurement Methods (see Table 1):**
**OSHA** PV2011

**Physical Description:** Clear, colorless liquid with a strong, fruity odor. [**Note:** Highly reactive; may contain an inhibitor to prevent spontaneous polymerization.]

**Chemical & Physical Properties:**
**MW:** 128.2
**BP:** 293°F
**Sol:** 0.1%
**Fl.P:** 103°F
**IP:** ?
**Sp.Gr:** 0.89
**VP:** 4 mmHg
**FRZ:** -83°F
**UEL:** 9.9%
**LEL:** 1.5%
Class II Combustible Liquid

**Personal Protection/Sanitation (see Table 2):**
**Skin:** Prevent skin contact
**Eyes:** Prevent eye contact
**Wash skin:** When contam
**Remove:** When wet or contam
**Change:** N.R.
**Provide:** Eyewash
Quick drench

**Respirator Recommendations (see Tables 3 and 4):**
Not available.

**Incompatibilities and Reactivities:** Strong acids & alkalis, amines, halogens, hydrogen compounds, oxidizers, heat, flame, sunlight [**Note:** Polymerizes readily on heating.]

**Exposure Routes, Symptoms, Target Organs (see Table 5):**
**ER:** Inh, Abs, Ing, Con
**SY:** Irrit eyes, skin, upper resp sys; sens derm; dysp
**TO:** Eyes, skin, resp sys

**First Aid (see Table 6):**
**Eye:** Irr immed
**Skin:** Soap wash immed
**Breath:** Resp support
**Swallow:** Medical attention immed

## n-Butyl alcohol

| n-Butyl alcohol | Formula: $CH_3CH_2CH_2CH_2OH$ | CAS#: 71-36-3 | RTECS#: EO1400000 | IDLH: 1400 ppm [10%LEL] |
|---|---|---|---|---|
| **Conversion:** 1 ppm = 3.03 mg/m$^3$ | **DOT:** 1120 129 | | | |

**Synonyms/Trade Names:** 1-Butanol, n-Butanol, Butyl alcohol, 1-Hydroxybutane, n-Propyl carbinol

**Exposure Limits:**
**NIOSH REL:** C 50 ppm (150 mg/m$^3$) [skin]
**OSHA PEL†:** TWA 100 ppm (300 mg/m$^3$)

**Measurement Methods (see Table 1):**
**NIOSH** 1401, 1405
**OSHA** 7

**Physical Description:** Colorless liquid with a strong, characteristic, mildly alcoholic odor.

**Chemical & Physical Properties:**
**MW:** 74.1
**BP:** 243°F
**Sol:** 9%
**Fl.P:** 84°F
**IP:** 10.04 eV
**Sp.Gr:** 0.81
**VP:** 6 mmHg
**FRZ:** -129°F
**UEL:** 11.2%
**LEL:** 1.4%
Class IC Flammable Liquid

**Personal Protection/Sanitation (see Table 2):**
**Skin:** Prevent skin contact
**Eyes:** Prevent eye contact
**Wash skin:** When contam
**Remove:** When wet (flamm)
**Change:** N.R.

**Respirator Recommendations (see Tables 3 and 4):**
**NIOSH**
**1250 ppm:** Sa:Cf£/PaprOv£
**1400 ppm:** CcrFOv/GmFOv/PaprTOv£/ScbaF/SaF
**§:** ScbaF:Pd,Pp/SaF:Pd,Pp:AScba
**Escape:** GmFOv/ScbaE

**Incompatibilities and Reactivities:** Strong oxidizers, strong mineral acids, alkali metals, halogens

**Exposure Routes, Symptoms, Target Organs (see Table 5):**
**ER:** Inh, Abs, Ing, Con
**SY:** Irrit eyes, nose, throat; head, dizz, drow; corn inflamm, blurred vision, lac, photo; derm; possible auditory nerve damage, hearing loss; CNS depres
**TO:** Eyes, skin, resp sys, CNS

**First Aid (see Table 6):**
**Eye:** Irr immed
**Skin:** Water flush prompt
**Breath:** Resp support
**Swallow:** Medical attention immed

## sec-Butyl alcohol

| Formula: | CAS#: | RTECS#: | IDLH: |
|---|---|---|---|
| $CH_3CH(OH)CH_2CH_3$ | 78-92-2 | EO1750000 | 2000 ppm |

**Conversion:** 1 ppm = 3.03 mg/m$^3$ **DOT:** 1120 129

**Synonyms/Trade Names:** 2-Butanol, Butylene hydrate, 2-Hydroxybutane, Methyl ethyl carbinol

**Exposure Limits:**
**NIOSH REL:** TWA 100 ppm (305 mg/m$^3$)
ST 150 ppm (455 mg/m$^3$)
**OSHA PEL†:** TWA 150 ppm (450 mg/m$^3$)

**Measurement Methods (see Table 1):**
**NIOSH** 1401, 1405
**OSHA** 7

**Physical Description:** Colorless liquid with a strong, pleasant odor.

**Chemical & Physical Properties:**
**MW:** 74.1
**BP:** 211°F
**Sol:** 16%
**Fl.P:** 75°F
**IP:** 10.10 eV
**Sp.Gr:** 0.81
**VP:** 12 mmHg
**FRZ:** -175°F
**UEL(212°F):** 9.8%
**LEL(212°F):** 1.7%
Class IC Flammable Liquid

**Personal Protection/Sanitation (see Table 2):**
**Skin:** Prevent skin contact
**Eyes:** Prevent eye contact
**Wash skin:** When contam
**Remove:** When wet (flamm)
**Change:** N.R.

**Respirator Recommendations (see Tables 3 and 4):**
**NIOSH**
**1000 ppm:** CcrOv*/Sa*
**2000 ppm:** Sa:Cf*/PaprOv*/CcrFOv/GmFOv/ScbaF/SaF
**§:** ScbaF:Pd,Pp/SaF:Pd,Pp:AScba
**Escape:** GmFOv/ScbaE

**Incompatibilities and Reactivities:** Strong oxidizers, organic peroxides, perchloric & permonosulfuric acids

**Exposure Routes, Symptoms, Target Organs (see Table 5):**
**ER:** Inh, Ing, Con
**SY:** Irrit eyes, skin, nose, throat; narco
**TO:** Eyes, skin, resp sys, CNS

**First Aid (see Table 6):**
**Eye:** Irr immed
**Skin:** Water flush prompt
**Breath:** Resp support
**Swallow:** Medical attention immed

## tert-Butyl alcohol

| Formula: | CAS#: | RTECS#: | IDLH: |
|---|---|---|---|
| $(CH_3)_3COH$ | 75-65-0 | EO1925000 | 1600 ppm |

**Conversion:** 1 ppm = 3.03 mg/m$^3$ **DOT:** 1120 129

**Synonyms/Trade Names:** 2-Methyl-2-propanol, Trimethyl carbinol

**Exposure Limits:**
**NIOSH REL:** TWA 100 ppm (300 mg/m$^3$)
ST 150 ppm (450 mg/m$^3$)
**OSHA PEL†:** TWA 100 ppm (300 mg/m$^3$)

**Measurement Methods (see Table 1):**
**NIOSH** 1400
**OSHA** 7

**Physical Description:** Colorless solid or liquid (above 77°F) with a camphor-like odor.
[**Note:** Often used in aqueous solutions.]

**Chemical & Physical Properties:**
**MW:** 74.1
**BP:** 180°F
**Sol:** Miscible
**Fl.P:** 52°F
**IP:** 9.70 eV
**Sp.Gr:** 0.79 (Solid)
**VP(77°F):** 42 mmHg
**FRZ:** 78°F
**UEL:** 8.0%
**LEL:** 2.4%
Combustible Solid
Class IB Flammable Liquid

**Personal Protection/Sanitation (see Table 2):**
**Skin:** Prevent skin contact
**Eyes:** Prevent eye contact
**Wash skin:** When contam
**Remove:** When wet (flamm)
**Change:** N.R.

**Respirator Recommendations (see Tables 3 and 4):**
**NIOSH/OSHA**
**1600 ppm:** Sa:Cf£/PaprOv£/CcrFOv/GmFOv/ScbaF/SaF
**§:** ScbaF:Pd,Pp/SaF:Pd,Pp:AScba
**Escape:** GmFOv/ScbaE

**Incompatibilities and Reactivities:** Strong mineral acids, strong hydrochloric acid, oxidizers

**Exposure Routes, Symptoms, Target Organs (see Table 5):**
**ER:** Inh, Ing, Con
**SY:** Irrit eyes, skin, nose, throat; drow, narco
**TO:** Eyes, skin, resp sys, CNS

**First Aid (see Table 6):**
**Eye:** Irr immed
**Skin:** Water flush prompt
**Breath:** Resp support
**Swallow:** Medical attention immed

## n-Butylamine

| Formula: | CAS#: | RTECS#: | IDLH: |
|---|---|---|---|
| $CH_3CH_2CH_2CH_2NH_2$ | 109-73-9 | EO2975000 | 300 ppm |

**Conversion:** 1 ppm = 2.99 mg/m$^3$
**DOT:** 1125 132

**Synonyms/Trade Names:** 1-Aminobutane, Butylamine

**Exposure Limits:**
**NIOSH REL:** C 5 ppm (15 mg/m$^3$) [skin]
**OSHA PEL:** C 5 ppm (15 mg/m$^3$) [skin]

**Measurement Methods (see Table 1):**
**NIOSH** 2012

**Physical Description:** Colorless liquid with a fishy, ammonia-like odor.

**Chemical & Physical Properties:**
**MW:** 73.2
**BP:** 172°F
**Sol:** Miscible
**Fl.P:** 10°F
**IP:** 8.71 eV
**Sp.Gr:** 0.74
**VP:** 82 mmHg
**FRZ:** -58°F
**UEL:** 9.8%
**LEL:** 1.7%
Class IB Flammable Liquid

**Personal Protection/Sanitation (see Table 2):**
**Skin:** Prevent skin contact
**Eyes:** Prevent eye contact
**Wash skin:** When contam
**Remove:** When wet (flamm)
**Change:** N.R.
**Provide:** Eyewash
Quick drench

**Respirator Recommendations (see Tables 3 and 4):**
**NIOSH/OSHA**
**50 ppm:** CcrS*/Sa*
**125 ppm:** Sa:Cf*/PaprS*
**250 ppm:** CcrFS/GmFS/PaprTS*/ ScbaF/SaF
**300 ppm:** SaF:Pd,Pp
**§:** ScbaF:Pd,Pp/SaF:Pd,Pp:AScba
**Escape:** GmFS/ScbaE

**Incompatibilities and Reactivities:** Strong oxidizers, strong acids
[**Note:** May corrode some metals in presence of water.]

**Exposure Routes, Symptoms, Target Organs (see Table 5):**
**ER:** Inh, Abs, Ing, Con
**SY:** Irrit eyes, skin, nose, throat; head; skin flush, burns
**TO:** Eyes, skin, resp sys

**First Aid (see Table 6):**
**Eye:** Irr immed
**Skin:** Water flush immed
**Breath:** Resp support
**Swallow:** Medical attention immed

## tert-Butyl chromate

| Formula: | CAS#: | RTECS#: | IDLH: |
|---|---|---|---|
| $[(CH_3)_3CO]_2CrO_2$ | 1189-85-1 | GB2900000 | Ca [15 mg/m$^3$ {as Cr(VI)}] |

**Conversion:**
**DOT:**

**Synonyms/Trade Names:** di-tert-Butyl ester of chromic acid

**Exposure Limits:**
**NIOSH REL:** Ca
TWA 0.001 mg Cr(VI)/m$^3$
See Appendix A
See Appendix C
**OSHA PEL:** TWA 0.005 mg $CrO_3$/m$^3$ [skin]
See Appendix C

**Measurement Methods (see Table 1):**
**NIOSH** 7604
**OSHA** ID103, ID215

**Physical Description:** Liquid. [**Note:** Solidifies at 32-23°F.]

**Chemical & Physical Properties:**
**MW:** 230.3
**BP:** ?
**Sol:** ?
**Fl.P:** ?
**IP:** ?
**Sp.Gr:** ?
**VP:** ?
**FRZ:** 32-23°F
**UEL:** ?
**LEL:** ?

**Personal Protection/Sanitation (see Table 2):**
**Skin:** Prevent skin contact
**Eyes:** Prevent eye contact
**Wash skin:** When contam/Daily
**Remove:** When wet or contam
**Change:** N.R.
**Provide:** Eyewash
Quick drench

**Respirator Recommendations (see Tables 3 and 4):**
**NIOSH**
**¥:** ScbaF:Pd,Pp/SaF:Pd,Pp:AScba
**Escape:** GmFOv100/ScbaE

**Incompatibilities and Reactivities:** Reducing agents, moisture, acids, alcohols, hydrazine, combustible materials

**Exposure Routes, Symptoms, Target Organs (see Table 5):**
**ER:** Inh, Abs, Ing, Con
**SY:** Irrit eyes, skin, resp sys; eye, skin burns; drow, musc weak; skin ulcers; lung changes; [carc]
**TO:** Eyes, skin, resp sys, CNS [lung cancer]

**First Aid (see Table 6):**
**Eye:** Irr immed
**Skin:** Soap wash immed
**Breath:** Resp support
**Swallow:** Medical attention immed

## n-Butyl glycidyl ether

**Formula:** $C_7H_{14}O_2$ | **CAS#:** 2426-08-6 | **RTECS#:** TX4200000 | **IDLH:** 250 ppm

**Conversion:** 1 ppm = 5.33 mg/m$^3$ | **DOT:**

**Synonyms/Trade Names:** BGE; 1,2-Epoxy-3-butoxypropane

**Exposure Limits:**
**NIOSH REL:** C 5.6 ppm (30 mg/m$^3$) [15-minute]
**OSHA PEL†:** TWA 50 ppm (270 mg/m$^3$)

**Measurement Methods (see Table 1):**
**NIOSH** 1616
**OSHA** 7

**Physical Description:** Colorless liquid with an irritating odor.

**Chemical & Physical Properties:**
**MW:** 130.2
**BP:** 327°F
**Sol:** 2%
**Fl.P:** 130°F
**IP:** ?
**Sp.Gr:** 0.91
**VP(77°F):** 3 mmHg
**FRZ:** ?
**UEL:** ?
**LEL:** ?
Class II Combustible Liquid

**Personal Protection/Sanitation (see Table 2):**
**Skin:** Prevent skin contact
**Eyes:** Prevent eye contact
**Wash skin:** When contam
**Remove:** When wet or contam
**Change:** N.R.

**Respirator Recommendations (see Tables 3 and 4):**
**NIOSH**
**56 ppm:** CcrOv*/Sa*
**140 ppm:** Sa:Cf*/PaprOv*
**250 ppm:** CcrFOv/GmFOv/PaprTOv*/ScbaF/SaF
**§:** ScbaF:Pd,Pp/SaF:Pd,Pp:AScba
**Escape:** GmFOv/ScbaE

**Incompatibilities and Reactivities:** Strong oxidizers, strong caustics

**Exposure Routes, Symptoms, Target Organs (see Table 5):**
**ER:** Inh, Ing, Con
**SY:** Irrit eyes, skin, nose; skin sens; narco; possible hemato effects; CNS depres
**TO:** Eyes, skin, resp sys, CNS, blood

**First Aid (see Table 6):**
**Eye:** Irr immed
**Skin:** Soap wash immed
**Breath:** Resp support
**Swallow:** Medical attention immed

## n-Butyl lactate

**Formula:** $CH_3CH(OH)COOC_4H_9$ | **CAS#:** 138-22-7 | **RTECS#:** OD4025000 | **IDLH:** N.D.

**Conversion:** 1 ppm = 5.98 mg/m$^3$ | **DOT:** 1993 128 (combustible liquid, n.o.s.)

**Synonyms/Trade Names:** Butyl ester of 2-hydroxypropanoic acid, Butyl ester of lactic acid, Butyl lactate

**Exposure Limits:**
**NIOSH REL:** TWA 5 ppm (25 mg/m$^3$)
**OSHA PEL†:** none

**Measurement Methods (see Table 1):**
None available

**Physical Description:** Clear, colorless to white liquid with a mild, transient odor.

**Chemical & Physical Properties:**
**MW:** 146.2
**BP:** 370°F
**Sol:** Slight
**Fl.P:** 160°F
**IP:** ?
**Sp.Gr:** 0.98
**VP:** 0.4 mmHg
**FRZ:** -45°F
**UEL:** ?
**LEL:** 1.15%
Class IIIA Combustible Liquid

**Personal Protection/Sanitation (see Table 2):**
**Skin:** Prevent skin contact
**Eyes:** Prevent eye contact
**Wash skin:** When contam
**Remove:** When wet or contam
**Change:** N.R.
**Provide:** Eyewash
Quick drench

**Respirator Recommendations (see Tables 3 and 4):**
Not available.

**Incompatibilities and Reactivities:** Strong acids & bases, strong oxidizers, heat, sparks, open flames

**Exposure Routes, Symptoms, Target Organs (see Table 5):**
**ER:** Inh, Ing, Con
**SY:** Irrit eyes, skin, nose, throat; drow, head, CNS depres; nau, vomit
**TO:** Eyes, skin, resp sys, CNS

**First Aid (see Table 6):**
**Eye:** Irr immed
**Skin:** Soap wash immed
**Breath:** Resp support
**Swallow:** Medical attention immed

| n-Butyl mercaptan | Formula: $CH_3CH_2CH_2CH_2SH$ | CAS#: 109-79-5 | RTECS#: EK6300000 | IDLH: 500 ppm |
|---|---|---|---|---|
| **Conversion:** 1 ppm = 3.69 mg/m$^3$ | **DOT:** 2347 130 | | | |

B

**Synonyms/Trade Names:** Butanethiol, 1-Butanethiol, n-Butanethiol, 1-Mercaptobutane

**Exposure Limits:**
**NIOSH REL:** C 0.5 ppm (1.8 mg/m$^3$) [15-minute]
**OSHA PEL†:** TWA 10 ppm (35 mg/m$^3$)

**Measurement Methods (see Table 1):**
**NIOSH** 2525, 2542

**Physical Description:** Colorless liquid with a strong, garlic-, cabbage-, or skunk-like odor.

| Chemical & Physical Properties: | Personal Protection/Sanitation (see Table 2): | Respirator Recommendations (see Tables 3 and 4): |
|---|---|---|
| **MW:** 90.2<br>**BP:** 209°F<br>**Sol:** 0.06%<br>**Fl.P:** 35°F<br>**IP:** 9.15 eV<br>**Sp.Gr:** 0.83<br>**VP:** 35 mmHg<br>**FRZ:** -176°F<br>**UEL:** ?<br>**LEL:** ?<br>Class IB Flammable Liquid | **Skin:** Prevent skin contact<br>**Eyes:** Prevent eye contact<br>**Wash skin:** When contam<br>**Remove:** When wet (flamm)<br>**Change:** N.R. | **NIOSH**<br>**5 ppm:** CcrOv/Sa<br>**12.5 ppm:** Sa:Cf/PaprOv<br>**25 ppm:** CcrFOv/GmFOv/PaprTOv/ScbaF/SaF<br>**500 ppm:** Sa:Pd,Pp*<br>**§:** ScbaF:Pd,Pp/SaF:Pd,Pp:AScba<br>**Escape:** GmFOv/ScbaE |

**Incompatibilities and Reactivities:** Strong oxidizers (such as dry bleaches), acids

**Exposure Routes, Symptoms, Target Organs (see Table 5):**
**ER:** Inh, Ing, Con
**SY:** Irrit eyes, skin; musc weak, mal, sweat, nau, vomit, head, conf; in animals: narco, inco, lass; cyan, pulm irrit; liver, kidney damage
**TO:** Eyes, skin, resp sys, CNS, liver, kidneys

**First Aid (see Table 6):**
**Eye:** Irr immed
**Skin:** Soap wash prompt
**Breath:** Resp support
**Swallow:** Medical attention immed

| o-sec-Butylphenol | Formula: $CH_3CH_2CH(CH_3)C_6H_4OH$ | CAS#: 89-72-5 | RTECS#: SJ8920000 | IDLH: N.D. |
|---|---|---|---|---|
| **Conversion:** 1 ppm = 6.14 mg/m$^3$ | **DOT:** | | | |

**Synonyms/Trade Names:** 2-sec-Butylphenol; 2-(1-Methylpropyl)phenol

**Exposure Limits:**
**NIOSH REL:** TWA 5 ppm (30 mg/m$^3$) [skin]
**OSHA PEL†:** none

**Measurement Methods (see Table 1):**
None available

**Physical Description:** Colorless liquid or solid (below 61°F).

| Chemical & Physical Properties: | Personal Protection/Sanitation (see Table 2): | Respirator Recommendations (see Tables 3 and 4): |
|---|---|---|
| **MW:** 150.2<br>**BP:** 227°F<br>**Sol:** Insoluble<br>**Fl.P:** 225°F<br>**IP:** ?<br>**Sp.Gr:** 0.89<br>**VP:** Low<br>**FRZ:** 61°F<br>**UEL:** ?<br>**LEL:** ?<br>Class IIB Combustible Liquid<br>Combustible Solid | **Skin:** Prevent skin contact<br>**Eyes:** Prevent eye contact<br>**Wash skin:** When contam<br>**Remove:** When wet or contam<br>**Change:** N.R.<br>**Provide:** Eyewash<br>Quick drench | Not available. |

**Incompatibilities and Reactivities:** None reported

**Exposure Routes, Symptoms, Target Organs (see Table 5):**
**ER:** Inh, Abs, Ing, Con
**SY:** Irrit eyes, skin, resp sys; skin burns
**TO:** Eyes, skin, resp sys

**First Aid (see Table 6):**
**Eye:** Irr immed
**Skin:** Soap flush immed
**Breath:** Resp support
**Swallow:** Medical attention immed

B

## p-tert-Butyltoluene

| Formula: | CAS#: | RTECS#: | IDLH: |
|---|---|---|---|
| $(CH_3)_3CC_6H_4CH_3$ | 98-51-1 | XS8400000 | 100 ppm |

**Conversion:** 1 ppm = 6.07 mg/m$^3$ | **DOT:** 2667 152

**Synonyms/Trade Names:** 4-tert-Butyltoluene, 1-Methyl-4-tert-butylbenzene

**Exposure Limits:**
**NIOSH REL:** TWA 10 ppm (60 mg/m$^3$)
ST 20 ppm (120 mg/m$^3$)
**OSHA PEL†:** TWA 10 ppm (60 mg/m$^3$)

**Measurement Methods (see Table 1):**
**NIOSH** 1501
**OSHA** 7

**Physical Description:** Colorless liquid with a distinct aromatic odor, somewhat like gasoline.

**Chemical & Physical Properties:**
**MW:** 148.3
**BP:** 379°F
**Sol:** Insoluble
**Fl.P:** 155°F
**IP:** 8.28 eV
**Sp.Gr:** 0.86
**VP(77°F):** 0.7 mmHg
**FRZ:** -62°F
**UEL:** ?
**LEL:** ?
Class IIIA Combustible Liquid

**Personal Protection/Sanitation (see Table 2):**
**Skin:** Prevent skin contact
**Eyes:** Prevent eye contact
**Wash skin:** When contam
**Remove:** When wet or contam
**Change:** N.R.

**Respirator Recommendations (see Tables 3 and 4):**
**NIOSH/OSHA**
**100 ppm:** Sa:Cf£/PaprOv£/CcrFOv/GmFOv/ScbaF/SaF
**§:** ScbaF:Pd,Pp/SaF:Pd,Pp:AScba
**Escape:** GmFOv/ScbaE

**Incompatibilities and Reactivities:** Oxidizers

**Exposure Routes, Symptoms, Target Organs (see Table 5):**
**ER:** Inh, Ing, Con
**SY:** Irrit eyes, skin; dry nose, throat; head; low BP, tacar, abnor CVS stress; CNS, hemato depres; metallic taste; liver, kidney inj
**TO:** Eyes, skin, resp sys, CVS, CNS, bone marrow, liver, kidneys

**First Aid (see Table 6):**
**Eye:** Irr immed
**Skin:** Water flush prompt
**Breath:** Resp support
**Swallow:** Medical attention immed

## n-Butyronitrile

| Formula: | CAS#: | RTECS#: | IDLH: |
|---|---|---|---|
| $CH_3CH_2CH_2CN$ | 109-74-0 | ET8750000 | N.D. |

**Conversion:** 1 ppm = 2.83 mg/m$^3$ | **DOT:** 2411 131

**Synonyms/Trade Names:** Butanenitrile, Butyronitrile, 1-Cyanopropane, Propyl cyanide, n-Propyl cyanide

**Exposure Limits:**
**NIOSH REL:** TWA 8 ppm (22 mg/m$^3$)
**OSHA PEL:** none

**Measurement Methods (see Table 1):**
**NIOSH** 1606 (adapt)

**Physical Description:** Colorless liquid with a sharp, suffocating odor.
[**Note:** Forms cyanide in the body.]

**Chemical & Physical Properties:**
**MW:** 69.1
**BP:** 244°F
**Sol(77°F):** 3%
**Fl.P:** 62°F
**IP:** 11.67 eV
**Sp.Gr:** 0.81
**VP:** 14 mmHg
**FRZ:** -170°F
**UEL:** ?
**LEL:** 1.65%
Class IB Flammable Liquid

**Personal Protection/Sanitation (see Table 2):**
**Skin:** Prevent skin contact
**Eyes:** Prevent eye contact
**Wash skin:** When contam
**Remove:** When wet (flamm)
**Change:** N.R.
**Provide:** Quick drench

**Respirator Recommendations (see Tables 3 and 4):**
**NIOSH**
**80 ppm:** CcrOv/Sa
**200 ppm:** Sa:Cf/PaprOv
**400 ppm:** CcrFOv/GmFOv/PaprTOv/ScbaF/SaF
**1000 ppm:** SaF:Pd,Pp
**§:** ScbaF:Pd,Pp/SaF:Pd,Pp:AScba
**Escape:** GmFOv/ScbaE

**Incompatibilities and Reactivities:** Strong oxidizers & reducing agents, strong acids & bases

**Exposure Routes, Symptoms, Target Organs (see Table 5):**
**ER:** Inh, Abs, Ing, Con
**SY:** Irrit eyes, skin, resp sys; head, dizz, lass, conf, convuls; dysp; abdom pain, nau, vomit
**TO:** Eyes, skin, resp sys, CNS, CVS

**First Aid (see Table 6):**
**Eye:** Irr immed
**Skin:** Soap wash immed
**Breath:** Resp support
**Swallow:** Medical attention immed

## Cadmium dust (as Cd)

| Formula: | CAS#: | RTECS#: | IDLH: |
|---|---|---|---|
| Cd (metal) | 7440-43-9 (metal) | EU9800000 (metal) | Ca [9 $mg/m^3$ (as Cd)] |

**Conversion:**
**DOT:** 2570 154 (cadmium compound)

**Synonyms/Trade Names: Cadmium metal:** Cadmium
Other synonyms vary depending upon the specific cadmium compound.

**Exposure Limits:**
**NIOSH REL*:** Ca
See Appendix A
**OSHA PEL*:** [1910.1027] TWA 0.005 $mg/m^3$
[***Note:** The REL and PEL apply to all Cadmium compounds (as Cd).]

**Measurement Methods (see Table 1):**
**NIOSH** 7048, 7300, 7301, 7303, 9102
**OSHA** ID121, ID125G, ID189, ID206

**Physical Description:** Metal: Silver-white, blue-tinged lustrous, odorless solid.

**Chemical & Physical Properties:**
**MW:** 112.4
**BP:** 1409°F
**Sol:** Insoluble
**Fl.P:** NA
**IP:** NA
**Sp.Gr:** 8.65 (metal)
**VP:** 0 mmHg (approx)
**MLT:** 610°F
**UEL:** NA
**LEL:** NA
Metal: Noncombustible Solid in bulk form, but will burn in powder form.

**Personal Protection/Sanitation (see Table 2):**
**Skin:** N.R.
**Eyes:** N.R.
**Wash skin:** Daily
**Remove:** N.R.
**Change:** Daily

**Respirator Recommendations (see Tables 3 and 4):**
**NIOSH**
**¥:** ScbaF:Pd,Pp/SaF:Pd,Pp:AScba
**Escape:** 100F/ScbaE

**See Appendix E** (page 351)

**Incompatibilities and Reactivities:** Strong oxidizers; elemental sulfur, selenium & tellurium

**Exposure Routes, Symptoms, Target Organs (see Table 5):**
**ER:** Inh, Ing
**SY:** Pulm edema, dysp, cough, chest tight, subs pain; head; chills, musc aches; nau, vomit, diarr; anos, emphy, prot, mild anemia; [carc]
**TO:** Resp sys, kidneys, prostate, blood [prostatic & lung cancer]

**First Aid (see Table 6):**
**Eye:** Irr immed
**Skin:** Soap wash
**Breath:** Resp support
**Swallow:** Medical attention immed

## Cadmium fume (as Cd)

| Formula: | CAS#: | RTECS#: | IDLH: |
|---|---|---|---|
| CdO/Cd | 1306-19-0 (CdO) | EV1930000 (CdO) | Ca [9 $mg/m^3$ (as Cd)] |

**Conversion:**
**DOT:**

**Synonyms/Trade Names: CdO:** Cadmium monoxide, Cadmium oxide fume **Cd:** Cadmium

**Exposure Limits:**
**NIOSH REL*:** Ca
See Appendix A
**OSHA PEL*:** [1910.1027] TWA 0.005 $mg/m^3$
[***Note:** The REL and PEL apply to all Cadmium compounds (as Cd).]

**Measurement Methods (see Table 1):**
**NIOSH** 7048, 7300, 7301, 7303
**OSHA** ID121, ID125G, ID189, ID206

**Physical Description:** Odorless, yellow-brown, finely divided particulate dispersed in air. [**Note:** See listing for Cadmium dust for properties of Cd.]

**Chemical & Physical Properties:**
**MW:** 128.4
**BP:** Decomposes
**Sol:** Insoluble
**Fl.P:** NA
**IP:** NA
**Sp.Gr:** 8.15 (crystalline form)
6.95 (amorphous form)
**VP:** 0 mmHg (approx)
**MLT:** 2599°F
**UEL:** NA
**LEL:** NA
Noncombustible Solid

**Personal Protection/Sanitation (see Table 2):**
**Skin:** N.R.
**Eyes:** N.R.
**Wash skin:** Daily
**Remove:** N.R.
**Change:** Daily

**Respirator Recommendations (see Tables 3 and 4):**
**NIOSH**
**¥:** ScbaF:Pd,Pp/SaF:Pd,Pp:AScba
**Escape:** 100F/ScbaE

**See Appendix E** (page 351)

**Incompatibilities and Reactivities:** Not applicable

**Exposure Routes, Symptoms, Target Organs (see Table 5):**
**ER:** Inh
**SY:** Pulm edema, dysp, cough, chest tight, subs pain; head; chills, musc aches; nau, vomit, diarr; emphy, prot, anos, mild anemia; [carc]
**TO:** Resp sys, kidneys, blood [prostatic & lung cancer]

**First Aid (see Table 6):**
**Breath:** Resp support

| **Calcium arsenate (as As)** | **Formula:** $Ca_3(AsO_4)_2$ | **CAS#:** 7778-44-1 | **RTECS#:** CG0830000 | **IDLH:** Ca [5 mg/m$^3$ (as As)] |
|---|---|---|---|---|
| **Conversion:** | **DOT:** 1573 151 | | | |

**Synonyms/Trade Names:** Calcium salt (2:3) of arsenic acid, Cucumber dust, Tricalcium arsenate, Tricalcium ortho-arsenate [**Note:** Also see specific listing for Arsenic (inorganic compounds, as As).]

**Exposure Limits:**
**NIOSH REL:** Ca
C 0.002 mg/m$^3$ [15-minute]
See Appendix A
**OSHA PEL:** [1910.1018] TWA 0.010 mg/m$^3$

**Measurement Methods (see Table 1):**
**NIOSH** 7900
**OSHA** ID105

**Physical Description:** Colorless to white, odorless solid. [insecticide/herbicide]

**Chemical & Physical Properties:**
**MW:** 398.1
**BP:** Decomposes
**Sol(77°F):** 0.01%
**Fl.P:** NA
**IP:** NA
**Sp.Gr:** 3.62
**VP:** 0 mmHg (approx)
**MLT:** ?
**UEL:** NA
**LEL:** NA
Noncombustible Solid

**Personal Protection/Sanitation (see Table 2):**
**Skin:** Prevent skin contact
**Eyes:** Prevent eye contact
**Wash skin:** When contam/Daily
**Remove:** When wet or contam
**Change:** Daily
**Provide:** Eyewash
Quick drench

**Respirator Recommendations (see Tables 3 and 4):**
**NIOSH**
**¥:** ScbaF:Pd,Pp/SaF:Pd,Pp:AScba
**Escape:** 100F/ScbaE

**Incompatibilities and Reactivities:** None reported
[**Note:** Produces toxic fumes of arsenic when heated to decomposition.]

**Exposure Routes, Symptoms, Target Organs (see Table 5):**
**ER:** Inh, Abs, Ing, Con
**SY:** Lass; GI dist; peri neur; skin hyperpig, palmar planter hyperkeratoses; derm; [carc]; in animals: liver damage
**TO:** Eyes, resp sys, liver, skin, CNS, lymphatic sys [lymphatic & lung cancer]

**First Aid (see Table 6):**
**Eye:** Irr immed
**Skin:** Soap wash prompt
**Breath:** Resp support
**Swallow:** Medical attention immed

| **Calcium carbonate** | **Formula:** $CaCO_3$ | **CAS#:** 471-34-1 (synthetic) 1317-65-3 (natural) | **RTECS#:** EV9580000 | **IDLH:** N.D. |
|---|---|---|---|---|
| **Conversion:** | **DOT:** | | | |

**Synonyms/Trade Names:** Calcium salt of carbonic acid
[**Note:** Occurs in nature as as limestone, chalk, marble, dolomite, aragonite, calcite & oyster shells.]

**Exposure Limits:**
**NIOSH REL:** TWA 10 mg/m$^3$ (total)
TWA 5 mg/m$^3$ (resp)
**OSHA PEL:** TWA 15 mg/m$^3$ (total)
TWA 5 mg/m$^3$ (resp)

**Measurement Methods (see Table 1):**
**NIOSH** 7020, 7303
**OSHA** ID121

**Physical Description:** White, odorless powder or colorless crystals.

**Chemical & Physical Properties:**
**MW:** 100.1
**BP:** Decomposes
**Sol:** 0.001%
**Fl.P:** NA
**IP:** NA
**Sp.Gr:** 2.7-2.95
**VP:** 0 mmHg (approx)
**MLT:** 1517-2442°F (Decomposes)
**UEL:** NA
**LEL:** NA
Noncombustible Solid

**Personal Protection/Sanitation (see Table 2):**
**Skin:** N.R.
**Eyes:** N.R.
**Wash skin:** N.R.
**Remove:** N.R.
**Change:** N.R.

**Respirator Recommendations (see Tables 3 and 4):**
Not available.

**Incompatibilities and Reactivities:** Acids, alum, ammonium salts, mercury & hydrogen, fluorine, magnesium

**Exposure Routes, Symptoms, Target Organs (see Table 5):**
**ER:** Inh, Con
**SY:** Irrit eyes, skin, resp sys; cough
**TO:** Eyes, skin, resp sys

**First Aid (see Table 6):**
**Eye:** Irr immed
**Skin:** Soap wash
**Breath:** Fresh air

C

| Calcium cyanamide | Formula: $CaCN_2$ | CAS#: 156-62-7 | RTECS#: GS6000000 | IDLH: N.D. |
|---|---|---|---|---|
| Conversion: | DOT: 1403 138 (with >0.1% calcium carbide) | | | |

**Synonyms/Trade Names:** Calcium carbimide, Cyanamide, Lime nitrogen, Nitrogen lime
[**Note:** Cyanamide is also a synonym for Hydrogen cyanamide, $NH_2CN$.]

**Exposure Limits:**
**NIOSH REL:** TWA 0.5 mg/m$^3$
**OSHA PEL†:** none

**Measurement Methods (see Table 1):**
**NIOSH** 0500

**Physical Description:** Colorless, gray, or black crystals or powder. [fertilizer]
[**Note:** Commercial grades may contain calcium carbide.]

**Chemical & Physical Properties:**
**MW:** 80.1
**BP:** Sublimes
**Sol:** Insoluble
**Fl.P:** NA
**IP:** NA
**Sp.Gr:** 2.29
**VP:** 0 mmHg (approx)
**MLT:** 2444°F
**UEL:** NA
**LEL:** NA
Noncombustible Solid, but a fire risk if it contains calcium carbide.

**Personal Protection/Sanitation (see Table 2):**
**Skin:** Prevent skin contact
**Eyes:** Prevent eye contact
**Wash skin:** When contam
**Remove:** When wet or contam
**Change:** Daily
**Provide:** Eyewash
Quick drench

**Respirator Recommendations (see Tables 3 and 4):**
Not available.

**Incompatibilities and Reactivities:** Water
[**Note:** May polymerize in water or alkaline solutions to dicyanamide. Decomposes in water to form acetylene & ammonia.]

**Exposure Routes, Symptoms, Target Organs (see Table 5):**
**ER:** Inh, Ing, Con
**SY:** Irrit eyes, skin, resp sys; head, dizz, rapid breath, low BP, nau, vomit; skin burns, sens; cough; Antabuse-like effects
**TO:** Eyes, skin, resp sys, vasomotor sys

**First Aid (see Table 6):**
**Eye:** Irr immed
**Skin:** Soap flush immed
**Breath:** Resp support
**Swallow:** Medical attention immed

| Calcium hydroxide | Formula: $Ca(OH)_2$ | CAS#: 1305-62-0 | RTECS#: EW2800000 | IDLH: N.D. |
|---|---|---|---|---|
| Conversion: | DOT: | | | |

**Synonyms/Trade Names:** Calcium hydrate, Caustic lime, Hydrated lime, Slaked lime

**Exposure Limits:**
**NIOSH REL:** TWA 5 mg/m$^3$
**OSHA PEL:** TWA 15 mg/m$^3$ (total)
TWA 5 mg/m$^3$ (resp)

**Measurement Methods (see Table 1):**
**NIOSH** 7020
**OSHA** ID121

**Physical Description:** White, odorless powder.
[**Note:** Readily absorbs $CO_2$ from the air to form calcium carbonate.]

**Chemical & Physical Properties:**
**MW:** 74.1
**BP:** Decomposes
**Sol(32°F):** 0.2%
**Fl.P:** NA
**IP:** NA
**Sp.Gr:** 2.24
**VP:** 0 mmHg (approx)
**MLT:** 1076°F (Decomposes) (Loses $H_2O$)
**UEL:** NA
**LEL:** NA
Noncombustible Solid

**Personal Protection/Sanitation (see Table 2):**
**Skin:** Prevent skin contact
**Eyes:** Prevent eye contact
**Wash skin:** When contam/Daily
**Remove:** When wet or contam
**Change:** Daily
**Provide:** Eyewash
Quick drench

**Respirator Recommendations (see Tables 3 and 4):**
Not available.

**Incompatibilities and Reactivities:** Maleic anhydride, phosphorus, nitroethane, nitromethane, nitroparaffins, nitropropane [**Note:** Attacks some metals.]

**Exposure Routes, Symptoms, Target Organs (see Table 5):**
**ER:** Inh, Ing, Con
**SY:** Irrit eyes, skin, upper resp sys; eye, skin burns; skin vesic; cough, bron, pneu
**TO:** Eyes, skin, resp sys

**First Aid (see Table 6):**
**Eye:** Irr immed
**Skin:** Soap flush immed
**Breath:** Resp support
**Swallow:** Medical attention immed

C

| **Calcium oxide** | **Formula:** CaO | **CAS#:** 1305-78-8 | **RTECS#:** EW3100000 | **IDLH:** 25 mg/m$^3$ |
|---|---|---|---|---|
| **Conversion:** | **DOT:** 1910 157 | | | |

**Synonyms/Trade Names:** Burned lime, Burnt lime, Lime, Pebble lime, Quick lime, Unslaked lime

**Exposure Limits:**
**NIOSH REL:** TWA 2 mg/m$^3$
**OSHA PEL:** TWA 5 mg/m$^3$

**Measurement Methods (see Table 1):**
**NIOSH** 7020, 7303
**OSHA** ID121

**Physical Description:** White or gray, odorless lumps or granular powder.

| **Chemical & Physical Properties:** | **Personal Protection/Sanitation (see Table 2):** | **Respirator Recommendations (see Tables 3 and 4):** |
|---|---|---|
| **MW:** 56.1<br>**BP:** 5162°F<br>**Sol:** Reacts<br>**Fl.P:** NA<br>**IP:** NA<br>**Sp.Gr:** 3.34<br>**VP:** 0 mmHg (approx)<br>**MLT:** 4662°F<br>**UEL:** NA<br>**LEL:** NA<br>Noncombustible Solid, but will support combustion by liberation of oxygen. | **Skin:** Prevent skin contact<br>**Eyes:** Prevent eye contact<br>**Wash skin:** When contam/Daily<br>**Remove:** When wet or contam<br>**Change:** Daily<br>**Provide:** Eyewash<br>Quick drench | **NIOSH**<br>**10 mg/m$^3$:** Qm<br>**20 mg/m$^3$:** 95XQ/Sa<br>**25 mg/m$^3$:** Sa:Cf/PaprHie/100F/ ScbaF/SaF<br>**§:** ScbaF:Pd,Pp/SaF:Pd,Pp:AScba<br>**Escape:** 100F/ScbaE |

**Incompatibilities and Reactivities:** Water (liberates heat), fluorine, ethanol
[**Note:** Reacts with water to form calcium hydroxide.]

| **Exposure Routes, Symptoms, Target Organs (see Table 5):** | **First Aid (see Table 6):** |
|---|---|
| **ER:** Inh, Ing, Con<br>**SY:** Irrit eyes, skin, upper resp tract; ulcer, perf nasal septum; pneu; derm<br>**TO:** Eyes, skin, resp sys | **Eye:** Irr immed<br>**Skin:** Water flush immed<br>**Breath:** Resp support<br>**Swallow:** Medical attention immed |

| **Calcium silicate** | **Formula:** $CaSiO_3$ | **CAS#:** 1344-95-2 | **RTECS#:** VV9150000 | **IDLH:** N.D. |
|---|---|---|---|---|
| **Conversion:** | **DOT:** | | | |

**Synonyms/Trade Names:** Calcium hydrosilicate, Calcium metasilicate, Calcium monosilicate, Calcium salt of silicic acid

**Exposure Limits:**
**NIOSH REL:** TWA 10 mg/m$^3$ (total)
TWA 5 mg/m$^3$ (resp)
**OSHA PEL:** TWA 15 mg/m$^3$ (total)
TWA 5 mg/m$^3$ (resp)

**Measurement Methods (see Table 1):**
**NIOSH** 7020
**OSHA** ID121

**Physical Description:** White or cream-colored, free-flowing powder.
[**Note:** The commercial product is prepared from diatomaceous earth & lime.]

| **Chemical & Physical Properties:** | **Personal Protection/Sanitation (see Table 2):** | **Respirator Recommendations (see Tables 3 and 4):** |
|---|---|---|
| **MW:** 116.2<br>**BP:** ?<br>**Sol:** 0.01%<br>**Fl.P:** NA<br>**IP:** NA<br>**Sp.Gr:** 2.9<br>**VP:** 0 mmHg (approx)<br>**MLT:** 2804°F<br>**UEL:** NA<br>**LEL:** NA<br>Noncombustible Solid | **Skin:** N.R.<br>**Eyes:** N.R.<br>**Wash skin:** N.R.<br>**Remove:** N.R.<br>**Change:** N.R. | Not available. |

**Incompatibilities and Reactivities:** None reported
[**Note:** After prolonged contact with water, solution reverts to soluble calcium salts & amorphous silica.]

| **Exposure Routes, Symptoms, Target Organs (see Table 5):** | **First Aid (see Table 6):** |
|---|---|
| **ER:** Inh, Con<br>**SY:** Irrit eyes, skin, upper resp sys<br>**TO:** Eyes, skin, resp sys | **Eye:** Irr immed<br>**Skin:** Soap wash<br>**Breath:** Fresh air |

C

## Calcium sulfate

| Formula: | CAS#: | RTECS#: | IDLH: |
|---|---|---|---|
| $CaSO_4$ | 7778-18-9 | WS6920000 | N.D. |

**Conversion:**

**DOT:**

**Synonyms/Trade Names:** Anhydrous calcium sulfate, Anhydrous gypsum, Anhydrous sulfate of lime, Calcium salt of sulfuric acid [**Note:** Gypsum is the dihydrate form & Plaster of Paris is the hemihydrate form.]

**Exposure Limits:**
**NIOSH REL:** TWA 10 mg/m$^3$ (total)
TWA 5 mg/m$^3$ (resp)
**OSHA PEL:** TWA 15 mg/m$^3$ (total)
TWA 5 mg/m$^3$ (resp)

**Measurement Methods (see Table 1):**
**NIOSH** 0500, 0600

**Physical Description:** Odorless, white powder or colorless, crystalline solid.
[**Note:** May have blue, gray, or reddish tinge.]

**Chemical & Physical Properties:**
**MW:** 136.1
**BP:** Decomposes
**Sol:** 0.3%
**Fl.P:** NA
**IP:** NA
**Sp.Gr:** 2.96
**VP:** 0 mmHg (approx)
**MLT:** 2840°F (Decomposes)
**UEL:** NA
**LEL:** NA
Noncombustible Solid

**Personal Protection/Sanitation (see Table 2):**
**Skin:** N.R.
**Eyes:** N.R.
**Wash skin:** N.R.
**Remove:** N.R.
**Change:** N.R.

**Respirator Recommendations (see Tables 3 and 4):**
Not available.

**Incompatibilities and Reactivities:** Diazomethane, aluminum, phosphorus, water
[**Note:** Hygroscopic (i.e., absorbs moisture from the air). Reacts with water to form Gypsum & Plaster of Paris.]

**Exposure Routes, Symptoms, Target Organs (see Table 5):**
**ER:** Inh, Con
**SY:** Irrit eyes, skin, upper resp sys; conj; rhinitis, epis
**TO:** Eyes, skin, resp sys

**First Aid (see Table 6):**
**Eye:** Irr immed
**Skin:** Soap wash
**Breath:** Fresh air

## Camphor (synthetic)

| Formula: | CAS#: | RTECS#: | IDLH: |
|---|---|---|---|
| $C_{10}H_{16}O$ | 76-22-2 | EX1225000 | 200 mg/m$^3$ |

**Conversion:**

**DOT:** 2717 133

**Synonyms/Trade Names:** 2-Camphonone, Gum camphor, Laurel camphor, Synthetic camphor

**Exposure Limits:**
**NIOSH REL:** TWA 2 mg/m$^3$
**OSHA PEL:** TWA 2 mg/m$^3$

**Measurement Methods (see Table 1):**
**NIOSH** 1301, 2553
**OSHA** 7

**Physical Description:** Colorless or white crystals with a penetrating, aromatic odor.

**Chemical & Physical Properties:**
**MW:** 152.3
**BP:** 399°F
**Sol:** Insoluble
**Fl.P:** 150°F
**IP:** 8.76 eV
**Sp.Gr:** 0.99
**VP:** 0.2 mmHg
**MLT:** 345°F
**UEL:** 3.5%
**LEL:** 0.6%
Combustible Solid

**Personal Protection/Sanitation (see Table 2):**
**Skin:** Prevent skin contact
**Eyes:** Prevent eye contact
**Wash skin:** When contam
**Remove:** When wet or contam
**Change:** Daily

**Respirator Recommendations (see Tables 3 and 4):**
**NIOSH/OSHA**
**50 mg/m$^3$:** Sa:Cf£/PaprOvHie£
**100 mg/m$^3$:** CcrFOv100/GmFOv100/PaprTOvHie£/ScbaF/SaF
**200 mg/m$^3$:** SaF:Pd,Pp
**§:** ScbaF:Pd,Pp/SaF:Pd,Pp:AScba
**Escape:** GmFOv100/ScbaE

**Incompatibilities and Reactivities:** Strong oxidizers (especially chromic anhydride & potassium permanganate)

**Exposure Routes, Symptoms, Target Organs (see Table 5):**
**ER:** Inh, Abs, Ing, Con
**SY:** Irrit eyes, skin, muc memb; nau, vomit, diarr; head, dizz, excitement, epilep convuls
**TO:** Eyes, skin, resp sys, CNS

**First Aid (see Table 6):**
**Eye:** Irr immed
**Skin:** Soap wash immed
**Breath:** Resp support
**Swallow:** Medical attention immed

C

| Caprolactam | Formula: $C_2H_{11}NO$ | CAS#: 105-60-2 | RTECS#: CM3675000 | IDLH: N.D. |
|---|---|---|---|---|
| **Conversion:** 1 ppm = 4.63 mg/m$^3$ | **DOT:** | | | |

**Synonyms/Trade Names:** Aminocaproic lactam, epsilon-Caprolactam, Hexahydro-2H-azepin-2-one, 2-Oxohexamethyleneimine

**Exposure Limits:**
**NIOSH REL:** Dust: TWA 1 mg/m$^3$
ST 3 mg/m$^3$
Vapor: TWA 0.22 ppm (1 mg/m$^3$)
ST 0.66 ppm (3 mg/m$^3$)
**OSHA PEL†:** none

**Measurement Methods (see Table 1):**
**OSHA** PV2012

**Physical Description:** White, crystalline solid or flakes with an unpleasant odor.
[**Note:** Significant vapor concentrations would be expected only at elevated temperatures.]

**Chemical & Physical Properties:**
**MW:** 113.2
**BP:** 515°F
**Sol:** 53%
**Fl.P:** 282°F
**IP:** ?
**Sp.Gr:** 1.01
**VP:** 0.00000008 mmHg
**MLT:** 156°F
**UEL:** 8.0%
**LEL:** 1.4%
Combustible Solid

**Personal Protection/Sanitation (see Table 2):**
**Skin:** Prevent skin contact
**Eyes:** Prevent eye contact
**Wash skin:** When contam
**Remove:** When wet or contam
**Change:** Daily

**Respirator Recommendations (see Tables 3 and 4):**
Not available.

**Incompatibilities and Reactivities:** Strong oxidizers, (acetic acid + dinitrogen trioxide)

**Exposure Routes, Symptoms, Target Organs (see Table 5):**
**ER:** Inh, Ing, Con
**SY:** Irrit skin, eyes, resp sys; epis; derm, skin sens; asthma; irrity, conf, dizz, head; abdom cramps, diarr, nau, vomit; liver, kidney inj
**TO:** Eyes, skin, resp sys, CNS, CVS, liver, kidneys

**First Aid (see Table 6):**
**Eye:** Irr immed
**Skin:** Water wash immed
**Breath:** Resp support
**Swallow:** Medical attention immed

| Captafol | Formula: $C_{10}H_9Cl_{14}NO_2S$ | CAS#: 2425-06-1 | RTECS#: GW4900000 | IDLH: Ca [N.D.] |
|---|---|---|---|---|
| **Conversion:** | **DOT:** | | | |

**Synonyms/Trade Names:** Captofol; Difolatan®; N-((1,1,2,2-Tetrachloroethyl)thio)-4-cyclohexene-1,2-dicarboximide

**Exposure Limits:**
**NIOSH REL:** Ca
TWA 0.1 mg/m$^3$ [skin]
See Appendix A
**OSHA PEL†:** none

**Measurement Methods (see Table 1):**
**NIOSH** 0500

**Physical Description:** White, crystalline solid with a slight, characteristic pungent odor. [fungicide]
[**Note:** Available commercially as a wettable powder or in liquid form.]

**Chemical & Physical Properties:**
**MW:** 349.1
**BP:** Decomposes
**Sol:** 0.0001%
**Fl.P:** NA
**IP:** NA
**Sp.Gr:** ?
**VP:** 0.000008 mmHg
**MLT:** 321°F (Decomposes)
**UEL:** NA
**LEL:** NA
Noncombustible Solid, but may be dissolved in flammable liquids.

**Personal Protection/Sanitation (see Table 2):**
**Skin:** Prevent skin contact
**Eyes:** Prevent eye contact
**Wash skin:** When contam/Daily
**Remove:** When wet or contam
**Change:** Daily
**Provide:** Eyewash
Quick drench

**Respirator Recommendations (see Tables 3 and 4):**
**NIOSH**
**¥:** ScbaF:Pd,Pp/SaF:Pd,Pp:AScba
**Escape:** GmFOv/ScbaE

**Incompatibilities and Reactivities:** Acids, acid vapors, strong oxidizers

**Exposure Routes, Symptoms, Target Organs (see Table 5):**
**ER:** Inh, Abs, Ing, Con
**SY:** Irrit eyes, skin, resp sys; derm, skin sens; conj; bron, wheez; diarr, vomit; liver, kidney inj; high BP; in animals: terato effects; [carc]
**TO:** Eyes, skin, resp sys, CNS, liver, kidneys, CVS [in animals: tumors at many sites]

**First Aid (see Table 6):**
**Eye:** Irr immed
**Skin:** Soap wash immed
**Breath:** Resp support
**Swallow:** Medical attention immed

## Captan

| Formula: | CAS#: | RTECS#: | IDLH: |
|---|---|---|---|
| $C_9H_8Cl_3NO_2S$ | 133-06-2 | GW5075000 | Ca [N.D.] |

**Conversion:**

**DOT:**

**Synonyms/Trade Names:** Captane; N-Trichloromethylmercapto-4-cyclohexene-1,2-dicarboximide

**Exposure Limits:**
**NIOSH REL:** Ca
TWA 5 $mg/m^3$
See Appendix A

**OSHA PEL†:** none

**Measurement Methods (see Table 1):**
**NIOSH** 5601, 9202, 9205

**Physical Description:** Odorless, white, crystalline powder. [fungicide]
[**Note:** Commercial product is a yellow powder with a pungent odor.]

**Chemical & Physical Properties:**
**MW:** 300.6
**BP:** Decomposes
**Sol(77°F):** 0.0003%
**Fl.P:** ?
**IP:** NA
**Sp.Gr:** 1.74
**VP:** 0 mmHg (approx)
**MLT:** 352°F (Decomposes)
**UEL:** ?
**LEL:** ?
Combustible Solid; may be dissolved in flammable liquids.

**Personal Protection/Sanitation (see Table 2):**
**Skin:** Prevent skin contact
**Eyes:** Prevent eye contact
**Wash skin:** When contam/Daily
**Remove:** When wet or contam
**Change:** Daily
**Provide:** Eyewash
Quick drench

**Respirator Recommendations (see Tables 3 and 4):**
**NIOSH**
**¥:** ScbaF:Pd,Pp/SaF:Pd,Pp:AScba
**Escape:** GmFOv/ScbaE

**Incompatibilities and Reactivities:** Strong alkaline materials (e.g., hydrated lime) [**Note:** Corrosive to metals.]

**Exposure Routes, Symptoms, Target Organs (see Table 5):**
**ER:** Inh, Abs, Ing, Con
**SY:** Irrit eyes, skin, upper resp sys; blurred vision; derm, skin sens; dysp; diarr, vomit; [carc]
**TO:** Eyes, skin, resp sys, GI tract, liver, kidneys [in animals: duodenal tumors]

**First Aid (see Table 6):**
**Eye:** Irr immed
**Skin:** Soap wash immed
**Breath:** Resp support
**Swallow:** Medical attention immed

## Carbaryl

| Formula: | CAS#: | RTECS#: | IDLH: |
|---|---|---|---|
| $CH_3NHCOOC_{10}H_7$ | 63-25-2 | EC5950000 | 100 $mg/m^3$ |

**Conversion:**

**DOT:** 2757 151

**Synonyms/Trade Names:** α-Naphthyl N-methyl-carbamate, 1-Naphthyl N-Methyl-carbamate, Sevin®

**Exposure Limits:**
**NIOSH REL:** TWA 5 $mg/m^3$
**OSHA PEL:** TWA 5 $mg/m^3$

**Measurement Methods (see Table 1):**
**NIOSH** 5006, 5601
**OSHA** 63

**Physical Description:** White or gray, odorless solid. [pesticide]

**Chemical & Physical Properties:**
**MW:** 201.2
**BP:** Decomposes
**Sol:** 0.01%
**Fl.P:** NA
**IP:** ?
**Sp.Gr:** 1.23
**VP(77°F):** <0.00004 mmHg
**MLT:** 293°F
**UEL:** NA
**LEL:** NA
Noncombustible Solid, but may be dissolved in flammable liquids.

**Personal Protection/Sanitation (see Table 2):**
**Skin:** Prevent skin contact
**Eyes:** Prevent eye contact
**Wash skin:** When contam
**Remove:** When wet or contam
**Change:** Daily

**Respirator Recommendations (see Tables 3 and 4):**
**NIOSH/OSHA**
**50 $mg/m^3$:** Sa*
**100 $mg/m^3$:** Sa:Cf*/ScbaF/SaF
**§:** ScbaF:Pd,Pp/SaF:Pd,Pp:AScba
**Escape:** GmFOv100/ScbaE

**Incompatibilities and Reactivities:** Strong oxidizers, strongly alkaline pesticides

**Exposure Routes, Symptoms, Target Organs (see Table 5):**
**ER:** Inh, Abs, Ing, Con
**SY:** Miosis, blurred vision, tear; rhin, salv; sweat; abdom cramps, nau, vomit, diarr; tremor; cyan; convuls; irrit skin; possible repro effects
**TO:** Resp sys, CNS, CVS, skin, blood chol, repro sys

**First Aid (see Table 6):**
**Eye:** Irr immed
**Skin:** Soap wash prompt
**Breath:** Resp support
**Swallow:** Medical attention immed

C

## Carbofuran

| | | | | |
|---|---|---|---|---|
| **Carbofuran** | **Formula:** $C_{12}H_{15}NO_3$ | **CAS#:** 1563-66-2 | **RTECS#:** FB9450000 | **IDLH:** N.D. |
| **Conversion:** | **DOT:** 2757 151 | | | |

**Synonyms/Trade Names:** 2,3-Dihydro-2,2-dimethyl-7-benzofuranyl methylcarbamate; Furacarb®; Furadan®

**Exposure Limits:**
**NIOSH REL:** TWA 0.1 mg/m$^3$
**OSHA PEL†:** none

**Measurement Methods (see Table 1):**
**NIOSH** 5601

**Physical Description:** Odorless, white or grayish, crystalline solid. [insecticide]
[**Note:** May be dissolved in a liquid carrier.]

**Chemical & Physical Properties:**
**MW:** 221.3
**BP:** ?
**Sol(77°F):** 0.07%
**Fl.P:** NA
**IP:** NA
**Sp.Gr:** 1.18
**VP(77°F):** 0.000003 mmHg
**MLT:** 304°F
**UEL:** NA
**LEL:** NA
Noncombustible Solid

**Personal Protection/Sanitation (see Table 2):**
**Skin:** Prevent skin contact
**Eyes:** Prevent eye contact
**Wash skin:** When contam
**Remove:** When wet or contam
**Change:** Daily
**Provide:** Eyewash
Quick drench

**Respirator Recommendations (see Tables 3 and 4):**
Not available.

**Incompatibilities and Reactivities:** Alkaline substances, acid, strong oxidizers (e.g., perchlorates, peroxides, chlorates, nitrates, permanganates)

**Exposure Routes, Symptoms, Target Organs (see Table 5):**
**ER:** Inh, Abs, Ing, Con
**SY:** Miosis, blurred vision; sweat, salv, abdom cramps, diarr, head, nau, vomit; lass, musc twitch, inco, convuls
**TO:** CNS, PNS, blood chol

**First Aid (see Table 6):**
**Eye:** Irr immed
**Skin:** Soap flush immed
**Breath:** Fresh air
**Swallow:** Medical attention immed

## Carbon black

| | | | | |
|---|---|---|---|---|
| **Carbon black** | **Formula:** C | **CAS#:** 1333-86-4 | **RTECS#:** FF5800000 | **IDLH:** 1750 mg/m$^3$ |
| **Conversion:** | **DOT:** | | | |

**Synonyms/Trade Names:** Acetylene black, Channel black, Furnace black, Lamp black, Thermal black

**Exposure Limits:**
**NIOSH REL:** TWA 3.5 mg/m$^3$
Carbon black in presence of polycyclic aromatic hydrocarbons (PAHs):
Ca
TWA 0.1 mg PAHs/m$^3$
See Appendix A
See Appendix C
**OSHA PEL:** TWA 3.5 mg/m$^3$

**Measurement Methods (see Table 1):**
**NIOSH** 5000
**OSHA** ID196

**Physical Description:** Black, odorless solid.

**Chemical & Physical Properties:**
**MW:** 12.0
**BP:** Sublimes
**Sol:** Insoluble
**Fl.P:** NA
**IP:** NA
**Sp.Gr:** 1.8-2.1
**VP:** 0 mmHg (approx)
**MLT:** Sublimes
**UEL:** NA
**LEL:** NA
Combustible Solid that may contain flammable hydrocarbons.

**Personal Protection/Sanitation (see Table 2):**
**Skin:** N.R.
**Eyes:** Prevent eye contact
**Wash skin:** Daily
**Remove:** N.R.
**Change:** N.R.

**Respirator Recommendations (see Tables 3 and 4):**
**NIOSH/OSHA**
**17.5 mg/m$^3$:** Qm
**35 mg/m$^3$:** 95XQ/Sa
**87.5 mg/m$^3$:** Sa:Cf/PaprHie
**175 mg/m$^3$:** 100F/PaprTHie/ScbaF/SaF
**1750 mg/m$^3$:** Sa:Pd,Pp
**§:** ScbaF:Pd,Pp/SaF:Pd,Pp:AScba
**Escape:** 100F/ScbaE

In presence of polycyclic aromatic hydrocarbons:
**NIOSH**
**¥:** ScbaF:Pd,Pp/SaF:Pd,Pp:AScba
**Escape:** 100F/ScbaE

**Incompatibilities and Reactivities:** Strong oxidizers such as chlorates, bromates & nitrates

**Exposure Routes, Symptoms, Target Organs (see Table 5):**
**ER:** Inh, Con
**SY:** Cough; irrit eyes; in presence of polycyclic aromatic hydrocarbons: [carc]
**TO:** Resp sys, eyes [lymphatic cancer (in presence of PAHs)]

**First Aid (see Table 6):**
**Eye:** Irr prompt
**Breath:** Fresh air

C

## Carbon dioxide

| Formula: | CAS#: | RTECS#: | IDLH: |
|---|---|---|---|
| $CO_2$ | 124-38-9 | FF6400000 | 40,000 ppm |

**Conversion:** 1 ppm = 1.80 mg/m$^3$

**DOT:** 1013 120; 1845 120 (dry ice); 2187 120 (liquid)

**Synonyms/Trade Names:** Carbonic acid gas, Dry ice [**Note:** Normal constituent of air (about 300 ppm)].

**Exposure Limits:**
**NIOSH REL:** TWA 5000 ppm (9000 mg/m$^3$)
ST 30,000 ppm (54,000 mg/m$^3$)
**OSHA PEL†:** TWA 5000 ppm (9000 mg/m$^3$)

**Measurement Methods (see Table 1):**
**NIOSH** 6603
**OSHA** ID172

**Physical Description:** Colorless, odorless gas.
[**Note:** Shipped as a liquefied compressed gas. Solid form is utilized as dry ice.]

**Chemical & Physical Properties:**
**MW:** 44.0
**BP:** Sublimes
**Sol(77°F):** 0.2%
**Fl.P:** NA
**IP:** 13.77 eV
**RGasD:** 1.53
**VP:** 56.5 atm
**MLT:** -109°F (Sublimes)
**UEL:** NA
**LEL:** NA
Nonflammable Gas

**Personal Protection/Sanitation (see Table 2):**
**Skin:** Frostbite
**Eyes:** Frostbite
**Wash skin:** N.R.
**Remove:** N.R.
**Change:** N.R.
**Provide:** Frostbite wash

**Respirator Recommendations (see Tables 3 and 4):**
**NIOSH/OSHA**
**40,000 ppm:** Sa/ScbaF
**§:** ScbaF:Pd,Pp/SaF:Pd,Pp:AScba
**Escape:** ScbaE

**Incompatibilities and Reactivities:** Dusts of various metals, such as magnesium, zirconium, titanium, aluminum, chromium & manganese are ignitable and explosive when suspended in carbon dioxide. Forms carbonic acid in water.

**Exposure Routes, Symptoms, Target Organs (see Table 5):**
**ER:** Inh, Con (liquid/solid)
**SY:** Head, dizz, restless, pares; dysp; sweat, mal; incr heart rate, card output, BP; coma; asphy; convuls; frostbite (liq, dry ice)
**TO:** Resp sys, CVS

**First Aid (see Table 6):**
**Eye:** Frostbite
**Skin:** Frostbite
**Breath:** Resp support

## Carbon disulfide

| Formula: | CAS#: | RTECS#: | IDLH: |
|---|---|---|---|
| $CS_2$ | 75-15-0 | FF6650000 | 500 ppm |

**Conversion:** 1 ppm = 3.11 mg/m$^3$

**DOT:** 1131 131

**Synonyms/Trade Names:** Carbon bisulfide

**Exposure Limits:**
**NIOSH REL:** TWA 1 ppm (3 mg/m$^3$)
ST 10 ppm (30 mg/m$^3$) [skin]
**OSHA PEL†:** TWA 20 ppm
C 30 ppm
100 ppm (30-minute maximum peak)

**Measurement Methods (see Table 1):**
**NIOSH** 1600, 3800

**Physical Description:** Colorless to faint-yellow liquid with a sweet ether-like odor.
[**Note:** Reagent grades are foul smelling.]

**Chemical & Physical Properties:**
**MW:** 76.1
**BP:** 116°F
**Sol:** 0.3%
**Fl.P:** -22°F
**IP:** 10.08 eV
**Sp.Gr:** 1.26
**VP:** 297 mmHg
**FRZ:** -169°F
**UEL:** 50.0%
**LEL:** 1.3%
Class IB Flammable Liquid

**Personal Protection/Sanitation (see Table 2):**
**Skin:** Prevent skin contact
**Eyes:** Prevent eye contact
**Wash skin:** When contam
**Remove:** When wet (flamm)
**Change:** N.R.

**Respirator Recommendations (see Tables 3 and 4):**
**NIOSH**
**10 ppm:** CcrOv/Sa
**25 ppm:** Sa:Cf/PaprOv
**50 ppm:** CcrFOv/GmFOv/PaprTOv/ScbaF/SaF
**500 ppm:** Sa:Pd,Pp
**§:** ScbaF:Pd,Pp/SaF:Pd,Pp:AScba
**Escape:** GmFOv/ScbaE

**Incompatibilities and Reactivities:** Strong oxidizers; chemically-active metals such as sodium, potassium & zinc; azides; rust; halogens; amines
[**Note:** Vapors may be ignited by contact with an ordinary light bulb.]

**Exposure Routes, Symptoms, Target Organs (see Table 5):**
**ER:** Inh, Abs, Ing, Con
**SY:** Dizz, head, poor sleep, lass, anxi, anor, low-wgt; psychosis; polyneur; Parkinson-like syndrome; ocular changes; coronary heart disease; gastritis; kidney, liver inj; eye, skin burns; derm; repro effects
**TO:** CNS, PNS, CVS, eyes, kidneys, liver, skin, repro sys

**First Aid (see Table 6):**
**Eye:** Irr immed
**Skin:** Soap wash immed
**Breath:** Resp support
**Swallow:** Medical attention immed

C

## Carbon monoxide

| Carbon monoxide | Formula: CO | CAS#: 630-08-0 | RTECS#: FG3500000 | IDLH: 1200 ppm |
|---|---|---|---|---|

**Conversion:** 1 ppm = 1.15 mg/m$^3$

**DOT:** 1016 119; 9202 168 (cryogenic liquid)

**Synonyms/Trade Names:** Carbon oxide, Flue gas, Monoxide

**Exposure Limits:**
**NIOSH REL:** TWA 35 ppm (40 mg/m$^3$)
C 200 ppm (229 mg/m$^3$)
**OSHA PEL†:** TWA 50 ppm (55 mg/m$^3$)

**Measurement Methods (see Table 1):**
**NIOSH** 6604
**OSHA** ID209, ID210

**Physical Description:** Colorless, odorless gas.
[**Note:** Shipped as a nonliquefied or liquefied compressed gas.]

**Chemical & Physical Properties:**
**MW:** 28.0
**BP:** -313°F
**Sol:** 2%
**Fl.P:** NA (Gas)
**IP:** 14.01 eV
**RGasD:** 0.97
**VP:** >35 atm
**MLT:** -337°F
**UEL:** 74%
**LEL:** 12.5%
Flammable Gas

**Personal Protection/Sanitation (see Table 2):**
**Skin:** Frostbite
**Eyes:** Frostbite
**Wash skin:** N.R.
**Remove:** When wet (flamm)
**Change:** N.R.
**Provide:** Frostbite wash

**Respirator Recommendations (see Tables 3 and 4):**
**NIOSH**
**350 ppm:** Sa
**875 ppm:** Sa:Cf
**1200 ppm:** GmFS†/ScbaF/SaF
**§:** ScbaF:Pd,Pp/SaF:Pd,Pp:AScba
**Escape:** GmFS†/ScbaE

**Incompatibilities and Reactivities:** Strong oxidizers, bromine trifluoride, chlorine trifluoride, lithium

**Exposure Routes, Symptoms, Target Organs (see Table 5):**
**ER:** Inh, Con (liquid)
**SY:** Head, tachypnea, nau, lass, dizz, conf, halu; cyan; depres S-T segment of electrocardiogram, angina, syncope
**TO:** CVS, lungs, blood, CNS

**First Aid (see Table 6):**
**Eye:** Frostbite
**Skin:** Frostbite
**Breath:** Resp support

## Carbon tetrabromide

| Carbon tetrabromide | Formula: $CBr_4$ | CAS#: 558-13-4 | RTECS#: FG4725000 | IDLH: N.D. |
|---|---|---|---|---|

**Conversion:** 1 ppm = 13.57 mg/m$^3$

**DOT:** 2516 151

**Synonyms/Trade Names:** Carbon bromide, Methane tetrabromide, Tetrabromomethane

**Exposure Limits:**
**NIOSH REL:** TWA 0.1 ppm (1.4 mg/m$^3$)
ST 0.3 ppm (4 mg/m$^3$)
**OSHA PEL†:** none

**Measurement Methods (see Table 1):**
None available

**Physical Description:** Colorless to yellow-brown crystals with a slight odor.

**Chemical & Physical Properties:**
**MW:** 331.7
**BP:** 374°F
**Sol:** 0.02%
**Fl.P:** NA
**IP:** 10.31 eV
**Sp.Gr:** 3.42
**VP(205°F):** 40 mmHg
**MLT:** 194°F
**UEL:** NA
**LEL:** NA
Noncombustible Solid

**Personal Protection/Sanitation (see Table 2):**
**Skin:** N.R.
**Eyes:** Prevent eye contact
**Wash skin:** Daily
**Remove:** N.R.
**Change:** Daily
**Provide:** Eyewash

**Respirator Recommendations (see Tables 3 and 4):**
Not available.

**Incompatibilities and Reactivities:** Strong oxidizers, hexacyclohexyldilead, lithium

**Exposure Routes, Symptoms, Target Organs (see Table 5):**
**ER:** Inh, Ing, Con
**SY:** Irrit eyes, skin, resp sys; lac; lung, liver, kidney inj; in animals: corn damage
**TO:** Eyes, skin, resp sys, liver, kidneys

**First Aid (see Table 6):**
**Eye:** Irr immed
**Skin:** Soap wash prompt
**Breath:** Resp support
**Swallow:** Medical attention immed

## Carbon tetrachloride

| Carbon tetrachloride | Formula: $CCl_4$ | CAS#: 56-23-5 | RTECS#: FG4900000 | IDLH: Ca [200 ppm] |
|---|---|---|---|---|
| **Conversion:** 1 ppm = 6.29 mg/m$^3$ | **DOT:** 1846 151 | | | |

**Synonyms/Trade Names:** Carbon chloride, Carbon tet, Freon® 10, Halon® 104, Tetrachloromethane

**Exposure Limits:**
**NIOSH REL:** Ca
ST 2 ppm (12.6 mg/m$^3$) [60-minute]
See Appendix A
**OSHA PEL†:** TWA 10 ppm
C 25 ppm
200 ppm (5-minute maximum peak in any 4 hours)

**Measurement Methods (see Table 1):**
**NIOSH** 1003
**OSHA** 7

**Physical Description:** Colorless liquid with a characteristic ether-like odor.

**Chemical & Physical Properties:**
**MW:** 153.8
**BP:** 170°F
**Sol:** 0.05%
**Fl.P:** NA
**IP:** 11.47 eV
**Sp.Gr:** 1.59
**VP:** 91 mmHg
**FRZ:** -9°F
**UEL:** NA
**LEL:** NA
Noncombustible Liquid

**Personal Protection/Sanitation (see Table 2):**
**Skin:** Prevent skin contact
**Eyes:** Prevent eye contact
**Wash skin:** When contam
**Remove:** When wet or contam
**Change:** N.R.
**Provide:** Eyewash
Quick drench

**Respirator Recommendations (see Tables 3 and 4):**
**NIOSH**
**¥:** ScbaF:Pd,Pp/SaF:Pd,Pp:AScba
**Escape:** GmFOv/ScbaE

**Incompatibilities and Reactivities:** Chemically-active metals such as sodium, potassium & magnesium; fluorine; aluminum [**Note:** Forms highly toxic phosgene gas when exposed to flames or welding arcs.]

**Exposure Routes, Symptoms, Target Organs (see Table 5):**
**ER:** Inh, Abs, Ing, Con
**SY:** Irrit eyes, skin; CNS depres; nau, vomit; liver, kidney inj; drow, dizz, inco; [carc]
**TO:** CNS, eyes, lungs, liver, kidneys, skin [in animals: liver cancer]

**First Aid (see Table 6):**
**Eye:** Irr immed
**Skin:** Soap wash immed
**Breath:** Resp support
**Swallow:** Medical attention immed

## Carbonyl fluoride

| Carbonyl fluoride | Formula: $COF_2$ | CAS#: 353-50-4 | RTECS#: FG6125000 | IDLH: N.D. |
|---|---|---|---|---|
| **Conversion:** 1 ppm = 2.70 mg/m$^3$ | **DOT:** 2417 125 | | | |

**Synonyms/Trade Names:** Carbon difluoride oxide, Carbon fluoride oxide, Carbon oxyfluoride, Carbonyl difluoride, Fluoroformyl fluoride, Fluorophosgene

**Exposure Limits:**
**NIOSH REL:** TWA 2 ppm (5 mg/m$^3$)
ST 5 ppm (15 mg/m$^3$)
**OSHA PEL†:** none

**Measurement Methods (see Table 1):**
None available

**Physical Description:** Colorless gas with a pungent and very irritating odor.
[**Note:** Shipped as a liquefied compressed gas.]

**Chemical & Physical Properties:**
**MW:** 66.0
**BP:** -118°F
**Sol:** Reacts
**Fl.P:** NA
**IP:** 13.02 eV
**RGasD:** 2.29
**VP:** 55.4 atm
**FRZ:** -173°F
**UEL:** NA
**LEL:** NA
Nonflammable Gas

**Personal Protection/Sanitation (see Table 2):**
**Skin:** Frostbite
**Eyes:** Frostbite
**Wash skin:** N.R.
**Remove:** N.R.
**Change:** N.R.
**Provide:** Frostbite wash

**Respirator Recommendations (see Tables 3 and 4):**
Not available.

**Incompatibilities and Reactivities:** Heat, moisture, hexafluoroisopropylideneamino-lithium
[**Note:** Reacts with water to form hydrogen fluoride & carbon dioxide.]

**Exposure Routes, Symptoms, Target Organs (see Table 5):**
**ER:** Inh, Con
**SY:** Irrit eyes, skin, muc memb, resp sys; eye, skin burns; lac; cough, pulm edema, dysp; chronic exposure: GI pain, musc fib, skeletal fluorosis; liquid: frostbite
**TO:** Eyes, skin, resp sys, bone

**First Aid (see Table 6):**
**Eye:** Frostbite
**Skin:** Frostbite
**Breath:** Resp support

## Catechol

| Formula: | CAS#: | RTECS#: | IDLH: |
|---|---|---|---|
| $C_6H_4(OH)_2$ | 120-80-9 | UX1050000 | N.D. |

**Conversion:** 1 ppm = 4.50 mg/m$^3$
**DOT:**

**Synonyms/Trade Names:** 1,2-Benzenediol; o-Benzenediol; 1,2-Dihydroxybenzene; o-Dihydroxybenzene; 2-Hydroxyphenol; Pyrocatechol

**Exposure Limits:**
**NIOSH REL:** TWA 5 ppm (20 mg/m$^3$) [skin]
**OSHA PEL†:** none

**Measurement Methods (see Table 1):**
**OSHA** PV2014

**Physical Description:** Colorless, crystalline solid with a faint odor.
[**Note:** Discolors to brown in air & light.]

**Chemical & Physical Properties:**
**MW:** 110.1
**BP:** 474°F
**Sol:** 44%
**Fl.P:** 261°F
**IP:** ?
**Sp.Gr:** 1.34
**VP(244°F):** 10 mmHg
**MLT:** 221°F
**UEL:** ?
**LEL:** 1.4%
Combustible Solid

**Personal Protection/Sanitation (see Table 2):**
**Skin:** Prevent skin contact
**Eyes:** Prevent eye contact
**Wash skin:** When contam
**Remove:** When wet or contam
**Change:** Daily
**Provide:** Eyewash

**Respirator Recommendations (see Tables 3 and 4):**
Not available.

**Incompatibilities and Reactivities:** Strong oxidizers, nitric acid

**Exposure Routes, Symptoms, Target Organs (see Table 5):**
**ER:** Inh, Abs, Ing, Con
**SY:** Irrit eyes, skin, resp sys; skin sens, derm; lac, burns eyes; convuls, incr BP, kidney inj
**TO:** Eyes, skin, resp sys, CNS, kidneys

**First Aid (see Table 6):**
**Eye:** Irr immed
**Skin:** Water wash immed
**Breath:** Resp support
**Swallow:** Medical attention immed

## Cellulose

| Formula: | CAS#: | RTECS#: | IDLH: |
|---|---|---|---|
| $(C_6H_{10}O_5)_n$ | 9004-34-6 | FJ5691460 | N.D. |

**Conversion:**
**DOT:**

**Synonyms/Trade Names:** Hydroxycellulose, Pyrocellulose

**Exposure Limits:**
**NIOSH REL:** TWA 10 mg/m$^3$ (total)
TWA 5 mg/m$^3$ (resp)
**OSHA PEL:** TWA 15 mg/m$^3$ (total)
TWA 5 mg/m$^3$ (resp)

**Measurement Methods (see Table 1):**
**NIOSH** 0500, 0600, 7404

**Physical Description:** Odorless, white substance.
[**Note:** The principal fiber cell wall material of vegetable tissues (wood, cotton, flax, grass, etc.).]

**Chemical & Physical Properties:**
**MW:** 160,000-560,000
**BP:** Decomposes
**Sol:** Insoluble
**Fl.P:** NA
**IP:** NA
**Sp.Gr:** 1.27-1.61
**VP:** 0 mmHg (approx)
**MLT:** 500-518°F (Decomposes)
**UEL:** NA
**LEL:** NA
Combustible Solid

**Personal Protection/Sanitation (see Table 2):**
**Skin:** N.R.
**Eyes:** N.R.
**Wash skin:** N.R.
**Remove:** N.R.
**Change:** N.R.

**Respirator Recommendations (see Tables 3 and 4):**
Not available.

**Incompatibilities and Reactivities:** Water, bromine pentafluoride, sodium nitrate, fluorine, strong oxidizers

**Exposure Routes, Symptoms, Target Organs (see Table 5):**
**ER:** Inh, Con
**SY:** Irrit eyes, skin, muc memb
**TO:** Eyes, skin, resp sys

**First Aid (see Table 6):**
**Eye:** Irr immed
**Skin:** Soap wash
**Breath:** Fresh air

C

| **Cesium hydroxide** | **Formula:** CsOH | **CAS#:** 21351-79-1 | **RTECS#:** FK9800000 | **IDLH:** N.D. |
|---|---|---|---|---|
| **Conversion:** | **DOT:** 2682 157; 2681 154 (solution) | | | |

**Synonyms/Trade Names:** Cesium hydrate, Cesium hydroxide dimer

**Exposure Limits:**
**NIOSH REL:** TWA 2 $mg/m^3$
**OSHA PEL†:** none

**Measurement Methods (see Table 1):**
None available

**Physical Description:** Colorless or yellowish, crystalline solid.
[**Note:** Hygroscopic (i.e., absorbs moisture from the air).]

**Chemical & Physical Properties:**
**MW:** 149.9
**BP:** ?
**Sol(59°F):** 395%
**Fl.P:** NA
**IP:** NA
**Sp.Gr:** 3.68
**VP:** 0 mmHg (approx)
**MLT:** 522°F
**UEL:** NA
**LEL:** NA
Noncombustible Solid

**Personal Protection/Sanitation (see Table 2):**
**Skin:** Prevent skin contact
**Eyes:** Prevent eye contact
**Wash skin:** When contam
**Remove:** When wet or contam
**Change:** Daily
**Provide:** Eyewash
Quick drench

**Respirator Recommendations (see Tables 3 and 4):**
Not available.

**Incompatibilities and Reactivities:** Water, acids, $CO_2$, metals (e.g., Al, Pb, Sn, Zn), oxygen
[**Note:** CsOH is a strong base, causing the generation of considerable heat in contact with water or moisture.]

**Exposure Routes, Symptoms, Target Organs (see Table 5):**
**ER:** Inh, Ing, Con
**SY:** Irrit eyes, skin, upper resp sys; eye, skin burns
**TO:** Eyes, skin, resp sys

**First Aid (see Table 6):**
**Eye:** Irr immed
**Skin:** Water flush immed
**Breath:** Resp support
**Swallow:** Medical attention immed

| **Chlordane** | **Formula:** $C_{10}H_6Cl_8$ | **CAS#:** 57-74-9 | **RTECS#:** PB9800000 | **IDLH:** Ca [100 $mg/m^3$] |
|---|---|---|---|---|
| **Conversion:** | **DOT:** 2996 151 | | | |

**Synonyms/Trade Names:** Chlordan; Chlordano;
1,2,4,5,6,7,8,8-Octachloro-3a,4,7,7a-tetrahydro-4,7-methanoindane

**Exposure Limits:**
**NIOSH REL:** Ca
TWA 0.5 $mg/m^3$ [skin]
See Appendix A
**OSHA PEL:** TWA 0.5 $mg/m^3$ [skin]

**Measurement Methods (see Table 1):**
**NIOSH** 5510
**OSHA** 67

**Physical Description:** Amber-colored, viscous liquid with a pungent, chlorine-like odor. [insecticide]

**Chemical & Physical Properties:**
**MW:** 409.8
**BP:** Decomposes
**Sol:** 0.0001%
**Fl.P:** NA
**IP:** ?
**Sp.Gr(77°F):** 1.6
**VP:** 0.00001 mmHg
**FRZ:** 217-228°F
**UEL:** NA
**LEL:** NA
Noncombustible Liquid, but may be utilized in flammable solutions.

**Personal Protection/Sanitation (see Table 2):**
**Skin:** Prevent skin contact
**Eyes:** Prevent eye contact
**Wash skin:** When contam
**Remove:** When wet or contam
**Change:** Daily
**Provide:** Eyewash
Quick drench

**Respirator Recommendations (see Tables 3 and 4):**
**NIOSH**
**¥:** ScbaF:Pd,Pp/SaF:Pd,Pp:AScba
**Escape:** GmFOv100/ScbaE

**Incompatibilities and Reactivities:** Strong oxidizers, alkaline reagents

**Exposure Routes, Symptoms, Target Organs (see Table 5):**
**ER:** Inh, Abs, Ing, Con
**SY:** Blurred vision; conf; ataxia, delirium; cough; abdom pain, nau, vomit, diarr; irrity, tremor, convuls; anuria; in animals: lung, liver, kidney damage; [carc]
**TO:** CNS, eyes, lungs, liver, kidneys [in animals: liver cancer]

**First Aid (see Table 6):**
**Eye:** Irr immed
**Skin:** Soap wash immed
**Breath:** Resp support
**Swallow:** Medical attention immed

C

## Chlorinated camphene

| Formula: | CAS#: | RTECS#: | IDLH: |
|---|---|---|---|
| $C_{10}H_{10}Cl_8$ | 8001-35-2 | XW5250000 | Ca [200 mg/m$^3$] |

**Conversion:**

**DOT:** 2761 151

**Synonyms/Trade Names:** Chlorocamphene, Octachlorocamphene, Polychlorocamphene, Toxaphene

**Exposure Limits:**
**NIOSH REL:** Ca [skin]
See Appendix A
**OSHA PEL†:** TWA 0.5 mg/m$^3$ [skin]

**Measurement Methods (see Table 1):**
**NIOSH** 5039

**Physical Description:** Amber, waxy solid with a mild, piney, chlorine- and camphor-like odor. [insecticide]

| Chemical & Physical Properties: | Personal Protection/Sanitation (see Table 2): | Respirator Recommendations (see Tables 3 and 4): |
|---|---|---|
| **MW:** 413.8<br>**BP:** Decomposes<br>**Sol:** 0.0003%<br>**Fl.P:** NA<br>**IP:** ?<br>**Sp.Gr:** 1.65<br>**VP(77°F):** 0.4 mmHg<br>**MLT:** 149-194°F<br>**UEL:** NA<br>**LEL:** NA<br>Noncombustible Solid, but may be dissolved in flammable liquids. | **Skin:** Prevent skin contact<br>**Eyes:** Prevent eye contact<br>**Wash skin:** When contam/Daily<br>**Remove:** When wet or contam<br>**Change:** Daily<br>**Provide:** Eyewash<br>Quick drench | **NIOSH**<br>**¥:** ScbaF:Pd,Pp/SaF:Pd,Pp:AScba<br>**Escape:** GmFOv100/ScbaE |

**Incompatibilities and Reactivities:** Strong oxidizers [**Note:** Slightly corrosive to metals under moist conditions.]

**Exposure Routes, Symptoms, Target Organs (see Table 5):**
**ER:** Inh, Abs, Ing, Con
**SY:** Nau, conf, agitation, tremor, convuls, uncon; dry, red skin; [carc]
**TO:** CNS, skin [in animals: liver cancer]

**First Aid (see Table 6):**
**Eye:** Irr immed
**Skin:** Soap wash prompt
**Breath:** Resp support
**Swallow:** Medical attention immed

## Chlorinated diphenyl oxide

| Formula: | CAS#: | RTECS#: | IDLH: |
|---|---|---|---|
| $C_{12}H_{10-n}Cl_nO$ | | | 5 mg/m$^3$ |

**Conversion:**

**DOT:**

**Synonyms/Trade Names:** Synonyms depend on the degree of chlorination of diphenyl oxide [$(C_6H_5)_2O$], ranging from monochlorodiphenyl oxide [$(C_6H_4Cl)O(C_6H_5)$] to decachlorodiphenyl oxide [$(C_6Cl_5)O(C_6Cl_5)$].

**Exposure Limits:**
**NIOSH REL:** TWA 0.5 mg/m$^3$
**OSHA PEL:** TWA 0.5 mg/m$^3$

**Measurement Methods (see Table 1):**
**NIOSH** 5025

**Physical Description:** Appearance and odor vary depending upon the specific compound.

| Chemical & Physical Properties: | Personal Protection/Sanitation (see Table 2): | Respirator Recommendations (see Tables 3 and 4): |
|---|---|---|
| Properties vary depending upon the specific compound. | **Skin:** Prevent skin contact<br>**Eyes:** Prevent eye contact<br>**Wash skin:** When contam<br>**Remove:** When wet or contam<br>**Change:** Daily | **NIOSH/OSHA**<br>**5 mg/m$^3$:** Sa/ScbaF<br>**§:** ScbaF:Pd,Pp/SaF:Pd,Pp:AScba<br>**Escape:** GmFOvAg100/ScbaE |

**Incompatibilities and Reactivities:** Strong oxidizers

**Exposure Routes, Symptoms, Target Organs (see Table 5):**
**ER:** Inh, Ing, Con
**SY:** Acne-form derm, liver damage
**TO:** Skin, liver

**First Aid (see Table 6):**
**Eye:** Irr immed
**Skin:** Soap wash prompt
**Breath:** Resp support
**Swallow:** Medical attention immed

## Chlorine

| Chlorine | Formula: $Cl_2$ | CAS#: 7782-50-5 | RTECS#: FO2100000 | IDLH: 10 ppm |
|---|---|---|---|---|

**Conversion:** 1 ppm = 2.90 mg/m³ **DOT:** 1017 124

**Synonyms/Trade Names:** Molecular chlorine

**Exposure Limits:**
**NIOSH REL:** C 0.5 ppm (1.45 mg/$m^3$) [15-minute]
**OSHA PEL†:** C 1 ppm (3 mg/$m^3$)

**Measurement Methods (see Table 1):**
**NIOSH** 6011
**OSHA** ID101, ID126SGX

**Physical Description:** Greenish-yellow gas with a pungent, irritating odor.
[**Note:** Shipped as a liquefied compressed gas.]

**Chemical & Physical Properties:**
**MW:** 70.9
**BP:** -29°F
**Sol:** 0.7%
**Fl.P:** NA
**IP:** 11.48 eV
**RGasD:** 2.47
**VP:** 6.8 atm
**FRZ:** -150°F
**UEL:** NA
**LEL:** NA
Nonflammable Gas, but a strong oxidizer.

**Personal Protection/Sanitation (see Table 2):**
**Skin:** Frostbite
**Eyes:** Frostbite
**Wash skin:** N.R.
**Remove:** N.R.
**Change:** N.R.
**Provide:** Frostbite wash

**Respirator Recommendations (see Tables 3 and 4):**
**NIOSH**
**5 ppm:** CcrS*/Sa*
**10 ppm:** Sa:Cf*/PaprS*/CcrFS/GmFS/ScbaF/SaF
**§:** ScbaF:Pd,Pp/SaF:Pd,Pp:AScba
**Escape:** GmFS/ScbaE

**Incompatibilities and Reactivities:** Reacts explosively or forms explosive compounds with many common substances such as acetylene, ether, turpentine, ammonia, fuel gas, hydrogen & finely divided metals.

**Exposure Routes, Symptoms, Target Organs (see Table 5):**
**ER:** Inh, Con
**SY:** Burning of eyes, nose, mouth; lac, rhin; cough, choking, subs pain; nau, vomit; head, dizz; syncope; pulm edema; pneu; hypox; derm; liquid: frostbite
**TO:** Eyes, skin, resp sys

**First Aid (see Table 6):**
**Eye:** Frostbite
**Skin:** Frostbite
**Breath:** Resp support

## Chlorine dioxide

| Chlorine dioxide | Formula: $ClO_2$ | CAS#: 10049-04-4 | RTECS#: FO3000000 | IDLH: 5 ppm |
|---|---|---|---|---|

**Conversion:** 1 ppm = 2.76 mg/m³ **DOT:** 9191 143 (hydrate, frozen)

**Synonyms/Trade Names:** Chlorine oxide, Chlorine peroxide

**Exposure Limits:**
**NIOSH REL:** TWA 0.1 ppm (0.3 mg/$m^3$)
ST 0.3 ppm (0.9 mg/$m^3$)
**OSHA PEL†:** TWA 0.1 ppm (0.3 mg/$m^3$)

**Measurement Methods (see Table 1):**
**OSHA** ID126SGX, ID202

**Physical Description:** Yellow to red gas or a red-brown liquid (below 52°F) with an unpleasant odor similar to chlorine and nitric acid.

**Chemical & Physical Properties:**
**MW:** 67.5
**BP:** 52°F
**Sol(77°F):** 0.3%
**Fl.P:** NA (Gas) ? (Liquid)
**IP:** 10.36 eV
**RGasD:** 2.33
**Sp.Gr:** 1.6 (Liquid at 32°F)
**VP:** >1 atm
**FRZ:** -74°F
**UEL:** ?
**LEL:** ?
Flammable Gas, Combustible Liquid

**Personal Protection/Sanitation (see Table 2):**
**Skin:** Prevent skin contact (liquid)
**Eyes:** Prevent eye contact (liquid)
**Wash skin:** When contam (liquid)
**Remove:** When wet (flamm)
**Change:** N.R.
**Provide:** Eyewash (liquid)
Quick drench (liquid)

**Respirator Recommendations (see Tables 3 and 4):**
**NIOSH/OSHA**
**1 ppm:** CcrS/Sa
**2.5 ppm:** Sa:Cf£/PaprS£
**5 ppm:** CcrFS/GmFS/ScbaF/SaF
**§:** ScbaF:Pd,Pp/SaF:Pd,Pp:AScba
**Escape:** GmFS¿/ScbaE

**Incompatibilities and Reactivities:** Organic materials, heat, phosphorus, potassium hydroxide, sulfur, mercury, carbon monoxide [**Note:** Unstable in light. A powerful oxidizer.]

**Exposure Routes, Symptoms, Target Organs (see Table 5):**
**ER:** Inh, Ing (liquid), Con
**SY:** Irrit eyes, nose, throat; cough, wheez, bron, pulm edema; chronic bron
**TO:** Eyes, resp sys

**First Aid (see Table 6):**
**Eye:** Irr immed (liquid)
**Skin:** Soap wash immed (liquid)
**Breath:** Resp support
**Swallow:** Medical attention immed (liquid)

C

## Chlorine trifluoride

| | | | | |
|---|---|---|---|---|
| **Chlorine trifluoride** | **Formula:** $ClF_3$ | **CAS#:** 7790-91-2 | **RTECS#:** FO2800000 | **IDLH:** 20 ppm |
| **Conversion:** 1 ppm = 3.78 mg/m$^3$ | **DOT:** 1749 124 | | | |

**Synonyms/Trade Names:** Chlorine fluoride, Chlorotrifluoride

**Exposure Limits:**
**NIOSH REL:** C 0.1 ppm (0.4 mg/m$^3$)
**OSHA PEL:** C 0.1 ppm (0.4 mg/m$^3$)

**Measurement Methods (see Table 1):**
None available

**Physical Description:** Colorless gas or a greenish-yellow liquid (below 53°F) with a somewhat sweet, suffocating odor. [**Note:** Shipped as a liquefied compressed gas.]

**Chemical & Physical Properties:**
**MW:** 92.5
**BP:** 53°F
**Sol:** Reacts
**Fl.P:** NA
**IP:** 13.00 eV
**RGasD:** 3.21
**Sp.Gr:** 1.77 (Liquid at 53°F)
**VP:** 1.4 atm
**FRZ:** -105°F
**UEL:** NA
**LEL:** NA
Nonflammable Gas
Noncombustible Liquid, but contact with organic materials may result in SPONTANEOUS ignition.

**Personal Protection/Sanitation (see Table 2):**
**Skin:** Prevent skin contact
**Eyes:** Prevent eye contact
**Wash skin:** When contam (liquid)
**Remove:** When wet or contam (liquid)
**Change:** N.R.
**Provide:** Eyewash (liquid)
Quick drench (liquid)

**Respirator Recommendations (see Tables 3 and 4):**
**NIOSH/OSHA**
**2.5 ppm:** Sa:Cf£
**5 ppm:** ScbaF/SaF
**20 ppm:** SaF:Pd,Pp
**§:** ScbaF:Pd,Pp/SaF:Pd,Pp:AScba
**Escape:** GmFS/ScbaE

**Incompatibilities and Reactivities:** Oxidizers, water, acids, combustible materials, sand, glass, metals (corrosive)
[**Note:** Reacts with water to form chlorine & hydrofluoric acid.]

**Exposure Routes, Symptoms, Target Organs (see Table 5):**
**ER:** Inh, Ing (liquid), Con
**SY:** Eye, skin burns (liq or high vap conc); resp irrit; in animals: lac, corn ulcer; pulm edema
**TO:** Skin, eyes, resp sys

**First Aid (see Table 6):**
**Eye:** Irr immed
**Skin:** Water flush immed
**Breath:** Resp support
**Swallow:** Medical attention immed (liquid)

## Chloroacetaldehyde

| | | | | |
|---|---|---|---|---|
| **Chloroacetaldehyde** | **Formula:** $ClCH_2CHO$ | **CAS#:** 107-20-0 | **RTECS#:** AB2450000 | **IDLH:** 45 ppm |
| **Conversion:** 1 ppm = 3.21 mg/m$^3$ | **DOT:** 2232 153 | | | |

**Synonyms/Trade Names:** Chloroacetaldehyde (40% aqueous solution), 2-Chloroacetaldehyde, 2-Chloroethanal

**Exposure Limits:**
**NIOSH REL:** C 1 ppm (3 mg/m$^3$)
**OSHA PEL:** C 1 ppm (3 mg/m$^3$)

**Measurement Methods (see Table 1):**
**NIOSH** 2015
**OSHA** 76

**Physical Description:** Colorless liquid with an acrid, penetrating odor.
[**Note:** Typically found as a 40% aqueous solution.]

**Chemical & Physical Properties:**
**MW:** 78.5
**BP:** 186°F
**Sol:** Miscible
**Fl.P:** 190°F (40% solution)
**IP:** 10.61 eV
**Sp.Gr:** 1.19 (40% solution)
**VP:** 100 mmHg
**FRZ:** -3°F (40% solution)
**UEL:** ?
**LEL:** ?
Class IIIA Combustible Liquid

**Personal Protection/Sanitation (see Table 2):**
**Skin:** Prevent skin contact
**Eyes:** Prevent eye contact
**Wash skin:** When contam
**Remove:** When wet or contam
**Change:** N.R.
**Provide:** Eyewash
Quick drench

**Respirator Recommendations (see Tables 3 and 4):**
**NIOSH/OSHA**
**10 ppm:** CcrOv*/Sa*
**25 ppm:** Sa:Cf*/PaprOv*
**45 ppm:** CcrFOv/GmFOv/PaprTOv*/ScbaF/SaF
**§:** ScbaF:Pd,Pp/SaF:Pd,Pp:AScba
**Escape:** GmFOv/ScbaE

**Incompatibilities and Reactivities:** Oxidizers, acids

**Exposure Routes, Symptoms, Target Organs (see Table 5):**
**ER:** Inh, Abs, Ing, Con
**SY:** Irrit skin, eyes, muc memb; skin burns; eye damage; pulm edema; skin, resp sys sens
**TO:** Eyes, skin, resp sys

**First Aid (see Table 6):**
**Eye:** Irr immed
**Skin:** Water flush immed
**Breath:** Resp support
**Swallow:** Medical attention immed

## α-Chloroacetophenone

**Formula:** $C_6H_5COCH_2Cl$ | **CAS#:** 532-27-4 | **RTECS#:** AM6300000 | **IDLH:** 15 mg/m³

**Conversion:** 1 ppm = 6.32 mg/m³ | **DOT:** 1697 153

**Synonyms/Trade Names:** 2-Chloroacetophenone, Chloromethyl phenyl ketone, Mace®, Phenacyl chloride, Phenyl chloromethyl ketone, Tear gas

**Exposure Limits:**
**NIOSH REL:** TWA 0.3 mg/m³ (0.05 ppm)
**OSHA PEL:** TWA 0.3 mg/m³ (0.05 ppm)

**Measurement Methods (see Table 1):**
**NIOSH** P&CAM291 (II-5)

**Physical Description:** Colorless to gray crystalline solid with a sharp, irritating odor.

**Chemical & Physical Properties:**
**MW:** 154.6
**BP:** 472°F
**Sol:** Insoluble
**Fl.P:** 244°F
**IP:** 9.44 eV
**Sp.Gr:** 1.32
**VP:** 0.005 mmHg
**MLT:** 134°F
**UEL:** ?
**LEL:** ?
Combustible Solid

**Personal Protection/Sanitation (see Table 2):**
**Skin:** Prevent skin contact
**Eyes:** Prevent eye contact
**Wash skin:** When contam
**Remove:** When wet or contam
**Change:** Daily
**Provide:** Eyewash

**Respirator Recommendations (see Tables 3 and 4):**
**NIOSH/OSHA**
**3 mg/m³:** CcrOv95/Sa
**7.5 mg/m³:** Sa:Cf£/PaprOvHie£
**15 mg/m³:** CcrFOv100/GmFS100/ScbaF/SaF
**§:** ScbaF:Pd,Pp/SaF:Pd,Pp:AScba
**Escape:** GmFS100/ScbaE

**Incompatibilities and Reactivities:** Water, steam, strong oxidizers [**Note:** Slowly corrodes metals.]

**Exposure Routes, Symptoms, Target Organs (see Table 5):**
**ER:** Inh, Ing, Con
**SY:** Irrit eyes, skin, resp sys; pulm edema
**TO:** Eyes, skin, resp sys

**First Aid (see Table 6):**
**Eye:** Irr immed
**Skin:** Soap wash immed
**Breath:** Resp support
**Swallow:** Medical attention immed

C

## Chloroacetyl chloride

**Formula:** $ClCH_2COCl$ | **CAS#:** 79-04-9 | **RTECS#:** AO6475000 | **IDLH:** N.D.

**Conversion:** 1 ppm = 4.62 mg/m³ | **DOT:** 1752 156

**Synonyms/Trade Names:** Chloroacetic acid chloride, Chloroacetic chloride, Monochloroacetyl chloride

**Exposure Limits:**
**NIOSH REL:** TWA 0.05 ppm (0.2 mg/m³)
**OSHA PEL†:** none

**Measurement Methods (see Table 1):**
None available

**Physical Description:** Colorless to yellowish liquid with a strong, pungent odor.

**Chemical & Physical Properties:**
**MW:** 112.9
**BP:** 223°F
**Sol:** Decomposes
**Fl.P:** NA
**IP:** 10.30 eV
**Sp.Gr:** 1.42
**VP:** 19 mmHg
**FRZ:** -7°F
**UEL:** NA
**LEL:** NA
Noncombustible Liquid

**Personal Protection/Sanitation (see Table 2):**
**Skin:** Prevent skin contact
**Eyes:** Prevent eye contact
**Wash skin:** When contam
**Remove:** When wet or contam
**Change:** N.R.
**Provide:** Eyewash
Quick drench

**Respirator Recommendations (see Tables 3 and 4):**
Not available.

**Incompatibilities and Reactivities:** Water, alcohols, bases, metals (corrosive), amines
[**Note:** Decomposes in water to form chloroacetic acid & hydrogen chloride gas.]

**Exposure Routes, Symptoms, Target Organs (see Table 5):**
**ER:** Inh, Abs, Ing, Con
**SY:** Irrit eyes, skin, resp sys; eye, skin burns; cough, wheez, dysp; lac
**TO:** Eyes, skin, resp sys

**First Aid (see Table 6):**
**Eye:** Irr immed
**Skin:** Water flush immed
**Breath:** Resp support
**Swallow:** Medical attention immed

C

| Chlorobenzene | Formula: $C_6H_5Cl$ | CAS#: 108-90-7 | RTECS#: CZ0175000 | IDLH: 1000 ppm |
|---|---|---|---|---|
| **Conversion:** 1 ppm = 4.61 mg/m$^3$ | **DOT:** 1134 130 | | | |

**Synonyms/Trade Names:** Benzene chloride, Chlorobenzol, MCB, Monochlorobenzene, Phenyl chloride

**Exposure Limits:**
**NIOSH REL:** See Appendix D
**OSHA PEL:** TWA 75 ppm (350 mg/m$^3$)

**Measurement Methods (see Table 1):**
**NIOSH** 1003
**OSHA** 7

**Physical Description:** Colorless liquid with an almond-like odor.

| Chemical & Physical Properties: | Personal Protection/Sanitation (see Table 2): | Respirator Recommendations (see Tables 3 and 4): |
|---|---|---|
| **MW:** 112.6<br>**BP:** 270°F<br>**Sol:** 0.05%<br>**Fl.P:** 82°F<br>**IP:** 9.07 eV<br>**Sp.Gr:** 1.11<br>**VP:** 9 mmHg<br>**FRZ:** -50°F<br>**UEL:** 9.6%<br>**LEL:** 1.3%<br>Class IC Flammable Liquid | **Skin:** Prevent skin contact<br>**Eyes:** Prevent eye contact<br>**Wash skin:** When contam<br>**Remove:** When wet (flamm)<br>**Change:** N.R. | **OSHA**<br>**1000 ppm:** Sa:Cf£/PaprOv£/CcrFOv/ GmFOv/ScbaF/SaF<br>**§:** ScbaF:Pd,Pp/SaF:Pd,Pp:AScba<br>**Escape:** GmFOv/ScbaE |

**Incompatibilities and Reactivities:** Strong oxidizers

| Exposure Routes, Symptoms, Target Organs (see Table 5): | First Aid (see Table 6): |
|---|---|
| **ER:** Inh, Ing, Con<br>**SY:** Irrit eyes, skin, nose; drow, inco; CNS depres; in animals: liver, lung, kidney inj<br>**TO:** Eyes, skin, resp sys, CNS, liver | **Eye:** Irr immed<br>**Skin:** Soap wash prompt<br>**Breath:** Resp support<br>**Swallow:** Medical attention immed |

| o-Chlorobenzylidene malononitrile | Formula: $ClC_6H_4CH{=}C(CN)_2$ | CAS#: 2698-41-1 | RTECS#: OO3675000 | IDLH: 2 mg/m$^3$ |
|---|---|---|---|---|
| **Conversion:** 1 ppm = 7.71 mg/m$^3$ | **DOT:** 2810 153 | | | |

**Synonyms/Trade Names:** 2-Chlorobenzalmalonitrile, CS, OCBM

**Exposure Limits:**
**NIOSH REL:** C 0.05 ppm (0.4 mg/m$^3$) [skin]
**OSHA PEL†:** TWA 0.05 ppm (0.4 mg/m$^3$)

**Measurement Methods (see Table 1):**
**NIOSH** P&CAM304 (II-5)

**Physical Description:** White crystalline solid with a pepper-like odor.

| Chemical & Physical Properties: | Personal Protection/Sanitation (see Table 2): | Respirator Recommendations (see Tables 3 and 4): |
|---|---|---|
| **MW:** 188.6<br>**BP:** 590-599°F<br>**Sol:** Insoluble<br>**Fl.P:** ?<br>**IP:** ?<br>**Sp.Gr:** ?<br>**VP:** 0.00003 mmHg<br>**MLT:** 203-205°F<br>**UEL:** ?<br>**LEL:** ?<br>**MEC:** 25 g/m$^3$<br>Combustible Solid | **Skin:** Prevent skin contact<br>**Eyes:** Prevent eye contact<br>**Wash skin:** When contam/Daily<br>**Remove:** When wet or contam<br>**Change:** Daily | **NIOSH/OSHA**<br>**2 mg/m$^3$:** Sa:Cf£/GmFS100/ScbaF/SaF<br>**§:** ScbaF:Pd,Pp/SaF:Pd,Pp:AScba<br>**Escape:** GmFS100/ScbaE |

**Incompatibilities and Reactivities:** Strong oxidizers

| Exposure Routes, Symptoms, Target Organs (see Table 5): | First Aid (see Table 6): |
|---|---|
| **ER:** Inh, Abs, Ing, Con<br>**SY:** Pain, burn eyes, lac, conj; eryt eyelids, blepharospasm; irrit throat, cough, chest tight; head; eryt, vesic skin<br>**TO:** Eyes, skin, resp sys | **Eye:** Irr immed<br>**Skin:** Soap wash immed<br>**Breath:** Resp support<br>**Swallow:** Medical attention immed |

## Chlorobromomethane

| Chlorobromomethane | Formula: | CAS#: | RTECS#: | IDLH: |
|---|---|---|---|---|
| | $CH_2BrCl$ | 74-97-5 | PA5250000 | 2000 ppm |
| **Conversion:** 1 ppm = 5.29 mg/m$^3$ | **DOT:** 1887 160 | | | |

**Synonyms/Trade Names:** Bromochloromethane, CB, CBM, Fluorocarbon 1011, Halon® 1011, Methyl chlorobromide

**Exposure Limits:**
**NIOSH REL:** TWA 200 ppm (1050 mg/m$^3$)
**OSHA PEL:** TWA 200 ppm (1050 mg/m$^3$)

**Measurement Methods (see Table 1):**
**NIOSH** 1003

**Physical Description:** Colorless to pale-yellow liquid with a chloroform-like odor.
[**Note:** May be used as a fire extinguishing agent.]

**Chemical & Physical Properties:**
**MW:** 129.4
**BP:** 155°F
**Sol:** Insoluble
**Fl.P:** NA
**IP:** 10.77 eV
**Sp.Gr:** 1.93
**VP:** 115 mmHg
**FRZ:** -124°F
**UEL:** NA
**LEL:** NA
Noncombustible Liquid

**Personal Protection/Sanitation (see Table 2):**
**Skin:** Prevent skin contact
**Eyes:** Prevent eye contact
**Wash skin:** When contam
**Remove:** When wet or contam
**Change:** N.R.

**Respirator Recommendations (see Tables 3 and 4):**
**NIOSH/OSHA**
**2000 ppm:** Sa:Cf£/PaprOv£/CcrFOv/GmFOv/ScbaF/SaF
**§:** ScbaF:Pd,Pp/SaF:Pd,Pp:AScba
**Escape:** GmFOv/ScbaE

**Incompatibilities and Reactivities:** Chemically-active metals such as calcium, powdered aluminum, zinc, and magnesium

**Exposure Routes, Symptoms, Target Organs (see Table 5):**
**ER:** Inh, Ing, Con
**SY:** Irrit eyes, skin, throat; conf, dizz, CNS depres; pulm edema
**TO:** Eyes, skin, resp sys, liver, kidneys, CNS

**First Aid (see Table 6):**
**Eye:** Irr immed
**Skin:** Soap wash prompt
**Breath:** Resp support
**Swallow:** Medical attention immed

## Chlorodifluoromethane

| Chlorodifluoromethane | Formula: | CAS#: | RTECS#: | IDLH: |
|---|---|---|---|---|
| | $CHClF_2$ | 75-45-6 | PA6390000 | N.D. |
| **Conversion:** 1 ppm = 3.54 mg/m$^3$ | **DOT:** 1018 126 | | | |

**Synonyms/Trade Names:** Difluorochloromethane, Fluorocarbon-22, Freon® 22, Genetron® 22, Monochlorodifluoromethane, Refrigerant 22

**Exposure Limits:**
**NIOSH REL:** TWA 1000 ppm (3500 mg/m$^3$)
ST 1250 ppm (4375 mg/m$^3$)
**OSHA PEL†:** none

**Measurement Methods (see Table 1):**
**NIOSH** 1018

**Physical Description:** Colorless gas with a faint, sweetish odor.
[**Note:** Shipped as a liquefied compressed gas.]

**Chemical & Physical Properties:**
**MW:** 86.5
**BP:** -41°F
**Sol(77°F):** 0.3%
**Fl.P:** NA
**IP:** 12.45 eV
**RGasD:** 3.11
**VP:** 9.4 atm
**FRZ:** -231°F
**UEL:** NA
**LEL:** NA
Nonflammable Gas

**Personal Protection/Sanitation (see Table 2):**
**Skin:** Frostbite
**Eyes:** Frostbite
**Wash skin:** N.R.
**Remove:** N.R.
**Change:** N.R.
**Provide:** Frostbite wash

**Respirator Recommendations (see Tables 3 and 4):**
Not available.

**Incompatibilities and Reactivities:** Alkalis, alkaline earth metals (e.g., powdered aluminum, sodium, potassium, zinc)

**Exposure Routes, Symptoms, Target Organs (see Table 5):**
**ER:** Inh, Con (liquid)
**SY:** Irrit resp sys; conf, drow, ringing in ears; heart palp, card arrhy; asphy; liver, kidney, spleen inj; liquid: frostbite
**TO:** Resp sys, CVS, CNS, liver, kidneys, spleen

**First Aid (see Table 6):**
**Eye:** Frostbite
**Skin:** Frostbite
**Breath:** Resp support

C

| Chlorodiphenyl (42% chlorine) | Formula: $C_6H_4ClC_6H_3Cl_2$ (approx) | CAS#: 53469-21-9 | RTECS#: TQ1356000 | IDLH: Ca [5 mg/m$^3$] |
|---|---|---|---|---|
| **Conversion:** | **DOT:** 2315 171 | | | |

**Synonyms/Trade Names:** Aroclor® 1242, PCB, Polychlorinated biphenyl

**Exposure Limits:**
**NIOSH REL*:** Ca
TWA 0.001 mg/m$^3$
See Appendix A
[***Note:** The REL also applies to other PCBs.]

**OSHA PEL:** TWA 1 mg/m$^3$ [skin]

**Measurement Methods (see Table 1):**
**NIOSH** 5503
**OSHA** PV2089

**Physical Description:** Colorless to light-colored, viscous liquid with a mild, hydrocarbon odor.

**Chemical & Physical Properties:**
**MW:** 258 (approx)
**BP:** 617-691°F
**Sol:** Insoluble
**Fl.P:** NA
**IP:** ?
**Sp.Gr(77°F):** 1.39
**VP:** 0.001 mmHg
**FRZ:** -2°F
**UEL:** NA
**LEL:** NA

**Personal Protection/Sanitation (see Table 2):**
**Skin:** Prevent skin contact
**Eyes:** Prevent eye contact
**Wash skin:** When contam
**Remove:** When wet or contam
**Change:** Daily
**Provide:** Eyewash
Quick drench

**Respirator Recommendations (see Tables 3 and 4):**
**NIOSH**
**¥:** ScbaF:Pd,Pp/SaF:Pd,Pp:AScba
**Escape:** GmFOv100/ScbaE

Nonflammable Liquid, but exposure in a fire results in the formation of a black soot containing PCBs, polychlorinated dibenzofurans & chlorinated dibenzo-p-dioxins.

**Incompatibilities and Reactivities:** Strong oxidizers

**Exposure Routes, Symptoms, Target Organs (see Table 5):**
**ER:** Inh, Abs, Ing, Con
**SY:** Irrit eyes; chloracne; liver damage; repro effects; [carc]
**TO:** Skin, eyes, liver, repro sys [in animals: tumors of the pituitary gland & liver, leukemia]

**First Aid (see Table 6):**
**Eye:** Irr immed
**Skin:** Soap wash immed
**Breath:** Resp support
**Swallow:** Medical attention immed

| Chlorodiphenyl (54% chlorine) | Formula: $C_6H_3Cl_2C_6H_2Cl_3$ (approx) | CAS#: 11097-69-1 | RTECS#: TQ1360000 | IDLH: Ca [5 mg/m$^3$] |
|---|---|---|---|---|
| **Conversion:** | **DOT:** 2315 171 | | | |

**Synonyms/Trade Names:** Aroclor® 1254, PCB, Polychlorinated biphenyl

**Exposure Limits:**
**NIOSH REL*:** Ca
TWA 0.001 mg/m$^3$
See Appendix A
[***Note:** The REL also applies to other PCBs.]

**OSHA PEL:** TWA 0.5 mg/m$^3$ [skin]

**Measurement Methods (see Table 1):**
**NIOSH** 5503
**OSHA** PV2088

**Physical Description:** Colorless to pale-yellow, viscous liquid or solid (below 50°F) with a mild, hydrocarbon odor.

**Chemical & Physical Properties:**
**MW:** 326 (approx)
**BP:** 689-734°F
**Sol:** Insoluble
**Fl.P:** NA
**IP:** ?
**Sp.Gr(77°F):** 1.38
**VP:** 0.00006 mmHg
**FRZ:** 50°F
**UEL:** NA
**LEL:** NA

**Personal Protection/Sanitation (see Table 2):**
**Skin:** Prevent skin contact
**Eyes:** Prevent eye contact
**Wash skin:** When contam
**Remove:** When wet or contam
**Change:** Daily
**Provide:** Eyewash
Quick drench

**Respirator Recommendations (see Tables 3 and 4):**
**NIOSH**
**¥:** ScbaF:Pd,Pp/SaF:Pd,Pp:AScba
**Escape:** GmFOv100/ScbaE

Nonflammable Liquid, but exposure in a fire results in the formation of a black soot containing PCBs, polychlorinated dibenzofurans, and chlorinated dibenzo-p-dioxins.

**Incompatibilities and Reactivities:** Strong oxidizers

**Exposure Routes, Symptoms, Target Organs (see Table 5):**
**ER:** Inh, Abs, Ing, Con
**SY:** Irrit eyes, chloracne; liver damage; repro effects; [carc]
**TO:** Skin, eyes, liver, repro sys [in animals: tumors of the pituitary gland & liver, leukemia]

**First Aid (see Table 6):**
**Eye:** Irr immed
**Skin:** Soap wash immed
**Breath:** Resp support
**Swallow:** Medical attention immed

## Chloroform

| Chloroform | Formula: $CHCl_3$ | CAS#: 67-66-3 | RTECS#: FS9100000 | IDLH: Ca [500 ppm] |
|---|---|---|---|---|
| **Conversion:** 1 ppm = 4.88 mg/$m^3$ | **DOT:** 1888 151 | | | |

**Synonyms/Trade Names:** Methane trichloride, Trichloromethane

**Exposure Limits:**
**NIOSH REL:** Ca
ST 2 ppm (9.78 mg/$m^3$) [60-minute]
See Appendix A
**OSHA PEL†:** C 50 ppm (240 mg/$m^3$)

**Measurement Methods (see Table 1):**
**NIOSH** 1003

**Physical Description:** Colorless liquid with a pleasant odor.

| Chemical & Physical Properties: | Personal Protection/Sanitation (see Table 2): | Respirator Recommendations (see Tables 3 and 4): |
|---|---|---|
| **MW:** 119.4<br>**BP:** 143°F<br>**Sol(77°F):** 0.5%<br>**Fl.P:** NA<br>**IP:** 11.42 eV<br>**Sp.Gr:** 1.48<br>**VP:** 160 mmHg<br>**FRZ:** -82°F<br>**UEL:** NA<br>**LEL:** NA<br>Noncombustible Liquid | **Skin:** Prevent skin contact<br>**Eyes:** Prevent eye contact<br>**Wash skin:** When contam<br>**Remove:** When wet or contam<br>**Change:** N.R.<br>**Provide:** Eyewash<br>Quick drench | **NIOSH**<br>**¥:** ScbaF:Pd,Pp/SaF:Pd,Pp:AScba<br>**Escape:** GmFOv/ScbaE |

**Incompatibilities and Reactivities:** Strong caustics; chemically-active metals such as aluminum or magnesium powder, sodium & potassium; strong oxidizers [**Note:** When heated to decomposition, forms phosgene gas.]

**Exposure Routes, Symptoms, Target Organs (see Table 5):**
**ER:** Inh, Abs, Ing, Con
**SY:** Irrit eyes, skin; dizz, mental dullness, nau, conf; head, lass; anes; enlarged liver; [carc]
**TO:** Liver, kidneys, heart, eyes, skin, CNS [in animals: liver & kidney cancer]

**First Aid (see Table 6):**
**Eye:** Irr immed
**Skin:** Soap wash prompt
**Breath:** Resp support
**Swallow:** Medical attention immed

## bis-Chloromethyl ether

| bis-Chloromethyl ether | Formula: $(CH_2Cl)_2O$ | CAS#: 542-88-1 | RTECS#: KN1575000 | IDLH: Ca [N.D.] |
|---|---|---|---|---|
| **Conversion:** | **DOT:** 2249 131 | | | |

**Synonyms/Trade Names:** BCME, bis-CME, Chloromethyl ether, Dichlorodimethyl ether, Dichloromethyl ether, Oxybis(chloromethane)

**Exposure Limits:**
**NIOSH REL:** Ca
See Appendix A
**OSHA PEL:** [1910.1008]
See Appendix B

**Measurement Methods (see Table 1):**
**OSHA** 10

**Physical Description:** Colorless liquid with a suffocating odor.

| Chemical & Physical Properties: | Personal Protection/Sanitation (see Table 2): | Respirator Recommendations (see Tables 3 and 4): |
|---|---|---|
| **MW:** 115.0<br>**BP:** 223°F<br>**Sol:** Reacts<br>**Fl.P:** <66°F<br>**IP:** ?<br>**Sp.Gr:** 1.32<br>**VP(72°F):** 30 mmHg<br>**FRZ:** -43°F<br>**UEL:** ?<br>**LEL:** ?<br>Class IB Flammable Liquid | **Skin:** Prevent skin contact<br>**Eyes:** Prevent eye contact<br>**Wash skin:** When contam/Daily<br>**Remove:** When wet (flamm)<br>**Change:** Daily<br>**Provide:** Eyewash<br>Quick drench | **NIOSH**<br>**¥:** ScbaF:Pd,Pp/SaF:Pd,Pp:AScba<br>**Escape:** GmFOv/ScbaE<br><br>**See Appendix E** (page 351) |

**Incompatibilities and Reactivities:** Acids, water
[**Note:** Reacts with water to form hydrochloric acid & formaldehyde.]

**Exposure Routes, Symptoms, Target Organs (see Table 5):**
**ER:** Inh, Abs, Ing, Con
**SY:** Irrit eyes, skin, muc memb, resp sys; pulm congestion, edema; corn damage, nec; decr pulm function, cough, dysp, wheez; blood-stained sputum, bronchial secretions; [carc]
**TO:** Eyes, skin, resp sys [lung cancer]

**First Aid (see Table 6):**
**Eye:** Irr immed
**Skin:** Soap wash immed
**Breath:** Resp support
**Swallow:** Medical attention immed

C

| Chloromethyl methyl ether | Formula: $CH_3OCH_2Cl$ | CAS#: 107-30-2 | RTECS#: KN6650000 | IDLH: Ca [N.D.] |
|---|---|---|---|---|
| Conversion: | DOT: 1239 131 | | | |

**Synonyms/Trade Names:** Chlorodimethyl ether, Chloromethoxymethane, CMME, Dimethylchloroether, Methylchloromethyl ether

| Exposure Limits: | Measurement Methods (see Table 1): |
|---|---|
| **NIOSH REL:** Ca<br>See Appendix A<br>**OSHA PEL:** [1910.1006]<br>See Appendix B | **NIOSH** P&CAM220 (II-1)<br>**OSHA** 10 |

**Physical Description:** Colorless liquid with an irritating odor.

| Chemical & Physical Properties: | Personal Protection/Sanitation (see Table 2): | Respirator Recommendations (see Tables 3 and 4): |
|---|---|---|
| **MW:** 80.5<br>**BP:** 138°F<br>**Sol:** Reacts<br>**Fl.P(oc):** 32°F<br>**IP:** 10.25 eV<br>**Sp.Gr:** 1.06<br>**VP(70°F):** 192 mmHg<br>**FRZ:** -154°F<br>**UEL:** ?<br>**LEL:** ?<br>Class IB Flammable Liquid | **Skin:** Prevent skin contact<br>**Eyes:** Prevent eye contact<br>**Wash skin:** When contam/Daily<br>**Remove:** When wet (flamm)<br>**Change:** Daily<br>**Provide:** Eyewash<br>Quick drench | **NIOSH**<br>**¥:** ScbaF:Pd,Pp/SaF:Pd,Pp:AScba<br>**Escape:** GmFOv/ScbaE<br><br>**See Appendix E** (page 351) |

**Incompatibilities and Reactivities:** Water [**Note:** Reacts with water to form hydrochloric acid & formaldehyde.]

| Exposure Routes, Symptoms, Target Organs (see Table 5): | First Aid (see Table 6): |
|---|---|
| **ER:** Inh, Abs, Ing, Con<br>**SY:** Irrit eyes, skin, muc memb; pulm edema, pulm congestion, pneu; skin burns, nec; cough, wheez, pulm congestion; blood stained-sputum; low-wgt; bronchial secretions; [carc]<br>**TO:** Eyes, skin, resp sys [in animals: skin & lung cancer] | **Eye:** Irr immed<br>**Skin:** Soap wash immed<br>**Breath:** Resp support<br>**Swallow:** Medical attention immed |

| 1-Chloro-1-nitropropane | Formula: $CH_3CH_2CHClNO_2$ | CAS#: 600-25-9 | RTECS#: TX5075000 | IDLH: 100 ppm |
|---|---|---|---|---|
| Conversion: 1 ppm = 5.06 mg/m$^3$ | DOT: | | | |

**Synonyms/Trade Names:** Korax®, Lanstan®

| Exposure Limits: | Measurement Methods (see Table 1): |
|---|---|
| **NIOSH REL:** TWA 2 ppm (10 mg/m$^3$)<br>**OSHA PEL†:** TWA 20 ppm (100 mg/m$^3$) | **NIOSH** S211 (II-5) |

**Physical Description:** Colorless liquid with an unpleasant odor. [fungicide]

| Chemical & Physical Properties: | Personal Protection/Sanitation (see Table 2): | Respirator Recommendations (see Tables 3 and 4): |
|---|---|---|
| **MW:** 123.6<br>**BP:** 289°F<br>**Sol:** 0.5%<br>**Fl.P(oc):** 144°F<br>**IP:** 9.90 eV<br>**Sp.Gr:** 1.21<br>**VP(77°F):** 6 mmHg<br>**FRZ:** ?<br>**UEL:** ?<br>**LEL:** ?<br>Class IIIA Combustible Liquid | **Skin:** Prevent skin contact<br>**Eyes:** Prevent eye contact<br>**Wash skin:** When contam<br>**Remove:** When wet or contam<br>**Change:** N.R. | **NIOSH**<br>**20 ppm:** Sa*<br>**50 ppm:** Sa:Cf*/PaprOv*<br>**100 ppm:** CcrFOv/GmFOv/PaprTOv*/<br>ScbaF/SaF<br>**§:** ScbaF:Pd,Pp/SaF:Pd,Pp:AScba<br>**Escape:** GmFOv/ScbaE |

**Incompatibilities and Reactivities:** Strong oxidizers, acids

| Exposure Routes, Symptoms, Target Organs (see Table 5): | First Aid (see Table 6): |
|---|---|
| **ER:** Inh, Ing, Con<br>**SY:** In animals: irrit eyes; pulm edema; liver, kidney, heart damage<br>**TO:** Resp sys, liver, kidneys, CVS, eyes | **Eye:** Irr immed<br>**Skin:** Soap wash<br>**Breath:** Resp support<br>**Swallow:** Medical attention immed |

C

| Chloropentafluoroethane | Formula: $CClF_2CF_3$ | CAS#: 76-15-3 | RTECS#: KH7877500 | IDLH: N.D. |
|---|---|---|---|---|
| **Conversion:** 1 ppm = 6.32 mg/$m^3$ | **DOT:** 1020 126 | | | |

**Synonyms/Trade Names:** Fluorocarbon-115, Freon® 115, Genetron® 115, Halocarbon 115, Monochloropentafluoroethane

**Exposure Limits:**
**NIOSH REL:** TWA 1000 ppm (6320 mg/$m^3$)
**OSHA PEL†:** none

**Measurement Methods (see Table 1):**
None available

**Physical Description:** Colorless gas with a slight, ethereal odor.
[**Note:** Shipped as a liquefied compressed gas.]

**Chemical & Physical Properties:**
**MW:** 154.5
**BP:** -38°F
**Sol(77°F):** 0.006%
**Fl.P:** NA
**IP:** 12.96 eV
**RGasD:** 5.55
**VP(70°F):** 7.9 atm
**FRZ:** -223°F
**UEL:** NA
**LEL:** NA
Nonflammable Gas

**Personal Protection/Sanitation (see Table 2):**
**Skin:** Frostbite
**Eyes:** Frostbite
**Wash skin:** N.R.
**Remove:** N.R.
**Change:** N.R.
**Provide:** Frostbite wash

**Respirator Recommendations (see Tables 3 and 4):**
Not available.

**Incompatibilities and Reactivities:** Alkalis, alkaline earth metals (e.g., aluminum powder, sodium, potassium, zinc)

**Exposure Routes, Symptoms, Target Organs (see Table 5):**
**ER:** Inh, Con (liquid)
**SY:** Dysp; dizz, inco, narco; nau, vomit; heart palp, card arrhy, asphy; liquid: frostbite, derm
**TO:** Skin, CNS, CVS

**First Aid (see Table 6):**
**Eye:** Frostbite
**Skin:** Frostbite
**Breath:** Resp support

| Chloropicrin | Formula: $CCl_3NO_2$ | CAS#: 76-06-2 | RTECS#: PB6300000 | IDLH: 2 ppm |
|---|---|---|---|---|
| **Conversion:** 1 ppm = 6.72 mg/$m^3$ | **DOT:** 1580 154; 1583 154 (mixture, n.o.s.) | | | |

**Synonyms/Trade Names:** Nitrochloroform, Nitrotrichloromethane, Trichloronitromethane

**Exposure Limits:**
**NIOSH REL:** TWA 0.1 ppm (0.7 mg/$m^3$)
**OSHA PEL:** TWA 0.1 ppm (0.7 mg/$m^3$)

**Measurement Methods (see Table 1):**
None available

**Physical Description:** Colorless to faint-yellow, oily liquid with an intensely irritating odor. [pesticide]

**Chemical & Physical Properties:**
**MW:** 164.4
**BP:** 234°F
**Sol:** 0.2%
**Fl.P:** NA
**IP:** ?
**Sp.Gr:** 1.66
**VP:** 18 mmHg
**FRZ:** -93°F
**UEL:** NA
**LEL:** NA
Noncombustible Liquid

**Personal Protection/Sanitation (see Table 2):**
**Skin:** Prevent skin contact
**Eyes:** Prevent eye contact
**Wash skin:** When contam
**Remove:** When wet or contam
**Change:** N.R.
**Provide:** Eyewash
Quick drench

**Respirator Recommendations (see Tables 3 and 4):**
**NIOSH/OSHA**
**2 ppm:** Sa:Cf£/PaprOv£/CcrFOv/GmFOv/ScbaF/SaF
**§:** ScbaF:Pd,Pp/SaF:Pd,Pp:AScba
**Escape:** GmFOv/ScbaE

**Incompatibilities and Reactivities:** Strong oxidizers
[**Note:** The material may explode when heated under confinement.]

**Exposure Routes, Symptoms, Target Organs (see Table 5):**
**ER:** Inh, Ing, Con
**SY:** Irrit eyes, skin, resp sys; lac; cough, pulm edema; nau, vomit
**TO:** Eyes, skin, resp sys

**First Aid (see Table 6):**
**Eye:** Irr immed
**Skin:** Soap wash immed
**Breath:** Resp support
**Swallow:** Medical attention immed

C

## β-Chloroprene

| | | | | |
|---|---|---|---|---|
| **β-Chloroprene** | **Formula:** $CH_2=CClCH=CH_2$ | **CAS#:** 126-99-8 | **RTECS#:** EI9625000 | **IDLH:** Ca [300 ppm] |
| **Conversion:** 1 ppm = 3.62 mg/m$^3$ | **DOT:** 1991 131P (inhibited) | | | |

**Synonyms/Trade Names:** 2-Chloro-1,3-butadiene; Chlorobutadiene; Chloroprene

**Exposure Limits:**
**NIOSH REL:** Ca
C 1 ppm (3.6 mg/m$^3$) [15-minute]
See Appendix A
**OSHA PEL†:** TWA 25 ppm (90 mg/m$^3$) [skin]

**Measurement Methods (see Table 1):**
**NIOSH** 1002
**OSHA** 112

**Physical Description:** Colorless liquid with a pungent, ether-like odor.

| Chemical & Physical Properties: | Personal Protection/Sanitation (see Table 2): | Respirator Recommendations (see Tables 3 and 4): |
|---|---|---|
| **MW:** 88.5<br>**BP:** 139°F<br>**Sol:** Slight<br>**Fl.P:** -4°F<br>**IP:** 8.79 eV<br>**Sp.Gr:** 0.96<br>**VP:** 188 mmHg<br>**FRZ:** -153°F<br>**UEL:** 11.3%<br>**LEL:** 1.9%<br>Class IB Flammable Liquid | **Skin:** Prevent skin contact<br>**Eyes:** Prevent eye contact<br>**Wash skin:** When contam<br>**Remove:** When wet (flamm)<br>**Change:** N.R.<br>**Provide:** Eyewash<br>Quick drench | **NIOSH**<br>**¥:** ScbaF:Pd,Pp/SaF:Pd,Pp:AScba<br>**Escape:** GmFOv/ScbaE |

**Incompatibilities and Reactivities:** Peroxides & other oxidizers
[**Note:** Polymerizes at room temperature unless inhibited with antioxidants.]

| Exposure Routes, Symptoms, Target Organs (see Table 5): | First Aid (see Table 6): |
|---|---|
| **ER:** Inh, Abs, Ing, Con<br>**SY:** Irrit eyes, skin, resp sys; anxi, irrity; derm; alopecia; repro effects; [carc]<br>**TO:** Eyes, skin, resp sys, repro sys [lung & skin cancer] | **Eye:** Irr immed<br>**Skin:** Soap wash immed<br>**Breath:** Resp support<br>**Swallow:** Medical attention immed |

## o-Chlorostyrene

| | | | | |
|---|---|---|---|---|
| **o-Chlorostyrene** | **Formula:** $ClC_6H_4CH=CH_2$ | **CAS#:** 2039-87-4 | **RTECS#:** WL4160000 | **IDLH:** N.D. |
| **Conversion:** 1 ppm = 5.67 mg/m$^3$ | **DOT:** | | | |

**Synonyms/Trade Names:** 2-Chlorostyrene, ortho-Chlorostyrene, 1-Chloro-2-ethenylbenzene

**Exposure Limits:**
**NIOSH REL:** TWA 50 ppm (285 mg/m$^3$)
ST 75 ppm (428 mg/m$^3$)
**OSHA PEL†:** none

**Measurement Methods (see Table 1):**
None available

**Physical Description:** Colorless liquid.

| Chemical & Physical Properties: | Personal Protection/Sanitation (see Table 2): | Respirator Recommendations (see Tables 3 and 4): |
|---|---|---|
| **MW:** 138.6<br>**BP:** 372°F<br>**Sol:** Insoluble<br>**Fl.P:** 138°F<br>**IP:** ?<br>**Sp.Gr:** 1.10<br>**VP(77°F):** 0.96 mmHg<br>**FRZ:** -82°F<br>**UEL:** ?<br>**LEL:** ?<br>Class II Combustible Liquid | **Skin:** Prevent skin contact<br>**Eyes:** Prevent eye contact<br>**Wash skin:** When contam<br>**Remove:** When wet or contam<br>**Change:** N.R. | Not available. |

**Incompatibilities and Reactivities:** None reported

| Exposure Routes, Symptoms, Target Organs (see Table 5): | First Aid (see Table 6): |
|---|---|
| **ER:** Inh, Ing, Con<br>**SY:** In animals: irrit eyes, skin; hema, prot, acidosis; enlarged liver, jaun<br>**TO:** Eyes, skin, liver, kidneys, CNS, PNS | **Eye:** Irr immed<br>**Skin:** Soap wash<br>**Breath:** Resp support<br>**Swallow:** Medical attention immed |

| **o-Chlorotoluene** | **Formula:** $ClC_6H_4CH_3$ | **CAS#:** 95-49-8 | **RTECS#:** XS9000000 | **IDLH:** N.D. |
|---|---|---|---|---|
| **Conversion:** 1 ppm = 5.18 mg/m$^3$ | **DOT:** 2238 129 | | | |

**Synonyms/Trade Names:** 1-Chloro-2-methylbenzene, 2-Chloro-1-methylbenzene, 2-Chlorotoluene, o-Tolyl chloride

**Exposure Limits:**
**NIOSH REL:** TWA 50 ppm (250 mg/m$^3$)
ST 75 ppm (375 mg/m$^3$)
**OSHA PEL†:** none

**Measurement Methods (see Table 1):**
None available

**Physical Description:** Colorless liquid with an aromatic odor.

| **Chemical & Physical Properties:** | **Personal Protection/Sanitation (see Table 2):** | **Respirator Recommendations (see Tables 3 and 4):** |
|---|---|---|
| **MW:** 126.6<br>**BP:** 320°F<br>**Sol(77°F):** 0.009%<br>**Fl.P:** 96°F<br>**IP:** 8.83 eV<br>**Sp.Gr:** 1.08<br>**VP(77°F):** 4 mmHg<br>**FRZ:** -31°F<br>**UEL:** ?<br>**LEL:** ?<br>Class IC Flammable Liquid | **Skin:** Prevent skin contact<br>**Eyes:** Prevent eye contact<br>**Wash skin:** When contam<br>**Remove:** When wet (flamm)<br>**Change:** N.R.<br>**Provide:** Eyewash | Not available. |

**Incompatibilities and Reactivities:** Acids, alkalis, oxidizers, reducing materials, water

| **Exposure Routes, Symptoms, Target Organs (see Table 5):** | **First Aid (see Table 6):** |
|---|---|
| **ER:** Inh, Abs, Ing, Con<br>**SY:** Irrit eyes, skin, muc memb; derm; drow, inco, anes; cough; liver, kidney inj<br>**TO:** Eyes, skin, resp sys, CNS, liver, kidneys | **Eye:** Irr immed<br>**Skin:** Soap wash immed<br>**Breath:** Resp support<br>**Swallow:** Medical attention immed |

| **2-Chloro-6-trichloromethyl pyridine** | **Formula:** $ClC_5H_3NCCl_3$ | **CAS#:** 1929-82-4 | **RTECS#:** US7525000 | **IDLH:** N.D. |
|---|---|---|---|---|
| **Conversion:** | **DOT:** | | | |

**Synonyms/Trade Names:** 2-Chloro-6-(trichloro-methyl)pyridine; Nitrapyrin; N-serve®; 2,2,2,6-Tetrachloro-2-picoline

**Exposure Limits:**
**NIOSH REL:** TWA 10 mg/m$^3$ (total)
ST 20 mg/m$^3$ (total)
TWA 5 mg/m$^3$ (resp)
**OSHA PEL:** TWA 15 mg/m$^3$ (total)
TWA 5 mg/m$^3$ (resp)

**Measurement Methods (see Table 1):**
None available

**Physical Description:** Colorless or white, crystalline solid with a mild, sweet odor.

| **Chemical & Physical Properties:** | **Personal Protection/Sanitation (see Table 2):** | **Respirator Recommendations (see Tables 3 and 4):** |
|---|---|---|
| **MW:** 230.9<br>**BP:** ?<br>**Sol:** Insoluble<br>**Fl.P:** ?<br>**IP:** ?<br>**Sp.Gr:** ?<br>**VP(73°F):** 0.003 mmHg<br>**MLT:** 145°F<br>**UEL:** ?<br>**LEL:** ?<br>Combustible Solid [Explosive] | **Skin:** Prevent skin contact<br>**Eyes:** Prevent eye contact<br>**Wash skin:** When contam<br>**Remove:** When wet or contam<br>**Change:** Daily | Not available. |

**Incompatibilities and Reactivities:** Aluminum, magnesium
[**Note:** Emits oxides of nitrogen and chloride ion when heated to decomposition.]

| **Exposure Routes, Symptoms, Target Organs (see Table 5):** | **First Aid (see Table 6):** |
|---|---|
| **ER:** Inh, Abs, Ing, Con<br>**SY:** No adverse effects noted in ingestion studies with animals.<br>**TO:** Eyes, skin | **Eye:** Irr immed<br>**Skin:** Soap wash immed<br>**Breath:** Resp support<br>**Swallow:** Medical attention immed |

C

## Chlorpyrifos

| Chlorpyrifos | Formula: $C_9H_{11}Cl_3NO_3PS$ | CAS#: 2921-88-2 | RTECS#: TF6300000 | IDLH: N.D. |
|---|---|---|---|---|

**Conversion:**

**DOT:** 2783 152

**Synonyms/Trade Names:** Chlorpyrifos-ethyl; O,O-Diethyl O-3,5,6-trichloro-2-pyridyl phosphorothioate; Dursban®

**Exposure Limits:**
**NIOSH REL:** TWA 0.2 mg/m$^3$
ST 0.6 mg/m$^3$ [skin]
**OSHA PEL†:** none

**Measurement Methods (see Table 1):**
**NIOSH** 5600 **OSHA** 62

**Physical Description:** Colorless to white, crystalline solid with a mild, mercaptan-like odor. [pesticide]
[**Note:** Commercial formulations may be combined with combustible liquids.]

**Chemical & Physical Properties:**
**MW:** 350.6
**BP:** 320°F (Decomposes)
**Sol:** 0.0002%
**Fl.P:** ?
**IP:** ?
**Sp.Gr:** 1.40 (Liquid at 110°F)
**VP:** 0.00002 mmHg
**MLT:** 108°F
**UEL:** ?
**LEL:** ?
Combustible Solid

**Personal Protection/Sanitation (see Table 2):**
**Skin:** Prevent skin contact
**Eyes:** Prevent eye contact
**Wash skin:** When contam
**Remove:** When wet or contam
**Change:** Daily

**Respirator Recommendations (see Tables 3 and 4):**
Not available.

**Incompatibilities and Reactivities:** Strong acids, caustics, amines
[**Note:** Corrosive to copper & brass.]

**Exposure Routes, Symptoms, Target Organs (see Table 5):**
**ER:** Inh, Abs, Ing, Con
**SY:** Wheez, lar spasms, salv; bluish lips, skin; miosis, blurred vision; nau, vomit, abdom cramps, diarr
**TO:** Resp sys, CNS, PNS, plasma chol

**First Aid (see Table 6):**
**Eye:** Irr immed
**Skin:** Soap wash immed
**Breath:** Resp support
**Swallow:** Medical attention immed

## Chromic acid and chromates

| Chromic acid and chromates | Formula: $CrO_3$ (acid) | CAS#: 1333-82-0 ($CrO_3$) | RTECS#: GB6650000 ($CrO_3$) | IDLH: Ca [15 mg/m$^3$ {as Cr(VI)}] |
|---|---|---|---|---|

**Conversion:**

**DOT:** 1755 154 (acid solution); 1463 141 (acid, solid)

**Synonyms/Trade Names: Chromic acid ($CrO_3$):** Chromic anhydride, Chromic oxide, Chromium(VI) oxide (1:3), Chromium trioxide. Synonyms of chromates (i.e., chromium(VI) compounds) such as zinc chromate vary depending upon the specific compound.

**Exposure Limits:**
**NIOSH REL (as Cr):** Ca
TWA 0.001 mg/m$^3$
See Appendix A
See Appendix C
**OSHA PEL (as $CrO_3$):** TWA 0.005 mg/m$^3$ See Appendix C

**Measurement Methods (see Table 1):**
**NIOSH** 7600, 7604, 7605
**OSHA** ID103, ID215, W4001

**Physical Description:** $CrO_3$: Dark-red, odorless flakes or powder.
[**Note:** Often used in an aqueous solution ($H_2CrO_4$).]

**Chemical & Physical Properties:**
**MW:** 100.0
**BP:** 482°F (Decomposes)
**Sol:** 63%
**Fl.P:** NA
**IP:** NA
**Sp.Gr:** 2.70 ($CrO_3$)
**VP:** Very low
**MLT:** 387°F (Decomposes)
**UEL:** NA
**LEL:** NA
$CrO_3$: Noncombustible Solid, but will accelerate the burning of combustible materials.

**Personal Protection/Sanitation (see Table 2):**
**Skin:** Prevent skin contact
**Eyes:** Prevent eye contact
**Wash skin:** When contam
**Remove:** When wet or contam
**Change:** Daily
**Provide:** Eyewash
Quick drench

**Respirator Recommendations (see Tables 3 and 4):**
**NIOSH**
**¥:** ScbaF:Pd,Pp/SaF:Pd,Pp:AScba
**Escape:** 100F/ScbaE

**Incompatibilities and Reactivities:** Combustible, organic, or other readily oxidizable materials (paper, wood, sulfur, aluminum, plastics, etc.); corrosive to metals

**Exposure Routes, Symptoms, Target Organs (see Table 5):**
**ER:** Inh, Ing, Con
**SY:** Irrit resp sys; nasal septum perf; liver, kidney damage; leucyt, leupen, eosin; eye inj, conj; skin ulcer, sens derm; [carc]
**TO:** Blood, resp sys, liver, kidneys, eyes, skin [lung cancer]

**First Aid (see Table 6):**
**Eye:** Irr immed
**Skin:** Soap flush immed
**Breath:** Resp support
**Swallow:** Medical attention immed

## Chromium(II) compounds (as Cr)

| Formula: | CAS#: | RTECS#: | IDLH: 250 mg/m$^3$ [as Cr(II)] |
|---|---|---|---|

**Conversion:**

**DOT:**

**Synonyms/Trade Names:** Synonyms vary depending upon the specific Chromium(II) compound.
[**Note:** Chromium(II) compounds include soluble chromous salts.]

**Exposure Limits:**
**NIOSH REL:** TWA 0.5 mg/m$^3$
See Appendix C
**OSHA PEL:** TWA 0.5 mg/m$^3$
See Appendix C

**Measurement Methods (see Table 1):**
**NIOSH** 7024, 7300, 7301, 7303, 9102
**OSHA** ID121, ID125G

**Physical Description:** Appearance and odor vary depending upon the specific compound.

**Chemical & Physical Properties:**
Properties vary depending upon the specific compound.

**Personal Protection/Sanitation (see Table 2):**
**Skin:** Prevent skin contact
**Eyes:** Prevent eye contact
**Wash skin:** When contam
**Remove:** When wet or contam
**Change:** N.R.

**Respirator Recommendations (see Tables 3 and 4):**
**NIOSH/OSHA**
**2.5 mg/m$^3$:** Qm*
**5 mg/m$^3$:** 95XQ*/Sa*
**12.5 mg/m$^3$:** Sa:Cf*/PaprHie*
**25 mg/m$^3$:** 100F/PaprTHie*/ScbaF/SaF
**250 mg/m$^3$:** SaF:Pd,Pp
**§:** ScbaF:Pd,Pp/SaF:Pd,Pp:AScba
**Escape:** 100F/ScbaE

**Incompatibilities and Reactivities:** Varies

**Exposure Routes, Symptoms, Target Organs (see Table 5):**
**ER:** Inh, Ing, Con
**SY:** Irrit eyes; sens derm
**TO:** Eyes, skin

**First Aid (see Table 6):**
**Eye:** Irr immed
**Skin:** Water flush prompt
**Breath:** Resp support
**Swallow:** Medical attention immed



## Chromium(III) compounds (as Cr)

| Formula: | CAS#: | RTECS#: | IDLH: 25 mg/m$^3$ [as Cr(III)] |
|---|---|---|---|

**Conversion:**

**DOT:**

**Synonyms/Trade Names:** Synonyms vary depending upon the specific Chromium(III) compound.
[**Note:** Chromium(III) compounds include soluble chromic salts.]

**Exposure Limits:**
**NIOSH REL:** TWA 0.5 mg/m$^3$
See Appendix C
**OSHA PEL:** TWA 0.5 mg/m$^3$
See Appendix C

**Measurement Methods (see Table 1):**
**NIOSH** 7024, 7300, 7301, 7303, 9102
**OSHA** ID121, ID125G

**Physical Description:** Appearance and odor vary depending upon the specific compound.

**Chemical & Physical Properties:**
Properties vary depending upon the specific compound.

**Personal Protection/Sanitation (see Table 2):**
**Skin:** Prevent skin contact
**Eyes:** Prevent eye contact
**Wash skin:** When contam
**Remove:** When wet or contam
**Change:** N.R.

**Respirator Recommendations (see Tables 3 and 4):**
**NIOSH/OSHA**
**2.5 mg/m$^3$:** Qm*
**5 mg/m$^3$:** 95XQ*/Sa*
**12.5 mg/m$^3$:** Sa:Cf*/PaprHie*
**25 mg/m$^3$:** 100F/PaprTHie*/ScbaF/SaF
**§:** ScbaF:Pd,Pp/SaF:Pd,Pp:AScba
**Escape:** 100F/ScbaE

**Incompatibilities and Reactivities:** Varies

**Exposure Routes, Symptoms, Target Organs (see Table 5):**
**ER:** Inh, Ing, Con
**SY:** Irrit eyes; sens derm
**TO:** Eyes, skin

**First Aid (see Table 6):**
**Eye:** Irr immed
**Skin:** Water flush prompt
**Breath:** Resp support
**Swallow:** Medical attention immed

C

## Chromium metal

| Formula: | CAS#: | RTECS#: | IDLH: |
|---|---|---|---|
| Cr | 7440-47-3 | GB4200000 | 250 mg/m$^3$ (as Cr) |

**Conversion:**

**DOT:**

**Synonyms/Trade Names:** Chrome, Chromium

**Exposure Limits:**
**NIOSH REL:** TWA 0.5 mg/m$^3$
See Appendix C
**OSHA PEL*:** TWA 1 mg/m$^3$
See Appendix C
[***Note:** The PEL also applies to insoluble chromium salts.]

**Measurement Methods (see Table 1):**
**NIOSH** 7024, 7300, 7301, 7303, 9102
**OSHA** ID121, ID125G

**Physical Description:** Blue-white to steel-gray, lustrous, brittle, hard, odorless solid.

**Chemical & Physical Properties:**
**MW:** 52.0
**BP:** 4788°F
**Sol:** Insoluble
**Fl.P:** NA
**IP:** NA
**Sp.Gr:** 7.14
**VP:** 0 mmHg (approx)
**MLT:** 3452°F
**UEL:** NA
**LEL:** NA
Noncombustible Solid in bulk form, but finely divided dust burns rapidly if heated in a flame.

**Personal Protection/Sanitation (see Table 2):**
**Skin:** N.R.
**Eyes:** N.R.
**Wash skin:** N.R.
**Remove:** N.R.
**Change:** N.R.

**Respirator Recommendations (see Tables 3 and 4):**
**NIOSH**
**2.5 mg/m$^3$:** Qm*
**5 mg/m$^3$:** 95XQ*/Sa*
**12.5 mg/m$^3$:** Sa:Cf*/PaprHie*
**25 mg/m$^3$:** 100F/PaprTHie*/ScbaF/SaF
**250 mg/m$^3$:** SaF:Pd,Pp
**§:** ScbaF:Pd,Pp/SaF:Pd,Pp:AScba
**Escape:** 100F/ScbaE

**Incompatibilities and Reactivities:** Strong oxidizers (such as hydrogen peroxide), alkalis

**Exposure Routes, Symptoms, Target Organs (see Table 5):**
**ER:** Inh, Ing, Con
**SY:** Irrit eyes, skin; lung fib (histologic)
**TO:** Eyes, skin, resp sys

**First Aid (see Table 6):**
**Eye:** Irr immed
**Skin:** Soap wash
**Breath:** Resp support
**Swallow:** Medical attention immed

## Chromyl chloride

| Formula: | CAS#: | RTECS#: | IDLH: |
|---|---|---|---|
| $Cr(OCl)_2$ | 14977-61-8 | GB5775000 | Ca [N.D.] |

**Conversion:**

**DOT:** 1758 137

**Synonyms/Trade Names:** Chlorochromic anhydride, Chromic oxychloride, Chromium chloride oxide, Chromium dichloride dioxide, Chromium dioxide dichloride, Chromium dioxychloride, Chromium oxychloride, Dichlorodioxochromium

**Exposure Limits:**
**NIOSH REL:** Ca
0.001 mg Cr(VI)/m$^3$
See Appendix A, See Appendix C
**OSHA PEL:** none

**Measurement Methods (see Table 1):**
None available

**Physical Description:** Deep-red liquid with a musty, burning, acrid odor. [**Note:** Fumes in moist air.]

**Chemical & Physical Properties:**
**MW:** 154.9
**BP:** 243°F
**Sol:** Reacts
**Fl.P:** NA
**IP:** 12.60 eV
**Sp.Gr(77°F):** 1.91
**VP:** 20 mmHg
**FRZ:** -142°F
**UEL:** NA
**LEL:** NA
Noncombustible Liquid, but a powerful oxidizer.

**Personal Protection/Sanitation (see Table 2):**
**Skin:** Prevent skin contact
**Eyes:** Prevent eye contact
**Wash skin:** When contam
**Remove:** When wet or contam
**Change:** N.R.
**Provide:** Eyewash
Quick drench

**Respirator Recommendations (see Tables 3 and 4):**
**NIOSH**
**¥:** ScbaF:Pd,Pp/SaF:Pd,Pp:AScba
**Escape:** GmFOv/ScbaE

**Incompatibilities and Reactivities:** Water, combustible substances, halides, phosphorus, turpentine [**Note:** Reacts violently in water; forms chromic acid, chromic chloride, hydrochloric acid & chlorine. Corrodes common metals.]

**Exposure Routes, Symptoms, Target Organs (see Table 5):**
**ER:** Inh, Abs, Ing, Con
**SY:** Irrit eyes, skin, upper resp sys; eye, skin burns [carc]
**TO:** Eyes, skin, resp sys [lung cancer]

**First Aid (see Table 6):**
**Eye:** Irr immed
**Skin:** Water flush immed
**Breath:** Resp support
**Swallow:** Medical attention immed

## Clopidol

**Formula:** $C_7H_7Cl_2NO$ | **CAS#:** 2971-90-6 | **RTECS#:** UU7711500 | **IDLH:** N.D.

**Conversion:**

**DOT:**

**Synonyms/Trade Names:** Coyden®; 3,5-Dichloro-2,6-dimethyl-4-pyridinol

**Exposure Limits:**
**NIOSH REL:** TWA 10 mg/m$^3$ (total)
ST 20 mg/m$^3$ (total)
TWA 5 mg/m$^3$ (resp)
**OSHA PEL:** TWA 15 mg/m$^3$ (total)
TWA 5 mg/m$^3$ (resp)

**Measurement Methods (see Table 1):**
**NIOSH** 0500, 0600

**Physical Description:** White to light-brown, crystalline solid.

**Chemical & Physical Properties:**
**MW:** 192.1
**BP:** ?
**Sol:** Insoluble
**Fl.P:** NA
**IP:** ?
**Sp.Gr:** ?
**VP:** ?
**MLT:** >608°F
**UEL:** NA
**LEL:** NA
Noncombustible Solid, but dust may explode in cloud form.

**Personal Protection/Sanitation (see Table 2):**
**Skin:** N.R.
**Eyes:** N.R.
**Wash skin:** N.R.
**Remove:** N.R.
**Change:** N.R.

**Respirator Recommendations (see Tables 3 and 4):**
Not available.

**Incompatibilities and Reactivities:** None reported

**Exposure Routes, Symptoms, Target Organs (see Table 5):**
**ER:** Inh, Con
**SY:** Irrit eyes, skin, nose, throat; cough
**TO:** Eyes, skin, resp sys

**First Aid (see Table 6):**
**Eye:** Irr immed
**Skin:** Soap wash
**Breath:** Fresh air

## Coal dust

**Formula:** | **CAS#:** | **RTECS#:** GF8281000 | **IDLH:** N.D.

**Conversion:**

**DOT:** 1361 133

**Synonyms/Trade Names:** Anthracite coal dust, Bituminous coal dust, Lignite coal dust

**Exposure Limits:**
**NIOSH REL:** TWA 1 mg/m$^3$ [measured according to MSHA method (CPSU)]
TWA 0.9 mg/m$^3$ [measured according to ISO/CEN/ACGIH criteria]
See Appendix C (Coal Dust and Coal Mine Dust)
**OSHA PEL†:** TWA 2.4 mg/m$^3$ [respirable, < 5% $SiO_2$]
TWA (10 mg/m$^3$)/(%$SiO_2$ + 2) [respirable, > 5% $SiO_2$]
See Appendix C (Mineral Dusts)
[**Note:** The Mine Safety and Health Administration (**MSHA**) PEL for respirable coal mine dust with < 5% silica is 2.0 mg/m$^3$, or (10 mg/m$^3$) / (% respirable quartz + 2) for coal dust with > 5% silica.]

**Measurement Methods (see Table 1):**
**NIOSH** 0600, 7500

**Physical Description:** Dark-brown to black solid dispersed in air.

**Chemical & Physical Properties:**
Properties vary depending upon the specific coal type.

Combustible Solid; slightly explosive when exposed to flame.

**Personal Protection/Sanitation (see Table 2):**
**Skin:** N.R.
**Eyes:** N.R.
**Wash skin:** N.R.
**Remove:** N.R.
**Change:** N.R.

**Respirator Recommendations (see Tables 3 and 4):**
Not available.

**Incompatibilities and Reactivities:** None reported

**Exposure Routes, Symptoms, Target Organs (see Table 5):**
**ER:** Inh
**SY:** Chronic bron, decr pulm func, emphy
**TO:** Resp sys

**First Aid (see Table 6):**
**Breath:** Fresh air

C

## Coal tar pitch volatiles

| | | | | |
|---|---|---|---|---|
| **Coal tar pitch volatiles** | **Formula:** | **CAS#:** 65996-93-2 | **RTECS#:** GF8655000 | **IDLH:** Ca [80 mg/m³] |
| **Conversion:** | **DOT:** 2713 153 (acridine) | | | |

**Synonyms/Trade Names:** Synonyms vary depending upon the specific compound (e.g., pyrene, phenanthrene, acridine, chrysene, anthracene & benzo(a)pyrene).
[**Note:** NIOSH considers coal tar, coal tar pitch, and creosote to be coal tar products.]

**Exposure Limits:**
**NIOSH REL:** Ca
TWA 0.1 mg/m³ (cyclohexane-extractable fraction)
See Appendix A
See Appendix C
**OSHA PEL:** TWA 0.2 mg/m³ (benzene-soluble fraction) [1910.1002]
See Appendix C

**Measurement Methods (see Table 1):**
**OSHA** 58

**Physical Description:** Black or dark-brown amorphous residue.

**Chemical & Physical Properties:**
Properties vary depending upon the specific compound.
Combustible Solids

**Personal Protection/Sanitation (see Table 2):**
**Skin:** Prevent skin contact
**Eyes:** Prevent eye contact
**Wash skin:** Daily
**Remove:** N.R.
**Change:** Daily

**Respirator Recommendations (see Tables 3 and 4):**
**NIOSH**
**¥:** ScbaF:Pd,Pp/SaF:Pd,Pp:AScba
**Escape:** GmFOv100/ScbaE

**Incompatibilities and Reactivities:** Strong oxidizers

**Exposure Routes, Symptoms, Target Organs (see Table 5):**
**ER:** Inh, Con
**SY:** Derm, bron, [carc]
**TO:** Resp sys, skin, bladder, kidneys [lung, kidney & skin cancer]

**First Aid (see Table 6):**
**Eye:** Irr immed
**Skin:** Soap wash immed
**Breath:** Resp support
**Swallow:** Medical attention immed

## Cobalt carbonyl (as Co)

| | | | | |
|---|---|---|---|---|
| **Cobalt carbonyl (as Co)** | **Formula:** $C_8Co_2O_8$ | **CAS#:** 10210-68-1 | **RTECS#:** GG0300000 | **IDLH:** N.D. |
| **Conversion:** | **DOT:** | | | |

**Synonyms/Trade Names:** di-mu-Carbonylhexacarbonyldicobalt, Cobalt octacarbonyl, Cobalt tetracarbonyl dimer, Dicobalt carbonyl, Dicobalt Octacarbonyl, Octacarbonyldicobalt

**Exposure Limits:**
**NIOSH REL:** TWA 0.1 mg/m³
**OSHA PEL†:** none

**Measurement Methods (see Table 1):**
None available

**Physical Description:** Orange to dark-brown, crystalline solid.
[**Note:** The pure substance is white.]

**Chemical & Physical Properties:**
**MW:** 341.9
**BP:** 126°F (Decomposes)
**Sol:** Insoluble
**Fl.P:** NA
**IP:** ?
**Sp.Gr:** 1.87
**VP:** 0.7 mmHg
**MLT:** 124°F
**UEL:** NA
**LEL:** NA
Noncombustible Solid, but flammable carbon monoxide is emitted during decomposition.

**Personal Protection/Sanitation (see Table 2):**
**Skin:** Prevent skin contact
**Eyes:** Prevent eye contact
**Wash skin:** When contam
**Remove:** When wet or contam
**Change:** Daily

**Respirator Recommendations (see Tables 3 and 4):**
Not available.

**Incompatibilities and Reactivities:** Air
[**Note:** Decomposes on exposure to air or heat; stable in atmosphere of hydrogen & carbon monoxide.]

**Exposure Routes, Symptoms, Target Organs (see Table 5):**
**ER:** Inh, Abs, Ing, Con
**SY:** Irrit eyes, skin, muc memb; cough, decr pulm func, wheez, dysp; in animals: liver, kidney inj, pulm edema
**TO:** Eyes, skin, resp sys, blood, CNS

**First Aid (see Table 6):**
**Eye:** Irr immed
**Skin:** Soap wash
**Breath:** Resp support
**Swallow:** Medical attention immed

C

## Cobalt hydrocarbonyl (as Co)

| Cobalt hydrocarbonyl (as Co) | Formula: $HCo(CO)_4$ | CAS#: 16842-03-8 | RTECS#: GG0900000 | IDLH: N.D. |
|---|---|---|---|---|

**Conversion:**

**DOT:**

**Synonyms/Trade Names:** Hydrocobalt tetracarbonyl, Tetracarbonylhydridocobalt, Tetracarbonylhydrocobalt

**Exposure Limits:**
**NIOSH REL:** TWA 0.1 mg/m$^3$
**OSHA PEL†:** none

**Measurement Methods (see Table 1):**
None available

**Physical Description:** Gas with an offensive odor.

**Chemical & Physical Properties:**
**MW:** 172.0
**BP:** ?
**Sol:** 0.05%
**Fl.P:** NA (Gas)
**IP:** ?
**RGasD:** 5.93
**VP:** >1 atm
**FRZ:** -15°F
**UEL:** ?
**LEL:** ?
Flammable Gas

**Personal Protection/Sanitation (see Table 2):**
**Skin:** Prevent skin contact
**Eyes:** Prevent eye contact
**Wash skin:** When contam
**Remove:** When wet or contam
**Change:** Daily

**Respirator Recommendations (see Tables 3 and 4):**
Not available.

**Incompatibilities and Reactivities:** Air
[**Note:** Unstable gas that decomposes rapidly in air at room temperature to cobalt carbonyl & hydrogen.]

**Exposure Routes, Symptoms, Target Organs (see Table 5):**
**ER:** Inh, Con
**SY:** In animals: irrit resp sys; dysp, cough, decr pulm func, pulm edema
**TO:** Eyes, skin, resp sys

**First Aid (see Table 6):**
**Eye:** Irr immed
**Skin:** Soap wash
**Breath:** Resp support

## Cobalt metal dust and fume (as Co)

| Cobalt metal dust and fume (as Co) | Formula: Co | CAS#: 7440-48-4 | RTECS#: GF8750000 | IDLH: 20 mg/m$^3$ (as Co) |
|---|---|---|---|---|

**Conversion:**

**DOT:**

**Synonyms/Trade Names:** Cobalt metal dust, Cobalt metal fume

**Exposure Limits:**
**NIOSH REL:** TWA 0.05 mg/m$^3$
**OSHA PEL†:** TWA 0.1 mg/m$^3$

**Measurement Methods (see Table 1):**
**NIOSH** 7027, 7300, 7301, 7303, 9102
**OSHA** ID121, ID125G, ID213

**Physical Description:** Odorless, silver-gray to black solid.

**Chemical & Physical Properties:**
**MW:** 58.9
**BP:** 5612°F
**Sol:** Insoluble
**Fl.P:** NA
**IP:** NA
**Sp.Gr:** 8.92
**VP:** 0 mmHg (approx)
**MLT:** 2719°F
**UEL:** NA
**LEL:** NA
Noncombustible Solid in bulk form, but finely divided dust will burn at high temperatures.

**Personal Protection/Sanitation (see Table 2):**
**Skin:** Prevent skin contact
**Eyes:** N.R.
**Wash skin:** When contam
**Remove:** When wet or contam
**Change:** Daily

**Respirator Recommendations (see Tables 3 and 4):**
**NIOSH**
**0.25 mg/m$^3$:** Qm
**0.5 mg/m$^3$:** 95XQ*/Sa*
**1.25 mg/m$^3$:** Sa:Cf*/PaprHie*
**2.5 mg/m$^3$:** 100F/ScbaF/SaF
**20 mg/m$^3$:** SaF:Pd,Pp
**§:** ScbaF:Pd,Pp/SaF:Pd,Pp:AScba
**Escape:** 100F/ScbaE

**Incompatibilities and Reactivities:** Strong oxidizers, ammonium nitrate

**Exposure Routes, Symptoms, Target Organs (see Table 5):**
**ER:** Inh, Ing, Con
**SY:** Cough, dysp, wheez, decr pulm func; low-wgt; derm; diffuse nodular fib; resp hypersensitivity, asthma
**TO:** Skin, resp sys

**First Aid (see Table 6):**
**Eye:** Irr immed
**Skin:** Soap wash
**Breath:** Resp support
**Swallow:** Medical attention immed

C

## Coke oven emissions

| Formula: | CAS#: | RTECS#: GH0346000 | IDLH: Ca [N.D.] |
|---|---|---|---|

**Conversion:** | **DOT:**

**Synonyms/Trade Names:** Synonyms vary depending upon the specific constituent.

**Exposure Limits:**
**NIOSH REL:** Ca
TWA 0.2 mg/m$^3$ (benzene-soluble fraction)
See Appendix A
See Appendix C
**OSHA PEL:** [1910.1029] TWA 0.150 mg/m$^3$ (benzene-soluble fraction)

**Measurement Methods (see Table 1):**
**OSHA** 58

**Physical Description:** Emissions released during the carbonization of bituminous coal for the production of coke. [**Note:** See Appendix C for more information.]

**Chemical & Physical Properties:**
Properties vary depending upon the constituent.

**Personal Protection/Sanitation (see Table 2):**
**Skin:** Prevent skin contact
**Eyes:** Prevent eye contact
**Wash skin:** Daily
**Remove:** N.R.
**Change:** Daily

**Respirator Recommendations (see Tables 3 and 4):**
**NIOSH**
**¥:** ScbaF:Pd,Pp/SaF:Pd,Pp:AScba
**Escape:** GmFOv100/ScbaE

**See Appendix E** (page 351)

**Incompatibilities and Reactivities:** None reported

**Exposure Routes, Symptoms, Target Organs (see Table 5):**
**ER:** Inh, Con
**SY:** Irrit eyes, resp sys; cough, dysp, wheez; [carc]
**TO:** Skin, resp sys, urinary sys [skin, lung, kidney & bladder cancer]

**First Aid (see Table 6):**
**Eye:** Irr immed
**Breath:** Resp support

## Copper (dusts and mists, as Cu)

| Formula: Cu | CAS#: 7440-50-8 | RTECS#: GL5325000 | IDLH: 100 mg/m$^3$ (as Cu) |
|---|---|---|---|

**Conversion:** | **DOT:**

**Synonyms/Trade Names:** Copper metal dusts, Copper metal fumes

**Exposure Limits:**
**NIOSH REL*:** TWA 1 mg/m$^3$
**OSHA PEL*:** TWA 1 mg/m$^3$
[***Note:** The REL and PEL also apply to other copper compounds (as Cu) except copper fume.]

**Measurement Methods (see Table 1):**
**NIOSH** 7029, 7300, 7301, 7303, 9102
**OSHA** ID121, ID125G

**Physical Description:** Reddish, lustrous, malleable, odorless solid.

**Chemical & Physical Properties:**
**MW:** 63.5
**BP:** 4703°F
**Sol:** Insoluble
**Fl.P:** NA
**IP:** NA
**Sp.Gr:** 8.94
**VP:** 0 mmHg (approx)
**MLT:** 1981°F
**UEL:** NA
**LEL:** NA
Noncombustible Solid in bulk form, but powdered form may ignite.

**Personal Protection/Sanitation (see Table 2):**
**Skin:** Prevent skin contact
**Eyes:** Prevent eye contact
**Wash skin:** When contam
**Remove:** When wet or contam
**Change:** Daily

**Respirator Recommendations (see Tables 3 and 4):**
**NIOSH/OSHA**
**5 mg/m$^3$:** Qm*
**10 mg/m$^3$:** 95XQ*/Sa*
**25 mg/m$^3$:** Sa:Cf*/PaprHie*
**50 mg/m$^3$:** 100F/PaprTHie*/ScbaF/SaF
**100 mg/m$^3$:** SaF:Pd,Pp
**§:** ScbaF:Pd,Pp/SaF:Pd,Pp:AScba
**Escape:** 100F/ScbaE

**Incompatibilities and Reactivities:** Oxidizers, alkalis, sodium azide, acetylene

**Exposure Routes, Symptoms, Target Organs (see Table 5):**
**ER:** Inh, Ing, Con
**SY:** Irrit eyes, nose, pharynx; nasal septum perf; metallic taste; derm; in animals: lung, liver, kidney damage; anemia
**TO:** Eyes, skin, resp sys, liver, kidneys (incr risk with Wilson's disease)

**First Aid (see Table 6):**
**Eye:** Irr immed
**Skin:** Soap wash prompt
**Breath:** Resp support
**Swallow:** Medical attention immed

## Copper fume (as Cu)

| Copper fume (as Cu) | Formula: | CAS#: | RTECS#: | IDLH: |
|---|---|---|---|---|
| | CuO/Cu | 1317-38-0 (CuO) | GL7900000 (CuO) | 100 mg/m³ (as Cu) |

**Conversion:**

**DOT:**

**Synonyms/Trade Names: Cu:** Copper fume **CuO:** Black copper oxide fume, Copper monoxide fume, Copper(II) oxide fume, Cupric oxide fume
[**Note:** Also see specific listing for Copper (dusts and mists).]

**Exposure Limits:**
**NIOSH REL:** TWA 0.1 mg/m³
**OSHA PEL:** TWA 0.1 mg/m³

**Measurement Methods (see Table 1):**
**NIOSH** 7029, 7300, 7301, 7303
**OSHA** ID121, ID125G, ID206

**Physical Description:** Finely divided black particulate dispersed in air.
[**Note:** Exposure may occur in copper & brass plants and during the welding of copper alloys.]

| Chemical & Physical Properties: | Personal Protection/Sanitation (see Table 2): | Respirator Recommendations (see Tables 3 and 4): |
|---|---|---|
| **MW:** 79.5<br>**BP:** Decomposes<br>**Sol:** Insoluble<br>**Fl.P:** NA<br>**IP:** NA<br>**Sp.Gr:** 6.4 (CuO)<br>**VP:** 0 mmHg (approx)<br>**MLT:** 1879°F (Decomposes)<br>**UEL:** NA<br>**LEL:** NA<br>CuO: Noncombustible Solid | **Skin:** N.R.<br>**Eyes:** N.R.<br>**Wash skin:** N.R.<br>**Remove:** N.R.<br>**Change:** N.R. | **NIOSH/OSHA**<br>**1 mg/m³:** 95XQ/Sa<br>**2.5 mg/m³:** Sa:Cf/PaprHie<br>**5 mg/m³:** 100F/SaT:Cf/PaprTHie/ScbaF/SaF<br>**100 mg/m³:** SaF:Pd,Pp<br>**§:** ScbaF:Pd,Pp/SaF:Pd,Pp:AScba<br>**Escape:** 100F/ScbaE |

**Incompatibilities and Reactivities: CuO:** Acetylene, zirconium
[**Note:** See Copper (dusts and mists) for properties of Copper metal.]

**Exposure Routes, Symptoms, Target Organs (see Table 5):**
**ER:** Inh, Con
**SY:** Irrit eyes, upper resp sys; metal fume fever: chills, musc ache, nau, fever, dry throat, cough, lass; metallic or sweet taste; discoloration skin, hair
**TO:** Eyes, skin, resp sys (incr risk with Wilson's disease)

**First Aid (see Table 6):**
**Breath:** Resp support

## Cotton dust (raw)

| Cotton dust (raw) | Formula: | CAS#: | RTECS#: | IDLH: |
|---|---|---|---|---|
| | | | GN2275000 | 100 mg/m³ |

**Conversion:**

**DOT:** 1365 133 (cotton)

**Synonyms/Trade Names:** Raw cotton dust

**Exposure Limits:**
**NIOSH REL:** TWA <0.200 mg/m³
See Appendix C
**OSHA PEL:** [Z-1-A & 1910.1043]
See Appendix C

**Measurement Methods (see Table 1):**
**OSHA** [1910.1043]

**Physical Description:** Colorless, odorless solid.

| Chemical & Physical Properties: | Personal Protection/Sanitation (see Table 2): | Respirator Recommendations (see Tables 3 and 4): |
|---|---|---|
| **MW:** ?<br>**BP:** Decomposes<br>**Sol:** Insoluble<br>**Fl.P:** NA<br>**IP:** NA<br>**Sp.Gr:** ?<br>**VP:** 0 mmHg (approx)<br>**MLT:** Decomposes<br>**UEL:** NA<br>**LEL:** NA<br>Combustible Solid | **Skin:** N.R.<br>**Eyes:** N.R.<br>**Wash skin:** N.R.<br>**Remove:** N.R.<br>**Change:** N.R. | **NIOSH**<br>**1 mg/m³:** Qm<br>**2 mg/m³:** 95XQ/Sa<br>**5 mg/m³:** Sa:Cf/PaprHie<br>**10 mg/m³:** 100F/SaT:Cf/PaprTHie/ScbaF/SaF<br>**100 mg/m³:** Sa:Pd,Pp<br>**§:** ScbaF:Pd,Pp/SaF:Pd,Pp:AScba<br>**Escape:** 100F/ScbaE<br>**See Appendix E** (page 351) |

**Incompatibilities and Reactivities:** Strong oxidizers

**Exposure Routes, Symptoms, Target Organs (see Table 5):**
**ER:** Inh
**SY:** Byssinosis: chest tight, cough, wheez, dysp; decr FEV; bron; mal; fever, chills, upper resp symptoms after initial exposure
**TO:** CVS, resp sys

**First Aid (see Table 6):**
**Breath:** Fresh air

C

## Crag® herbicide

| Formula: | CAS#: | RTECS#: | IDLH: |
|---|---|---|---|
| $C_6H_3Cl_2OCH_2CH_2OSO_3Na$ | 136-78-7 | KK4900000 | 500 mg/m$^3$ |

**Conversion:**
**DOT:**

**Synonyms/Trade Names:** Crag® herbicide No. 1; 2-(2,4-Dichlorophenoxy)ethyl sodium sulfate; Sesone

**Exposure Limits:**
**NIOSH REL:** TWA 10 mg/m$^3$ (total)
TWA 5 mg/m$^3$ (resp)
**OSHA PEL†:** TWA 15 mg/m$^3$ (total)
TWA 5 mg/m$^3$ (resp)

**Measurement Methods (see Table 1):**
**NIOSH** S356 (II-5)

**Physical Description:** Colorless to white crystalline, odorless solid. [herbicide]

**Chemical & Physical Properties:**
**MW:** 309.1
**BP:** Decomposes
**Sol(77°F):** 26%
**Fl.P:** NA
**IP:** ?
**Sp.Gr:** 1.70
**VP:** 0.1 mmHg
**MLT:** 473°F (Decomposes)
**UEL:** NA
**LEL:** NA
Noncombustible Solid

**Personal Protection/Sanitation (see Table 2):**
**Skin:** Prevent skin contact
**Eyes:** Prevent eye contact
**Wash skin:** When contam
**Remove:** When wet or contam
**Change:** Daily

**Respirator Recommendations (see Tables 3 and 4):**
**NIOSH**
**50 mg/m$^3$:** Qm
**100 mg/m$^3$:** 95XQ/Sa
**250 mg/m$^3$:** Sa:Cf/PaprHie
**500 mg/m$^3$:** 100F/PaprTHie*/SaT:Cf*/ScbaF/SaF
**§:** ScbaF:Pd,Pp/SaF:Pd,Pp:AScba
**Escape:** 100F/ScbaE

**Incompatibilities and Reactivities:** Strong oxidizers, acids

**Exposure Routes, Symptoms, Target Organs (see Table 5):**
**ER:** Inh, Ing, Con
**SY:** Irrit eyes, skin; liver, kidney damage; in animals: CNS effects, convuls
**TO:** Eyes, skin, CNS, liver, kidneys

**First Aid (see Table 6):**
**Eye:** Irr immed
**Skin:** Water wash prompt
**Breath:** Resp support
**Swallow:** Medical attention immed

## m-Cresol

| Formula: | CAS#: | RTECS#: | IDLH: |
|---|---|---|---|
| $CH_3C_6H_4OH$ | 108-39-4 | GO6125000 | 250 ppm |

**Conversion:** 1 ppm = 4.43 mg/m$^3$
**DOT:** 2076 153

**Synonyms/Trade Names:** meta-Cresol, 3-Cresol, m-Cresylic acid, 1-Hydroxy-3-methylbenzene, 3-Hydroxytoluene, 3-Methyl phenol

**Exposure Limits:**
**NIOSH REL:** TWA 2.3 ppm (10 mg/m$^3$)
**OSHA PEL:** TWA 5 ppm (22 mg/m$^3$) [skin]

**Measurement Methods (see Table 1):**
**NIOSH** 2546
**OSHA** 32

**Physical Description:** Colorless to yellowish liquid with a sweet, tarry odor.
[**Note:** A solid below 54°F.]

**Chemical & Physical Properties:**
**MW:** 108.2
**BP:** 397°F
**Sol:** 2%
**Fl.P:** 187°F
**IP:** 8.98 eV
**Sp.Gr:** 1.03
**VP(77°F):** 0.14 mmHg
**FRZ:** 54°F
**UEL:** ?
**LEL(300°F):** 1.1%
Class IIIA Combustible Liquid

**Personal Protection/Sanitation (see Table 2):**
**Skin:** Prevent skin contact
**Eyes:** Prevent eye contact
**Wash skin:** When contam
**Remove:** When wet or contam
**Change:** Daily
**Provide:** Eyewash
Quick drench

**Respirator Recommendations (see Tables 3 and 4):**
**NIOSH**
**23 ppm:** CcrOv95/Sa
**57.5 ppm:** Sa:Cf/PaprOvHie
**115 ppm:** CcrFOv100/GmFOv100/PaprTOvHie*/SaT:Cf*/ScbaF/SaF
**250 ppm:** SaF:Pd,Pp
**§:** ScbaF:Pd,Pp/SaF:Pd,Pp:AScba
**Escape:** GmFOv100/ScbaE

**Incompatibilities and Reactivities:** Strong oxidizers, acids

**Exposure Routes, Symptoms, Target Organs (see Table 5):**
**ER:** Inh, Abs, Ing, Con
**SY:** Irrit eyes, skin, muc memb; CNS effects: conf, depres, resp fail; dysp, irreg rapid resp, weak pulse; eye, skin burns; derm; lung, liver, kidney, pancreas damage
**TO:** Eyes, skin, resp sys, CNS, liver, kidneys, pancreas, CVS

**First Aid (see Table 6):**
**Eye:** Irr immed
**Skin:** Soap wash immed
**Breath:** Resp support
**Swallow:** Medical attention immed

## o-Cresol

**Formula:** $CH_3C_6H_4OH$ | **CAS#:** 95-48-7 | **RTECS#:** GO6300000 | **IDLH:** 250 ppm

**Conversion:** 1 ppm = 4.43 mg/m$^3$ | **DOT:** 2076 153

C

**Synonyms/Trade Names:** ortho-Cresol, 2-Cresol, o-Cresylic acid, 1-Hydroxy-2-methylbenzene, 2-Hydroxytoluene, 2-Methyl phenol

**Exposure Limits:**
**NIOSH REL:** TWA 2.3 ppm (10 mg/m$^3$)
**OSHA PEL:** TWA 5 ppm (22 mg/m$^3$) [skin]

**Measurement Methods (see Table 1):**
NIOSH 2546
OSHA 32

**Physical Description:** White crystals with a sweet, tarry odor.
[**Note:** A liquid above 88°F.]

**Chemical & Physical Properties:**
**MW:** 108.2
**BP:** 376°F
**Sol:** 2%
**Fl.P:** 178°F
**IP:** 8.93 eV
**Sp.Gr:** 1.05
**VP(77°F):** 0.29 mmHg
**MLT:** 88°F
**UEL:** ?
**LEL(300°F):** 1.4%
Combustible Solid
Class IIIA Combustible Liquid

**Personal Protection/Sanitation (see Table 2):**
**Skin:** Prevent skin contact
**Eyes:** Prevent eye contact
**Wash skin:** When contam
**Remove:** When wet or contam
**Change:** Daily
**Provide:** Eyewash
Quick drench

**Respirator Recommendations (see Tables 3 and 4):**
**NIOSH**
**23 ppm:** CcrOv95/Sa
**57.5 ppm:** Sa:Cf/PaprOvHie
**115 ppm:** CcrFOv100/GmFOv100/PaprTOvHie*/SaT:Cf*/ScbaF/SaF
**250 ppm:** SaF:Pd,Pp
**§:** ScbaF:Pd,Pp/SaF:Pd,Pp:AScba
**Escape:** GmFOv100/ScbaE

**Incompatibilities and Reactivities:** Strong oxidizers, acids

**Exposure Routes, Symptoms, Target Organs (see Table 5):**
**ER:** Inh, Abs, Ing, Con
**SY:** Irrit eyes, skin, muc memb; CNS effects: conf, depres, resp fail; dysp, irreg rapid resp, weak pulse; eye, skin burns; derm; lung, liver, kidney, pancreas damage
**TO:** Eyes, skin, resp sys, CNS, liver, kidneys, pancreas, CVS

**First Aid (see Table 6):**
**Eye:** Irr immed
**Skin:** Soap wash immed
**Breath:** Resp support
**Swallow:** Medical attention immed

## p-Cresol

**Formula:** $CH_3C_6H_4OH$ | **CAS#:** 106-44-5 | **RTECS#:** GO6475000 | **IDLH:** 250 ppm

**Conversion:** 1 ppm = 4.43 mg/m$^3$ | **DOT:** 2076 153

**Synonyms/Trade Names:** para-Cresol, 4-Cresol, p-Cresylic acid, 1-Hydroxy-4-methylbenzene, 4-Hydroxytoluene, 4-Methyl phenol

**Exposure Limits:**
**NIOSH REL:** TWA 2.3 ppm (10 mg/m$^3$)
**OSHA PEL:** TWA 5 ppm (22 mg/m$^3$) [skin]

**Measurement Methods (see Table 1):**
NIOSH 2546
OSHA 32

**Physical Description:** Crystalline solid with a sweet, tarry odor.
[**Note:** A liquid above 95°F.]

**Chemical & Physical Properties:**
**MW:** 108.2
**BP:** 396°F
**Sol:** 2%
**Fl.P:** 187°F
**IP:** 8.97 eV
**Sp.Gr:** 1.04
**VP(77°F):** 0.11 mmHg
**MLT:** 95°F
**UEL:** ?
**LEL(300°F):** 1.1%
Combustible Solid
Class IIIA Combustible Liquid

**Personal Protection/Sanitation (see Table 2):**
**Skin:** Prevent skin contact
**Eyes:** Prevent eye contact
**Wash skin:** When contam
**Remove:** When wet or contam
**Change:** Daily
**Provide:** Eyewash
Quick drench

**Respirator Recommendations (see Tables 3 and 4):**
**NIOSH**
**23 ppm:** CcrOv95/Sa
**57.5 ppm:** Sa:Cf/PaprOvHie
**115 ppm:** CcrFOv100/GmFOv100/PaprTOvHie*/SaT:Cf*/ScbaF/SaF
**250 ppm:** SaF:Pd,Pp
**§:** ScbaF:Pd,Pp/SaF:Pd,Pp:AScba
**Escape:** GmFOv100/ScbaE

**Incompatibilities and Reactivities:** Strong oxidizers, acids

**Exposure Routes, Symptoms, Target Organs (see Table 5):**
**ER:** Inh, Abs, Ing, Con
**SY:** Irrit eyes, skin, muc memb; CNS effects: conf, depres, resp fail; dysp, irreg rapid resp, weak pulse; eye, skin burns; derm; lung, liver, kidney, pancreas damage
**TO:** Eyes, skin, resp sys, CNS, liver, kidneys, pancreas, CVS

**First Aid (see Table 6):**
**Eye:** Irr immed
**Skin:** Soap wash immed
**Breath:** Resp support
**Swallow:** Medical attention immed

C

## Crotonaldehyde

| Crotonaldehyde | Formula: $CH_3CH{=}CHCHO$ | CAS#: 4170-30-3 | RTECS#: GP9499000 | IDLH: 50 ppm |
|---|---|---|---|---|

**Conversion:** 1 ppm = 2.87 mg/$m^3$ | **DOT:** 1143 131P (inhibited)

**Synonyms/Trade Names:** 2-Butenal, β-Methyl acrolein, Propylene aldehyde

**Exposure Limits:**
**NIOSH REL:** TWA 2 ppm (6 mg/$m^3$)
See Appendix C (Aldehydes)
**OSHA PEL:** TWA 2 ppm (6 mg/$m^3$)

**Measurement Methods (see Table 1):**
**NIOSH** 3516
**OSHA** 81

**Physical Description:** Water-white liquid with a suffocating odor.
[**Note:** Turns pale-yellow on contact with air.]

**Chemical & Physical Properties:**
**MW:** 70.1
**BP:** 219°F
**Sol:** 18%
**Fl.P:** 45°F
**IP:** 9.73 eV
**Sp.Gr:** 0.87
**VP:** 19 mmHg
**FRZ:** -101°F
**UEL:** 15.5%
**LEL:** 2.1%
Class IB Flammable Liquid

**Personal Protection/Sanitation (see Table 2):**
**Skin:** Prevent skin contact
**Eyes:** Prevent eye contact
**Wash skin:** When contam
**Remove:** When wet (flamm)
**Change:** N.R.
**Provide:** Eyewash
Quick drench

**Respirator Recommendations (see Tables 3 and 4):**
**NIOSH/OSHA**
**20 ppm:** CcrOv*/Sa*
**50 ppm:** Sa:Cf*/PaprOv*/CcrFOv/GmFOv/ScbaF/SaF
**§:** ScbaF:Pd,Pp/SaF:Pd,Pp:AScba
**Escape:** GmFOv/ScbaE

**Incompatibilities and Reactivities:** Caustics, ammonia, strong oxidizers, nitric acid, amines
[**Note:** Polymerization may occur at elevated temperatures, such as in fire conditions.]

**Exposure Routes, Symptoms, Target Organs (see Table 5):**
**ER:** Inh, Ing, Con
**SY:** Irrit eyes, resp sys; in animals: dysp, pulm edema, irrit skin
**TO:** Eyes, skin, resp sys

**First Aid (see Table 6):**
**Eye:** Irr immed
**Skin:** Water flush immed
**Breath:** Resp support
**Swallow:** Medical attention immed

## Crufomate

| Crufomate | Formula: $C_{12}H_{19}ClNO_3P$ | CAS#: 299-86-5 | RTECS#: TB3850000 | IDLH: N.D. |
|---|---|---|---|---|

**Conversion:** | **DOT:**

**Synonyms/Trade Names:** 4-t-Butyl-2-chlorophenylmethyl methylphosphoramidate, Dowco® 132, Ruelene®

**Exposure Limits:**
**NIOSH REL:** TWA 5 mg/$m^3$
ST 20 mg/$m^3$
**OSHA PEL†:** none

**Measurement Methods (see Table 1):**
**NIOSH** 0500
**OSHA** PV2015

**Physical Description:** White, crystalline solid in pure form. [pesticide]
[**Note:** Commercial product is a yellow oil.]

**Chemical & Physical Properties:**
**MW:** 291.7
**BP:** Decomposes
**Sol:** Insoluble
**Fl.P:** ?
**IP:** ?
**Sp.Gr:** 1.16
**VP(243°F):** 0.01 mmHg
**MLT:** 140°F
**UEL:** ?
**LEL:** ?
Combustible Solid

**Personal Protection/Sanitation (see Table 2):**
**Skin:** Prevent skin contact
**Eyes:** Prevent eye contact
**Wash skin:** When contam
**Remove:** When wet or contam
**Change:** Daily

**Respirator Recommendations (see Tables 3 and 4):**
Not available.

**Incompatibilities and Reactivities:** Strongly alkaline & strongly acidic media
[**Note:** Unstable over long periods in aqueous preparations or above 140°F.]

**Exposure Routes, Symptoms, Target Organs (see Table 5):**
**ER:** Inh, Abs, Ing, Con
**SY:** Irrit eyes, skin, resp sys; wheez, dysp; blurred vision, lac; sweat; abdom cramps, diarr, nau, anor
**TO:** Eyes, skin, resp sys, blood chol

**First Aid (see Table 6):**
**Eye:** Irr immed
**Skin:** Soap wash immed
**Breath:** Resp support
**Swallow:** Medical attention immed

| Cumene | Formula: $C_6H_5CH(CH_3)_2$ | CAS#: 98-82-8 | RTECS#: GR8575000 | IDLH: 900 ppm [10%LEL] |
|---|---|---|---|---|
| **Conversion:** 1 ppm = 4.92 mg/m$^3$ | **DOT:** 1918 130 | | | |

**Synonyms/Trade Names:** Cumol, Isopropyl benzene, 2-Phenyl propane

**Exposure Limits:**
**NIOSH REL:** TWA 50 ppm (245 mg/m$^3$) [skin]
**OSHA PEL:** TWA 50 ppm (245 mg/m$^3$) [skin]

**Measurement Methods (see Table 1):**
**NIOSH** 1501

**Physical Description:** Colorless liquid with a sharp, penetrating, aromatic odor.

| Chemical & Physical Properties: | Personal Protection/Sanitation (see Table 2): | Respirator Recommendations (see Tables 3 and 4): |
|---|---|---|
| **MW:** 120.2<br>**BP:** 306°F<br>**Sol:** Insoluble<br>**Fl.P:** 96°F<br>**IP:** 8.75 eV<br>**Sp.Gr:** 0.86<br>**VP:** 8 mmHg<br>**FRZ:** -141°F<br>**UEL:** 6.5%<br>**LEL:** 0.9%<br>Class IC Flammable Liquid | **Skin:** Prevent skin contact<br>**Eyes:** Prevent eye contact<br>**Wash skin:** When contam<br>**Remove:** When wet (flamm)<br>**Change:** N.R. | **NIOSH/OSHA**<br>**500 ppm:** CcrOv*/Sa*<br>**900 ppm:** Sa:Cf*/PaprOv*/CcrFOv/GmFOv/ScbaF/SaF<br>**§:** ScbaF:Pd,Pp/SaF:Pd,Pp:AScba<br>**Escape:** GmFOv/ScbaE |

**Incompatibilities and Reactivities:** Oxidizers, nitric acid, sulfur acid
[**Note:** Forms cumene hydroperoxide upon long exposure to air.]

| Exposure Routes, Symptoms, Target Organs (see Table 5): | First Aid (see Table 6): |
|---|---|
| **ER:** Inh, Abs, Ing, Con<br>**SY:** Irrit eyes, skin, muc memb; derm; head, narco, coma<br>**TO:** Eyes, skin, resp sys, CNS | **Eye:** Irr immed<br>**Skin:** Water flush prompt<br>**Breath:** Resp support<br>**Swallow:** Medical attention immed |

| Cyanamide | Formula: $NH_2CN$ | CAS#: 420-04-2 | RTECS#: GS5950000 | IDLH: N.D. |
|---|---|---|---|---|
| **Conversion:** | **DOT:** | | | |

**Synonyms/Trade Names:** Amidocyanogen, Carbimide, Carbodiimide, Cyanogen nitride, Hydrogen cyanamide
[**Note:** Cyanamide is also a synonym for Calcium cyanamide.]

**Exposure Limits:**
**NIOSH REL:** TWA 2 mg/m$^3$
**OSHA PEL†:** none

**Measurement Methods (see Table 1):**
**NIOSH** 0500

**Physical Description:** Crystalline solid.

| Chemical & Physical Properties: | Personal Protection/Sanitation (see Table 2): | Respirator Recommendations (see Tables 3 and 4): |
|---|---|---|
| **MW:** 42.1<br>**BP:** 500°F (Decomposes)<br>**Sol(59°F):** 78%<br>**Fl.P:** 286°F<br>**IP:** 10.65 eV<br>**Sp.Gr:** 1.28<br>**VP:** ?<br>**MLT:** 113°F<br>**UEL:** ?<br>**LEL:** ?<br>Combustible Solid | **Skin:** Prevent skin contact<br>**Eyes:** Prevent eye contact<br>**Wash skin:** When contam<br>**Remove:** When wet or contam<br>**Change:** Daily<br>**Provide:** Eyewash<br>Quick drench | Not available. |

**Incompatibilities and Reactivities:** Above 104°F: Moisture, acids, or alkalis; 1,2-phenylene diamine salts
[**Note:** Polymerization may occur on evaporation of aqueous solutions.]

| Exposure Routes, Symptoms, Target Organs (see Table 5): | First Aid (see Table 6): |
|---|---|
| **ER:** Inh, Abs, Ing, Con<br>**SY:** Irrit eyes, skin, resp sys; eye, skin burns; miosis, salv, lac, twitch; Antabuse-like effects<br>**TO:** Eyes, skin, resp sys, CNS | **Eye:** Irr immed<br>**Skin:** Water flush immed<br>**Breath:** Resp support<br>**Swallow:** Medical attention immed |

C

## Cyanogen

| Cyanogen | Formula: NCCN | CAS#: 460-19-5 | RTECS#: GT1925000 | IDLH: N.D. |
|---|---|---|---|---|
| **Conversion:** 1 ppm = 2.13 mg/m$^3$ | **DOT:** 1026 119 | | | |

**Synonyms/Trade Names:** Carbon nitride, Dicyan, Dicyanogen, Ethanedinitrile, Oxalonitrile

**Exposure Limits:**
**NIOSH REL:** TWA 10 ppm (20 mg/m$^3$)
**OSHA PEL†:** none

**Measurement Methods (see Table 1):**
**OSHA** PV2104

**Physical Description:** Colorless gas with a pungent, almond-like odor.
[**Note:** Shipped as a liquefied compressed gas. Forms cyanide in the body.]

**Chemical & Physical Properties:**
**MW:** 52.0
**BP:** -6°F
**Sol:** 1%
**Fl.P:** NA (Gas)
**IP:** 13.57 eV
**RGasD:** 1.82
**Sp.Gr:** 0.95 (Liquid at -6°F)
**VP(70°F):** 5.1 atm
**FRZ:** -18°F
**UEL:** 32%
**LEL:** 6.6%
Flammable Gas

**Personal Protection/Sanitation (see Table 2):**
**Skin:** Frostbite
**Eyes:** Prevent eye contact/Frostbite
**Wash skin:** N.R.
**Remove:** When wet (flamm)
**Change:** N.R.
**Provide:** Frostbite wash

**Respirator Recommendations (see Tables 3 and 4):**
Not available.

**Incompatibilities and Reactivities:** Acids, water, strong oxidizers (e.g., dichlorine oxide, fluorine)
[**Note:** Slowly hydrolyzed in water to form hydrogen cyanide, oxalic acid, or ammonia.]

**Exposure Routes, Symptoms, Target Organs (see Table 5):**
**ER:** Inh, Con
**SY:** Irrit eyes, nose, upper resp sys; lac; cherry red lips, tachypnea, hypernea, bradycardia; head, convuls; dizz, loss of appetite, low-wgt; liquid: frostbite
**TO:** Eyes, resp sys, CNS, CVS

**First Aid (see Table 6):**
**Eye:** Frostbite
**Skin:** Frostbite
**Breath:** Resp support

## Cyanogen chloride

| Cyanogen chloride | Formula: ClCN | CAS#: 506-77-4 | RTECS#: GT2275000 | IDLH: N.D. |
|---|---|---|---|---|
| **Conversion:** 1 ppm = 2.52 mg/m$^3$ | **DOT:** 1589 125 (inhibited) | | | |

**Synonyms/Trade Names:** Chlorcyan, Chlorine cyanide, Chlorocyanide, Chlorocyanogen

**Exposure Limits:**
**NIOSH REL:** C 0.3 ppm (0.6 mg/m$^3$)
**OSHA PEL†:** none

**Measurement Methods (see Table 1):**
None available

**Physical Description:** Colorless gas or liquid (below 55°F) with an irritating odor.
[**Note:** Shipped as a liquefied gas. A solid below 20°F. Forms cyanide in the body.]

**Chemical & Physical Properties:**
**MW:** 61.5
**BP:** 55°F
**Sol:** 7%
**Fl.P:** NA
**IP:** 12.49 eV
**RGasD:** 2.16
**Sp.Gr:** 1.22 (Liquid at 32°F)
**VP:** 1010 mmHg
**FRZ:** 20°F
**UEL:** NA
**LEL:** NA
Nonflammable Gas

**Personal Protection/Sanitation (see Table 2):**
**Skin:** Prevent skin contact (liquid)
**Eyes:** Prevent eye contact (liquid)
**Wash skin:** When contam (liquid)
**Remove:** When wet or contam (liquid)
**Change:** N.R.
**Provide:** Eyewash (liquid)
Quick drench (liquid)

**Respirator Recommendations (see Tables 3 and 4):**
Not available.

**Incompatibilities and Reactivities:** Water, acids, alkalis, ammonia, alcohols
[**Note:** Can react very slowly with water to form hydrogen cyanide. May be stabilized to prevent polymerization.]

**Exposure Routes, Symptoms, Target Organs (see Table 5):**
**ER:** Inh, Abs (liquid), Ing (liquid), Con (liquid)
**SY:** Irrit eyes, upper resp sys; cough, delayed pulm edema; lass, head, dizz, conf, nau, vomit; irreg heartbeat; irrit skin (liquid)
**TO:** Eyes, skin, resp sys, CNS, CVS

**First Aid (see Table 6):**
**Eye:** Irr immed
**Skin:** Water wash immed (liquid)
**Breath:** Resp support
**Swallow:** Medical attention immed (liquid)

## Cyclohexane

| Formula: | CAS#: | RTECS#: | IDLH: |
|---|---|---|---|
| $C_6H_{12}$ | 110-82-7 | GU6300000 | 1300 ppm [10%LEL] |

**Conversion:** 1 ppm = 3.44 mg/m$^3$ | **DOT:** 1145 128

**Synonyms/Trade Names:** Benzene hexahydride, Hexahydrobenzene, Hexamethylene, Hexanaphthene

C

**Exposure Limits:**
**NIOSH REL:** TWA 300 ppm (1050 mg/m$^3$)
**OSHA PEL:** TWA 300 ppm (1050 mg/m$^3$)

**Measurement Methods (see Table 1):**
**NIOSH** 1500
**OSHA** 7

**Physical Description:** Colorless liquid with a sweet, chloroform-like odor.
[**Note:** A solid below 44°F.]

**Chemical & Physical Properties:**
**MW:** 84.2
**BP:** 177°F
**Sol:** Insoluble
**Fl.P:** 0°F
**IP:** 9.88 eV
**Sp.Gr:** 0.78
**VP:** 78 mmHg
**FRZ:** 44°F
**UEL:** 8%
**LEL:** 1.3%
Class IB Flammable Liquid

**Personal Protection/Sanitation (see Table 2):**
**Skin:** Prevent skin contact
**Eyes:** Prevent eye contact
**Wash skin:** When contam
**Remove:** When wet (flamm)
**Change:** N.R.

**Respirator Recommendations (see Tables 3 and 4):**
**NIOSH/OSHA**
**1300 ppm:** Sa:Cf£/PaprOv£/CcrFOv/GmFOv/ScbaF/SaF
**§:** ScbaF:Pd,Pp/SaF:Pd,Pp:AScba
**Escape:** GmFOv/ScbaE

**Incompatibilities and Reactivities:** Oxidizers

**Exposure Routes, Symptoms, Target Organs (see Table 5):**
**ER:** Inh, Ing, Con
**SY:** Irrit eyes, skin, resp sys; drow; derm; narco, coma
**TO:** Eyes, skin, resp sys, CNS

**First Aid (see Table 6):**
**Eye:** Irr immed
**Skin:** Water flush prompt
**Breath:** Resp support
**Swallow:** Medical attention immed

## Cyclohexanethiol

| Formula: | CAS#: | RTECS#: | IDLH: |
|---|---|---|---|
| $C_6H_{11}SH$ | 1569-69-3 | GV7525000 | N.D. |

**Conversion:** 1 ppm = 4.75 mg/m$^3$ | **DOT:** 3054 129

**Synonyms/Trade Names:** Cyclohexylmercaptan, Cyclohexylthiol

**Exposure Limits:**
**NIOSH REL:** C 0.5 ppm (2.4 mg/m$^3$) [15-minute]
**OSHA PEL:** none

**Measurement Methods (see Table 1):**
None available

**Physical Description:** Colorless liquid with a strong, offensive odor.

**Chemical & Physical Properties:**
**MW:** 116.2
**BP:** 316°F
**Sol:** Insoluble
**Fl.P:** 110°F
**IP:** ?
**Sp.Gr:** 0.98
**VP:** 10 mmHg
**FRZ:** -181°F
**UEL:** ?
**LEL:** ?
Class II Combustible Liquid

**Personal Protection/Sanitation (see Table 2):**
**Skin:** Prevent skin contact
**Eyes:** Prevent eye contact
**Wash skin:** When contam
**Remove:** When wet or contam
**Change:** N.R.
**Provide:** Eyewash
Quick drench

**Respirator Recommendations (see Tables 3 and 4):**
**NIOSH**
**5 ppm:** CcrOv/Sa
**12.5 ppm:** Sa:Cf/PaprOv
**25 ppm:** CcrFOv/GmFOv/PaprTOv/ScbaF/SaF
**§:** ScbaF:Pd,Pp/SaF:Pd,Pp:AScba
**Escape:** GmFOv/ScbaE

**Incompatibilities and Reactivities:** Oxidizers, reducing agents, strong acids, alkali metals

**Exposure Routes, Symptoms, Target Organs (see Table 5):**
**ER:** Inh, Abs, Ing, Con
**SY:** Irrit eyes, skin, resp sys; head, dizz, lass, nau, vomit, convuls; cough, wheez, laryngitis, dysp
**TO:** Eyes, skin, resp sys, CNS

**First Aid (see Table 6):**
**Eye:** Irr immed
**Skin:** Soap flush immed
**Breath:** Resp support
**Swallow:** Medical attention immed

C

## Cyclohexanol

| Cyclohexanol | Formula: $C_6H_{11}OH$ | CAS#: 108-93-0 | RTECS#: GV7875000 | IDLH: 400 ppm |
|---|---|---|---|---|

**Conversion:** 1 ppm = 4.10 mg/m$^3$

**DOT:** 1993 128 (combustible liquid, n.o.s.)

**Synonyms/Trade Names:** Anol, Cyclohexyl alcohol, Hexahydrophenol, Hexalin, Hydralin, Hydroxycyclohexane

**Exposure Limits:**
**NIOSH REL:** TWA 50 ppm (200 mg/m$^3$) [skin]
**OSHA PEL†:** TWA 50 ppm (200 mg/m$^3$)

**Measurement Methods (see Table 1):**
**NIOSH** 1402, 1405
**OSHA** 7

**Physical Description:** Sticky solid or colorless to light-yellow liquid (above 77°F) with a camphor-like odor.

| Chemical & Physical Properties: | Personal Protection/Sanitation (see Table 2): | Respirator Recommendations (see Tables 3 and 4): |
|---|---|---|
| **MW:** 100.2<br>**BP:** 322°F<br>**Sol:** 4%<br>**Fl.P:** 154°F<br>**IP:** 10.00 eV<br>**Sp.Gr:** 0.96<br>**VP:** 1 mmHg<br>**MLT:** 77°F<br>**UEL:** ?<br>**LEL:** ?<br>Class IIIA Combustible Liquid | **Skin:** Prevent skin contact<br>**Eyes:** Prevent eye contact<br>**Wash skin:** When contam<br>**Remove:** When wet or contam<br>**Change:** Daily | **NIOSH/OSHA**<br>**400 ppm:** CcrOv*/PaprOv*/GmFOv/Sa*/ScbaF<br>**§:** ScbaF:Pd,Pp/SaF:Pd,Pp:AScba<br>**Escape:** GmFOv/ScbaE |

**Incompatibilities and Reactivities:** Strong oxidizers (such as hydrogen peroxide & nitric acid)

| Exposure Routes, Symptoms, Target Organs (see Table 5): | First Aid (see Table 6): |
|---|---|
| **ER:** Inh, Abs, Ing, Con<br>**SY:** Irrit eyes, skin, nose, throat; narco<br>**TO:** Eyes, skin, resp sys | **Eye:** Irr immed<br>**Skin:** Water wash prompt<br>**Breath:** Resp support<br>**Swallow:** Medical attention immed |

## Cyclohexanone

| Cyclohexanone | Formula: $C_6H_{10}O$ | CAS#: 108-94-1 | RTECS#: GW1050000 | IDLH: 700 ppm |
|---|---|---|---|---|

**Conversion:** 1 ppm = 4.02 mg/m$^3$

**DOT:** 1915 127

**Synonyms/Trade Names:** Anone, Cyclohexyl ketone, Pimelic ketone

**Exposure Limits:**
**NIOSH REL:** TWA 25 ppm (100 mg/m$^3$) [skin]
**OSHA PEL†:** TWA 50 ppm (200 mg/m$^3$)

**Measurement Methods (see Table 1):**
**NIOSH** 1300, 2555
**OSHA** 1

**Physical Description:** Water-white to pale-yellow liquid with a peppermint- or acetone-like odor.

| Chemical & Physical Properties: | Personal Protection/Sanitation (see Table 2): | Respirator Recommendations (see Tables 3 and 4): |
|---|---|---|
| **MW:** 98.2<br>**BP:** 312°F<br>**Sol:** 15%<br>**Fl.P:** 146°F<br>**IP:** 9.14 eV<br>**Sp.Gr:** 0.95<br>**VP:** 5 mmHg<br>**FRZ:** -49°F<br>**UEL:** 9.4%<br>**LEL(212°F):** 1.1%<br>Class IIIA Combustible Liquid | **Skin:** Prevent skin contact<br>**Eyes:** Prevent eye contact<br>**Wash skin:** When contam<br>**Remove:** When wet or contam<br>**Change:** N.R. | **NIOSH**<br>**625 ppm:** Sa:Cf£/PaprOv£<br>**700 ppm:** CcrFOv/GmFOv/PaprTOv£/ScbaF/SaF<br>**§:** ScbaF:Pd,Pp/SaF:Pd,Pp:AScba<br>**Escape:** GmFOv/ScbaE |

**Incompatibilities and Reactivities:** Oxidizers, nitric acid

| Exposure Routes, Symptoms, Target Organs (see Table 5): | First Aid (see Table 6): |
|---|---|
| **ER:** Inh, Abs, Ing, Con<br>**SY:** Irrit eyes, skin, muc memb; head; narco, coma; derm; in animals: liver, kidney damage<br>**TO:** Eyes, skin, resp sys, CNS, liver, kidneys | **Eye:** Irr immed<br>**Skin:** Water flush prompt<br>**Breath:** Resp support<br>**Swallow:** Medical attention immed |

## Cyclohexene

| Formula: | CAS#: | RTECS#: | IDLH: |
|---|---|---|---|
| $C_6H_{10}$ | 110-83-8 | GW2500000 | 2000 ppm |

**Conversion:** 1 ppm = 3.36 mg/m$^3$ **DOT:** 2256 130

**Synonyms/Trade Names:** Benzene tetrahydride, Tetrahydrobenzene

**Exposure Limits:**
**NIOSH REL:** TWA 300 ppm (1015 mg/m$^3$)
**OSHA PEL:** TWA 300 ppm (1015 mg/m$^3$)

**Measurement Methods (see Table 1):**
**NIOSH** 1500
**OSHA** 7

**Physical Description:** Colorless liquid with a sweet odor.

**Chemical & Physical Properties:**
**MW:** 82.2
**BP:** 181°F
**Sol:** Insoluble
**Fl.P:** 11°F
**IP:** 8.95 eV
**Sp.Gr:** 0.81
**VP:** 67 mmHg
**FRZ:** -154°F
**UEL:** ?
**LEL:** ?
Class IB Flammable Liquid

**Personal Protection/Sanitation (see Table 2):**
**Skin:** Prevent skin contact
**Eyes:** Prevent eye contact
**Wash skin:** When contam
**Remove:** When wet (flamm)
**Change:** N.R.

**Respirator Recommendations (see Tables 3 and 4):**
**NIOSH/OSHA**
**2000 ppm:** Sa:Cf£/PaprOv£/CcrFOv/GmFOv/ScbaF/SaF
**§:** ScbaF:Pd,Pp/SaF:Pd,Pp:AScba
**Escape:** GmFOv/ScbaE

**Incompatibilities and Reactivities:** Strong oxidizers [**Note:** Forms explosive peroxides with oxygen upon storage.]

**Exposure Routes, Symptoms, Target Organs (see Table 5):**
**ER:** Inh, Ing, Con
**SY:** Irrit eyes, skin, resp sys; drow
**TO:** Eyes, skin, resp sys, CNS

**First Aid (see Table 6):**
**Eye:** Irr immed
**Skin:** Soap wash prompt
**Breath:** Resp support
**Swallow:** Medical attention immed

## Cyclohexylamine

| Formula: | CAS#: | RTECS#: | IDLH: |
|---|---|---|---|
| $C_6H_{11}NH_2$ | 108-91-8 | GX0700000 | N.D. |

**Conversion:** 1 ppm = 4.06 mg/m$^3$ **DOT:** 2357 132

**Synonyms/Trade Names:** Aminocyclohexane, Aminohexahydrobenzene, Hexahydroaniline, Hexahydrobenzenamine

**Exposure Limits:**
**NIOSH REL:** TWA 10 ppm (40 mg/m$^3$)
**OSHA PEL†:** none

**Measurement Methods (see Table 1):**
**NIOSH** 2010
**OSHA** PV2016

**Physical Description:** Colorless or yellow liquid with a strong, fishy, amine-like odor.

**Chemical & Physical Properties:**
**MW:** 99.2
**BP:** 274°F
**Sol:** Miscible
**Fl.P:** 88°F
**IP:** 8.37 eV
**Sp.Gr:** 0.87
**VP:** 11 mmHg
**FRZ:** 0°F
**UEL:** 9.4%
**LEL:** 1.5%
Class IC Flammable Liquid

**Personal Protection/Sanitation (see Table 2):**
**Skin:** Prevent skin contact
**Eyes:** Prevent eye contact
**Wash skin:** When contam
**Remove:** When wet (flamm)
**Change:** N.R.
**Provide:** Eyewash
Quick drench

**Respirator Recommendations (see Tables 3 and 4):**
Not available.

**Incompatibilities and Reactivities:** Oxidizers, organic compounds, acid anhydrides, acid chlorides, acids, lead [**Note:** Corrosive to copper, aluminum, zinc & galvanized steel.]

**Exposure Routes, Symptoms, Target Organs (see Table 5):**
**ER:** Inh, Abs, Ing, Con
**SY:** Irrit eyes, skin, muc memb, resp sys; eye, skin burns; skin sens; cough, pulm edema; drow, dizz; diarr, nau, vomit
**TO:** Eyes, skin, resp sys, CNS

**First Aid (see Table 6):**
**Eye:** Irr immed
**Skin:** Water flush immed
**Breath:** Resp support
**Swallow:** Medical attention immed

C

## Cyclonite

| Formula: | CAS#: | RTECS#: | IDLH: |
|---|---|---|---|
| $C_3H_6N_6O_6$ | 121-82-4 | XY9450000 | N.D. |

**Conversion:**

**DOT:**

**Synonyms/Trade Names:** Cyclotrimethylenetrinitramine; Hexahydro-1,3,5-trinitro-s-triazine; RDX; Trimethylenetrinitramine; 1,3,5-Trinitro-1,3,5-triazacyclohexane

**Exposure Limits:**
**NIOSH REL:** TWA 1.5 mg/m$^3$
ST 3 mg/m$^3$ [skin]
**OSHA PEL†:** none

**Measurement Methods (see Table 1):**
**NIOSH** 0500

**Physical Description:** White, crystalline powder. [**Note:** A powerful high explosive.]

**Chemical & Physical Properties:**
**MW:** 222.2
**BP:** ?
**Sol:** Insoluble
**Fl.P:** Explodes
**IP:** ?
**Sp.Gr:** 1.82
**VP(230°F):** 0.0004 mmHg
**MLT:** 401°F
**UEL:** ?
**LEL:** ?
Combustible Solid [EXPLOSIVE!]

**Personal Protection/Sanitation (see Table 2):**
**Skin:** Prevent skin contact
**Eyes:** Prevent eye contact
**Wash skin:** When contam/Daily
**Remove:** When wet or contam
**Change:** Daily
**Provide:** Eyewash
Quick drench

**Respirator Recommendations (see Tables 3 and 4):**
Not available.

**Incompatibilities and Reactivities:** Strong oxidizers, combustible materials, heat
[**Note:** Detonates on contact with mercury fulminate.]

**Exposure Routes, Symptoms, Target Organs (see Table 5):**
**ER:** Inh, Abs, Ing, Con
**SY:** Irrit eyes, skin; head, irrity, lass, tremor, nau, dizz, vomit, insom, convuls
**TO:** Eyes, skin, CNS

**First Aid (see Table 6):**
**Eye:** Irr immed
**Skin:** Soap flush immed
**Breath:** Resp support
**Swallow:** Medical attention immed

## Cyclopentadiene

| Formula: | CAS#: | RTECS#: | IDLH: |
|---|---|---|---|
| $C_5H_6$ | 542-92-7 | GY1000000 | 750 ppm |

**Conversion:** 1 ppm = 2.70 mg/m$^3$

**DOT:**

**Synonyms/Trade Names:** 1,3-Cyclopentadiene

**Exposure Limits:**
**NIOSH REL:** TWA 75 ppm (200 mg/m$^3$)
**OSHA PEL:** TWA 75 ppm (200 mg/m$^3$)

**Measurement Methods (see Table 1):**
**NIOSH** 2523

**Physical Description:** Colorless liquid with an irritating, terpene-like odor.

**Chemical & Physical Properties:**
**MW:** 66.1
**BP:** 107°F
**Sol:** Insoluble
**Fl.P(oc):** 77°F
**IP:** 8.56 eV
**Sp.Gr:** 0.80
**VP:** 400 mmHg
**FRZ:** -121°F
**UEL:** ?
**LEL:** ?
Class IC Flammable Liquid

**Personal Protection/Sanitation (see Table 2):**
**Skin:** Prevent skin contact
**Eyes:** Prevent eye contact
**Wash skin:** When contam
**Remove:** When wet (flamm)
**Change:** N.R.

**Respirator Recommendations (see Tables 3 and 4):**
**NIOSH/OSHA**
**750 ppm:** CcrOv/GmFOv/PaprOv/Sa/ScbaF
**§:** ScbaF:Pd,Pp/SaF:Pd,Pp:AScba
**Escape:** GmFOv/ScbaE

**Incompatibilities and Reactivities:** Strong oxidizers, fuming nitric acid, sulfuric acid
[**Note:** Polymerizes to dicyclopentadiene upon standing.]

**Exposure Routes, Symptoms, Target Organs (see Table 5):**
**ER:** Inh, Ing, Con
**SY:** Irrit eyes, nose
**TO:** Eyes, resp sys

**First Aid (see Table 6):**
**Eye:** Irr immed
**Skin:** Soap wash prompt
**Breath:** Resp support
**Swallow:** Medical attention immed

C

| **Cyclopentane** | **Formula:** $C_5H_{10}$ | **CAS#:** 287-92-3 | **RTECS#:** GY2390000 | **IDLH:** N.D. |
|---|---|---|---|---|
| **Conversion:** 1 ppm = 2.87 mg/$m^3$ | **DOT:** 1146 128 | | | |

**Synonyms/Trade Names:** Pentamethylene

**Exposure Limits:**
**NIOSH REL:** TWA 600 ppm (1720 mg/$m^3$)
**OSHA PEL†:** none

**Measurement Methods (see Table 1):**
None available

**Physical Description:** Colorless liquid with a mild, sweet odor.

**Chemical & Physical Properties:**
**MW:** 70.2
**BP:** 121°F
**Sol:** Insoluble
**Fl.P:** -35°F
**IP:** 10.52 eV
**Sp.Gr:** 0.75
**VP(88°F):** 400 mmHg
**FRZ:** -137°F
**UEL:** 8.7%
**LEL:** 1.1%
Class IB Flammable Liquid

**Personal Protection/Sanitation (see Table 2):**
**Skin:** Prevent skin contact
**Eyes:** Prevent eye contact
**Wash skin:** Daily
**Remove:** When wet (flamm)
**Change:** N.R.

**Respirator Recommendations (see Tables 3 and 4):**
Not available.

**Incompatibilities and Reactivities:** Strong oxidizers (e.g., chlorine, bromine, fluorine)

**Exposure Routes, Symptoms, Target Organs (see Table 5):**
**ER:** Inh, Ing, Con
**SY:** Irrit eyes, skin, nose, throat; dizz, euph, inco, nau, vomit, stupor; dry, cracking skin
**TO:** Eyes, skin, resp sys, CNS

**First Aid (see Table 6):**
**Eye:** Irr immed
**Skin:** Soap wash
**Breath:** Resp support
**Swallow:** Medical attention immed

| **Cyhexatin** | **Formula:** $(C_6H_{11})_3SnOH$ | **CAS#:** 13121-70-5 | **RTECS#:** WH8750000 | **IDLH:** 80 mg/$m^3$ [25 mg/$m^3$ (as Sn)] |
|---|---|---|---|---|
| **Conversion:** | **DOT:** | | | |

**Synonyms/Trade Names:** TCHH, Tricyclohexylhydroxystannane, Tricyclohexylhydroxytin, Tricyclohexylstannium hydroxide, Tricyclohexyltin hydroxide

**Exposure Limits:**
**NIOSH REL:** TWA 5 mg/$m^3$
**OSHA PEL†:** TWA 0.32 mg/$m^3$ [0.1 mg/$m^3$ (as Sn)]

**Measurement Methods (see Table 1):**
**NIOSH** 5504

**Physical Description:** Colorless to white, nearly odorless, crystalline powder. [insecticide]

**Chemical & Physical Properties:**
**MW:** 385.2
**BP:** 442°F (Decomposes)
**Sol:** Insoluble
**Fl.P:** NA
**IP:** NA
**Sp.Gr:** ?
**VP:** 0 mmHg (approx)
**MLT:** 383°F
**UEL:** NA
**LEL:** NA

**Personal Protection/Sanitation (see Table 2):**
**Skin:** Prevent skin contact
**Eyes:** N.R.
**Wash skin:** When contam
**Remove:** When wet or contam
**Change:** Daily

**Respirator Recommendations (see Tables 3 and 4):**
**OSHA**
**3.2 mg/$m^3$:** CcrOv95/Sa
**8 mg/$m^3$:** Sa:Cf/PaprOvHie
**16 mg/$m^3$:** CcrFOv100/GmFOv100/PaprTOvHie/SaT:Cf/ScbaF/SaF
**80 mg/$m^3$:** SaF:Pd,Pp
**§:** ScbaF:Pd,Pp/SaF:Pd,Pp:AScba
**Escape:** GmFOv100/ScbaE

**Incompatibilities and Reactivities:** Strong oxidizers, ultraviolet light

**Exposure Routes, Symptoms, Target Organs (see Table 5):**
**ER:** Inh, Abs, Ing, Con
**SY:** Irrit eyes, skin, resp sys; head, dizz; sore throat, cough; abdom pain, vomit; skin burns, pruritus; in animals: liver, kidney damage
**TO:** Eyes, skin, resp sys, liver, kidneys

**First Aid (see Table 6):**
**Eye:** Irr immed
**Skin:** Soap wash immed
**Breath:** Resp support
**Swallow:** Medical attention immed

| 2,4-D | Formula: $Cl_2C_6H_3OCH_2COOH$ | CAS#: 94-75-7 | RTECS#: AG6825000 | IDLH: 100 mg/$m^3$ |
|---|---|---|---|---|
| **Conversion:** | **DOT:** 2765 152 | | | |

**Synonyms/Trade Names:** Dichlorophenoxyacetic acid; 2,4-Dichlorophenoxyacetic acid

D

**Exposure Limits:**
**NIOSH REL:** TWA 10 mg/$m^3$
**OSHA PEL:** TWA 10 mg/$m^3$

**Measurement Methods (see Table 1):**
**NIOSH** 5001

**Physical Description:** White to yellow, crystalline, odorless powder. [herbicide]

| Chemical & Physical Properties: | Personal Protection/Sanitation (see Table 2): | Respirator Recommendations (see Tables 3 and 4): |
|---|---|---|
| **MW:** 221.0<br>**BP:** Decomposes<br>**Sol:** 0.05%<br>**Fl.P:** NA<br>**IP:** ?<br>**Sp.Gr:** 1.57<br>**VP(320°F):** 0.4 mmHg<br>**MLT:** 280°F<br>**UEL:** NA<br>**LEL:** NA<br>Noncombustible Solid, but may be dissolved in flammable liquids. | **Skin:** Prevent skin contact<br>**Eyes:** Prevent eye contact<br>**Wash skin:** When contam<br>**Remove:** When wet or contam<br>**Change:** Daily | **NIOSH/OSHA**<br>**100 mg/$m^3$:** CcrOv95/GmFOv100/ PaprOvHie/Sa/ScbaF<br>**§:** ScbaF:Pd,Pp/SaF:Pd,Pp:AScba<br>**Escape:** GmFOv100/ScbaE |

**Incompatibilities and Reactivities:** Strong oxidizers

| Exposure Routes, Symptoms, Target Organs (see Table 5): | First Aid (see Table 6): |
|---|---|
| **ER:** Inh, Abs, Ing, Con<br>**SY:** Lass, stupor, hyporeflexia, musc twitch; convuls; derm; in animals: liver, kidney inj<br>**TO:** Skin, CNS, liver, kidneys | **Eye:** Irr immed<br>**Skin:** Soap wash prompt<br>**Breath:** Resp support<br>**Swallow:** Medical attention immed |

| DDT | Formula: $(C_6H_4Cl)_2CHCCl_3$ | CAS#: 50-29-3 | RTECS#: KJ3325000 | IDLH: Ca [500 mg/$m^3$] |
|---|---|---|---|---|
| **Conversion:** | **DOT:** 2761 151 | | | |

**Synonyms/Trade Names:** p,p'-DDT; Dichlorodiphenyltrichloroethane; 1,1,1-Trichloro-2,2-bis(p-chlorophenyl)ethane

**Exposure Limits:**
**NIOSH REL:** Ca
TWA 0.5 mg/$m^3$
See Appendix A
**OSHA PEL:** TWA 1 mg/$m^3$ [skin]

**Measurement Methods (see Table 1):**
**NIOSH** S274 (II-3)

**Physical Description:** Colorless crystals or off-white powder with a slight, aromatic odor. [pesticide]

| Chemical & Physical Properties: | Personal Protection/Sanitation (see Table 2): | Respirator Recommendations (see Tables 3 and 4): |
|---|---|---|
| **MW:** 354.5<br>**BP:** 230°F (Decomposes)<br>**Sol:** Insoluble<br>**Fl.P:** 162-171°F<br>**IP:** ?<br>**Sp.Gr:** 0.99<br>**VP:** 0.0000002 mmHg<br>**MLT:** 227°F<br>**UEL:** ?<br>**LEL:** ?<br>Combustible Solid | **Skin:** Prevent skin contact<br>**Eyes:** Prevent eye contact<br>**Wash skin:** When contam/Daily<br>**Remove:** When wet or contam<br>**Change:** Daily<br>**Provide:** Eyewash<br>Quick drench | **NIOSH**<br>**¥:** ScbaF:Pd,Pp/SaF:Pd,Pp:AScba<br>**Escape:** GmFOv100/ScbaE |

**Incompatibilities and Reactivities:** Strong oxidizers, alkalis

| Exposure Routes, Symptoms, Target Organs (see Table 5): | First Aid (see Table 6): |
|---|---|
| **ER:** Inh, Abs, Ing, Con<br>**SY:** Irrit eyes, skin; pares tongue, lips, face; tremor; anxi, dizz, conf, mal, head, lass; convuls; paresis hands; vomit; [carc]<br>**TO:** Eyes, skin, CNS, kidneys, liver, PNS [in animals: liver, lung & lymphatic tumors] | **Eye:** Irr immed<br>**Skin:** Soap wash prompt<br>**Breath:** Resp support<br>**Swallow:** Medical attention immed |

## Decaborane

| Decaborane | Formula: $B_{10}H_{14}$ | CAS#: 17702-41-9 | RTECS#: HD1400000 | IDLH: 15 mg/m$^3$ |
|---|---|---|---|---|
| **Conversion:** 1 ppm = 5.00 mg/m$^3$ | **DOT:** 1868 134 | | | |

**Synonyms/Trade Names:** Decaboron tetradecahydride

**Exposure Limits:**
**NIOSH REL:** TWA 0.3 mg/m$^3$ (0.05 ppm) [skin]
ST 0.9 mg/m$^3$ (0.15 ppm)
**OSHA PEL†:** TWA 0.3 mg/m$^3$ (0.05 ppm) [skin]

**Measurement Methods (see Table 1):**
None available

**Physical Description:** Colorless to white crystalline solid with an intense, bitter, chocolate-like odor.

| Chemical & Physical Properties: | Personal Protection/Sanitation (see Table 2): | Respirator Recommendations (see Tables 3 and 4): |
|---|---|---|
| **MW:** 122.2<br>**BP:** 415°F<br>**Sol:** Slight<br>**Fl.P:** 176°F<br>**IP:** 9.88 eV<br>**Sp.Gr:** 0.94<br>**VP:** 0.2 mmHg<br>**MLT:** 211°F<br>**UEL:** ?<br>**LEL:** ?<br>Combustible Solid | **Skin:** Prevent skin contact<br>**Eyes:** Prevent eye contact<br>**Wash skin:** When contam/Daily<br>**Remove:** When wet or contam<br>**Change:** Daily<br>**Provide:** Eyewash<br>Quick drench | **NIOSH/OSHA**<br>**3 mg/m$^3$:** Sa<br>**7.5 mg/m$^3$:** Sa:Cf<br>**15 mg/m$^3$:** SaT:Cf/ScbaF/SaF<br>**§:** ScbaF:Pd,Pp/SaF:Pd,Pp:AScba<br>**Escape:** GmFOv100/ScbaE |

**Incompatibilities and Reactivities:** Oxidizers, water, halogenated compounds (especially carbon tetrachloride) [**Note:** May ignite SPONTANEOUSLY on exposure to air. Decomposes slowly in hot water.]

| Exposure Routes, Symptoms, Target Organs (see Table 5): | First Aid (see Table 6): |
|---|---|
| **ER:** Inh, Abs, Ing, Con<br>**SY:** Dizz, head, nau, drow; inco, local musc spasm, tremor, convuls; lass; in animals: dysp; lass; liver, kidney damage<br>**TO:** CNS, liver, kidneys | **Eye:** Irr immed<br>**Skin:** Soap wash immed<br>**Breath:** Resp support<br>**Swallow:** Medical attention immed |

## 1-Decanethiol

| 1-Decanethiol | Formula: $CH_3(CH_2)_9SH$ | CAS#: 143-10-2 | RTECS#: | IDLH: N.D. |
|---|---|---|---|---|
| **Conversion:** 1 ppm = 7.13 mg/m$^3$ | **DOT:** 1228 131 | | | |

**Synonyms/Trade Names:** Decylmercaptan, n-Decylmercaptan, 1-Mercaptodecane

**Exposure Limits:**
**NIOSH REL:** C 0.5 ppm (3.6 mg/m$^3$) [15-minute]
**OSHA PEL:** none

**Measurement Methods (see Table 1):**
None available

**Physical Description:** Colorless liquid with a strong odor.

| Chemical & Physical Properties: | Personal Protection/Sanitation (see Table 2): | Respirator Recommendations (see Tables 3 and 4): |
|---|---|---|
| **MW:** 174.4<br>**BP:** 465°F<br>**Sol:** Insoluble<br>**Fl.P:** 209°F<br>**IP:** ?<br>**Sp.Gr:** 0.84<br>**VP:** ?<br>**FRZ:** -15°F<br>**UEL:** ?<br>**LEL:** ?<br>Class IIIB Combustible Liquid | **Skin:** Prevent skin contact<br>**Eyes:** Prevent eye contact<br>**Wash skin:** When contam<br>**Remove:** When wet or contam<br>**Change:** N.R. | **NIOSH**<br>**5 ppm:** CcrOv/Sa<br>**12.5 ppm:** Sa:Cf/PaprOv<br>**25 ppm:** CcrFOv/GmFOv/PaprTOv/ScbaF/SaF<br>**§:** ScbaF:Pd,Pp/SaF:Pd,Pp:AScba<br>**Escape:** GmFOv/ScbaE |

**Incompatibilities and Reactivities:** Oxidizers, strong acids & bases, alkali metals, nitric acid

| Exposure Routes, Symptoms, Target Organs (see Table 5): | First Aid (see Table 6): |
|---|---|
| **ER:** Inh, Abs, Ing, Con<br>**SY:** Irrit eyes, skin, resp sys; conf, dizz, head, drow, nau, vomit, lass, convuls<br>**TO:** Eyes, skin, resp sys, CNS | **Eye:** Irr immed<br>**Skin:** Soap wash<br>**Breath:** Resp support<br>**Swallow:** Medical attention immed |

| Demeton | Formula: $(C_2H_5O)_2PSOC_2H_4SC_2H_5$ | CAS#: 8065-48-3 | RTECS#: TF3150000 | IDLH: 10 mg/m$^3$ |
|---|---|---|---|---|
| **Conversion:** | **DOT:** | | | |

**Synonyms/Trade Names:** O-O-Diethyl-O(and S)-2-(ethylthio)ethyl phosphorothioate mixture, Systox®

D

**Exposure Limits:**
**NIOSH REL:** TWA 0.1 mg/m$^3$ [skin]
**OSHA PEL:** TWA 0.1 mg/m$^3$ [skin]

**Measurement Methods (see Table 1):**
**NIOSH** 5514

**Physical Description:** Amber, oily liquid with a sulfur-like odor. [insecticide]

| Chemical & Physical Properties: | Personal Protection/Sanitation (see Table 2): | Respirator Recommendations (see Tables 3 and 4): |
|---|---|---|
| **MW:** 258.3<br>**BP:** Decomposes<br>**Sol:** 0.01%<br>**Fl.P:** 113°F<br>**IP:** ?<br>**Sp.Gr:** 1.12<br>**VP:** 0.0003 mmHg<br>**FRZ:** <-13°F<br>**UEL:** ?<br>**LEL:** ?<br>Class II Combustible Liquid | **Skin:** Prevent skin contact<br>**Eyes:** Prevent eye contact<br>**Wash skin:** When contam<br>**Remove:** When wet or contam<br>**Change:** Daily<br>**Provide:** Eyewash<br>Quick drench | **NIOSH/OSHA**<br>**1 mg/m$^3$:** Sa<br>**2.5 mg/m$^3$:** Sa:Cf<br>**5 mg/m$^3$:** SaT:Cf/ScbaF/SaF<br>**10 mg/m$^3$:** Sa:Pd,Pp<br>**§:** ScbaF:Pd,Pp/SaF:Pd,Pp:AScba<br>**Escape:** GmFOv100/ScbaE |

**Incompatibilities and Reactivities:** Strong oxidizers, alkalis, water

| Exposure Routes, Symptoms, Target Organs (see Table 5): | First Aid (see Table 6): |
|---|---|
| **ER:** Inh, Abs, Ing, Con<br>**SY:** Irrit eyes, skin; miosis, ache eyes, rhin, head; chest tight, wheez, lar spasm, salv, cyan; anor, nau, vomit, abdom cramps, diarr; local sweat; musc fasc, lass, para; dizz, conf, ataxia; convuls, coma; low BP; card irreg<br>**TO:** Eyes, skin, resp sys, CVS, CNS, blood chol | **Eye:** Irr immed<br>**Skin:** Soap wash immed<br>**Breath:** Resp support<br>**Swallow:** Medical attention immed |

| Diacetone alcohol | Formula: $CH_3COCH_2C(CH_3)_2OH$ | CAS#: 123-42-2 | RTECS#: SA9100000 | IDLH: 1800 ppm [10%LEL] |
|---|---|---|---|---|
| **Conversion:** 1 ppm = 4.75 mg/m$^3$ | **DOT:** 1148 129 | | | |

**Synonyms/Trade Names:** Diacetone, 4-Hydroxy-4-methyl-2-pentanone, 2-Methyl-2-pentanol-4-one

**Exposure Limits:**
**NIOSH REL:** TWA 50 ppm (240 mg/m$^3$)
**OSHA PEL:** TWA 50 ppm (240 mg/m$^3$)

**Measurement Methods (see Table 1):**
**NIOSH** 1402, 1405
**OSHA** 7

**Physical Description:** Colorless liquid with a faint, minty odor.

| Chemical & Physical Properties: | Personal Protection/Sanitation (see Table 2): | Respirator Recommendations (see Tables 3 and 4): |
|---|---|---|
| **MW:** 116.2<br>**BP:** 334°F<br>**Sol:** Miscible<br>**Fl.P:** 125°F<br>**IP:** ?<br>**Sp.Gr:** 0.94<br>**VP:** 1 mmHg<br>**FRZ:** -47°F<br>**UEL:** 6.9%<br>**LEL:** 1.8%<br>Class II Combustible Liquid | **Skin:** Prevent skin contact<br>**Eyes:** Prevent eye contact<br>**Wash skin:** When contam<br>**Remove:** When wet or contam<br>**Change:** N.R. | **NIOSH/OSHA**<br>**1250 ppm:** Sa:Cf£/PaprOv£<br>**1800 ppm:** CcrFOv/GmFOv/PaprTOv£/ ScbaF/SaF<br>**§:** ScbaF:Pd,Pp/SaF:Pd,Pp:AScba<br>**Escape:** GmFOv/ScbaE |

**Incompatibilities and Reactivities:** Strong oxidizers, strong alkalis

| Exposure Routes, Symptoms, Target Organs (see Table 5): | First Aid (see Table 6): |
|---|---|
| **ER:** Inh, Ing, Con<br>**SY:** Irrit eyes, skin, nose, throat; corn damage; in animals: narco, liver damage<br>**TO:** Eyes, skin, resp sys, CNS, liver | **Eye:** Irr immed<br>**Skin:** Water flush prompt<br>**Breath:** Resp support<br>**Swallow:** Medical attention immed |

D

## 2,4-Diaminoanisole (and its salts)

| 2,4-Diaminoanisole (and its salts) | Formula: $(NH_2)_2C_6H_3OCH_3$ | CAS#: 615-05-4 | RTECS#: BZ8580500 | IDLH: Ca [N.D.] |
|---|---|---|---|---|
| **Conversion:** | **DOT:** | | | |

**Synonyms/Trade Names:** 1,3-Diamino-4-methoxybenzene; 4-Methoxy-1,3-benzene-diamine; 4-Methoxy-m-phenylene-diamine (Synonyms of salts vary depending upon the specific compound.)

**Exposure Limits:**
**NIOSH REL:** Ca
Minimize occupational exposure (especially skin exposures)
See Appendix A
**OSHA PEL:** none

**Measurement Methods (see Table 1):**
None available

**Physical Description:** Colorless solid (needles).
[**Note:** The primary use (including its salts such as 2,4-diaminoanisole sulfate) is a component of hair & fur dye formulations.]

| Chemical & Physical Properties: | Personal Protection/Sanitation (see Table 2): | Respirator Recommendations (see Tables 3 and 4): |
|---|---|---|
| **MW:** 138.2<br>**BP:** ?<br>**Sol:** ?<br>**Fl.P:** ?<br>**IP:** ?<br>**Sp.Gr:** ?<br>**VP:** ?<br>**MLT:** 153°F<br>**UEL:** ?<br>**LEL:** ?<br>Combustible Solid | **Skin:** Prevent skin contact<br>**Eyes:** Prevent eye contact<br>**Wash skin:** When contam/Daily<br>**Remove:** When wet or contam<br>**Change:** Daily<br>**Provide:** Eyewash<br>Quick drench | **NIOSH**<br>**¥:** ScbaF:Pd,Pp/SaF:Pd,Pp:AScba<br>**Escape:** GmFOv100/ScbaE |

**Incompatibilities and Reactivities:** Strong oxidizers

| Exposure Routes, Symptoms, Target Organs (see Table 5): | First Aid (see Table 6): |
|---|---|
| **ER:** Inh, Abs, Ing, Con<br>**SY:** In animals: irrit skin; thyroid, liver changes; terato effects; [carc]<br>**TO:** Skin, thyroid, liver, repro sys [in animals: thyroid, liver, skin & lymphatic sys tumors] | **Eye:** Irr immed<br>**Skin:** Soap wash immed<br>**Breath:** Resp support<br>**Swallow:** Medical attention immed |

## o-Dianisidine

| o-Dianisidine | Formula: $(NH_2C_6H_3OCH_3)_2$ | CAS#: 119-90-4 | RTECS#: DD0875000 | IDLH: Ca [N.D.] |
|---|---|---|---|---|
| **Conversion:** | **DOT:** | | | |

**Synonyms/Trade Names:** Dianisidine; 3,3'-Dianisidine; 3,3'-Dimethoxybenzidine

**Exposure Limits:**
**NIOSH REL:** Ca
See Appendix A
See Appendix C
**OSHA PEL:** See Appendix C

**Measurement Methods (see Table 1):**
**NIOSH** 5013
**OSHA** 71

**Physical Description:** Colorless crystals that turn a violet color on standing.
[**Note:** Used as a basis for many dyes.]

| Chemical & Physical Properties: | Personal Protection/Sanitation (see Table 2): | Respirator Recommendations (see Tables 3 and 4): |
|---|---|---|
| **MW:** 244.3<br>**BP:** ?<br>**Sol:** Insoluble<br>**Fl.P:** 403°F<br>**IP:** ?<br>**Sp.Gr:** ?<br>**VP:** ?<br>**MLT:** 279°F<br>**UEL:** ?<br>**LEL:** ?<br>Combustible Solid | **Skin:** Prevent skin contact<br>**Eyes:** Prevent eye contact<br>**Wash skin:** When contam/Daily<br>**Remove:** When wet or contam<br>**Change:** Daily<br>**Provide:** Eyewash<br>Quick drench | **NIOSH**<br>**¥:** ScbaF:Pd,Pp/SaF:Pd,Pp:AScba<br>**Escape:** GmFOv100/ScbaE |

**Incompatibilities and Reactivities:** Oxidizers

| Exposure Routes, Symptoms, Target Organs (see Table 5): | First Aid (see Table 6): |
|---|---|
| **ER:** Inh, Abs, Ing, Con<br>**SY:** Irrit skin; in animals: kidney, liver damage; thyroid, spleen changes; [carc]<br>**TO:** Skin, kidneys, liver, thyroid, liver [in animals: bladder, liver, stomach & mammary gland tumors] | **Eye:** Irr immed<br>**Skin:** Soap wash immed<br>**Breath:** Resp support<br>**Swallow:** Medical attention immed |

D

## Diazinon®

| Diazinon® | Formula: $C_{12}H_{21}N_2O_3PS$ | CAS#: 333-41-5 | RTECS#: TF3325000 | IDLH: N.D. |
|---|---|---|---|---|
| Conversion: | DOT: 2783 152 | | | |

**Synonyms/Trade Names:** Basudin®; Diazide®; O,O-Diethyl-O-2-isopropyl-4-methyl-6-pyrimidinyl-phosphorothioate; Spectracide®

**Exposure Limits:**
**NIOSH REL:** TWA 0.1 mg/m$^3$ [skin]
**OSHA PEL†:** none

**Measurement Methods (see Table 1):**
**NIOSH** 5600
**OSHA** 62

**Physical Description:** Colorless liquid with a faint ester-like odor. [insecticide]
[**Note:** Technical grade is pale to dark brown.]

| Chemical & Physical Properties: | Personal Protection/Sanitation (see Table 2): | Respirator Recommendations (see Tables 3 and 4): |
|---|---|---|
| **MW:** 304.4<br>**BP:** Decomposes<br>**Sol:** 0.004%<br>**Fl.P:** 180°F<br>**IP:** ?<br>**Sp.Gr:** 1.12<br>**VP:** 0.0001 mmHg<br>**FRZ:** ?<br>**UEL:** ?<br>**LEL:** ?<br>Class IIIA Combustible Liquid | **Skin:** Prevent skin contact<br>**Eyes:** Prevent eye contact<br>**Wash skin:** When contam<br>**Remove:** When wet or contam<br>**Change:** Daily<br>**Provide:** Eyewash<br>Quick drench | Not available. |

**Incompatibilities and Reactivities:** Strong acids & alkalis, copper-containing compounds
[**Note:** Hydrolyzes slowly in water & dilute acid.]

| Exposure Routes, Symptoms, Target Organs (see Table 5): | First Aid (see Table 6): |
|---|---|
| **ER:** Inh, Abs, Ing, Con<br>**SY:** Irrit eyes; miosis, blurred vision; dizz, conf, lass, convuls; dysp; salv, abdom cramps, nau, vomit<br>**TO:** Eyes, resp sys, CNS, CVS, blood chol | **Eye:** Irr immed<br>**Skin:** Soap wash immed<br>**Breath:** Resp support<br>**Swallow:** Medical attention immed |

## Diazomethane

| Diazomethane | Formula: $CH_2N_2$ | CAS#: 334-88-3 | RTECS#: PA7000000 | IDLH: 2 ppm |
|---|---|---|---|---|
| Conversion: 1 ppm = 1.72 mg/m$^3$ | DOT: | | | |

**Synonyms/Trade Names:** Azimethylene, Azomethylene, Diazirine

**Exposure Limits:**
**NIOSH REL:** TWA 0.2 ppm (0.4 mg/m$^3$)
**OSHA PEL:** TWA 0.2 ppm (0.4 mg/m$^3$)

**Measurement Methods (see Table 1):**
**NIOSH** 2515

**Physical Description:** Yellow gas with a musty odor. [**Note:** Shipped as a liquefied compressed gas.]

| Chemical & Physical Properties: | Personal Protection/Sanitation (see Table 2): | Respirator Recommendations (see Tables 3 and 4): |
|---|---|---|
| **MW:** 42.1<br>**BP:** -9°F<br>**Sol:** Reacts<br>**Fl.P:** NA (Gas)<br>**IP:** 9.00 eV<br>**RGasD:** 1.45<br>**VP:** >1 atm<br>**FRZ:** -229°F<br>**UEL:** ?<br>**LEL:** ?<br>Flammable Gas [EXPLOSIVE!] | **Skin:** Frostbite<br>**Eyes:** Frostbite<br>**Wash skin:** N.R.<br>**Remove:** When wet (flamm)<br>**Change:** N.R.<br>**Provide:** Frostbite wash | **NIOSH/OSHA**<br>**2 ppm:** Sa*/ScbaF<br>**§:** ScbaF:Pd,Pp/SaF:Pd,Pp:AScba<br>**Escape:** GmFOv/ScbaE |

**Incompatibilities and Reactivities:** Alkali metals, water, drying agents such as calcium arsenate
[**Note:** May explode violently on heating, exposure to sunlight, or contact with rough edges such as ground glass.]

| Exposure Routes, Symptoms, Target Organs (see Table 5): | First Aid (see Table 6): |
|---|---|
| **ER:** Inh, Con (liquid)<br>**SY:** Irrit eyes; cough, short breath; head, lass; flush skin, fever; chest pain, pulm edema, pneu; asthma; liquid: frostbite<br>**TO:** Eyes, resp sys | **Eye:** Frostbite<br>**Skin:** Frostbite<br>**Breath:** Resp support |

## Diborane

| Diborane | Formula: $B_2H_6$ | CAS#: 19287-45-7 | RTECS#: HQ9275000 | IDLH: 15 ppm |
|---|---|---|---|---|
| **Conversion:** 1 ppm = 1.13 mg/m$^3$ | **DOT:** 1911 119 | | | |

**Synonyms/Trade Names:** Boroethane, Boron hydride, Diboron hexahydride

**Exposure Limits:**
**NIOSH REL:** TWA 0.1 ppm (0.1 mg/m$^3$)
**OSHA PEL:** TWA 0.1 ppm (0.1 mg/m$^3$)

**Measurement Methods (see Table 1):**
**NIOSH** 6006

D

**Physical Description:** Colorless gas with a repulsive, sweet odor.
[**Note:** Usually shipped in pressurized cylinders diluted with hydrogen, argon, nitrogen, or helium.]

**Chemical & Physical Properties:**
**MW:** 27.7
**BP:** -135°F
**Sol:** Reacts
**Fl.P:** NA (Gas)
**IP:** 11.38 eV
**RGasD:** 0.97
**VP(62°F):** 39.5 atm
**FRZ:** -265°F
**UEL:** 88%
**LEL:** 0.8%
Flammable Gas

**Personal Protection/Sanitation (see Table 2):**
**Skin:** N.R.
**Eyes:** N.R.
**Wash skin:** N.R.
**Remove:** N.R.
**Change:** N.R.

**Respirator Recommendations (see Tables 3 and 4):**
**NIOSH/OSHA**
**1 ppm:** Sa
**2.5 ppm:** Sa:Cf
**5 ppm:** SaT:Cf/ScbaF/SaF
**15 ppm:** Sa:Pd,Pp
**§:** ScbaF:Pd,Pp/SaF:Pd,Pp:AScba
**Escape:** GmFS/ScbaE

**Incompatibilities and Reactivities:** Water, halogenated compounds, aluminum, lithium, oxidized surfaces, acids
[**Note:** Will ignite spontaneously in moist air at room temperature. Reacts with water to form hydrogen & boric acid.]

**Exposure Routes, Symptoms, Target Organs (see Table 5):**
**ER:** Inh
**SY:** Chest tight, precordial pain, short breath, nonproductive cough, nau; head, dizz, chills, fever, lass, tremor, musc fasc; in animals: liver, kidney damage; pulm edema; hemorr
**TO:** Resp sys, CNS, liver, kidneys

**First Aid (see Table 6):**
**Breath:** Resp support

## 1,2-Dibromo-3-chloropropane

| 1,2-Dibromo-3-chloropropane | Formula: $CH_2BrCHBrCH_2Cl$ | CAS#: 96-12-8 | RTECS#: TX8750000 | IDLH: Ca [N.D.] |
|---|---|---|---|---|
| **Conversion:** 1 ppm = 9.67 mg/m$^3$ | **DOT:** 2872 159 | | | |

**Synonyms/Trade Names:** 1-Chloro-2,3-dibromopropane; DBCP; Dibromochloropropane

**Exposure Limits:**
**NIOSH REL:** Ca
See Appendix A
**OSHA PEL:** [1910.1044] TWA 0.001 ppm

**Measurement Methods (see Table 1):**
None available

**Physical Description:** Dense yellow or amber liquid with a pungent odor at high concentrations. [pesticide]
[**Note:** A solid below 43°F.]

**Chemical & Physical Properties:**
**MW:** 236.4
**BP:** 384°F
**Sol:** 0.1%
**Fl.P(oc):** 170°F
**IP:** ?
**Sp.Gr:** 2.05
**VP:** 0.8 mmHg
**FRZ:** 43°F
**UEL:** ?
**LEL:** ?
Class IIIA Combustible Liquid

**Personal Protection/Sanitation (see Table 2):**
**Skin:** Prevent skin contact
**Eyes:** Prevent eye contact
**Wash skin:** When contam/Daily
**Remove:** When wet or contam
**Change:** Daily
**Provide:** Eyewash
Quick drench

**Respirator Recommendations (see Tables 3 and 4):**
**NIOSH**
**¥:** ScbaF:Pd,Pp/SaF:Pd,Pp:AScba
**Escape:** GmFOv100/ScbaE

**See Appendix E** (page 351)

**Incompatibilities and Reactivities:** Chemically-active metals such as aluminum, magnesium & tin alloys [**Note:** Corrosive to metals.]

**Exposure Routes, Symptoms, Target Organs (see Table 5):**
**ER:** Inh, Abs, Ing, Con
**SY:** Irrit eyes, skin, nose, throat; drow; nau, vomit; pulm edema; liver, kidney inj; sterility; [carc]
**TO:** Eyes, skin, resp sys, CNS, liver, kidneys, spleen, repro sys, digestive sys [in animals: cancer of the nasal cavity, tongue, pharynx, lungs, stomach, adrenal & mammary glands]

**First Aid (see Table 6):**
**Eye:** Irr immed
**Skin:** Soap wash immed
**Breath:** Resp support
**Swallow:** Medical attention immed

D

| 2-N-Dibutylaminoethanol | Formula: $(C_4H_9)_2NCH_2CH_2OH$ | CAS#: 102-81-8 | RTECS#: KK3850000 | IDLH: N.D. |
|---|---|---|---|---|
| **Conversion:** 1 ppm = 7.09 mg/m³ | **DOT:** 2873 153 | | | |

**Synonyms/Trade Names:** Dibutylaminoethanol; 2-Dibutylaminoethanol; 2-Di-N-butylaminoethanol; 2-Di-N-butylaminoethyl alcohol; N,N-Dibutylethanolamine

**Exposure Limits:**
**NIOSH REL:** TWA 2 ppm (14 mg/m³) [skin]
**OSHA PEL†:** none

**Measurement Methods (see Table 1):**
**NIOSH** 2007

**Physical Description:** Colorless liquid with a faint, amine-like odor.

| Chemical & Physical Properties: | Personal Protection/Sanitation (see Table 2): | Respirator Recommendations (see Tables 3 and 4): |
|---|---|---|
| **MW:** 173.3<br>**BP:** 446°F<br>**Sol:** 0.4%<br>**Fl.P:** 195°F<br>**IP:** ?<br>**Sp.Gr:** 0.86<br>**VP:** 0.1 mmHg<br>**FRZ:** ?<br>**UEL:** ?<br>**LEL:** ?<br>Class IIIA Combustible Liquid | **Skin:** Prevent skin contact<br>**Eyes:** Prevent eye contact<br>**Wash skin:** When contam<br>**Remove:** When wet or contam<br>**Change:** N.R.<br>**Provide:** Eyewash<br>Quick drench | Not available. |

**Incompatibilities and Reactivities:** Oxidizers

| Exposure Routes, Symptoms, Target Organs (see Table 5): | First Aid (see Table 6): |
|---|---|
| **ER:** Inh, Abs, Ing, Con<br>**SY:** In animals: irrit eyes, skin, nose; derm; skin, corn nec; low-wgt<br>**TO:** Eyes, skin, resp sys | **Eye:** Irr immed<br>**Skin:** Soap flush immed<br>**Breath:** Resp support<br>**Swallow:** Medical attention immed |

| 2,6-Di-tert-butyl-p-cresol | Formula: $[C(CH_3)_3]_2CH_3C_6H_2OH$ | CAS#: 128-37-0 | RTECS#: GO7875000 | IDLH: N.D. |
|---|---|---|---|---|
| **Conversion:** | **DOT:** | | | |

**Synonyms/Trade Names:** BHT; Butylated hydroxytoluene; Dibutylated hydroxytoluene; 4-Methyl-2,6-di-tert-butyl phenol

**Exposure Limits:**
**NIOSH REL:** TWA 10 mg/m³
**OSHA PEL†:** none

**Measurement Methods (see Table 1):**
**NIOSH** P&CAM226 (II-1)
**OSHA** PV2108

**Physical Description:** White to pale-yellow, crystalline solid with a slight, phenolic odor. [food preservative]

| Chemical & Physical Properties: | Personal Protection/Sanitation (see Table 2): | Respirator Recommendations (see Tables 3 and 4): |
|---|---|---|
| **MW:** 220.4<br>**BP:** 509°F<br>**Sol:** 0.00004%<br>**Fl.P:** 261°F<br>**IP:** ?<br>**Sp.Gr:** 1.05<br>**VP:** 0.01 mmHg<br>**MLT:** 158°F<br>**UEL:** ?<br>**LEL:** ?<br>Class IIIB Combustible Liquid | **Skin:** Prevent skin contact<br>**Eyes:** Prevent eye contact<br>**Wash skin:** When contam<br>**Remove:** When wet or contam<br>**Change:** Daily | Not available. |

**Incompatibilities and Reactivities:** Oxidizers

| Exposure Routes, Symptoms, Target Organs (see Table 5): | First Aid (see Table 6): |
|---|---|
| **ER:** Inh, Ing, Con<br>**SY:** Irrit eyes, skin; in animals: decr growth rate, incr liver weight<br>**TO:** Eyes, skin | **Eye:** Irr immed<br>**Skin:** Soap wash<br>**Breath:** Fresh air<br>**Swallow:** Medical attention immed |

D

## Dibutyl phosphate

| Formula: | CAS#: | RTECS#: | IDLH: |
|---|---|---|---|
| $(C_4H_9O)_2(OH)PO$ | 107-66-4 | TB9605000 | 30 ppm |

**Conversion:** 1 ppm = 8.60 mg/m$^3$ | **DOT:**

**Synonyms/Trade Names:** Dibutyl acid o-phosphate, Di-n-butyl hydrogen phosphate, Dibutyl phosphoric acid

**Exposure Limits:**
**NIOSH REL:** TWA 1 ppm (5 mg/m$^3$)
ST 2 ppm (10 mg/m$^3$)
**OSHA PEL†:** TWA 1 ppm (5 mg/m$^3$)

**Measurement Methods (see Table 1):**
**NIOSH** 5017

**Physical Description:** Pale-amber, odorless liquid.

**Chemical & Physical Properties:**
**MW:** 210.2
**BP:** 212°F (Decomposes)
**Sol:** Insoluble
**Fl.P:** ?
**IP:** ?
**Sp.Gr:** 1.06
**VP:** 1 mmHg (approx)
**FRZ:** ?
**UEL:** ?
**LEL:** ?
Combustible Liquid

**Personal Protection/Sanitation (see Table 2):**
**Skin:** Prevent skin contact
**Eyes:** Prevent eye contact
**Wash skin:** When contam
**Remove:** When wet or contam
**Change:** N.R.
**Provide:** Quick drench

**Respirator Recommendations (see Tables 3 and 4):**
**NIOSH/OSHA**
**10 ppm:** Sa
**25 ppm:** Sa:Cf
**30 ppm:** SaT:Cf/ScbaF/SaF
**§:** ScbaF:Pd,Pp/SaF:Pd,Pp:AScba
**Escape:** GmFOv100/ScbaE

**Incompatibilities and Reactivities:** Strong oxidizers

**Exposure Routes, Symptoms, Target Organs (see Table 5):**
**ER:** Inh, Ing, Con
**SY:** Irrit eyes, skin, resp sys; head
**TO:** Eyes, skin, resp sys

**First Aid (see Table 6):**
**Eye:** Irr immed
**Skin:** Soap wash prompt
**Breath:** Resp support
**Swallow:** Medical attention immed

## Dibutyl phthalate

| Formula: | CAS#: | RTECS#: | IDLH: |
|---|---|---|---|
| $C_6H_4(COOC_4H_9)_2$ | 84-74-2 | TI0875000 | 4000 mg/m$^3$ |

**Conversion:** 1 ppm = 11.57 mg/m$^3$ | **DOT:**

**Synonyms/Trade Names:** DBP; Dibutyl-1,2-benzene-dicarboxylate; Di-n-butyl phthalate

**Exposure Limits:**
**NIOSH REL:** TWA 5 mg/m$^3$
**OSHA PEL:** TWA 5 mg/m$^3$

**Measurement Methods (see Table 1):**
**NIOSH** 5020
**OSHA** 104

**Physical Description:** Colorless to faint-yellow, oily liquid with a slight, aromatic odor.

**Chemical & Physical Properties:**
**MW:** 278.3
**BP:** 644°F
**Sol(77°F):** 0.001%
**Fl.P:** 315°F
**IP:** ?
**Sp.Gr:** 1.05
**VP:** 0.00007 mmHg
**FRZ:** -31°F
**UEL:** ?
**LEL(456°F):** 0.5%
Class IIIB Combustible Liquid

**Personal Protection/Sanitation (see Table 2):**
**Skin:** N.R.
**Eyes:** Prevent eye contact
**Wash skin:** N.R.
**Remove:** N.R.
**Change:** N.R.

**Respirator Recommendations (see Tables 3 and 4):**
**NIOSH/OSHA**
**50 mg/m$^3$:** 95F
**125 mg/m$^3$:** Sa:Cf£/PaprHie£
**250 mg/m$^3$:** 100F/ScbaF/SaF
**4000 mg/m$^3$:** SaF:Pd,Pp
**§:** ScbaF:Pd,Pp/SaF:Pd,Pp:AScba
**Escape:** 100F/ScbaE

**Incompatibilities and Reactivities:** Nitrates; strong oxidizers, alkalis & acids; liquid chlorine

**Exposure Routes, Symptoms, Target Organs (see Table 5):**
**ER:** Inh, Ing, Con
**SY:** Irrit eyes, upper resp sys, stomach
**TO:** Eyes, resp sys, GI tract

**First Aid (see Table 6):**
**Eye:** Irr immed
**Skin:** Wash regularly
**Breath:** Resp support
**Swallow:** Medical attention immed

D

| Dichloroacetylene | Formula: $C_2Cl_2$ | CAS#: 7572-29-4 | RTECS#: AP1080000 | IDLH: Ca [N.D.] |
|---|---|---|---|---|
| **Conversion:** 1 ppm = 3.88 mg/m$^3$ | **DOT:** | | | |

**Synonyms/Trade Names:** DCA, Dichloroethyne
[**Note:** DCA is a possible decomposition product of trichloroethylene or trichloroethane.]

**Exposure Limits:**
**NIOSH REL:** Ca
C 0.1 ppm (0.4 mg/m$^3$)
See Appendix A
**OSHA PEL†:** none

**Measurement Methods (see Table 1):**
None available

**Physical Description:** Volatile oil with a disagreeable, sweetish odor.
[**Note:** A gas above 90°F. DCA is not produced commercially.]

| Chemical & Physical Properties: | Personal Protection/Sanitation (see Table 2): | Respirator Recommendations (see Tables 3 and 4): |
|---|---|---|
| **MW:** 94.9<br>**BP:** 90°F (Explodes)<br>**Sol:** ?<br>**Fl.P:** ?<br>**IP:** ?<br>**Sp.Gr:** 1.26<br>**VP:** ?<br>**FRZ:** -58 to -87°F<br>**UEL:** ?<br>**LEL:** ?<br>Combustible Liquid | **Skin:** Prevent skin contact<br>**Eyes:** Prevent eye contact<br>**Wash skin:** When contam<br>**Remove:** When wet (flamm)<br>**Change:** N.R.<br>**Provide:** Eyewash<br>Quick drench | **NIOSH**<br>**¥:** ScbaF:Pd,Pp/SaF:Pd,Pp:AScba<br>**Escape:** GmFOv/ScbaE |

**Incompatibilities and Reactivities:** Oxidizers, heat, shock

**Exposure Routes, Symptoms, Target Organs (see Table 5):**
**ER:** Inh, Abs, Ing, Con
**SY:** Head, loss of appetite, nau, vomit, intense jaw pain, cranial nerve palsy; in animals: kidney, liver, brain inj; low-wgt; [carc]
**TO:** CNS [in animals: kidney tumors]

**First Aid (see Table 6):**
**Eye:** Irr immed
**Skin:** Soap flush immed
**Breath:** Resp support
**Swallow:** Medical attention immed

| o-Dichlorobenzene | Formula: $C_6H_4Cl_2$ | CAS#: 95-50-1 | RTECS#: CZ4500000 | IDLH: 200 ppm |
|---|---|---|---|---|
| **Conversion:** 1 ppm = 6.01 mg/m$^3$ | **DOT:** 1591 152 | | | |

**Synonyms/Trade Names:** o-DCB; 1,2-Dichlorobenzene; ortho-Dichlorobenzene; o-Dichlorobenzol

**Exposure Limits:**
**NIOSH REL:** C 50 ppm (300 mg/m$^3$)
**OSHA PEL:** C 50 ppm (300 mg/m$^3$)

**Measurement Methods (see Table 1):**
**NIOSH** 1003
**OSHA** 7

**Physical Description:** Colorless to pale-yellow liquid with a pleasant, aromatic odor. [herbicide]

| Chemical & Physical Properties: | Personal Protection/Sanitation (see Table 2): | Respirator Recommendations (see Tables 3 and 4): |
|---|---|---|
| **MW:** 147.0<br>**BP:** 357°F<br>**Sol:** 0.01%<br>**Fl.P:** 151°F<br>**IP:** 9.06 eV<br>**Sp.Gr:** 1.30<br>**VP:** 1 mmHg<br>**FRZ:** 1°F<br>**UEL:** 9.2%<br>**LEL:** 2.2%<br>Class IIIA Combustible Liquid | **Skin:** Prevent skin contact<br>**Eyes:** Prevent eye contact<br>**Wash skin:** When contam<br>**Remove:** When wet or contam<br>**Change:** N.R. | **NIOSH/OSHA**<br>**200 ppm:** CcrFOv/PaprOv£/<br>ScbaF/SaF<br>**§:** ScbaF:Pd,Pp/SaF:Pd,Pp:AScba<br>**Escape:** GmFOv/ScbaE |

**Incompatibilities and Reactivities:** Strong oxidizers, aluminum, chlorides, acids, acid fumes

**Exposure Routes, Symptoms, Target Organs (see Table 5):**
**ER:** Inh, Abs, Ing, Con
**SY:** Irrit eyes, nose; liver, kidney damage; skin blisters
**TO:** Eyes, skin, resp sys, liver, kidneys

**First Aid (see Table 6):**
**Eye:** Irr immed
**Skin:** Soap wash prompt
**Breath:** Resp support
**Swallow:** Medical attention immed

D

| p-Dichlorobenzene | Formula: $C_6H_4Cl_2$ | CAS#: 106-46-7 | RTECS#: CZ4550000 | IDLH: Ca [150 ppm] |
|---|---|---|---|---|
| **Conversion:** 1 ppm = 6.01 mg/m$^3$ | **DOT:** | | | |

**Synonyms/Trade Names:** p-DCB; 1,4-Dichlorobenzene; para-Dichlorobenzene; Dichlorocide

**Exposure Limits:**
**NIOSH REL:** Ca
See Appendix A
**OSHA PEL†:** TWA 75 ppm (450 mg/m$^3$)

**Measurement Methods (see Table 1):**
**NIOSH** 1003
**OSHA** 7

**Physical Description:** Colorless or white crystalline solid with a mothball-like odor. [insecticide]

| Chemical & Physical Properties: | Personal Protection/Sanitation (see Table 2): | Respirator Recommendations (see Tables 3 and 4): |
|---|---|---|
| **MW:** 147.0<br>**BP:** 345°F<br>**Sol:** 0.008%<br>**Fl.P:** 150°F<br>**IP:** 8.98 eV<br>**Sp.Gr:** 1.25<br>**VP:** 1.3 mmHg<br>**MLT:** 128°F<br>**UEL:** ?<br>**LEL:** 2.5%<br>Combustible Solid, but may take some effort to ignite. | **Skin:** Prevent skin contact<br>**Eyes:** Prevent eye contact<br>**Wash skin:** When contam/Daily<br>**Remove:** When wet or contam<br>**Change:** Daily<br>**Provide:** Eyewash<br>Quick drench | **NIOSH**<br>**¥:** ScbaF:Pd,Pp/SaF:Pd,Pp:AScba<br>**Escape:** GmFOv/ScbaE |

**Incompatibilities and Reactivities:** Strong oxidizers (such as chlorine or permanganate)

| Exposure Routes, Symptoms, Target Organs (see Table 5): | First Aid (see Table 6): |
|---|---|
| **ER:** Inh, Abs, Ing, Con<br>**SY:** Eye irrit, swell periorb; profuse rhinitis; head, anor, nau, vomit; low-wgt, jaun, cirr; in animals: liver, kidney inj; [carc]<br>**TO:** Liver, resp sys, eyes, kidneys, skin [in animals: liver & kidney cancer] | **Eye:** Irr immed<br>**Skin:** Soap wash<br>**Breath:** Resp support<br>**Swallow:** Medical attention immed |

| 3,3'-Dichlorobenzidine (and its salts) | Formula: $NH_2ClC_6H_3C_6H_3ClNH_2$ | CAS#: 91-94-1 | RTECS#: DD0525000 | IDLH: Ca [N.D.] |
|---|---|---|---|---|
| **Conversion:** | **DOT:** | | | |

**Synonyms/Trade Names:** 4,4'-Diamino-3,3'-dichlorobiphenyl; Dichlorobenzidine base; o,o'-Dichlorobenzidine; 3,3'-Dichlorobiphenyl-4,4'-diamine; 3,3'-Dichloro-4,4'-biphenyldiamine; 3,3'-Dichloro-4,4'-diaminobiphenyl

**Exposure Limits:**
**NIOSH REL:** Ca
See Appendix A
**OSHA PEL:** [1910.1007] See Appendix B

**Measurement Methods (see Table 1):**
**NIOSH** 5509
**OSHA** 65

**Physical Description:** Gray to purple, crystalline solid.

| Chemical & Physical Properties: | Personal Protection/Sanitation (see Table 2): | Respirator Recommendations (see Tables 3 and 4): |
|---|---|---|
| **MW:** 253.1<br>**BP:** 788°F<br>**Sol(59°F):** 0.07%<br>**Fl.P:** ?<br>**IP:** ?<br>**Sp.Gr:** ?<br>**VP:** ?<br>**MLT:** 271°F<br>**UEL:** ?<br>**LEL:** ? | **Skin:** Prevent skin contact<br>**Eyes:** Prevent eye contact<br>**Wash skin:** When contam/Daily<br>**Remove:** When wet or contam<br>**Change:** Daily<br>**Provide:** Eyewash<br>Quick drench | **NIOSH**<br>**¥:** ScbaF:Pd,Pp/SaF:Pd,Pp:AScba<br>**Escape:** 100F/ScbaE<br><br>**See Appendix E** (page 351) |

**Incompatibilities and Reactivities:** None reported

| Exposure Routes, Symptoms, Target Organs (see Table 5): | First Aid (see Table 6): |
|---|---|
| **ER:** Inh, Abs, Ing, Con<br>**SY:** Skin sens, derm; head, dizz; caustic burns; frequent urination, dysuria; hema; GI upset; upper resp infection; [carc]<br>**TO:** Bladder, liver, lung, skin, GI tract [in animals: liver & bladder cancer] | **Eye:** Irr immed<br>**Skin:** Soap wash immed<br>**Breath:** Resp support<br>**Swallow:** Medical attention immed |

D

## Dichlorodifluoromethane

| | | | | |
|---|---|---|---|---|
| **Dichlorodifluoromethane** | **Formula:** $CCl_2F_2$ | **CAS#:** 75-71-8 | **RTECS#:** PA8200000 | **IDLH:** 15,000 ppm |
| **Conversion:** 1 ppm = 4.95 mg/m$^3$ | **DOT:** 1028 126 | | | |

**Synonyms/Trade Names:** Difluorodichloromethane, Fluorocarbon 12, Freon® 12, Genetron® 12, Halon® 122, Propellant 12, Refrigerant 12

**Exposure Limits:**
**NIOSH REL:** TWA 1000 ppm (4950 mg/m$^3$)
**OSHA PEL:** TWA 1000 ppm (4950 mg/m$^3$)

**Measurement Methods (see Table 1):**
**NIOSH** 1018

**Physical Description:** Colorless gas with an ether-like odor at extremely high concentrations. [**Note:** Shipped as a liquefied compressed gas.]

| Chemical & Physical Properties: | Personal Protection/Sanitation (see Table 2): | Respirator Recommendations (see Tables 3 and 4): |
|---|---|---|
| **MW:** 120.9<br>**BP:** -22°F<br>**Sol(77°F):** 0.03%<br>**Fl.P:** NA<br>**IP:** 11.75 eV<br>**RGasD:** 4.2<br>**VP:** 5.7 atm<br>**FRZ:** -252°F<br>**UEL:** NA<br>**LEL:** NA<br>Nonflammable Gas | **Skin:** Frostbite<br>**Eyes:** Frostbite<br>**Wash skin:** N.R.<br>**Remove:** N.R.<br>**Change:** N.R.<br>**Provide:** Frostbite wash | **NIOSH/OSHA**<br>**10,000 ppm:** Sa<br>**15,000 ppm:** Sa:Cf/ScbaF/SaF<br>**§:** ScbaF:Pd,Pp/SaF:Pd,Pp:AScba<br>**Escape:** GmFOv/ScbaE |

**Incompatibilities and Reactivities:** Chemically-active metals such as sodium, potassium, calcium, powdered aluminum, zinc & magnesium

| Exposure Routes, Symptoms, Target Organs (see Table 5): | First Aid (see Table 6): |
|---|---|
| **ER:** Inh, Con (liquid)<br>**SY:** Dizz, tremor, asphy, uncon, card arrhy, card arrest; liquid: frostbite<br>**TO:** CVS, PNS | **Eye:** Frostbite<br>**Skin:** Frostbite<br>**Breath:** Resp support |

## 1,3-Dichloro-5,5-dimethylhydantoin

| | | | | |
|---|---|---|---|---|
| **1,3-Dichloro-5,5-dimethylhydantoin** | **Formula:** $C_5H_6Cl_2N_2O_2$ | **CAS#:** 118-52-5 | **RTECS#:** MU0700000 | **IDLH:** 5 mg/m$^3$ |
| **Conversion:** | **DOT:** | | | |

**Synonyms/Trade Names:** Dactin, DDH, Halane

**Exposure Limits:**
**NIOSH REL:** TWA 0.2 mg/m$^3$
ST 0.4 mg/m$^3$
**OSHA PEL†:** TWA 0.2 mg/m$^3$

**Measurement Methods (see Table 1):**
None available

**Physical Description:** White powder with a chlorine-like odor.

| Chemical & Physical Properties: | Personal Protection/Sanitation (see Table 2): | Respirator Recommendations (see Tables 3 and 4): |
|---|---|---|
| **MW:** 197.0<br>**BP:** ?<br>**Sol:** 0.2%<br>**Fl.P:** 346°F<br>**IP:** ?<br>**Sp.Gr:** 1.5<br>**VP:** ?<br>**MLT:** 270°F<br>**UEL:** ?<br>**LEL:** ?<br>Combustible Solid | **Skin:** Prevent skin contact<br>**Eyes:** Prevent eye contact<br>**Wash skin:** When contam<br>**Remove:** When wet or contam<br>**Change:** Daily<br>**Provide:** Eyewash | **NIOSH/OSHA**<br>**2 mg/m$^3$:** Sa<br>**5 mg/m$^3$:** Sa:Cf/ScbaF/SaF<br>**§:** ScbaF:Pd,Pp/SaF:Pd,Pp:AScba<br>**Escape:** GmFS100/ScbaE |

**Incompatibilities and Reactivities:** Water, strong acids, easily oxidized materials such as ammonia salts & sulfides

| Exposure Routes, Symptoms, Target Organs (see Table 5): | First Aid (see Table 6): |
|---|---|
| **ER:** Inh, Ing, Con<br>**SY:** Irrit eyes, muc memb, resp sys<br>**TO:** Eyes, resp sys | **Eye:** Irr immed<br>**Skin:** Soap wash prompt<br>**Breath:** Resp support<br>**Swallow:** Medical attention immed |

D

## 1,1-Dichloroethane

| Formula: | CAS#: | RTECS#: | IDLH: |
|---|---|---|---|
| $CHCl_2CH_3$ | 75-34-3 | KI0175000 | 3000 ppm |

**Conversion:** 1 ppm = 4.05 mg/m$^3$ | **DOT:** 2362 130

**Synonyms/Trade Names:** Asymmetrical dichloroethane; Ethylidene chloride; 1,1-Ethylidene dichloride

**Exposure Limits:**
**NIOSH REL:** TWA 100 ppm (400 mg/m$^3$)
See Appendix C (Chloroethanes)
**OSHA PEL:** TWA 100 ppm (400 mg/m$^3$)

**Measurement Methods (see Table 1):**
**NIOSH** 1003
**OSHA** 7

**Physical Description:** Colorless, oily liquid with a chloroform-like odor.

**Chemical & Physical Properties:**
**MW:** 99.0
**BP:** 135°F
**Sol:** 0.6%
**Fl.P:** 2°F
**IP:** 11.06 eV
**Sp.Gr:** 1.18
**VP:** 182 mmHg
**FRZ:** -143°F
**UEL:** 11.4%
**LEL:** 5.4%
Class IB Flammable Liquid

**Personal Protection/Sanitation (see Table 2):**
**Skin:** Prevent skin contact
**Eyes:** Prevent eye contact
**Wash skin:** When contam
**Remove:** When wet (flamm)
**Change:** N.R.

**Respirator Recommendations (see Tables 3 and 4):**
**NIOSH/OSHA**
**1000 ppm:** Sa
**2500 ppm:** Sa:Cf
**3000 ppm:** ScbaF/SaF
**§:** ScbaF:Pd,Pp/SaF:Pd,Pp:AScba
**Escape:** GmFOv/ScbaE

**Incompatibilities and Reactivities:** Strong oxidizers, strong caustics

**Exposure Routes, Symptoms, Target Organs (see Table 5):**
**ER:** Inh, Ing, Con
**SY:** Irrit skin; CNS depres; liver, kidney, lung damage
**TO:** Skin, liver, kidneys, lungs, CNS

**First Aid (see Table 6):**
**Eye:** Irr immed
**Skin:** Soap flush prompt
**Breath:** Resp support
**Swallow:** Medical attention immed

## 1,2-Dichloroethylene

| Formula: | CAS#: | RTECS#: | IDLH: |
|---|---|---|---|
| ClCH=CHCl | 540-59-0 | KV9360000 | 1000 ppm |

**Conversion:** 1 ppm = 3.97 mg/m$^3$ | **DOT:** 1150 130P

**Synonyms/Trade Names:** Acetylene dichloride, cis-Acetylene dichloride, trans-Acetylene dichloride, sym-Dichloroethylene

**Exposure Limits:**
**NIOSH REL:** TWA 200 ppm (790 mg/m$^3$)
**OSHA PEL:** TWA 200 ppm (790 mg/m$^3$)

**Measurement Methods (see Table 1):**
**NIOSH** 1003
**OSHA** 7

**Physical Description:** Colorless liquid (usually a mixture of the cis & trans isomers) with a slightly acrid, chloroform-like odor.

**Chemical & Physical Properties:**
**MW:** 97.0
**BP:** 118-140°F
**Sol:** 0.4%
**Fl.P:** 36-39°F
**IP:** 9.65 eV
**Sp.Gr(77°F):** 1.27
**VP:** 180-265 mmHg
**FRZ:** -57 to -115°F
**UEL:** 12.8%
**LEL:** 5.6%
Class IB Flammable Liquid

**Personal Protection/Sanitation (see Table 2):**
**Skin:** Prevent skin contact
**Eyes:** Prevent eye contact
**Wash skin:** When contam
**Remove:** When wet (flamm)
**Change:** N.R.

**Respirator Recommendations (see Tables 3 and 4):**
**NIOSH/OSHA**
**1000 ppm:** Sa:Cf£/PaprOv£/CcrFOv/GmFOv/ScbaF/SaF
**§:** ScbaF:Pd,Pp/SaF:Pd,Pp:AScba
**Escape:** GmFOv/ScbaE

**Incompatibilities and Reactivities:** Strong oxidizers, strong alkalis, potassium hydroxide, copper [**Note:** Usually contains inhibitors to prevent polymerization.]

**Exposure Routes, Symptoms, Target Organs (see Table 5):**
**ER:** Inh, Ing, Con
**SY:** Irrit eyes, resp sys; CNS depres
**TO:** Eyes, resp sys, CNS

**First Aid (see Table 6):**
**Eye:** Irr immed
**Skin:** Soap wash prompt
**Breath:** Resp support
**Swallow:** Medical attention immed

D

## Dichloroethyl ether

| Formula: | CAS#: | RTECS#: | IDLH: |
|---|---|---|---|
| $(ClCH_2CH_2)_2O$ | 111-44-4 | KN0875000 | Ca [100 ppm] |

**Conversion:** 1 ppm = 5.85 mg/m$^3$ | **DOT:** 1916 152

**Synonyms/Trade Names:** bis(2-Chloroethyl)ether; 2,2'-Dichlorodiethyl ether, 2,2'-Dichloroethyl ether

**Exposure Limits:**
**NIOSH REL:** Ca
TWA 5 ppm (30 mg/m$^3$)
ST 10 ppm (60 mg/m$^3$) [skin]
See Appendix A
**OSHA PEL†:** TWA 15 ppm (90 mg/m$^3$) [skin]

**Measurement Methods (see Table 1):**
**NIOSH** 1004
**OSHA** 7

**Physical Description:** Colorless liquid with a chlorinated solvent-like odor.

**Chemical & Physical Properties:**
**MW:** 143.0
**BP:** 352°F
**Sol:** 1%
**Fl.P:** 131°F
**IP:** ?
**Sp.Gr:** 1.22
**VP:** 0.7 mmHg
**FRZ:** -58°F
**UEL:** ?
**LEL:** 2.7%
Class II Combustible Liquid

**Personal Protection/Sanitation (see Table 2):**
**Skin:** Prevent skin contact
**Eyes:** Prevent eye contact
**Wash skin:** When contam
**Remove:** When wet or contam
**Change:** N.R.
**Provide:** Eyewash
Quick drench

**Respirator Recommendations (see Tables 3 and 4):**
**NIOSH**
**¥:** ScbaF:Pd,Pp/SaF:Pd,Pp:AScba
**Escape:** GmFOv/ScbaE

**Incompatibilities and Reactivities:** Strong oxidizers
[**Note:** Decomposes in presence of moisture to form hydrochloric acid.]

**Exposure Routes, Symptoms, Target Organs (see Table 5):**
**ER:** Inh, Abs, Ing, Con
**SY:** Irrit nose, throat, resp sys; lac; cough; nau, vomit; in animals: pulm edema; liver damage; [carc]
**TO:** Eyes, resp sys, liver [in animals: liver tumors]

**First Aid (see Table 6):**
**Eye:** Irr immed
**Skin:** Soap wash
**Breath:** Resp support
**Swallow:** Medical attention immed

## Dichloromonofluoromethane

| Formula: | CAS#: | RTECS#: | IDLH: |
|---|---|---|---|
| $CHCl_2F$ | 75-43-4 | PA8400000 | 5000 ppm |

**Conversion:** 1 ppm = 4.21 mg/m$^3$ | **DOT:** 1029 126

**Synonyms/Trade Names:** Dichlorofluoromethane, Fluorodichloromethane, Freon® 21, Genetron® 21, Halon® 112, Refrigerant 21

**Exposure Limits:**
**NIOSH REL:** TWA 10 ppm (40 mg/m$^3$)
**OSHA PEL†:** TWA 1000 ppm (4200 mg/m$^3$)

**Measurement Methods (see Table 1):**
**NIOSH** 2516

**Physical Description:** Colorless gas with a slight, ether-like odor.
[**Note:** A liquid below 48°F. Shipped as a liquefied compressed gas.]

**Chemical & Physical Properties:**
**MW:** 102.9
**BP:** 48°F
**Sol(86°F):** 0.7%
**Fl.P:** NA
**IP:** 12.39 eV
**RGasD:** 3.57
**VP(70°F):** 1.6 atm
**FRZ:** -211°F
**UEL:** NA
**LEL:** NA
Nonflammable Gas

**Personal Protection/Sanitation (see Table 2):**
**Skin:** Frostbite
**Eyes:** Frostbite
**Wash skin:** N.R.
**Remove:** N.R.
**Change:** N.R.
**Provide:** Frostbite wash

**Respirator Recommendations (see Tables 3 and 4):**
**NIOSH**
**100 ppm:** Sa
**250 ppm:** Sa:Cf
**500 ppm:** ScbaF/SaF
**5000 ppm:** Sa:Pd,Pp
**§:** ScbaF:Pd,Pp/SaF:Pd,Pp:AScba
**Escape:** GmFOv/ScbaE

**Incompatibilities and Reactivities:** Chemically-active metals such as sodium, potassium, calcium, powdered aluminum, zinc & magnesium; acid; acid fumes

**Exposure Routes, Symptoms, Target Organs (see Table 5):**
**ER:** Inh, Con (liquid)
**SY:** Asphy, card arrhy, card arrest; liquid: frostbite
**TO:** Resp sys, CVS

**First Aid (see Table 6):**
**Eye:** Frostbite
**Skin:** Frostbite
**Breath:** Resp support

## 1,1-Dichloro-1-nitroethane

**Formula:** $CH_3CCl_2NO_2$ | **CAS#:** 594-72-9 | **RTECS#:** KI0500000 | **IDLH:** 25 ppm

**Conversion:** 1 ppm = 5.89 mg/m$^3$ | **DOT:** 2650 153

**Synonyms/Trade Names:** Dichloronitroethane

**Exposure Limits:**
**NIOSH REL:** TWA 2 ppm (10 mg/m$^3$)
**OSHA PEL†:** C 10 ppm (60 mg/m$^3$)

**Measurement Methods (see Table 1):**
NIOSH 1601
OSHA 7

**Physical Description:** Colorless liquid with an unpleasant odor. [fumigant]

D

**Chemical & Physical Properties:**
**MW:** 143.9
**BP:** 255°F
**Sol:** 0.3%
**Fl.P:** 136°F
**IP:** ?
**Sp.Gr:** 1.43
**VP:** 15 mmHg
**FRZ:** ?
**UEL:** ?
**LEL:** ?
Class II Combustible Liquid

**Personal Protection/Sanitation (see Table 2):**
**Skin:** Prevent skin contact
**Eyes:** Prevent eye contact
**Wash skin:** When contam
**Remove:** When wet or contam
**Change:** N.R.

**Respirator Recommendations (see Tables 3 and 4):**
**NIOSH**
**20 ppm:** Sa
**25 ppm:** Sa:Cf/ScbaF/SaF
**§:** ScbaF:Pd,Pp/SaF:Pd,Pp:AScba
**Escape:** GmFOv/ScbaE

**Incompatibilities and Reactivities:** Strong oxidizers [**Note:** Corrosive to iron in presence of moisture.]

**Exposure Routes, Symptoms, Target Organs (see Table 5):**
**ER:** Inh, Ing, Con
**SY:** In animals: irrit eyes, skin; liver, heart, kidney damage; pulm edema, hemorr
**TO:** Eyes, skin, resp sys, liver, kidneys, CVS

**First Aid (see Table 6):**
**Eye:** Irr immed
**Skin:** Soap wash immed
**Breath:** Resp support
**Swallow:** Medical attention immed

## 1,3-Dichloropropene

**Formula:** $ClHC{=}CHCH_2Cl$ | **CAS#:** 542-75-6 | **RTECS#:** UC8310000 | **IDLH:** Ca [N.D.]

**Conversion:** 1 ppm = 4.54 mg/m$^3$ | **DOT:** 2047 129

**Synonyms/Trade Names:** 3-Chloroallyl chloride; DCP; 1,3-Dichloro-1-propene; 1,3-Dichloropropylene; Telone®

**Exposure Limits:**
**NIOSH REL:** Ca
TWA 1 ppm (5 mg/m$^3$) [skin]
See Appendix A
**OSHA PEL†:** none

**Measurement Methods (see Table 1):**
None available

**Physical Description:** Colorless to straw-colored liquid with a sharp, sweet, irritating, chloroform-like odor. [insecticide] [**Note:** Exists as mixture of cis- & trans-isomers.]

**Chemical & Physical Properties:**
**MW:** 111.0
**BP:** 226°F
**Sol:** 0.2%
**Fl.P:** 77°F
**IP:** ?
**Sp.Gr:** 1.21
**VP:** 28 mmHg
**FRZ:** -119°F
**UEL:** 14.5%
**LEL:** 5.3%
Class IC Flammable Liquid

**Personal Protection/Sanitation (see Table 2):**
**Skin:** Prevent skin contact
**Eyes:** Prevent eye contact
**Wash skin:** When contam
**Remove:** When wet (flamm)
**Change:** N.R.
**Provide:** Eyewash
Quick drench

**Respirator Recommendations (see Tables 3 and 4):**
**NIOSH**
**¥:** ScbaF:Pd,Pp/SaF:Pd,Pp:AScba
**Escape:** GmFOv/ScbaE

**Incompatibilities and Reactivities:** Aluminum, magnesium, halogens, oxidizers [**Note:** Epichlorohydrin may be added as a stabilizer.]

**Exposure Routes, Symptoms, Target Organs (see Table 5):**
**ER:** Inh, Abs, Ing, Con
**SY:** Irrit eyes, skin, resp sys; eye, skin burns; lac; head, dizz; in animals; liver, kidney damage; [carc]
**TO:** Eyes, skin, resp sys, CNS, liver, kidneys [in animals: cancer of the bladder, liver, lung & forestomach]

**First Aid (see Table 6):**
**Eye:** Irr immed
**Skin:** Soap flush immed
**Breath:** Resp support
**Swallow:** Medical attention immed

D

| 2,2-Dichloropropionic acid | Formula: $CH_3CCl_2COOH$ | CAS#: 75-99-0 | RTECS#: UF0690000 | IDLH: N.D. |
|---|---|---|---|---|

**Conversion:** 1 ppm = 5.85 mg/$m^3$ | **DOT:**

**Synonyms/Trade Names:** Dalapon; 2,2-Dichloropropanoic acid; α,α-Dichloropropionic acid

**Exposure Limits:**
**NIOSH REL:** TWA 1 ppm (6 mg/$m^3$)
**OSHA PEL†:** none

**Measurement Methods (see Table 1):**
**OSHA** PV2017

**Physical Description:** Colorless liquid with an acrid odor. [herbicide] [**Note:** A white to tan powder below 46°F. The sodium salt, a white powder, is often used.]

| Chemical & Physical Properties: | Personal Protection/Sanitation (see Table 2): | Respirator Recommendations (see Tables 3 and 4): |
|---|---|---|
| **MW:** 143.0<br>**BP:** 374°F<br>**Sol:** 50%<br>**Fl.P:** NA<br>**IP:** ?<br>**Sp.Gr:** 1.40<br>**VP:** ?<br>**FRZ:** 46°F<br>**UEL:** NA<br>**LEL:** NA<br>Noncombustible Liquid | **Skin:** Prevent skin contact<br>**Eyes:** Prevent eye contact<br>**Wash skin:** When contam<br>**Remove:** When wet or contam<br>**Change:** N.R.<br>**Provide:** Eyewash<br>Quick drench | Not available. |

**Incompatibilities and Reactivities:** Metals
[**Note:** Very corrosive to aluminum & copper alloys. Reacts slowly in water to form hydrochloric & pyruvic acids.]

| Exposure Routes, Symptoms, Target Organs (see Table 5): | First Aid (see Table 6): |
|---|---|
| **ER:** Inh, Ing, Con<br>**SY:** Irrit eyes, skin, upper resp sys; skin burns; lass, loss of appetite, diarr, vomit, slowing of pulse; CNS depres<br>**TO:** Eyes, skin, resp sys, GI tract, CNS | **Eye:** Irr immed<br>**Skin:** Water wash immed<br>**Breath:** Resp support<br>**Swallow:** Medical attention immed |

| Dichlorotetrafluoroethane | Formula: $CClF_2CClF_2$ | CAS#: 76-14-2 | RTECS#: KI1101000 | IDLH: 15,000 ppm |
|---|---|---|---|---|

**Conversion:** 1 ppm = 6.99 mg/$m^3$ | **DOT:** 1958 126

**Synonyms/Trade Names:** 1,2-Dichlorotetrafluoroethane; Freon® 114; Genetron® 114; Halon® 242; Refrigerant 114

**Exposure Limits:**
**NIOSH REL:** TWA 1000 ppm (7000 mg/$m^3$)
**OSHA PEL:** TWA 1000 ppm (7000 mg/$m^3$)

**Measurement Methods (see Table 1):**
**NIOSH** 1018

**Physical Description:** Colorless gas with a faint, ether-like odor at high concentrations. [**Note:** A liquid below 38°F. Shipped as a liquefied compressed gas.]

| Chemical & Physical Properties: | Personal Protection/Sanitation (see Table 2): | Respirator Recommendations (see Tables 3 and 4): |
|---|---|---|
| **MW:** 170.9<br>**BP:** 38°F<br>**Sol:** 0.01%<br>**Fl.P:** NA<br>**IP:** 12.20 eV<br>**RGasD:** 5.93<br>**VP(70°F):** 1.9 atm<br>**FRZ:** -137°F<br>**UEL:** NA<br>**LEL:** NA<br>Nonflammable Gas | **Skin:** Frostbite<br>**Eyes:** Frostbite<br>**Wash skin:** N.R.<br>**Remove:** N.R.<br>**Change:** N.R.<br>**Provide:** Frostbite wash | **NIOSH/OSHA**<br>**10,000 ppm:** Sa<br>**15,000 ppm:** Sa:Cf/ScbaF/SaF<br>**§:** ScbaF:Pd,Pp/SaF:Pd,Pp:AScba<br>**Escape:** GmFOv/ScbaE |

**Incompatibilities and Reactivities:** Chemically-active metals such as sodium, potassium, calcium, powdered aluminum, zinc & magnesium; acids; acid fumes

| Exposure Routes, Symptoms, Target Organs (see Table 5): | First Aid (see Table 6): |
|---|---|
| **ER:** Inh, Con (liquid)<br>**SY:** Irrit resp sys; asphy; card arrhy, card arrest; liquid: frostbite<br>**TO:** Resp sys, CVS | **Eye:** Frostbite<br>**Skin:** Frostbite<br>**Breath:** Resp support |

D

## Dichlorvos

| Dichlorvos | Formula: $(CH_3O)_2P(O)OCH=CCl_2$ | CAS#: 62-73-7 | RTECS#: TC0350000 | IDLH: 100 mg/m$^3$ |
|---|---|---|---|---|
| **Conversion:** 1 ppm = 9.04 mg/m$^3$ | **DOT:** 2783 152 | | | |

**Synonyms/Trade Names:** DDVP; 2,2-Dichlorovinyl dimethyl phosphate

**Exposure Limits:**
**NIOSH REL:** TWA 1 mg/m$^3$ [skin]
**OSHA PEL:** TWA 1 mg/m$^3$ [skin]

**Measurement Methods (see Table 1):**
**NIOSH** P&CAM295 (II-5)
**OSHA** 62

**Physical Description:** Colorless to amber liquid with a mild, chemical odor.
[**Note:** Insecticide that may be absorbed on a dry carrier.]

**Chemical & Physical Properties:**
**MW:** 221.0
**BP:** Decomposes
**Sol:** 0.5%
**Fl.P:** >175°F
**IP:** ?
**Sp.Gr(77°F):** 1.42
**VP:** 0.01 mmHg
**FRZ:** ?
**UEL:** ?
**LEL:** ?
Class III Combustible Liquid

**Personal Protection/Sanitation (see Table 2):**
**Skin:** Prevent skin contact
**Eyes:** Prevent eye contact
**Wash skin:** When contam
**Remove:** When wet or contam
**Change:** N.R.

**Respirator Recommendations (see Tables 3 and 4):**
**NIOSH/OSHA**
**10 mg/m$^3$:** Sa
**25 mg/m$^3$:** Sa:Cf
**50 mg/m$^3$:** SaT:Cf/ScbaF/SaF
**100 mg/m$^3$:** Sa:Pd,Pp
**§:** ScbaF:Pd,Pp/SaF:Pd,Pp:AScba
**Escape:** GmFOv100/ScbaE

**Incompatibilities and Reactivities:** Strong acids, strong alkalis [**Note:** Corrosive to iron & mild steel.]

**Exposure Routes, Symptoms, Target Organs (see Table 5):**
**ER:** Inh, Abs, Ing, Con
**SY:** Irrit eyes, skin; miosis, ache eyes; rhin; head; chest tight, wheez, lar spasm, salv; cyan; anor, nau, vomit, diarr; sweat; musc fasc, para, dizz, ataxia; convuls; low BP, card irreg
**TO:** Eyes, skin, resp sys, CVS, CNS, blood chol

**First Aid (see Table 6):**
**Eye:** Irr immed
**Skin:** Soap wash immed
**Breath:** Resp support
**Swallow:** Medical attention immed

## Dicrotophos

| Dicrotophos | Formula: $C_8H_{16}NO_5P$ | CAS#: 141-66-2 | RTECS#: TC3850000 | IDLH: N.D. |
|---|---|---|---|---|
| **Conversion:** 1 ppm = 9.70 mg/m$^3$ | **DOT:** | | | |

**Synonyms/Trade Names:** Bidrin®, Carbicron®, 2-Dimethyl-cis-2-dimethylcarbamoyl-1-methylvinylphosphate

**Exposure Limits:**
**NIOSH REL:** TWA 0.25 mg/m$^3$ [skin]
**OSHA PEL†:** none

**Measurement Methods (see Table 1):**
**NIOSH** 5600

**Physical Description:** Yellow-brown liquid with a mild, ester odor. [insecticide]

**Chemical & Physical Properties:**
**MW:** 237.2
**BP:** 752°F
**Sol:** Miscible
**Fl.P:** >200°F
**IP:** ?
**Sp.Gr(59°F):** 1.22
**VP:** 0.0001 mmHg
**FRZ:** ?
**UEL:** ?
**LEL:** ?
Class IIIB Combustible Liquid

**Personal Protection/Sanitation (see Table 2):**
**Skin:** Prevent skin contact
**Eyes:** Prevent eye contact
**Wash skin:** When contam
**Remove:** When wet or contam
**Change:** Daily
**Provide:** Quick drench

**Respirator Recommendations (see Tables 3 and 4):**
Not available.

**Incompatibilities and Reactivities:** Metals [**Note:** Corrosive to cast iron, mild steel, brass & stainless steel.]

**Exposure Routes, Symptoms, Target Organs (see Table 5):**
**ER:** Inh, Abs, Ing, Con
**SY:** Head, nau, dizz, anxi, restless, musc twitch, lass, tremor, inco, vomit, abdom cramps, diarr; salv, sweat, lac, rhinitis; anor, mal
**TO:** CNS, blood chol

**First Aid (see Table 6):**
**Eye:** Irr immed
**Skin:** Water wash immed
**Breath:** Resp support
**Swallow:** Medical attention immed

D

| Dicyclopentadiene | Formula: $C_{10}H_{12}$ | CAS#: 77-73-6 | RTECS#: PC1050000 | IDLH: N.D. |
|---|---|---|---|---|
| **Conversion:** 1 ppm = 5.41 mg/m$^3$ | **DOT:** 2048 130 | | | |

**Synonyms/Trade Names:** Bicyclopentadiene; DCPD; 1,3-Dicyclopentadiene dimer; 3a,4,7,7a-Tetrahydro-4,7-methanoindene [**Note:** Exists in two stereoisomeric forms.]

**Exposure Limits:**
**NIOSH REL:** TWA 5 ppm (30 mg/m$^3$)
**OSHA PEL†:** none

**Measurement Methods (see Table 1):**
**OSHA** PV2098

**Physical Description:** Colorless, crystalline solid with a disagreeable, camphor-like odor. [**Note:** A liquid above 90°F.]

**Chemical & Physical Properties:**
**MW:** 132.2
**BP:** 342°F
**Sol:** 0.02%
**Fl.P(oc):** 90°F
**IP:** ?
**Sp.Gr:** 0.98 (Liquid at 95°F)
**VP:** 1.4 mmHg
**FRZ:** 90°F
**UEL:** 6.3%
**LEL:** 0.8%
Class IC Flammable Liquid
Combustible Solid

**Personal Protection/Sanitation (see Table 2):**
**Skin:** Prevent skin contact
**Eyes:** Prevent eye contact
**Wash skin:** When contam
**Remove:** When wet or contam
**Change:** Daily
**Provide:** Eyewash
Quick drench

**Respirator Recommendations (see Tables 3 and 4):**
Not available.

**Incompatibilities and Reactivities:** Oxidizers
[**Note:** Depolymerizes at boiling point and forms two molecules of cyclopentadiene. Must be inhibited and maintained under an inert atmosphere to prevent polymerization.]

**Exposure Routes, Symptoms, Target Organs (see Table 5):**
**ER:** Inh, Ing, Con
**SY:** Irrit eyes, skin, nose, throat; inco, head; sneez, cough; skin blisters; in animals: kidney, lung damage
**TO:** Eyes, skin, resp sys, CNS, kidneys

**First Aid (see Table 6):**
**Eye:** Irr immed
**Skin:** Soap flush immed
**Breath:** Resp support
**Swallow:** Medical attention immed

| Dicyclopentadienyl iron | Formula: $(C_5H_5)_2Fe$ | CAS#: 102-54-5 | RTECS#: LK0700000 | IDLH: N.D. |
|---|---|---|---|---|
| **Conversion:** | **DOT:** | | | |

**Synonyms/Trade Names:** bis(Cyclopentadienyl)iron, Ferrocene, Iron dicyclopentadienyl

**Exposure Limits:**
**NIOSH REL:** TWA 10 mg/m$^3$ (total)
TWA 5 mg/m$^3$ (resp)
**OSHA PEL†:** TWA 15 mg/m$^3$ (total)
TWA 5 mg/m$^3$ (resp)

**Measurement Methods (see Table 1):**
**OSHA** ID125G

**Physical Description:** Orange, crystalline solid with a camphor-like odor.

**Chemical & Physical Properties:**
**MW:** 186.1
**BP:** 480°F
**Sol:** Insoluble
**Fl.P:** ?
**IP:** 6.88 eV
**Sp.Gr:** ?
**VP:** ?
**MLT:** 343°F
**UEL:** ?
**LEL:** ?
Combustible Solid

**Personal Protection/Sanitation (see Table 2):**
**Skin:** N.R.
**Eyes:** N.R.
**Wash skin:** N.R.
**Remove:** N.R.
**Change:** Daily

**Respirator Recommendations (see Tables 3 and 4):**
Not available.

**Incompatibilities and Reactivities:** Ammonium perchlorate, tetranitromethane, mercury(II) nitrate

**Exposure Routes, Symptoms, Target Organs (see Table 5):**
**ER:** Inh, Ing, Con
**SY:** Possible irrit eyes, skin, resp sys; in animals: liver, RBC, testicular changes
**TO:** Eyes, skin, resp sys, liver, blood, repro sys

**First Aid (see Table 6):**
**Eye:** Irr immed
**Skin:** Soap wash
**Breath:** Resp support
**Swallow:** Medical attention immed

## Dieldrin

| Dieldrin | Formula: $C_{12}H_8Cl_6O$ | CAS#: 60-57-1 | RTECS#: IO1750000 | IDLH: Ca [50 mg/m$^3$] |
|---|---|---|---|---|
| **Conversion:** | **DOT:** 2761 151 | | | |

**Synonyms/Trade Names:** HEOD; 1,2,3,4,10,10-Hexachloro-6,7-epoxy-1,4,4a,5,6,7,8,8a-octahydro-1,4-endo,exo-5,8-dimethanonaphthalene

**Exposure Limits:**
**NIOSH REL:** Ca
TWA 0.25 mg/m$^3$ [skin]
See Appendix A
**OSHA PEL:** TWA 0.25 mg/m$^3$ [skin]

**Measurement Methods (see Table 1):**
**NIOSH** S283 (II-3)

**Physical Description:** Colorless to light-tan crystals with a mild, chemical odor. [insecticide]

| Chemical & Physical Properties: | Personal Protection/Sanitation (see Table 2): | Respirator Recommendations (see Tables 3 and 4): |
|---|---|---|
| **MW:** 380.9<br>**BP:** Decomposes<br>**Sol:** 0.02%<br>**Fl.P:** NA<br>**IP:** ?<br>**Sp.Gr:** 1.75<br>**VP(77°F):** 8 x 10$^{-7}$ mmHg<br>**MLT:** 349°F<br>**UEL:** NA<br>**LEL:** NA<br>Noncombustible Solid | **Skin:** Prevent skin contact<br>**Eyes:** Prevent eye contact<br>**Wash skin:** When contam/Daily<br>**Remove:** When wet or contam<br>**Change:** Daily<br>**Provide:** Eyewash<br>Quick drench | **NIOSH**<br>**¥:** ScbaF:Pd,Pp/SaF:Pd,Pp:AScba<br>**Escape:** GmFOv100/ScbaE |

**Incompatibilities and Reactivities:** Strong oxidizers, active metals such as sodium, strong acids, phenols

**Exposure Routes, Symptoms, Target Organs (see Table 5):**
**ER:** Inh, Abs, Ing, Con
**SY:** Head, dizz; nau, vomit, mal, sweat; myoclonic limb jerks; clonic, tonic convuls; coma; [carc]; in animals: liver, kidney damage
**TO:** CNS, liver, kidneys, skin [in animals: lung, liver, thyroid & adrenal gland tumors]

**First Aid (see Table 6):**
**Eye:** Irr immed
**Skin:** Soap wash immed
**Breath:** Resp support
**Swallow:** Medical attention immed

## Diesel exhaust

| Diesel exhaust | Formula: | CAS#: | RTECS#: HZ1755000 | IDLH: Ca [N.D.] |
|---|---|---|---|---|
| **Conversion:** | **DOT:** | | | |

**Synonyms/Trade Names:** Synonyms vary depending upon the specific diesel exhaust component.

**Exposure Limits:**
**NIOSH REL:** Ca
See Appendix A
**OSHA PEL:** none

**Measurement Methods (see Table 1):**
**NIOSH** 2560, 5040

**Physical Description:** Appearance and odor vary depending upon the specific diesel exhaust component.

| Chemical & Physical Properties: | Personal Protection/Sanitation (see Table 2): | Respirator Recommendations (see Tables 3 and 4): |
|---|---|---|
| Properties vary depending upon the specific component diesel exhaust component. | **Skin:** N.R.<br>**Eyes:** N.R.<br>**Wash skin:** N.R.<br>**Remove:** N.R.<br>**Change:** N.R. | **NIOSH**<br>**¥:** ScbaF:Pd,Pp/SaF:Pd,Pp:AScba<br>**Escape:** GmFOv100/ScbaE |

**Incompatibilities and Reactivities:** Varies

**Exposure Routes, Symptoms, Target Organs (see Table 5):**
**ER:** Inh, Con
**SY:** Eye irrit, pulm func changes; [carc]
**TO:** Eyes, resp sys [in animals: lung tumors]

**First Aid (see Table 6):**
**Breath:** Resp support

## Diethanolamine

| Formula: | CAS#: | RTECS#: | IDLH: |
|---|---|---|---|
| $(HOCH_2CH_2)_2NH$ | 111-42-2 | KL2975000 | N.D. |

**Conversion:** 1 ppm = 4.30 mg/m$^3$

**DOT:**

**Synonyms/Trade Names:** DEA; Di(2-hydroxyethyl)amine; 2,2'-Dihydroxydiethyamine; Diolamine; bis(2-Hydroxyethyl)amine; 2,2'-Iminodiethanol

**Exposure Limits:**
**NIOSH REL:** TWA 3 ppm (15 mg/m$^3$)
**OSHA PEL†:** none

**Measurement Methods (see Table 1):**
**NIOSH** 3509
**OSHA** PV2018

**Physical Description:** Colorless crystals or a syrupy, white liquid (above 82°F) with a mild, ammonia-like odor.

**Chemical & Physical Properties:**
**MW:** 105.2
**BP:** 516°F (Decomposes)
**Sol:** 95%
**Fl.P:** 279°F
**IP:** ?
**Sp.Gr:** 1.10
**VP:** <0.01 mmHg
**MLT:** 82°F
**UEL:** 9.8%
**LEL:** 1.6%
Class IIIB Combustible Liquid
Combustible Solid

**Personal Protection/Sanitation (see Table 2):**
**Skin:** Prevent skin contact
**Eyes:** Prevent eye contact
**Wash skin:** When contam
**Remove:** When wet or contam
**Change:** Daily
**Provide:** Eyewash
Quick drench

**Respirator Recommendations (see Tables 3 and 4):**
Not available.

**Incompatibilities and Reactivities:** Oxidizers, strong acids, acid anhydrides, halides
[**Note:** Reacts with $CO_2$ in the air. Hygroscopic (i.e., absorbs moisture from the air). Corrosive to copper, zinc, and galvanized iron.]

**Exposure Routes, Symptoms, Target Organs (see Table 5):**
**ER:** Inh, Ing, Con
**SY:** Irrit eyes, skin, nose, throat; eye burns, corn nec; skin burns; lac, cough, sneez
**TO:** Eyes, skin, resp sys

**First Aid (see Table 6):**
**Eye:** Irr immed
**Skin:** Water flush immed
**Breath:** Resp support
**Swallow:** Medical attention immed

## Diethylamine

| Formula: | CAS#: | RTECS#: | IDLH: |
|---|---|---|---|
| $(C_2H_5)_2NH$ | 109-89-7 | HZ8750000 | 200 ppm |

**Conversion:** 1 ppm = 2.99 mg/m$^3$

**DOT:** 1154 132

**Synonyms/Trade Names:** Diethamine; N,N-Diethylamine; N-Ethylethanamine

**Exposure Limits:**
**NIOSH REL:** TWA 10 ppm (30 mg/m$^3$)
ST 25 ppm (75 mg/m$^3$)
**OSHA PEL†:** TWA 25 ppm (75 mg/m$^3$)

**Measurement Methods (see Table 1):**
**NIOSH** 2010
**OSHA** 41

**Physical Description:** Colorless liquid with a fishy, ammonia-like odor.

**Chemical & Physical Properties:**
**MW:** 73.1
**BP:** 132°F
**Sol:** Miscible
**Fl.P:** -15°F
**IP:** 8.01 eV
**Sp.Gr:** 0.71
**VP:** 192 mmHg
**FRZ:** -58°F
**UEL:** 10.1%
**LEL:** 1.8%
Class IB Flammable Liquid

**Personal Protection/Sanitation (see Table 2):**
**Skin:** Prevent skin contact
**Eyes:** Prevent eye contact
**Wash skin:** When contam
**Remove:** When wet (flamm)
**Change:** N.R.
**Provide:** Eyewash (>0.5%)
Quick drench (liquid)

**Respirator Recommendations (see Tables 3 and 4):**
**NIOSH**
**200 ppm:** Sa:Cf£/PaprS£/CcrFS/GmFS/ScbaF/SaF
**§:** ScbaF:Pd,Pp/SaF:Pd,Pp:AScba
**Escape:** GmFS/ScbaE

**Incompatibilities and Reactivities:** Strong oxidizers, strong acids, cellulose nitrate

**Exposure Routes, Symptoms, Target Organs (see Table 5):**
**ER:** Inh, Abs, Ing, Con
**SY:** Irrit eyes, skin, resp sys; in animals; myocardial degeneration
**TO:** Eyes, skin, resp sys, CVS

**First Aid (see Table 6):**
**Eye:** Irr immed
**Skin:** Water flush immed
**Breath:** Resp support
**Swallow:** Medical attention immed

D

## 2-Diethylaminoethanol

| | | | | |
|---|---|---|---|---|
| **2-Diethylaminoethanol** | **Formula:** $(C_2H_5)_2NCH_2CH_2OH$ | **CAS#:** 100-37-8 | **RTECS#:** KK5075000 | **IDLH:** 100 ppm |
| **Conversion:** 1 ppm = 4.79 mg/m$^3$ | **DOT:** 2686 132 | | | |

**Synonyms/Trade Names:** Diethylaminoethanol; 2-Diethylaminoethyl alcohol; N,N-Diethylethanolamine; Diethyl-(2-hydroxyethyl)amine; 2-Hydroxytriethylamine

**Exposure Limits:**
**NIOSH REL:** TWA 10 ppm (50 mg/m$^3$) [skin]
**OSHA PEL:** TWA 10 ppm (50 mg/m$^3$) [skin]

**Measurement Methods (see Table 1):**
**NIOSH** 2007

**Physical Description:** Colorless liquid with a nauseating, ammonia-like odor.

| **Chemical & Physical Properties:** | **Personal Protection/Sanitation (see Table 2):** | **Respirator Recommendations (see Tables 3 and 4):** |
|---|---|---|
| **MW:** 117.2<br>**BP:** 325°F<br>**Sol:** Miscible<br>**Fl.P:** 126°F<br>**IP:** ?<br>**Sp.Gr:** 0.89<br>**VP:** 1 mmHg<br>**FRZ:** -94°F<br>**UEL:** 11.7%<br>**LEL:** 6.7%<br>Class II Combustible Liquid | **Skin:** Prevent skin contact<br>**Eyes:** Prevent eye contact<br>**Wash skin:** When contam<br>**Remove:** When wet or contam<br>**Change:** N.R.<br>**Provide:** Eyewash (>5%)<br>Quick drench | **NIOSH/OSHA**<br>**100 ppm:** CcrOv*/GmFOv/PaprOv*/Sa*/ScbaF<br>**§:** ScbaF:Pd,Pp/SaF:Pd,Pp:AScba<br>**Escape:** GmFOv/ScbaE |

**Incompatibilities and Reactivities:** Strong oxidizers, strong acids

| **Exposure Routes, Symptoms, Target Organs (see Table 5):** | **First Aid (see Table 6):** |
|---|---|
| **ER:** Inh, Abs, Ing, Con<br>**SY:** Irrit eyes, skin, resp sys; nau, vomit<br>**TO:** Eyes, skin, resp sys | **Eye:** Irr immed<br>**Skin:** Water flush immed<br>**Breath:** Resp support<br>**Swallow:** Medical attention immed |

## Diethylenetriamine

| | | | | |
|---|---|---|---|---|
| **Diethylenetriamine** | **Formula:** $(NH_2CH_2CH_2)_2NH$ | **CAS#:** 111-40-0 | **RTECS#:** IE1225000 | **IDLH:** N.D. |
| **Conversion:** 1 ppm = 4.22 mg/m$^3$ | **DOT:** 2079 154 | | | |

**Synonyms/Trade Names:** N-(2-Aminoethyl)-1,2-ethanediamine; bis(2-Aminoethyl)amine; DETA; 2,2'-Diaminodiethylamine

**Exposure Limits:**
**NIOSH REL:** TWA 1 ppm (4 mg/m$^3$) [skin]
**OSHA PEL†:** none

**Measurement Methods (see Table 1):**
**NIOSH** 2540
**OSHA** 60

**Physical Description:** Colorless to yellow liquid with a strong, ammonia-like odor.
[**Note:** Hygroscopic (i.e., absorbs moisture from the air).]

| **Chemical & Physical Properties:** | **Personal Protection/Sanitation (see Table 2):** | **Respirator Recommendations (see Tables 3 and 4):** |
|---|---|---|
| **MW:** 103.2<br>**BP:** 405°F<br>**Sol:** Miscible<br>**Fl.P:** 208°F<br>**IP:** ?<br>**Sp.Gr:** 0.96<br>**VP:** 0.4 mmHg<br>**FRZ:** -38°F<br>**UEL:** 6.7%<br>**LEL:** 2%<br>Class IIIB Combustible Liquid | **Skin:** Prevent skin contact<br>**Eyes:** Prevent eye contact<br>**Wash skin:** When contam<br>**Remove:** When wet or contam<br>**Change:** N.R.<br>**Provide:** Eyewash<br>Quick drench | Not available. |

**Incompatibilities and Reactivities:** Oxidizers, strong acids, cellulose nitrate
[**Note:** May form explosive complexes with silver, cobalt, or chromium compounds. Corrosive to aluminum, copper, brass & zinc.]

| **Exposure Routes, Symptoms, Target Organs (see Table 5):** | **First Aid (see Table 6):** |
|---|---|
| **ER:** Inh, Abs, Ing, Con<br>**SY:** Irrit eyes, skin, muc memb, upper resp sys; derm, skin sens; eye, skin nec; cough, dysp, pulm sens<br>**TO:** Eyes, skin, resp sys | **Eye:** Irr immed<br>**Skin:** Water flush immed<br>**Breath:** Resp support<br>**Swallow:** Medical attention immed |

D

| Diethyl ketone | Formula: $CH_3CH_2COCH_2CH_3$ | CAS#: 96-22-0 | RTECS#: SA8050000 | IDLH: N.D. |
|---|---|---|---|---|
| **Conversion:** 1 ppm = 3.53 mg/m$^3$ | **DOT:** 1156 127 | | | |

**Synonyms/Trade Names:** DEK, Dimethylacetone, Ethyl ketone, Metacetone, 3-Pentanone, Propione

**Exposure Limits:**
**NIOSH REL:** TWA 200 ppm (705 mg/m$^3$)
**OSHA PEL†:** none

**Measurement Methods (see Table 1):**
None available

**Physical Description:** Colorless liquid with an acetone-like odor.

| Chemical & Physical Properties: | Personal Protection/Sanitation (see Table 2): | Respirator Recommendations (see Tables 3 and 4): |
|---|---|---|
| **MW:** 86.2<br>**BP:** 215°F<br>**Sol:** 5%<br>**Fl.P(oc):** 55°F<br>**IP:** 9.32 eV<br>**Sp.Gr:** 0.81<br>**VP(77°F):** 35 mmHg<br>**FRZ:** -44°F<br>**UEL:** 6.4%<br>**LEL:** 1.6%<br>Class IB Flammable Liquid | **Skin:** N.R.<br>**Eyes:** Prevent eye contact<br>**Wash skin:** Daily<br>**Remove:** When wet (flamm)<br>**Change:** N.R. | Not available. |

**Incompatibilities and Reactivities:** Strong oxidizers, alkalis, mineral acids, (hydrogen peroxide + nitric acid)

| Exposure Routes, Symptoms, Target Organs (see Table 5): | First Aid (see Table 6): |
|---|---|
| **ER:** Inh, Ing, Con<br>**SY:** Irrit eyes, skin, muc memb, resp sys; cough, sneez<br>**TO:** Eyes, skin, resp sys | **Eye:** Irr immed<br>**Skin:** Soap wash<br>**Breath:** Resp support<br>**Swallow:** Medical attention immed |

| Diethyl phthalate | Formula: $C_6H_4(COOC_2H_5)_2$ | CAS#: 84-66-2 | RTECS#: TI1050000 | IDLH: N.D. |
|---|---|---|---|---|
| **Conversion:** | **DOT:** | | | |

**Synonyms/Trade Names:** DEP, Diethyl ester of phthalic acid, Ethyl phthalate

**Exposure Limits:**
**NIOSH REL:** TWA 5 mg/m$^3$
**OSHA PEL†:** none

**Measurement Methods (see Table 1):**
OSHA 104

**Physical Description:** Colorless to water-white, oily liquid with a very slight, aromatic odor. [pesticide]

| Chemical & Physical Properties: | Personal Protection/Sanitation (see Table 2): | Respirator Recommendations (see Tables 3 and 4): |
|---|---|---|
| **MW:** 222.3<br>**BP:** 563°F<br>**Sol(77°F):** 0.1%<br>**Fl.P(oc):** 322°F<br>**IP:** ?<br>**Sp.Gr:** 1.12<br>**VP(77°F):** 0.002 mmHg<br>**FRZ:** -41°F<br>**UEL:** ?<br>**LEL(368°F):** 0.7%<br>Class IIIB Combustible Liquid; however, ignition is difficult. | **Skin:** N.R.<br>**Eyes:** N.R.<br>**Wash skin:** N.R.<br>**Remove:** N.R.<br>**Change:** N.R. | Not available. |

**Incompatibilities and Reactivities:** Strong oxidizers, strong acids, nitric acid, permanganates, water

| Exposure Routes, Symptoms, Target Organs (see Table 5): | First Aid (see Table 6): |
|---|---|
| **ER:** Inh, Ing, Con<br>**SY:** Irrit eyes, skin, nose, throat; head, dizz, nau; lac; possible polyneur, vestibular dysfunc; pain, numb, lass, spasms in arms & legs; in animals: possible repro effects<br>**TO:** Eyes, skin, resp sys, CNS, PNS, repro sys | **Eye:** Irr immed<br>**Skin:** Wash regularly<br>**Breath:** Resp support<br>**Swallow:** Medical attention immed |

## Difluorodibromomethane

| Formula: | CAS#: | RTECS#: | IDLH: |
|---|---|---|---|
| $CBr_2F_2$ | 75-61-6 | PA7525000 | 2000 ppm |

**Conversion:** 1 ppm = 8.58 mg/m$^3$ **DOT:** 1941 171

**Synonyms/Trade Names:** Dibromodifluoromethane, Freon® 12B2, Halon® 1202

**Exposure Limits:**
**NIOSH REL:** TWA 100 ppm (860 mg/m$^3$)
**OSHA PEL:** TWA 100 ppm (860 mg/m$^3$)

**Measurement Methods (see Table 1):**
**NIOSH** 1012
**OSHA** 7

**Physical Description:** Colorless, heavy liquid or gas (above 76°F) with a characteristic odor.

**Chemical & Physical Properties:**
**MW:** 209.8
**BP:** 76°F
**Sol:** Insoluble
**Fl.P:** NA
**IP:** 11.07 eV
**Sp.Gr(59°F):** 2.29
**VP:** 620 mmHg
**FRZ:** -231°F
**UEL:** NA
**LEL:** NA
Noncombustible Liquid
Nonflammable Gas

**Personal Protection/Sanitation (see Table 2):**
**Skin:** Prevent skin contact
**Eyes:** Prevent eye contact
**Wash skin:** N.R.
**Remove:** When wet or contam
**Change:** N.R.

**Respirator Recommendations (see Tables 3 and 4):**
**NIOSH/OSHA**
**1000 ppm:** Sa
**2000 ppm:** Sa:Cf/ScbaF/SaF
**§:** ScbaF:Pd,Pp/SaF:Pd,Pp:AScba
**Escape:** GmFOv/ScbaE

**Incompatibilities and Reactivities:** Chemically-active metals such as sodium, potassium, calcium, powdered aluminum, zinc & magnesium

**Exposure Routes, Symptoms, Target Organs (see Table 5):**
**ER:** Inh, Ing, Con
**SY:** In animals: irrit resp sys; CNS symptoms; liver damage
**TO:** Resp sys, CNS, liver

**First Aid (see Table 6):**
**Eye:** Irr immed
**Skin:** Water flush immed
**Breath:** Resp support
**Swallow:** Medical attention immed

D

## Diglycidyl ether

| Formula: | CAS#: | RTECS#: | IDLH: |
|---|---|---|---|
| $C_6H_{10}O_3$ | 2238-07-5 | KN2350000 | Ca [10 ppm] |

**Conversion:** 1 ppm = 5.33 mg/m$^3$ **DOT:**

**Synonyms/Trade Names:** Diallyl ether dioxide; DGE; Di(2,3-epoxypropyl) ether; 2-Epoxypropyl ether; bis(2,3-Epoxypropyl) ether

**Exposure Limits:**
**NIOSH REL:** Ca
TWA 0.1 ppm (0.5 mg/m$^3$)
See Appendix A
**OSHA PEL†:** C 0.5 ppm (2.8 mg/m$^3$)

**Measurement Methods (see Table 1):**
None available

**Physical Description:** Colorless liquid with a strong, irritating odor.

**Chemical & Physical Properties:**
**MW:** 130.2
**BP:** 500°F
**Sol:** ?
**Fl.P:** 147°F
**IP:** ?
**Sp.Gr:** 1.12
**VP(77°F):** 0.09 mmHg
**FRZ:** ?
**UEL:** ?
**LEL:** ?
Class IIIA Combustible Liquid

**Personal Protection/Sanitation (see Table 2):**
**Skin:** Prevent skin contact
**Eyes:** Prevent eye contact
**Wash skin:** When contam/Daily
**Remove:** When wet or contam
**Change:** Daily
**Provide:** Eyewash
Quick drench

**Respirator Recommendations (see Tables 3 and 4):**
**NIOSH**
**¥:** ScbaF:Pd,Pp/SaF:Pd,Pp:AScba
**Escape:** GmFOv/ScbaE

**Incompatibilities and Reactivities:** Strong oxidizers

**Exposure Routes, Symptoms, Target Organs (see Table 5):**
**ER:** Inh, Abs, Ing, Con
**SY:** Irrit eyes, skin, resp sys; skin burns; in animals: hemato sys, lung, liver, kidney damage; repro effects; [carc]
**TO:** Eyes, skin, resp sys, repro sys [in animals: skin tumors]

**First Aid (see Table 6):**
**Eye:** Irr immed
**Skin:** Soap wash immed
**Breath:** Resp support
**Swallow:** Medical attention immed

D

## Diisobutyl ketone

**Formula:** $[(CH_3)_2CHCH_2]_2CO$ | **CAS#:** 108-83-8 | **RTECS#:** MJ5775000 | **IDLH:** 500 ppm

**Conversion:** 1 ppm = 5.82 mg/m$^3$ | **DOT:** 1157 128

**Synonyms/Trade Names:** DIBK; sym-Diisopropyl acetone; 2,6-Dimethyl-4-heptanone; Isovalerone; Valerone

**Exposure Limits:**
**NIOSH REL:** TWA 25 ppm (150 mg/m$^3$)
**OSHA PEL†:** TWA 50 ppm (290 mg/m$^3$)

**Measurement Methods (see Table 1):**
**NIOSH** 1300, 2555
**OSHA** 7

**Physical Description:** Colorless liquid with a mild, sweet odor.

**Chemical & Physical Properties:**
**MW:** 142.3
**BP:** 334°F
**Sol:** 0.05%
**Fl.P:** 120°F
**IP:** 9.04 eV
**Sp.Gr:** 0.81
**VP:** 2 mmHg
**FRZ:** -43°F
**UEL(200°F):** 7.1%
**LEL(200°F):** 0.8%
Class II Combustible Liquid

**Personal Protection/Sanitation (see Table 2):**
**Skin:** Prevent skin contact
**Eyes:** N.R.
**Wash skin:** When contam
**Remove:** When wet or contam
**Change:** N.R.

**Respirator Recommendations (see Tables 3 and 4):**
**NIOSH**
**500 ppm:** Sa:Cf£/PaprOv£/CcrFOv/GmFOv/ScbaF/SaF
**§:** ScbaF:Pd,Pp/SaF:Pd,Pp:AScba
**Escape:** GmFOv/ScbaE

**Incompatibilities and Reactivities:** Strong oxidizers

**Exposure Routes, Symptoms, Target Organs (see Table 5):**
**ER:** Inh, Ing, Con
**SY:** Irrit eyes, skin, nose, throat; head, dizz; derm; liver, kidney damage
**TO:** Eyes, skin, resp sys, CNS, liver, kidneys

**First Aid (see Table 6):**
**Eye:** Irr immed
**Skin:** Soap wash prompt
**Breath:** Resp support
**Swallow:** Medical attention immed

## Diisopropylamine

**Formula:** $[(CH_3)_2CH]_2NH$ | **CAS#:** 108-18-9 | **RTECS#:** IM4025000 | **IDLH:** 200 ppm

**Conversion:** 1 ppm = 4.14 mg/m$^3$ | **DOT:** 1158 132

**Synonyms/Trade Names:** DIPA, N-(1-Methylethyl)-2-propanamine

**Exposure Limits:**
**NIOSH REL:** TWA 5 ppm (20 mg/m$^3$) [skin]
**OSHA PEL:** TWA 5 ppm (20 mg/m$^3$) [skin]

**Measurement Methods (see Table 1):**
**NIOSH** S141 (II-4)

**Physical Description:** Colorless liquid with an ammonia- or fish-like odor.

**Chemical & Physical Properties:**
**MW:** 101.2
**BP:** 183°F
**Sol:** Miscible
**Fl.P:** 20°F
**IP:** 7.73 eV
**Sp.Gr:** 0.72
**VP:** 70 mmHg
**FRZ:** -141°F
**UEL:** 7.1%
**LEL:** 1.1%
Class IB Flammable Liquid

**Personal Protection/Sanitation (see Table 2):**
**Skin:** Prevent skin contact
**Eyes:** Prevent eye contact (>5%)
**Wash skin:** When contam
**Remove:** When wet (flamm)
**Change:** N.R.
**Provide:** Eyewash (>5%)

**Respirator Recommendations (see Tables 3 and 4):**
**NIOSH/OSHA**
**125 ppm:** Sa:Cf£/PaprOv£
**200 ppm:** CcrFOv/GmFOv/PaprTOv£/ScbaF/SaF
**§:** ScbaF:Pd,Pp/SaF:Pd,Pp:AScba
**Escape:** GmFOv/ScbaE

**Incompatibilities and Reactivities:** Strong oxidizers, strong acids

**Exposure Routes, Symptoms, Target Organs (see Table 5):**
**ER:** Inh, Abs, Ing, Con
**SY:** Irrit eyes, skin, resp sys; nau, vomit; head; vis dist
**TO:** Eyes, skin, resp sys

**First Aid (see Table 6):**
**Eye:** Irr immed
**Skin:** Water wash immed
**Breath:** Resp support
**Swallow:** Medical attention immed

## Dimethyl acetamide

| Dimethyl acetamide | Formula: $CH_3CON(CH_3)_2$ | CAS#: 127-19-5 | RTECS#: AB7700000 | IDLH: 300 ppm |
|---|---|---|---|---|
| **Conversion:** 1 ppm = 3.56 mg/m$^3$ | **DOT:** | | | |

**Synonyms/Trade Names:** N,N-Dimethyl acetamide; DMAC

**Exposure Limits:**
**NIOSH REL:** TWA 10 ppm (35 mg/m$^3$) [skin]
**OSHA PEL:** TWA 10 ppm (35 mg/m$^3$) [skin]

**Measurement Methods (see Table 1):**
**NIOSH** 2004

**Physical Description:** Colorless liquid with a weak, ammonia- or fish-like odor.

| Chemical & Physical Properties: | Personal Protection/Sanitation (see Table 2): | Respirator Recommendations (see Tables 3 and 4): |
|---|---|---|
| **MW:** 87.1<br>**BP:** 329°F<br>**Sol:** Miscible<br>**Fl.P(oc):** 158°F<br>**IP:** 8.81 eV<br>**Sp.Gr:** 0.94<br>**VP:** 2 mmHg<br>**FRZ:** -4°F<br>**UEL(320°F):** 11.5%<br>**LEL(212°F):** 1.8%<br>Class IIIA Combustible Liquid | **Skin:** Prevent skin contact<br>**Eyes:** Prevent eye contact<br>**Wash skin:** When contam<br>**Remove:** When wet or contam<br>**Change:** N.R.<br>**Provide:** Quick drench | **NIOSH/OSHA**<br>**100 ppm:** Sa<br>**250 ppm:** Sa:Cf£<br>**300 ppm:** ScbaF/SaF<br>**§:** ScbaF:Pd,Pp/SaF:Pd,Pp:AScba<br>**Escape:** GmFOv/ScbaE |

**Incompatibilities and Reactivities:** Carbon tetrachloride, other halogenated compounds when in contact with iron, oxidizers

| Exposure Routes, Symptoms, Target Organs (see Table 5): | First Aid (see Table 6): |
|---|---|
| **ER:** Inh, Abs, Ing, Con<br>**SY:** Irrit skin; jaun, liver damage; depres, drow, halu, delusions<br>**TO:** Skin, liver, CNS | **Eye:** Irr immed<br>**Skin:** Water flush immed<br>**Breath:** Resp support<br>**Swallow:** Medical attention immed |

## Dimethylamine

| Dimethylamine | Formula: $(CH_3)_2NH$ | CAS#: 124-40-3 | RTECS#: IP8750000 | IDLH: 500 ppm |
|---|---|---|---|---|
| **Conversion:** 1 ppm = 1.85 mg/m$^3$ | **DOT:** 1032 118 (anhydrous); 1160 132 (solution) | | | |

**Synonyms/Trade Names:** Dimethylamine (anhydrous), N-Methylmethanamine

**Exposure Limits:**
**NIOSH REL:** TWA 10 ppm (18 mg/m$^3$)
**OSHA PEL:** TWA 10 ppm (18 mg/m$^3$)

**Measurement Methods (see Table 1):**
**NIOSH** 2010
**OSHA** 34

**Physical Description:** Colorless gas with an ammonia- or fish-like odor.
[**Note:** A liquid below 44°F. Shipped as a liquefied compressed gas.]

| Chemical & Physical Properties: | Personal Protection/Sanitation (see Table 2): | Respirator Recommendations (see Tables 3 and 4): |
|---|---|---|
| **MW:** 45.1<br>**BP:** 44°F<br>**Sol(140°F):** 24%<br>**Fl.P:** NA (Gas) 20°F (Liquid)<br>**IP:** 8.24 eV<br>**RGasD:** 1.56<br>**Sp.Gr:** 0.67 (Liquid at 44°F)<br>**VP:** 1.7 atm<br>**FRZ:** -134°F<br>**UEL:** 14.4%<br>**LEL:** 2.8%<br>Flammable Gas | **Skin:** Prevent skin contact (liquid)<br>Frostbite<br>**Eyes:** Prevent eye contact (liquid)<br>Frostbite<br>**Wash skin:** When contam (liquid)<br>**Remove:** When wet (flamm)<br>**Change:** N.R.<br>**Provide:** Eyewash (liquid)<br>Quick drench (liquid)<br>Frostbite wash | **NIOSH/OSHA**<br>**250 ppm:** Sa:Cf£<br>**500 ppm:** ScbaF/SaF<br>**§:** ScbaF:Pd,Pp/SaF:Pd,Pp:AScba<br>**Escape:** GmFS/ScbaE |

**Incompatibilities and Reactivities:** Strong oxidizers, chlorine, mercury, acraldehyde, fluorides, maleic anhydride, aluminum, brass, copper, zinc

| Exposure Routes, Symptoms, Target Organs (see Table 5): | First Aid (see Table 6): |
|---|---|
| **ER:** Inh, Con (liquid)<br>**SY:** Irrit nose, throat; sneez, cough, dysp; pulm edema; conj; derm; liquid: frostbite<br>**TO:** Eyes, skin, resp sys | **Eye:** Irr immed (liquid)/Frostbite<br>**Skin:** Water flush immed (liquid)/Frostbite<br>**Breath:** Resp support |

## 4-Dimethylaminoazobenzene

| Formula: | CAS#: | RTECS#: | IDLH: |
|---|---|---|---|
| $C_6H_5NNC_6H_4N(CH_3)_2$ | 60-11-7 | BX7350000 | Ca [N.D.] |

**Conversion:**

**DOT:**

**Synonyms/Trade Names:** Butter yellow; DAB; p-Dimethylaminoazobenzene; N,N-Dimethyl-4-aminoazobenzene; Methyl yellow

**Exposure Limits:**
**NIOSH REL:** Ca
See Appendix A
**OSHA PEL:** [1910.1015] See Appendix B

**Measurement Methods (see Table 1):**
**NIOSH** P&CAM284 (II-4)

**Physical Description:** Yellow, leaf-shaped crystals.

**Chemical & Physical Properties:**
**MW:** 225.3
**BP:** Sublimes
**Sol:** 0.001%
**Fl.P:** ?
**IP:** ?
**Sp.Gr:** ?
**VP:** 0.0000003 mmHg (est.)
**MLT:** 237°F
**UEL:** ?
**LEL:** ?

**Personal Protection/Sanitation (see Table 2):**
**Skin:** Prevent skin contact
**Eyes:** Prevent eye contact
**Wash skin:** When contam/Daily
**Remove:** When wet or contam
**Change:** Daily
**Provide:** Eyewash
Quick drench

**Respirator Recommendations (see Tables 3 and 4):**
**NIOSH**
**¥:** ScbaF:Pd,Pp/SaF:Pd,Pp:AScba
**Escape:** 100F/ScbaE

**See Appendix E** (page 351)

**Incompatibilities and Reactivities:** None reported

**Exposure Routes, Symptoms, Target Organs (see Table 5):**
**ER:** Inh, Abs, Ing, Con
**SY:** Enlarged liver; liver, kidney dist; contact derm; cough, wheez, dysp; bloody sputum; bronchial secretions; frequent urination, hema, dysuria; [carc]
**TO:** Skin, resp sys, liver, kidneys, bladder [in animals: liver & bladder tumors]

**First Aid (see Table 6):**
**Eye:** Irr immed
**Skin:** Soap wash immed
**Breath:** Resp support
**Swallow:** Medical attention immed

## bis(2-(Dimethylamino)ethyl)ether

| Formula: | CAS#: | RTECS#: | IDLH: |
|---|---|---|---|
| $C_8H_{20}N_2O$ | 3033-62-3 | KR9460000 | N.D. |

**Conversion:**

**DOT:**

**Synonyms/Trade Names:** NIAX® A99; NIAX® Catalyst A1; 2,2'-Oxybis(N,N-dimethyl ethylamine)
[**Note:** A component (5%) of NIAX® Catalyst ESN, along with dimethylaminopropionitrile (95%).]

**Exposure Limits:**
**NIOSH REL:** See Appendix C (NIAX® Catalyst ESN)
**OSHA PEL:** See Appendix C (NIAX® Catalyst ESN)

**Measurement Methods (see Table 1):**
None available

**Physical Description:** Liquid.

**Chemical & Physical Properties:**
**MW:** 160.3
**BP:** 372°F
**Sol:** ?
**Fl.P:** ?
**IP:** ?
**Sp.Gr:** ?
**VP:** ?
**FRZ:** ?
**UEL:** ?
**LEL:** ?

**Personal Protection/Sanitation (see Table 2):**
**Skin:** Prevent skin contact
**Eyes:** Prevent eye contact
**Wash skin:** When contam
**Remove:** When wet or contam
**Change:** N.R.
**Provide:** Eyewash
Quick drench

**Respirator Recommendations (see Tables 3 and 4):**
**NIOSH**
**¥:** ScbaF:Pd,Pp/SaF:Pd,Pp:AScba
**Escape:** GmFOv/ScbaE

**Incompatibilities and Reactivities:** None reported

**Exposure Routes, Symptoms, Target Organs (see Table 5):**
**ER:** Inh, Abs, Ing, Con
**SY:** Possible urinary dist, neurological disorders; in animals: irrit eyes, skin
**TO:** Eyes, skin, urinary tract, PNS

**First Aid (see Table 6):**
**Eye:** Irr immed
**Skin:** Water flush immed
**Breath:** Resp support
**Swallow:** Medical attention immed

## Dimethylaminopropionitrile

| Formula: | CAS#: | RTECS#: | IDLH: |
|---|---|---|---|
| $(CH_3)_2NCH_2CH_2CN$ | 1738-25-6 | UG1575000 | N.D. |

**Conversion:**

**DOT:**

**Synonyms/Trade Names:** 3-(Dimethylamino)propionitrile; N,N-Dimethylamino-3-propionitrile
[**Note:** A component (95%) of NIAX® Catalyst ESN, along with bis(2-(dimethylamino)ethyl) ether (5%).]

**Exposure Limits:**
**NIOSH REL:** See Appendix C (NIAX® Catalyst ESN)
**OSHA PEL:** See Appendix C (NIAX® Catalyst ESN)

**Measurement Methods (see Table 1):**
None available

**Physical Description:** Colorless liquid.

**Chemical & Physical Properties:**
**MW:** 98.2
**BP:** 342°F
**Sol:** Miscible
**Fl.P:** 147°F
**IP:** ?
**Sp.Gr(86°F):** 0.86
**VP(135°F):** 10 mmHg
**FRZ:** -48°F
**UEL:** ?
**LEL:** ?
Class IIIA Combustible Liquid

**Personal Protection/Sanitation (see Table 2):**
**Skin:** Prevent skin contact
**Eyes:** Prevent eye contact
**Wash skin:** When contam
**Remove:** When wet or contam
**Change:** N.R.
**Provide:** Eyewash
Quick drench

**Respirator Recommendations (see Tables 3 and 4):**
**NIOSH**
**¥:** ScbaF:Pd,Pp/SaF:Pd,Pp:AScba
**Escape:** GmFOv/ScbaE

**Incompatibilities and Reactivities:** Oxidizers
[**Note:** Emits toxic oxides of nitrogen and cyanide fumes when heated to decomposition.]

**Exposure Routes, Symptoms, Target Organs (see Table 5):**
**ER:** Inh, Abs, Ing, Con
**SY:** Irrit eyes, skin; urinary dist; neurological disorders; pins & needles in hands & feet; musc weak, lass, nau, vomit; decr nerve conduction in lower legs
**TO:** Eyes, skin, CNS, urinary tract

**First Aid (see Table 6):**
**Eye:** Irr immed
**Skin:** Water flush immed
**Breath:** Resp support
**Swallow:** Medical attention immed

D

## N,N-Dimethylaniline

| Formula: | CAS#: | RTECS#: | IDLH: |
|---|---|---|---|
| $C_6H_5N(CH_3)_2$ | 121-69-7 | BX4725000 | 100 ppm |

**Conversion:** 1 ppm = 4.96 mg/m$^3$

**DOT:** 2253 153

**Synonyms/Trade Names:** N,N-Dimethylbenzeneamine; N,N-Dimethylphenylamine
[**Note:** Also known as Dimethylaniline which is a correct synonym for Xylidine.]

**Exposure Limits:**
**NIOSH REL:** TWA 5 ppm (25 mg/m$^3$)
ST 10 ppm (50 mg/m$^3$) [skin]
**OSHA PEL†:** TWA 5 ppm (25 mg/m$^3$) [skin]

**Measurement Methods (see Table 1):**
**NIOSH** 2002
**OSHA** PV2064

**Physical Description:** Pale yellow, oily liquid with an amine-like odor. [**Note:** A solid below 36°F.]

**Chemical & Physical Properties:**
**MW:** 121.2
**BP:** 378°F
**Sol:** 2%
**Fl.P:** 142°F
**IP:** 7.14 eV
**Sp.Gr:** 0.96
**VP:** 1 mmHg
**FRZ:** 36°F
**UEL:** ?
**LEL:** ?
Class IIIA Combustible Liquid

**Personal Protection/Sanitation (see Table 2):**
**Skin:** Prevent skin contact
**Eyes:** Prevent eye contact
**Wash skin:** When contam
**Remove:** When wet or contam
**Change:** N.R.
**Provide:** Quick drench

**Respirator Recommendations (see Tables 3 and 4):**
**NIOSH**
**50 ppm:** Sa
**100 ppm:** Sa:Cf/ScbaF/SaF
**§:** ScbaF:Pd,Pp/SaF:Pd,Pp:AScba
**Escape:** GmFOv/ScbaE

**Incompatibilities and Reactivities:** Strong oxidizers, strong acids, benzoyl peroxide

**Exposure Routes, Symptoms, Target Organs (see Table 5):**
**ER:** Inh, Abs, Ing, Con
**SY:** Anoxia symptoms: cyan, lass, dizz, ataxia; methemo
**TO:** Blood, kidneys, liver, CVS

**First Aid (see Table 6):**
**Eye:** Irr immed
**Skin:** Soap wash immed
**Breath:** Resp support
**Swallow:** Medical attention immed

| Dimethyl carbamoyl chloride | Formula: $(CH_3)_2NCOCl$ | CAS#: 79-44-7 | RTECS#: FD4200000 | IDLH: Ca [N.D.] |
|---|---|---|---|---|
| **Conversion:** | **DOT:** 2262 156 | | | |

**Synonyms/Trade Names:** Chloroformic acid dimethylamide; Dimethylcarbamic chloride; N,N-Dimethylcarbamoyl chloride; DMCC

**Exposure Limits:**
**NIOSH REL:** Ca
See Appendix A
**OSHA PEL:** none

**Measurement Methods (see Table 1):**
None available

**Physical Description:** Clear, colorless liquid.

**Chemical & Physical Properties:**
**MW:** 107.6
**BP:** 329°F
**Sol:** Reacts
**Fl.P:** 155°F
**IP:** ?
**Sp.Gr:** 1.17
**VP:** ?
**FRZ:** -27°F
**UEL:** ?
**LEL:** ?
Class IIIA Combustible Liquid

**Personal Protection/Sanitation (see Table 2):**
**Skin:** Prevent skin contact
**Eyes:** Prevent eye contact
**Wash skin:** When contam
**Remove:** When wet or contam
**Change:** N.R.
**Provide:** Eyewash
Quick drench

**Respirator Recommendations (see Tables 3 and 4):**
**NIOSH**
**¥:** ScbaF:Pd,Pp/SaF:Pd,Pp:AScba
**Escape:** GmFOv/ScbaE

**Incompatibilities and Reactivities:** Acids, water
[**Note:** Rapidly hydrolyzes in water to dimethylamine, carbon dioxide, and hydrogen chloride.]

**Exposure Routes, Symptoms, Target Organs (see Table 5):**
**ER:** Inh, Abs, Ing, Con
**SY:** Irrit eyes, skin, nose, throat, resp sys; eye, skin burns; cough, wheez, larnygitis, dysp; head, nau, vomit; liver inj; [carc]
**TO:** Eyes, skin, resp sys, liver [in animals: nasal cancer]

**First Aid (see Table 6):**
**Eye:** Irr immed
**Skin:** Water flush immed
**Breath:** Resp support
**Swallow:** Medical attention immed

| Dimethyl-1,2-dibromo-2,2-dichlorethyl phosphate | Formula: $(CH_3O)_2P(O)OCHBrCBrCl_2$ | CAS#: 300-76-5 | RTECS#: TB9450000 | IDLH: 200 mg/m$^3$ |
|---|---|---|---|---|
| **Conversion:** | **DOT:** | | | |

**Synonyms/Trade Names:** Dibrom®; 1,2-Dibromo-2,2-dichloroethyl dimethyl phosphate; Naled

**Exposure Limits:**
**NIOSH REL:** TWA 3 mg/m$^3$ [skin]
**OSHA PEL†:** TWA 3 mg/m$^3$

**Measurement Methods (see Table 1):**
None available

**Physical Description:** Colorless to white solid or straw-colored liquid (above 80°F) with a slightly pungent odor. [insecticide]

**Chemical & Physical Properties:**
**MW:** 380.8
**BP:** Decomposes
**Sol:** Insoluble
**Fl.P:** NA
**IP:** ?
**Sp.Gr(77°F):** 1.96
**VP:** 0.0002 mmHg
**MLT:** 80°F
**UEL:** NA
**LEL:** NA
Noncombustible Solid

**Personal Protection/Sanitation (see Table 2):**
**Skin:** Prevent skin contact
**Eyes:** Prevent eye contact
**Wash skin:** When contam
**Remove:** When wet or contam
**Change:** Daily
**Provide:** Eyewash

**Respirator Recommendations (see Tables 3 and 4):**
**NIOSH/OSHA**
**30 mg/m$^3$:** 95XQ/Sa
**75 mg/m$^3$:** Sa:Cf/PaprHie
**150 mg/m$^3$:** 100F/SaT:Cf/PaprTHie/ScbaF/SaF
**200 mg/m$^3$:** Sa:Pd,Pp
**§:** ScbaF:Pd,Pp/SaF:Pd,Pp:AScba
**Escape:** 100F/ScbaE

**Incompatibilities and Reactivities:** Strong oxidizers, acids, sunlight, water
[**Note:** Corrosive to metals. Hydrolyzed in presence of water.]

**Exposure Routes, Symptoms, Target Organs (see Table 5):**
**ER:** Inh, Abs, Ing, Con
**SY:** Irrit eyes, skin; miosis, lac; head; chest tight, wheez, lar spasm; salv; cyan; anor, nau, vomit, abdom cramp, diarr; lass, twitch, para; dizz, ataxia, convuls; low BP; card irreg
**TO:** Eyes, skin, resp sys, CNS, CVS, blood chol

**First Aid (see Table 6):**
**Eye:** Irr immed
**Skin:** Soap wash immed
**Breath:** Resp support
**Swallow:** Medical attention immed

| **Dimethylformamide** | **Formula:** $HCON(CH_3)_2$ | **CAS#:** 68-12-2 | **RTECS#:** LQ2100000 | **IDLH:** 500 ppm |
|---|---|---|---|---|
| **Conversion:** 1 ppm = 2.99 $mg/m^3$ | **DOT:** 2265 129 | | | |

**Synonyms/Trade Names:** Dimethyl formamide; N,N-Dimethylformamide; DMF

| **Exposure Limits:** | **Measurement Methods (see Table 1):** |
|---|---|
| **NIOSH REL:** TWA 10 ppm (30 $mg/m^3$) [skin]<br>**OSHA PEL:** TWA 10 ppm (30 $mg/m^3$) [skin] | **NIOSH** 2004<br>**OSHA** 66 |
| **Physical Description:** Colorless to pale-yellow liquid with a faint, amine-like odor. | |

| **Chemical & Physical Properties:** | **Personal Protection/Sanitation (see Table 2):** | **Respirator Recommendations (see Tables 3 and 4):** |
|---|---|---|
| **MW:** 73.1<br>**BP:** 307°F<br>**Sol:** Miscible<br>**Fl.P:** 136°F<br>**IP:** 9.12 eV<br>**Sp.Gr:** 0.95<br>**VP:** 3 mmHg<br>**FRZ:** -78°F<br>**UEL:** 15.2%<br>**LEL(212°F):** 2.2%<br>Class II Combustible Liquid | **Skin:** Prevent skin contact<br>**Eyes:** Prevent eye contact<br>**Wash skin:** When contam<br>**Remove:** When wet or contam<br>**Change:** N.R. | **NIOSH**<br>**100 ppm:** Sa*<br>**250 ppm:** Sa:Cf*<br>**500 ppm:** SaT:Cf*/ScbaF/SaF<br>**§:** ScbaF:Pd,Pp/SaF:Pd,Pp:AScba<br>**Escape:** GmFOv/ScbaE |

**Incompatibilities and Reactivities:** Carbon tetrachloride; other halogenated compounds when in contact with iron; strong oxidizers; alkyl aluminums; inorganic nitrates

| **Exposure Routes, Symptoms, Target Organs (see Table 5):** | **First Aid (see Table 6):** |
|---|---|
| **ER:** Inh, Abs, Ing, Con<br>**SY:** Irrit eyes, skin, resp sys; nau, vomit, colic; liver damage, enlarged liver; high BP; face flush; derm; in animals: kidney, heart damage<br>**TO:** Eyes, skin, resp sys, liver, kidneys, CVS | **Eye:** Irr immed<br>**Skin:** Water flush prompt<br>**Breath:** Resp support<br>**Swallow:** Medical attention immed |

| **1,1-Dimethylhydrazine** | **Formula:** $(CH_3)_2NNH_2$ | **CAS#:** 57-14-7 | **RTECS#:** MV2450000 | **IDLH:** Ca [15 ppm] |
|---|---|---|---|---|
| **Conversion:** 1 ppm = 2.46 $mg/m^3$ | **DOT:** 1163 131 | | | |

**Synonyms/Trade Names:** Dimazine, DMH, UDMH, Unsymmetrical dimethylhydrazine

| **Exposure Limits:** | **Measurement Methods (see Table 1):** |
|---|---|
| **NIOSH REL:** Ca<br>C 0.06 ppm (0.15 $mg/m^3$) [2-hr]<br>See Appendix A<br>**OSHA PEL:** TWA 0.5 ppm (1 $mg/m^3$) [skin] | **NIOSH** 3515 |
| **Physical Description:** Colorless liquid with an ammonia- or fish-like odor. | |

| **Chemical & Physical Properties:** | **Personal Protection/Sanitation (see Table 2):** | **Respirator Recommendations (see Tables 3 and 4):** |
|---|---|---|
| **MW:** 60.1<br>**BP:** 147°F<br>**Sol:** Miscible<br>**Fl.P:** 5°F<br>**IP:** 8.05 eV<br>**Sp.Gr:** 0.79<br>**VP:** 103 mmHg<br>**FRZ:** -72°F<br>**UEL:** 95%<br>**LEL:** 2%<br>Class IB Flammable Liquid | **Skin:** Prevent skin contact<br>**Eyes:** Prevent eye contact<br>**Wash skin:** When contam<br>**Remove:** When wet (flamm)<br>**Change:** N.R.<br>**Provide:** Eyewash<br>Quick drench | **NIOSH**<br>**¥:** ScbaF:Pd,Pp/SaF:Pd,Pp:AScba<br>**Escape:** GmFS/ScbaE |

**Incompatibilities and Reactivities:** Oxidizers, halogens, metallic mercury, fuming nitric acid, hydrogen peroxide [**Note:** May ignite SPONTANEOUSLY in contact with oxidizers.]

| **Exposure Routes, Symptoms, Target Organs (see Table 5):** | **First Aid (see Table 6):** |
|---|---|
| **ER:** Inh, Abs, Ing, Con<br>**SY:** Irrit eyes, skin; choking, chest pain, dysp; drow; nau; anoxia; convuls; liver inj; [carc]<br>**TO:** CNS, liver, GI tract, blood, resp sys, eyes, skin [in animals: tumors of the lungs, liver, blood vessels & intestines] | **Eye:** Irr immed<br>**Skin:** Water flush immed<br>**Breath:** Resp support<br>**Swallow:** Medical attention immed |

## Dimethylphthalate

**Formula:** $C_6H_4(COOCH_3)_2$ | **CAS#:** 131-11-3 | **RTECS#:** TI1575000 | **IDLH:** 2000 mg/m$^3$

**Conversion:** | **DOT:**

**Synonyms/Trade Names:** Dimethyl ester of 1,2-benzenedicarboxylic acid; DMP

**Exposure Limits:**
**NIOSH REL:** TWA 5 mg/m$^3$
**OSHA PEL:** TWA 5 mg/m$^3$

**Measurement Methods (see Table 1):**
**OSHA** 104

**Physical Description:** Colorless, oily liquid with a slight, aromatic odor.
[**Note:** A solid below 42°F.]

**Chemical & Physical Properties:**
**MW:** 194.2
**BP:** 543°F
**Sol:** 0.4%
**Fl.P:** 295°F
**IP:** 9.64 eV
**Sp.Gr:** 1.19
**VP:** 0.01 mmHg
**FRZ:** 42°F
**UEL:** ?
**LEL(358°F):** 0.9%
Class IIIB Combustible Liquid; however, ignition is difficult.

**Personal Protection/Sanitation (see Table 2):**
**Skin:** N.R.
**Eyes:** Prevent eye contact
**Wash skin:** N.R.
**Remove:** N.R.
**Change:** N.R.

**Respirator Recommendations (see Tables 3 and 4):**
**NIOSH/OSHA**
**50 mg/m$^3$:** 95F
**125 mg/m$^3$:** Sa:Cf£/PaprHie£
**250 mg/m$^3$:** 100F/ScbaF/SaF
**2000 mg/m$^3$:** SaF:Pd,Pp
**§:** ScbaF:Pd,Pp/SaF:Pd,Pp:AScba
**Escape:** 100F/ScbaE

**Incompatibilities and Reactivities:** Nitrates; strong oxidizers, alkalis & acids

**Exposure Routes, Symptoms, Target Organs (see Table 5):**
**ER:** Inh, Ing, Con
**SY:** Irrit eyes, upper resp sys; stomach pain
**TO:** Eyes, resp sys, GI tract

**First Aid (see Table 6):**
**Eye:** Irr prompt
**Skin:** Wash regularly
**Breath:** Resp support
**Swallow:** Medical attention immed

## Dimethyl sulfate

**Formula:** $(CH_3)_2SO_4$ | **CAS#:** 77-78-1 | **RTECS#:** WS8225000 | **IDLH:** Ca [7 ppm]

**Conversion:** 1 ppm = 5.16 mg/m$^3$ | **DOT:** 1595 156

**Synonyms/Trade Names:** Dimethyl ester of sulfuric acid, Dimethylsulfate, Methyl sulfate

**Exposure Limits:**
**NIOSH REL:** Ca
TWA 0.1 ppm (0.5 mg/m$^3$) [skin]
See Appendix A
**OSHA PEL†:** TWA 1 ppm (5 mg/m$^3$) [skin]

**Measurement Methods (see Table 1):**
**NIOSH** 2524

**Physical Description:** Colorless, oily liquid with a faint, onion-like odor.

**Chemical & Physical Properties:**
**MW:** 126.1
**BP:** 370°F (Decomposes)
**Sol(64°F):** 3%
**Fl.P:** 182°F
**IP:** ?
**Sp.Gr:** 1.33
**VP:** 0.1 mmHg
**FRZ:** -25°F
**UEL:** ?
**LEL:** ?
Class IIIA Combustible Liquid

**Personal Protection/Sanitation (see Table 2):**
**Skin:** Prevent skin contact
**Eyes:** Prevent eye contact
**Wash skin:** When contam
**Remove:** When wet or contam
**Change:** N.R.
**Provide:** Eyewash
Quick drench

**Respirator Recommendations (see Tables 3 and 4):**
**NIOSH**
**¥:** ScbaF:Pd,Pp/SaF:Pd,Pp:AScba
**Escape:** GmFS/ScbaE

**Incompatibilities and Reactivities:** Strong oxidizers, ammonia solutions
[**Note:** Decomposes in water to sulfuric acid; corrosive to metals.]

**Exposure Routes, Symptoms, Target Organs (see Table 5):**
**ER:** Inh, Abs, Ing, Con
**SY:** Irrit eyes, nose; head; dizz; conj; photo; periorb edema; dysphonia, aphonia, dysphagia, productive cough; chest pain; dysp, cyan; vomit, diarr; dysuria; analgesia; fever; prot, hema; eye, skin burns; delirium; [carc]
**TO:** Eyes, skin, resp sys, liver, kidneys, CNS [in animals: nasal & lung cancer]

**First Aid (see Table 6):**
**Eye:** Irr immed
**Skin:** Water flush immed
**Breath:** Resp support
**Swallow:** Medical attention immed

| Dinitolmide | Formula: $(NO_2)_2C_6H_2(CH_3)CONH_2$ | CAS#: 148-01-6 | RTECS#: XS4200000 | IDLH: N.D. |
|---|---|---|---|---|
| **Conversion:** | **DOT:** | | | |

**Synonyms/Trade Names:** 3,5-Dinitro-o-toluamide; 2-Methyl-3,5-dinitrobenzamide; Zoalene

**Exposure Limits:**
**NIOSH REL:** TWA 5 mg/m³
**OSHA PEL†:** none

**Measurement Methods (see Table 1):**
**NIOSH** 0500

D

**Physical Description:** Yellowish, crystalline solid.

| Chemical & Physical Properties: | Personal Protection/Sanitation (see Table 2): | Respirator Recommendations (see Tables 3 and 4): |
|---|---|---|
| **MW:** 225.2<br>**BP:** ?<br>**Sol:** Slight<br>**Fl.P:** NA<br>**IP:** ?<br>**Sp.Gr:** ?<br>**VP:** ?<br>**MLT:** 351°F<br>**UEL:** NA<br>**LEL:** NA<br>Noncombustible Solid | **Skin:** Prevent skin contact<br>**Eyes:** Prevent eye contact<br>**Wash skin:** When contam<br>**Remove:** When wet or contam<br>**Change:** Daily | Not available. |

**Incompatibilities and Reactivities:** None reported

| Exposure Routes, Symptoms, Target Organs (see Table 5): | First Aid (see Table 6): |
|---|---|
| **ER:** Inh, Ing, Con<br>**SY:** Contact eczema; in animals: methemo, liver changes<br>**TO:** Skin, liver, blood | **Eye:** Irr immed<br>**Skin:** Soap wash immed<br>**Breath:** Resp support<br>**Swallow:** Medical attention immed |

| m-Dinitrobenzene | Formula: $C_6H_4(NO_2)_2$ | CAS#: 99-65-0 | RTECS#: CZ7350000 | IDLH: 50 mg/m³ |
|---|---|---|---|---|
| **Conversion:** | **DOT:** 1597 152 | | | |

**Synonyms/Trade Names:** meta-Dinitrobenzene; 1,3-Dinitrobenzene

**Exposure Limits:**
**NIOSH REL:** TWA 1 mg/m³ [skin]
**OSHA PEL:** TWA 1 mg/m³ [skin]

**Measurement Methods (see Table 1):**
**NIOSH** S214 (II-4)

**Physical Description:** Pale-white or yellow solid.

| Chemical & Physical Properties: | Personal Protection/Sanitation (see Table 2): | Respirator Recommendations (see Tables 3 and 4): |
|---|---|---|
| **MW:** 168.1<br>**BP:** 572°F<br>**Sol:** 0.02%<br>**Fl.P:** 302°F<br>**IP:** 10.43 eV<br>**Sp.Gr:** 1.58<br>**VP:** ?<br>**MLT:** 192°F<br>**UEL:** ?<br>**LEL:** ?<br>Combustible Solid | **Skin:** Prevent skin contact<br>**Eyes:** Prevent eye contact<br>**Wash skin:** When contam<br>**Remove:** When wet or contam<br>**Change:** Daily<br>**Provide:** Quick drench | **NIOSH/OSHA**<br>**5 mg/m³:** Qm<br>**10 mg/m³:** 95XQ/Sa<br>**25 mg/m³:** Sa:Cf/PaprHie<br>**50 mg/m³:** 100F/SaT:Cf/PaprTHie/ ScbaF/SaF<br>**§:** ScbaF:Pd,Pp/SaF:Pd,Pp:AScba<br>**Escape:** 100F/ScbaE |

**Incompatibilities and Reactivities:** Strong oxidizers, caustics, metals such as tin & zinc
[**Note:** Prolonged exposure to fire and heat may result in an explosion due to SPONTANEOUS decomposition.]

| Exposure Routes, Symptoms, Target Organs (see Table 5): | First Aid (see Table 6): |
|---|---|
| **ER:** Inh, Abs, Ing, Con<br>**SY:** Anoxia, cyan; vis dist, central scotomas; bad taste, burning mouth, dry throat, thirst; yellowing hair, eyes, skin; anemia; liver damage<br>**TO:** Eyes, skin, blood, liver, CVS, CNS | **Eye:** Irr immed<br>**Skin:** Soap wash immed<br>**Breath:** Resp support<br>**Swallow:** Medical attention immed |

D

## o-Dinitrobenzene

| o-Dinitrobenzene | Formula: $C_6H_4(NO_2)_2$ | CAS#: 528-29-0 | RTECS#: CZ7450000 | IDLH: 50 $mg/m^3$ |
|---|---|---|---|---|
| **Conversion:** | **DOT:** 1597 152 | | | |

**Synonyms/Trade Names:** ortho-Dinitrobenzene; 1,2-Dinitrobenzene

**Exposure Limits:**
**NIOSH REL:** TWA 1 $mg/m^3$ [skin]
**OSHA PEL:** TWA 1 $mg/m^3$ [skin]

**Measurement Methods (see Table 1):**
**NIOSH** S214 (II-4)

**Physical Description:** Pale-white or yellow solid.

| Chemical & Physical Properties: | Personal Protection/Sanitation (see Table 2): | Respirator Recommendations (see Tables 3 and 4): |
|---|---|---|
| **MW:** 168.1<br>**BP:** 606°F<br>**Sol:** 0.05%<br>**Fl.P:** 302°F<br>**IP:** 10.71 eV<br>**Sp.Gr:** 1.57<br>**VP:** ?<br>**MLT:** 244°F<br>**UEL:** ?<br>**LEL:** ?<br>Combustible Solid | **Skin:** Prevent skin contact<br>**Eyes:** Prevent eye contact<br>**Wash skin:** When contam<br>**Remove:** When wet or contam<br>**Change:** Daily<br>**Provide:** Quick drench | **NIOSH/OSHA**<br>**5 $mg/m^3$:** Qm<br>**10 $mg/m^3$:** 95XQ/Sa<br>**25 $mg/m^3$:** Sa:Cf/PaprHie<br>**50 $mg/m^3$:** 100F/SaT:Cf/PaprTHie/ScbaF/SaF<br>**§:** ScbaF:Pd,Pp/SaF:Pd,Pp:AScba<br>**Escape:** 100F/ScbaE |

**Incompatibilities and Reactivities:** Strong oxidizers, caustics, metals such as tin & zinc
[**Note:** Prolonged exposure to fire and heat may result in an explosion due to SPONTANEOUS decomposition.]

| Exposure Routes, Symptoms, Target Organs (see Table 5): | First Aid (see Table 6): |
|---|---|
| **ER:** Inh, Abs, Ing, Con<br>**SY:** Anoxia, cyan; vis dist, central scotomas; bad taste, burning mouth, dry throat, thirst; yellowing hair, eyes, skin; anemia; liver damage<br>**TO:** Eyes, skin, blood, liver, CVS, CNS | **Eye:** Irr immed<br>**Skin:** Soap wash immed<br>**Breath:** Resp support<br>**Swallow:** Medical attention immed |

## p-Dinitrobenzene

| p-Dinitrobenzene | Formula: $C_6H_4(NO_2)_2$ | CAS#: 100-25-4 | RTECS#: CZ7525000 | IDLH: 50 $mg/m^3$ |
|---|---|---|---|---|
| **Conversion:** | **DOT:** 1597 152 | | | |

**Synonyms/Trade Names:** para-Dinitrobenzene; 1,4-Dinitrobenzene

**Exposure Limits:**
**NIOSH REL:** TWA 1 $mg/m^3$ [skin]
**OSHA PEL:** TWA 1 $mg/m^3$ [skin]

**Measurement Methods (see Table 1):**
**NIOSH** S214 (II-4)

**Physical Description:** Pale-white or yellow solid.

| Chemical & Physical Properties: | Personal Protection/Sanitation (see Table 2): | Respirator Recommendations (see Tables 3 and 4): |
|---|---|---|
| **MW:** 168.1<br>**BP:** 570°F<br>**Sol:** 0.01%<br>**Fl.P:** ?<br>**IP:** 10.50 eV<br>**Sp.Gr:** 1.63<br>**VP:** ?<br>**MLT:** 343°F<br>**UEL:** ?<br>**LEL:** ?<br>Combustible Solid | **Skin:** Prevent skin contact<br>**Eyes:** Prevent eye contact<br>**Wash skin:** When contam<br>**Remove:** When wet or contam<br>**Change:** Daily<br>**Provide:** Quick drench | **NIOSH/OSHA**<br>**5 $mg/m^3$:** Qm<br>**10 $mg/m^3$:** 95XQ/Sa<br>**25 $mg/m^3$:** Sa:Cf/PaprHie<br>**50 $mg/m^3$:** 100F/SaT:Cf/PaprTHie/ScbaF/SaF<br>**§:** ScbaF:Pd,Pp/SaF:Pd,Pp:AScba<br>**Escape:** 100F/ScbaE |

**Incompatibilities and Reactivities:** Strong oxidizers, caustics, metals such as tin & zinc
[**Note:** Prolonged exposure to fire and heat may result in an explosion due to SPONTANEOUS decomposition.]

| Exposure Routes, Symptoms, Target Organs (see Table 5): | First Aid (see Table 6): |
|---|---|
| **ER:** Inh, Abs, Ing, Con<br>**SY:** Anoxia, cyan; vis dist, central scotomas; bad taste, burning mouth, dry throat, thirst; yellowing hair, eyes, skin; anemia; liver damage<br>**TO:** Eyes, skin, blood, liver, CVS, CNS | **Eye:** Irr immed<br>**Skin:** Soap wash immed<br>**Breath:** Resp support<br>**Swallow:** Medical attention immed |

## Dinitro-o-cresol

| Dinitro-o-cresol | Formula: $CH_3C_6H_2OH(NO_2)_2$ | CAS#: 534-52-1 | RTECS#: GO9625000 | IDLH: 5 mg/m$^3$ |
|---|---|---|---|---|
| **Conversion:** | **DOT:** 1598 153 | | | |

**Synonyms/Trade Names:** 4,6-Dinitro-o-cresol; 3,5-Dinitro-2-hydroxytoluene; 4,6-Dinitro-2-methyl phenol; DNC; DNOC

**Exposure Limits:**
**NIOSH REL:** TWA 0.2 mg/m$^3$ [skin]
**OSHA PEL:** TWA 0.2 mg/m$^3$ [skin]

**Measurement Methods (see Table 1):**
**NIOSH** S166 (II-5)

**Physical Description:** Yellow, odorless solid. [insecticide]

**Chemical & Physical Properties:**
**MW:** 198.1
**BP:** 594°F
**Sol:** 0.01%
**Fl.P:** NA
**IP:** ?
**Sp.Gr:** 1.1 (estimated)
**VP:** 0.00005 mmHg
**MLT:** 190°F
**UEL:** NA
**LEL:** NA
**MEC:** 30 g/m$^3$
Noncombustible Solid

**Personal Protection/Sanitation (see Table 2):**
**Skin:** Prevent skin contact
**Eyes:** Prevent eye contact
**Wash skin:** When contam/Daily
**Remove:** When wet or contam
**Change:** Daily

**Respirator Recommendations (see Tables 3 and 4):**
**NIOSH/OSHA**
**2 mg/m$^3$:** 95F
**5 mg/m$^3$:** 100F/Sa:Cf£/PaprHie£/ScbaF/SaF
**§:** ScbaF:Pd,Pp/SaF:Pd,Pp:AScba
**Escape:** 100F/ScbaE

**Incompatibilities and Reactivities:** Strong oxidizers

**Exposure Routes, Symptoms, Target Organs (see Table 5):**
**ER:** Inh, Abs, Ing, Con
**SY:** Sense of well being; head, fever, lass, profuse sweat, excess thirst, tacar, hyperpnea, cough, short breath, coma
**TO:** CVS, endocrine sys

**First Aid (see Table 6):**
**Eye:** Irr immed
**Skin:** Soap wash immed
**Breath:** Resp support
**Swallow:** Medical attention immed

## Dinitrotoluene

| Dinitrotoluene | Formula: $CH_3C_6H_3(NO_2)_2$ | CAS#: 25321-14-6 | RTECS#: XT1300000 | IDLH: Ca [50 mg/m$^3$] |
|---|---|---|---|---|
| **Conversion:** | **DOT:** 1600 152 (molten); 2038 152 (solid) | | | |

**Synonyms/Trade Names:** Dinitrotoluol, DNT, Methyldinitrobenzene [**Note:** Various isomers of DNT exist.]

**Exposure Limits:**
**NIOSH REL:** Ca
TWA 1.5 mg/m$^3$ [skin]
See Appendix A
**OSHA PEL:** TWA 1.5 mg/m$^3$ [skin]

**Measurement Methods (see Table 1):**
**OSHA** 44

**Physical Description:** Orange-yellow crystalline solid with a characteristic odor. [**Note:** Often shipped molten.]

**Chemical & Physical Properties:**
**MW:** 182.2
**BP:** 572°F
**Sol:** Insoluble
**Fl.P:** 404°F
**IP:** ?
**Sp.Gr:** 1.32
**VP:** 1 mmHg
**MLT:** 158°F
**UEL:** ?
**LEL:** ?
Combustible Solid, but difficult to ignite.

**Personal Protection/Sanitation (see Table 2):**
**Skin:** Prevent skin contact
**Eyes:** Prevent eye contact
**Wash skin:** When contam/Daily
**Remove:** When wet or contam
**Change:** Daily
**Provide:** Quick drench

**Respirator Recommendations (see Tables 3 and 4):**
**NIOSH**
**¥:** ScbaF:Pd,Pp/SaF:Pd,Pp:AScba
**Escape:** GmFOv100/ScbaE

**Incompatibilities and Reactivities:** Strong oxidizers, caustics, metals such as tin & zinc
[**Note:** Commercial grades will decompose at 482°F, with self-sustaining decomposition at 536°F.]

**Exposure Routes, Symptoms, Target Organs (see Table 5):**
**ER:** Inh, Abs, Ing, Con
**SY:** Anoxia, cyan; anemia, jaun; repro effects; [carc]
**TO:** Blood, liver, CVS, repro sys [in animals: liver, skin & kidney tumors]

**First Aid (see Table 6):**
**Eye:** Irr immed
**Skin:** Soap wash immed
**Breath:** Resp support
**Swallow:** Medical attention immed

## Di-sec octyl phthalate

| Formula: | CAS#: | RTECS#: | IDLH: |
|---|---|---|---|
| $C_{24}H_{38}O_4$ | 117-81-7 | TI0350000 | Ca [5000 mg/m$^3$] |

**Conversion:**

**DOT:**

**Synonyms/Trade Names:** DEHP, Di(2-ethylhexyl)phthalate, DOP, bis-(2-Ethylhexyl)phthalate, Octyl phthalate

**Exposure Limits:**
**NIOSH REL:** Ca
TWA 5 mg/m$^3$
ST 10 mg/m$^3$
See Appendix A
**OSHA PEL†:** TWA 5 mg/m$^3$

**Measurement Methods (see Table 1):**
**NIOSH** 5020

**Physical Description:** Colorless, oily liquid with a slight odor.

**Chemical & Physical Properties:**
**MW:** 390.5
**BP:** 727°F
**Sol(75°F):** 0.00003%
**Fl.P(oc):** 420°F
**IP:** ?
**Sp.Gr:** 0.99
**VP:** <0.01 mmHg
**FRZ:** -58°F
**UEL:** ?
**LEL(474°F):** 0.3%
Class IIIB Combustible Liquid

**Personal Protection/Sanitation (see Table 2):**
**Skin:** N.R.
**Eyes:** N.R.
**Wash skin:** N.R.
**Remove:** N.R.
**Change:** N.R.

**Respirator Recommendations (see Tables 3 and 4):**
**NIOSH**
**¥:** ScbaF:Pd,Pp/SaF:Pd,Pp:AScba
**Escape:** 100F/ScbaE

**Incompatibilities and Reactivities:** Nitrates; strong oxidizers, acids & alkalis

**Exposure Routes, Symptoms, Target Organs (see Table 5):**
**ER:** Inh, Ing, Con
**SY:** Irrit eyes, muc memb; in animals: liver damage; terato effects; [carc]
**TO:** Eyes, resp sys, CNS, liver, repro sys, GI tract [in animals: liver tumors]

**First Aid (see Table 6):**
**Eye:** Irr immed
**Breath:** Resp support
**Swallow:** Medical attention immed

## Dioxane

| Formula: | CAS#: | RTECS#: | IDLH: |
|---|---|---|---|
| $C_4H_8O_2$ | 123-91-1 | JG8225000 | Ca [500 ppm] |

**Conversion:** 1 ppm = 3.60 mg/m$^3$

**DOT:** 1165 127

**Synonyms/Trade Names:** Diethylene dioxide; Diethylene ether; Dioxan; p-Dioxane; 1,4-Dioxane

**Exposure Limits:**
**NIOSH REL:** Ca
C 1 ppm (3.6 mg/m$^3$) [30-minute]
See Appendix A
**OSHA PEL†:** TWA 100 ppm (360 mg/m$^3$) [skin]

**Measurement Methods (see Table 1):**
**NIOSH** 1602
**OSHA** 7

**Physical Description:** Colorless liquid or solid (below 53°F) with a mild, ether-like odor.

**Chemical & Physical Properties:**
**MW:** 88.1
**BP:** 214°F
**Sol:** Miscible
**Fl.P:** 55°F
**IP:** 9.13 eV
**Sp.Gr:** 1.03
**VP:** 29 mmHg
**FRZ:** 53°F
**UEL:** 22%
**LEL:** 2.0%
Class IB Flammable Liquid

**Personal Protection/Sanitation (see Table 2):**
**Skin:** Prevent skin contact
**Eyes:** Prevent eye contact
**Wash skin:** When contam
**Remove:** When wet (flamm)
**Change:** N.R.
**Provide:** Eyewash
Quick drench

**Respirator Recommendations (see Tables 3 and 4):**
**NIOSH**
**¥:** ScbaF:Pd,Pp/SaF:Pd,Pp:AScba
**Escape:** GmFOv/ScbaE

**Incompatibilities and Reactivities:** Strong oxidizers, decaborane, triethynyl aluminum

**Exposure Routes, Symptoms, Target Organs (see Table 5):**
**ER:** Inh, Abs, Ing, Con
**SY:** Irrit eyes, skin, nose, throat; drow, head; nau, vomit; liver damage; kidney failure; [carc]
**TO:** Eyes, skin, resp sys, liver, kidneys [in animals: lung, liver & nasal cavity tumors]

**First Aid (see Table 6):**
**Eye:** Irr immed
**Skin:** Water wash prompt
**Breath:** Resp support
**Swallow:** Medical attention immed

| **Dioxathion** | **Formula:** $C_4H_6O_2[SPS(OC_2H_5)_2]_2$ | **CAS#:** 78-34-2 | **RTECS#:** TE3350000 | **IDLH:** N.D. |
|---|---|---|---|---|
| **Conversion:** | **DOT:** | | | |

**Synonyms/Trade Names:** Delnav®; p-Dioxane-2,3-diyl ethyl phosphorodithioate; Dioxane phosphate; 2,3-p-Dioxanethiol-S,S-bis(O,O-diethyl phosphoro-dithioate); Navadel®

**Exposure Limits:**
**NIOSH REL:** TWA 0.2 mg/m³ [skin]
**OSHA PEL†:** none

**Measurement Methods (see Table 1):**
None available

**Physical Description:** Viscous, brown, tan, or dark-amber liquid. [insecticide]
[**Note:** Technical product is a mixture of cis- & trans-isomers.]

| Chemical & Physical Properties: | Personal Protection/Sanitation (see Table 2): | Respirator Recommendations (see Tables 3 and 4): |
|---|---|---|
| **MW:** 456.6<br>**BP:** ?<br>**Sol:** Insoluble<br>**Fl.P:** NA<br>**IP:** ?<br>**Sp.Gr(79°F):** 1.26<br>**VP:** ?<br>**FRZ:** -4°F<br>**UEL:** NA<br>**LEL:** NA<br>Noncombustible Liquid | **Skin:** Prevent skin contact<br>**Eyes:** Prevent eye contact<br>**Wash skin:** When contam<br>**Remove:** When wet or contam<br>**Change:** N.R.<br>**Provide:** Eyewash<br>Quick drench | Not available. |

**Incompatibilities and Reactivities:** Alkalis, iron or tin surfaces, heat

| Exposure Routes, Symptoms, Target Organs (see Table 5): | First Aid (see Table 6): |
|---|---|
| **ER:** Inh, Abs, Ing, Con<br>**SY:** Irrit eyes, skin; head, dizz, lass; rhin, chest tight; miosis; nau, vomit, abdom cramps, diarr, salv; musc fasc; conf, drow<br>**TO:** Eyes, skin, resp sys, CNS, CVS, blood chol | **Eye:** Irr immed<br>**Skin:** Soap flush immed<br>**Breath:** Resp support<br>**Swallow:** Medical attention immed |

| **Diphenyl** | **Formula:** $C_6H_5C_6H_5$ | **CAS#:** 92-52-4 | **RTECS#:** DU8050000 | **IDLH:** 100 mg/m³ |
|---|---|---|---|---|
| **Conversion:** 1 ppm = 6.31 mg/m³ | **DOT:** | | | |

**Synonyms/Trade Names:** Biphenyl, Phenyl benzene

**Exposure Limits:**
**NIOSH REL:** TWA 1 mg/m³ (0.2 ppm)
**OSHA PEL:** TWA 1 mg/m³ (0.2 ppm)

**Measurement Methods (see Table 1):**
**NIOSH** 2530
**OSHA** PV2022

**Physical Description:** Colorless to pale-yellow solid with a pleasant, characteristic odor. [fungicide]

| Chemical & Physical Properties: | Personal Protection/Sanitation (see Table 2): | Respirator Recommendations (see Tables 3 and 4): |
|---|---|---|
| **MW:** 154.2<br>**BP:** 489°F<br>**Sol:** Insoluble<br>**Fl.P:** 235°F<br>**IP:** 7.95 eV<br>**Sp.Gr:** 1.04<br>**VP:** 0.005 mmHg<br>**MLT:** 156°F<br>**UEL(311°F):** 5.8%<br>**LEL(232°F):** 0.6%<br>Combustible Solid | **Skin:** Prevent skin contact<br>**Eyes:** Prevent eye contact<br>**Wash skin:** When contam<br>**Remove:** When wet or contam<br>**Change:** Daily<br>**Provide:** Eyewash (molt)<br>Quick drench (molt) | **NIOSH/OSHA**<br>**10 mg/m³:** CcrOv95/Sa<br>**25 mg/m³:** Sa:Cf*/PaprOvHie*<br>**50 mg/m³:** CcrFOv100/GmFOv100/PaprTOvHie*/ScbaF/SaF<br>**100 mg/m³:** SaF:Pd,Pp<br>**§:** ScbaF:Pd,Pp/SaF:Pd,Pp:AScba<br>**Escape:** GmFOv100/ScbaE |

**Incompatibilities and Reactivities:** Oxidizers

| Exposure Routes, Symptoms, Target Organs (see Table 5): | First Aid (see Table 6): |
|---|---|
| **ER:** Inh, Abs, Ing, Con<br>**SY:** Irrit eyes, throat; head, nau, lass, numb limbs; liver damage<br>**TO:** Eyes, resp sys, liver, CNS | **Eye:** Irr immed<br>**Skin:** Water flush immed<br>**Breath:** Resp support<br>**Swallow:** Medical attention immed |

D

## Diphenylamine

| Diphenylamine | Formula: $(C_6H_5)_2NH$ | CAS#: 122-39-4 | RTECS#: JJ7800000 | IDLH: N.D. |
|---|---|---|---|---|
| Conversion: | DOT: | | | |

**Synonyms/Trade Names:** Anilinobenzene, DPA, Phenylaniline, N-Phenylaniline, N-Phenylbenzenamine [**Note:** The carcinogen 4-Aminodiphenyl may be present as an impurity in the commercial product.]

**Exposure Limits:**
**NIOSH REL:** TWA 10 mg/m$^3$
**OSHA PEL†:** none

**Measurement Methods (see Table 1):**
**OSHA** 22, 78

**Physical Description:** Colorless, tan, amber, or brown crystalline solid with a pleasant, floral odor. [fungicide]

| Chemical & Physical Properties: | Personal Protection/Sanitation (see Table 2): | Respirator Recommendations (see Tables 3 and 4): |
|---|---|---|
| **MW:** 169.2<br>**BP:** 576°F<br>**Sol:** 0.03%<br>**Fl.P:** 307°F<br>**IP:** 7.40 eV<br>**Sp.Gr:** 1.16<br>**VP(227°F):** 1 mmHg<br>**MLT:** 127°F<br>**UEL:** ?<br>**LEL:** ?<br>Combustible Solid; explosive if a cloud of dust is exposed to a source of ignition. | **Skin:** Prevent skin contact<br>**Eyes:** Prevent eye contact<br>**Wash skin:** Daily<br>**Remove:** When wet or contam<br>**Change:** Daily | Not available. |

**Incompatibilities and Reactivities:** Oxidizers, hexachloromelamine, trichloromelamine

| Exposure Routes, Symptoms, Target Organs (see Table 5): | First Aid (see Table 6): |
|---|---|
| **ER:** Inh, Abs, Ing, Con<br>**SY:** Irrit eyes, skin, muc memb; eczema; tacar, hypertension; cough, sneez; methemo; incr BP, heart rate; prot, hema, bladder inj; in animals: terato effects<br>**TO:** Eyes, skin, resp sys, CVS, blood, bladder, repro sys | **Eye:** Irr immed<br>**Skin:** Soap wash prompt<br>**Breath:** Resp support<br>**Swallow:** Medical attention immed |

## Dipropylene glycol methyl ether

| Dipropylene glycol methyl ether | Formula: $CH_3OC_3H_6OC_3H_6OH$ | CAS#: 34590-94-8 | RTECS#: JM1575000 | IDLH: 600 ppm |
|---|---|---|---|---|
| Conversion: 1 ppm = 6.06 mg/m$^3$ | DOT: | | | |

**Synonyms/Trade Names:** Dipropylene glycol monomethyl ether, Dowanol® 50B

**Exposure Limits:**
**NIOSH REL:** TWA 100 ppm (600 mg/m$^3$)
ST 150 ppm (900 mg/m$^3$) [skin]
**OSHA PEL†:** TWA 100 ppm (600 mg/m$^3$) [skin]

**Measurement Methods (see Table 1):**
**NIOSH** 2554, S69 (II-2)

**Physical Description:** Colorless liquid with a mild, ether-like odor.

| Chemical & Physical Properties: | Personal Protection/Sanitation (see Table 2): | Respirator Recommendations (see Tables 3 and 4): |
|---|---|---|
| **MW:** 148.2<br>**BP:** 408°F<br>**Sol:** Miscible<br>**Fl.P:** 180°F<br>**IP:** ?<br>**Sp.Gr:** 0.95<br>**VP:** 0.5 mmHg<br>**FRZ:** -112°F<br>**UEL:** 3.0%<br>**LEL(392°F):** 1.1%<br>Class IIIA Combustible Liquid | **Skin:** N.R.<br>**Eyes:** N.R.<br>**Wash skin:** N.R.<br>**Remove:** N.R.<br>**Change:** N.R. | **NIOSH/OSHA**<br>**600 ppm:** Sa/ScbaF<br>**§:** ScbaF:Pd,Pp/SaF:Pd,Pp:AScba<br>**Escape:** GmFOv100/ScbaE |

**Incompatibilities and Reactivities:** Strong oxidizers

| Exposure Routes, Symptoms, Target Organs (see Table 5): | First Aid (see Table 6): |
|---|---|
| **ER:** Inh, Abs, Ing, Con<br>**SY:** Irrit eyes, nose, throat; lass, dizz, head<br>**TO:** Eyes, resp sys, CNS | **Eye:** Irr immed<br>**Skin:** Water wash prompt<br>**Breath:** Resp support<br>**Swallow:** Medical attention immed |

| Dipropyl ketone | Formula: $(CH_3CH_2CH_2)_2CO$ | CAS#: 123-19-3 | RTECS#: MJ5600000 | IDLH: N.D. |
|---|---|---|---|---|
| **Conversion:** 1 ppm = 4.67 mg/m$^3$ | **DOT:** 2710 128 | | | |

**Synonyms/Trade Names:** Butyrone, DPK, 4-Heptanone, Heptan-4-one, Propyl ketone

**Exposure Limits:**
**NIOSH REL:** TWA 50 ppm (235 mg/m$^3$)
**OSHA PEL†:** none

**Measurement Methods (see Table 1):**
**OSHA** 7

D

**Physical Description:** Colorless liquid with a pleasant odor.

**Chemical & Physical Properties:**
**MW:** 114.2
**BP:** 291°F
**Sol:** Insoluble
**Fl.P:** 120°F
**IP:** 9.10 eV
**Sp.Gr:** 0.82
**VP:** 5 mmHg
**FRZ:** -27°F
**UEL:** ?
**LEL:** ?
Class II Combustible Liquid

**Personal Protection/Sanitation (see Table 2):**
**Skin:** Prevent skin contact
**Eyes:** Prevent eye contact
**Wash skin:** Daily
**Remove:** When wet or contam
**Change:** N.R.

**Respirator Recommendations (see Tables 3 and 4):**
Not available.

**Incompatibilities and Reactivities:** Oxidizers

**Exposure Routes, Symptoms, Target Organs (see Table 5):**
**ER:** Inh, Ing, Con
**SY:** Irrit eyes, skin; CNS depres, dizz, drow, decr breath; in animals: liver inj; narco
**TO:** Eyes, skin, CNS, liver

**First Aid (see Table 6):**
**Eye:** Irr immed
**Skin:** Soap wash
**Breath:** Resp support
**Swallow:** Medical attention immed

| Diquat (Diquat dibromide) | Formula: $C_{12}H_{12}N_2Br_2$ | CAS#: 85-00-7 | RTECS#: JM5690000 | IDLH: N.D. |
|---|---|---|---|---|
| **Conversion:** | **DOT:** 2781 151 (solid); 2782 131 (liquid) | | | |

**Synonyms/Trade Names:** Diquat dibromide; 1,1'-Ethylene-2,2'-bipyridyllium dibromide
[**Note:** Diquat is a cation ($C_{12}H_{12}N_2^{++}$; 1,1'-Ethylene-2,2-bipyridyllium ion). Various diquat salts are commercially available.]

**Exposure Limits:**
**NIOSH REL:** TWA 0.5 mg/m$^3$
**OSHA PEL†:** none

**Measurement Methods (see Table 1):**
None available

**Physical Description:** Dibromide salt: Yellow crystals. [herbicide]
[**Note:** Commercial product may be found in a liquid concentrate or a solution.]

**Chemical & Physical Properties:**
**MW:** 344.1
**BP:** Decomposes
**Sol:** 70%
**Fl.P:** ?
**IP:** ?
**Sp.Gr:** 1.22-1.27
**VP:** <0.00001 mmHg
**MLT:** 635°F
**UEL:** ?
**LEL:** ?
Combustible Solid, but does not readily ignite and burns with difficulty.

**Personal Protection/Sanitation (see Table 2):**
**Skin:** Prevent skin contact
**Eyes:** Prevent eye contact
**Wash skin:** When contam
**Remove:** When wet or contam
**Change:** Daily
**Provide:** Quick drench

**Respirator Recommendations (see Tables 3 and 4):**
Not available.

**Incompatibilities and Reactivities:** Alkalis, UV light, basic solutions
[**Note:** Concentrated diquat solutions corrode aluminum.]

**Exposure Routes, Symptoms, Target Organs (see Table 5):**
**ER:** Inh, Abs, Ing, Con
**SY:** Irrit eyes, skin, muc memb, resp sys; rhin, epis; skin burns; nau, vomit, diarr, mal; kidney, liver inj; cough, chest pain, dysp, pulm edema; tremor, convuls; delayed healing of wounds
**TO:** Eyes, skin, resp sys, kidneys, liver, CNS

**First Aid (see Table 6):**
**Eye:** Irr immed
**Skin:** Water flush immed
**Breath:** Resp support
**Swallow:** Medical attention immed

## Disulfiram

| Disulfiram | Formula: $[(C_2H_5)_2NCS]_2S_2$ | CAS#: 97-77-8 | RTECS#: JO1225000 | IDLH: N.D. |
|---|---|---|---|---|
| **Conversion:** | **DOT:** | | | |

**Synonyms/Trade Names:** Antabuse®, bis(Diethylthiocarbamoyl) disulfide, Ro-Sulfiram®, TETD, Tetraethylthiuram disulfide

**Exposure Limits:**
**NIOSH REL:** TWA 2 mg/m$^3$
[Precautions should be taken to avoid concurrent exposure to ethylene dibromide.]
**OSHA PEL†:** none

**Measurement Methods (see Table 1):**
None available

**Physical Description:** White, yellowish, or light-gray powder with a slight odor. [fungicide]

**Chemical & Physical Properties:**
**MW:** 296.6
**BP:** ?
**Sol:** 0.02%
**Fl.P:** NA
**IP:** ?
**Sp.Gr:** 1.30
**VP:** ?
**MLT:** 158°F
**UEL:** NA
**LEL:** NA
Noncombustible Solid

**Personal Protection/Sanitation (see Table 2):**
**Skin:** Prevent skin contact
**Eyes:** Prevent eye contact
**Wash skin:** When contam
**Remove:** When wet or contam
**Change:** Daily

**Respirator Recommendations (see Tables 3 and 4):**
Not available.

**Incompatibilities and Reactivities:** None reported

**Exposure Routes, Symptoms, Target Organs (see Table 5):**
**ER:** Inh, Ing, Con
**SY:** Irrit eyes, skin, resp sys; sens derm; lass, tremor, restless, head, dizz; metallic taste; peri neur; liver damage
**TO:** Eyes, skin, resp sys, CNS, PNS, liver

**First Aid (see Table 6):**
**Eye:** Irr immed
**Skin:** Soap wash immed
**Breath:** Resp support
**Swallow:** Medical attention immed

## Disulfoton

| Disulfoton | Formula: $C_8H_{19}O_2PS_3$ | CAS#: 298-04-4 | RTECS#: TD9275000 | IDLH: N.D. |
|---|---|---|---|---|
| **Conversion:** | **DOT:** 2783 152 | | | |

**Synonyms/Trade Names:** O,O-Diethyl S-2-(ethylthio)-ethyl phosphorodithioate; Di-Syston®; Thiodemeton

**Exposure Limits:**
**NIOSH REL:** TWA 0.1 mg/m$^3$ [skin]
**OSHA PEL†:** none

**Measurement Methods (see Table 1):**
**NIOSH** 5600

**Physical Description:** Oily, colorless to yellow liquid with a characteristic, sulfur odor. [insecticide] [**Note:** Technical product is a brown liquid.]

**Chemical & Physical Properties:**
**MW:** 274.4
**BP:** ?
**Sol(73°F):** 0.003%
**Fl.P:** >180°F
**IP:** ?
**Sp.Gr:** 1.14
**VP:** 0.0002 mmHg
**FRZ:** >-13°F
**UEL:** ?
**LEL:** ?
Combustible Liquid, but will not ignite easily.

**Personal Protection/Sanitation (see Table 2):**
**Skin:** Prevent skin contact
**Eyes:** Prevent eye contact
**Wash skin:** When contam
**Remove:** When wet or contam
**Change:** Daily
**Provide:** Eyewash
Quick drench

**Respirator Recommendations (see Tables 3 and 4):**
Not available.

**Incompatibilities and Reactivities:** Alkalis

**Exposure Routes, Symptoms, Target Organs (see Table 5):**
**ER:** Inh, Abs, Ing, Con
**SY:** Irrit eyes, skin; nau, vomit, abdom cramps, diarr, salv; head, dizz, lass; rhin, chest tight; blurred vision, miosis; card irreg; musc fasc; dysp; eye, skin burns
**TO:** Eyes, skin, resp sys, CNS, CVS, blood chol

**First Aid (see Table 6):**
**Eye:** Irr immed
**Skin:** Soap flush immed
**Breath:** Resp support
**Swallow:** Medical attention immed

## Diuron

| | | | | |
|---|---|---|---|---|
| **Diuron** | **Formula:** $C_6H_3Cl_2NHCON(CH_3)_2$ | **CAS#:** 330-54-1 | **RTECS#:** YS8925000 | **IDLH:** N.D. |
| **Conversion:** | **DOT:** | | | |

**Synonyms/Trade Names:** 3-(3,4-Dichlorophenyl)-1,1-dimethylurea; Direx®; Karmex®

**Exposure Limits:**
**NIOSH REL:** TWA 10 mg/m³
**OSHA PEL†:** none

**Measurement Methods (see Table 1):**
**NIOSH** 5601
**OSHA** PV2097

**Physical Description:** White, odorless, crystalline solid. [herbicide]

| **Chemical & Physical Properties:** | **Personal Protection/Sanitation (see Table 2):** | **Respirator Recommendations (see Tables 3 and 4):** |
|---|---|---|
| **MW:** 233.1<br>**BP:** 356°F (Decomposes)<br>**Sol:** 0.004%<br>**Fl.P:** NA<br>**IP:**?<br>**Sp.Gr:** ?<br>**VP:** 0.000000002 mmHg<br>**MLT:** 316°F<br>**UEL:** NA<br>**LEL:** NA<br>Noncombustible Solid | **Skin:** Prevent skin contact<br>**Eyes:** Prevent eye contact<br>**Wash skin:** Daily<br>**Remove:** N.R.<br>**Change:** Daily | Not available. |

**Incompatibilities and Reactivities:** Strong acids

| **Exposure Routes, Symptoms, Target Organs (see Table 5):** | **First Aid (see Table 6):** |
|---|---|
| **ER:** Inh, Ing, Con<br>**SY:** Irrit eyes, skin, nose, throat; in animals: anemia, methemo<br>**TO:** Eyes, skin, resp sys, blood | **Eye:** Irr immed<br>**Skin:** Water flush immed<br>**Breath:** Resp support<br>**Swallow:** Medical attention immed |

## Divinyl benzene

| | | | | |
|---|---|---|---|---|
| **Divinyl benzene** | **Formula:** $C_6H_4(HC{=}CH_2)_2$ | **CAS#:** 1321-74-0 (mixed isomers) | **RTECS#:** CZ9370000 | **IDLH:** N.D. |
| **Conversion:** 1 ppm = 5.33 mg/m³ | **DOT:** 2049 130 | | | |

**Synonyms/Trade Names:** Diethyl benzene, DVB, Vinylstyrene
[**Note:** Commercial product contains all 3 isomers, but m-isomer predominates. Usually contains an inhibitor to prevent polymerization.]

**Exposure Limits:**
**NIOSH REL:** TWA 10 ppm (50 mg/m³)
**OSHA PEL†:** none

**Measurement Methods (see Table 1):**
**OSHA** 89

**Physical Description:** Pale, straw-colored liquid.

| **Chemical & Physical Properties:** | **Personal Protection/Sanitation (see Table 2):** | **Respirator Recommendations (see Tables 3 and 4):** |
|---|---|---|
| **MW:** 130.2<br>**BP:** 392°F<br>**Sol:** 0.005%<br>**Fl.P(oc):** 169°F<br>**IP:** ?<br>**Sp.Gr:** 0.93<br>**VP:** 0.7 mmHg<br>**FRZ:** -88°F<br>**UEL:** 6.2%<br>**LEL:** 1.1%<br>Class IIIA Combustible Liquid | **Skin:** Prevent skin contact<br>**Eyes:** Prevent eye contact<br>**Wash skin:** When contam<br>**Remove:** When wet or contam<br>**Change:** N.R.<br>**Provide:** Eyewash<br>Quick drench | Not available. |

**Incompatibilities and Reactivities:** None reported

| **Exposure Routes, Symptoms, Target Organs (see Table 5):** | **First Aid (see Table 6):** |
|---|---|
| **ER:** Inh, Ing, Con<br>**SY:** Irrit eyes, skin, resp sys; skin burns; in animals: CNS depres<br>**TO:** Eyes, skin, resp sys, CNS | **Eye:** Irr immed<br>**Skin:** Soap flush immed<br>**Breath:** Resp support<br>**Swallow:** Medical attention immed |

E

| 1-Dodecanethiol | Formula: $CH_3(CH_2)_{11}SH$ | CAS#: 112-55-0 | RTECS#: JR3155000 | IDLH: N.D. |
|---|---|---|---|---|
| **Conversion:** 1 ppm = 8.28 mg/m$^3$ | **DOT:** 1228 131 | | | |

**Synonyms/Trade Names:** Dodecyl mercaptan, 1-Dodecyl mercaptan, n-Dodecyl mercaptan, Lauryl mercaptan, n-Lauryl mercaptan, 1-Mercaptododecane

| Exposure Limits | Measurement Methods |
|---|---|
| **Exposure Limits:**<br>**NIOSH REL:** C 0.5 ppm (4.1 mg/m$^3$) [15-minute]<br>**OSHA PEL:** none<br>**Physical Description:** Colorless, water-white, or pale-yellow, oily liquid with a mild, skunk-like odor. [**Note:** A solid below 15°F.] | **Measurement Methods (see Table 1):**<br>None available |

| Chemical & Physical Properties | Personal Protection/Sanitation | Respirator Recommendations |
|---|---|---|
| **Chemical & Physical Properties:**<br>**MW:** 202.4<br>**BP:** 441-478°F<br>**Sol:** Insoluble<br>**Fl.P(oc):** 190°F<br>**IP:** ?<br>**Sp.Gr:** 0.85<br>**VP(77°F):** 3 mmHg<br>**FRZ:** 15°F<br>**UEL:** ?<br>**LEL:** ?<br>Class IIIA Combustible Liquid | **Personal Protection/Sanitation (see Table 2):**<br>**Skin:** Prevent skin contact<br>**Eyes:** Prevent eye contact<br>**Wash skin:** When contam<br>**Remove:** When wet or contam<br>**Change:** N.R.<br>**Provide:** Eyewash | **Respirator Recommendations (see Tables 3 and 4):**<br>**NIOSH**<br>**5 ppm:** CcrOv/Sa<br>**12.5 ppm:** Sa:Cf/PaprOv<br>**25 ppm:** CcrFOv/GmFOv/PaprTOv/ ScbaF/SaF<br>**§:** ScbaF:Pd,Pp/SaF:Pd,Pp:AScba<br>**Escape:** GmFOv/ScbaE |

**Incompatibilities and Reactivities:** Strong oxidizers & acids, strong bases, reducing agents, alkali metals, water, steam

| Exposure Routes, Symptoms, Target Organs | First Aid |
|---|---|
| **Exposure Routes, Symptoms, Target Organs (see Table 5):**<br>**ER:** Inh, Ing, Con<br>**SY:** Irrit eyes, skin, resp sys; cough; dizz, dysp, lass, conf, cyan; abdom pain, nau; skin sens<br>**TO:** Eyes, skin, resp sys, CNS, blood | **First Aid (see Table 6):**<br>**Eye:** Irr immed<br>**Skin:** Soap wash immed<br>**Breath:** Resp support<br>**Swallow:** Medical attention immed |

| Emery | Formula: $Al_2O_3$ | CAS#: 1302-74-5 (corundum) | RTECS#: GN2310000 (corundum) | IDLH: N.D. |
|---|---|---|---|---|
| **Conversion:** | **DOT:** | | | |

**Synonyms/Trade Names:** Aluminum oxide, Aluminum trioxide, Corundum, Impure corundum, Natural aluminum oxide [**Note:** Emery is an impure variety of $Al_2O_3$ which may contain small impurities of iron, magnesium & silica Corundum is natural $Al_2O_3$.]

| Exposure Limits | Measurement Methods |
|---|---|
| **Exposure Limits:**<br>**NIOSH REL:** See Appendix D<br>**OSHA PEL†:** TWA 15 mg/m$^3$ (total)<br>TWA 5 mg/m$^3$ (resp)<br>**Physical Description:** Odorless, white, crystalline powder. | **Measurement Methods (see Table 1):**<br>**NIOSH** 0500, 0600 |

| Chemical & Physical Properties | Personal Protection/Sanitation | Respirator Recommendations |
|---|---|---|
| **Chemical & Physical Properties:**<br>See $\alpha$-Alumina for physical & chemical properties. | **Personal Protection/Sanitation (see Table 2):**<br>**Skin:** N.R.<br>**Eyes:** N.R.<br>**Wash skin:** N.R.<br>**Remove:** N.R.<br>**Change:** N.R. | **Respirator Recommendations (see Tables 3 and 4):**<br>Not available. |

**Incompatibilities and Reactivities:**

| Exposure Routes, Symptoms, Target Organs | First Aid |
|---|---|
| **Exposure Routes, Symptoms, Target Organs (see Table 5):**<br>**ER:** Inh, Ing, Con<br>**SY:** Irrit eyes, skin, resp sys<br>**TO:** Eyes, skin, resp sys | **First Aid (see Table 6):**<br>**Eye:** Irr immed<br>**Breath:** Fresh air<br>**Swallow:** Medical attention immed |

## Endosulfan

| Endosulfan | Formula: $C_9H_6Cl_6O_3S$ | CAS#: 115-29-7 | RTECS#: RB9275000 | IDLH: N.D. |
|---|---|---|---|---|
| Conversion: | DOT: 2761 151 | | | |

**Synonyms/Trade Names:** Benzoepin; Endosulphan; 6,7,8,9,10-Hexachloro-1,5,5a,6,9,9a-hexachloro-6,9-methano-2,4,3-benzo-dioxathiepin-3-oxide; Thiodan®

**Exposure Limits:**
**NIOSH REL:** TWA 0.1 mg/m³ [skin] **OSHA PEL†:** none

**Measurement Methods (see Table 1):**
**OSHA** PV2023

**Physical Description:** Brown crystals with a slight, sulfur dioxide odor. [insecticide]
[**Note:** Technical product is a tan, waxy, isomer mixture.]

E

**Chemical & Physical Properties:**
**MW:** 406.9
**BP:** Decomposes
**Sol:** 0.00001%
**Fl.P:** NA
**IP:** ?
**Sp.Gr:** 1.74
**VP(77°F):** 0.00001 mmHg
**MLT:** 223°F
**UEL:** NA
**LEL:** NA Noncombustible Solid, but may be dissolved in flammable liquids.

**Personal Protection/Sanitation (see Table 2):**
**Skin:** Prevent skin contact
**Eyes:** Prevent eye contact
**Wash skin:** When contam
**Remove:** When wet or contam
**Change:** Daily
**Provide:** Eyewash
Quick drench

**Respirator Recommendations (see Tables 3 and 4):**
Not available.

**Incompatibilities and Reactivities:** Alkalis, acids, water
[**Note:** Corrosive to iron. Hydrolyzes slowly on contact with water or decomposes in presence of alkalis and acids to form sulfur dioxide.]

**Exposure Routes, Symptoms, Target Organs (see Table 5):**
**ER:** Inh, Abs, Ing, Con
**SY:** Irrit skin; nau, conf, agitation, flushing, dry mouth, tremor, convuls, head; in animals: kidney, liver inj; decr testis weight
**TO:** Skin, CNS, liver, kidneys, repro sys

**First Aid (see Table 6):**
**Eye:** Irr immed
**Skin:** Soap flush immed
**Breath:** Resp support
**Swallow:** Medical attention immed

## Endrin

| Endrin | Formula: $C_{12}H_8Cl_6O$ | CAS#: 72-20-8 | RTECS#: IO1575000 | IDLH: 2 mg/m³ |
|---|---|---|---|---|
| Conversion: | DOT: 2761 151 | | | |

**Synonyms/Trade Names:** Hexadrin®, 1,2,3,4,10,10-Hexachloro-6,7-epoxy-1,4,4a,5,6,7,8,8a-octahydro-1,4-endo,endo-5,8-dimethanonaphthalene

**Exposure Limits:**
**NIOSH REL:** TWA 0.1 mg/m³ [skin]
**OSHA PEL:** TWA 0.1 mg/m³ [skin]

**Measurement Methods (see Table 1):**
**NIOSH** 5519

**Physical Description:** Colorless to tan, crystalline solid with a mild, chemical odor. [insecticide]

**Chemical & Physical Properties:**
**MW:** 380.9
**BP:** Decomposes
**Sol:** Insoluble
**Fl.P:** NA
**IP:** ?
**Sp.Gr:** 1.70
**VP:** Low
**MLT:** 392°F (Decomposes)
**UEL:** NA
**LEL:** NA Noncombustible Solid, but may be dissolved in flammable liquids.

**Personal Protection/Sanitation (see Table 2):**
**Skin:** Prevent skin contact
**Eyes:** Prevent eye contact
**Wash skin:** When contam
**Remove:** When wet or contam
**Change:** Daily
**Provide:** Eyewash
Quick drench

**Respirator Recommendations (see Tables 3 and 4):**
**NIOSH/OSHA**
**1 mg/m³:** CcrOv95/Sa
**2 mg/m³:** Sa:Cf/PaprOvHie/CcrFOv100/GmFOv100/ScbaF/SaF
**§:** ScbaF:Pd,Pp/SaF:Pd,Pp:AScba
**Escape:** GmFOv100/ScbaE

**Incompatibilities and Reactivities:** Strong oxidizers, strong acids, parathion
[**Note:** May emit hydrogen chloride & phosgene when heated or burned.]

**Exposure Routes, Symptoms, Target Organs (see Table 5):**
**ER:** Inh, Abs, Ing, Con
**SY:** Epilep convuls; stupor, head, dizz; abdom discomfort, nau, vomit; insom; aggressiveness, conf; drow, lass; anor; in animals: liver damage
**TO:** CNS, liver

**First Aid (see Table 6):**
**Eye:** Irr immed
**Skin:** Soap wash immed
**Breath:** Resp support
**Swallow:** Medical attention immed

E

| Enflurane | Formula: $CHF_2OCF_2CHClF$ | CAS#: 13838-16-9 | RTECS#: KN6800000 | IDLH: N.D. |
|---|---|---|---|---|
| **Conversion:** 1 ppm = 7.55 mg/m$^3$ | **DOT:** | | | |

**Synonyms/Trade Names:** 2-Chloro-1-(difluoromethoxy)-1,1,2-trifluoroethane; 2-Chloro-1,1,2-trifluoroethyl difluoromethyl ether; Ethrane®

**Exposure Limits:**
**NIOSH REL*:** C 2 ppm (15.1 mg/m$^3$) [60-minute]
[***Note:** REL for exposure to waste anesthetic gas.]
**OSHA PEL:** none

**Measurement Methods (see Table 1):**
**OSHA** 29, 103

**Physical Description:** Clear, colorless liquid with a mild, sweet odor. [inhalation anesthetic]

**Chemical & Physical Properties:**
**MW:** 184.5
**BP:** 134°F
**Sol:** Low
**Fl.P:** NA
**IP:** ?
**Sp.Gr(77°F):** 1.52
**VP:** 175 mmHg
**FRZ:** ?
**UEL:** NA
**LEL:** NA
Noncombustible Liquid

**Personal Protection/Sanitation (see Table 2):**
**Skin:** N.R.
**Eyes:** Prevent eye contact
**Wash skin:** N.R.
**Remove:** N.R.
**Change:** N.R.

**Respirator Recommendations (see Tables 3 and 4):**
Not available.

**Incompatibilities and Reactivities:** None reported

**Exposure Routes, Symptoms, Target Organs (see Table 5):**
**ER:** Inh, Ing, Con
**SY:** Irrit eyes; CNS depres, analgesia, anes, convuls, resp depres
**TO:** Eyes, CNS

**First Aid (see Table 6):**
**Eye:** Irr immed
**Skin:** Soap wash
**Breath:** Resp support
**Swallow:** Medical attention immed

| Epichlorohydrin | Formula: $C_3H_5OCl$ | CAS#: 106-89-8 | RTECS#: TX4900000 | IDLH: Ca [75 ppm] |
|---|---|---|---|---|
| **Conversion:** 1 ppm = 3.78 mg/m$^3$ | **DOT:** 2023 131P | | | |

**Synonyms/Trade Names:** 1-Chloro-2,3-epoxypropane; 2-Chloropropylene oxide; γ-Chloropropylene oxide

**Exposure Limits:**
**NIOSH REL:** Ca
See Appendix A
**OSHA PEL†:** TWA 5 ppm (19 mg/m$^3$) [skin]

**Measurement Methods (see Table 1):**
**NIOSH** 1010
**OSHA** 7

**Physical Description:** Colorless liquid with a slightly irritating, chloroform-like odor.

**Chemical & Physical Properties:**
**MW:** 92.5
**BP:** 242°F
**Sol:** 7%
**Fl.P:** 93°F
**IP:** 10.60 eV
**Sp.Gr:** 1.18
**VP:** 13 mmHg
**FRZ:** -54°F
**UEL:** 21.0%
**LEL:** 3.8%
Class IC Flammable Liquid

**Personal Protection/Sanitation (see Table 2):**
**Skin:** Prevent skin contact
**Eyes:** Prevent eye contact
**Wash skin:** When contam
**Remove:** When wet (flamm)
**Change:** N.R.
**Provide:** Eyewash
Quick drench

**Respirator Recommendations (see Tables 3 and 4):**
**NIOSH**
**¥:** ScbaF:Pd,Pp/SaF:Pd,Pp:AScba
**Escape:** GmFOvAg/ScbaE

**Incompatibilities and Reactivities:** Strong oxidizers, strong acids, certain salts, caustics, zinc, aluminum, water
[**Note:** May polymerize in presence of strong acids and bases, particularly when hot.]

**Exposure Routes, Symptoms, Target Organs (see Table 5):**
**ER:** Inh, Abs, Ing, Con
**SY:** Irrit eyes, skin with deep pain; nau, vomit; abdom pain; resp distress, cough; cyan; repro effects; [carc]
**TO:** Eyes, skin, resp sys, kidneys, liver, repro sys [in animals: nasal cancer]

**First Aid (see Table 6):**
**Eye:** Irr immed
**Skin:** Soap wash immed
**Breath:** Resp support
**Swallow:** Medical attention immed

| EPN | Formula: $C_{14}H_{14}O_4NSP$ | CAS#: 2104-64-5 | RTECS#: TB1925000 | IDLH: 5 mg/m$^3$ |
|---|---|---|---|---|
| **Conversion:** | **DOT:** | | | |

**Synonyms/Trade Names:** Ethyl p-nitrophenyl benzenethionophosphonate, O-Ethyl O-(4-nitrophenyl) phenylphosphonothioate

**Exposure Limits:**
**NIOSH REL:** TWA 0.5 mg/m$^3$ [skin]
**OSHA PEL:** TWA 0.5 mg/m$^3$ [skin]

**Measurement Methods (see Table 1):**
**NIOSH** 5012

**Physical Description:** Yellow solid with an aromatic odor. [pesticide]
[**Note:** A brown liquid above 97°F.]

**Chemical & Physical Properties:**
**MW:** 323.3
**BP:** ?
**Sol:** Insoluble
**Fl.P:** NA
**IP:** ?
**Sp.Gr(77°F):** 1.27
**VP(212°F):** 0.0003 mmHg
**MLT:** 97°F
**UEL:** NA
**LEL:** NA
Noncombustible Solid

**Personal Protection/Sanitation (see Table 2):**
**Skin:** Prevent skin contact
**Eyes:** Prevent eye contact
**Wash skin:** When contam
**Remove:** When wet or contam
**Change:** Daily
**Provide:** Eyewash
Quick drench

**Respirator Recommendations (see Tables 3 and 4):**
**NIOSH/OSHA**
**5 mg/m$^3$:** Sa/ScbaF
**§:** ScbaF:Pd,Pp/SaF:Pd,Pp:AScba
**Escape:** GmFOv100/ScbaE

**Incompatibilities and Reactivities:** Strong oxidizers

**Exposure Routes, Symptoms, Target Organs (see Table 5):**
**ER:** Inh, Abs, Ing, Con
**SY:** Irrit eyes, skin; miosis, lac; rhin; head; chest tight, wheez, lar spasm; salv; cyan; anor, nau, abdom cramps, diarr; para, convuls; low BP, card irreg
**TO:** Eyes, skin, resp sys, CVS, CNS, blood chol

**First Aid (see Table 6):**
**Eye:** Irr immed
**Skin:** Soap wash immed
**Breath:** Resp support
**Swallow:** Medical attention immed

E

| Ethanolamine | Formula: $NH_2CH_2CH_2OH$ | CAS#: 141-43-5 | RTECS#: KJ5775000 | IDLH: 30 ppm |
|---|---|---|---|---|
| **Conversion:** 1 ppm = 2.50 mg/m$^3$ | **DOT:** 2491 153 | | | |

**Synonyms/Trade Names:** 2-Aminoethanol, β-Aminoethyl alcohol, Ethylolamine, 2-Hydroxyethylamine, Monoethanolamine

**Exposure Limits:**
**NIOSH REL:** TWA 3 ppm (8 mg/m$^3$)
ST 6 ppm (15 mg/m$^3$)
**OSHA PEL†:** TWA 3 ppm (6 mg/m$^3$)

**Measurement Methods (see Table 1):**
**NIOSH** 2007

**Physical Description:** Colorless, viscous liquid or solid (below 51°F) with an unpleasant, ammonia-like odor.

**Chemical & Physical Properties:**
**MW:** 61.1
**BP:** 339°F
**Sol:** Miscible
**Fl.P:** 186°F
**IP:** 8.96 eV
**Sp.Gr:** 1.02
**VP:** 0.4 mmHg
**FRZ:** 51°F
**UEL:** 23.5%
**LEL(284°F):** 3.0%
Class IIIA Combustible Liquid

**Personal Protection/Sanitation (see Table 2):**
**Skin:** Prevent skin contact
**Eyes:** Prevent eye contact
**Wash skin:** When contam
**Remove:** When wet or contam
**Change:** Daily
**Provide:** Eyewash

**Respirator Recommendations (see Tables 3 and 4):**
**NIOSH/OSHA**
**30 ppm:** CcrS*/GmFS/PaprS*/Sa*/ScbaF
**§:** ScbaF:Pd,Pp/SaF:Pd,Pp:AScba
**Escape:** GmFS/ScbaE

**Incompatibilities and Reactivities:** Strong oxidizers, strong acids, iron
[**Note:** May attack copper, brass, and rubber.]

**Exposure Routes, Symptoms, Target Organs (see Table 5):**
**ER:** Inh, Ing, Con
**SY:** Irrit eyes, skin, resp sys; drow
**TO:** Eyes, skin, resp sys, CNS

**First Aid (see Table 6):**
**Eye:** Irr immed
**Skin:** Water flush prompt
**Breath:** Resp support
**Swallow:** Medical attention immed

## Ethion

| Formula: | CAS#: | RTECS#: | IDLH: |
|---|---|---|---|
| $[(C_2H_5O)_2P(S)S]_2CH_2$ | 563-12-2 | TE4550000 | N.D. |

**Conversion:**
**DOT:**

**Synonyms/Trade Names:** O,O,O',O'-Tetraethyl S,S'-methylene di(phosphorodithioate)

**Exposure Limits:**
**NIOSH REL:** 0.4 mg/m$^3$ [skin]
**OSHA PEL†:** none

**Measurement Methods (see Table 1):**
**NIOSH** 5600

**Physical Description:** Colorless to amber-colored, odorless liquid. [insecticide]
[**Note:** A solid below 10°F. The technical product has a very disagreeable odor.]

**Chemical & Physical Properties:**
**MW:** 384.5
**BP:** >302°F (Decomposes)
**Sol:** 0.0001%
**Fl.P:** 349°F
**IP:** ?
**Sp.Gr:** 1.22
**VP:** 0.0000015 mmHg
**FRZ:** 10°F
**UEL:** ?
**LEL:** ?
Class IIIB Combustible Liquid

**Personal Protection/Sanitation (see Table 2):**
**Skin:** Prevent skin contact
**Eyes:** Prevent eye contact
**Wash skin:** When contam
**Remove:** When wet or contam
**Change:** Daily
**Provide:** Eyewash
Quick drench

**Respirator Recommendations (see Tables 3 and 4):**
Not available.

**Incompatibilities and Reactivities:** Acids, alkalis

**Exposure Routes, Symptoms, Target Organs (see Table 5):**
**ER:** Inh, Abs, Ing, Con
**SY:** Irrit eyes, skin; nau, vomit, abdom cramps, diarr, salv; head, dizz, lass; rhin, chest tight; blurred vision, miosis; card irreg; musc fasc; dysp
**TO:** Eyes, skin, resp sys, CNS, CVS, blood chol

**First Aid (see Table 6):**
**Eye:** Irr immed
**Skin:** Soap wash immed
**Breath:** Resp support
**Swallow:** Medical attention immed

## 2-Ethoxyethanol

| Formula: | CAS#: | RTECS#: | IDLH: |
|---|---|---|---|
| $C_2H_5OCH_2CH_2OH$ | 110-80-5 | KK8050000 | 500 ppm |

**Conversion:** 1 ppm = 3.69 mg/m$^3$
**DOT:** 1171 127

**Synonyms/Trade Names:** Cellosolve®, EGEE, Ethylene glycol monoethyl ether

**Exposure Limits:**
**NIOSH REL:** TWA 0.5 ppm (1.8 mg/m$^3$) [skin]
**OSHA PEL:** TWA 200 ppm (740 mg/m$^3$) [skin]

**Measurement Methods (see Table 1):**
**NIOSH** 1403
**OSHA** 53, 79

**Physical Description:** Colorless liquid with a sweet, pleasant, ether-like odor.

**Chemical & Physical Properties:**
**MW:** 90.1
**BP:** 275°F
**Sol:** Miscible
**Fl.P:** 110°F
**IP:** ?
**Sp.Gr:** 0.93
**VP:** 4 mmHg
**FRZ:** -130°F
**UEL(200°F):** 15.6%
**LEL(200°F):** 1.7%
Class II Combustible Liquid

**Personal Protection/Sanitation (see Table 2):**
**Skin:** Prevent skin contact
**Eyes:** Prevent eye contact
**Wash skin:** When contam
**Remove:** When wet or contam
**Change:** N.R.

**Respirator Recommendations (see Tables 3 and 4):**
**NIOSH**
**5 ppm:** Sa*
**12.5 ppm:** Sa:Cf*
**25 ppm:** ScbaF/SaF
**500 ppm:** Sa:Pd,Pp*
**§:** ScbaF:Pd,Pp/SaF:Pd,Pp:AScba
**Escape:** GmFOv/ScbaE

**Incompatibilities and Reactivities:** Strong oxidizers

**Exposure Routes, Symptoms, Target Organs (see Table 5):**
**ER:** Inh, Abs, Ing, Con
**SY:** In animals: irrit eyes, resp sys; blood changes; liver, kidney, lung damage; repro, terato effects
**TO:** Eyes, resp sys, blood, kidneys, liver, repro sys, hemato sys

**First Aid (see Table 6):**
**Eye:** Irr immed
**Skin:** Water flush prompt
**Breath:** Resp support
**Swallow:** Medical attention immed

## 2-Ethoxyethyl acetate

| 2-Ethoxyethyl acetate | Formula: $CH_3COOCH_2CH_2OC_2H_5$ | CAS#: 111-15-9 | RTECS#: KK8225000 | IDLH: 500 ppm |
|---|---|---|---|---|
| **Conversion:** 1 ppm = 5.41 mg/m$^3$ | **DOT:** 1172 129 | | | |

**Synonyms/Trade Names:** Cellosolve® acetate, EGEEA, Ethylene glycol monoethyl ether acetate, Glycol monoethyl ether acetate

**Exposure Limits:**
**NIOSH REL:** TWA 0.5 ppm (2.7 mg/m$^3$) [skin]
**OSHA PEL:** TWA 100 ppm (540 mg/m$^3$) [skin]

**Measurement Methods (see Table 1):**
**NIOSH** 1450
**OSHA** 53

**Physical Description:** Colorless liquid with a mild odor.

**Chemical & Physical Properties:**
**MW:** 132.2
**BP:** 313°F
**Sol:** 23%
**Fl.P:** 124°F
**IP:** ?
**Sp.Gr:** 0.98
**VP:** 2 mmHg
**FRZ:** -79°F
**UEL:** ?
**LEL:** 1.7%
Class II Combustible Liquid

**Personal Protection/Sanitation (see Table 2):**
**Skin:** Prevent skin contact
**Eyes:** Prevent eye contact
**Wash skin:** When contam
**Remove:** When wet or contam
**Change:** N.R.

**Respirator Recommendations (see Tables 3 and 4):**
**NIOSH**
**5 ppm:** CcrOv*/Sa*
**12.5 ppm:** Sa:Cf*/PaprOv*
**25 ppm:** CcrFOv/GmFOv/PaprTOv*/ScbaF/SaF
**500 ppm:** Sa:Pd,Pp*
**§:** ScbaF:Pd,Pp/SaF:Pd,Pp:AScba
**Escape:** GmFOv/ScbaE

**Incompatibilities and Reactivities:** Nitrates; strong oxidizers, alkalis & acids

**Exposure Routes, Symptoms, Target Organs (see Table 5):**
**ER:** Inh, Abs, Ing, Con
**SY:** Irrit eyes, nose; vomit; kidney damage; para; in animals: repro, terato effects
**TO:** Eyes, resp sys, GI tract, repro sys, hemato sys

**First Aid (see Table 6):**
**Eye:** Irr immed
**Skin:** Water flush prompt
**Breath:** Resp support
**Swallow:** Medical attention immed

## Ethyl acetate

| Ethyl acetate | Formula: $CH_3COOC_2H_5$ | CAS#: 141-78-6 | RTECS#: AH5425000 | IDLH: 2000 ppm [10%LEL] |
|---|---|---|---|---|
| **Conversion:** 1 ppm = 3.60 mg/m$^3$ | **DOT:** 1173 129 | | | |

**Synonyms/Trade Names:** Acetic ester, Acetic ether, Ethyl ester of acetic acid, Ethyl ethanoate

**Exposure Limits:**
**NIOSH REL:** TWA 400 ppm (1400 mg/m$^3$)
**OSHA PEL:** TWA 400 ppm (1400 mg/m$^3$)

**Measurement Methods (see Table 1):**
**NIOSH** 1457
**OSHA** 7

**Physical Description:** Colorless liquid with an ether-like, fruity odor.

**Chemical & Physical Properties:**
**MW:** 88.1
**BP:** 171°F
**Sol(77°F):** 10%
**Fl.P:** 24°F
**IP:** 10.01 eV
**Sp.Gr:** 0.90
**VP:** 73 mmHg
**FRZ:** -117°F
**UEL:** 11.5%
**LEL:** 2.0%
Class IB Flammable Liquid

**Personal Protection/Sanitation (see Table 2):**
**Skin:** Prevent skin contact
**Eyes:** Prevent eye contact
**Wash skin:** When contam
**Remove:** When wet (flamm)
**Change:** N.R.

**Respirator Recommendations (see Tables 3 and 4):**
**NIOSH/OSHA**
**2000 ppm:** Sa:Cf£/PaprOv£/CcrFOv/GmFOv/ScbaF/SaF
**§:** ScbaF:Pd,Pp/SaF:Pd,Pp:AScba
**Escape:** GmFOv/ScbaE

**Incompatibilities and Reactivities:** Nitrates; strong oxidizers, alkalis & acids

**Exposure Routes, Symptoms, Target Organs (see Table 5):**
**ER:** Inh, Ing, Con
**SY:** Irrit eyes, skin, nose, throat; narco; derm
**TO:** Eyes, skin, resp sys

**First Aid (see Table 6):**
**Eye:** Irr immed
**Skin:** Water flush prompt
**Breath:** Resp support
**Swallow:** Medical attention immed

## Ethyl acrylate

| Formula: | CAS#: | RTECS#: | IDLH: |
|---|---|---|---|
| $CH_2=CHCOOC_2H_5$ | 140-88-5 | AT0700000 | Ca [300 ppm] |

**Conversion:** 1 ppm = 4.09 mg/m$^3$ | **DOT:** 1917 129P (inhibited)

**Synonyms/Trade Names:** Ethyl acrylate (inhibited), Ethyl ester of acrylic acid, Ethyl propenoate

**Exposure Limits:**
**NIOSH REL:** Ca
See Appendix A
**OSHA PEL†:** TWA 25 ppm (100 mg/m$^3$) [skin]

**Measurement Methods (see Table 1):**
**NIOSH** 1450
**OSHA** 92

**Physical Description:** Colorless liquid with an acrid odor.

**Chemical & Physical Properties:**
**MW:** 100.1
**BP:** 211°F
**Sol:** 2%
**Fl.P:** 48°F
**IP:** 10.30 eV
**Sp.Gr:** 0.92
**VP:** 29 mmHg
**FRZ:** -96°F
**UEL:** 14%
**LEL:** 1.4%
Class IB Flammable Liquid

**Personal Protection/Sanitation (see Table 2):**
**Skin:** Prevent skin contact
**Eyes:** Prevent eye contact
**Wash skin:** When contam
**Remove:** When wet (flamm)
**Change:** N.R.
**Provide:** Eyewash
Quick drench

**Respirator Recommendations (see Tables 3 and 4):**
**NIOSH**
**¥:** ScbaF:Pd,Pp/SaF:Pd,Pp:AScba
**Escape:** GmFOv/ScbaE

**Incompatibilities and Reactivities:** Oxidizers, peroxides, polymerizers, strong alkalis, moisture, chlorosulfonic acid [**Note:** Polymerizes readily unless an inhibitor such as hydroquinone is added.]

**Exposure Routes, Symptoms, Target Organs (see Table 5):**
**ER:** Inh, Abs, Ing, Con
**SY:** Irrit eyes, skin, resp sys; [carc]
**TO:** Eyes, skin, resp sys [in animals: tumors of the forestomach]

**First Aid (see Table 6):**
**Eye:** Irr immed
**Skin:** Water flush immed
**Breath:** Resp support
**Swallow:** Medical attention immed

## Ethyl alcohol

| Formula: | CAS#: | RTECS#: | IDLH: |
|---|---|---|---|
| $CH_3CH_2OH$ | 64-17-5 | KQ6300000 | 3300 ppm [10%LEL] |

**Conversion:** 1 ppm = 1.89 mg/m$^3$ | **DOT:** 1170 127

**Synonyms/Trade Names:** Alcohol, Cologne spirit, Ethanol, EtOH, Grain alcohol

**Exposure Limits:**
**NIOSH REL:** TWA 1000 ppm (1900 mg/m$^3$)
**OSHA PEL:** TWA 1000 ppm (1900 mg/m$^3$)

**Measurement Methods (see Table 1):**
**NIOSH** 1400
**OSHA** 100

**Physical Description:** Clear, colorless liquid with a weak, ethereal, vinous odor.

**Chemical & Physical Properties:**
**MW:** 46.1
**BP:** 173°F
**Sol:** Miscible
**Fl.P:** 55°F
**IP:** 10.47 eV
**Sp.Gr:** 0.79
**VP:** 44 mmHg
**FRZ:** -173°F
**UEL:** 19%
**LEL:** 3.3%
Class IB Flammable Liquid

**Personal Protection/Sanitation (see Table 2):**
**Skin:** Prevent skin contact
**Eyes:** Prevent eye contact
**Wash skin:** When contam
**Remove:** When wet (flamm)
**Change:** N.R.

**Respirator Recommendations (see Tables 3 and 4):**
**NIOSH/OSHA**
**3300 ppm:** Sa/ScbaF
**§:** ScbaF:Pd,Pp/SaF:Pd,Pp:AScba
**Escape:** ScbaE

**Incompatibilities and Reactivities:** Strong oxidizers, potassium dioxide, bromine pentafluoride, acetyl bromide, acetyl chloride, platinum, sodium

**Exposure Routes, Symptoms, Target Organs (see Table 5):**
**ER:** Inh, Ing, Con
**SY:** Irrit eyes, skin, nose; head, drow, lass, narco; cough; liver damage; anemia; repro, terato effects
**TO:** Eyes, skin, resp sys, CNS, liver, blood, repro sys

**First Aid (see Table 6):**
**Eye:** Irr immed
**Skin:** Water flush prompt
**Breath:** Fresh air
**Swallow:** Medical attention immed

## Ethylamine

**Formula:** $CH_3CH_2NH_2$ | **CAS#:** 75-04-7 | **RTECS#:** KH2100000 | **IDLH:** 600 ppm

**Conversion:** 1 ppm = 1.85 mg/m$^3$ | **DOT:** 1036 118

**Synonyms/Trade Names:** Aminoethane, Ethylamine (anhydrous), Monoethylamine

**Exposure Limits:**
**NIOSH REL:** TWA 10 ppm (18 mg/m$^3$)
**OSHA PEL:** TWA 10 ppm (18 mg/m$^3$)

**Measurement Methods (see Table 1):**
**NIOSH** S144 (II-3)
**OSHA** 36

**Physical Description:** Colorless gas or water-white liquid (below 62°F) with an ammonia-like odor. [**Note:** Shipped as a liquefied compressed gas.]

**Chemical & Physical Properties:**
**MW:** 45.1
**BP:** 62°F
**Sol:** Miscible
**Fl.P:** 1°F
**IP:** 8.86 eV
**RGasD:** 1.61
**Sp.Gr:** 0.69 (Liquid)
**VP:** 874 mmHg
**FRZ:** -114°F
**UEL:** 14.0%
**LEL:** 3.5%
Flammable Gas

**Personal Protection/Sanitation (see Table 2):**
**Skin:** Prevent skin contact (liquid)
**Eyes:** Prevent eye contact (liquid)
**Wash skin:** When contam (liquid)
**Remove:** When wet or contam (liquid)
**Change:** N.R.
**Provide:** Eyewash (liquid)
Quick drench (liquid)

**Respirator Recommendations (see Tables 3 and 4):**
**NIOSH/OSHA**
**250 ppm:** Sa:Cf£/PaprS£
**500 ppm:** CcrFS/GmFS/ScbaF/SaF
**600 ppm:** SaF:Pd,Pp
**§:** ScbaF:Pd,Pp/SaF:Pd,Pp:AScba
**Escape:** GmFS/ScbaE

**Incompatibilities and Reactivities:** Strong acids; strong oxidizers; copper, tin & zinc in presence of moisture; cellulose nitrate; chlorine; hypochlorites

**Exposure Routes, Symptoms, Target Organs (see Table 5):**
**ER:** Inh, Abs (liquid), Ing (liquid), Con (liquid)
**SY:** Irrit eyes, skin, resp sys; skin burns, derm
**TO:** Eyes, skin, resp sys

**First Aid (see Table 6):**
**Eye:** Irr immed (liquid)
**Skin:** Water flush immed (liquid)
**Breath:** Resp support
**Swallow:** Medical attention immed (liquid)

## Ethyl benzene

**Formula:** $CH_3CH_2C_6H_5$ | **CAS#:** 100-41-4 | **RTECS#:** DA0700000 | **IDLH:** 800 ppm [10%LEL]

**Conversion:** 1 ppm = 4.34 mg/m$^3$ | **DOT:** 1175 130

**Synonyms/Trade Names:** Ethylbenzol, Phenylethane

**Exposure Limits:**
**NIOSH REL:** TWA 100 ppm (435 mg/m$^3$)
ST 125 ppm (545 mg/m$^3$)
**OSHA PEL†:** TWA 100 ppm (435 mg/m$^3$)

**Measurement Methods (see Table 1):**
**NIOSH** 1501
**OSHA** 7, 1002

**Physical Description:** Colorless liquid with an aromatic odor.

**Chemical & Physical Properties:**
**MW:** 106.2
**BP:** 277°F
**Sol:** 0.01%
**Fl.P:** 55°F
**IP:** 8.76 eV
**Sp.Gr:** 0.87
**VP:** 7 mmHg
**FRZ:** -139°F
**UEL:** 6.7%
**LEL:** 0.8%
Class IB Flammable Liquid

**Personal Protection/Sanitation (see Table 2):**
**Skin:** Prevent skin contact
**Eyes:** Prevent eye contact
**Wash skin:** When contam
**Remove:** When wet (flamm)
**Change:** N.R.

**Respirator Recommendations (see Tables 3 and 4):**
**NIOSH/OSHA**
**800 ppm:** CcrOv*/GmFOv/PaprOv*/Sa*/ScbaF
**§:** ScbaF:Pd,Pp/SaF:Pd,Pp:AScba
**Escape:** GmFOv/ScbaE

**Incompatibilities and Reactivities:** Strong oxidizers

**Exposure Routes, Symptoms, Target Organs (see Table 5):**
**ER:** Inh, Ing, Con
**SY:** Irrit eyes, skin, muc memb; head; derm; narco, coma
**TO:** Eyes, skin, resp sys, CNS

**First Aid (see Table 6):**
**Eye:** Irr immed
**Skin:** Water flush prompt
**Breath:** Resp support
**Swallow:** Medical attention immed

E

| Ethyl bromide | Formula: $CH_3CH_2Br$ | CAS#: 74-96-4 | RTECS#: KH6475000 | IDLH: 2000 ppm |
|---|---|---|---|---|
| **Conversion:** 1 ppm = 4.46 mg/m$^3$ | **DOT:** 1891 131 | | | |

**Synonyms/Trade Names:** Bromoethane, Monobromoethane

**Exposure Limits:**
**NIOSH REL:** See Appendix D
**OSHA PEL†:** TWA 200 ppm (890 mg/m$^3$)

**Measurement Methods (see Table 1):**
**NIOSH** 1011
**OSHA** 7

**Physical Description:** Colorless to yellow liquid with an ether-like odor. [**Note:** A gas above 101°F.]

| Chemical & Physical Properties: | Personal Protection/Sanitation (see Table 2): | Respirator Recommendations (see Tables 3 and 4): |
|---|---|---|
| **MW:** 109.0<br>**BP:** 101°F<br>**Sol:** 0.9%<br>**Fl.P:** <4°F<br>**IP:** 10.29 eV<br>**Sp.Gr:** 1.46<br>**VP:** 375 mmHg<br>**FRZ:** -182°F<br>**UEL:** 8.0%<br>**LEL:** 6.8%<br>Class IB Flammable Liquid | **Skin:** Prevent skin contact<br>**Eyes:** Prevent eye contact<br>**Wash skin:** When contam<br>**Remove:** When wet (flamm)<br>**Change:** N.R. | **OSHA**<br>**2000 ppm:** Sa/ScbaF<br>**§:** ScbaF:Pd,Pp/SaF:Pd,Pp:AScba<br>**Escape:** GmFOv/ScbaE |

**Incompatibilities and Reactivities:** Chemically-active metals such as sodium, potassium, calcium, powdered aluminum, zinc & magnesium

| Exposure Routes, Symptoms, Target Organs (see Table 5): | First Aid (see Table 6): |
|---|---|
| **ER:** Inh, Ing, Con<br>**SY:** Irrit eyes, skin, resp sys; CNS depres; pulm edema; liver, kidney disease; card arrhy, card arrest<br>**TO:** Eyes, skin, resp sys, liver, kidneys, CVS, CNS | **Eye:** Irr immed<br>**Skin:** Soap flush prompt<br>**Breath:** Resp support<br>**Swallow:** Medical attention immed |

| Ethyl butyl ketone | Formula: $CH_3CH_2CO[CH_2]_3CH_3$ | CAS#: 106-35-4 | RTECS#: MJ5250000 | IDLH: 1000 ppm |
|---|---|---|---|---|
| **Conversion:** 1 ppm = 4.67 mg/m$^3$ | **DOT:** 1224 127 | | | |

**Synonyms/Trade Names:** Butyl ethyl ketone, 3-Heptanone

**Exposure Limits:**
**NIOSH REL:** TWA 50 ppm (230 mg/m$^3$)
**OSHA PEL:** TWA 50 ppm (230 mg/m$^3$)

**Measurement Methods (see Table 1):**
**NIOSH** 1301, 2553
**OSHA** 7

**Physical Description:** Colorless liquid with a powerful, fruity odor.

| Chemical & Physical Properties: | Personal Protection/Sanitation (see Table 2): | Respirator Recommendations (see Tables 3 and 4): |
|---|---|---|
| **MW:** 114.2<br>**BP:** 298°F<br>**Sol:** 1%<br>**Fl.P(oc):** 115°F<br>**IP:** 9.02 eV<br>**Sp.Gr:** 0.82<br>**VP:** 4 mmHg<br>**FRZ:** -38°F<br>**UEL:** ?<br>**LEL:** ?<br>Class II Combustible Liquid | **Skin:** Prevent skin contact<br>**Eyes:** Prevent eye contact<br>**Wash skin:** When contam<br>**Remove:** When wet or contam<br>**Change:** N.R. | **NIOSH/OSHA**<br>**500 ppm:** CcrOv*/Sa*<br>**1000 ppm:** Sa:Cf*/PaprOv*/CcrFOv/GmFOv/ScbaF/SaF<br>**§:** ScbaF:Pd,Pp/SaF:Pd,Pp:AScba<br>**Escape:** GmFOv/ScbaE |

**Incompatibilities and Reactivities:** Oxidizers, acetaldehyde, perchloric acid

| Exposure Routes, Symptoms, Target Organs (see Table 5): | First Aid (see Table 6): |
|---|---|
| **ER:** Inh, Ing, Con<br>**SY:** Irrit eyes, skin, muc memb; head, narco, coma; derm<br>**TO:** Eyes, skin, resp sys, CNS | **Eye:** Irr immed<br>**Skin:** Water flush<br>**Breath:** Resp support<br>**Swallow:** Medical attention immed |

E

## Ethyl chloride

| Ethyl chloride | Formula: $CH_3CH_2Cl$ | CAS#: 75-00-3 | RTECS#: KH7525000 | IDLH: 3800 ppm [10%LEL] |
|---|---|---|---|---|
| **Conversion:** 1 ppm = 2.64 mg/$m^3$ | **DOT:** 1037 115 | | | |

**Synonyms/Trade Names:** Chloroethane, Hydrochloric ether, Monochloroethane, Muriatic ether

**Exposure Limits:**
**NIOSH REL:** Handle with caution in the workplace.
See Appendix C (Chloroethanes)
**OSHA PEL:** TWA 1000 ppm (2600 mg/$m^3$)

**Measurement Methods (see Table 1):**
**NIOSH** 2519

**Physical Description:** Colorless gas or liquid (below 54°F) with a pungent, ether-like odor.
[**Note:** Shipped as a liquefied compressed gas.]

**Chemical & Physical Properties:**
**MW:** 64.5
**BP:** 54°F
**Sol:** 0.6%
**Fl.P:** NA (Gas)
-58°F (Liquid)
**IP:** 10.97 eV
**RGasD:** 2.23
**Sp.Gr:** 0.92 (Liquid at 32°F)
**VP:** 1000 mmHg
**FRZ:** -218°F
**UEL:** 15.4%
**LEL:** 3.8%
Flammable Gas

**Personal Protection/Sanitation (see Table 2):**
**Skin:** Prevent skin contact (liquid)
**Eyes:** Prevent eye contact (liquid)
**Wash skin:** N.R.
**Remove:** When wet (flamm)
**Change:** N.R.

**Respirator Recommendations (see Tables 3 and 4):**
**OSHA**
**3800 ppm:** Sa*/ScbaF
**§:** ScbaF:Pd,Pp/SaF:Pd,Pp:AScba
**Escape:** GmFOv/ScbaE

**Incompatibilities and Reactivities:** Chemically-active metals such as sodium, potassium, calcium, powdered aluminum, zinc & magnesium; oxidizers; water or steam [**Note:** Reacts with water to form hydrochloric acid.]

**Exposure Routes, Symptoms, Target Organs (see Table 5):**
**ER:** Inh, Abs (liquid), Ing (liquid), Con
**SY:** Inco, inebri; abdom cramps; card arrhy, card arrest; liver, kidney damage
**TO:** Liver, kidneys, resp sys, CVS, CNS

**First Aid (see Table 6):**
**Eye:** Irr immed (liquid)
**Skin:** Water flush prompt (liquid)
**Breath:** Resp support
**Swallow:** Medical attention immed (liquid)

## Ethylene chlorohydrin

| Ethylene chlorohydrin | Formula: $CH_2ClCH_2OH$ | CAS#: 107-07-3 | RTECS#: KK0875000 | IDLH: 7 ppm |
|---|---|---|---|---|
| **Conversion:** 1 ppm = 3.29 mg/$m^3$ | **DOT:** 1135 131 | | | |

**Synonyms/Trade Names:** 2-Chloroethanol, 2-Chloroethyl alcohol, Ethylene chlorhydrin

**Exposure Limits:**
**NIOSH REL:** C 1 ppm (3 mg/$m^3$) [skin]
**OSHA PEL†:** TWA 5 ppm (16 mg/$m^3$) [skin]

**Measurement Methods (see Table 1):**
**NIOSH** 2513
**OSHA** 7

**Physical Description:** Colorless liquid with a faint, ether-like odor.

**Chemical & Physical Properties:**
**MW:** 80.5
**BP:** 262°F
**Sol:** Miscible
**Fl.P:** 140°F
**IP:** 10.90 eV
**Sp.Gr:** 1.20
**VP:** 5 mmHg
**FRZ:** -90°F
**UEL:** 15.9%
**LEL:** 4.9%
Class IIIA Combustible Liquid

**Personal Protection/Sanitation (see Table 2):**
**Skin:** Prevent skin contact
**Eyes:** Prevent eye contact
**Wash skin:** When contam
**Remove:** When wet or contam
**Change:** N.R.
**Provide:** Eyewash
Quick drench

**Respirator Recommendations (see Tables 3 and 4):**
**NIOSH**
**7 ppm:** Sa*/ScbaF
**§:** ScbaF:Pd,Pp/SaF:Pd,Pp:AScba
**Escape:** GmFOv/ScbaE

**Incompatibilities and Reactivities:** Strong oxidizers, strong caustics, water or steam

**Exposure Routes, Symptoms, Target Organs (see Table 5):**
**ER:** Inh, Abs, Ing, Con
**SY:** Irrit muc memb; nau, vomit; dizz, inco; numb; vis dist; head; thirst; delirium; low BP; collapse, shock, coma; liver, kidney damage
**TO:** Resp sys, liver, kidneys, CNS, CVS, eyes

**First Aid (see Table 6):**
**Eye:** Irr immed
**Skin:** Water flush immed
**Breath:** Resp support
**Swallow:** Medical attention immed

## Ethylenediamine

| Formula: | CAS#: | RTECS#: | IDLH: |
|---|---|---|---|
| $NH_2CH_2CH_2NH_2$ | 107-15-3 | KH8575000 | 1000 ppm |

**Conversion:** 1 ppm = 2.46 mg/m$^3$

**DOT:** 1604 132

**Synonyms/Trade Names:** 1,2-Diaminoethane; 1,2-Ethanediamine; Ethylenediamine (anhydrous)

**Exposure Limits:**
**NIOSH REL:** TWA 10 ppm (25 mg/m$^3$)
**OSHA PEL:** TWA 10 ppm (25 mg/m$^3$)

**Measurement Methods (see Table 1):**
**NIOSH** 2540
**OSHA** 60

**Physical Description:** Colorless, viscous liquid with an ammonia-like odor. [fungicide] [**Note:** A solid below 47°F.]

**Chemical & Physical Properties:**
**MW:** 60.1
**BP:** 241°F
**Sol:** Miscible
**Fl.P:** 93°F
**IP:** 8.60 eV
**Sp.Gr:** 0.91
**VP:** 11 mmHg
**FRZ:** 47°F
**UEL(212°F):** 12%
**LEL(212°F):** 2.5%
Class IC Flammable Liquid

**Personal Protection/Sanitation (see Table 2):**
**Skin:** Prevent skin contact
**Eyes:** Prevent eye contact
**Wash skin:** When contam/Daily
**Remove:** When wet (flamm)
**Change:** N.R.
**Provide:** Eyewash (>5%)
Quick drench

**Respirator Recommendations (see Tables 3 and 4):**
**NIOSH/OSHA**
**250 ppm:** Sa:Cf£/PaprS£
**500 ppm:** CcrFS/GmFS/PaprTS£/ScbaF/SaF
**1000 ppm:** SaF:Pd,Pp
**§:** ScbaF:Pd,Pp/SaF:Pd,Pp:AScba
**Escape:** GmFS/ScbaE

**Incompatibilities and Reactivities:** Strong acids & oxidizers, carbon tetrachloride & other chlorinated organic compounds, carbon disulfide [**Note:** Corrosive to metals.]

**Exposure Routes, Symptoms, Target Organs (see Table 5):**
**ER:** Inh, Abs, Ing, Con
**SY:** Irrit nose, resp sys; sens derm; asthma; liver, kidney damage
**TO:** Skin, resp sys, liver, kidneys

**First Aid (see Table 6):**
**Eye:** Irr immed
**Skin:** Water flush immed
**Breath:** Resp support
**Swallow:** Medical attention immed

## Ethylene dibromide

| Formula: | CAS#: | RTECS#: | IDLH: |
|---|---|---|---|
| $BrCH_2CH_2Br$ | 106-93-4 | KH9275000 | Ca [100 ppm] |

**Conversion:** 1 ppm = 7.69 mg/m$^3$

**DOT:** 1605 154

**Synonyms/Trade Names:** 1,2-Dibromoethane; Ethylene bromide; Glycol dibromide

**Exposure Limits:**
**NIOSH REL:** Ca
TWA 0.045 ppm
C 0.13 ppm [15-minute]
See Appendix A
**OSHA PEL:** TWA 20 ppm
C 30 ppm
50 ppm [5-minute maximum peak]

**Measurement Methods (see Table 1):**
**NIOSH** 1008
**OSHA** 2

**Physical Description:** Colorless liquid or solid (below 50°F) with a sweet odor. [fumigant]

**Chemical & Physical Properties:**
**MW:** 187.9
**BP:** 268°F
**Sol:** 0.4%
**Fl.P:** NA
**IP:** 9.45 eV
**Sp.Gr:** 2.17
**VP:** 12 mmHg
**FRZ:** 50°F
**UEL:** NA
**LEL:** NA
Noncombustible Liquid

**Personal Protection/Sanitation (see Table 2):**
**Skin:** Prevent skin contact
**Eyes:** Prevent eye contact
**Wash skin:** When contam
**Remove:** When wet or contam
**Change:** N.R.
**Provide:** Eyewash
Quick drench

**Respirator Recommendations (see Tables 3 and 4):**
**NIOSH**
**¥:** ScbaF:Pd,Pp/SaF:Pd,Pp:AScba
**Escape:** GmFOv/ScbaE

**Incompatibilities and Reactivities:** Chemically-active metals such as sodium, potassium, calcium, hot aluminum & magnesium; liquid ammonia; strong oxidizers

**Exposure Routes, Symptoms, Target Organs (see Table 5):**
**ER:** Inh, Abs, Ing, Con
**SY:** Irrit eyes, skin, resp sys; derm with vesic; liver, heart, spleen, kidney damage; repro effects; [carc]
**TO:** Eyes, skin, resp sys, liver, kidneys, repro sys [in animals: skin & lung tumors]

**First Aid (see Table 6):**
**Eye:** Irr immed
**Skin:** Soap wash immed
**Breath:** Resp support
**Swallow:** Medical attention immed

## Ethylene dichloride

| Ethylene dichloride | Formula: $ClCH_2CH_2Cl$ | CAS#: 107-06-2 | RTECS#: KI0525000 | IDLH: Ca [50 ppm] |
|---|---|---|---|---|
| **Conversion:** 1 ppm = 4.05 mg/m$^3$ | **DOT:** 1184 131 | | | |

**Synonyms/Trade Names:** 1,2-Dichloroethane; Ethylene chloride; Glycol dichloride

**Exposure Limits:**
**NIOSH REL:** Ca
TWA 1 ppm (4 mg/m$^3$)
ST 2 ppm (8 mg/m$^3$)
See Appendix A,
See Appendix C (Chloroethanes)
**OSHA PEL†:** TWA 50 ppm
C 100 ppm
200 ppm [5-minute maximum peak in any 3 hours]

**Measurement Methods (see Table 1):**
**NIOSH** 1003
**OSHA** 3

**Physical Description:** Colorless liquid with a pleasant, chloroform-like odor. [**Note:** Decomposes slowly, becomes acidic & darkens in color.]

**Chemical & Physical Properties:**
**MW:** 99.0
**BP:** 182°F
**Sol:** 0.9%
**Fl.P:** 56°F
**IP:** 11.05 eV
**Sp.Gr:** 1.24
**VP:** 64 mmHg
**FRZ:** -32°F
**UEL:** 16%
**LEL:** 6.2%
Class IB Flammable Liquid

**Personal Protection/Sanitation (see Table 2):**
**Skin:** Prevent skin contact
**Eyes:** Prevent eye contact
**Wash skin:** When contam
**Remove:** When wet (flamm)
**Change:** N.R.
**Provide:** Eyewash
Quick drench

**Respirator Recommendations (see Tables 3 and 4):**
**NIOSH**
**¥:** ScbaF:Pd,Pp/SaF:Pd,Pp:AScba
**Escape:** GmFOv/ScbaE

**Incompatibilities and Reactivities:** Strong oxidizers & caustics; chemically-active metals such as magnesium or aluminum powder, sodium & potassium; liquid ammonia [**Note:** Decomposes to vinyl chloride & HCl above 1112°F.]

**Exposure Routes, Symptoms, Target Organs (see Table 5):**
**ER:** Inh, Ing, Abs, Con
**SY:** Irrit eyes, corn opac; CNS depres; nau, vomit; derm; liver, kidney, CVS damage; [carc]
**TO:** Eyes, skin, kidneys, liver, CNS, CVS [in animals: forestomach, mammary gland & circulatory sys cancer]

**First Aid (see Table 6):**
**Eye:** Irr immed
**Skin:** Soap wash prompt
**Breath:** Resp support
**Swallow:** Medical attention immed

## Ethylene glycol

| Ethylene glycol | Formula: $HOCH_2CH_2OH$ | CAS#: 107-21-1 | RTECS#: KW2975000 | IDLH: N.D. |
|---|---|---|---|---|
| **Conversion:** | **DOT:** | | | |

**Synonyms/Trade Names:** 1,2-Dihydroxyethane; 1,2-Ethanediol; Glycol; Glycol alcohol; Monoethylene glycol

**Exposure Limits:**
**NIOSH REL:** See Appendix D
**OSHA PEL†:** none

**Measurement Methods (see Table 1):**
**NIOSH** 5523
**OSHA** PV2024

**Physical Description:** Clear, colorless, syrupy, odorless liquid. [antifreeze]
[**Note:** A solid below 9°F.]

**Chemical & Physical Properties:**
**MW:** 62.1
**BP:** 388°F
**Sol:** Miscible
**Fl.P:** 232°F
**IP:** ?
**Sp.Gr:** 1.11
**VP:** 0.06 mmHg
**FRZ:** 9°F
**UEL:** 15.3%
**LEL:** 3.2%
Class IIIB Combustible Liquid

**Personal Protection/Sanitation (see Table 2):**
**Skin:** Prevent skin contact
**Eyes:** Prevent eye contact
**Wash skin:** When contam
**Remove:** When wet or contam
**Change:** Daily

**Respirator Recommendations (see Tables 3 and 4):**
Not available.

**Incompatibilities and Reactivities:** Strong oxidizers, chromium trioxide, potassium permanganate, sodium peroxide
[**Note:** Hygroscopic (i.e., absorbs moisture from the air).]

**Exposure Routes, Symptoms, Target Organs (see Table 5):**
**ER:** Inh, Ing, Con
**SY:** Irrit eyes, skin, nose, throat; nau, vomit, abdom pain, lass; dizz, stupor, convuls, CNS depres; skin sens
**TO:** Eyes, skin, resp sys, CNS

**First Aid (see Table 6):**
**Eye:** Irr immed
**Skin:** Water wash immed
**Breath:** Resp support
**Swallow:** Medical attention immed

E

## Ethylene glycol dinitrate

**Formula:** $O_2NOCH_2CH_2ONO_2$
**CAS#:** 628-96-6
**RTECS#:** KW5600000
**IDLH:** 75 mg/m$^3$

**Conversion:** 1 ppm = 6.22 mg/m$^3$
**DOT:**

**Synonyms/Trade Names:** EGDN; 1,2-Ethanediol dinitrate; Ethylene dinitrate; Ethylene nitrate; Glycol dinitrate; Nitroglycol

**Exposure Limits:**
**NIOSH REL:** ST 0.1 mg/m$^3$ [skin]
**OSHA PEL†:** C 0.2 ppm (1 mg/m$^3$) [skin]

**Measurement Methods (see Table 1):**
**NIOSH** 2507
**OSHA** 43

**Physical Description:** Colorless to yellow, oily, odorless liquid. [**Note:** An explosive ingredient (60-80%) in dynamite along with nitroglycerine (40-20%).]

**Chemical & Physical Properties:**
**MW:** 152.1
**BP:** 387°F
**Sol:** Insoluble
**Fl.P:** 419°F
**IP:** ?
**Sp.Gr:** 1.49
**VP:** 0.05 mmHg
**FRZ:** -8°F
**UEL:** ?
**LEL:** ?
Explosive Liquid

**Personal Protection/Sanitation (see Table 2):**
**Skin:** Prevent skin contact
**Eyes:** Prevent eye contact
**Wash skin:** When contam
**Remove:** When wet (flamm)
**Change:** Daily
**Provide:** Quick drench

**Respirator Recommendations (see Tables 3 and 4):**
**NIOSH**
**1 mg/m$^3$:** Sa*
**2.5 mg/m$^3$:** Sa:Cf*
**5 mg/m$^3$:** SaT:Cf*/ScbaF/SaF
**75 mg/m$^3$:** SaF:Pd,Pp
**§:** ScbaF:Pd,Pp/SaF:Pd,Pp:AScba
**Escape:** GmFOv100/ScbaE

**Incompatibilities and Reactivities:** Acids, alkalis

**Exposure Routes, Symptoms, Target Organs (see Table 5):**
**ER:** Inh, Abs, Ing, Con
**SY:** Throb head; dizz; nau, vomit, abdom pain; hypotension, flush, palp, angina; methemo; delirium, CNS depres; irrit skin; in animals: anemia; liver, kidney damage
**TO:** Skin, CVS, blood, liver, kidneys

**First Aid (see Table 6):**
**Eye:** Irr immed
**Skin:** Soap wash immed
**Breath:** Resp support
**Swallow:** Medical attention immed

## Ethyleneimine

**Formula:** $C_2H_5N$
**CAS#:** 151-56-4
**RTECS#:** KX5075000
**IDLH:** Ca [100 ppm]

**Conversion:** 1 ppm = 1.76 mg/m$^3$
**DOT:** 1185 131P (inhibited)

**Synonyms/Trade Names:** Aminoethylene, Azirane, Aziridine, Dimethyleneimine, Dimethylenimine, Ethylenimine, Ethylimine

**Exposure Limits:**
**NIOSH REL:** Ca
See Appendix A
**OSHA PEL:** [1910.1012] See Appendix B

**Measurement Methods (see Table 1):**
**NIOSH** 3514

**Physical Description:** Colorless liquid with an ammonia-like odor.
[**Note:** Usually contains inhibitors to prevent polymerization.]

**Chemical & Physical Properties:**
**MW:** 43.1
**BP:** 133°F
**Sol:** Miscible
**Fl.P:** 12°F
**IP:** 9.20 eV
**Sp.Gr:** 0.83
**VP:** 160 mmHg
**FRZ:** -97°F
**UEL:** 54.8%
**LEL:** 3.3%
Class IB Flammable Liquid

**Personal Protection/Sanitation (see Table 2):**
**Skin:** Prevent skin contact
**Eyes:** Prevent eye contact
**Wash skin:** When contam/Daily
**Remove:** When wet or contam
**Change:** Daily
**Provide:** Eyewash
Quick drench

**Respirator Recommendations (see Tables 3 and 4):**
**NIOSH**
**¥:** ScbaF:Pd,Pp/SaF:Pd,Pp:AScba
**Escape:** GmFOv/ScbaE

**See Appendix E** (page 351)

**Incompatibilities and Reactivities:** Polymerizes explosively in presence of acids [**Note:** Explosive silver derivatives may be formed with silver alloys (e.g., silver solder).]

**Exposure Routes, Symptoms, Target Organs (see Table 5):**
**ER:** Inh, Abs, Ing, Con
**SY:** Irrit eyes, skin, nose, throat; nau, vomit; head, dizz; pulm edema; liver, kidney damage; eye burns; skin sens; [carc]
**TO:** Eyes, skin, resp sys, liver, kidneys [in animals: lung & liver tumors]

**First Aid (see Table 6):**
**Eye:** Irr immed
**Skin:** Soap wash immed
**Breath:** Resp support
**Swallow:** Medical attention immed

## Ethylene oxide

| Ethylene oxide | Formula: $C_2H_4O$ | CAS#: 75-21-8 | RTECS#: KX2450000 | IDLH: Ca [800 ppm] |
|---|---|---|---|---|
| **Conversion:** 1 ppm = 1.80 mg/m$^3$ | **DOT:** 1040 119P | | | |

**Synonyms/Trade Names:** Dimethylene oxide; 1,2-Epoxy ethane; Oxirane

**Exposure Limits:**
**NIOSH REL:** Ca
TWA <0.1 ppm (0.18 mg/m$^3$)
C 5 ppm (9 mg/m$^3$) [10-min/day]
See Appendix A
**OSHA PEL:** [1910.1047] TWA 1 ppm
5 ppm [15-minute Excursion]

**Measurement Methods (see Table 1):**
**NIOSH** 1614, 3800
**OSHA** 30, 49, 50

**Physical Description:** Colorless gas or liquid (below 51°F) with an ether-like odor.

**Chemical & Physical Properties:**
**MW:** 44.1
**BP:** 51°F
**Sol:** Miscible
**Fl.P:** NA (Gas)
-20°F (Liquid)
**IP:** 10.56 eV
**RGasD:** 1.49
**Sp.Gr:** 0.82 (Liquid at 50°F)
**VP:** 1.46 atm
**FRZ:** -171°F
**UEL:** 100%
**LEL:** 3.0%
Flammable Gas

**Personal Protection/Sanitation (see Table 2):**
**Skin:** Prevent skin contact (liquid)
**Eyes:** Prevent eye contact (liquid)
**Wash skin:** When contam (liquid)
**Remove:** When wet (flamm)
**Change:** N.R.
**Provide:** Quick drench (liquid)

**Respirator Recommendations (see Tables 3 and 4):**
**NIOSH**
**5 ppm:** GmFS†/ScbaF/SaF
**§:** ScbaF:Pd,Pp/SaF:Pd,Pp:AScba
**Escape:** GmFS†/ScbaE

**See Appendix E** (page 351)

**Incompatibilities and Reactivities:** Strong acids, alkalis & oxidizers; chlorides of iron, aluminum & tin; oxides of iron & aluminum; water

**Exposure Routes, Symptoms, Target Organs (see Table 5):**
**ER:** Inh, Ing, (liquid), Con
**SY:** Irrit eyes, skin, nose, throat; peculiar taste; head; nau, vomit, diarr; dysp, cyan, pulm edema; drow, lass, inco; EKG abnor; eye, skin burns (liq or high vap conc); liquid: frostbite; repro effects; [carc]; in animals: convuls; liver, kidney damage
**TO:** Eyes, skin, resp sys, liver, CNS, blood, kidneys, repro sys [peritoneal cancer, leukemia]

**First Aid (see Table 6):**
**Eye:** Irr immed
**Skin:** Water flush immed
**Breath:** Resp support
**Swallow:** Medical attention immed (liquid)

## Ethylene thiourea

| Ethylene thiourea | Formula: $C_3H_6N_2S$ | CAS#: 96-45-7 | RTECS#: NI9625000 | IDLH: Ca [N.D.] |
|---|---|---|---|---|
| **Conversion:** | **DOT:** | | | |

**Synonyms/Trade Names:** 1,3-Ethylene-2-thiourea; N,N-Ethylenethiourea; ETU; 2-Imidazolidine-2-thione

**Exposure Limits:**
**NIOSH REL:** Ca
Use encapsulated form.
See Appendix A
**OSHA PEL:** none

**Measurement Methods (see Table 1):**
**NIOSH** 5011
**OSHA** 95

**Physical Description:** White to pale-green, crystalline solid with a faint, amine odor.
[**Note:** Used as an accelerator in the curing of polychloroprene & other elastomers.]

**Chemical & Physical Properties:**
**MW:** 102.2
**BP:** 446-595°F
**Sol(86°F):** 2%
**Fl.P:** 486°F
**IP:** 8.15 eV
**Sp.Gr:** ?
**VP:** 16 mmHg
**MLT:** 392°F
**UEL:** ?
**LEL:** ?
Combustible Solid

**Personal Protection/Sanitation (see Table 2):**
**Skin:** Prevent skin contact
**Eyes:** Prevent eye contact
**Wash skin:** When contam/Daily
**Remove:** When wet or contam
**Change:** Daily

**Respirator Recommendations (see Tables 3 and 4):**
**NIOSH**
**¥:** ScbaF:Pd,Pp/SaF:Pd,Pp:AScba
**Escape:** GmFOv100/ScbaE

**Incompatibilities and Reactivities:** Acrolein

**Exposure Routes, Symptoms, Target Organs (see Table 5):**
**ER:** Inh, Ing, Con
**SY:** Irrit eyes; in animals: thickening of the skin; goiter; terato effects; [carc]
**TO:** Eyes, skin, thyroid, repro sys [in animals: liver, thyroid & lymphatic sys tumors]

**First Aid (see Table 6):**
**Eye:** Irr immed
**Skin:** Soap wash immed
**Breath:** Resp support
**Swallow:** Medical attention immed

E

## Ethyl ether

**Formula:** $C_2H_5OC_2H_5$ | **CAS#:** 60-29-7 | **RTECS#:** KI5775000 | **IDLH:** 1900 ppm [10%LEL]

**Conversion:** 1 ppm = 3.03 mg/m$^3$ | **DOT:** 1155 127

**Synonyms/Trade Names:** Diethyl ether, Diethyl oxide, Ethyl oxide, Ether, Solvent ether

**Exposure Limits:**
**NIOSH REL:** See Appendix D
**OSHA PEL†:** TWA 400 ppm (1200 mg/m$^3$)

**Measurement Methods (see Table 1):**
**NIOSH** 1610
**OSHA** 7

**Physical Description:** Colorless liquid with a pungent, sweetish odor.
[**Note:** A gas above 94°F.]

**Chemical & Physical Properties:**
**MW:** 74.1
**BP:** 94°F
**Sol:** 8%
**Fl.P:** -49°F
**IP:** 9.53 eV
**Sp.Gr:** 0.71
**VP:** 440 mmHg
**FRZ:** -177°F
**UEL:** 36.0%
**LEL:** 1.9%
Class IA Flammable Liquid

**Personal Protection/Sanitation (see Table 2):**
**Skin:** Prevent skin contact
**Eyes:** Prevent eye contact
**Wash skin:** N.R.
**Remove:** When wet (flamm)
**Change:** N.R.

**Respirator Recommendations (see Tables 3 and 4):**
**OSHA**
**1900 ppm:** CcrOv*/GmFOv/PaprOv*/Sa*/ScbaF
**§:** ScbaF:Pd,Pp/SaF:Pd,Pp:AScba
**Escape:** GmFOv/ScbaE

**Incompatibilities and Reactivities:** Strong oxidizers, halogens, sulfur, sulfur compounds
[**Note:** Tends to form explosive peroxides under influence of air and light.]

**Exposure Routes, Symptoms, Target Organs (see Table 5):**
**ER:** Inh, Ing, Con
**SY:** Irrit eyes, skin, upper resp sys; dizz, drow, head, excited, narco; nau, vomit
**TO:** Eyes, skin, resp sys, CNS

**First Aid (see Table 6):**
**Eye:** Irr immed
**Skin:** Water wash prompt
**Breath:** Resp support
**Swallow:** Medical attention immed

## Ethyl formate

**Formula:** $CH_3CH_2OCHO$ | **CAS#:** 109-94-4 | **RTECS#:** LQ8400000 | **IDLH:** 1500 ppm

**Conversion:** 1 ppm = 3.03 mg/m$^3$ | **DOT:** 1190 129

**Synonyms/Trade Names:** Ethyl ester of formic acid, Ethyl methanoate

**Exposure Limits:**
**NIOSH REL:** TWA 100 ppm (300 mg/m$^3$)
**OSHA PEL:** TWA 100 ppm (300 mg/m$^3$)

**Measurement Methods (see Table 1):**
**NIOSH** 1452
**OSHA** 7

**Physical Description:** Colorless liquid with a fruity odor.

**Chemical & Physical Properties:**
**MW:** 74.1
**BP:** 130°F
**Sol(64°F):** 9%
**Fl.P:** -4°F
**IP:** 10.61 eV
**Sp.Gr:** 0.92
**VP:** 200 mmHg
**FRZ:** -113°F
**UEL:** 16.0%
**LEL:** 2.8%
Class IB Flammable Liquid

**Personal Protection/Sanitation (see Table 2):**
**Skin:** Prevent skin contact
**Eyes:** Prevent eye contact
**Wash skin:** When contam
**Remove:** When wet (flamm)
**Change:** N.R.

**Respirator Recommendations (see Tables 3 and 4):**
**NIOSH/OSHA**
**1500 ppm:** Sa:Cf£/PaprOv£/CcrFOv/GmFOv/ScbaF/SaF
**§:** ScbaF:Pd,Pp/SaF:Pd,Pp:AScba
**Escape:** GmFOv/ScbaE

**Incompatibilities and Reactivities:** Nitrates; strong oxidizers, alkalis & acids
[**Note:** Decomposes slowly in water to form ethyl alcohol and formic acid.]

**Exposure Routes, Symptoms, Target Organs (see Table 5):**
**ER:** Inh, Ing, Con
**SY:** Irrit eyes, upper resp sys; in animals: narco
**TO:** Eyes, resp sys, CNS

**First Aid (see Table 6):**
**Eye:** Irr immed
**Skin:** Water flush immed
**Breath:** Resp support
**Swallow:** Medical attention immed

## Ethylidene norbornene

**Formula:** $C_9H_{12}$ | **CAS#:** 16219-75-3 | **RTECS#:** RB9450000 | **IDLH:** N.D.

**Conversion:** 1 ppm = 4.92 mg/m$^3$ | **DOT:**

**Synonyms/Trade Names:** ENB, 5-Ethylidenebicyclo(2.2.1)hept-2-ene, 5-Ethylidene-2-norbornene
[**Note:** Due to its reactivity, ENB may be stabilized with tert-butyl catechol.]

**Exposure Limits:**
**NIOSH REL:** C 5 ppm (25 mg/m$^3$)
**OSHA PEL†:** none

**Measurement Methods (see Table 1):**
None available

**Physical Description:** Colorless to white liquid with a turpentine-like odor.

**Chemical & Physical Properties:**
**MW:** 120.2
**BP:** 298°F
**Sol:** ?
**Fl.P(oc):** 101°F
**IP:** ?
**Sp.Gr:** 0.90
**VP:** 4 mmHg
**FRZ:** -112°F
**UEL:** ?
**LEL:** ?
Class II Combustible Liquid

**Personal Protection/Sanitation (see Table 2):**
**Skin:** Prevent skin contact
**Eyes:** Prevent eye contact
**Wash skin:** Daily
**Remove:** When wet or contam
**Change:** N.R.

**Respirator Recommendations (see Tables 3 and 4):**
Not available.

**Incompatibilities and Reactivities:** Oxygen
[**Note:** ENB should be stored in a nitrogen atmosphere since it reacts with oxygen.]

**Exposure Routes, Symptoms, Target Organs (see Table 5):**
**ER:** Inh, Abs, Ing, Con
**SY:** Irrit eyes, skin, nose, throat; head; cough, dysp; nau, vomit; olfactory, taste changes; chemical pneu (aspir liquid); in animals: liver, kidney, urogenital inj; bone marrow effects
**TO:** Eyes, skin, resp sys, CNS, liver, kidneys, urogenital system, bone marrow

**First Aid (see Table 6):**
**Eye:** Irr immed
**Skin:** Soap wash immed
**Breath:** Resp support
**Swallow:** Medical attention immed

## Ethyl mercaptan

**Formula:** $CH_3CH_2SH$ | **CAS#:** 75-08-1 | **RTECS#:** KI9625000 | **IDLH:** 500 ppm

**Conversion:** 1 ppm = 2.54 mg/m$^3$ | **DOT:** 2363 129

**Synonyms/Trade Names:** Ethanethiol, Ethyl sulfhydrate, Mercaptoethane

**Exposure Limits:**
**NIOSH REL:** C 0.5 ppm (1.3 mg/m$^3$) [15-minute]
**OSHA PEL†:** C 10 ppm (25 mg/m$^3$)

**Measurement Methods (see Table 1):**
**NIOSH** 2542

**Physical Description:** Colorless liquid with a strong, skunk-like odor.
[**Note:** A gas above 95°F.]

**Chemical & Physical Properties:**
**MW:** 62.1
**BP:** 95°F
**Sol:** 0.7%
**Fl.P:** -55°F
**IP:** 9.29 eV
**Sp.Gr:** 0.84
**VP:** 442 mmHg
**FRZ:** -228°F
**UEL:** 18.0%
**LEL:** 2.8%
Class IA Flammable Liquid

**Personal Protection/Sanitation (see Table 2):**
**Skin:** Prevent skin contact
**Eyes:** Prevent eye contact
**Wash skin:** When contam
**Remove:** When wet (flamm)
**Change:** N.R.

**Respirator Recommendations (see Tables 3 and 4):**
**NIOSH**
**5 ppm:** CcrOv/Sa
**12.5 ppm:** Sa:Cf/PaprOv
**25 ppm:** CcrFOv/GmFOv/SaT:Cf/PaprTOv/ScbaF/SaF
**500 ppm:** Sa:Pd,Pp
**§:** ScbaF:Pd,Pp/SaF:Pd,Pp:AScba
**Escape:** GmFOv/ScbaE

**Incompatibilities and Reactivities:** Strong oxidizers [**Note:** Reacts violently with calcium hypochlorite.]

**Exposure Routes, Symptoms, Target Organs (see Table 5):**
**ER:** Inh, Ing, Con
**SY:** Irrit muc memb; head, nau; in animals: inco, lass; liver, kidney damage; cyan; narco
**TO:** Eyes, resp sys, liver, kidneys, blood

**First Aid (see Table 6):**
**Eye:** Irr immed
**Skin:** Soap wash immed
**Breath:** Resp support
**Swallow:** Medical attention immed

E

## N-Ethylmorpholine

| | | | | |
|---|---|---|---|---|
| **N-Ethylmorpholine** | **Formula:** $C_4H_8ONCH_2CH_3$ | **CAS#:** 100-74-3 | **RTECS#:** QE4025000 | **IDLH:** 100 ppm |
| **Conversion:** 1 ppm = 4.71 mg/m$^3$ | **DOT:** | | | |

**Synonyms/Trade Names:** 4-Ethylmorpholine

**Exposure Limits:**
**NIOSH REL:** TWA 5 ppm (23 mg/m$^3$) [skin]
**OSHA PEL†:** TWA 20 ppm (94 mg/m$^3$) [skin]

**Measurement Methods (see Table 1):**
**NIOSH** S146 (II-3)

**Physical Description:** Colorless liquid with an ammonia-like odor.

| **Chemical & Physical Properties:** | **Personal Protection/Sanitation (see Table 2):** | **Respirator Recommendations (see Tables 3 and 4):** |
|---|---|---|
| **MW:** 115.2<br>**BP:** 281°F<br>**Sol:** Miscible<br>**Fl.P(oc):** 90°F<br>**IP:** ?<br>**Sp.Gr:** 0.90<br>**VP:** 6 mmHg<br>**FRZ:** -81°F<br>**UEL:** ?<br>**LEL:** ?<br>Class IC Flammable Liquid | **Skin:** Prevent skin contact<br>**Eyes:** Prevent eye contact<br>**Wash skin:** When contam<br>**Remove:** When wet (flamm)<br>**Change:** N.R.<br>**Provide:** Eyewash (>15%)<br>Quick drench | **NIOSH**<br>**50 ppm:** CcrOv*/Sa*<br>**100 ppm:** Sa:Cf*/PaprOv*/CcrFOv/GmFOv/ScbaF/SaF<br>**§:** ScbaF:Pd,Pp/SaF:Pd,Pp:AScba<br>**Escape:** GmFOv/ScbaE |

**Incompatibilities and Reactivities:** Strong acids, strong oxidizers

**Exposure Routes, Symptoms, Target Organs (see Table 5):**
**ER:** Inh, Abs, Ing, Con
**SY:** Irrit eyes, nose, throat; vis dist: corn edema, blue-gray vision, colored haloes
**TO:** Eyes, resp sys

**First Aid (see Table 6):**
**Eye:** Irr immed
**Skin:** Water flush prompt
**Breath:** Resp support
**Swallow:** Medical attention immed

## Ethyl silicate

| | | | | |
|---|---|---|---|---|
| **Ethyl silicate** | **Formula:** $(C_2H_5)_4SiO_4$ | **CAS#:** 78-10-4 | **RTECS#:** VV9450000 | **IDLH:** 700 ppm |
| **Conversion:** 1 ppm = 8.52 mg/m$^3$ | **DOT:** 1292 129 | | | |

**Synonyms/Trade Names:** Ethyl orthosilicate, Ethyl silicate (condensed), Tetraethoxysilane, Tetraethyl orthosilicate, Tetraethyl silicate

**Exposure Limits:**
**NIOSH REL:** TWA 10 ppm (85 mg/m$^3$)
**OSHA PEL†:** TWA 100 ppm (850 mg/m$^3$)

**Measurement Methods (see Table 1):**
**NIOSH** S264 (II-3)

**Physical Description:** Colorless liquid with a sharp, alcohol-like odor.

| **Chemical & Physical Properties:** | **Personal Protection/Sanitation (see Table 2):** | **Respirator Recommendations (see Tables 3 and 4):** |
|---|---|---|
| **MW:** 208.3<br>**BP:** 336°F<br>**Sol:** Reacts<br>**Fl.P:** 99°F<br>**IP:** 9.77 eV<br>**Sp.Gr:** 0.93<br>**VP:** 1 mmHg<br>**FRZ:** -117°F<br>**UEL:** ?<br>**LEL:** ?<br>Class IC Flammable Liquid | **Skin:** Prevent skin contact<br>**Eyes:** Prevent eye contact<br>**Wash skin:** When contam<br>**Remove:** When wet (flamm)<br>**Change:** N.R. | **NIOSH**<br>**100 ppm:** Sa*<br>**250 ppm:** Sa:Cf*<br>**500 ppm:** ScbaF/SaF<br>**700 ppm:** SaF:Pd,Pp<br>**§:** ScbaF:Pd,Pp/SaF:Pd,Pp:AScba<br>**Escape:** GmFOv/ScbaE |

**Incompatibilities and Reactivities:** Strong oxidizers, water
[**Note:** Reacts with water to form a silicone adhesive (a milky-white mass).]

**Exposure Routes, Symptoms, Target Organs (see Table 5):**
**ER:** Inh, Ing, Con
**SY:** Irrit eyes, nose; in animals: lac; dysp, pulm edema; tremor, narco; liver, kidney damage; anemia
**TO:** Eyes, resp sys, liver, kidneys, blood, skin

**First Aid (see Table 6):**
**Eye:** Irr immed
**Skin:** Soap wash prompt
**Breath:** Resp support
**Swallow:** Medical attention immed

## Fenamiphos

| Fenamiphos | Formula: $C_{13}H_{22}NO_3PS$ | CAS#: 22224-92-6 | RTECS#: TB3675000 | IDLH: N.D. |
|---|---|---|---|---|

**Conversion:**

**DOT:**

**Synonyms/Trade Names:** Ethyl 3-methyl-4-(methylthio)phenyl-(1-methylethyl)phosphoramidate, Nemacur®, Phenamiphos

**Exposure Limits:**
**NIOSH REL:** TWA 0.1 mg/m³ [skin]
**OSHA PEL†:** none

**Measurement Methods (see Table 1):**
**NIOSH** 5600

**Physical Description:** Off-white to tan, waxy solid. [insecticide]
[**Note:** Found commercially as a granular ingredient (5-15%) or in an emulsifiable concentrate (400 g/l).]

**Chemical & Physical Properties:**
**MW:** 303.4
**BP:** ?
**Sol:** 0.03%
**Fl.P:** ?
**IP:** ?
**Sp.Gr:** 1.14
**VP:** 0.00005 mmHg
**MLT:** 121°F
**UEL:** ?
**LEL:** ?

**Personal Protection/Sanitation (see Table 2):**
**Skin:** Prevent skin contact
**Eyes:** Prevent eye contact
**Wash skin:** When contam/Daily
**Remove:** When wet or contam
**Change:** Daily
**Provide:** Quick drench

**Respirator Recommendations (see Tables 3 and 4):**
Not available.

**Incompatibilities and Reactivities:** None reported [**Note:** May hydrolyze under alkaline conditions.]

**Exposure Routes, Symptoms, Target Organs (see Table 5):**
**ER:** Inh, Abs, Ing, Con
**SY:** Nau, vomit, abdom cramps, diarr, salv; head, dizz, lass; rhin, chest tight; blurred vision, miosis; card irreg; musc fasc; dysp
**TO:** Resp sys, CNS, CVS, blood chol

**First Aid (see Table 6):**
**Eye:** Irr immed
**Skin:** Soap flush immed
**Breath:** Resp support
**Swallow:** Medical attention immed

## Fensulfothion

| Fensulfothion | Formula: $C_{11}H_{17}O_4PS_2$ | CAS#: 115-90-2 | RTECS#: TF3850000 | IDLH: N.D. |
|---|---|---|---|---|

**Conversion:**

**DOT:**

**Synonyms/Trade Names:** Dasanit®; O,O-Diethyl O-(p-methylsulfinyl)phenyl)phosphorothioate; Terracur P®

**Exposure Limits:**
**NIOSH REL:** TWA 0.1 mg/m³
**OSHA PEL†:** none

**Measurement Methods (see Table 1):**
None available

**Physical Description:** Brown liquid or yellow oil. [pesticide]

**Chemical & Physical Properties:**
**MW:** 308.4
**BP:** ?
**Sol(77°F):** 0.2%
**Fl.P:** ?
**IP:** ?
**Sp.Gr:** 1.20
**VP:** ?
**FRZ:** ?
**UEL:** ?
**LEL:** ?
Combustible Liquid

**Personal Protection/Sanitation (see Table 2):**
**Skin:** Prevent skin contact
**Eyes:** Prevent eye contact
**Wash skin:** When contam
**Remove:** When wet or contam
**Change:** N.R.
**Provide:** Eyewash
Quick drench

**Respirator Recommendations (see Tables 3 and 4):**
Not available.

**Incompatibilities and Reactivities:** Alkalis

**Exposure Routes, Symptoms, Target Organs (see Table 5):**
**ER:** Inh, Abs, Ing, Con
**SY:** Irrit skin; nau, vomit, abdom cramps, diarr, salv; head, dizz, lass; rhin, chest tight; blurred vision, miosis; card irreg; musc fasc; dys
**TO:** Skin, resp sys, CNS, CVS, blood chol

**First Aid (see Table 6):**
**Eye:** Irr immed
**Skin:** Soap flush immed
**Breath:** Resp support
**Swallow:** Medical attention immed

F

## Fenthion

| Fenthion | Formula: $C_{10}H_{15}O_3PS$ | CAS#: 55-38-9 | RTECS#: TF9625000 | IDLH: N.D. |
|---|---|---|---|---|
| **Conversion:** | **DOT:** | | | |

**Synonyms/Trade Names:** Baytex; Entex; O,O-Dimethyl O-3-methyl-4-methylthiophenyl phosphorothioate

**Exposure Limits:**
**NIOSH REL:** See Appendix D
**OSHA PEL†:** none

**Measurement Methods (see Table 1):**
None available

**Physical Description:** Colorless to brown liquid with a slight, garlic-like odor. [insecticide]

**Chemical & Physical Properties:**
**MW:** 278.3
**BP:** ?
**Sol:** 0.006%
**Fl.P:** NA
**IP:** ?
**Sp.Gr:** 1.25
**VP:** 0.0003 mmHg
**FRZ:** 43°F
**UEL:** NA
**LEL:** NA
Noncombustible Liquid

**Personal Protection/Sanitation (see Table 2):**
**Skin:** Prevent skin contact
**Eyes:** Prevent eye contact
**Wash skin:** When contam
**Remove:** When wet or contam
**Change:** Daily

**Respirator Recommendations (see Tables 3 and 4):**
Not available.

**Incompatibilities and Reactivities:** Oxidizers

**Exposure Routes, Symptoms, Target Organs (see Table 5):**
**ER:** Inh, Abs, Ing, Con
**SY:** Nau, vomit, abdom cramps, diarr, salv; head, dizz, lass; rhin, chest tight; blurred vision, miosis; card irregularities; musc fasc; dysp
**TO:** Resp sys, CNS, CVS, plasma chol

**First Aid (see Table 6):**
**Eye:** Irr immed
**Skin:** Soap flush immed
**Breath:** Resp support
**Swallow:** Medical attention immed

## Ferbam

| Ferbam | Formula: $[(CH_3)_2NCS_2]_3Fe$ | CAS#: 14484-64-1 | RTECS#: NO8750000 | IDLH: 800 mg/m$^3$ |
|---|---|---|---|---|
| **Conversion:** | **DOT:** | | | |

**Synonyms/Trade Names:** tris(Dimethyldithiocarbamato)iron, Ferric dimethyl dithiocarbamate

**Exposure Limits:**
**NIOSH REL:** TWA 10 mg/m$^3$
**OSHA PEL†:** TWA 15 mg/m$^3$

**Measurement Methods (see Table 1):**
NIOSH 0500

**Physical Description:** Dark brown to black, odorless solid. [fungicide]

**Chemical & Physical Properties:**
**MW:** 416.5
**BP:** Decomposes
**Sol:** 0.01%
**Fl.P:** ?
**IP:** 7.72 eV
**Sp.Gr:** ?
**VP:** 0 mmHg (approx)
**MLT:** >356°F (Decomposes)
**UEL:** ?
**LEL:** ?
**MEC:** 55 g/m$^3$
Combustible Solid

**Personal Protection/Sanitation (see Table 2):**
**Skin:** Prevent skin contact
**Eyes:** Prevent eye contact
**Wash skin:** When contam
**Remove:** When wet or contam
**Change:** Daily

**Respirator Recommendations (see Tables 3 and 4):**
**NIOSH**
**50 mg/m$^3$:** Qm
**100 mg/m$^3$:** 95XQ*/Sa*
**250 mg/m$^3$:** Sa:Cf*/PaprHie*
**500 mg/m$^3$:** 100F/SaT:Cf*/PaprTHie*/ScbaF/SaF
**800 mg/m$^3$:** SaF:Pd,Pp
**§:** ScbaF:Pd,Pp/SaF:Pd,Pp:AScba
**Escape:** 100F/ScbaE

**Incompatibilities and Reactivities:** Strong oxidizers, moisture

**Exposure Routes, Symptoms, Target Organs (see Table 5):**
**ER:** Inh, Ing, Con
**SY:** Irrit eyes, resp tract; derm; GI dist
**TO:** Eyes, skin, resp sys, GI tract

**First Aid (see Table 6):**
**Eye:** Irr immed
**Skin:** Soap wash prompt
**Breath:** Resp support
**Swallow:** Medical attention immed

| Ferrovanadium dust | Formula: FeV | CAS#: 12604-58-9 | RTECS#: LK2900000 | IDLH: 500 mg/m$^3$ |
|---|---|---|---|---|
| **Conversion:** | **DOT:** | | | |

**Synonyms/Trade Names:** Ferrovanadium

**Exposure Limits:**
**NIOSH REL*:** TWA 1 mg/m$^3$
ST 3 mg/m$^3$
[***Note:** The REL also applies to Vanadium metal and Vanadium carbide.]
**OSHA PEL†:** TWA 1 mg/m$^3$

**Measurement Methods (see Table 1):**
**OSHA** ID121, ID125G

**Physical Description:** Dark, odorless particulate dispersed in air.
[**Note:** Ferrovanadium metal is an alloy usually containing 50-80% vanadium.]

**Chemical & Physical Properties:**
**MW:** 106.8
**BP:** ?
**Sol:** Insoluble
**Fl.P:** NA
**IP:** NA
**Sp.Gr:** ?
**VP:** 0 mmHg (approx)
**MLT:** 2696-2768°F
**UEL:** NA
**LEL:** NA
**MEC:** 1.3 g/m$^3$
Metal: Noncombustible Solid, but dust may be an explosion hazard.

**Personal Protection/Sanitation (see Table 2):**
**Skin:** N.R.
**Eyes:** N.R.
**Wash skin:** N.R.
**Remove:** N.R.
**Change:** N.R.

**Respirator Recommendations (see Tables 3 and 4):**
**NIOSH/OSHA**
**5 mg/m$^3$:** Qm*
**10 mg/m$^3$:** 95XQ*/Sa*
**25 mg/m$^3$:** Sa:Cf*/PaprHie*
**50 mg/m$^3$:** 100F/SaT:Cf*/PaprTHie*/ScbaF/SaF
**500 mg/m$^3$:** SaF:Pd,Pp
**§:** ScbaF:Pd,Pp/SaF:Pd,Pp:AScba
**Escape:** 100F/ScbaE

**Incompatibilities and Reactivities:** Strong oxidizers

**Exposure Routes, Symptoms, Target Organs (see Table 5):**
**ER:** Inh, Con
**SY:** Irrit eyes, resp sys; in animals: bron, pneu
**TO:** Eyes, resp sys

**First Aid (see Table 6):**
**Eye:** Irr immed
**Breath:** Resp support

F

| Fibrous glass dust | Formula: | CAS#: | RTECS#: LK3651000 | IDLH: N.D. |
|---|---|---|---|---|
| **Conversion:** | **DOT:** | | | |

**Synonyms/Trade Names:** Fiber glas®, Fiberglass, Glass fibers, Glass wool
[**Note:** Usually produced from borosilicate & low alkali silicate glasses.]

**Exposure Limits:**
**NIOSH REL:** TWA 3 fibers/cm$^3$ (fibers ≤ 3.5 µm in diameter & ≥ 10 µm in length)
TWA 5 mg/m$^3$ (total)
**OSHA PEL:** TWA 15 mg/m3 (total)
TWA 5 mg/m$^3$ (resp)

**Measurement Methods (see Table 1):**
**NIOSH** 7400

**Physical Description:** Typically, glass filaments >3 µm in diameter or glass "wool" with diameters down to 0.05 µm & >1 µm in length.

**Chemical & Physical Properties:**
**MW:** NA
**BP:** NA
**Sol:** Insoluble
**Fl.P:** NA
**IP:** NA
**Sp.Gr:** 2.5
**VP:** 0 mmHg (approx)
**MLT:** ?
**UEL:** NA
**LEL:** NA
Noncombustible Fibers

**Personal Protection/Sanitation (see Table 2):**
**Skin:** Prevent skin contact
**Eyes:** Prevent eye contact
**Wash skin:** Daily
**Remove:** N.R.
**Change:** Daily

**Respirator Recommendations (see Tables 3 and 4):**
**NIOSH**
**5X REL:** Qm
**10X REL:** 95XQ/Sa
**25X REL:** Sa:Cf/PaprHie
**50X REL:** 100F/PaprTHie/ScbaF/SaF
**1000X REL:** SaF:Pd,Pp
**§:** ScbaF:Pd,Pp/SaF:Pd,Pp:AScba
**Escape:** 100F/ScbaE

**Incompatibilities and Reactivities:** None reported

**Exposure Routes, Symptoms, Target Organs (see Table 5):**
**ER:** Inh, Con
**SY:** Irrit eyes, skin, nose, throat; dysp
**TO:** Eyes, skin, resp sys

**First Aid (see Table 6):**
**Eye:** Irr immed
**Breath:** Fresh air

F

| Fluorine | Formula: $F_2$ | CAS#: 7782-41-4 | RTECS#: LM6475000 | IDLH: 25 ppm |
|---|---|---|---|---|
| **Conversion:** 1 ppm = 1.55 mg/m$^3$ | **DOT:** 1045 124; 9192 167 (cryogenic liquid) | | | |

**Synonyms/Trade Names:** Fluorine-19

**Exposure Limits:**
**NIOSH REL:** TWA 0.1 ppm (0.2 mg/m$^3$)
**OSHA PEL:** TWA 0.1 ppm (0.2 mg/m$^3$)

**Measurement Methods (see Table 1):**
None available

**Physical Description:** Pale-yellow to greenish gas with a pungent, irritating odor.

**Chemical & Physical Properties:**
**MW:** 38.0
**BP:** -307°F
**Sol:** Reacts
**Fl.P:** NA
**IP:** 15.70 eV
**RGasD:** 1.31
**VP:** >1 atm
**FRZ:** -363°F
**UEL:** NA
**LEL:** NA
Nonflammable Gas, but an extremely strong oxidizer.

**Personal Protection/Sanitation (see Table 2):**
**Skin:** Prevent skin contact (liquid)
**Eyes:** Prevent eye contact (liquid)
**Wash skin:** When contam (liquid)
**Remove:** When wet or contam (liquid)
**Change:** N.R.
**Provide:** Eyewash (liquid)
Quick drench (liquid)

**Respirator Recommendations (see Tables 3 and 4):**
**NIOSH/OSHA**
**1 ppm:** Sa*
**2.5 ppm:** Sa:Cf*
**5 ppm:** ScbaF/SaF
**25 ppm:** SaF:Pd,Pp
**§:** ScbaF:Pd,Pp/SaF:Pd,Pp:AScba
**Escape:** GmFS¿/ScbaE

**Incompatibilities and Reactivities:** Water, nitric acid, oxidizers, organic compounds
[**Note:** Reacts violently with all combustible materials, except the metal containers in which it is shipped. Reacts with $H_2O$ to form hydrofluoric acid.]

**Exposure Routes, Symptoms, Target Organs (see Table 5):**
**ER:** Inh, Con
**SY:** Irrit eyes, nose, resp sys; lar spasm, wheez; pulm edema; eye, skin burns; in animals: liver, kidney damage
**TO:** Eyes, skin, resp sys, liver, kidneys

**First Aid (see Table 6):**
**Eye:** Irr immed
**Skin:** Water flush immed
**Breath:** Resp support

| Fluorotrichloromethane | Formula: $CCl_3F$ | CAS#: 75-69-4 | RTECS#: PB6125000 | IDLH: 2000 ppm |
|---|---|---|---|---|
| **Conversion:** 1 ppm = 5.62 mg/m$^3$ | **DOT:** | | | |

**Synonyms/Trade Names:** Freon® 11, Monofluorotrichloromethane, Refrigerant 11, Trichlorofluoromethane, Trichloromonofluoromethane

**Exposure Limits:**
**NIOSH REL:** C 1000 ppm (5600 mg/m$^3$)
**OSHA PEL†:** TWA 1000 ppm (5600 mg/m$^3$)

**Measurement Methods (see Table 1):**
**NIOSH** 1006

**Physical Description:** Colorless to water-white, nearly odorless liquid or gas (above 75°F).

**Chemical & Physical Properties:**
**MW:** 137.4
**BP:** 75°F
**Sol(75°F):** 0.1%
**Fl.P:** NA
**IP:** 11.77 eV
**RGasD:** 4.74
**Sp.Gr:** 1.47 (Liquid at 75°F)
**VP:** 690 mmHg
**FRZ:** -168°F
**UEL:** NA
**LEL:** NA
Noncombustible Liquid
Nonflammable Gas

**Personal Protection/Sanitation (see Table 2):**
**Skin:** Prevent skin contact
**Eyes:** Prevent eye contact
**Wash skin:** N.R.
**Remove:** When wet or contam
**Change:** N.R.
**Provide:** Eyewash
Quick drench

**Respirator Recommendations (see Tables 3 and 4):**
**NIOSH/OSHA**
**2000 ppm:** Sa/ScbaF
**§:** ScbaF:Pd,Pp/SaF:Pd,Pp:AScba
**Escape:** GmFOv/ScbaE

**Incompatibilities and Reactivities:** Chemically-active metals such as sodium, potassium, calcium, powdered aluminum, zinc, magnesium & lithium shavings; granular barium

**Exposure Routes, Symptoms, Target Organs (see Table 5):**
**ER:** Inh, Ing, Con
**SY:** Inco, tremor; derm; card arrhy, card arrest; asphy; liquid: frostbite
**TO:** Skin, resp sys, CVS

**First Aid (see Table 6):**
**Eye:** Irr immed
**Skin:** Water flush immed
**Breath:** Resp support
**Swallow:** Medical attention immed

| Fluoroxene | Formula: $CF_3CH_2OCH{=}CH_2$ | CAS#: 406-90-6 | RTECS#: KO4250000 | IDLH: N.D. |
|---|---|---|---|---|
| **Conversion:** 1 ppm = 5.16 mg/m$^3$ | **DOT:** | | | |

**Synonyms/Trade Names:** 2,2,2-Trifluoroethoxyethene; 2,2,2-Trifluoroethyl vinyl ether

**Exposure Limits:**
**NIOSH REL*:** C 2 ppm (10.3 mg/m$^3$) [60-minute]
[***Note:** REL for exposure to waste anesthetic gas.]
**OSHA PEL:** none

**Measurement Methods (see Table 1):**
None available

**Physical Description:** Liquid. [inhalation anesthetic] [**Note:** A gas above 109°F.]

| Chemical & Physical Properties: | Personal Protection/Sanitation (see Table 2): | Respirator Recommendations (see Tables 3 and 4): |
|---|---|---|
| **MW:** 126.1<br>**BP:** 109°F<br>**Sol:** ?<br>**Fl.P:** ?<br>**IP:** ?<br>**Sp.Gr:** 1.14<br>**VP:** 286 mmHg<br>**FRZ:** ?<br>**UEL:** ?<br>**LEL:** ?<br>Combustible Liquid<br>[Potentially EXPLOSIVE!] | **Skin:** N.R.<br>**Eyes:** Prevent eye contact<br>**Wash skin:** N.R.<br>**Remove:** N.R.<br>**Change:** N.R. | Not available. |

**Incompatibilities and Reactivities:** None reported

| Exposure Routes, Symptoms, Target Organs (see Table 5): | First Aid (see Table 6): |
|---|---|
| **ER:** Inh, Ing, Con<br>**SY:** Irrit eyes; CNS depres, analgesia, anes, convuls, resp depres<br>**TO:** Eyes, CNS | **Eye:** Irr immed<br>**Skin:** Soap wash<br>**Breath:** Resp support<br>**Swallow:** Medical attention immed |

| Fonofos | Formula: $C_{10}H_{15}OPS_2$ | CAS#: 944-22-9 | RTECS#: TA5950000 | IDLH: N.D. |
|---|---|---|---|---|
| **Conversion:** 1 ppm = 10.07 mg/m$^3$ | **DOT:** | | | |

**Synonyms/Trade Names:** Dyfonate®, Dyphonate, O-Ethyl-S-phenyl ethylphosphorothioate, Fonophos

**Exposure Limits:**
**NIOSH REL:** TWA 0.1 mg/m$^3$ [skin]
**OSHA PEL†:** none

**Measurement Methods (see Table 1):**
**NIOSH** 5600
**OSHA** PV2027

**Physical Description:** Light-yellow liquid with an aromatic odor. [insecticide]

| Chemical & Physical Properties: | Personal Protection/Sanitation (see Table 2): | Respirator Recommendations (see Tables 3 and 4): |
|---|---|---|
| **MW:** 246.3<br>**BP:** ?<br>**Sol:** 0.001%<br>**Fl.P:** >201°F<br>**IP:** ?<br>**Sp.Gr:** 1.15<br>**VP(77°F):** 0.0002 mmHg<br>**FRZ:** ?<br>**UEL:** ?<br>**LEL:** ?<br>Class IIIB Combustible Liquid | **Skin:** Prevent skin contact<br>**Eyes:** Prevent eye contact<br>**Wash skin:** When contam<br>**Remove:** When wet or contam<br>**Change:** Daily<br>**Provide:** Eyewash<br>Quick drench | Not available. |

**Incompatibilities and Reactivities:** None reported

| Exposure Routes, Symptoms, Target Organs (see Table 5): | First Aid (see Table 6): |
|---|---|
| **ER:** Inh, Abs, Ing, Con<br>**SY:** Nau, vomit, abdom cramps, diarr, salv; head, dizz, lass; rhin, chest tight; blurred vision, miosis; card irreg; musc fasc; dysp<br>**TO:** Resp sys, CNS, CVS, blood chol | **Eye:** Irr immed<br>**Skin:** Soap flush immed<br>**Breath:** Resp support<br>**Swallow:** Medical attention immed |

F

## Formaldehyde

| Formaldehyde | Formula: HCHO | CAS#: 50-00-0 | RTECS#: LP8925000 | IDLH: Ca [20 ppm] |
|---|---|---|---|---|
| Conversion: 1 ppm = 1.23 mg/m$^3$ | DOT: | | | |

**Synonyms/Trade Names:** Methanal, Methyl aldehyde, Methylene oxide

**Exposure Limits:**
**NIOSH REL:** Ca
TWA 0.016 ppm
C 0.1 ppm [15-minute]
See Appendix A
**OSHA PEL:** [1910.1048] TWA 0.75 ppm
ST 2 ppm

**Measurement Methods (see Table 1):**
**NIOSH** 2016, 2541, 3500, 3800
**OSHA** ID205, 52

**Physical Description:** Nearly colorless gas with a pungent, suffocating odor. [**Note:** Often used in an aqueous solution (see specific listing for Formalin).]

**Chemical & Physical Properties:**
**MW:** 30.0
**BP:** -6°F
**Sol:** Miscible
**Fl.P:** NA (Gas)
**IP:** 10.88 eV
**RGasD:** 1.04
**VP:** >1 atm
**FRZ:** -134°F
**UEL:** 73%
**LEL:** 7.0%
Flammable Gas

**Personal Protection/Sanitation (see Table 2):**
**Skin:** N.R.
**Eyes:** Prevent eye contact
**Wash skin:** N.R.
**Remove:** N.R.
**Change:** N.R.

**Respirator Recommendations (see Tables 3 and 4):**
**NIOSH**
**¥:** ScbaF:Pd,Pp/SaF:Pd,Pp:AScba
**Escape:** GmFS/ScbaE

**See Appendix E** (page 351)

**Incompatibilities and Reactivities:** Strong oxidizers, alkalis & acids; phenols; urea [**Note:** Pure formaldehyde has a tendency to polymerize. Reacts with HCl to form bis-Chloromethyl ether.]

**Exposure Routes, Symptoms, Target Organs (see Table 5):**
**ER:** Inh, Con
**SY:** Irrit eyes, nose, throat, resp sys; lac; cough; wheez; [carc]
**TO:** Eyes, resp sys [nasal cancer]

**First Aid (see Table 6):**
**Eye:** Irr immed
**Breath:** Resp support

## Formalin (as formaldehyde)

| Formalin (as formaldehyde) | Formula: | CAS#: | RTECS#: | IDLH: Ca [20 ppm] |
|---|---|---|---|---|
| Conversion: | DOT: 1198 132; 2209 132 | | | |

**Synonyms/Trade Names:** Formaldehyde solution
[**Note:** Formalin is an aqueous solution that is 37% formaldehyde by weight; inhibited solutions usually contain 6-12% methyl alcohol. Also see specific listings for Formaldehyde and Methyl alcohol.]

**Exposure Limits:**
**NIOSH REL:** Ca
TWA 0.016 ppm
C 0.1 ppm [15-minute]
See Appendix A
**OSHA PEL:** [1910.1048] TWA 0.75 ppm
ST 2 ppm

**Measurement Methods (see Table 1):**
**NIOSH** 2016, 2541, 3500, 3800
**OSHA** ID205, 52

**Physical Description:** Colorless liquid with a pungent odor.

**Chemical & Physical Properties:**
**MW:** Varies
**BP:** 214°F
**Sol:** Miscible
**Fl.P:** 185°F
**IP:** ?
**Sp.Gr(77°F):** 1.08
**VP:** 1 mmHg
**FRZ:** ?
**UEL:** 73%
**LEL:** 7%
Class IIIA Combustible Liquid

**Personal Protection/Sanitation (see Table 2):**
**Skin:** Prevent skin contact
**Eyes:** Prevent eye contact
**Wash skin:** When contam
**Remove:** When wet or contam
**Change:** N.R.
**Provide:** Eyewash
Quick drench

**Respirator Recommendations (see Tables 3 and 4):**
**NIOSH**
**¥:** ScbaF:Pd,Pp/SaF:Pd,Pp:AScba
**Escape:** GmFS/ScbaE

**See Appendix E** (page 351)

**Incompatibilities and Reactivities:** Strong oxidizers, alkalis & acids; phenols; urea; oxides; isocyanates; caustics; anhydrides

**Exposure Routes, Symptoms, Target Organs (see Table 5):**
**ER:** Inh, Ing, Con
**SY:** Irrit eyes, nose, throat, resp sys; lac; cough; wheez, derm; [carc]
**TO:** Eyes, skin, resp sys [nasal cancer]

**First Aid (see Table 6):**
**Eye:** Irr immed
**Skin:** Water flush prompt
**Breath:** Resp support
**Swallow:** Medical attention immed

## Formamide

| Formamide | Formula: $HCONH_2$ | CAS#: 75-12-7 | RTECS#: LQ0525000 | IDLH: N.D. |
|---|---|---|---|---|

**Conversion:** 1 ppm = 1.85 mg/m$^3$ | **DOT:**

**Synonyms/Trade Names:** Carbamaldehyde, Methanamide

**Exposure Limits:**
**NIOSH REL:** TWA 10 ppm (15 mg/m$^3$) [skin]
**OSHA PEL†:** none

**Measurement Methods (see Table 1):**
None available

**Physical Description:** Colorless, oily liquid. [**Note:** A solid below 37°F.]

**Chemical & Physical Properties:**
**MW:** 45.1
**BP:** 411°F (Decomposes)
**Sol:** Miscible
**Fl.P(oc):** 310°F
**IP:** 10.20 eV
**Sp.Gr:** 1.13
**VP(86°F):** 0.1 mmHg
**FRZ:** 37°F
**UEL:** ?
**LEL:** ?
Class IIIB Combustible Liquid

**Personal Protection/Sanitation (see Table 2):**
**Skin:** N.R.
**Eyes:** Prevent eye contact
**Wash skin:** N.R.
**Remove:** N.R.
**Change:** N.R.

**Respirator Recommendations (see Tables 3 and 4):**
Not available.

**Incompatibilities and Reactivities:** Oxidizers, iodine, pyridine, sulfur trioxide, copper, brass, lead
[**Note:** Hygroscopic (i.e., absorbs moisture from the air).]

**Exposure Routes, Symptoms, Target Organs (see Table 5):**
**ER:** Inh, Ing, Con
**SY:** Irrit eyes, skin, muc memb; drow, lass; nau; acidosis; skin eruptions; in animals: repro effects
**TO:** Eyes, skin, resp sys, CNS, repro sys

**First Aid (see Table 6):**
**Eye:** Irr immed
**Skin:** Water wash
**Breath:** Resp support
**Swallow:** Medical attention immed

F

## Formic acid

| Formic acid | Formula: HCOOH | CAS#: 64-18-6 | RTECS#: LQ4900000 | IDLH: 30 ppm |
|---|---|---|---|---|

**Conversion:** 1 ppm = 1.88 mg/m$^3$ | **DOT:** 1779 153

**Synonyms/Trade Names:** Formic acid (85-95% in aqueous solution); Hydrogen carboxylic acid; Methanoic acid

**Exposure Limits:**
**NIOSH REL:** TWA 5 ppm (9 mg/m$^3$)
**OSHA PEL:** TWA 5 ppm (9 mg/m$^3$)

**Measurement Methods (see Table 1):**
**NIOSH** 2011
**OSHA** ID186SG

**Physical Description:** Colorless liquid with a pungent, penetrating odor.
[**Note:** Often used in an aqueous solution.]

**Chemical & Physical Properties:**
**MW:** 46.0
**BP:** 224°F (90% solution)
**Sol:** Miscible
**Fl.P(oc):** 122°F (90% solution)
**IP:** 11.05 eV
**Sp.Gr:** 1.22 (90% solution)
**VP:** 35 mmHg
**FRZ:** 20°F (90% solution)
**UEL:** 57% (90% solution)
**LEL:** 18% (90% solution)
Class II Combustible Liquid (90% solution)

**Personal Protection/Sanitation (see Table 2):**
**Skin:** Prevent skin contact
**Eyes:** Prevent eye contact
**Wash skin:** When contam
**Remove:** When wet or contam
**Change:** N.R.
**Provide:** Eyewash
Quick drench

**Respirator Recommendations (see Tables 3 and 4):**
**NIOSH/OSHA**
**30 ppm:** Sa*/ScbaF
**§:** ScbaF:Pd,Pp/SaF:Pd,Pp:AScba
**Escape:** GmFOv100/ScbaE

**Incompatibilities and Reactivities:** Strong oxidizers, strong caustics, concentrated sulfuric acid
[**Note:** Corrosive to metals.]

**Exposure Routes, Symptoms, Target Organs (see Table 5):**
**ER:** Inh, Ing, Con
**SY:** Irrit eyes; skin, throat; skin burns, derm; lac; rhin; cough, dysp; nau
**TO:** Eyes, skin, resp sys

**First Aid (see Table 6):**
**Eye:** Irr immed
**Skin:** Water flush immed
**Breath:** Resp support
**Swallow:** Medical attention immed

## Furfural

**Formula:** $C_5H_4O_2$ | **CAS#:** 98-01-1 | **RTECS#:** LT7000000 | **IDLH:** 100 ppm

**Conversion:** 1 ppm = 3.93 mg/m$^3$ | **DOT:** 1199 132P

**Synonyms/Trade Names:** Fural, 2-Furancarboxaldehyde, Furfuraldehyde, 2-Furfuraldehyde

**Exposure Limits:**
**NIOSH REL:** See Appendix D
**OSHA PEL†:** TWA 5 ppm (20 mg/m$^3$) [skin]

**Measurement Methods (see Table 1):**
**NIOSH** 2529
**OSHA** 72

**Physical Description:** Colorless to amber liquid with an almond-like odor.
[**Note:** Darkens in light and air.]

**Chemical & Physical Properties:**
**MW:** 96.1
**BP:** 323°F
**Sol:** 8%
**Fl.P:** 140°F
**IP:** 9.21 eV
**Sp.Gr:** 1.16
**VP:** 2 mmHg
**FRZ:** -34°F
**UEL:** 19.3%
**LEL:** 2.1%
Class IIIA Combustible Liquid

**Personal Protection/Sanitation (see Table 2):**
**Skin:** Prevent skin contact
**Eyes:** Prevent eye contact
**Wash skin:** When contam
**Remove:** When wet or contam
**Change:** N.R.

**Respirator Recommendations (see Tables 3 and 4):**
**OSHA**
**50 ppm:** CcrOv*/Sa*
**100 ppm:** Sa:Cf*/CcrFOv/PaprOv*/GmFOv/ScbaF/SaF
**§:** ScbaF:Pd,Pp/SaF:Pd,Pp:AScba
**Escape:** GmFOv/ScbaE

**Incompatibilities and Reactivities:** Strong acids, oxidizers, strong alkalis
[**Note:** May polymerize on contact with strong acids or strong alkalis.]

**Exposure Routes, Symptoms, Target Organs (see Table 5):**
**ER:** Inh, Abs, Ing, Con
**SY:** Irrit eyes, skin, upper resp sys; head; derm
**TO:** Eyes, skin, resp sys

**First Aid (see Table 6):**
**Eye:** Irr immed
**Skin:** Water flush prompt
**Breath:** Resp support
**Swallow:** Medical attention immed

## Furfuryl alcohol

**Formula:** $C_5H_6O_2$ | **CAS#:** 98-00-0 | **RTECS#:** LU9100000 | **IDLH:** 75 ppm

**Conversion:** 1 ppm = 4.01 mg/m$^3$ | **DOT:** 2874 153

**Synonyms/Trade Names:** 2-Furylmethanol, 2-Hydroxymethylfuran

**Exposure Limits:**
**NIOSH REL:** TWA 10 ppm (40 mg/m$^3$) [skin]
ST 15 ppm (60 mg/m$^3$)
**OSHA PEL†:** TWA 50 ppm (200 mg/m$^3$)

**Measurement Methods (see Table 1):**
**NIOSH** 2505

**Physical Description:** Colorless to amber liquid with a faint, burning odor.
[**Note:** Darkens on exposure to light.]

**Chemical & Physical Properties:**
**MW:** 98.1
**BP:** 338°F
**Sol:** Miscible
**Fl.P:** 149°F
**IP:** ?
**Sp.Gr:** 1.13
**VP(77°F):** 0.6 mmHg
**FRZ:** 6°F
**UEL:** 16.3%
**LEL:** 1.8%
Class IIIA Combustible Liquid

**Personal Protection/Sanitation (see Table 2):**
**Skin:** Prevent skin contact
**Eyes:** Prevent eye contact
**Wash skin:** When contam
**Remove:** When wet or contam
**Change:** N.R.
**Provide:** Quick drench

**Respirator Recommendations (see Tables 3 and 4):**
**NIOSH**
**75 ppm:** CcrOv*/GmFOv/PaprOv*/Sa*/ScbaF
**§:** ScbaF:Pd,Pp/SaF:Pd,Pp:AScba
**Escape:** GmFOv/ScbaE

**Incompatibilities and Reactivities:** Strong oxidizers & acids
[**Note:** Contact with organic acids may lead to polymerization.]

**Exposure Routes, Symptoms, Target Organs (see Table 5):**
**ER:** Inh, Abs, Ing, Con
**SY:** Irrit eyes, muc memb; dizz; nau, diarr; diuresis; resp, body temperature depres; vomit; derm
**TO:** Eyes, skin, resp sys, CNS

**First Aid (see Table 6):**
**Eye:** Irr immed
**Skin:** Water flush immed
**Breath:** Resp support
**Swallow:** Medical attention immed

| **Gasoline** | **Formula:** | **CAS#:** 8006-61-9 | **RTECS#:** LX3300000 | **IDLH:** Ca [N.D.] |
|---|---|---|---|---|
| **Conversion:** 1 ppm = 4.5 $mg/m^3$ (approx) | **DOT:** 1203 128 | | | |

**Synonyms/Trade Names:** Motor fuel, Motor spirits, Natural gasoline, Petrol
[**Note:** A complex mixture of volatile hydrocarbons (paraffins, cycloparaffins & aromatics).]

**Exposure Limits:**
**NIOSH REL:** Ca
See Appendix A
**OSHA PEL†:** none

**Measurement Methods (see Table 1):**
**OSHA** PV2028

**Physical Description:** Clear liquid with a characteristic odor.

**Chemical & Physical Properties:**
**MW:** 110 (approx)
**BP:** 102°F
**Sol:** Insoluble
**Fl.P:** -45°F
**IP:** ?
**Sp.Gr(60°F):** 0.72-0.76
**VP:** 38-300 mmHg
**FRZ:** ?
**UEL:** 7.6%
**LEL:** 1.4%
Class IB Flammable Liquid

**Personal Protection/Sanitation (see Table 2):**
**Skin:** Prevent skin contact
**Eyes:** Prevent eye contact
**Wash skin:** When contam
**Remove:** When wet (flamm)
**Change:** N.R.
**Provide:** Eyewash
Quick drench

**Respirator Recommendations (see Tables 3 and 4):**
**NIOSH**
**¥:** ScbaF:Pd,Pp/SaF:Pd,Pp:AScba
**Escape:** GmFOv/ScbaE

**Incompatibilities and Reactivities:** Strong oxidizers such as peroxides, nitric acid & perchlorates

**Exposure Routes, Symptoms, Target Organs (see Table 5):**
**ER:** Inh, Abs, Ing, Con
**SY:** Irrit eyes, skin, muc memb; derm; head, lass, blurred vision, dizz, slurred speech, conf, convuls; chemical pneu (aspir liquid); possible liver, kidney damage; [carc]
**TO:** Eyes, skin, resp sys, CNS, liver, kidneys [in animals: liver & kidney cancer]

**First Aid (see Table 6):**
**Eye:** Irr immed
**Skin:** Soap flush immed
**Breath:** Resp support
**Swallow:** Medical attention immed

G

| **Germanium tetrahydride** | **Formula:** $GeH_4$ | **CAS#:** 7782-65-2 | **RTECS#:** LY4900000 | **IDLH:** N.D. |
|---|---|---|---|---|
| **Conversion:** 1 ppm = 3.13 $mg/m^3$ | **DOT:** 2192 119 | | | |

**Synonyms/Trade Names:** Germane, Germanium hydride, Germanomethane, Monogermane
[**Note:** Used chiefly for the production of high purity germanium for use in semiconductors.]

**Exposure Limits:**
**NIOSH REL:** TWA 0.2 ppm (0.6 $mg/m^3$)
**OSHA PEL†:** none

**Measurement Methods (see Table 1):**
None available

**Physical Description:** Colorless gas with a pungent odor.
[**Note:** Shipped as a compressed gas.]

**Chemical & Physical Properties:**
**MW:** 76.6
**BP:** -127°F
**Sol:** Insoluble
**Fl.P:** NA (Gas)
**IP:** 11.34 eV
**RGasD:** 2.65
**VP:** >1 atm
**FRZ:** -267°F
**UEL:** ?
**LEL:** ?
Flammable Gas (may ignite SPONTANEOUSLY in air).

**Personal Protection/Sanitation (see Table 2):**
**Skin:** N.R.
**Eyes:** N.R.
**Wash skin:** N.R.
**Remove:** N.R.
**Change:** N.R.

**Respirator Recommendations (see Tables 3 and 4):**
Not available.

**Incompatibilities and Reactivities:** Bromine

**Exposure Routes, Symptoms, Target Organs (see Table 5):**
**ER:** Inh
**SY:** Mal, head, dizz, fainting; dysp; nau, vomit; kidney inj; hemolytic effects
**TO:** CNS, kidneys, blood

**First Aid (see Table 6):**
**Breath:** Resp support

G

| **Glutaraldehyde** | **Formula:** $OCH(CH_2)_3CHO$ | **CAS#:** 111-30-8 | **RTECS#:** MA2450000 | **IDLH:** N.D. |
|---|---|---|---|---|
| **Conversion:** 1 ppm = 4.09 mg/m$^3$ | **DOT:** | | | |

**Synonyms/Trade Names:** Glutaric dialdehyde; 1,5-Pentanedial

| | |
|---|---|
| **Exposure Limits:**<br>**NIOSH REL:** C 0.2 ppm (0.8 mg/m$^3$)<br>See Appendix C (Aldehydes)<br>**OSHA PEL†:** none | **Measurement Methods (see Table 1):**<br>**NIOSH** 2532<br>**OSHA** 64 |

**Physical Description:** Colorless liquid with a pungent odor.

| **Chemical & Physical Properties:** | **Personal Protection/Sanitation (see Table 2):** | **Respirator Recommendations (see Tables 3 and 4):** |
|---|---|---|
| **MW:** 100.1<br>**BP:** 212°F<br>**Sol:** Miscible<br>**Fl.P:** NA<br>**IP:** ?<br>**Sp.Gr:** 1.10<br>**VP:** 17 mmHg<br>**FRZ:** 7°F<br>**UEL:** NA<br>**LEL:** NA<br>Noncombustible Liquid | **Skin:** Prevent skin contact<br>**Eyes:** Prevent eye contact<br>**Wash skin:** When contam<br>**Remove:** When wet or contam<br>**Change:** N.R.<br>**Provide:** Eyewash<br>Quick drench | Not available. |

**Incompatibilities and Reactivities:** Strong oxidizers, strong bases [**Note:** Alkaline solutions of glutaraldehyde (i.e., activated glutaraldehyde) react with alcohol, ketones, amines, hydrazines & proteins.]

| **Exposure Routes, Symptoms, Target Organs (see Table 5):** | **First Aid (see Table 6):** |
|---|---|
| **ER:** Inh, Abs, Ing, Con<br>**SY:** Irrit eyes, skin, resp sys; derm, sens skin; cough, asthma; nau, vomit<br>**TO:** Eyes, skin, resp sys | **Eye:** Irr immed<br>**Skin:** Water flush immed<br>**Breath:** Resp support<br>**Swallow:** Medical attention immed |

| **Glycerin (mist)** | **Formula:** $HOCH_2CH(OH)CH_2OH$ | **CAS#:** 56-81-5 | **RTECS#:** MA8050000 | **IDLH:** N.D. |
|---|---|---|---|---|
| **Conversion:** | **DOT:** | | | |

**Synonyms/Trade Names:** Glycerin (anhydrous); Glycerol; Glycyl alcohol; 1,2,3-Propanetriol; Trihydroxypropane

| | |
|---|---|
| **Exposure Limits:**<br>**NIOSH REL:** See Appendix D<br>**OSHA PEL†:** TWA 15 mg/m$^3$ (total)<br>TWA 5 mg/m$^3$ (resp) | **Measurement Methods (see Table 1):**<br>**NIOSH** 0500, 0600 |

**Physical Description:** Clear, colorless, odorless, syrupy liquid or solid (below 64°F).
[**Note:** The solid form melts above 64°F but the liquid form freezes at a much lower temperature.]

| **Chemical & Physical Properties:** | **Personal Protection/Sanitation (see Table 2):** | **Respirator Recommendations (see Tables 3 and 4):** |
|---|---|---|
| **MW:** 92.1<br>**BP:** 554°F (Decomposes)<br>**Sol:** Miscible<br>**Fl.P:** 320°F<br>**IP:** ?<br>**Sp.Gr:** 1.26<br>**VP(122°F):** 0.003 mmHg<br>**MLT:** 64°F<br>**UEL:** ?<br>**LEL:** ?<br>Class IIIB Combustible Liquid | **Skin:** N.R.<br>**Eyes:** N.R.<br>**Wash skin:** N.R.<br>**Remove:** N.R.<br>**Change:** N.R. | Not available. |

**Incompatibilities and Reactivities:** Strong oxidizers (e.g., chromium trioxide, potassium chlorate, potassium permanganate) [**Note:** Hygroscopic (i.e., absorbs moisture from the air).]

| **Exposure Routes, Symptoms, Target Organs (see Table 5):** | **First Aid (see Table 6):** |
|---|---|
| **ER:** Inh, Con<br>**SY:** Irrit eyes, skin, resp sys; head, nau, vomit; kidney inj<br>**TO:** Eyes, skin, resp sys, kidneys | **Eye:** Irr immed<br>**Skin:** Water wash<br>**Breath:** Fresh air |

## Glycidol

| Glycidol | Formula: $C_3H_6O_2$ | CAS#: 556-52-5 | RTECS#: UB4375000 | IDLH: 150 ppm |
|---|---|---|---|---|
| **Conversion:** 1 ppm = 3.03 mg/m$^3$ | **DOT:** | | | |

**Synonyms/Trade Names:** 2,3-Epoxy-1-propanol; Epoxypropyl alcohol; Glycide; Hydroxymethyl ethylene oxide; 2-Hydroxymethyl oxiran; 3-Hydroxypropylene oxide

**Exposure Limits:**
**NIOSH REL:** TWA 25 ppm (75 mg/m$^3$)
**OSHA PEL†:** TWA 50 ppm (150 mg/m$^3$)

**Measurement Methods (see Table 1):**
**NIOSH** 1608
**OSHA** 7

**Physical Description:** Colorless liquid.

| Chemical & Physical Properties: | Personal Protection/Sanitation (see Table 2): | Respirator Recommendations (see Tables 3 and 4): |
|---|---|---|
| **MW:** 74.1<br>**BP:** 320°F (Decomposes)<br>**Sol:** Miscible<br>**Fl.P:** 162°F<br>**IP:** ?<br>**Sp.Gr:** 1.12<br>**VP(77°F):** 0.9 mmHg<br>**FRZ:** -49°F<br>**UEL:** ?<br>**LEL:** ?<br>Class IIIA Combustible Liquid | **Skin:** Prevent skin contact<br>**Eyes:** Prevent eye contact<br>**Wash skin:** When contam<br>**Remove:** When wet or contam<br>**Change:** N.R. | **NIOSH**<br>**150 ppm:** Sa*/ScbaF<br>**§:** ScbaF:Pd,Pp/SaF:Pd,Pp:AScba<br>**Escape:** GmFOv/ScbaE |

**Incompatibilities and Reactivities:** Strong oxidizers, nitrates

| Exposure Routes, Symptoms, Target Organs (see Table 5): | First Aid (see Table 6): |
|---|---|
| **ER:** Inh, Ing, Con<br>**SY:** Irrit eyes, skin, nose, throat; narco<br>**TO:** Eyes, skin, resp sys, CNS | **Eye:** Irr immed<br>**Skin:** Water wash prompt<br>**Breath:** Resp support<br>**Swallow:** Medical attention immed |

G

## Glycolonitrile

| Glycolonitrile | Formula: $HOCH_2CN$ | CAS#: 107-16-4 | RTECS#: AM0350000 | IDLH: N.D. |
|---|---|---|---|---|
| **Conversion:** 1 ppm = 2.34 mg/m$^3$ | **DOT:** | | | |

**Synonyms/Trade Names:** Cyanomethanol, Formaldehyde cyanohydrin, Glycolic nitrile, Glyconitrile, Hydroxyacetonitrile

**Exposure Limits:**
**NIOSH REL:** C 2 ppm (5 mg/m$^3$) [15-minute]
**OSHA PEL:** none

**Measurement Methods (see Table 1):**
None available

**Physical Description:** Colorless, odorless, oily liquid.
[**Note:** Forms cyanide in the body.]

| Chemical & Physical Properties: | Personal Protection/Sanitation (see Table 2): | Respirator Recommendations (see Tables 3 and 4): |
|---|---|---|
| **MW:** 57.1<br>**BP:** 361°F (Decomposes)<br>**Sol:** Soluble<br>**Fl.P:** ?<br>**IP:** ?<br>**Sp.Gr(66°F):** 1.10<br>**VP(145°F):** 1 mmHg<br>**FRZ:** <-98°F<br>**UEL:** ?<br>**LEL:** ?<br>Combustible Liquid | **Skin:** Prevent skin contact<br>**Eyes:** Prevent eye contact<br>**Wash skin:** When contam<br>**Remove:** When wet or contam<br>**Change:** Daily<br>**Provide:** Eyewash<br>Quick drench | **NIOSH**<br>**20 ppm:** Sa<br>**50 ppm:** Sa:Cf<br>**100 ppm:** ScbaF/SaF<br>**250 ppm:** SaF:Pd,Pp<br>**§:** ScbaF:Pd,Pp/SaF:Pd,Pp:AScba<br>**Escape:** GmFOv/ScbaE |

**Incompatibilities and Reactivities:** Traces of alkalis (promote violent polymerization)

| Exposure Routes, Symptoms, Target Organs (see Table 5): | First Aid (see Table 6): |
|---|---|
| **ER:** Inh, Abs, Ing, Con<br>**SY:** Irrit eyes, skin, resp sys; head, dizz, lass, conf, convuls; dysp; abdom pain, nau, vomit<br>**TO:** Eyes, skin, resp sys, CNS, CVS | **Eye:** Irr immed<br>**Skin:** Water wash immed<br>**Breath:** Resp support<br>**Swallow:** Medical attention immed |

G

## Grain dust (oat, wheat, barley)

| Grain dust (oat, wheat, barley) | Formula: | CAS#: | RTECS#: MD7900000 | IDLH: N.D. |
|---|---|---|---|---|
| Conversion: | DOT: | | | |

**Synonyms/Trade Names:** None
[**Note:** Grain dust consists of 60-75% organic materials (cereal grains) & 25-40% inorganic materials (soil), and includes fertilizers, pesticides & microorganisms.]

**Exposure Limits:**
**NIOSH REL:** TWA 4 mg/m$^3$
**OSHA PEL:** TWA 10 mg/m$^3$

**Measurement Methods (see Table 1):**
**NIOSH** 0500

**Physical Description:** Mixture of grain and all the other substances associated with its cultivation & harvesting.

**Chemical & Physical Properties:**
Properties depend upon the specific component of the grain dust.

**Personal Protection/Sanitation (see Table 2):**
**Skin:** N.R.
**Eyes:** N.R.
**Wash skin:** N.R.
**Remove:** N.R.
**Change:** Daily

**Respirator Recommendations (see Tables 3 and 4):**
Not available.

**Incompatibilities and Reactivities:** None reported

**Exposure Routes, Symptoms, Target Organs (see Table 5):**
**ER:** Inh, Con
**SY:** Irrit eyes, skin, upper resp sys; cough, dysp, wheez, asthma, bron, chronic obstructive pulm disease; conj, derm, rhinitis, grain fever
**TO:** Eyes, skin, resp sys

**First Aid (see Table 6):**
**Eye:** Irr immed
**Breath:** Fresh air

## Graphite (natural)

| Graphite (natural) | Formula: C | CAS#: 7782-42-5 | RTECS#: MD9659600 | IDLH: 1250 mg/m$^3$ |
|---|---|---|---|---|
| Conversion: | DOT: | | | |

**Synonyms/Trade Names:** Black lead, Mineral carbon, Plumbago, Silver graphite, Stove black
[**Note:** Also see specific listing for Graphite (synthetic).]

**Exposure Limits:**
**NIOSH REL:** TWA 2.5 mg/m$^3$ (resp)
**OSHA PEL†:** TWA 15 mppcf

**Measurement Methods (see Table 1):**
**NIOSH** 0500, 0600

**Physical Description:** Steel gray to black, greasy feeling, odorless solid.

**Chemical & Physical Properties:**
**MW:** 12.0
**BP:** Sublimes
**Sol:** Insoluble
**Fl.P:** NA
**IP:** NA
**Sp.Gr:** 2.0-2.25
**VP:** 0 mmHg (approx)
**MLT:** 6602°F (Sublimes)
**UEL:** NA
**LEL:** NA
Combustible Solid

**Personal Protection/Sanitation (see Table 2):**
**Skin:** N.R.
**Eyes:** N.R.
**Wash skin:** N.R.
**Remove:** N.R.
**Change:** N.R.

**Respirator Recommendations (see Tables 3 and 4):**
**NIOSH**
**12.5 mg/m$^3$:** Qm
**25 mg/m$^3$:** 95XQ/Sa
**62.5 mg/m$^3$:** PaprHie/Sa:Cf
**125 mg/m$^3$:** 100F/PaprTHie/SaT:Cf/ScbaF/SaF
**1250 mg/m$^3$:** SaF:Pd,Pp
**§:** ScbaF:Pd,Pp/SaF:Pd,Pp:AScba
**Escape:** 100F/ScbaE

**Incompatibilities and Reactivities:** Very strong oxidizers such as fluorine, chlorine trifluoride & potassium peroxide

**Exposure Routes, Symptoms, Target Organs (see Table 5):**
**ER:** Inh, Con
**SY:** Cough, dysp, black sputum, decr pulm func, lung fib
**TO:** Resp sys, CVS

**First Aid (see Table 6):**
**Eye:** Irr immed
**Breath:** Fresh air

## Graphite (synthetic)

**Formula:** C | **CAS#:** 7440-44-0 (synthetic) | **RTECS#:** FF5250100 (synthetic) | **IDLH:** N.D.

**Conversion:** | **DOT:**

**Synonyms/Trade Names:** Acheson graphite, Artificial graphite
[**Note:** Also see specific listing for Graphite (natural).]

**Exposure Limits:**
**NIOSH REL:** See Appendix D
**OSHA PEL†:** TWA 15 mg/m$^3$ (total)
TWA 5 mg/m$^3$ (resp)

**Measurement Methods (see Table 1):**
**NIOSH** 0500, 0600

**Physical Description:** Steel gray to black, greasy feeling, odorless solid.

**Chemical & Physical Properties:**
**MW:** 12.0
**BP:** Sublimes
**Sol:** Insoluble
**Fl.P:** NA
**IP:** NA
**Sp.Gr:** 1.5-1.8
**VP:** 0 mmHg (approx)
**MLT:** 6602°F (Sublimes)
**UEL:** NA
**LEL:** NA
Combustible Solid

**Personal Protection/Sanitation (see Table 2):**
**Skin:** N.R.
**Eyes:** N.R.
**Wash skin:** N.R.
**Remove:** N.R.
**Change:** N.R.

**Respirator Recommendations (see Tables 3 and 4):**
Not available.

**Incompatibilities and Reactivities:** Very strong oxidizers such as fluorine, chlorine trifluoride & potassium peroxide

**Exposure Routes, Symptoms, Target Organs (see Table 5):**
**ER:** Inh, Con
**SY:** Cough, dysp, black sputum, decr pulm func, lung fib
**TO:** Resp sys, CVS

**First Aid (see Table 6):**
**Eye:** Irr immed
**Breath:** Fresh air

## Gypsum

**Formula:** $CaSO_4 \times 2H_2O$ | **CAS#:** 13397-24-5 | **RTECS#:** MG2360000 | **IDLH:** N.D.

**Conversion:** | **DOT:**

**Synonyms/Trade Names:** Calcium(II) sulfate dihydrate, Gypsum stone, Hydrated calcium sulfate, Mineral white
[**Note:** Gypsum is the dihydrate form of calcium sulfate; Plaster of Paris is the hemihydrate form.]

**Exposure Limits:**
**NIOSH REL:** TWA 10 mg/m$^3$ (total)
TWA 5 mg/m$^3$ (resp)
**OSHA PEL:** TWA 15 mg/m$^3$ (total)
TWA 5 mg/m$^3$ (resp)

**Measurement Methods (see Table 1):**
**NIOSH** 0500, 0600

**Physical Description:** White or nearly white, odorless, crystalline solid.

**Chemical & Physical Properties:**
**MW:** 172.2
**BP:** ?
**Sol(77°F):** 0.2%
**Fl.P:** NA
**IP:** NA
**Sp.Gr:** 2.32
**VP:** 0 mmHg (approx)
**MLT:** 262-325°F (Loses $H_2O$)
**UEL:** NA
**LEL:** NA
Noncombustible Solid

**Personal Protection/Sanitation (see Table 2):**
**Skin:** N.R.
**Eyes:** N.R.
**Wash skin:** N.R.
**Remove:** N.R.
**Change:** N.R.

**Respirator Recommendations (see Tables 3 and 4):**
Not available.

**Incompatibilities and Reactivities:** Aluminum (at high temperatures), diazomethane

**Exposure Routes, Symptoms, Target Organs (see Table 5):**
**ER:** Inh, Con
**SY:** Irrit eyes, skin, muc memb, upper resp sys; cough, sneez, rhin
**TO:** Eyes, skin, resp sys

**First Aid (see Table 6):**
**Eye:** Irr immed
**Breath:** Fresh air

H

## Hafnium

| Formula: | CAS#: | RTECS#: | IDLH: |
|---|---|---|---|
| Hf | 7440-58-6 | MG4600000 | 50 mg/m$^3$ (as Hf) |

**Conversion:**

**DOT:** 1326 170 (powder, wet); 2545 135 (powder, dry)

**Synonyms/Trade Names:** Celtium, Elemental hafnium, Hafnium metal

**Exposure Limits:**
**NIOSH REL*:** TWA 0.5 mg/m$^3$
**OSHA PEL*:** TWA 0.5 mg/m$^3$
[***Note:** The REL and PEL also apply to other hafnium compounds (as Hf).]

**Measurement Methods (see Table 1):**
**NIOSH** S194 (II-5)
**OSHA** ID121

**Physical Description:** Highly lustrous, ductile, grayish solid.

**Chemical & Physical Properties:**
**MW:** 178.5
**BP:** 8316°F
**Sol:** Insoluble
**Fl.P:** NA
**IP:** NA
**Sp.Gr:** 13.31
**VP:** 0 mmHg (approx)
**MLT:** 4041°F
**UEL:** NA
**LEL:** NA
Explosive in powder form (either dry or with <25% water); finely divided powder can be ignited by static electricity or even SPONTANEOUSLY.

**Personal Protection/Sanitation (see Table 2):**
**Skin:** Prevent skin contact
**Eyes:** Prevent eye contact
**Wash skin:** When contam/Daily
**Remove:** When wet or contam
**Change:** Daily
**Provide:** Eyewash
Quick drench

**Respirator Recommendations (see Tables 3 and 4):**
**NIOSH/OSHA**
**2.5 mg/m$^3$:** Qm
**5 mg/m$^3$:** 95XQ/Sa
**12.5 mg/m$^3$:** Sa:Cf*/PaprHie*
**25 mg/m$^3$:** 100F/SaT:Cf*/PaprTHie*/ScbaF/SaF
**50 mg/m$^3$:** SaF:Pd,Pp
**§:** ScbaF:Pd,Pp/SaF:Pd,Pp:AScba
**Escape:** 100F/ScbaE

**Incompatibilities and Reactivities:** Strong oxidizers, chlorine

**Exposure Routes, Symptoms, Target Organs (see Table 5):**
**ER:** Inh, Ing, Con
**SY:** In animals: irrit eyes, skin, muc memb; liver damage
**TO:** Eyes, skin, muc memb, liver

**First Aid (see Table 6):**
**Eye:** Irr immed
**Skin:** Soap wash prompt
**Breath:** Resp support
**Swallow:** Medical attention immed

## Halothane

| Formula: | CAS#: | RTECS#: | IDLH: |
|---|---|---|---|
| $CF_3CHBrCl$ | 151-67-7 | KH6550000 | N.D. |

**Conversion:** 1 ppm = 8.07 mg/m$^3$

**DOT:**

**Synonyms/Trade Names:** 1-Bromo-1-chloro-2,2,2-trifluoroethane; 2-Bromo-2-chloro-1,1,1-trifluoroethane; 1,1,1-Trifluoro-2-bromo-2-chloroethane; 2,2,2-Trifluoro-1-bromo-1-chloroethane

**Exposure Limits:**
**NIOSH REL*:** C 2 ppm (16.2 mg/m$^3$) [60-minute]
[***Note:** REL for exposure to waste anesthetic gas.]
**OSHA PEL:** none

**Measurement Methods (see Table 1):**
**OSHA** 29

**Physical Description:** Clear, colorless liquid with a sweetish, pleasant odor. [inhalation anesthetic]

**Chemical & Physical Properties:**
**MW:** 197.4
**BP:** 122°F
**Sol:** 0.3%
**Fl.P:** NA
**IP:** ?
**Sp.Gr:** 1.87
**VP:** 243 mmHg
**FRZ:** -180°F
**UEL:** NA
**LEL:** NA
Noncombustible Liquid

**Personal Protection/Sanitation (see Table 2):**
**Skin:** Prevent skin contact
**Eyes:** Prevent eye contact
**Wash skin:** When contam
**Remove:** When wet or contam
**Change:** N.R.
**Provide:** Eyewash

**Respirator Recommendations (see Tables 3 and 4):**
Not available.

**Incompatibilities and Reactivities:** May attack rubber & some plastics; sensitive to light.
[**Note:** Light causes decomposition. May be stabilized with 0.01% thymol.]

**Exposure Routes, Symptoms, Target Organs (see Table 5):**
**ER:** Inh, Abs, Ing, Con
**SY:** Irrit eyes, skin, resp sys; conf, drow, dizz, nau, analgesia, anes; card arrhy; liver, kidney damage; decr audio-visual performance; in animals: repro effects
**TO:** Eyes, skin, resp sys, CVS, CNS, liver, kidneys, repro sys

**First Aid (see Table 6):**
**Eye:** Irr immed
**Skin:** Soap wash prompt
**Breath:** Resp support
**Swallow:** Medical attention immed

## Heptachlor

| Heptachlor | Formula: $C_{10}H_5Cl_7$ | CAS#: 76-44-8 | RTECS#: PC0700000 | IDLH: Ca [35 mg/m$^3$] |
|---|---|---|---|---|

**Conversion:**

**DOT:** 2761 151 (organochlorine pesticide, solid)

**Synonyms/Trade Names:** 1,4,5,6,7,8,8-Heptachloro-3a,4,7,7a-tetrahydro-4,7-methanoindene

**Exposure Limits:**
**NIOSH REL:** Ca
TWA 0.5 mg/m$^3$ [skin]
See Appendix A
**OSHA PEL:** TWA 0.5 mg/m$^3$ [skin]

**Measurement Methods (see Table 1):**
**NIOSH** S287 (II-5)
**OSHA** PV2029

**Physical Description:** White to light-tan crystals with a camphor-like odor. [insecticide]

**Chemical & Physical Properties:**
**MW:** 373.4
**BP:** 293°F (Decomposes)
**Sol:** 0.0006%
**Fl.P:** NA
**IP:** ?
**Sp.Gr:** 1.66
**VP(77°F):** 0.0003 mmHg
**MLT:** 203°F
**UEL:** NA
**LEL:** NA
Noncombustible Solid, but may be dissolved in flammable liquids.

**Personal Protection/Sanitation (see Table 2):**
**Skin:** Prevent skin contact
**Eyes:** Prevent eye contact
**Wash skin:** When contam/Daily
**Remove:** When wet or contam
**Change:** Daily
**Provide:** Eyewash
Quick drench

**Respirator Recommendations (see Tables 3 and 4):**
**NIOSH**
**¥:** ScbaF:Pd,Pp/SaF:Pd,Pp:AScba
**Escape:** GmFOv100/ScbaE

**Incompatibilities and Reactivities:** Iron, rust

**Exposure Routes, Symptoms, Target Organs (see Table 5):**
**ER:** Inh, Abs, Ing, Con
**SY:** In animals: tremor, convuls; liver damage; [carc]
**TO:** CNS,liver [in animals: liver cancer]

**First Aid (see Table 6):**
**Eye:** Irr immed
**Skin:** Soap wash immed
**Breath:** Resp support
**Swallow:** Medical attention immed

## n-Heptane

| n-Heptane | Formula: $CH_3[CH_2]_5CH_3$ | CAS#: 142-82-5 | RTECS#: MI7700000 | IDLH: 750 ppm |
|---|---|---|---|---|

**Conversion:** 1 ppm = 4.10 mg/m$^3$

**DOT:** 1206 128

**Synonyms/Trade Names:** Heptane, normal-Heptane

**Exposure Limits:**
**NIOSH REL:** TWA 85 ppm (350 mg/m$^3$)
C 440 ppm (1800 mg/m$^3$) [15-minute]
**OSHA PEL†:** TWA 500 ppm (2000 mg/m$^3$)

**Measurement Methods (see Table 1):**
**NIOSH** 1500
**OSHA** 7

**Physical Description:** Colorless liquid with a gasoline-like odor.

**Chemical & Physical Properties:**
**MW:** 100.2
**BP:** 209°F
**Sol:** 0.0003%
**Fl.P:** 25°F
**IP:** 9.90 eV
**Sp.Gr:** 0.68
**VP(72°F):** 40 mmHg
**FRZ:** -131°F
**UEL:** 6.7%
**LEL:** 1.05%
Class IB Flammable Liquid

**Personal Protection/Sanitation (see Table 2):**
**Skin:** Prevent skin contact
**Eyes:** Prevent eye contact
**Wash skin:** When contam
**Remove:** When wet (flamm)
**Change:** N.R.

**Respirator Recommendations (see Tables 3 and 4):**
**NIOSH**
**750 ppm:** CcrOv/GmFOv/PaprOv/Sa/ScbaF
**§:** ScbaF:Pd,Pp/SaF:Pd,Pp:AScba
**Escape:** GmFOv/ScbaE

**Incompatibilities and Reactivities:** Strong oxidizers

**Exposure Routes, Symptoms, Target Organs (see Table 5):**
**ER:** Inh, Ing, Con
**SY:** Dizz, stupor, inco; loss of appetite, nau; derm; chemical pneu (aspir liquid); uncon
**TO:** Skin, resp sys, CNS

**First Aid (see Table 6):**
**Eye:** Irr immed
**Skin:** Soap wash prompt
**Breath:** Resp support
**Swallow:** Medical attention immed

| 1-Heptanethiol | Formula: $CH_3[CH_2]_6SH$ | CAS#: 1639-09-4 | RTECS#: MJ1400000 | IDLH: N.D. |
|---|---|---|---|---|

**Conversion:** 1 ppm = 5.41 mg/m$^3$ | **DOT:** 1228 131

**Synonyms/Trade Names:** Heptyl mercaptan, n-Heptyl mercaptan

**Exposure Limits:**
**NIOSH REL:** C 0.5 ppm (2.7 mg/m$^3$) [15-minute]
**OSHA PEL:** none

**Measurement Methods (see Table 1):**
None available

**Physical Description:** Colorless liquid with a strong odor.

| Chemical & Physical Properties: | Personal Protection/Sanitation (see Table 2): | Respirator Recommendations (see Tables 3 and 4): |
|---|---|---|
| **MW:** 132.3<br>**BP:** 351°F<br>**Sol:** Insoluble<br>**Fl.P:** 115°F<br>**IP:** ?<br>**Sp.Gr:** 0.84<br>**VP:** ?<br>**FRZ:** -46°F<br>**UEL:** ?<br>**LEL:** ?<br>Class II Combustible Liquid | **Skin:** Prevent skin contact<br>**Eyes:** Prevent eye contact<br>**Wash skin:** When contam<br>**Remove:** When wet or contam<br>**Change:** N.R. | **NIOSH**<br>**5 ppm:** CcrOv/Sa<br>**12.5 ppm:** Sa:Cf/PaprOv<br>**25 ppm:** CcrFOv/GmFOv/PaprTOv/ ScbaF/SaF<br>**§:** ScbaF:Pd,Pp/SaF:Pd,Pp:AScba<br>**Escape:** GmFOv/ScbaE |

**Incompatibilities and Reactivities:** Oxidizers, reducing agents, strong acids & bases, alkali metals

| Exposure Routes, Symptoms, Target Organs (see Table 5): | First Aid (see Table 6): |
|---|---|
| **ER:** Inh, Ing, Con<br>**SY:** Irrit eyes, skin, nose, throat; lass, cyan, incr respiration, nau, drow, head, vomit<br>**TO:** Eyes, skin, resp sys, CNS, blood | **Eye:** Irr immed<br>**Skin:** Soap wash<br>**Breath:** Resp support<br>**Swallow:** Medical attention immed |

H

| Hexachlorobutadiene | Formula: $Cl_2C{=}CClCCl{=}CCl_2$ | CAS#: 87-68-3 | RTECS#: EJ07000 | IDLH: Ca [N.D.] |
|---|---|---|---|---|

**Conversion:** 1 ppm = 10.66 mg/m$^3$ | **DOT:** 2279 151

**Synonyms/Trade Names:** HCBD; Hexachloro-1,3-butadiene; 1,3-Hexachlorobutadiene; Perchlorobutadiene

**Exposure Limits:**
**NIOSH REL:** Ca
TWA 0.02 ppm (0.24 mg/m$^3$) [skin]
See Appendix A
**OSHA PEL†:** none

**Measurement Methods (see Table 1):**
**NIOSH** 2543

**Physical Description:** Clear, colorless liquid with a mild, turpentine-like odor.

| Chemical & Physical Properties: | Personal Protection/Sanitation (see Table 2): | Respirator Recommendations (see Tables 3 and 4): |
|---|---|---|
| **MW:** 260.7<br>**BP:** 419°F<br>**Sol:** Insoluble<br>**Fl.P:** ?<br>**IP:** ?<br>**Sp.Gr:** 1.55<br>**VP:** 0.2 mmHg<br>**FRZ:** -6°F<br>**UEL:** ?<br>**LEL:** ?<br>Combustible Liquid | **Skin:** Prevent skin contact<br>**Eyes:** Prevent eye contact<br>**Wash skin:** When contam<br>**Remove:** When wet or contam<br>**Change:** N.R.<br>**Provide:** Eyewash<br>Quick drench | **NIOSH**<br>**¥:** ScbaF:Pd,Pp/SaF:Pd,Pp:AScba<br>**Escape:** GmFOv/ScbaE |

**Incompatibilities and Reactivities:** Oxidizers

| Exposure Routes, Symptoms, Target Organs (see Table 5): | First Aid (see Table 6): |
|---|---|
| **ER:** Inh, Abs, Ing, Con<br>**SY:** In animals: irrit eyes, skin, resp sys; kidney damage; [carc]<br>**TO:** Eyes, skin, resp sys, kidneys [in animals: kidney tumors] | **Eye:** Irr immed<br>**Skin:** Soap wash immed<br>**Breath:** Resp support<br>**Swallow:** Medical attention immed |

## Hexachlorocyclopentadiene

| Hexachlorocyclopentadiene | Formula: $C_5Cl_6$ | CAS#: 77-47-4 | RTECS#: GY1225000 | IDLH: N.D. |
|---|---|---|---|---|
| **Conversion:** 1 ppm = 11.16 mg/m$^3$ | **DOT:** 2646 151 | | | |

**Synonyms/Trade Names:** HCCPD; Hexachloro-1,3-cyclopentadiene; 1,2,3,4,5,5-Hexachloro-1,3-cyclopentadiene; Perchlorocyclopentadiene

**Exposure Limits:**
**NIOSH REL:** TWA 0.01 ppm (0.1 mg/m$^3$)
**OSHA PEL†:** none

**Measurement Methods (see Table 1):**
**NIOSH** 2518

**Physical Description:** Pale-yellow to amber-colored liquid with a pungent, unpleasant odor.
[**Note:** A solid below 16°F.]

**Chemical & Physical Properties:**
**MW:** 272.8
**BP:** 462°F
**Sol(77°F):** 0.0002% (Reacts)
**Fl.P:** NA
**IP:** ?
**Sp.Gr:** 1.71
**VP(77°F):** 0.08 mmHg
**FRZ:** 16°F
**UEL:** NA
**LEL:** NA
Noncombustible Liquid

**Personal Protection/Sanitation (see Table 2):**
**Skin:** Prevent skin contact
**Eyes:** Prevent eye contact
**Wash skin:** When contam
**Remove:** When wet or contam
**Change:** N.R.
**Provide:** Eyewash
Quick drench

**Respirator Recommendations (see Tables 3 and 4):**
Not available.

**Incompatibilities and Reactivities:** Water, light
[**Note:** Reacts slowly with water to form hydrochloric acid; will corrode iron & most metals in presence of moisture. Explosive hydrogen gas may collect in enclosed spaces in the presence of moisture.]

**Exposure Routes, Symptoms, Target Organs (see Table 5):**
**ER:** Inh, Abs, Ing, Con
**SY:** Irrit eyes, skin, resp sys; eye, skin burns; lac; sneez, cough, dysp, salv, pulm edema; nau, vomit, diarr; in animals: liver, kidney inj
**TO:** Eyes, skin, resp sys, liver, kidneys

**First Aid (see Table 6):**
**Eye:** Irr immed
**Skin:** Soap flush immed
**Breath:** Resp support
**Swallow:** Medical attention immed

## Hexachloroethane

| Hexachloroethane | Formula: $Cl_3CCCl_3$ | CAS#: 67-72-1 | RTECS#: KI4025000 | IDLH: Ca [300 ppm] |
|---|---|---|---|---|
| **Conversion:** 1 ppm = 9.68 mg/m$^3$ | **DOT:** | | | |

**Synonyms/Trade Names:** Carbon hexachloride, Ethane hexachloride, Perchloroethane

**Exposure Limits:**
**NIOSH REL:** Ca
TWA 1 ppm (10 mg/m$^3$) [skin]
See Appendix A
See Appendix C (Chloroethanes)
**OSHA PEL:** TWA 1 ppm (10 mg/m$^3$) [skin]

**Measurement Methods (see Table 1):**
**NIOSH** 1003
**OSHA** 7

**Physical Description:** Colorless crystals with a camphor-like odor.

**Chemical & Physical Properties:**
**MW:** 236.7
**BP:** Sublimes
**Sol(72°F):** 0.005%
**Fl.P:** NA
**IP:** 11.22 eV
**Sp.Gr:** 2.09
**VP:** 0.2 mmHg
**MLT:** 368°F (Sublimes)
**UEL:** NA
**LEL:** NA
Noncombustible Solid

**Personal Protection/Sanitation (see Table 2):**
**Skin:** Prevent skin contact
**Eyes:** Prevent eye contact
**Wash skin:** When contam/Daily
**Remove:** When wet or contam
**Change:** Daily
**Provide:** Eyewash
Quick drench

**Respirator Recommendations (see Tables 3 and 4):**
**NIOSH**
**¥:** ScbaF:Pd,Pp/SaF:Pd,Pp:AScba
**Escape:** GmFOv/ScbaE

**Incompatibilities and Reactivities:** Alkalis; metals such as zinc, cadmium, aluminum, hot iron & mercury

**Exposure Routes, Symptoms, Target Organs (see Table 5):**
**ER:** Inh, Abs, Ing, Con
**SY:** Irrit eyes, skin, muc memb; in animals: kidney damage; [carc]
**TO:** Eyes, skin, resp sys, kidneys [in animals: liver cancer]

**First Aid (see Table 6):**
**Eye:** Irr immed
**Skin:** Soap wash immed
**Breath:** Resp support
**Swallow:** Medical attention immed

H

## Hexachloronaphthalene

| Hexachloronaphthalene | Formula: $C_{10}H_2Cl_6$ | CAS#: 1335-87-1 | RTECS#: QJ7350000 | IDLH: 2 mg/m³ |
|---|---|---|---|---|
| Conversion: | DOT: | | | |

**Synonyms/Trade Names:** Halowax® 1014

**Exposure Limits:**
**NIOSH REL:** TWA 0.2 mg/m³ [skin]
**OSHA PEL:** TWA 0.2 mg/m³ [skin]

**Measurement Methods (see Table 1):**
**NIOSH** S100 (II-2)

**Physical Description:** White to light-yellow solid with an aromatic odor.

| Chemical & Physical Properties: | Personal Protection/Sanitation (see Table 2): | Respirator Recommendations (see Tables 3 and 4): |
|---|---|---|
| **MW:** 334.9<br>**BP:** 650-730°F<br>**Sol:** Insoluble<br>**Fl.P:** NA<br>**IP:** ?<br>**Sp.Gr:** 1.78<br>**VP:** <1 mmHg<br>**MLT:** 279°F<br>**UEL:** NA<br>**LEL:** NA<br>Noncombustible Solid | **Skin:** Prevent skin contact<br>**Eyes:** Prevent eye contact<br>**Wash skin:** When contam/Daily<br>**Remove:** When wet or contam<br>**Change:** Daily | **NIOSH/OSHA**<br>**2 mg/m³:** Sa*/ScbaF<br>**§:** ScbaF:Pd,Pp/SaF:Pd,Pp:AScba<br>**Escape:** GmFOv/ScbaE |

**Incompatibilities and Reactivities:** Strong oxidizers

**Exposure Routes, Symptoms, Target Organs (see Table 5):**
**ER:** Inh, Abs, Ing, Con
**SY:** Acne-form derm, nau, conf, jaun, coma
**TO:** Skin, liver

**First Aid (see Table 6):**
**Eye:** Irr immed
**Skin:** Soap wash prompt
**Breath:** Resp support
**Swallow:** Medical attention immed

## 1-Hexadecanethiol

| 1-Hexadecanethiol | Formula: $CH_3[CH_2]_{15}SH$ | CAS#: 2917-26-2 | RTECS#: | IDLH: N.D. |
|---|---|---|---|---|
| Conversion: 1 ppm = 10.59 mg/m³ | DOT: 1228 131 (liquid) | | | |

**Synonyms/Trade Names:** Cetyl mercaptan, Hexadecanethiol-1, n-Hexadecanethiol, Hexadecyl mercaptan

**Exposure Limits:**
**NIOSH REL:** C 0.5 ppm (5.3 mg/m³) [15-minute]
**OSHA PEL:** none

**Measurement Methods (see Table 1):**
None available

**Physical Description:** Colorless liquid or solid (below 64-68°F) with a strong odor.

| Chemical & Physical Properties: | Personal Protection/Sanitation (see Table 2): | Respirator Recommendations (see Tables 3 and 4): |
|---|---|---|
| **MW:** 258.5<br>**BP:** ?<br>**Sol:** Insoluble<br>**Fl.P:** 215°F<br>**IP:** ?<br>**Sp.Gr:** 0.85<br>**VP:** 0.1 mmHg<br>**FRZ:** 64-68°F<br>**UEL:** ?<br>**LEL:** ?<br>Class IIIB Combustible Liquid | **Skin:** Prevent skin contact<br>**Eyes:** Prevent eye contact<br>**Wash skin:** When contam<br>**Remove:** When wet or contam<br>**Change:** Daily | **NIOSH**<br>**5 ppm:** CcrOv/Sa<br>**12.5 ppm:** Sa:Cf/PaprOv<br>**25 ppm:** CcrFOv/GmFOv/PaprTOv/ScbaF/SaF<br>**§:** ScbaF:Pd,Pp/SaF:Pd,Pp:AScba<br>**Escape:** GmFOv/ScbaE |

**Incompatibilities and Reactivities:** Oxidizers, strong acids & bases, alkali metals, reducing agents

**Exposure Routes, Symptoms, Target Organs (see Table 5):**
**ER:** Inh, Abs, Ing, Con
**SY:** Irrit eyes, skin, resp sys; head, dizz, lass, cyan, nau, convuls
**TO:** Eyes, skin, resp sys, CNS, blood

**First Aid (see Table 6):**
**Eye:** Irr immed
**Skin:** Soap wash immed
**Breath:** Resp support
**Swallow:** Medical attention immed

## Hexafluoroacetone

**Formula:** $(CF_3)_2CO$ | **CAS#:** 684-16-2 | **RTECS#:** UC2450000 | **IDLH:** N.D.

**Conversion:** 1 ppm = 6.79 mg/m$^3$ | **DOT:** 2420 125

**Synonyms/Trade Names:** Hexafluoro-2-propanone; 1,1,1,3,3,3-Hexafluoro-2-propanone; HFA; Perfluoroacetone

**Exposure Limits:**
**NIOSH REL:** TWA 0.1 ppm (0.7 mg/m$^3$) [skin]
**OSHA PEL†:** none

**Measurement Methods (see Table 1):**
None available

**Physical Description:** Colorless gas with a musty odor.
[**Note:** Shipped as a liquefied compressed gas.]

**Chemical & Physical Properties:**
**MW:** 166.0
**BP:** -18°F
**Sol:** Reacts
**Fl.P:** NA
**IP:** 11.81 eV
**RGasD:** 5.76
**VP:** 5.8 atm
**FRZ:** -188°F
**UEL:** NA
**LEL:** NA
Nonflammable Gas, but highly reactive with water & other substances, releasing heat.

**Personal Protection/Sanitation (see Table 2):**
**Skin:** Prevent skin contact/Frostbite
**Eyes:** Prevent eye contact/Frostbite
**Wash skin:** N.R.
**Remove:** N.R.
**Change:** N.R.
**Provide:** Frostbite wash

**Respirator Recommendations (see Tables 3 and 4):**
Not available.

**Incompatibilities and Reactivities:** Water, acids
[**Note:** Hygroscopic (i.e., absorbs moisture from the air); reacts with moisture to form a highly acidic sesquihydrate.]

**Exposure Routes, Symptoms, Target Organs (see Table 5):**
**ER:** Inh, Abs, Con
**SY:** Irrit eyes, skin, muc memb, resp sys; pulm edema; liquid: frostbite; in animals: terato, repro effects; kidney inj
**TO:** Eyes, skin, resp sys, kidneys, repro sys

**First Aid (see Table 6):**
**Eye:** Frostbite
**Skin:** Frostbite
**Breath:** Resp support

H

## Hexamethylene diisocyanate

**Formula:** $OCN[CH_2]_6NCO$ | **CAS#:** 822-06-0 | **RTECS#:** MO1740000 | **IDLH:** N.D.

**Conversion:** 1 ppm = 6.88 mg/m$^3$ | **DOT:** 2281 156

**Synonyms/Trade Names:** 1,6-Diisocyanatohexane; HDI; Hexamethylene-1,6-diisocyanate; 1,6-Hexamethylene diisocyanate; HMDI

**Exposure Limits:**
**NIOSH REL:** TWA 0.005 ppm (0.035 mg/m$^3$)
C 0.020 ppm (0.140 mg/m$^3$) [10-minute]
**OSHA PEL:** none

**Measurement Methods (see Table 1):**
**NIOSH** 5521, 5522, 5525
**OSHA** 42

**Physical Description:** Clear, colorless to slightly yellow liquid with a sharp, pungent odor.

**Chemical & Physical Properties:**
**MW:** 168.2
**BP:** 415°F
**Sol:** Low (Reacts)
**Fl.P:** 284°F
**IP:**?
**Sp.Gr(77°F):** 1.04
**VP(77°F):** 0.05 mmHg
**FRZ:** -89°F
**UEL:** ?
**LEL:** ?
Class IIIB Combustible Liquid

**Personal Protection/Sanitation (see Table 2):**
**Skin:** Prevent skin contact
**Eyes:** Prevent eye contact
**Wash skin:** When contam
**Remove:** When wet or contam
**Change:** N.R.
**Provide:** Eyewash
Quick drench

**Respirator Recommendations (see Tables 3 and 4):**
**NIOSH**
**0.05 ppm:** Sa*
**0.125 ppm:** Sa:Cf*
**0.25 ppm:** ScbaF/SaF
**1 ppm:** SaF:Pd,Pp
**§:** ScbaF:Pd,Pp/SaF:Pd,Pp:AScba
**Escape:** GmFOv/ScbaE

**Incompatibilities and Reactivities:** Water, alcohols, strong bases, amines, carboxylic acids, organotin catalysts
[**Note:** Reacts slowly with water to form carbon dioxide. Avoid heating above 392°F (polymerizes).]

**Exposure Routes, Symptoms, Target Organs (see Table 5):**
**ER:** Inh, Ing, Con
**SY:** Irrit eyes, skin, resp sys; cough, dysp, bron, wheez, pulm edema, asthma; corn damage, skin blisters
**TO:** Eyes, skin, resp sys

**First Aid (see Table 6):**
**Eye:** Irr immed
**Skin:** Soap flush immed
**Breath:** Resp support
**Swallow:** Medical attention immed

H

| Hexamethyl phosphoramide | Formula: $[(CH_3)_2N]_3PO$ | CAS#: 680-31-9 | RTECS#: TD0875000 | IDLH: Ca [N.D.] |
|---|---|---|---|---|
| **Conversion:** | **DOT:** | | | |

**Synonyms/Trade Names:** Hexamethylphosphoric triamide, Hexamethylphosphorotriamide, HMPA, Tris(dimethylamino)phosphine oxide

**Exposure Limits:**
**NIOSH REL:** Ca
See Appendix A
**OSHA PEL:** none

**Measurement Methods (see Table 1):**
None available

**Physical Description:** Clear, colorless liquid with an aromatic or mild, amine-like odor. [**Note:** A solid below 43°F.]

| Chemical & Physical Properties: | Personal Protection/Sanitation (see Table 2): | Respirator Recommendations (see Tables 3 and 4): |
|---|---|---|
| **MW:** 179.2<br>**BP:** 451°F<br>**Sol:** Miscible<br>**Fl.P:** 220°F<br>**IP:** ?<br>**Sp.Gr:** 1.03<br>**VP:** 0.03 mmHg<br>**FRZ:** 43°F<br>**UEL:** ?<br>**LEL:** ?<br>Class IIIB Combustible Liquid | **Skin:** Prevent skin contact<br>**Eyes:** Prevent eye contact<br>**Wash skin:** When contam<br>**Remove:** When wet or contam<br>**Change:** N.R.<br>**Provide:** Eyewash<br>Quick drench | **NIOSH**<br>**¥:** ScbaF:Pd,Pp/SaF:Pd,Pp:AScba<br>**Escape:** GmFOv/ScbaE |

**Incompatibilities and Reactivities:** Oxidizers, strong acids, chemically-active metals (e.g., potassium, sodium, magnesium, zinc)

| Exposure Routes, Symptoms, Target Organs (see Table 5): | First Aid (see Table 6): |
|---|---|
| **ER:** Inh, Abs, Ing, Con<br>**SY:** Irrit eyes, skin, resp sys; dysp; abdom pain; [carc]<br>**TO:** Eyes, skin, resp sys, CNS, GI tract [in animals: cancer of the nasal cavity] | **Eye:** Irr immed<br>**Skin:** Water flush immed<br>**Breath:** Resp support<br>**Swallow:** Medical attention immed |

| n-Hexane | Formula: $CH_3[CH_2]_4CH_3$ | CAS#: 110-54-3 | RTECS#: MN9275000 | IDLH: 1100 ppm [10%LEL] |
|---|---|---|---|---|
| **Conversion:** 1 ppm = 3.53 mg/$m^3$ | **DOT:** 1208 128 | | | |

**Synonyms/Trade Names:** Hexane, Hexyl hydride, normal-Hexane

**Exposure Limits:**
**NIOSH REL:** TWA 50 ppm (180 mg/$m^3$)
**OSHA PEL†:** TWA 500 ppm (1800 mg/$m^3$)

**Measurement Methods (see Table 1):**
**NIOSH** 1500, 3800
**OSHA** 7

**Physical Description:** Colorless liquid with a gasoline-like odor.

| Chemical & Physical Properties: | Personal Protection/Sanitation (see Table 2): | Respirator Recommendations (see Tables 3 and 4): |
|---|---|---|
| **MW:** 86.2<br>**BP:** 156°F<br>**Sol:** 0.002%<br>**Fl.P:** -7°F<br>**IP:** 10.18 eV<br>**Sp.Gr:** 0.66<br>**VP:** 124 mmHg<br>**FRZ:** -219°F<br>**UEL:** 7.5%<br>**LEL:** 1.1%<br>Class IB Flammable Liquid | **Skin:** Prevent skin contact<br>**Eyes:** Prevent eye contact<br>**Wash skin:** When contam<br>**Remove:** When wet (flamm)<br>**Change:** N.R. | **NIOSH**<br>**500 ppm:** Sa*<br>**1100 ppm:** Sa:Cf*/ScbaF/SaF<br>**§:** ScbaF:Pd,Pp/SaF:Pd,Pp:AScba<br>**Escape:** GmFOv/ScbaE |

**Incompatibilities and Reactivities:** Strong oxidizers

| Exposure Routes, Symptoms, Target Organs (see Table 5): | First Aid (see Table 6): |
|---|---|
| **ER:** Inh, Ing, Con<br>**SY:** Irrit eyes, nose; nau, head; peri neur: numb extremities, musc weak; derm; dizz; chemical pneu (aspir liquid)<br>**TO:** Eyes, skin, resp sys, CNS, PNS | **Eye:** Irr immed<br>**Skin:** Soap wash immed<br>**Breath:** Resp support<br>**Swallow:** Medical attention immed |

## Hexane isomers (excluding n-Hexane)

| Hexane isomers (excluding n-Hexane) | Formula: $C_6H_{14}$ | CAS#: | RTECS#: | IDLH: N.D. |
|---|---|---|---|---|
| **Conversion:** 1 ppm = 3.53 mg/m$^3$ | **DOT:** 1208 128 | | | |

**Synonyms/Trade Names:** Diethylmethylmethane; Diisopropyl; 2,2-Dimethylbutane; 2,3-Dimethylbutane; Isohexane; 2-Methylpentane; 3-Methylpentane [**Note:** Also see specific listing for n-Hexane.]

**Exposure Limits:**
**NIOSH REL:** TWA 100 ppm (350 mg/m$^3$)
C 510 ppm (1800 mg/m$^3$) [15-minute]
**OSHA PEL†:** none

**Measurement Methods (see Table 1):**
None available

**Physical Description:** Clear liquids with mild, gasoline-like odors.
[**Note:** Includes all the isomers of hexane except n-hexane.]

| Chemical & Physical Properties: | Personal Protection/Sanitation (see Table 2): | Respirator Recommendations (see Tables 3 and 4): |
|---|---|---|
| **MW:** 86.2<br>**BP:** 122-145°F<br>**Sol:** Insoluble<br>**Fl.P:** -54 to 19°F<br>**IP:** ?<br>**Sp.Gr:** 0.65-0.66<br>**VP:** ?<br>**FRZ:** -245 to -148°F<br>**UEL:** ?<br>**LEL:** ?<br>Class IB Flammable Liquids | **Skin:** Prevent skin contact<br>**Eyes:** Prevent eye contact<br>**Wash skin:** When contam<br>**Remove:** When wet (flamm)<br>**Change:** N.R. | **NIOSH**<br>**1000 ppm:** Sa*<br>**2500 ppm:** Sa:Cf*<br>**5000 ppm:** SaT:Cf*/ScbaF/SaF<br>**§:** ScbaF:Pd,Pp/SaF:Pd,Pp:AScba<br>**Escape:** GmFOv/ScbaE |

**Incompatibilities and Reactivities:** Strong oxidizers

| Exposure Routes, Symptoms, Target Organs (see Table 5): | First Aid (see Table 6): |
|---|---|
| **ER:** Inh, Ing, Con<br>**SY:** Irrit eyes, skin, resp sys; head, dizz; nau; chemical pneu (aspir liquid); derm<br>**TO:** Eyes, skin, resp sys, CNS | **Eye:** Irr immed<br>**Skin:** Soap wash immed<br>**Breath:** Resp support<br>**Swallow:** Medical attention immed |

H

## n-Hexanethiol

| n-Hexanethiol | Formula: $CH_3[CH_2]_5SH$ | CAS#: 111-31-9 | RTECS#: MO4550000 | IDLH: N.D. |
|---|---|---|---|---|
| **Conversion:** 1 ppm = 4.83 mg/m$^3$ | **DOT:** 1228 131 | | | |

**Synonyms/Trade Names:** 1-Hexanethiol, Hexyl mercaptan, n-Hexyl mercaptan, n-Hexylthiol

**Exposure Limits:**
**NIOSH REL:** C 0.5 ppm (2.7 mg/m$^3$) [15-minute]
**OSHA PEL:** none

**Measurement Methods (see Table 1):**
None available

**Physical Description:** Colorless liquid with an unpleasant odor.

| Chemical & Physical Properties: | Personal Protection/Sanitation (see Table 2): | Respirator Recommendations (see Tables 3 and 4): |
|---|---|---|
| **MW:** 118.2<br>**BP:** 304°F<br>**Sol:** Insoluble<br>**Fl.P:** 68°F<br>**IP:** ?<br>**Sp.Gr:** 0.84<br>**VP:** ?<br>**FRZ:** -113°F<br>**UEL:** ?<br>**LEL:** ?<br>Class IB Flammable Liquid | **Skin:** Prevent skin contact<br>**Eyes:** Prevent eye contact<br>**Wash skin:** When contam<br>**Remove:** When wet (flamm)<br>**Change:** N.R. | **NIOSH**<br>**5 ppm:** CcrOv/Sa<br>**12.5 ppm:** Sa:Cf/PaprOv<br>**25 ppm:** CcrFOv/GmFOv/PaprTOv/ScbaF/SaF<br>**§:** ScbaF:Pd,Pp/SaF:Pd,Pp:AScba<br>**Escape:** GmFOv/ScbaE |

**Incompatibilities and Reactivities:** Oxidizers, reducing agents, strong acids & bases, alkali metals

| Exposure Routes, Symptoms, Target Organs (see Table 5): | First Aid (see Table 6): |
|---|---|
| **ER:** Inh, Ing, Con<br>**SY:** Irrit eyes, skin, nose, throat; lass, cyan, incr respiration, nau, drow, head, vomit<br>**TO:** Eyes, skin, resp sys, CNS, blood | **Eye:** Irr immed<br>**Skin:** Soap wash immed<br>**Breath:** Resp support<br>**Swallow:** Medical attention immed |

H

## 2-Hexanone

**Formula:** $CH_3CO[CH_2]_3CH_3$ | **CAS#:** 591-78-6 | **RTECS#:** MP1400000 | **IDLH:** 1600 ppm

**Conversion:** 1 ppm = 4.10 mg/m$^3$ | **DOT:**

**Synonyms/Trade Names:** Butyl methyl ketone, MBK, Methyl butyl ketone, Methyl n-butyl ketone

**Exposure Limits:**
**NIOSH REL:** TWA 1 ppm (4 mg/m$^3$)
**OSHA PEL†:** TWA 100 ppm (410 mg/m$^3$)

**Measurement Methods (see Table 1):**
**NIOSH** 1300, 2555
**OSHA** PV2031

**Physical Description:** Colorless liquid with an acetone-like odor.

**Chemical & Physical Properties:**
**MW:** 100.2
**BP:** 262°F
**Sol:** 2%
**Fl.P:** 77°F
**IP:** 9.34 eV
**Sp.Gr:** 0.81
**VP:** 11 mmHg
**FRZ:** -71°F
**UEL:** 8%
**LEL:** ?
Class IC Flammable Liquid

**Personal Protection/Sanitation (see Table 2):**
**Skin:** Prevent skin contact
**Eyes:** Prevent eye contact
**Wash skin:** When contam
**Remove:** When wet (flamm)
**Change:** N.R.

**Respirator Recommendations (see Tables 3 and 4):**
**NIOSH**
**10 ppm:** Sa
**25 ppm:** Sa:Cf
**50 ppm:** SaT:Cf/ScbaF/SaF
**1600 ppm:** SaF:Pd,Pp
**§:** ScbaF:Pd,Pp/SaF:Pd,Pp:AScba
**Escape:** GmFOv/ScbaE

**Incompatibilities and Reactivities:** Strong oxidizers

**Exposure Routes, Symptoms, Target Organs (see Table 5):**
**ER:** Inh, Abs, Ing, Con
**SY:** Irrit eyes, nose; peri neur: lass, pares; derm; head, drow
**TO:** Eyes, skin, resp sys, CNS, PNS

**First Aid (see Table 6):**
**Eye:** Irr immed
**Skin:** Soap wash immed
**Breath:** Resp support
**Swallow:** Medical attention immed

## Hexone

**Formula:** $CH_3COCH_2CH(CH_3)_2$ | **CAS#:** 108-10-1 | **RTECS#:** SA9275000 | **IDLH:** 500 ppm

**Conversion:** 1 ppm = 4.10 mg/m$^3$ | **DOT:** 1245 127

**Synonyms/Trade Names:** Isobutyl methyl ketone, Methyl isobutyl ketone, 4-Methyl 2-pentanone, MIBK

**Exposure Limits:**
**NIOSH REL:** TWA 50 ppm (205 mg/m$^3$)
ST 75 ppm (300 mg/m$^3$)
**OSHA PEL†:** TWA 100 ppm (410 mg/m$^3$)

**Measurement Methods (see Table 1):**
**NIOSH** 1300, 2555
**OSHA** 1004

**Physical Description:** Colorless liquid with a pleasant odor.

**Chemical & Physical Properties:**
**MW:** 100.2
**BP:** 242°F
**Sol:** 2%
**Fl.P:** 64°F
**IP:** 9.30 eV
**Sp.Gr:** 0.80
**VP:** 16 mmHg
**FRZ:** -120°F
**UEL(200°F):** 8.0%
**LEL(200°F):** 1.2%
Class IB Flammable Liquid

**Personal Protection/Sanitation (see Table 2):**
**Skin:** Prevent skin contact
**Eyes:** Prevent eye contact
**Wash skin:** When contam
**Remove:** When wet (flamm)
**Change:** N.R.

**Respirator Recommendations (see Tables 3 and 4):**
**NIOSH**
**500 ppm:** CcrOv*/GmFOv/PaprTOv*/Sa*/ScbaF
**§:** ScbaF:Pd,Pp/SaF:Pd,Pp:AScba
**Escape:** GmFOv/ScbaE

**Incompatibilities and Reactivities:** Strong oxidizers, potassium tert-butoxide

**Exposure Routes, Symptoms, Target Organs (see Table 5):**
**ER:** Inh, Ing, Con
**SY:** Irrit eyes, skin, muc memb; head, narco, coma; derm; in animals: liver, kidney damage
**TO:** Eyes, skin, resp sys, CNS, liver, kidneys

**First Aid (see Table 6):**
**Eye:** Irr immed
**Skin:** Water flush prompt
**Breath:** Resp support
**Swallow:** Medical attention immed

## sec-Hexyl acetate

**Formula:** $C_8H_{16}O_2$ | **CAS#:** 108-84-9 | **RTECS#:** SA7525000 | **IDLH:** 500 ppm

**Conversion:** 1 ppm = 5.90 mg/m$^3$ | **DOT:** 1233 130

**Synonyms/Trade Names:** 1,3-Dimethylbutyl acetate; Methylisoamyl acetate

**Exposure Limits:**
**NIOSH REL:** TWA 50 ppm (300 mg/m$^3$)
**OSHA PEL:** TWA 50 ppm (300 mg/m$^3$)

**Measurement Methods (see Table 1):**
**NIOSH** 1450
**OSHA** 7

**Physical Description:** Colorless liquid with a mild, pleasant, fruity odor.

**Chemical & Physical Properties:**
**MW:** 144.2
**BP:** 297°F
**Sol:** 0.08%
**Fl.P:** 113°F
**IP:** ?
**Sp.Gr:** 0.86
**VP:** 3 mmHg
**FRZ:** -83°F
**UEL:** ?
**LEL:** ?
Class II Combustible Liquid

**Personal Protection/Sanitation (see Table 2):**
**Skin:** Prevent skin contact
**Eyes:** Prevent eye contact
**Wash skin:** When contam
**Remove:** When wet or contam
**Change:** N.R.

**Respirator Recommendations (see Tables 3 and 4):**
**NIOSH/OSHA**
**500 ppm:** CcrOv*/GmFOv/PaprOv*/Sa*/ScbaF
**§:** ScbaF:Pd,Pp/SaF:Pd,Pp:AScba
**Escape:** GmFOv/ScbaE

**Incompatibilities and Reactivities:** Nitrates; strong oxidizers, alkalis & acids

**Exposure Routes, Symptoms, Target Organs (see Table 5):**
**ER:** Inh, Ing, Con
**SY:** Irrit eyes, skin, nose, throat; head; in animals: narco
**TO:** Eyes, skin, resp sys, CNS

**First Aid (see Table 6):**
**Eye:** Irr immed
**Skin:** Water flush prompt
**Breath:** Resp support
**Swallow:** Medical attention immed

H

## Hexylene glycol

**Formula:** $(CH_3)_2COHCH_2CHOHCH_3$ | **CAS#:** 107-41-5 | **RTECS#:** SA0810000 | **IDLH:** N.D.

**Conversion:** 1 ppm = 4.83 mg/m$^3$ | **DOT:**

**Synonyms/Trade Names:** 2,4-Dihydroxy-2-methylpentane; 2-Methyl-2,4-pentanediol; 4-Methyl-2,4-pentanediol; 2-Methylpentane-2,4-diol

**Exposure Limits:**
**NIOSH REL:** C 25 ppm (125 mg/m$^3$)
**OSHA PEL†:** none

**Measurement Methods (see Table 1):**
**OSHA** PV2101

**Physical Description:** Colorless liquid with a mild, sweetish odor.

**Chemical & Physical Properties:**
**MW:** 118.2
**BP:** 388°F
**Sol:** Miscible
**Fl.P:** 209°F
**IP:** ?
**Sp.Gr:** 0.92
**VP:** 0.05 mmHg
**FRZ:** -58°F (Sets to glass)
**UEL(est):** 7.4%
**LEL(calc):** 1.3%
Class IIIB Combustible Liquid

**Personal Protection/Sanitation (see Table 2):**
**Skin:** Prevent skin contact
**Eyes:** Prevent eye contact
**Wash skin:** When contam
**Remove:** When wet or contam
**Change:** N.R.
**Provide:** Eyewash

**Respirator Recommendations (see Tables 3 and 4):**
Not available.

**Incompatibilities and Reactivities:** Strong oxidizers, strong acids
[**Note:** Hygroscopic (i.e., absorbs moisture from the air).]

**Exposure Routes, Symptoms, Target Organs (see Table 5):**
**ER:** Inh, Ing, Con
**SY:** Irrit eyes, skin, resp sys; head, dizz, nau, inco, CNS depres; derm, skin sens
**TO:** Eyes, skin, resp sys, CNS

**First Aid (see Table 6):**
**Eye:** Irr immed
**Skin:** Water wash immed
**Breath:** Resp support
**Swallow:** Medical attention immed

H

## Hydrazine

| Hydrazine | Formula: $H_2NNH_2$ | CAS#: 302-01-2 | RTECS#: MU7175000 | IDLH: Ca [50 ppm] |
|---|---|---|---|---|

**Conversion:** 1 ppm = 1.31 mg/m$^3$

**DOT:** 2029 132 (anhydrous); 3293 152 (≤ 37% solution); 2030 153 (37-64% solution); 2029 132 (>64% solution)

**Synonyms/Trade Names:** Diamine, Hydrazine (anhydrous), Hydrazine base

**Exposure Limits:**
**NIOSH REL:** Ca
C 0.03 ppm (0.04 mg/m$^3$) [2-hour]
See Appendix A
**OSHA PEL†:** TWA 1 ppm (1.3 mg/m$^3$) [skin]

**Measurement Methods (see Table 1):**
**NIOSH** 3503
**OSHA** 20, 108

**Physical Description:** Colorless, fuming, oily liquid with an ammonia-like odor. [**Note:** A solid below 36°F.]

**Chemical & Physical Properties:**
**MW:** 32.1
**BP:** 236°F
**Sol:** Miscible
**Fl.P:** 99°F
**IP:** 8.93 eV
**Sp.Gr:** 1.01
**VP:** 10 mmHg
**FRZ:** 36°F
**UEL:** 98%
**LEL:** 2.9%
Class IC Flammable Liquid

**Personal Protection/Sanitation (see Table 2):**
**Skin:** Prevent skin contact
**Eyes:** Prevent eye contact
**Wash skin:** When contam
**Remove:** When wet (flamm)
**Change:** N.R.
**Provide:** Eyewash
Quick drench

**Respirator Recommendations (see Tables 3 and 4):**
**NIOSH**
**¥:** ScbaF:Pd,Pp/SaF:Pd,Pp:AScba
**Escape:** ScbaE

**Incompatibilities and Reactivities:** Oxidizers, hydrogen peroxide, nitric acid, metallic oxides, acids
[**Note:** Can ignite SPONTANEOUSLY on contact with oxidizers or porous materials such as earth, wood & cloth.]

**Exposure Routes, Symptoms, Target Organs (see Table 5):**
**ER:** Inh, Abs, Ing, Con
**SY:** Irrit eyes, skin, nose, throat; temporary blindness; dizz, nau; derm; eye, skin burns; in animals: bron, pulm edema; liver, kidney damage; convuls; [carc]
**TO:** Eyes, skin, resp sys, CNS, liver, kidneys [in animals: tumors of the lungs, liver, blood vessels & intestine]

**First Aid (see Table 6):**
**Eye:** Irr immed
**Skin:** Water flush immed
**Breath:** Resp support
**Swallow:** Medical attention immed

## Hydrogenated terphenyls

| Hydrogenated terphenyls | Formula: $(C_6H_n)_3$ | CAS#: 61788-32-7 | RTECS#: WZ6535000 | IDLH: N.D. |
|---|---|---|---|---|

**Conversion:** 1 ppm = 12.19 mg/m$^3$ (40% hydrogenated)

**DOT:**

**Synonyms/Trade Names:** Hydrogenated diphenylbenzenes, Hydrogenated phenylbiphenyls, Hydrogenated triphenyls [**Note:** Complex mixture of terphenyl isomers that are partially hydrogenated.]

**Exposure Limits:**
**NIOSH REL:** TWA 0.5 ppm (5 mg/m$^3$)
**OSHA PEL†:** none

**Measurement Methods (see Table 1):**
None available

**Physical Description:** Clear, oily, pale-yellow liquids with a faint odor. [plasticizer/heat-transfer media]

**Chemical & Physical Properties:**
**MW:** 298 (40% hydrogenated)
**BP:** 644°F (40% hydrogenated)
**Sol:** Insoluble
**Fl.P:** 315°F (40% hydrogenated)
**IP:** ?
**Sp.Gr(77°F):** 1.003-1.009 (40% hydrogenated)
**VP:** ?
**FRZ:** ?
**UEL:** ?
**LEL:** ?
Class IIIB Combustible Liquids

**Personal Protection/Sanitation (see Table 2):**
**Skin:** Prevent skin contact
**Eyes:** Prevent eye contact
**Wash skin:** When contam
**Remove:** When wet or contam
**Change:** Daily

**Respirator Recommendations (see Tables 3 and 4):**
Not available.

**Incompatibilities and Reactivities:** None reported [**Note:** When heated, irritating vapors will be released.]

**Exposure Routes, Symptoms, Target Organs (see Table 5):**
**ER:** Inh, Ing, Con
**SY:** Irrit eyes, skin, resp sys; liver, kidney, hemato damage
**TO:** Eyes, skin, resp sys, liver, kidneys, hemato sys

**First Aid (see Table 6):**
**Eye:** Irr immed
**Skin:** Soap wash immed
**Breath:** Resp support
**Swallow:** Medical attention immed

## Hydrogen bromide

**Formula:** HBr | **CAS#:** 10035-10-6 | **RTECS#:** MW3850000 | **IDLH:** 30 ppm

**Conversion:** 1 ppm = 3.31 mg/m$^3$ | **DOT:** 1048 125 (anhydrous); 1788 154 (solution)

**Synonyms/Trade Names:** Anhydrous hydrogen bromide; Aqueous hydrogen bromide (i.e., Hydrobromic acid)

**Exposure Limits:**
**NIOSH REL:** C 3 ppm (10 mg/m$^3$)
**OSHA PEL†:** TWA 3 ppm (10 mg/m$^3$)

**Measurement Methods (see Table 1):**
**NIOSH** 7903
**OSHA** ID165SG

**Physical Description:** Colorless gas with a sharp, irritating odor.
[**Note:** Shipped as a liquefied compressed gas. Often used in an aqueous solution.]

**Chemical & Physical Properties:**
**MW:** 80.9
**BP:** -88°F
**Sol:** 49%
**Fl.P:** NA
**IP:** 11.62 eV
**RGasD:** 2.81
**VP:** 20 atm
**FRZ:** -124°F
**UEL:** NA
**LEL:** NA
Nonflammable Gas

**Personal Protection/Sanitation (see Table 2):**
**Skin:** Prevent skin contact (solution)/Frostbite
**Eyes:** Prevent eye contact (solution)/Frostbite
**Wash skin:** When contam (solution)
**Remove:** When wet or contam (solution)
**Change:** N.R.
**Provide:** Eyewash (liquid)
Quick drench (solution)
Frostbite wash

**Respirator Recommendations (see Tables 3 and 4):**
**NIOSH/OSHA**
**30 ppm:** Sa:Cf£/PaprAg£/GmFAg/ScbaF/SaF
**§:** ScbaF:Pd,Pp/SaF:Pd,Pp:AScba
**Escape:** GmFAg/ScbaE

**Incompatibilities and Reactivities:** Strong oxidizers, strong caustics, moisture, copper, brass, zinc
[**Note:** Hydrobromic acid is highly corrosive to most metals.]

**Exposure Routes, Symptoms, Target Organs (see Table 5):**
**ER:** Inh, Ing (solution), Con
**SY:** Irrit eyes, skin, nose, throat; solution: eye, skin burns; liquid: frostbite
**TO:** Eyes, skin, resp sys

**First Aid (see Table 6):**
**Eye:** Irr immed (solution)/Frostbite
**Skin:** Water flush immed (solution)/Frostbite
**Breath:** Resp support
**Swallow:** Medical attention immed (solution)

## Hydrogen chloride

**Formula:** HCl | **CAS#:** 7647-01-0 | **RTECS#:** MW4025000 | **IDLH:** 50 ppm

**Conversion:** 1 ppm = 1.49 mg/m$^3$ | **DOT:** 1050 125 (anhydrous); 1789 157 (solution)

**Synonyms/Trade Names:** Anhydrous hydrogen chloride;
Aqueous hydrogen chloride (i.e., Hydrochloric acid, Muriatic acid) [**Note:** Often used in an aqueous solution.]

**Exposure Limits:**
**NIOSH REL:** C 5 ppm (7 mg/m$^3$)
**OSHA PEL:** C 5 ppm (7 mg/m$^3$)

**Measurement Methods (see Table 1):**
**NIOSH** 7903
**OSHA** ID174SG

**Physical Description:** Colorless to slightly yellow gas with a pungent, irritating odor.
[**Note:** Shipped as a liquefied compressed gas.]

**Chemical & Physical Properties:**
**MW:** 36.5
**BP:** -121°F
**Sol(86°F):** 67%
**Fl.P:** NA
**IP:** 12.74 eV
**RGasD:** 1.27
**VP:** 40.5 atm
**FRZ:** -174°F
**UEL:** NA
**LEL:** NA
Nonflammable Gas

**Personal Protection/Sanitation (see Table 2):**
**Skin:** Prevent skin contact (solution)/Frostbite
**Eyes:** Prevent eye contact/Frostbite
**Wash skin:** When contam (solution)
**Remove:** When wet or contam (solution)
**Change:** N.R.
**Provide:** Eyewash (solution)
Quick drench (solution)
Frostbite wash

**Respirator Recommendations (see Tables 3 and 4):**
**NIOSH/OSHA**
**50 ppm:** CcrS*/GmFS/PaprS*/Sa*/ScbaF
**§:** ScbaF:Pd,Pp/SaF:Pd,Pp:AScba
**Escape:** GmFAg/ScbaE

**Incompatibilities and Reactivities:** Hydroxides, amines, alkalis, copper, brass, zinc
[**Note:** Hydrochloric acid is highly corrosive to most metals.]

**Exposure Routes, Symptoms, Target Organs (see Table 5):**
**ER:** Inh, Ing (solution), Con
**SY:** Irrit nose, throat, larynx; cough, choking; derm; solution: eye, skin burns; liquid: frostbite; in animals: lar spasm; pulm edema
**TO:** Eyes, skin, resp sys

**First Aid (see Table 6):**
**Eye:** Irr immed (solution)/Frostbite
**Skin:** Water flush immed (solution)/Frostbite
**Breath:** Resp support
**Swallow:** Medical attention immed (solution)

H

## Hydrogen cyanide

| Hydrogen cyanide | Formula: HCN | CAS#: 74-90-8 | RTECS#: MW6825000 | IDLH: 50 ppm |
|---|---|---|---|---|

**Conversion:** 1 ppm = 1.10 mg/m$^3$

**DOT:** 1051 117 (>20% solution); 1051 117 (anhydrous); 1613 154 (20% solution)

**Synonyms/Trade Names:** Formonitrile, Hydrocyanic acid, Prussic acid

**Exposure Limits:**
**NIOSH REL:** ST 4.7 ppm (5 mg/m$^3$) [skin]
**OSHA PEL†:** TWA 10 ppm (11 mg/m$^3$) [skin]

**Measurement Methods (see Table 1):**
**NIOSH** 6010, 6017

**Physical Description:** Colorless or pale-blue liquid or gas (above 78°F) with a bitter, almond-like odor.
[**Note:** Often used as a 96% solution in water.]

**Chemical & Physical Properties:**
**MW:** 27.0
**BP:** 78°F (96%)
**Sol:** Miscible
**Fl.P:** 0°F (96%)
**IP:** 13.60 eV
**Sp.Gr:** 0.69
**VP:** 630 mmHg
**FRZ:** 7°F (96%)
**UEL:** 40.0%
**LEL:** 5.6%
Class IA Flammable Liquid
Flammable Gas

**Personal Protection/Sanitation (see Table 2):**
**Skin:** Prevent skin contact
**Eyes:** Prevent eye contact
**Wash skin:** When contam
**Remove:** When wet (flamm)
**Change:** N.R.
**Provide:** Eyewash
Quick drench

**Respirator Recommendations (see Tables 3 and 4):**
**NIOSH**
**47 ppm:** Sa
**50 ppm:** Sa:Cf/ScbaF/SaF
**§:** ScbaF:Pd,Pp/SaF:Pd,Pp:AScba
**Escape:** GmFS/ScbaE

**Incompatibilities and Reactivities:** Amines, oxidizers, acids, sodium hydroxide, calcium hydroxide, sodium carbonate, caustics, ammonia [**Note:** Can polymerize at 122-140°F.]

**Exposure Routes, Symptoms, Target Organs (see Table 5):**
**ER:** Inh, Abs, Ing, Con
**SY:** Asphy; lass, head, conf; nau, vomit; incr rate and depth of respiration or respiration slow and gasping; thyroid, blood changes
**TO:** CNS, CVS, thyroid, blood

**First Aid (see Table 6):**
**Eye:** Irr immed
**Skin:** Water flush immed
**Breath:** Resp support
**Swallow:** Medical attention immed

## Hydrogen fluoride

| Hydrogen fluoride | Formula: HF | CAS#: 7664-39-3 | RTECS#: MW7875000 | IDLH: 30 ppm |
|---|---|---|---|---|

**Conversion:** 1 ppm = 0.82 mg/m$^3$

**DOT:** 1052 125 (anhydrous); 1790 157 (solution)

**Synonyms/Trade Names:** Anhydrous hydrogen fluoride; Aqueous hydrogen fluoride (i.e., Hydrofluoric acid); HF-A

**Exposure Limits:**
**NIOSH REL:** TWA 3 ppm (2.5 mg/m$^3$)
C 6 ppm (5 mg/m$^3$) [15-minute]
**OSHA PEL†:** TWA 3 ppm

**Measurement Methods (see Table 1):**
**NIOSH** 3800, 7902, 7903, 7906
**OSHA** ID110

**Physical Description:** Colorless gas or fuming liquid (below 67°F) with a strong, irritating odor. [**Note:** Shipped in cylinders.]

**Chemical & Physical Properties:**
**MW:** 20.0
**BP:** 67°F
**Sol:** Miscible
**Fl.P:** NA
**IP:** 15.98 eV
**RGasD:** 0.69
**Sp.Gr:** 1.00 (Liquid at 67°F)
**VP:** 783 mmHg
**FRZ:** -118°F
**UEL:** NA
**LEL:** NA
Nonflammable Gas

**Personal Protection/Sanitation (see Table 2):**
**Skin:** Prevent skin contact (liquid)
**Eyes:** Prevent eye contact (liquid)
**Wash skin:** When contam (liquid)
**Remove:** When wet or contam (liquid)
**Change:** N.R.
**Provide:** Eyewash (liquid)
Quick drench (liquid)

**Respirator Recommendations (see Tables 3 and 4):**
**NIOSH/OSHA**
**30 ppm:** CcrS*/PaprS*/GmFS/Sa*/ScbaF
**§:** ScbaF:Pd,Pp/SaF:Pd,Pp:AScba
**Escape:** GmFS/ScbaE

**Incompatibilities and Reactivities:** Metals, water or steam
[**Note:** Corrosive to metals. Will attack glass and concrete.]

**Exposure Routes, Symptoms, Target Organs (see Table 5):**
**ER:** Inh, Abs (liquid), Ing (solution), Con
**SY:** Irrit eyes, skin, nose, throat; pulm edema; eye, skin burns; rhinitis; bron; bone changes
**TO:** Eyes, skin, resp sys, bones

**First Aid (see Table 6):**
**Eye:** Irr immed (solution/liquid)
**Skin:** Water flush immed (solution/liquid)
**Breath:** Resp support
**Swallow:** Medical attention immed (solution)

## Hydrogen peroxide

| Hydrogen peroxide | Formula: $H_2O_2$ | CAS#: 7722-84-1 | RTECS#: MX0900000 | IDLH: 75 ppm |
|---|---|---|---|---|
| **Conversion:** 1 ppm = 1.39 mg/m³ | **DOT:** 2984 140 (8-20% solution); 2014 140 (20-60% solution); 2015 143 (>60% solution) | | | |

**Synonyms/Trade Names:** High-strength hydrogen peroxide, Hydrogen dioxide, Hydrogen peroxide (aqueous), Hydroperoxide, Peroxide

**Exposure Limits:**
**NIOSH REL:** TWA 1 ppm (1.4 mg/m³)
**OSHA PEL:** TWA 1 ppm (1.4 mg/m³)

**Measurement Methods (see Table 1):**
**OSHA** ID126SG

**Physical Description:** Colorless liquid with a slightly sharp odor.
[**Note:** The pure compound is a crystalline solid below 12°F. Often used in an aqueous solution.]

**Chemical & Physical Properties:**
**MW:** 34.0
**BP:** 286°F
**Sol:** Miscible
**Fl.P:** NA
**IP:** 10.54 eV
**Sp.Gr:** 1.39
VP(86°F): 5 mmHg
**FRZ:** 12°F
**UEL:** NA
**LEL:** NA
Noncombustible Liquid, but a powerful oxidizer.

**Personal Protection/Sanitation (see Table 2):**
**Skin:** Prevent skin contact
**Eyes:** Prevent eye contact
**Wash skin:** When contam
**Remove:** When wet or contam
**Change:** N.R.
**Provide:** Eyewash
Quick drench

**Respirator Recommendations (see Tables 3 and 4):**
**NIOSH/OSHA**
**10 ppm:** Sa*
**25 ppm:** Sa:Cf*
**50 ppm:** ScbaF/SaF
**75 ppm:** SaF:Pd,Pp
**§:** ScbaF:Pd,Pp/SaF:Pd,Pp:AScba
**Escape:** GmFS/ScbaE

**Incompatibilities and Reactivities:** Oxidizable materials, iron, copper, brass, bronze, chromium, zinc, lead, silver, manganese [**Note:** Contact with combustible material may result in SPONTANEOUS combustion.]

**Exposure Routes, Symptoms, Target Organs (see Table 5):**
**ER:** Inh, Ing, Con
**SY:** Irrit eyes, nose, throat; corn ulcer; eryt, vesic skin; bleaching hair
**TO:** Eyes, skin, resp sys

**First Aid (see Table 6):**
**Eye:** Irr immed
**Skin:** Water flush immed
**Breath:** Resp support
**Swallow:** Medical attention immed

## Hydrogen selenide

| Hydrogen selenide | Formula: $H_2Se$ | CAS#: 7783-07-5 | RTECS#: MX1050000 | IDLH: 1 ppm |
|---|---|---|---|---|
| **Conversion:** 1 ppm = 3.31 mg/m³ | **DOT:** 2202 117 (anhydrous) | | | |

**Synonyms/Trade Names:** Selenium dihydride, Selenium hydride

**Exposure Limits:**
**NIOSH REL:** TWA 0.05 ppm (0.2 mg/m³)
**OSHA PEL:** TWA 0.05 ppm (0.2 mg/m³)

**Measurement Methods (see Table 1):**
None available

**Physical Description:** Colorless gas with an odor resembling decayed horseradish.
[**Note:** Shipped as a liquefied compressed gas.]

**Chemical & Physical Properties:**
**MW:** 81.0
**BP:** -42°F
**Sol(73°F):** 0.9%
**Fl.P:** NA (Gas)
**IP:** 9.88 eV
**RGasD:** 2.80
**VP(70°F):** 9.5 atm
**FRZ:** -87°F
**UEL:** ?
**LEL:** ?
Flammable Gas

**Personal Protection/Sanitation (see Table 2):**
**Skin:** Frostbite
**Eyes:** Frostbite
**Wash skin:** N.R.
**Remove:** When wet (flamm)
**Change:** N.R.
**Provide:** Frostbite wash

**Respirator Recommendations (see Tables 3 and 4):**
**NIOSH/OSHA**
**0.5 ppm:** Sa
**1 ppm:** Sa:Cf*/ScbaF/SaF
**§:** ScbaF:Pd,Pp/SaF:Pd,Pp:AScba
**Escape:** GmFS¿/ScbaE

**Incompatibilities and Reactivities:** Strong oxidizers, acids, water, halogenated hydrocarbons

**Exposure Routes, Symptoms, Target Organs (see Table 5):**
**ER:** Inh, Con
**SY:** Irrit eyes, nose, throat; nau, vomit, diarr; metallic taste, garlic breath; dizz, lass; liquid: frostbite; in animals: pneu; liver damage
**TO:** Eyes, resp sys, liver

**First Aid (see Table 6):**
**Eye:** Frostbite
**Skin:** Frostbite
**Breath:** Resp support

## Hydrogen sulfide

| Hydrogen sulfide | Formula: $H_2S$ | CAS#: 7783-06-4 | RTECS#: MX1225000 | IDLH: 100 ppm |
|---|---|---|---|---|
| **Conversion:** 1 ppm = 1.40 mg/m$^3$ | **DOT:** 1053 117 | | | |

**Synonyms/Trade Names:** Hydrosulfuric acid, Sewer gas, Sulfuretted hydrogen

**Exposure Limits:**
**NIOSH REL:** C 10 ppm (15 mg/m$^3$) [10-minute]
**OSHA PEL†:** C 20 ppm
50 ppm [10-minute maximum peak]

**Measurement Methods (see Table 1):**
**NIOSH** 6013
**OSHA** ID141

**Physical Description:** Colorless gas with a strong odor of rotten eggs. [**Note:** Sense of smell becomes rapidly fatigued & can NOT be relied upon to warn of the continuous presence of $H_2S$. Shipped as a liquefied compressed gas.]

**Chemical & Physical Properties:**
**MW:** 34.1
**BP:** -77°F
**Sol:** 0.4%
**Fl.P:** NA (Gas)
**IP:** 10.46 eV
**RGasD:** 1.19
**VP:** 17.6 atm
**FRZ:** -122°F
**UEL:** 44.0%
**LEL:** 4.0%
Flammable Gas

**Personal Protection/Sanitation (see Table 2):**
**Skin:** Frostbite
**Eyes:** Frostbite
**Wash skin:** N.R.
**Remove:** When wet (flamm)
**Change:** N.R.
**Provide:** Frostbite wash

**Respirator Recommendations (see Tables 3 and 4):**
**NIOSH**
**100 ppm:** PaprS/GmFS/Sa*/ScbaF
**§:** ScbaF:Pd,Pp/SaF:Pd,Pp:AScba
**Escape:** GmFS/ScbaE

**Incompatibilities and Reactivities:** Strong oxidizers, strong nitric acid, metals

**Exposure Routes, Symptoms, Target Organs (see Table 5):**
**ER:** Inh, Con
**SY:** Irrit eyes, resp sys; apnea, coma, convuls; conj, eye pain, lac, photo, corn vesic; dizz, head, lass, irrity, insom; GI dist; liquid: frostbite
**TO:** Eyes, resp sys, CNS

**First Aid (see Table 6):**
**Eye:** Frostbite
**Skin:** Frostbite
**Breath:** Resp support

## Hydroquinone

| Hydroquinone | Formula: $C_6H_4(OH)_2$ | CAS#: 123-31-9 | RTECS#: MX3500000 | IDLH: 50 mg/m$^3$ |
|---|---|---|---|---|
| **Conversion:** | **DOT:** 2662 153 | | | |

**Synonyms/Trade Names:** p-Benzenediol; 1,4-Benzenediol; Dihydroxybenzene; 1,4-Dihydroxybenzene; Quinol

**Exposure Limits:**
**NIOSH REL:** C 2 mg/m$^3$ [15-minute]
**OSHA PEL:** TWA 2 mg/m$^3$

**Measurement Methods (see Table 1):**
**NIOSH** 5004
**OSHA** PV2094

**Physical Description:** Light-tan, light-gray, or colorless crystals.

**Chemical & Physical Properties:**
**MW:** 110.1
**BP:** 545°F
**Sol:** 7%
**Fl.P:** 329°F (Molten)
**IP:** 7.95 eV
**Sp.Gr:** 1.33
**VP:** 0.00001 mmHg
**MLT:** 338°F
**UEL:** ?
**LEL:** ?
Combustible Solid; dust cloud may explode if ignited in an enclosed area.

**Personal Protection/Sanitation (see Table 2):**
**Skin:** Prevent skin contact
**Eyes:** Prevent eye contact
**Wash skin:** When contam
**Remove:** When wet or contam
**Change:** Daily
**Provide:** Eyewash (>7%)

**Respirator Recommendations (see Tables 3 and 4):**
**NIOSH/OSHA**
**50 mg/m$^3$:** PaprHie£/100F/SaT:Cf£/ScbaF/SaF
**§:** ScbaF:Pd,Pp/SaF:Pd,Pp:AScba
**Escape:** 100F/ScbaE

**Incompatibilities and Reactivities:** Strong oxidizers, alkalis

**Exposure Routes, Symptoms, Target Organs (see Table 5):**
**ER:** Inh, Ing, Con
**SY:** Irrit eyes; conj; kera; CNS excitement; colored urine, nau, dizz, suffocation, rapid breath; musc twitch, delirium; collapse; skin irrit, sens, derm
**TO:** Eyes, skin, resp sys, CNS

**First Aid (see Table 6):**
**Eye:** Irr immed
**Skin:** Water flush
**Breath:** Resp support
**Swallow:** Medical attention immed

## 2-Hydroxypropyl acrylate

| 2-Hydroxypropyl acrylate | Formula: $CH_2$=$CHCOOCH_2CHOHCH_3$ | CAS#: 999-61-1 | RTECS#: AT1925000 | IDLH: N.D. |
|---|---|---|---|---|
| **Conversion:** 1 ppm = 5.33 mg/$m^3$ | **DOT:** | | | |

**Synonyms/Trade Names:** HPA, β-Hydroxypropyl acrylate, Propylene glycol monoacrylate

**Exposure Limits:**
**NIOSH REL:** TWA 0.5 ppm (3 mg/$m^3$) [skin]
**OSHA PEL†:** none

**Measurement Methods (see Table 1):**
None available

**Physical Description:** Clear to light-yellow liquid wiith a sweetish, solvent odor.

| Chemical & Physical Properties: | Personal Protection/Sanitation (see Table 2): | Respirator Recommendations (see Tables 3 and 4): |
|---|---|---|
| **MW:** 130.2<br>**BP:** 376°F<br>**Sol:** ?<br>**Fl.P:** 149°F<br>**IP:** ?<br>**Sp.Gr:** 1.05<br>**VP:** ?<br>**FRZ:** ?<br>**UEL:** ?<br>**LEL:** 1.8%<br>Class IIIA Combustible Liquid | **Skin:** Prevent skin contact<br>**Eyes:** Prevent eye contact<br>**Wash skin:** When contam<br>**Remove:** When wet or contam<br>**Change:** N.R.<br>**Provide:** Eyewash<br>Quick drench | Not available. |

**Incompatibilities and Reactivities:** Water [**Note:** Can become unstable at high temperatures & pressures or may react with water with some release of energy, but not violently.]

| Exposure Routes, Symptoms, Target Organs (see Table 5): | First Aid (see Table 6): |
|---|---|
| **ER:** Inh, Abs, Ing, Con<br>**SY:** Irrit eyes, skin, resp sys; eye, skin burns; cough, dysp<br>**TO:** Eyes, skin, resp sys | **Eye:** Irr immed<br>**Skin:** Soap flush immed<br>**Breath:** Resp support<br>**Swallow:** Medical attention immed |

I

## Indene

| Indene | Formula: $C_9H_8$ | CAS#: 95-13-6 | RTECS#: NK8225000 | IDLH: N.D. |
|---|---|---|---|---|
| **Conversion:** 1 ppm = 4.75 mg/$m^3$ | **DOT:** | | | |

**Synonyms/Trade Names:** Indonaphthene

**Exposure Limits:**
**NIOSH REL:** TWA 10 ppm (45 mg/$m^3$)
**OSHA PEL†:** none

**Measurement Methods (see Table 1):**
None available

**Physical Description:** Colorless liquid. [**Note:** A solid below 29°F.]

| Chemical & Physical Properties: | Personal Protection/Sanitation (see Table 2): | Respirator Recommendations (see Tables 3 and 4): |
|---|---|---|
| **MW:** 116.2<br>**BP:** 359°F<br>**Sol:** Insoluble<br>**Fl.P:** 173°F<br>**IP:** 8.81 eV<br>**Sp.Gr:** 0.997<br>**VP:** ?<br>**FRZ:** 29°F<br>**UEL:** ?<br>**LEL:** ?<br>Class IIIA Combustible Liquid | **Skin:** Prevent skin contact<br>**Eyes:** Prevent eye contact<br>**Wash skin:** Daily<br>**Remove:** When wet or contam<br>**Change:** N.R. | Not available. |

**Incompatibilities and Reactivities:** None reported
[**Note:** Polymerizes & oxidizes on standing. It has exploded during nitration with ($H_2SO_4$ + $HNO_3$).]

| Exposure Routes, Symptoms, Target Organs (see Table 5): | First Aid (see Table 6): |
|---|---|
| **ER:** Inh, Ing, Con<br>**SY:** In animals: irrit eyes, skin, muc memb; derm, skin sens; chemical pneu (aspir liquid); liver, kidney, spleen inj<br>**TO:** Eyes, skin, resp sys, liver, kidneys, spleen | **Eye:** Irr immed<br>**Skin:** Soap wash<br>**Breath:** Resp support<br>**Swallow:** Medical attention immed |

| Indium | Formula: In | CAS#: 7440-74-6 | RTECS#: NL1050000 | IDLH: N.D. |
|---|---|---|---|---|
| **Conversion:** | **DOT:** | | | |

**Synonyms/Trade Names:** Indium metal

**Exposure Limits:**
**NIOSH REL*:** TWA 0.1 mg/m$^3$
[***Note:** The REL also applies to other indium compounds (as In).]
**OSHA PEL†:** none

**Measurement Methods (see Table 1):**
**NIOSH** 7303, P&CAM173 (II-5)
**OSHA** ID121

**Physical Description:** Ductile, shiny, silver-white metal that is softer than lead.

**Chemical & Physical Properties:**
**MW:** 114.8
**BP:** 3767°F
**Sol:** Insoluble
**Fl.P:** NA
**IP:** NA
**Sp.Gr:** 7.31
**VP:** 0 mmHg (approx)
**MLT:** 314°F
**UEL:** NA
**LEL:** NA
Noncombustible Solid in bulk form, but may ignite in powdered or dust form.

**Personal Protection/Sanitation (see Table 2):**
**Skin:** N.R.
**Eyes:** N.R.
**Wash skin:** N.R.
**Remove:** N.R.
**Change:** N.R.

**Respirator Recommendations (see Tables 3 and 4):**
Not available.

**Incompatibilities and Reactivities:** (Dinitrogen tetraoxide + acetonitrile), mercury(II) bromide (at 662°F), sulfur (mixtures ignite when heated) [**Note:** oxidizes readily at higher temperatures.]

**Exposure Routes, Symptoms, Target Organs (see Table 5):**
**ER:** Inh, Ing, Con
**SY:** Irrit eyes, skin, resp sys; possible liver, kidney, heart, blood effects; pulm edema
**TO:** Eyes, skin, resp sys, liver, kidneys, heart, blood

**First Aid (see Table 6):**
**Eye:** Irr immed
**Skin:** Soap wash
**Breath:** Resp support
**Swallow:** Medical attention immed

I

| Iodine | Formula: $I_2$ | CAS#: 7553-56-2 | RTECS#: NN1575000 | IDLH: 2 ppm |
|---|---|---|---|---|
| **Conversion:** 1 ppm = 10.38 mg/m$^3$ | **DOT:** | | | |

**Synonyms/Trade Names:** Iodine crystals, Molecular iodine

**Exposure Limits:**
**NIOSH REL:** C 0.1 ppm (1 mg/m$^3$)
**OSHA PEL:** C 0.1 ppm (1 mg/m$^3$)

**Measurement Methods (see Table 1):**
**NIOSH** 6005
**OSHA** ID212

**Physical Description:** Violet solid with a sharp, characteristic odor.

**Chemical & Physical Properties:**
**MW:** 253.8
**BP:** 365°F
**Sol:** 0.01%
**Fl.P:** NA
**IP:** 9.31 eV
**Sp.Gr:** 4.93
**VP(77°F):** 0.3 mmHg
**MLT:** 236°F
**UEL:** NA
**LEL:** NA
Noncombustible Solid

**Personal Protection/Sanitation (see Table 2):**
**Skin:** Prevent skin contact
**Eyes:** Prevent eye contact
**Wash skin:** When contam
**Remove:** When wet or contam
**Change:** Daily
**Provide:** Eyewash (>7%)
Quick drench (>7%)

**Respirator Recommendations (see Tables 3 and 4):**
**NIOSH/OSHA**
**1 ppm:** Sa*
**2 ppm:** Sa:Cf*/ScbaF/SaF
**§:** ScbaF:Pd,Pp/SaF:Pd,Pp:AScba
**Escape:** GmFAg100/ScbaE

**Incompatibilities and Reactivities:** Ammonia, acetylene, acetaldehyde, powdered aluminum, active metals, liquid chlorine

**Exposure Routes, Symptoms, Target Organs (see Table 5):**
**ER:** Inh, Ing, Con
**SY:** Irrit eyes, skin, nose; lac; head; chest tight; skin burns, rash; cutaneous hypersensitivity
**TO:** Eyes, skin, resp sys, CNS, CVS

**First Aid (see Table 6):**
**Eye:** Irr immed
**Skin:** Soap wash immed
**Breath:** Resp support
**Swallow:** Medical attention immed

| Iodoform | Formula: $CHI_3$ | CAS#: 75-47-8 | RTECS#: PB7000000 | IDLH: N.D. |
|---|---|---|---|---|
| **Conversion:** 1 ppm = 16.10 mg/m$^3$ | **DOT:** | | | |
| **Synonyms/Trade Names:** Triiodomethane | | | | |
| **Exposure Limits:**<br>**NIOSH REL:** TWA 0.6 ppm (10 mg/m$^3$)<br>**OSHA PEL†:** none | | | **Measurement Methods (see Table 1):**<br>None available | |
| **Physical Description:** Yellow to greenish-yellow powder or crystalline solid with a pungent, disagreeable odor. [antiseptic for external use] | | | | |
| **Chemical & Physical Properties:**<br>**MW:** 393.7<br>**BP:** 410°F (Decomposes)<br>**Sol:** 0.01%<br>**Fl.P:** NA<br>**IP:** ?<br>**Sp.Gr:** 4.01<br>**VP:** ?<br>**MLT:** 246°F<br>**UEL:** NA<br>**LEL:** NA<br>Noncombustible Solid | **Personal Protection/Sanitation (see Table 2):**<br>**Skin:** Prevent skin contact<br>**Eyes:** Prevent eye contact<br>**Wash skin:** When contam<br>**Remove:** When wet or contam<br>**Change:** Daily | | **Respirator Recommendations (see Tables 3 and 4):**<br>Not available. | |
| **Incompatibilities and Reactivities:** Strong oxidizers, lithium, metallic salts (e.g., mercuric oxide, silver nitrate), strong bases, calomel, tannin | | | | |
| **Exposure Routes, Symptoms, Target Organs (see Table 5):**<br>**ER:** Inh, Abs, Ing, Con<br>**SY:** Irrit eyes, skin; lass, dizz, nau, inco, CNS depres; dysp; liver, kidney, heart damage; vis dist<br>**TO:** Eyes, skin, resp sys, liver, kidneys, heart | | **First Aid (see Table 6):**<br>**Eye:** Irr immed<br>**Skin:** Soap wash immed<br>**Breath:** Resp support<br>**Swallow:** Medical attention immed | | |

I

| Iron oxide dust and fume (as Fe) | Formula: $Fe_2O_3$ | CAS#: 1309-37-1 | RTECS#: NO7400000 NO7525000 (fume) | IDLH: 2500 mg/m$^3$ (as Fe) |
|---|---|---|---|---|
| **Conversion:** | **DOT:** 1376 135 (spent) | | | |
| **Synonyms/Trade Names:** Ferric oxide, Iron(III) oxide | | | | |
| **Exposure Limits:**<br>**NIOSH REL:** TWA 5 mg/m$^3$<br>**OSHA PEL:** TWA 10 mg/m$^3$ | | | **Measurement Methods (see Table 1):**<br>**NIOSH** 7300, 7301, 7303, 9102<br>**OSHA** ID121, ID125G | |
| **Physical Description:** Reddish-brown solid.<br>[**Note:** Exposure to fume may occur during the arc-welding of iron.] | | | | |
| **Chemical & Physical Properties:**<br>**MW:** 159.7<br>**BP:** ?<br>**Sol:** Insoluble<br>**Fl.P:** NA<br>**IP:** NA<br>**Sp.Gr:** 5.24<br>**VP:** 0 mmHg (approx)<br>**MLT:** 2664°F<br>**UEL:** NA<br>**LEL:** NA<br>Noncombustible Solid | **Personal Protection/Sanitation (see Table 2):**<br>**Skin:** N.R.<br>**Eyes:** N.R.<br>**Wash skin:** N.R.<br>**Remove:** N.R.<br>**Change:** N.R. | | **Respirator Recommendations (see Tables 3 and 4):**<br>**NIOSH**<br>**50 mg/m$^3$:** 95XQ/Sa<br>**125 mg/m$^3$:** Sa:Cf/PaprHie<br>**250 mg/m$^3$:** 100F/SaT:Cf/PaprTHie/ ScbaF/SaF<br>**2500 mg/m$^3$:** Sa:Pd,Pp<br>**§:** ScbaF:Pd,Pp/SaF:Pd,Pp:AScba<br>**Escape:** 100F/ScbaE | |
| **Incompatibilities and Reactivities:** Calcium hypochlorite | | | | |
| **Exposure Routes, Symptoms, Target Organs (see Table 5):**<br>**ER:** Inh<br>**SY:** Benign pneumoconiosis with X-ray shadows indistinguishable from fibrotic pneumoconiosis (siderosis)<br>**TO:** Resp sys | | **First Aid (see Table 6):**<br>**Breath:** Resp support | | |

| Iron pentacarbonyl (as Fe) | Formula: $Fe(CO)_5$ | CAS#: 13463-40-6 | RTECS#: NO4900000 | IDLH: N.D. |
|---|---|---|---|---|
| **Conversion:** 1 ppm = 2.28 mg/m$^3$ (as Fe) | **DOT:** 1994 131 | | | |

**Synonyms/Trade Names:** Iron carbonyl, Pentacarbonyl iron

**Exposure Limits:**
**NIOSH REL:** TWA 0.1 ppm (0.23 mg/m$^3$)
ST 0.2 ppm (0.45 mg/m$^3$)
**OSHA PEL†:** none

**Measurement Methods (see Table 1):**
None available

**Physical Description:** Colorless to yellow to dark-red, oily liquid.

**Chemical & Physical Properties:**
**MW:** 195.9
**BP(749 mmHg):** 217°F
**Sol:** Insoluble
**Fl.P:** 5°F
**IP:** ?
**Sp.Gr:** 1.46-1.52
**VP(87°F):** 40 mmHg
**FRZ:** -6°F
**UEL:** ?
**LEL:** ?
Class IB Flammable Liquid

**Personal Protection/Sanitation (see Table 2):**
**Skin:** Prevent skin contact
**Eyes:** Prevent eye contact
**Wash skin:** When contam
**Remove:** When wet (flamm)
**Change:** N.R.
**Provide:** Quick drench

**Respirator Recommendations (see Tables 3 and 4):**
Not available.

**Incompatibilities and Reactivities:** Oxidizers, nitrogen oxide, (zinc + cobalt halides)
[**Note:** Pyrophoric (i.e., ignites spontaneously in air). Decomposed by light or air, releasing carbon monoxide.]

**Exposure Routes, Symptoms, Target Organs (see Table 5):**
**ER:** Inh, Abs, Ing, Con
**SY:** Irrit eyes, muc memb, resp sys; head, dizz, nau, vomit; fever, cyan, cough, dysp; liver, kidney, lung inj; degenerative changes in CNS
**TO:** Eyes, resp sys, CNS, liver, kidneys

**First Aid (see Table 6):**
**Eye:** Irr immed
**Skin:** Soap flush immed
**Breath:** Resp support
**Swallow:** Medical attention immed

I

| Iron salts (soluble, as Fe) | Formula: | CAS#: | RTECS#: | IDLH: N.D. |
|---|---|---|---|---|
| **Conversion:** | **DOT:** | | | |

**Synonyms/Trade Names:** **$FeSO_4$:** Ferrous sulfate, Iron(II) sulfate; **$FeCl_2$:** Ferrous chloride, Iron(II) chloride; **$Fe(NO_3)_3$:** Ferric nitrate, Iron(III) nitrate; **$Fe(SO_4)_3$:** Ferric sulfate, Iron(III) sulfate; **$FeCl_3$:** Ferric chloride, Iron (III) chloride

**Exposure Limits:**
**NIOSH REL:** TWA 1 mg/m$^3$
**OSHA PEL†:** none

**Measurement Methods (see Table 1):**
**NIOSH** 7300, 7301, 7303, 9102
**OSHA** ID121, ID125G

**Physical Description:** Appearance and odor vary depending upon the specific soluble iron salt.

**Chemical & Physical Properties:**
Properties vary depending upon the specific soluble iron salt.

Noncombustible Solids

**Personal Protection/Sanitation (see Table 2):**
**Skin:** Prevent skin contact
**Eyes:** Prevent eye contact
**Wash skin:** Daily
**Remove:** N.R.
**Change:** Daily

**Respirator Recommendations (see Tables 3 and 4):**
Not available.

**Incompatibilities and Reactivities:** Varies

**Exposure Routes, Symptoms, Target Organs (see Table 5):**
**ER:** Inh, Ing, Con
**SY:** Irrit eyes, skin, muc memb; abdom pain, diarr, vomit; possible liver damage
**TO:** Eyes, skin, resp sys, liver, GI tract

**First Aid (see Table 6):**
**Eye:** Irr immed
**Skin:** Soap wash
**Breath:** Resp support
**Swallow:** Medical attention immed

## Isoamyl acetate

| | | | | |
|---|---|---|---|---|
| **Isoamyl acetate** | **Formula:** $CH_3COOCH_2CH_2CH(CH_3)_2$ | **CAS#:** 123-92-2 | **RTECS#:** NS9800000 | **IDLH:** 1000 ppm |
| **Conversion:** 1 ppm = 5.33 mg/m$^3$ | **DOT:** 1104 129 | | | |

**Synonyms/Trade Names:** Banana oil, Isopentyl acetate, 3-Methyl-1-butanol acetate, 3-Methylbutyl ester of acetic acid, 3-Methylbutyl ethanoate

**Exposure Limits:**
**NIOSH REL:** TWA 100 ppm (525 mg/m$^3$)
**OSHA PEL:** TWA 100 ppm (525 mg/m$^3$)

**Measurement Methods (see Table 1):**
**NIOSH** 1450
**OSHA** 7

**Physical Description:** Colorless liquid with a banana-like odor.

**Chemical & Physical Properties:**
**MW:** 130.2
**BP:** 288°F
**Sol:** 0.3%
**Fl.P:** 77°F
**IP:** ?
**Sp.Gr:** 0.87
**VP:** 4 mmHg
**FRZ:** -109°F
**UEL:** 7.5%
**LEL(212°F):** 1.0%
Class IC Flammable Liquid

**Personal Protection/Sanitation (see Table 2):**
**Skin:** Prevent skin contact
**Eyes:** Prevent eye contact
**Wash skin:** When contam
**Remove:** When wet (flamm)
**Change:** N.R.

**Respirator Recommendations (see Tables 3 and 4):**
**NIOSH/OSHA**
**1000 ppm:** CcrOv/PaprOv/GmFOv/Sa/ScbaF
**§:** ScbaF:Pd,Pp/SaF:Pd,Pp:AScba
**Escape:** GmFOv/ScbaE

**Incompatibilities and Reactivities:** Nitrates; strong oxidizers, alkalis & acids

**Exposure Routes, Symptoms, Target Organs (see Table 5):**
**ER:** Inh, Ing, Con
**SY:** Irrit eyes, skin, nose, throat; derm; in animals: narco
**TO:** Eyes, skin, resp sys, CNS

**First Aid (see Table 6):**
**Eye:** Irr immed
**Skin:** Water flush prompt
**Breath:** Resp support
**Swallow:** Medical attention immed

I

## Isoamyl alcohol (primary)

| | | | | |
|---|---|---|---|---|
| **Isoamyl alcohol (primary)** | **Formula:** $(CH_3)_2CHCH_2CH_2OH$ | **CAS#:** 123-51-3 | **RTECS#:** EL5425000 | **IDLH:** 500 ppm |
| **Conversion:** 1 ppm = 3.61 mg/m$^3$ | **DOT:** 1105 129 | | | |

**Synonyms/Trade Names:** Fermentation amyl alcohol, Fusel oil, Isobutyl carbinol, Isopentyl alcohol, 3-Methyl-1-butanol, Primary isoamyl alcohol

**Exposure Limits:**
**NIOSH REL:** TWA 100 ppm (360 mg/m$^3$)
ST 125 ppm (450 mg/m$^3$)
**OSHA PEL†:** TWA 100 ppm (360 mg/m$^3$)

**Measurement Methods (see Table 1):**
**NIOSH** 1402, 1405

**Physical Description:** Colorless liquid with a disagreeable odor.

**Chemical & Physical Properties:**
**MW:** 88.2
**BP:** 270°F
**Sol(57°F):** 2%
**Fl.P:** 109°F
**IP:** ?
**Sp.Gr(57°F):** 0.81
**VP:** 28 mmHg
**FRZ:** -179°F
**UEL(212°F):** 9.0%
**LEL:** 1.2%
Class II Combustible Liquid

**Personal Protection/Sanitation (see Table 2):**
**Skin:** Prevent skin contact
**Eyes:** Prevent eye contact
**Wash skin:** When contam
**Remove:** When wet or contam
**Change:** N.R.

**Respirator Recommendations (see Tables 3 and 4):**
**NIOSH/OSHA**
**500 ppm:** Sa:Cf£/CcrFOv/GmFOv/PaprOv£/ScbaF/SaF
**§:** ScbaF:Pd,Pp/SaF:Pd,Pp:AScba
**Escape:** GmFOv/ScbaE

**Incompatibilities and Reactivities:** Strong oxidizers

**Exposure Routes, Symptoms, Target Organs (see Table 5):**
**ER:** Inh, Ing, Con
**SY:** Irrit eyes, skin, nose, throat; head, dizz; cough, dysp, nau, vomit, diarr; skin cracking; in animals: narco
**TO:** Eyes, skin, resp sys, CNS

**First Aid (see Table 6):**
**Eye:** Irr immed
**Skin:** Water flush prompt
**Breath:** Resp support
**Swallow:** Medical attention immed

| Isoamyl alcohol (secondary) | Formula: $(CH_3)_2CHCH(OH)CH_3$ | CAS#: 6032-29-7 | RTECS#: SA4900000 | IDLH: 500 ppm |
|---|---|---|---|---|
| **Conversion:** 1 ppm = 3.61 mg/m³ | **DOT:** 1105 129 | | | |

**Synonyms/Trade Names:** 3-Methyl-2-butanol, Secondary isoamyl alcohol

**Exposure Limits:**
**NIOSH REL:** TWA 100 ppm (360 mg/m³)
ST 125 ppm (450 mg/m³)
**OSHA PEL†:** TWA 100 ppm (360 mg/m³)

**Measurement Methods (see Table 1):**
**NIOSH** 1402

**Physical Description:** Colorless liquid with a disagreeable odor.

| Chemical & Physical Properties: | Personal Protection/Sanitation (see Table 2): | Respirator Recommendations (see Tables 3 and 4): |
|---|---|---|
| **MW:** 88.2<br>**BP:** 234°F<br>**Sol:** ?<br>**Fl.P(oc):** 95°F<br>**IP:** ?<br>**Sp.Gr:** 0.82<br>**VP:** 1 mmHg<br>**FRZ:** ?<br>**UEL:** ?<br>**LEL:** ?<br>Class IC Flammable Liquid | **Skin:** Prevent skin contact<br>**Eyes:** Prevent eye contact<br>**Wash skin:** When contam<br>**Remove:** When wet (flamm)<br>**Change:** N.R. | **NIOSH/OSHA**<br>**500 ppm:** Sa:Cf£/CcrFOv/GmFOv/PaprOv£/ScbaF/SaF<br>**§:** ScbaF:Pd,Pp/SaF:Pd,Pp:AScba<br>**Escape:** GmFOv/ScbaE |

**Incompatibilities and Reactivities:** Strong oxidizers

| Exposure Routes, Symptoms, Target Organs (see Table 5): | First Aid (see Table 6): |
|---|---|
| **ER:** Inh, Ing, Con<br>**SY:** Irrit eyes, skin, nose, throat; head, dizz; cough, dysp, nau, vomit, diarr; skin cracking; in animals: narco<br>**TO:** Eyes, skin, resp sys, CNS | **Eye:** Irr immed<br>**Skin:** Water flush prompt<br>**Breath:** Resp support<br>**Swallow:** Medical attention immed |

| Isobutane | Formula: $CH_3CH(CH_3)_2$ | CAS#: 75-28-5 | RTECS#: TZ4300000 | IDLH: N.D. |
|---|---|---|---|---|
| **Conversion:** 1 ppm = 2.38 mg/m³ | **DOT:** 1075 115; 1969 115 | | | |

**Synonyms/Trade Names:** 2-Methylpropane [**Note:** Also see specific listing for n-Butane.]

**Exposure Limits:**
**NIOSH REL:** TWA 800 ppm (1900 mg/m³)
**OSHA PEL†:** none

**Measurement Methods (see Table 1):**
None available

**Physical Description:** Colorless gas with a gasoline-like or natural gas odor.
[**Note:** Shipped as a liquefied compressed gas. A liquid below 11°F.]

| Chemical & Physical Properties: | Personal Protection/Sanitation (see Table 2): | Respirator Recommendations (see Tables 3 and 4): |
|---|---|---|
| **MW:** 58.1<br>**BP:** 11°F<br>**Sol:** Slight<br>**Fl.P:** NA (Gas)<br>**IP:** 10.74 eV<br>**RGasD:** 2.06<br>**VP(70°F):** 3.1 atm<br>**FRZ:** -255°F<br>**UEL:** 8.4%<br>**LEL:** 1.6%<br>Flammable Gas | **Skin:** Frostbite<br>**Eyes:** Frostbite<br>**Wash skin:** N.R.<br>**Remove:** When wet (flamm)<br>**Change:** N.R.<br>**Provide:** Frostbite wash | Not available. |

**Incompatibilities and Reactivities:** Strong oxidizers (e.g., nitrates & perchlorates), chlorine, fluorine, (nickel carbonyl + oxygen)

| Exposure Routes, Symptoms, Target Organs (see Table 5): | First Aid (see Table 6): |
|---|---|
| **ER:** Inh, Con (liquid)<br>**SY:** Drow, narco, asphy; liquid: frostbite<br>**TO:** CNS | **Eye:** Frostbite<br>**Skin:** Frostbite<br>**Breath:** Resp support |

I

## Isobutyl acetate

| Isobutyl acetate | Formula: $CH_3COOCH_2CH(CH_3)_2$ | CAS#: 110-19-0 | RTECS#: AI4025000 | IDLH: 1300 ppm [10%LEL] |
|---|---|---|---|---|
| **Conversion:** 1 ppm = 4.75 mg/m³ | **DOT:** 1213 129 | | | |

**Synonyms/Trade Names:** Isobutyl ester of acetic acid, 2-Methylpropyl acetate, 2-Methylpropyl ester of acetic acid, β-Methylpropyl ethanoate

**Exposure Limits:**
**NIOSH REL:** TWA 150 ppm (700 mg/m³)
**OSHA PEL:** TWA 150 ppm (700 mg/m³)

**Measurement Methods (see Table 1):**
**NIOSH** 1450
**OSHA** 7

**Physical Description:** Colorless liquid with a fruity, floral odor.

| Chemical & Physical Properties: | Personal Protection/Sanitation (see Table 2): | Respirator Recommendations (see Tables 3 and 4): |
|---|---|---|
| **MW:** 116.2<br>**BP:** 243°F<br>**Sol(77°F):** 0.6%<br>**Fl.P:** 64°F<br>**IP:** 9.97 eV<br>**Sp.Gr:** 0.87<br>**VP:** 13 mmHg<br>**FRZ:** -145°F<br>**UEL:** 10.5%<br>**LEL:** 1.3%<br>Class IB Flammable Liquid | **Skin:** Prevent skin contact<br>**Eyes:** Prevent eye contact<br>**Wash skin:** When contam<br>**Remove:** When wet (flamm)<br>**Change:** N.R. | **NIOSH/OSHA**<br>**1300 ppm:** Sa:Cf£/CcrFOv/GmFOv/PaprOv£/ScbaF/SaF<br>**§:** ScbaF:Pd,Pp/SaF:Pd,Pp:AScba<br>**Escape:** GmFOv/ScbaE |

**Incompatibilities and Reactivities:** Nitrates; strong oxidizers, alkalis & acids

**Exposure Routes, Symptoms, Target Organs (see Table 5):**
**ER:** Inh, Ing, Con
**SY:** Irrit eyes, skin, upper resp sys; head, drow, anes; in animals: narco
**TO:** Eyes, skin, resp sys, CNS

**First Aid (see Table 6):**
**Eye:** Irr immed
**Skin:** Water flush prompt
**Breath:** Resp support
**Swallow:** Medical attention immed

## Isobutyl alcohol

| Isobutyl alcohol | Formula: $(CH_3)_2CHCH_2OH$ | CAS#: 78-83-1 | RTECS#: NP9625000 | IDLH: 1600 ppm |
|---|---|---|---|---|
| **Conversion:** 1 ppm = 3.03 mg/m³ | **DOT:** 1212 129 | | | |

**Synonyms/Trade Names:** IBA, Isobutanol, Isopropylcarbinol, 2-Methyl-1-propanol

**Exposure Limits:**
**NIOSH REL:** TWA 50 ppm (150 mg/m³)
**OSHA PEL†:** TWA 100 ppm (300 mg/m³)

**Measurement Methods (see Table 1):**
**NIOSH** 1401, 1405
**OSHA** 7

**Physical Description:** Colorless, oily liquid with a sweet, musty odor.

| Chemical & Physical Properties: | Personal Protection/Sanitation (see Table 2): | Respirator Recommendations (see Tables 3 and 4): |
|---|---|---|
| **MW:** 74.1<br>**BP:** 227°F<br>**Sol:** 10%<br>**Fl.P:** 82°F<br>**IP:** 10.12 eV<br>**Sp.Gr:** 0.80<br>**VP:** 9 mmHg<br>**FRZ:** -162°F<br>**UEL(202°F):** 10.6%<br>**LEL(123°F):** 1.7%<br>Class IC Flammable Liquid | **Skin:** Prevent skin contact<br>**Eyes:** Prevent eye contact<br>**Wash skin:** When contam<br>**Remove:** When wet (flamm)<br>**Change:** N.R. | **NIOSH**<br>**500 ppm:** CcrOv*/Sa*<br>**1250 ppm:** Sa:Cf*/PaprOv*<br>**1600 ppm:** CcrFOv/GmFOv/PaprTOv*/ScbaF/SaF<br>**§:** ScbaF:Pd,Pp/SaF:Pd,Pp:AScba<br>**Escape:** GmFOv/ScbaE |

**Incompatibilities and Reactivities:** Strong oxidizers

**Exposure Routes, Symptoms, Target Organs (see Table 5):**
**ER:** Inh, Ing, Con
**SY:** Irrit eyes, skin, throat; head, drow; skin cracking; in animals: narco
**TO:** Eyes, skin, resp sys, CNS

**First Aid (see Table 6):**
**Eye:** Irr immed
**Skin:** Water flush prompt
**Breath:** Resp support
**Swallow:** Medical attention immed

## Isobutyronitrile

| Formula: | CAS#: | RTECS#: | IDLH: |
|---|---|---|---|
| $(CH_3)_2CHCN$ | 78-82-0 | TZ4900000 | N.D. |

**Conversion:** 1 ppm = 2.83 mg/$m^3$ | **DOT:** 2284 131

**Synonyms/Trade Names:** Isopropyl cyanide, 2-Methylpropanenitrile, 2-Methylpropionitrile

**Exposure Limits:**
**NIOSH REL:** TWA 8 ppm (22 mg/$m^3$)
**OSHA PEL:** none

**Measurement Methods (see Table 1):**
**NIOSH** 1606 (adapt)

**Physical Description:** Colorless liquid with an almond-like odor.
[**Note:** Forms cyanide in the body.]

**Chemical & Physical Properties:**
**MW:** 69.1
**BP:** 219°F
**Sol:** Slight
**Fl.P:** 47°F
**IP:** ?
**Sp.Gr:** 0.76
**VP(130°F):** 100 mmHg
**FRZ:** -97°F
**UEL:** ?
**LEL:** ?
Class IB Flammable Liquid

**Personal Protection/Sanitation (see Table 2):**
**Skin:** Prevent skin contact
**Eyes:** Prevent eye contact
**Wash skin:** When contam
**Remove:** When wet (flamm)
**Change:** N.R.
**Provide:** Eyewash
Quick drench

**Respirator Recommendations (see Tables 3 and 4):**
**NIOSH**
**80 ppm:** CcrOv/Sa
**200 ppm:** Sa:Cf/PaprOv
**400 ppm:** CcrFOv/GmFOv/PaprTOv/ScbaF/SaF
**1000 ppm:** SaF:Pd,Pp
**§:** ScbaF:Pd,Pp/SaF:Pd,Pp:AScba
**Escape:** GmFOv/ScbaE

**Incompatibilities and Reactivities:** Oxidizers, reducing agents, strong acids & bases

**Exposure Routes, Symptoms, Target Organs (see Table 5):**
**ER:** Inh, Abs, Ing, Con
**SY:** Irrit eyes, skin, nose, throat; head, dizz, lass, conf, convuls; dysp; abdom pain, nau, vomit
**TO:** Eyes, skin, resp sys, CNS, CVS

**First Aid (see Table 6):**
**Eye:** Irr immed
**Skin:** Soap flush immed
**Breath:** Resp support
**Swallow:** Medical attention immed

## Isooctyl alcohol

| Formula: | CAS#: | RTECS#: | IDLH: |
|---|---|---|---|
| $C_7H_{15}CH_2OH$ | 26952-21-6 | NS7700000 | N.D. |

**Conversion:** 1 ppm = 5.33 mg/$m^3$ | **DOT:**

**Synonyms/Trade Names:** Isooctanol, Oxooctyl alcohol [**Note:** A mixture of closely related isomeric, primary alcohols with branched chains such as 2-Ethylhexanol, $CH_3(CH_2)_3CH(CH_2CH_3)CH_2OH$.]

**Exposure Limits:**
**NIOSH REL:** TWA 50 ppm (270 mg/$m^3$) [skin]
**OSHA PEL†:** none

**Measurement Methods (see Table 1):**
**OSHA** PV2033

**Physical Description:** Clear, colorless liquid.

**Chemical & Physical Properties:**
**MW:** 130.3
**BP:** 367°F
**Sol:** Insoluble
**Fl.P(oc):** 180°F
**IP:** ?
**Sp.Gr:** 0.83
**VP:** 0.4 mmHg
**FRZ:** <-105°F
**UEL(est.):** 5.7%
**LEL(calc.):** 0.9%
Class IIIA Combustible Liquid

**Personal Protection/Sanitation (see Table 2):**
**Skin:** Prevent skin contact
**Eyes:** Prevent eye contact
**Wash skin:** When contam/Daily
**Remove:** When wet or contam
**Change:** N.R.
**Provide:** Eyewash

**Respirator Recommendations (see Tables 3 and 4):**
Not available.

**Incompatibilities and Reactivities:** None reported

**Exposure Routes, Symptoms, Target Organs (see Table 5):**
**ER:** Inh, Abs, Ing, Con
**SY:** Irrit eyes, skin, nose, throat; eye, skin burns
**TO:** Eyes, skin, resp sys

**First Aid (see Table 6):**
**Eye:** Irr immed
**Skin:** Soap wash immed
**Breath:** Resp support
**Swallow:** Medical attention immed

## Isophorone

| Isophorone | Formula: $C_9H_{14}O$ | CAS#: 78-59-1 | RTECS#: GW7700000 | IDLH: 200 ppm |
|---|---|---|---|---|

**Conversion:** 1 ppm = 5.65 mg/m$^3$ | **DOT:** 1993 128 (combustible liquid, n.o.s.)

**Synonyms/Trade Names:** Isoacetophorone; 3,5,5-Trimethyl-2-cyclohexenone; 3,5,5-Trimethyl-2-cyclo-hexen-1-one

**Exposure Limits:**
**NIOSH REL:** TWA 4 ppm (23 mg/m$^3$)
**OSHA PEL†:** TWA 25 ppm (140 mg/m$^3$)

**Measurement Methods (see Table 1):**
**NIOSH** 2508, 2556
**OSHA** 7

**Physical Description:** Colorless to white liquid with a peppermint-like odor.

| Chemical & Physical Properties: | Personal Protection/Sanitation (see Table 2): | Respirator Recommendations (see Tables 3 and 4): |
|---|---|---|
| **MW:** 138.2<br>**BP:** 419°F<br>**Sol:** 1%<br>**Fl.P:** 184°F<br>**IP:** 9.07 eV<br>**Sp.Gr:** 0.92<br>**VP:** 0.3 mmHg<br>**FRZ:** 17°F<br>**UEL:** 3.8%<br>**LEL:** 0.8%<br>Class IIIA Combustible Liquid | **Skin:** Prevent skin contact<br>**Eyes:** Prevent eye contact<br>**Wash skin:** When contam<br>**Remove:** When wet or contam<br>**Change:** N.R.<br>**Provide:** Eyewash | **NIOSH**<br>**40 ppm:** CcrOv*/Sa*<br>**100 ppm:** Sa:Cf*/PaprOv*<br>**200 ppm:** CcrFOv/GmFOv/PaprTOv*/ SaT:Cf*/ScbaF/SaF<br>**§:** ScbaF:Pd,Pp/SaF:Pd,Pp:AScba<br>**Escape:** GmFOv/ScbaE |

**Incompatibilities and Reactivities:** Oxidizers, strong alkalis, amines

| Exposure Routes, Symptoms, Target Organs (see Table 5): | First Aid (see Table 6): |
|---|---|
| **ER:** Inh, Ing, Con<br>**SY:** Irrit eyes, nose, throat; head, nau, dizz, lass, mal, narco; derm; in animals: kidney, liver damage<br>**TO:** Eyes, skin, resp sys, CNS, liver, kidneys | **Eye:** Irr immed<br>**Skin:** Soap wash prompt<br>**Breath:** Resp support<br>**Swallow:** Medical attention immed |

I

## Isophorone diisocyanate

| Isophorone diisocyanate | Formula: $C_{12}H_{18}N_2O_2$ | CAS#: 4098-71-9 | RTECS#: NQ9370000 | IDLH: N.D. |
|---|---|---|---|---|

**Conversion:** 1 ppm = 9.09 mg/m$^3$ | **DOT:** 2290 156

**Synonyms/Trade Names:** IPDI; 3-Isocyanatomethyl-3,5,5-trimethylcyclohexyl-isocyanate; Isophorone diamine diisocyanate

**Exposure Limits:**
**NIOSH REL:** TWA 0.005 ppm (0.045 mg/m$^3$) [skin]
ST 0.02 ppm (0.180 mg/m$^3$)
**OSHA PEL†:** none

**Measurement Methods (see Table 1):**
**NIOSH** 5525
**OSHA** PV2034

**Physical Description:** Colorless to slightly yellow liquid with a pungent odor.

| Chemical & Physical Properties: | Personal Protection/Sanitation (see Table 2): | Respirator Recommendations (see Tables 3 and 4): |
|---|---|---|
| **MW:** 222.3<br>**BP:** ?<br>**Sol:** Decomposes<br>**Fl.P:** 311°F<br>**IP:** ?<br>**Sp.Gr:** 1.06<br>**VP:** 0.0003 mmHg<br>**FRZ:** -76°F<br>**UEL:** ?<br>**LEL:** ?<br>Class IIIB Combustible Liquid | **Skin:** Prevent skin contact<br>**Eyes:** Prevent eye contact<br>**Wash skin:** When contam<br>**Remove:** When wet or contam<br>**Change:** Daily<br>**Provide:** Quick drench | **NIOSH**<br>**0.05 ppm:** Sa*<br>**0.125 ppm:** Sa:Cf*<br>**0.25 ppm:** ScbaF/SaF<br>**1 ppm:** SaF:Pd,Pp<br>**§:** ScbaF:Pd,Pp/SaF:Pd,Pp:AScba<br>**Escape:** GmFOv/ScbaE |

**Incompatibilities and Reactivities:** Water, alcohols, phenols, amines, mercaptans, amides, urethanes, ureas [**Note:** Reacts with water to form carbon dioxide.]

| Exposure Routes, Symptoms, Target Organs (see Table 5): | First Aid (see Table 6): |
|---|---|
| **ER:** Inh, Abs, Ing, Con<br>**SY:** Irrit eyes, skin, resp sys; chest tight, dysp, cough, sore throat; bron, wheez, pulm edema; possible resp sens, asthma<br>**TO:** Eyes, skin, resp sys | **Eye:** Irr immed<br>**Skin:** Water flush immed<br>**Breath:** Resp support<br>**Swallow:** Medical attention immed |

## 2-Isopropoxyethanol

**Formula:** $(CH_3)_2CHOCH_2CH_2OH$
**CAS#:** 109-59-1
**RTECS#:** KL5075000
**IDLH:** N.D.

**Conversion:**
**DOT:**

**Synonyms/Trade Names:** Ethylene glycol isopropyl ether, β-Hydroxyethyl isopropyl ether, Isopropyl Cellosolve®, Isopropyl glycol

**Exposure Limits:**
**NIOSH REL:** See Appendix D
**OSHA PEL†:** none

**Measurement Methods (see Table 1):**
None available

**Physical Description:** Colorless liquid with a mild, ethereal odor.

**Chemical & Physical Properties:**
**MW:** 104.2
**BP:** 283°F
**Sol:** Miscible
**Fl.P(oc):** 92°F
**IP:** ?
**Sp.Gr:** 0.90
**VP:** 3 mmHg
**FRZ:** ?
**UEL:** ?
**LEL:** ?
Class IC Flammable Liquid

**Personal Protection/Sanitation (see Table 2):**
**Skin:** Prevent skin contact
**Eyes:** Prevent eye contact
**Wash skin:** When contam
**Remove:** When wet (flamm)
**Change:** N.R.

**Respirator Recommendations (see Tables 3 and 4):**
Not available.

**Incompatibilities and Reactivities:** Oxidizers

**Exposure Routes, Symptoms, Target Organs (see Table 5):**
**ER:** Inh, Abs, Ing, Con
**SY:** In animals: irrit eyes, skin; hema, anemia, pulm edema
**TO:** Eyes, skin, resp sys, blood

**First Aid (see Table 6):**
**Eye:** Irr immed
**Skin:** Water wash immed
**Breath:** Resp support
**Swallow:** Medical attention immed

## Isopropyl acetate

**Formula:** $CH_3COOCH(CH_3)_2$
**CAS#:** 108-21-4
**RTECS#:** AI4930000
**IDLH:** 1800 ppm

**Conversion:** 1 ppm = 4.18 mg/m$^3$
**DOT:** 1220 129

**Synonyms/Trade Names:** Isopropyl ester of acetic acid, 1-Methylethyl ester of acetic acid, 2-Propyl acetate

**Exposure Limits:**
**NIOSH REL:** See Appendix D
**OSHA PEL†:** TWA 250 ppm (950 mg/m$^3$)

**Measurement Methods (see Table 1):**
**NIOSH** 1454, 1460
**OSHA** 7

**Physical Description:** Colorless liquid with a fruity odor.

**Chemical & Physical Properties:**
**MW:** 102.2
**BP:** 194°F
**Sol:** 3%
**Fl.P:** 36°F
**IP:** 9.95 eV
**Sp.Gr:** 0.87
**VP:** 42 mmHg
**FRZ:** -92°F
**UEL:** 8%
**LEL(100°F):** 1.8%
Class IB Flammable Liquid

**Personal Protection/Sanitation (see Table 2):**
**Skin:** Prevent skin contact
**Eyes:** Prevent eye contact
**Wash skin:** When contam
**Remove:** When wet (flamm)
**Change:** N.R.

**Respirator Recommendations (see Tables 3 and 4):**
**OSHA**
**1800 ppm:** Sa:Cf£/ScbaF/SaF
**§:** ScbaF:Pd,Pp/SaF:Pd,Pp:AScba
**Escape:** GmFOv/ScbaE

**Incompatibilities and Reactivities:** Nitrates; strong oxidizers, alkalis & acids

**Exposure Routes, Symptoms, Target Organs (see Table 5):**
**ER:** Inh, Ing, Con
**SY:** Irrit eyes, skin, nose; derm; in animals: narco
**TO:** Eyes, skin, resp sys, CNS

**First Aid (see Table 6):**
**Eye:** Irr immed
**Skin:** Water flush prompt
**Breath:** Resp support
**Swallow:** Medical attention immed

## Isopropyl alcohol

| Isopropyl alcohol | Formula: $(CH_3)_2CHOH$ | CAS#: 67-63-0 | RTECS#: NT8050000 | IDLH: 2000 ppm [10%LEL] |
|---|---|---|---|---|
| **Conversion:** 1 ppm = 2.46 mg/m$^3$ | **DOT:** 1219 129 | | | |

**Synonyms/Trade Names:** Dimethyl carbinol, IPA, Isopropanol, 2-Propanol, sec-Propyl alcohol, Rubbing alcohol

**Exposure Limits:**
**NIOSH REL:** TWA 400 ppm (980 mg/m$^3$)
ST 500 ppm (1225 mg/m$^3$)
**OSHA PEL†:** TWA 400 ppm (980 mg/m$^3$)

**Measurement Methods (see Table 1):**
**NIOSH** 1400
**OSHA** 109

**Physical Description:** Colorless liquid with the odor of rubbing alcohol.

**Chemical & Physical Properties:**
**MW:** 60.1
**BP:** 181°F
**Sol:** Miscible
**Fl.P:** 53°F
**IP:** 10.10 eV
**Sp.Gr:** 0.79
**VP:** 33 mmHg
**FRZ:** -127°F
**UEL(200°F):** 12.7%
**LEL:** 2.0%
Class IB Flammable Liquid

**Personal Protection/Sanitation (see Table 2):**
**Skin:** Prevent skin contact
**Eyes:** Prevent eye contact
**Wash skin:** When contam
**Remove:** When wet (flamm)
**Change:** N.R.

**Respirator Recommendations (see Tables 3 and 4):**
**NIOSH/OSHA**
**2000 ppm:** Sa:Cf£/CcrFOv/GmFOv/PaprOv£/ScbaF/SaF
**§:** ScbaF:Pd,Pp/SaF:Pd,Pp:AScba
**Escape:** GmFOv/ScbaE

**Incompatibilities and Reactivities:** Strong oxidizers, acetaldehyde, chlorine, ethylene oxide, acids, isocyanates

**Exposure Routes, Symptoms, Target Organs (see Table 5):**
**ER:** Inh, Ing, Con
**SY:** Irrit eyes, nose, throat; drow, dizz, head; dry cracking skin; in animals: narco
**TO:** Eyes, skin, resp sys

**First Aid (see Table 6):**
**Eye:** Irr immed
**Skin:** Water flush
**Breath:** Resp support
**Swallow:** Medical attention immed

I

## Isopropylamine

| Isopropylamine | Formula: $(CH_3)_2CHNH_2$ | CAS#: 75-31-0 | RTECS#: NT8400000 | IDLH: 750 ppm |
|---|---|---|---|---|
| **Conversion:** 1 ppm = 2.42 mg/m$^3$ | **DOT:** 1221 132 | | | |

**Synonyms/Trade Names:** 2-Aminopropane, Monoisopropylamine, 2-Propylamine, sec-Propylamine

**Exposure Limits:**
**NIOSH REL:** See Appendix D
**OSHA PEL†:** TWA 5 ppm (12 mg/m$^3$)

**Measurement Methods (see Table 1):**
**NIOSH** S147 (II-3)

**Physical Description:** Colorless liquid with an ammonia-like odor.
[**Note:** A gas above 91°F.]

**Chemical & Physical Properties:**
**MW:** 59.1
**BP:** 91°F
**Sol:** Miscible
**Fl.P(oc):** -35°F
**IP:** 8.72 eV
**Sp.Gr:** 0.69
**VP:** 460 mmHg
**FRZ:** -150°F
**UEL:** ?
**LEL:** ?
Class IA Flammable Liquid

**Personal Protection/Sanitation (see Table 2):**
**Skin:** Prevent skin contact
**Eyes:** Prevent eye contact
**Wash skin:** When contam
**Remove:** When wet (flamm)
**Change:** N.R.
**Provide:** Eyewash
Quick drench

**Respirator Recommendations (see Tables 3 and 4):**
**OSHA**
**125 ppm:** Sa:Cf£/PaprS£
**250 ppm:** CcrFS/GmFS/PaprTS£/ScbaF/SaF
**750 ppm:** SaF:Pd,Pp
**§:** ScbaF:Pd,Pp/SaF:Pd,Pp:AScba
**Escape:** GmFS/ScbaE

**Incompatibilities and Reactivities:** Strong acids, strong oxidizers, aldehydes, ketones, epoxides

**Exposure Routes, Symptoms, Target Organs (see Table 5):**
**ER:** Inh, Abs, Ing, Con
**SY:** Irrit eyes, skin, nose, throat; pulm edema; vis dist; eye, skin burns; derm
**TO:** Eyes, skin, resp sys

**First Aid (see Table 6):**
**Eye:** Irr immed
**Skin:** Water flush immed
**Breath:** Resp support
**Swallow:** Medical attention immed

## N-Isopropylaniline

| Formula: | CAS#: | RTECS#: | IDLH: |
|---|---|---|---|
| $C_6H_5NHCH(CH_3)_2$ | 768-52-5 | BY4190000 | N.D. |

**Conversion:** 1 ppm = 5.53 mg/m$^3$

**DOT:**

**Synonyms/Trade Names:** N-IPA, Isopropylaniline, N-(1-Methylethyl)-benzenamine, N-Phenylisopropylamine

**Exposure Limits:**
**NIOSH REL:** TWA 2 ppm (10 mg/m$^3$) [skin]
**OSHA PEL†:** none

**Measurement Methods (see Table 1):**
**OSHA** 78

**Physical Description:** Clear, yellowish liquid with a sweet, aromatic odor.

**Chemical & Physical Properties:**
**MW:** 135.2
**BP:** 397°F
**Sol:** ?
**Fl.P(oc):** 190°F
**IP:** ?
**Sp.Gr(60°F):** 0.93
**VP(77°F):** 0.03 mmHg
**FRZ:** -58°F
**UEL:** ?
**LEL:** ?
Class IIIB Combustible Liquid

**Personal Protection/Sanitation (see Table 2):**
**Skin:** Prevent skin contact
**Eyes:** Prevent eye contact
**Wash skin:** When contam
**Remove:** When wet or contam
**Change:** N.R.
**Provide:** Quick drench

**Respirator Recommendations (see Tables 3 and 4):**
Not available.

**Incompatibilities and Reactivities:** None reported

**Exposure Routes, Symptoms, Target Organs (see Table 5):**
**ER:** Inh, Abs, Ing, Con
**SY:** Irrit eyes, skin; head, lass, dizz; cyan; ataxia; dysp on effort; tacar; methemo
**TO:** Eyes, skin, resp sys, blood, CVS, liver, kidneys

**First Aid (see Table 6):**
**Eye:** Irr immed
**Skin:** Soap wash prompt
**Breath:** Resp support
**Swallow:** Medical attention immed

## Isopropyl ether

| Formula: | CAS#: | RTECS#: | IDLH: |
|---|---|---|---|
| $(CH_3)_2CHOCH(CH_3)_2$ | 108-20-3 | TZ5425000 | 1400 ppm [10%LEL] |

**Conversion:** 1 ppm = 4.18 mg/m$^3$

**DOT:** 1159 127

**Synonyms/Trade Names:** Diisopropyl ether, Diisopropyl oxide, 2-Isopropoxy propane

**Exposure Limits:**
**NIOSH REL:** TWA 500 ppm (2100 mg/m$^3$)
**OSHA PEL:** TWA 500 ppm (2100 mg/m$^3$)

**Measurement Methods (see Table 1):**
**NIOSH** 1618
**OSHA** 7

**Physical Description:** Colorless liquid with a sharp, sweet, ether-like odor.

**Chemical & Physical Properties:**
**MW:** 102.2
**BP:** 154°F
**Sol:** 0.2%
**Fl.P:** -18°F
**IP:** 9.20 eV
**Sp.Gr:** 0.73
**VP:** 119 mmHg
**FRZ:** -76°F
**UEL:** 7.9%
**LEL:** 1.4%
Class IB Flammable Liquid

**Personal Protection/Sanitation (see Table 2):**
**Skin:** Prevent skin contact
**Eyes:** Prevent eye contact
**Wash skin:** When contam
**Remove:** When wet (flamm)
**Change:** N.R.

**Respirator Recommendations (see Tables 3 and 4):**
**NIOSH/OSHA**
**1400 ppm:** CcrOv*/PaprOv*/GmFOv/Sa*/ScbaF
**§:** ScbaF:Pd,Pp/SaF:Pd,Pp:AScba
**Escape:** GmFOv/ScbaE

**Incompatibilities and Reactivities:** Strong oxidizers, acids
[**Note:** Unstable peroxides may form on long contact with air.]

**Exposure Routes, Symptoms, Target Organs (see Table 5):**
**ER:** Inh, Ing, Con
**SY:** Irrit eyes, skin, nose; resp discomfort; derm; in animals: drow, dizz, uncon, narco
**TO:** Eyes, skin, resp sys, CNS

**First Aid (see Table 6):**
**Eye:** Irr immed
**Skin:** Soap wash prompt
**Breath:** Resp support
**Swallow:** Medical attention immed

## Isopropyl glycidyl ether

| Isopropyl glycidyl ether | Formula: $C_6H_{12}O_2$ | CAS#: 4016-14-2 | RTECS#: TZ3500000 | IDLH: 400 ppm |
|---|---|---|---|---|

**Conversion:** 1 ppm = 4.75 mg/m$^3$ | **DOT:**

**Synonyms/Trade Names:** 1,2-Epoxy-3-isopropoxypropane; IGE; Isopropoxymethyl oxirane

**Exposure Limits:**
**NIOSH REL:** C 50 ppm (240 mg/m$^3$) [15-minute]
**OSHA PEL†:** TWA 50 ppm (240 mg/m$^3$)

**Measurement Methods (see Table 1):**
**NIOSH** 1620
**OSHA** 7

**Physical Description:** Colorless liquid.

**Chemical & Physical Properties:**
**MW:** 116.2
**BP:** 279°F
**Sol:** 19%
**Fl.P:** 92°F
**IP:** ?
**Sp.Gr:** 0.92
**VP(77°F):** 9 mmHg
**FRZ:** ?
**UEL:** ?
**LEL:** ?
Class IC Flammable Liquid

**Personal Protection/Sanitation (see Table 2):**
**Skin:** Prevent skin contact
**Eyes:** Prevent eye contact
**Wash skin:** When contam
**Remove:** When wet (flamm)
**Change:** N.R.

**Respirator Recommendations (see Tables 3 and 4):**
**NIOSH/OSHA**
**400 ppm:** Sa:Cf£/ScbaF
**§:** ScbaF:Pd,Pp/SaF:Pd,Pp:AScba
**Escape:** GmFOv/ScbaE

**Incompatibilities and Reactivities:** Strong oxidizers, strong caustics
[**Note:** May form explosive peroxides upon exposure to air or light.]

**Exposure Routes, Symptoms, Target Organs (see Table 5):**
**ER:** Inh, Ing, Con
**SY:** Irrit eyes, skin, upper resp sys; skin sens; possible hemato, repro effects
**TO:** Eyes, skin, resp sys, blood, repro sys

**First Aid (see Table 6):**
**Eye:** Irr immed
**Skin:** Soap wash immed
**Breath:** Resp support
**Swallow:** Medical attention immed

K

## Kaolin

| Kaolin | Formula: | CAS#: 1332-58-7 | RTECS#: GF1670500 | IDLH: N.D. |
|---|---|---|---|---|

**Conversion:** | **DOT:**

**Synonyms/Trade Names:** China clay, Clay, Hydrated aluminum silicate, Hydrite, Porcelain clay
[**Note:** Main constituent of Kaolin is Kaolinite ($Al_2Si_2O_5(OH)_4$).]

**Exposure Limits:**
**NIOSH REL:** TWA 10 mg/m$^3$ (total)
TWA 5 mg/m$^3$ (resp)
**OSHA PEL†:** TWA 15 mg/m$^3$ (total)
TWA 5 mg/m$^3$ (resp)

**Measurement Methods (see Table 1):**
**NIOSH** 0500, 0600

**Physical Description:** White to yellowish or grayish powder.
[**Note:** When moistened, darkens & develops a clay-like odor.]

**Chemical & Physical Properties:**
**MW:** varies
**BP:** ?
**Sol:** Insoluble
**Fl.P:** NA
**IP:** NA
**Sp.Gr:** 1.8-2.6
**VP:** 0 mmHg (approx)
**MLT:** ?
**UEL:** NA
**LEL:** NA
Noncombustible Solid

**Personal Protection/Sanitation (see Table 2):**
**Skin:** N.R.
**Eyes:** N.R.
**Wash skin:** N.R.
**Remove:** N.R.
**Change:** N.R.

**Respirator Recommendations (see Tables 3 and 4):**
Not available.

**Incompatibilities and Reactivities:** None reported

**Exposure Routes, Symptoms, Target Organs (see Table 5):**
**ER:** Inh, Con
**SY:** Chronic pulm fib, stomach granuloma
**TO:** Resp sys, stomach

**First Aid (see Table 6):**
**Eye:** Irr immed
**Breath:** Fresh air

K

## Kepone

| Formula: | CAS#: | RTECS#: | IDLH: |
|---|---|---|---|
| $C_{10}Cl_{10}O$ | 143-50-0 | PC8575000 | Ca [N.D.] |

**Conversion:**

**DOT:**

**Synonyms/Trade Names:** Chlordecone; Decachlorooctahydro-1,3,4-metheno-2H-cyclobuta(cd)-pentalen-2-one; Decachlorooctahydro-kepone-2-one; Decachlorotetrahydro-4,7-methanoindeneone

**Exposure Limits:**
**NIOSH REL:** Ca
TWA 0.001 mg/m$^3$
See Appendix A
**OSHA PEL:** none

**Measurement Methods (see Table 1):**
**NIOSH** 5508

**Physical Description:** Tan to white, crystalline, odorless solid. [insecticide]

**Chemical & Physical Properties:**
**MW:** 490.6
**BP:** Sublimes
**Sol(212°F):** 0.5%
**Fl.P:** NA
**IP:** ?
**Sp.Gr:** ?
**VP(77°F):** 3 x 10$^{-7}$ mmHg
**MLT:** 662°F (Sublimes)
**UEL:** NA
**LEL:** NA
Noncombustible Solid

**Personal Protection/Sanitation (see Table 2):**
**Skin:** Prevent skin contact
**Eyes:** Prevent eye contact
**Wash skin:** When contam/Daily
**Remove:** When wet or contam
**Change:** Daily
**Provide:** Eyewash
Quick drench

**Respirator Recommendations (see Tables 3 and 4):**
**NIOSH**
**¥:** ScbaF:Pd,Pp/SaF:Pd,Pp:AScba
**Escape:** GmFOv100/ScbaE

**Incompatibilities and Reactivities:** Acids, acid fumes

**Exposure Routes, Symptoms, Target Organs (see Table 5):**
**ER:** Inh, Abs, Ing, Con
**SY:** Head, anxi, tremor; liver, kidney damage; vis dist; ataxia, chest pain, skin eryt; testicular atrophy, low sperm count; [carc]
**TO:** Eyes, skin, resp sys, CNS, liver, kidneys, repro sys [in animal: liver cancer]

**First Aid (see Table 6):**
**Eye:** Irr immed
**Skin:** Soap wash immed
**Breath:** Resp support
**Swallow:** Medical attention immed

## Kerosene

| Formula: | CAS#: | RTECS#: | IDLH: |
|---|---|---|---|
| | 8008-20-6 | OA5500000 | N.D. |

**Conversion:**

**DOT:** 1223 128

**Synonyms/Trade Names:** Fuel Oil No. 1, Range oil
[**Note:** A refined petroleum solvent (predominantly $C_9$-$C_{16}$), which typically is 25% normal paraffins, 11% branched paraffins, 30% monocycloparaffins, 12% dicycloparaffins, 1% tricycloparaffins, 16% mononuclear aromatics, and 5% dinuclear aromatics.]

**Exposure Limits:**
**NIOSH REL:** TWA 100 mg/m$^3$
**OSHA PEL:** none

**Measurement Methods (see Table 1):**
**NIOSH** 1550

**Physical Description:** Colorless to yellowish, oily liquid with a strong, characteristic odor.

**Chemical & Physical Properties:**
**MW:** 170 (approx)
**BP:** 347-617°F
**Sol:** Insoluble
**Fl.P:** 100-162°F
**IP:** ?
**Sp.Gr:** 0.81
**VP(100°F):** 5 mmHg
**FRZ:** -50°F
**UEL:** 5%
**LEL:** 0.7%
Class II Combustible Liquid

**Personal Protection/Sanitation (see Table 2):**
**Skin:** Prevent skin contact
**Eyes:** Prevent eye contact
**Wash skin:** When contam
**Remove:** When wet or contam
**Change:** N.R.
**Provide:** Quick drench

**Respirator Recommendations (see Tables 3 and 4):**
**NIOSH**
**1000 mg/m$^3$:** CcrOv/Sa
**2500 mg/m$^3$:** Sa:Cf/PaprOv
**5000 mg/m$^3$:** CcrFOv/GmFOv/PaprTOv/ScbaF/SaF
**§:** ScbaF:Pd,Pp/SaF:Pd,Pp:AScba
**Escape:** GmFOv/ScbaE

**Incompatibilities and Reactivities:** Strong oxidizers

**Exposure Routes, Symptoms, Target Organs (see Table 5):**
**ER:** Inh, Ing, Con
**SY:** Irrit eyes, skin, nose, throat; burning sensation in chest; head, nau, lass, restless, inco, conf, drow; vomit, diarr; derm; chemical pneu (aspir liquid)
**TO:** Eyes, skin, resp sys, CNS

**First Aid (see Table 6):**
**Eye:** Irr immed
**Skin:** Soap flush immed
**Breath:** Resp support
**Swallow:** Medical attention immed

## Ketene

| Ketene | Formula: $CH_2=CO$ | CAS#: 463-51-4 | RTECS#: OA7700000 | IDLH: 5 ppm |
|---|---|---|---|---|
| **Conversion:** 1 ppm = 1.72 mg/m$^3$ | **DOT:** | | | |

**Synonyms/Trade Names:** Carbomethene, Ethenone, Keto-ethylene

**Exposure Limits:**
**NIOSH REL:** TWA 0.5 ppm (0.9 mg/m$^3$)
ST 1.5 ppm (3 mg/m$^3$)
**OSHA PEL†:** TWA 0.5 ppm (0.9 mg/m$^3$)

**Measurement Methods (see Table 1):**
**NIOSH** S92 (II-2)

**Physical Description:** Colorless gas with a penetrating odor.

**Chemical & Physical Properties:**
**MW:** 42.0
**BP:** -69°F
**Sol:** Reacts
**Fl.P:** NA (Gas)
**IP:** 9.61 eV
**RGasD:** 1.45
**VP:** >1 atm
**FRZ:** -238°F
**UEL:** ?
**LEL:** ?
Flammable Gas

**Personal Protection/Sanitation (see Table 2):**
**Skin:** N.R.
**Eyes:** N.R.
**Wash skin:** N.R.
**Remove:** N.R.
**Change:** N.R.

**Respirator Recommendations (see Tables 3 and 4):**
**NIOSH/OSHA**
**5 ppm:** Sa*/ScbaF
**§:** ScbaF:Pd,Pp/SaF:Pd,Pp:AScba
**Escape:** GmFOv/ScbaE

**Incompatibilities and Reactivities:** Water, alcohols, ammonia
[**Note:** Readily polymerizes. Reacts with water to form acetic acid.]

**Exposure Routes, Symptoms, Target Organs (see Table 5):**
**ER:** Inh, Con
**SY:** Irrit eyes, skin, nose, throat, resp sys; pulm edema
**TO:** Eyes, skin, resp sys

**First Aid (see Table 6):**
**Breath:** Resp support

L

## Lead

| Lead | Formula: Pb | CAS#: 7439-92-1 | RTECS#: OF7525000 | IDLH: 100 mg/m$^3$ (as Pb) |
|---|---|---|---|---|
| **Conversion:** | **DOT:** | | | |

**Synonyms/Trade Names:** Lead metal, Plumbum

**Exposure Limits:**
**NIOSH REL*:** TWA 0.050 mg/m$^3$
See Appendix C
**OSHA PEL*:** [1910.1025] TWA 0.050 mg/m$^3$
See Appendix C
[***Note:** The REL and PEL also apply to other lead compounds (as Pb) -- see Appendix C.]

**Measurement Methods (see Table 1):**
**NIOSH** 7082, 7105, 7300, 7301, 7303, 7700, 7701, 7702, 9102, 9105
**OSHA** ID121, ID125G, ID206

**Physical Description:** A heavy, ductile, soft, gray solid.

**Chemical & Physical Properties:**
**MW:** 207.2
**BP:** 3164°F
**Sol:** Insoluble
**Fl.P:** NA
**IP:** NA
**Sp.Gr:** 11.34
**VP:** 0 mmHg (approx)
**MLT:** 621°F
**UEL:** NA
**LEL:** NA
Noncombustible Solid in bulk form.

**Personal Protection/Sanitation (see Table 2):**
**Skin:** Prevent skin contact
**Eyes:** Prevent eye contact
**Wash skin:** Daily
**Remove:** When wet or contam
**Change:** Daily

**Respirator Recommendations (see Tables 3 and 4):**
**NIOSH/OSHA**
**0.5 mg/m$^3$:** 100XQ/Sa
**1.25 mg/m$^3$:** Sa:Cf/PaprHie
**2.5 mg/m$^3$:** 100F/SaT:Cf/PaprTHie/ScbaF/SaF
**50 mg/m$^3$:** Sa:Pd,Pp
**100 mg/m$^3$:** SaF:Pd,Pp
**§:** ScbaF:Pd,Pp/SaF:Pd,Pp:AScba
**Escape:** 100F/ScbaE

**See Appendix E** (page 351)

**Incompatibilities and Reactivities:** Strong oxidizers, hydrogen peroxide, acids

**Exposure Routes, Symptoms, Target Organs (see Table 5):**
**ER:** Inh, Ing, Con
**SY:** Lass, insom; facial pallor; anor, low-wgt, malnut; constip, abdom pain, colic; anemia; gingival lead line; tremor; para wrist, ankles; encephalopathy; kidney disease; irrit eyes; hypotension
**TO:** Eyes, GI tract, CNS, kidneys, blood, gingival tissue

**First Aid (see Table 6):**
**Eye:** Irr immed
**Skin:** Soap flush prompt
**Breath:** Resp support
**Swallow:** Medical attention immed

L

## Limestone

| | | | | |
|---|---|---|---|---|
| **Limestone** | **Formula:** $CaCO_3$ | **CAS#:** 1317-65-3 | **RTECS#:** | **IDLH:** N.D. |
| **Conversion:** | **DOT:** | | | |

**Synonyms/Trade Names:** Calcium carbonate, Natural calcium carbonate
[**Note:** Calcite & aragonite are commercially important natural calcium carbonates.]

**Exposure Limits:**
**NIOSH REL:** TWA 10 mg/m$^3$ (total)
TWA 5 mg/m$^3$ (resp)
**OSHA PEL:** TWA 15 mg/m$^3$ (total)
TWA 5 mg/m$^3$ (resp)

**Measurement Methods (see Table 1):**
**NIOSH** 0500, 0600

**Physical Description:** Odorless, white to tan powder.

**Chemical & Physical Properties:**
**MW:** 100.1
**BP:** Decomposes
**Sol:** 0.001%
**Fl.P:** NA
**IP:** NA
**Sp.Gr:** 2.7-2.9
**VP:** 0 mmHg (approx)
**MLT:** 1517-2442°F (Decomposes)
**UEL:** NA
**LEL:** NA
Noncombustible Solid

**Personal Protection/Sanitation (see Table 2):**
**Skin:** N.R.
**Eyes:** N.R.
**Wash skin:** N.R.
**Remove:** N.R.
**Change:** N.R.

**Respirator Recommendations (see Tables 3 and 4):**
Not available.

**Incompatibilities and Reactivities:** Fluorine, magnesium, acids, alum, ammonium salts

**Exposure Routes, Symptoms, Target Organs (see Table 5):**
**ER:** Inh, Con
**SY:** Irrit eyes, skin, muc memb; cough, sneez, rhin; lac
**TO:** Eyes, skin, resp sys

**First Aid (see Table 6):**
**Eye:** Irr immed
**Skin:** Soap wash
**Breath:** Fresh air

## Lindane

| | | | | |
|---|---|---|---|---|
| **Lindane** | **Formula:** $C_6H_6Cl_6$ | **CAS#:** 58-89-9 | **RTECS#:** GV4900000 | **IDLH:** 50 mg/m$^3$ |
| **Conversion:** | **DOT:** 2761 151 | | | |

**Synonyms/Trade Names:** BHC; HCH; γ-Hexachlorocyclohexane;
gamma isomer of 1,2,3,4,5,6-Hexachlorocyclohexane

**Exposure Limits:**
**NIOSH REL:** TWA 0.5 mg/m$^3$ [skin]
**OSHA PEL:** TWA 0.5 mg/m$^3$ [skin]

**Measurement Methods (see Table 1):**
**NIOSH** 5502

**Physical Description:** White to yellow, crystalline powder with a slight, musty odor.
[pesticide]

**Chemical & Physical Properties:**
**MW:** 290.8
**BP:** 614°F
**Sol:** 0.001%
**Fl.P:** NA
**IP:** ?
**Sp.Gr:** 1.85
**VP:** 0.00001 mmHg
**MLT:** 235°F
**UEL:** NA
**LEL:** NA
Noncombustible Solid, but may be dissolved in flammable liquids.

**Personal Protection/Sanitation (see Table 2):**
**Skin:** Prevent skin contact
**Eyes:** N.R.
**Wash skin:** When contam
**Remove:** When wet or contam
**Change:** Daily
**Provide:** Quick drench

**Respirator Recommendations (see Tables 3 and 4):**
**NIOSH/OSHA**
**5 mg/m$^3$:** CcrOv95/Sa
**12.5 mg/m$^3$:** Sa:Cf*/PaprOvHie*
**25 mg/m$^3$:** CcrFOv100/GmFOv100/
PaprTOvHie*/ScbaF/SaF
**50 mg/m$^3$:** SaF:Pd,Pp
**§:** ScbaF:Pd,Pp/SaF:Pd,Pp:AScba
**Escape:** GmFOv100/ScbaE

**Incompatibilities and Reactivities:** Corrosive to metals

**Exposure Routes, Symptoms, Target Organs (see Table 5):**
**ER:** Inh, Abs, Ing, Con
**SY:** Irrit eyes, skin, nose, throat; head; nau; clonic convuls; resp difficulty; cyan; aplastic anemia; musc spasm; in animals: liver, kidney damage
**TO:** Eyes, skin, resp sys, CNS, blood, liver, kidneys

**First Aid (see Table 6):**
**Eye:** Irr immed
**Skin:** Soap wash prompt
**Breath:** Resp support
**Swallow:** Medical attention immed

## Lithium hydride

| Lithium hydride | Formula: LiH | CAS#: 7580-67-8 | RTECS#: OJ6300000 | IDLH: 0.5 mg/m$^3$ |
|---|---|---|---|---|

**Conversion:** | **DOT:** 1414 138; 2805 138 (fused, solid)

**Synonyms/Trade Names:** Lithium monohydride

**Exposure Limits:**
**NIOSH REL:** TWA 0.025 mg/m$^3$
**OSHA PEL:** TWA 0.025 mg/m$^3$

**Measurement Methods (see Table 1):**
**OSHA** ID121

**Physical Description:** Odorless, off-white to gray, translucent, crystalline mass or white powder.

**Chemical & Physical Properties:**
**MW:** 7.95
**BP:** Decomposes
**Sol:** Reacts
**Fl.P:** NA
**IP:** NA
**Sp.Gr:** 0.78
**VP:** 0 mmHg (approx)
**MLT:** 1256°F
**UEL:** NA
**LEL:** NA
Combustible Solid that can form airborne dust clouds which may explode on contact with flame, heat, or oxidizers.

**Personal Protection/Sanitation (see Table 2):**
**Skin:** Prevent skin contact
**Eyes:** Prevent eye contact
**Wash skin:** Brush (DO NOT WASH)
**Remove:** When wet or contam
**Change:** Daily
**Provide:** Eyewash
Quick drench (>0.5 mg/m$^3$)

**Respirator Recommendations (see Tables 3 and 4):**
**NIOSH/OSHA**
**0.25 mg/m$^3$:** 100XQ/Sa
**0.5 mg/m$^3$:** Sa:Cf*/100F/PaprHie*/ScbaF/SaF
**§:** ScbaF:Pd,Pp/SaF:Pd,Pp:AScba
**Escape:** 100F/ScbaE

**Incompatibilities and Reactivities:** Strong oxidizers, halogenated hydrocarbons, acids, water
[**Note:** May ignite SPONTANEOUSLY in air and may reignite after fire is extinguished. Reacts with water to form hydrogen & lithium hydroxide.]

**Exposure Routes, Symptoms, Target Organs (see Table 5):**
**ER:** Inh, Ing, Con
**SY:** Irrit eyes, skin; eye, skin burns; mouth, esophagus burns (if ingested); nau; musc twitches; mental conf; blurred vision
**TO:** Eyes, skin, resp sys, CNS

**First Aid (see Table 6):**
**Eye:** Irr immed
**Skin:** Brush (DO NOT WASH)
**Breath:** Resp support
**Swallow:** Medical attention immed

## L.P.G.

| L.P.G. | Formula: $C_3H_8/C_3H_6/C_4H_{10}/C_4H_8$ | CAS#: 68476-85-7 | RTECS#: SE7545000 | IDLH: 2000 ppm [10%LEL] |
|---|---|---|---|---|

**Conversion:** 1 ppm = 1.72-2.37 mg/m$^3$ | **DOT:** 1075 115

**Synonyms/Trade Names:** Bottled gas, Compressed petroleum gas, Liquefied hydrocarbon gas, Liquefied petroleum gas, LPG [**Note:** A fuel mixture of propane, propylene, butanes & butylenes.]

**Exposure Limits:**
**NIOSH REL:** TWA 1000 ppm (1800 mg/m$^3$)
**OSHA PEL:** TWA 1000 ppm (1800 mg/m$^3$)

**Measurement Methods (see Table 1):**
**NIOSH** S93 (II-2)

**Physical Description:** Colorless, noncorrosive, odorless gas when pure.
[**Note:** A foul-smelling odorant is usually added. Shipped as a liquefied compressed gas.]

**Chemical & Physical Properties:**
**MW:** 42-58
**BP:** >-44°F
**Sol:** Insoluble
**Fl.P:** NA (Gas)
**IP:** 10.95 eV
**RGasD:** 1.45-2.00
**VP:** >1 atm
**FRZ:** ?
**UEL:** 9.5% (Propane) 8.5% (Butane)
**LEL:** 2.1% (Propane) 1.9% (Butane)
Flammable Gas

**Personal Protection/Sanitation (see Table 2):**
**Skin:** Frostbite
**Eyes:** Frostbite
**Wash skin:** N.R.
**Remove:** When wet (flamm)
**Change:** N.R.
**Provide:** Frostbite wash

**Respirator Recommendations (see Tables 3 and 4):**
**NIOSH/OSHA**
**2000 ppm:** Sa/ScbaF
**§:** ScbaF:Pd,Pp/SaF:Pd,Pp:AScba
**Escape:** ScbaE

**Incompatibilities and Reactivities:** Strong oxidizers, chlorine dioxide

**Exposure Routes, Symptoms, Target Organs (see Table 5):**
**ER:** Inh, Con (liquid)
**SY:** Dizz, drow, asphy; liquid: frostbite
**TO:** Resp sys, CNS

**First Aid (see Table 6):**
**Eye:** Irr immed (liquid)
**Skin:** Water flush immed (liquid)
**Breath:** Resp support

| Magnesite | Formula: $MgCO_3$ | CAS#: 546-93-0 | RTECS#: OM2470000 | IDLH: N.D. |
|---|---|---|---|---|
| Conversion: | DOT: | | | |

**Synonyms/Trade Names:** Carbonate magnesium, Hydromagnesite, Magnesium carbonate, Magnesium(II) carbonate [**Note:** Magnesite is a naturally-occurring form of magnesium carbonate.]

**Exposure Limits:**
**NIOSH REL:** TWA 10 mg/m$^3$ (total)
TWA 5 mg/m$^3$ (resp)
**OSHA PEL:** TWA 15 mg/m$^3$ (total)
TWA 5 mg/m$^3$ (resp)

**Measurement Methods (see Table 1):**
**NIOSH** 0500, 0600

**Physical Description:** White, odorless, crystalline powder.

**Chemical & Physical Properties:**
**MW:** 84.3
**BP:** Decomposes
**Sol:** 0.01%
**Fl.P:** NA
**IP:** NA
**Sp.Gr:** 2.96
**VP:** 0 mmHg (approx)
**MLT:** 662°F (Decomposes)
**UEL:** NA
**LEL:** NA
Noncombustible Solid

**Personal Protection/Sanitation (see Table 2):**
**Skin:** N.R.
**Eyes:** N.R.
**Wash skin:** N.R.
**Remove:** N.R.
**Change:** N.R.

**Respirator Recommendations (see Tables 3 and 4):**
Not available.

**Incompatibilities and Reactivities:** Acids, formaldehyde

M

**Exposure Routes, Symptoms, Target Organs (see Table 5):**
**ER:** Inh, Con
**SY:** Irrit eyes, skin, resp sys; cough
**TO:** Eyes, skin, resp sys

**First Aid (see Table 6):**
**Eye:** Irr immed
**Breath:** Fresh air

| Magnesium oxide fume | Formula: MgO | CAS#: 1309-48-4 | RTECS#: OM3850000 | IDLH: 750 mg/m$^3$ |
|---|---|---|---|---|
| Conversion: | DOT: | | | |

**Synonyms/Trade Names:** Magnesia fume

**Exposure Limits:**
**NIOSH REL:** See Appendix D
**OSHA PEL†:** TWA 15 mg/m$^3$

**Measurement Methods (see Table 1):**
**NIOSH** 7300, 7301, 7303
**OSHA** ID121

**Physical Description:** Finely divided white particulate dispersed in air.
[**Note:** Exposure may occur when magnesium is burned, thermally cut, or welded upon.]

**Chemical & Physical Properties:**
**MW:** 40.3
**BP:** 6512°F
**Sol(86°F):** 0.009%
**Fl.P:** NA
**IP:** NA
**Sp.Gr:** 3.58
**VP:** 0 mmHg (approx)
**MLT:** 5072°F
**UEL:** NA
**LEL:** NA
Noncombustible Solid

**Personal Protection/Sanitation (see Table 2):**
**Skin:** N.R.
**Eyes:** N.R.
**Wash skin:** N.R.
**Remove:** N.R.
**Change:** N.R.

**Respirator Recommendations (see Tables 3 and 4):**
**OSHA**
**150 mg/m$^3$:** 95XQ/Sa
**375 mg/m$^3$:** Sa:Cf/PaprHie
**750 mg/m$^3$:** 100F/PaprTHie*/ScbaF/SaF
**§:** ScbaF:Pd,Pp/SaF:Pd,Pp:AScba
**Escape:** 100F/ScbaE

**Incompatibilities and Reactivities:** Chlorine trifluoride, phosphorus pentachloride

**Exposure Routes, Symptoms, Target Organs (see Table 5):**
**ER:** Inh, Con
**SY:** Irrit eyes, nose; metal fume fever: cough, chest pain, flu-like fever
**TO:** Eyes, resp sys

**First Aid (see Table 6):**
**Breath:** Resp support

## Malathion

| | | | | |
|---|---|---|---|---|
| **Malathion** | **Formula:** $C_{10}H_{19}O_6PS_2$ | **CAS#:** 121-75-5 | **RTECS#:** WM8400000 | **IDLH:** 250 mg/m$^3$ |
| **Conversion:** | **DOT:** 2783 152 | | | |

**Synonyms/Trade Names:** S-[1,2-bis(ethoxycarbonyl) ethyl]O,O-dimethyl-phosphorodithioate; Diethyl (dimethoxyphosphinothioylthio) succinate

**Exposure Limits:**
**NIOSH REL:** TWA 10 mg/m$^3$ [skin]
**OSHA PEL†:** TWA 15 mg/m$^3$ [skin]

**Measurement Methods (see Table 1):**
**NIOSH** 5600
**OSHA** 62

**Physical Description:** Deep-brown to yellow liquid with a garlic-like odor. [insecticide] [**Note:** A solid below 37°F.]

**Chemical & Physical Properties:**
**MW:** 330.4
**BP:** 140°F (Decomposes)
**Sol:** 0.02%
**Fl.P(oc):** >325°F
**IP:** ?
**Sp.Gr:** 1.21
**VP:** 0.00004 mmHg
**FRZ:** 37°F
**UEL:** ?
**LEL:** ?
Class IIIB Combustible Liquid, but may be difficult to ignite.

**Personal Protection/Sanitation (see Table 2):**
**Skin:** Prevent skin contact
**Eyes:** Prevent eye contact
**Wash skin:** When contam
**Remove:** When wet or contam
**Change:** Daily

**Respirator Recommendations (see Tables 3 and 4):**
**NIOSH**
**100 mg/m$^3$:** CcrOv95/Sa
**250 mg/m$^3$:** Sa:Cf*/CcrFOv100/GmFOv100/PaprOvHie*/ScbaF/SaF
**§:** ScbaF:Pd,Pp/SaF:Pd,Pp:AScba
**Escape:** GmFOv100/ScbaE

**Incompatibilities and Reactivities:** Strong oxidizers, magnesium, alkaline pesticides [**Note:** Corrosive to metals.]

**Exposure Routes, Symptoms, Target Organs (see Table 5):**
**ER:** Inh, Abs, Ing, Con
**SY:** Irrit eyes, skin; miosis, aching eyes, blurred vision, lac; salv; anor, nau, vomit, abdom cramps, diarr, dizz, conf, ataxia; rhin, head; chest tight, wheez, lar spasm
**TO:** Eyes, skin, resp sys, liver, blood chol, CNS, CVS, GI tract

**First Aid (see Table 6):**
**Eye:** Irr immed
**Skin:** Soap wash prompt
**Breath:** Resp support
**Swallow:** Medical attention immed

M

## Maleic anhydride

| | | | | |
|---|---|---|---|---|
| **Maleic anhydride** | **Formula:** $C_4H_2O_3$ | **CAS#:** 108-31-6 | **RTECS#:** ON3675000 | **IDLH:** 10 mg/m$^3$ |
| **Conversion:** 1 ppm = 4.01 mg/m$^3$ | **DOT:** 2215 156 | | | |

**Synonyms/Trade Names:** cis-Butenedioic anhydride; 2,5-Furanedione; Maleic acid anhydride; Toxilic anhydride

**Exposure Limits:**
**NIOSH REL:** TWA 1 mg/m$^3$ (0.25 ppm)
**OSHA PEL:** TWA 1 mg/m$^3$ (0.25 ppm)

**Measurement Methods (see Table 1):**
**NIOSH** 3512
**OSHA** 25, 86

**Physical Description:** Colorless needles, white lumps, or pellets with an irritating, choking odor.

**Chemical & Physical Properties:**
**MW:** 98.1
**BP:** 396°F
**Sol:** Reacts
**Fl.P:** 218°F
**IP:** 9.90 eV
**Sp.Gr:** 1.48
**VP:** 0.2 mmHg
**MLT:** 127°F
**UEL:** 7.1%
**LEL:** 1.4%
Combustible Solid, but may be difficult to ignite.

**Personal Protection/Sanitation (see Table 2):**
**Skin:** Prevent skin contact
**Eyes:** Prevent eye contact
**Wash skin:** When contam
**Remove:** When wet or contam
**Change:** N.R.
**Provide:** Eyewash

**Respirator Recommendations (see Tables 3 and 4):**
**NIOSH/OSHA**
**10 mg/m$^3$:** Sa:Cf£/ScbaF/SaF
**§:** ScbaF:Pd,Pp/SaF:Pd,Pp:AScba
**Escape:** GmFOv100/ScbaE

**Incompatibilities and Reactivities:** Strong oxidizers, water, alkalis, metals, caustics, and amines above 150°F
[**Note:** Reacts slowly with water (hydrolyzes) to form maleic acid.]

**Exposure Routes, Symptoms, Target Organs (see Table 5):**
**ER:** Inh, Ing, Con
**SY:** Irrit nose, upper resp sys; conj; photo, double vision; bronchial asthma; derm
**TO:** Eyes, skin, resp sys

**First Aid (see Table 6):**
**Eye:** Irr immed
**Skin:** Soap wash immed
**Breath:** Resp support
**Swallow:** Medical attention immed

## Malonaldehyde

| Malonaldehyde | Formula: $CHOCH_2CHO$ | CAS#: 542-78-9 | RTECS#: TX6475000 | IDLH: Ca [N.D.] |
|---|---|---|---|---|
| Conversion: | DOT: | | | |

**Synonyms/Trade Names:** Malonic aldehyde; Malonodialdehyde; Propanedial; 1,3-Propanedial
[**Note:** Pure Malonaldehyde is unstable and may be used as its sodium salt.]

**Exposure Limits:**
**NIOSH REL:** Ca
See Appendix A
See Appendix C (Aldehydes)
**OSHA PEL:** none

**Measurement Methods (see Table 1):**
None available

**Physical Description:** Solid (needles).

**Chemical & Physical Properties:**
**MW:** 72.1
**BP:** ?
**Sol:** ?
**Fl.P:** ?
**IP:** ?
**Sp.Gr:** ?
**VP:** ?
**MLT:** 161°F
**UEL:** ?
**LEL:** ?

**Personal Protection/Sanitation (see Table 2):**
**Skin:** Prevent skin contact
**Eyes:** Prevent eye contact
**Wash skin:** When contam/Daily
**Remove:** When wet or contam
**Change:** Daily
**Provide:** Eyewash
Quick drench

**Respirator Recommendations (see Tables 3 and 4):**
**NIOSH**
**¥:** ScbaF:Pd,Pp/SaF:Pd,Pp:AScba
**Escape:** GmFOv/ScbaE

**Incompatibilities and Reactivities:** Proteins
[**Note:** Pure compound is stable under neutral conditions, but not under acidic conditions.]

**Exposure Routes, Symptoms, Target Organs (see Table 5):**
**ER:** Inh, Abs, Ing, Con
**SY:** Irrit eyes, skin, resp sys; CNS depres; [carc]
**TO:** Eyes, skin, resp sys, CNS [in animals: thyroid gland tumors]

**First Aid (see Table 6):**
**Eye:** Irr immed
**Skin:** Water flush immed
**Breath:** Resp support
**Swallow:** Medical attention immed

M

## Malononitrile

| Malononitrile | Formula: $NCCH_2CN$ | CAS#: 109-77-3 | RTECS#: OO3150000 | IDLH: N.D. |
|---|---|---|---|---|
| Conversion: 1 ppm = 2.70 $mg/m^3$ | DOT: 2647 153 | | | |

**Synonyms/Trade Names:** Cyanoacetonitrile, Dicyanomethane, Malonic dinitrile

**Exposure Limits:**
**NIOSH REL:** TWA 3 ppm (8 $mg/m^3$)
**OSHA PEL:** none

**Measurement Methods (see Table 1):**
**NIOSH** Nitriles Criteria Document

**Physical Description:** White powder or colorless crystals.
[**Note:** Melts above 90°F. Forms cyanide in the body.]

**Chemical & Physical Properties:**
**MW:** 66.1
**BP:** 426°F
**Sol:** 13%
**Fl.P(oc):** 266°F
**IP:** 12.88 eV
**Sp.Gr:** 1.19
**VP:** ?
**MLT:** 90°F
**UEL:** ?
**LEL:** ?
Combustible Solid

**Personal Protection/Sanitation (see Table 2):**
**Skin:** Prevent skin contact
**Eyes:** Prevent eye contact
**Wash skin:** When contam
**Remove:** When wet or contam
**Change:** Daily
**Provide:** Eyewash
Quick drench

**Respirator Recommendations (see Tables 3 and 4):**
**NIOSH**
**80 $mg/m^3$:** Sa
**200 $mg/m^3$:** Sa:Cf
**400 $mg/m^3$:** ScbaF/SaF
**667 $mg/m^3$:** SaF:Pd,Pp
**§:** ScbaF:Pd,Pp/SaF:Pd,Pp:AScba
**Escape:** GmFOv/ScbaE

**Incompatibilities and Reactivities:** Strong bases [**Note:** May polymerize violently on prolonged heating at 265°F, or in contact with strong bases at lower temperatures.]

**Exposure Routes, Symptoms, Target Organs (see Table 5):**
**ER:** Inh, Abs, Ing, Con
**SY:** Irrit eyes, skin, nose, throat; head, dizz, lass, conf, convuls; dysp; abdom pain, nau, vomit
**TO:** Eyes, skin, resp sys, CNS, CVS

**First Aid (see Table 6):**
**Eye:** Irr immed
**Skin:** Water wash immed
**Breath:** Resp support
**Swallow:** Medical attention immed

| **Manganese compounds and fume (as Mn)** | **Formula:** Mn (metal) | **CAS#:** 7439-96-5 (metal) | **RTECS#:** OO9275000 (metal) | **IDLH:** 500 mg/m$^3$ (as Mn) |
|---|---|---|---|---|
| **Conversion:** | **DOT:** | | | |

**Synonyms/Trade Names: Manganese metal:** Colloidal manganese, Manganese-55
Synonyms of other compounds vary depending upon the specific manganese compound.

**Exposure Limits:**
**NIOSH REL*:** TWA 1 mg/m$^3$
ST 3 mg/m$^3$
[***Note:** Also see specific listings for Manganese cyclopentadienyl tricarbonyl, Methyl cyclopentadienyl manganese tricarbonyl, and Manganese tetroxide.]
**OSHA PEL*:** C 5 mg/m$^3$
[***Note:** Also see specific listings for Manganese cyclopentadienyl tricarbonyl and Methyl cyclopentadienyl manganese tricarbonyl.]

**Measurement Methods (see Table 1):**
**NIOSH** 7300, 7301, 7303, 9102
**OSHA** ID121, ID125G

**Physical Description:** A lustrous, brittle, silvery solid.

**Chemical & Physical Properties:**
**MW:** 54.9
**BP:** 3564°F
**Sol:** Insoluble
**Fl.P:** NA
**IP:** NA
**Sp.Gr:** 7.20 (metal)
**VP:** 0 mmHg (approx)
**MLT:** 2271°F
**UEL:** NA
**LEL:** NA
Metal: Combustible Solid

**Personal Protection/Sanitation (see Table 2):**
**Skin:** N.R.
**Eyes:** N.R.
**Wash skin:** N.R.
**Remove:** N.R.
**Change:** N.R.

**Respirator Recommendations (see Tables 3 and 4):**
**NIOSH**
**10 mg/m$^3$:** 95XQ/Sa
**25 mg/m$^3$:** Sa:Cf/PaprHie
**50 mg/m$^3$:** 100F/SaT:Cf/PaprTHie/ScbaF/SaF
**500 mg/m$^3$:** Sa:Pd,Pp
**§:** ScbaF:Pd,Pp/SaF:Pd,Pp:AScba
**Escape:** 100F/ScbaE

**Incompatibilities and Reactivities:** Oxidizers
[**Note:** Will react with water or steam to produce hydrogen.]

**Exposure Routes, Symptoms, Target Organs (see Table 5):**
**ER:** Inh, Ing
**SY:** Parkinson's; asthenia, insom, mental conf; metal fume fever: dry throat, cough, chest tight, dysp, rales, flu-like fever; low-back pain; vomit; mal; lass; kidney damage
**TO:** Resp sys, CNS, blood, kidneys

**First Aid (see Table 6):**
**Breath:** Resp support
**Swallow:** Medical attention immed

M

| **Manganese cyclopentadienyl tricarbonyl (as Mn)** | **Formula:** $C_5H_5Mn(CO)_3$ | **CAS#:** 12079-65-1 | **RTECS#:** OO9720000 | **IDLH:** N.D. |
|---|---|---|---|---|
| **Conversion:** | **DOT:** | | | |

**Synonyms/Trade Names:** Cyclopentadienylmanganese tricarbonyl, Cyclopentadienyl tricarbonyl manganese, MCT

**Exposure Limits:**
**NIOSH REL:** TWA 0.1 mg/m$^3$ [skin]
**OSHA PEL†:** C 5 mg/m$^3$

**Measurement Methods (see Table 1):**
None available

**Physical Description:** Yellow, crystalline solid with a characteristic odor.
[**Note:** An antiknock additive for gasoline. May be found in an oil & gaseous solution.]

**Chemical & Physical Properties:**
**MW:** 204.1
**BP:** Sublimes
**Sol:** Slight
**Fl.P:** ?
**IP:** ?
**Sp.Gr:** ?
**VP:** ?
**MLT:** 167°F (Sublimes)
**UEL:** ?
**LEL:** ?
Combustible Solid

**Personal Protection/Sanitation (see Table 2):**
**Skin:** Prevent skin contact
**Eyes:** Prevent eye contact
**Wash skin:** When contam
**Remove:** When wet or contam
**Change:** Daily

**Respirator Recommendations (see Tables 3 and 4):**
Not available.

**Incompatibilities and Reactivities:** Oxygen

**Exposure Routes, Symptoms, Target Organs (see Table 5):**
**ER:** Inh, Abs, Ing, Con
**SY:** In animals: irrit skin; pulm edema; convuls; CNS, resp sys, kidney changes; decr resistance to infection
**TO:** Skin, resp sys, CNS, kidneys

**First Aid (see Table 6):**
**Eye:** Irr immed
**Skin:** Soap wash
**Breath:** Resp support
**Swallow:** Medical attention immed

## Manganese tetroxide (as Mn)

**Formula:** $Mn_3O_4$ | **CAS#:** 1317-35-7 | **RTECS#:** OP0895000 | **IDLH:** N.D.

**Conversion:**

**DOT:**

**Synonyms/Trade Names:** Manganese oxide, Manganomanganic oxide, Trimanganese tetraoxide, Trimanganese tetroxide

**Exposure Limits:**
**NIOSH REL:** See Appendix D
**OSHA PEL†:** C 5 mg/m$^3$

**Measurement Methods (see Table 1):**
**NIOSH** 7300, 7301, 7303, 9102
**OSHA** ID121, ID125G

**Physical Description:** Brownish-black powder.
[**Note:** Fumes are generated whenever manganese oxides are heated strongly in air.]

**Chemical & Physical Properties:**
**MW:** 228.8
**BP:** ?
**Sol:** Insoluble
**Fl.P:** NA
**IP:** NA
**Sp.Gr:** 4.88
**VP:** 0 mmHg (approx)
**MLT:** 2847°F
**UEL:** NA
**LEL:** NA

**Personal Protection/Sanitation (see Table 2):**
**Skin:** N.R.
**Eyes:** N.R.
**Wash skin:** N.R.
**Remove:** N.R.
**Change:** Daily

**Respirator Recommendations (see Tables 3 and 4):**
Not available.

**Incompatibilities and Reactivities:** Soluble in hydrochloric acid (liberates chlorine gas)

M

**Exposure Routes, Symptoms, Target Organs (see Table 5):**
**ER:** Inh, Ing, Con
**SY:** Asthenia, insom, mental conf; low-back pain; vomit; mal, lass; kidney damage; pneu
**TO:** Resp sys, CNS, blood, kidneys

**First Aid (see Table 6):**
**Eye:** Irr immed
**Skin:** Soap wash
**Breath:** Resp support
**Swallow:** Medical attention immed

## Marble

**Formula:** $CaCO_3$ | **CAS#:** 1317-65-3 | **RTECS#:** EV9580000 | **IDLH:** N.D.

**Conversion:**

**DOT:**

**Synonyms/Trade Names:** Calcium carbonate, Natural calcium carbonate
[**Note:** Marble is a metamorphic form of calcium carbonate.]

**Exposure Limits:**
**NIOSH REL:** TWA 10 mg/m$^3$ (total)
TWA 5 mg/m$^3$ (resp)
**OSHA PEL:** TWA 15 mg/m$^3$ (total)
TWA 5 mg/m$^3$ (resp)

**Measurement Methods (see Table 1):**
**NIOSH** 0500, 0600

**Physical Description:** Odorless, white powder.

**Chemical & Physical Properties:**
**MW:** 100.1
**BP:** Decomposes
**Sol:** 0.001%
**Fl.P:** NA
**IP:** NA
**Sp.Gr:** 2.7-2.9
**VP:** 0 mmHg (approx)
**MLT:** 1517-2442°F (Decomposes)
**UEL:** NA
**LEL:** NA
Noncombustible Solid

**Personal Protection/Sanitation (see Table 2):**
**Skin:** N.R.
**Eyes:** N.R.
**Wash skin:** N.R.
**Remove:** N.R.
**Change:** N.R.

**Respirator Recommendations (see Tables 3 and 4):**
Not available.

**Incompatibilities and Reactivities:** Fluorine, magnesium, acids, alum, ammonium salts

**Exposure Routes, Symptoms, Target Organs (see Table 5):**
**ER:** Inh, Con
**SY:** Irrit eyes, skin, muc memb, upper resp sys; cough, sneez, rhin; lac
**TO:** Eyes, skin, resp sys

**First Aid (see Table 6):**
**Eye:** Irr immed
**Skin:** Soap wash
**Breath:** Fresh air

M

| Mercury compounds [except (organo) alkyls] (as Hg) | Formula: Hg (metal) | CAS#: 7439-97-6 (metal) | RTECS#: OV4550000 (metal) | IDLH: 10 mg/m$^3$ (as Hg) |
|---|---|---|---|---|
| **Conversion:** | **DOT:** 2809 172 (metal) | | | |

**Synonyms/Trade Names: Mercury metal:** Colloidal mercury, Metallic mercury, Quicksilver
Synonyms of "other" Hg compounds vary depending upon the specific compound.

**Exposure Limits:**
**NIOSH REL:** Hg Vapor: TWA 0.05 mg/m$^3$ [skin]
Other: C 0.1 mg/m$^3$ [skin]
**OSHA PEL†:** C 0.1 mg/m$^3$

**Measurement Methods (see Table 1):**
**NIOSH** 6009 **OSHA** ID140

**Physical Description:** Metal: Silver-white, heavy, odorless liquid.
[**Note:** "Other" Hg compounds include all inorganic & aryl Hg compounds except (organo) alkyls.]

**Chemical & Physical Properties:**
**MW:** 200.6
**BP:** 674°F
**Sol:** Insoluble
**Fl.P:** NA
**IP:** ?
**Sp.Gr:** 13.6 (metal)
**VP:** 0.0012 mmHg
**FRZ:** -38°F
**UEL:** NA
**LEL:** NA
Metal: Noncombustible Liquid

**Personal Protection/Sanitation (see Table 2):**
**Skin:** Prevent skin contact
**Eyes:** N.R.
**Wash skin:** When contam
**Remove:** When wet or contam
**Change:** Daily

**Respirator Recommendations (see Tables 3 and 4):**
**Mercury vapor:**
**NIOSH**
**0.5 mg/m$^3$:** CcrS†/Sa
**1.25 mg/m$^3$:** Sa:Cf/PaprS†(canister)
**2.5 mg/m$^3$:** CcrFS†/GmFS†/SaT:Cf/PaprTS(canister)/ScbaF/SaF
**10 mg/m$^3$:** Sa:Pd,Pp
**§:** ScbaF:Pd,Pp/SaF:Pd,Pp:AScba
**Escape:** GmFS/ScbaE

**Other mercury compounds:**
**NIOSH/OSHA**
**1 mg/m$^3$:** CcrS†/Sa
**2.5 mg/m$^3$:** Sa:Cf/PaprS†(canister)
**5 mg/m$^3$:** CcrFS†/GmFS†/SaT:Cf/PaprTS(canister)/ScbaF/SaF
**10 mg/m$^3$:** Sa:Pd,Pp
**§:** ScbaF:Pd,Pp/SaF:Pd,Pp:AScba
**Escape:** GmFS/ScbaE

**Incompatibilities and Reactivities:** Acetylene, ammonia, chlorine dioxide, azides, calcium (amalgam formation), sodium carbide, lithium, rubidium, copper

**Exposure Routes, Symptoms, Target Organs (see Table 5):**
**ER:** Inh, Abs, Ing, Con
**SY:** Irrit eyes, skin; cough, chest pain, dysp, bron, pneu; tremor, insom, irrity, indecision, head, lass; stomatitis, salv; GI dist, anor, low-wgt; prot
**TO:** Eyes, skin, resp sys, CNS, kidneys

**First Aid (see Table 6):**
**Eye:** Irr immed
**Skin:** Soap wash prompt
**Breath:** Resp support
**Swallow:** Medical attention immed

| Mercury (organo) alkyl compounds (as Hg) | Formula: | CAS#: | RTECS#: | IDLH: 2 mg/m$^3$ (as Hg) |
|---|---|---|---|---|
| **Conversion:** | **DOT:** | | | |

**Synonyms/Trade Names:** Synonyms vary depending upon the specific (organo) alkyl mercury compound.

**Exposure Limits:**
**NIOSH REL:** TWA 0.01 mg/m$^3$
ST 0.03 mg/m$^3$ [skin]
**OSHA PEL†:** TWA 0.01 mg/m$^3$
C 0.04 mg/m$^3$

**Measurement Methods (see Table 1):**
None available

**Physical Description:** Appearance and odor vary depending upon the specific (organo) alkyl mercury compound.

**Chemical & Physical Properties:**
Properties vary depending upon the specific (organo) alkyl mercury compound.

**Personal Protection/Sanitation (see Table 2):**
**Skin:** Prevent skin contact
**Eyes:** Prevent eye contact
**Wash skin:** When contam
**Remove:** When wet or contam
**Change:** Daily
**Provide:** Eyewash
Quick drench

**Respirator Recommendations (see Tables 3 and 4):**
**NIOSH/OSHA**
**0.1 mg/m$^3$:** Sa
**0.25 mg/m$^3$:** Sa:Cf
**0.5 mg/m$^3$:** SaT:Cf/ScbaF/SaF
**2 mg/m$^3$:** Sa:Pd,Pp
**§:** ScbaF:Pd,Pp/SaF:Pd,Pp:AScba
**Escape:** ScbaE

**Incompatibilities and Reactivities:** Strong oxidizers such as chlorine

**Exposure Routes, Symptoms, Target Organs (see Table 5):**
**ER:** Inh, Abs, Ing, Con
**SY:** Pares; ataxia, dysarthria; vision, hearing dist; spasticity, jerking limbs; dizz; salv; lac; nau, vomit, diarr, constip; skin burns; emotional dist; kidney inj; possible terato effects
**TO:** Eyes, skin, CNS, PNS, kidneys

**First Aid (see Table 6):**
**Eye:** Irr immed
**Skin:** Soap wash immed
**Breath:** Resp support
**Swallow:** Medical attention immed

| Mesityl oxide | Formula: $(CH_3)_2C{=}CHCOCH_3$ | CAS#: 141-79-7 | RTECS#: SB4200000 | IDLH: 1400 ppm [10%LEL] |
|---|---|---|---|---|
| **Conversion:** 1 ppm = 4.02 mg/m$^3$ | **DOT:** 1229 129 | | | |

**Synonyms/Trade Names:** Isobutenyl methyl ketone, Isopropylideneacetone, Methyl isobutenyl ketone, 4-Methyl-3-penten-2-one

**Exposure Limits:**
**NIOSH REL:** TWA 10 ppm (40 mg/m$^3$)
**OSHA PEL†:** TWA 25 ppm (100 mg/m$^3$)

**Measurement Methods (see Table 1):**
**NIOSH** 1301, 2553
**OSHA** 7

**Physical Description:** Oily, colorless to light-yellow liquid with a peppermint- or honey-like odor.

**Chemical & Physical Properties:**
**MW:** 98.2
**BP:** 266°F
**Sol:** 3%
**Fl.P:** 87°F
**IP:** 9.08 eV
**Sp.Gr(59°F):** 0.86
**VP:** 9 mmHg
**FRZ:** -52°F
**UEL:** 7.2%
**LEL:** 1.4%
Class IC Flammable Liquid

**Personal Protection/Sanitation (see Table 2):**
**Skin:** Prevent skin contact
**Eyes:** Prevent eye contact
**Wash skin:** When contam
**Remove:** When wet (flamm)
**Change:** N.R.
**Provide:** Quick drench

**Respirator Recommendations (see Tables 3 and 4):**
**NIOSH**
**250 ppm:** Sa:Cf£/PaprOv£
**500 ppm:** CcrFOv/GmFOv/PaprTOv£/ScbaF/SaF
**1400 ppm:** SaF:Pd,Pp
**§:** ScbaF:Pd,Pp/SaF:Pd,Pp:AScba
**Escape:** GmFOv/ScbaE

**Incompatibilities and Reactivities:** Oxidizers, acids

**Exposure Routes, Symptoms, Target Organs (see Table 5):**
**ER:** Inh, Ing, Con
**SY:** Irrit eyes, skin, muc memb; narco, coma; in animals: liver, kidney damage; CNS effects
**TO:** Eyes, skin, resp sys, CNS, liver, kidneys

**First Aid (see Table 6):**
**Eye:** Irr immed
**Skin:** Water flush immed
**Breath:** Resp support
**Swallow:** Medical attention immed

| Methacrylic acid | Formula: $CH_2{=}C(CH_3)COOH$ | CAS#: 79-41-4 | RTECS#: OZ2975000 | IDLH: N.D. |
|---|---|---|---|---|
| **Conversion:** 1 ppm = 3.52 mg/m$^3$ | **DOT:** 2531 153P (inhibited) | | | |

**Synonyms/Trade Names:** Methacrylic acid (glacial), Methacrylic acid (inhibited), α-Methacrylic acid, 2-Methylacrylic acid, 2-Methylpropenoic acid

**Exposure Limits:**
**NIOSH REL:** TWA 20 ppm (70 mg/m$^3$) [skin]
**OSHA PEL†:** none

**Measurement Methods (see Table 1):**
**OSHA** PV2005

**Physical Description:** Colorless liquid or solid (below 61°F) with an acrid, repulsive odor.

**Chemical & Physical Properties:**
**MW:** 86.1
**BP:** 325°F
**Sol(77°F):** 9%
**Fl.P(oc):** 171°F
**IP:** ?
**Sp.Gr:** 1.02 (Liquid)
**VP:** 0.7 mmHg
**FRZ:** 61°F
**UEL:** ?
**LEL:** ?
Class IIIA Combustible Liquid

**Personal Protection/Sanitation (see Table 2):**
**Skin:** Prevent skin contact
**Eyes:** Prevent eye contact
**Wash skin:** When contam
**Remove:** When wet or contam
**Change:** Daily
**Provide:** Eyewash
Quick drench

**Respirator Recommendations (see Tables 3 and 4):**
Not available.

**Incompatibilities and Reactivities:** Oxidizers, elevated temperatures, hydrochloric acid
[**Note:** Typically contains 100 ppm of the monomethyl ether of hydroquinone to prevent polymerization.]

**Exposure Routes, Symptoms, Target Organs (see Table 5):**
**ER:** Inh, Abs, Ing, Con
**SY:** Irrit eyes, skin, muc memb; eye, skin burns
**TO:** Eyes, skin, resp sys

**First Aid (see Table 6):**
**Eye:** Irr immed
**Skin:** Water flush immed
**Breath:** Resp support
**Swallow:** Medical attention immed

| Methomyl | Formula: $CH_3C(SCH_3)NOC(O)NHCH_3$ | CAS#: 16752-77-5 | RTECS#: AK2975000 | IDLH: N.D. |
|---|---|---|---|---|
| **Conversion:** | **DOT:** 2757 151 (carbamate pesticide, solid, toxic) | | | |

**Synonyms/Trade Names:** Lannate®, Methyl N-((methylamino)carbonyl)oxy)ethanimidothioate, S-Methyl-N-(methylcarbamoyloxy)thioacetimidate

**Exposure Limits:**
**NIOSH REL:** TWA 2.5 mg/$m^3$
**OSHA PEL†:** none

**Measurement Methods (see Table 1):**
**NIOSH** 5601

**Physical Description:** White, crystalline solid with a slight, sulfur-like odor. [insecticide]

| Chemical & Physical Properties: | Personal Protection/Sanitation (see Table 2): | Respirator Recommendations (see Tables 3 and 4): |
|---|---|---|
| **MW:** 162.2<br>**BP:** ?<br>**Sol(77°F):** 6%<br>**Fl.P:** NA<br>**IP:** ?<br>**Sp.Gr(75°F):** 1.29<br>**VP(77°F):** 0.00005 mmHg<br>**MLT:** 172°F<br>**UEL:** NA<br>**LEL:** NA<br>Noncombustible Solid, but may be dissolved in flammable liquids. | **Skin:** Prevent skin contact<br>**Eyes:** Prevent eye contact<br>**Wash skin:** When contam<br>**Remove:** When wet or contam<br>**Change:** Daily<br>**Provide:** Quick drench | Not available. |

**Incompatibilities and Reactivities:** Strong bases

| Exposure Routes, Symptoms, Target Organs (see Table 5): | First Aid (see Table 6): |
|---|---|
| **ER:** Inh, Ing, Con<br>**SY:** Irrit eyes; blurred vision, miosis; salv; abdom cramps, nau, vomit; dysp; lass, musc twitch; liver, kidney damage<br>**TO:** Eyes, resp sys, CNS, CVS, liver, kidneys, blood chol | **Eye:** Irr immed<br>**Skin:** Water flush immed<br>**Breath:** Resp support<br>**Swallow:** Medical attention immed |

M

| Methoxychlor | Formula: $(C_6H_4OCH_3)_2CHCCl_3$ | CAS#: 72-43-5 | RTECS#: KJ3675000 | IDLH: Ca [5000 mg/$m^3$] |
|---|---|---|---|---|
| **Conversion:** | **DOT:** 2761 151 (organochlorine pesticide, solid, toxic) | | | |

**Synonyms/Trade Names:** p,p'-Dimethoxydiphenyltrichloroethane; DMDT; Methoxy-DDT; 2,2-bis(p-Methoxyphenyl)-1,1,1-trichloroethane; 1,1,1-Trichloro-2,2-bis-(p-methoxyphenyl)ethane

**Exposure Limits:**
**NIOSH REL:** Ca
See Appendix A
**OSHA PEL†:** TWA 15 mg/$m^3$

**Measurement Methods (see Table 1):**
**NIOSH** S371 (II-4)
**OSHA** PV2038

**Physical Description:** Colorless to light-yellow crystals with a slight, fruity odor. [insecticide]

| Chemical & Physical Properties: | Personal Protection/Sanitation (see Table 2): | Respirator Recommendations (see Tables 3 and 4): |
|---|---|---|
| **MW:** 345.7<br>**BP:** Decomposes<br>**Sol:** 0.00001%<br>**Fl.P:** ?<br>**IP:** ?<br>**Sp.Gr:** 1.41<br>**VP:** Very low<br>**MLT:** 171°F<br>**UEL:** ?<br>**LEL:** ?<br>Combustible Solid, but difficult to burn. | **Skin:** Prevent skin contact<br>**Eyes:** N.R.<br>**Wash skin:** When contam/Daily<br>**Remove:** When wet or contam<br>**Change:** Daily | **NIOSH**<br>**¥:** ScbaF:Pd,Pp/SaF:Pd,Pp:AScba<br>**Escape:** GmFOv100/ScbaE |

**Incompatibilities and Reactivities:** Oxidizers

| Exposure Routes, Symptoms, Target Organs (see Table 5): | First Aid (see Table 6): |
|---|---|
| **ER:** Inh, Ing<br>**SY:** In animals: fasc, trembling, convuls; kidney, liver damage; [carc]<br>**TO:** CNS, liver, kidneys [in animals: liver & ovarian cancer] | **Skin:** Soap wash<br>**Breath:** Fresh air<br>**Swallow:** Medical attention immed |

| Methoxyflurane | Formula: $CHCl_2CF_2OCH_3$ | CAS#: 76-38-0 | RTECS#: KN7820000 | IDLH: N.D. |
|---|---|---|---|---|
| Conversion: 1 ppm = 6.75 mg/m³ | DOT: | | | |

**Synonyms/Trade Names:** 2,2-Dichloro-1,1-difluoroethyl methyl ether; 2,2-Dichloro-1,1-difluoro-1-methoxyethane; Methoflurane; Methoxyfluorane; Penthrane

**Exposure Limits:**
**NIOSH REL*:** C 2 ppm (13.5 mg/m³) [60-minute]
[***Note:** REL for exposure to waste anesthetic gas.]
**OSHA PEL:** none

**Measurement Methods (see Table 1):**
None available

**Physical Description:** Colorless liquid with a fruity odor. [inhalation anesthetic]

| Chemical & Physical Properties: | Personal Protection/Sanitation (see Table 2): | Respirator Recommendations (see Tables 3 and 4): |
|---|---|---|
| **MW:** 165.0<br>**BP:** 220°F<br>**Sol:** Slight<br>**Fl.P:** ?<br>**IP:** ?<br>**Sp.Gr(77°F):** 1.42<br>**VP:** 23 mmHg<br>**FRZ:** -31°F<br>**UEL:** ?<br>**LEL(176°F):** 7%<br>Combustible Liquid | **Skin:** N.R.<br>**Eyes:** Prevent eye contact<br>**Wash skin:** N.R.<br>**Remove:** When wet or contam<br>**Change:** N.R. | Not available. |

**Incompatibilities and Reactivities:** None reported

| Exposure Routes, Symptoms, Target Organs (see Table 5): | First Aid (see Table 6): |
|---|---|
| **ER:** Inh, Ing, Con<br>**SY:** Irrit eyes; CNS depres, analgesia, anes, convuls, resp depres; liver, kidney inj; in animals: repro, terato effects<br>**TO:** Eyes, CNS, liver, kidneys, repro sys | **Eye:** Irr immed<br>**Skin:** Soap wash<br>**Breath:** Resp support<br>**Swallow:** Medical attention immed |

M

| 4-Methoxyphenol | Formula: $CH_3OC_6H_4OH$ | CAS#: 150-76-5 | RTECS#: SL7700000 | IDLH: N.D. |
|---|---|---|---|---|
| Conversion: | DOT: | | | |

**Synonyms/Trade Names:** Hydroquinone monomethyl ether, p-Hydroxyanisole, Mequinol, p-Methoxyphenol, Monomethyl ether hydroquinone

**Exposure Limits:**
**NIOSH REL:** TWA 5 mg/m³
**OSHA PEL†:** none

**Measurement Methods (see Table 1):**
None available

**Physical Description:** Colorless to white, waxy solid with an odor of caramel & phenol.

| Chemical & Physical Properties: | Personal Protection/Sanitation (see Table 2): | Respirator Recommendations (see Tables 3 and 4): |
|---|---|---|
| **MW:** 124.2<br>**BP:** 469°F<br>**Sol(77°F):** 4%<br>**Fl.P(oc):** 270°F<br>**IP:** 7.50 eV<br>**Sp.Gr:** 1.55<br>**VP:** <0.01 mmHg<br>**MLT:** 135°F<br>**UEL:** ?<br>**LEL:** ?<br>Combustible Solid; under certain conditions, a dust cloud can probably explode if ignited by a spark or flame. | **Skin:** Prevent skin contact<br>**Eyes:** Prevent eye contact<br>**Wash skin:** When contam<br>**Remove:** When wet or contam<br>**Change:** Daily<br>**Provide:** Eyewash<br>Quick drench | Not available. |

**Incompatibilities and Reactivities:** Strong oxidizers, strong bases, acid chlorides, acid anhydrides

| Exposure Routes, Symptoms, Target Organs (see Table 5): | First Aid (see Table 6): |
|---|---|
| **ER:** Inh, Abs, Ing, Con<br>**SY:** Irrit eyes, skin, nose, throat, upper resp sys; eye, skin burns; CNS depres<br>**TO:** Eyes, skin, resp sys, CNS | **Eye:** Irr immed<br>**Skin:** Soap flush immed<br>**Breath:** Resp support<br>**Swallow:** Medical attention immed |

| Methyl acetate | Formula: $CH_3COOCH_3$ | CAS#: 79-20-9 | RTECS#: AI9100000 | IDLH: 3100 ppm [10%LEL] |
|---|---|---|---|---|
| **Conversion:** 1 ppm = 3.03 mg/m$^3$ | **DOT:** 1231 129 | | | |

**Synonyms/Trade Names:** Methyl ester of acetic acid, Methyl ethanoate

**Exposure Limits:**
**NIOSH REL:** TWA 200 ppm (610 mg/m$^3$)
ST 250 ppm (760 mg/m$^3$)
**OSHA PEL†:** TWA 200 ppm (610 mg/m$^3$)

**Measurement Methods (see Table 1):**
**NIOSH** 1458
**OSHA** 7

**Physical Description:** Colorless liquid with a fragrant, fruity odor.

| Chemical & Physical Properties: | Personal Protection/Sanitation (see Table 2): | Respirator Recommendations (see Tables 3 and 4): |
|---|---|---|
| **MW:** 74.1<br>**BP:** 135°F<br>**Sol:** 25%<br>**Fl.P:** 14°F<br>**IP:** 10.27 eV<br>**Sp.Gr:** 0.93<br>**VP:** 173 mmHg<br>**FRZ:** -145°F<br>**UEL:** 16%<br>**LEL:** 3.1%<br>Class IB Flammable Liquid | **Skin:** Prevent skin contact<br>**Eyes:** Prevent eye contact<br>**Wash skin:** When contam<br>**Remove:** When wet (flamm)<br>**Change:** N.R. | **NIOSH/OSHA**<br>**2000 ppm:** CcrOv*/Sa*<br>**3100 ppm:** Sa:Cf*/CcrFOv/GmFOv/PaprOv*/ScbaF/SaF<br>**§:** ScbaF:Pd,Pp/SaF:Pd,Pp:AScba<br>**Escape:** GmFOv/ScbaE |

**Incompatibilities and Reactivities:** Nitrates; strong oxidizers, alkalis & acids; water
[**Note:** Reacts slowly with water to form acetic acid & methanol.]

| Exposure Routes, Symptoms, Target Organs (see Table 5): | First Aid (see Table 6): |
|---|---|
| **ER:** Inh, Ing, Con<br>**SY:** Irrit eyes, skin, nose, throat; head, drow; optic nerve atrophy; chest tight; in animals: narco<br>**TO:** Eyes, skin, resp sys, CNS | **Eye:** Irr immed<br>**Skin:** Water flush prompt<br>**Breath:** Resp support<br>**Swallow:** Medical attention immed |

M

| Methyl acetylene | Formula: $CH_3C{\equiv}CH$ | CAS#: 74-99-7 | RTECS#: UK4250000 | IDLH: 1700 ppm [10%LEL] |
|---|---|---|---|---|
| **Conversion:** 1 ppm = 1.64 mg/m$^3$ | **DOT:** | | | |

**Synonyms/Trade Names:** Allylene, Propine, Propyne, 1-Propyne

**Exposure Limits:**
**NIOSH REL:** TWA 1000 ppm (1650 mg/m$^3$)
**OSHA PEL:** TWA 1000 ppm (1650 mg/m$^3$)

**Measurement Methods (see Table 1):**
**NIOSH** S84 (II-5)

**Physical Description:** Colorless gas with a sweet odor.
[**Note:** A fuel that is shipped as a liquefied compressed gas.]

| Chemical & Physical Properties: | Personal Protection/Sanitation (see Table 2): | Respirator Recommendations (see Tables 3 and 4): |
|---|---|---|
| **MW:** 40.1<br>**BP:** -10°F<br>**Sol:** Insoluble<br>**Fl.P:** NA (Gas)<br>**IP:** 10.36 eV<br>**RGasD:** 1.41<br>**VP:** 5.2 atm<br>**FRZ:** -153°F<br>**UEL:** ?<br>**LEL:** 1.7%<br>Flammable Gas | **Skin:** Frostbite<br>**Eyes:** Frostbite<br>**Wash skin:** N.R.<br>**Remove:** When wet (flamm)<br>**Change:** N.R.<br>**Provide:** Frostbite wash | **NIOSH/OSHA**<br>**1700 ppm:** Sa/ScbaF<br>**§:** ScbaF:Pd,Pp/SaF:Pd,Pp:AScba<br>**Escape:** GmFOv/ScbaE |

**Incompatibilities and Reactivities:** Strong oxidizers (such as chlorine), copper alloys
[**Note:** Can decompose explosively at 4.5 to 5.6 atmospheres of pressure.]

| Exposure Routes, Symptoms, Target Organs (see Table 5): | First Aid (see Table 6): |
|---|---|
| **ER:** Inh, Con (liquid)<br>**SY:** Irrit resp sys; tremor, hyperexcitability, anes; liquid: frostbite<br>**TO:** Resp sys, CNS | **Eye:** Frostbite<br>**Skin:** Frostbite<br>**Breath:** Resp support |

| Methyl acetylene-propadiene mixture | Formula: $CH_3{\equiv}CH/CH_2{=}C{=}CH_2$ | CAS#: 59355-75-8 | RTECS#: UK4920000 | IDLH: 3400 ppm [10%LEL] |
|---|---|---|---|---|
| **Conversion:** 1 ppm = 1.64 mg/m$^3$ | **DOT:** 1060 116P (stabilized) | | | |

**Synonyms/Trade Names:** MAPP gas, Methyl acetylene-allene mixture, Propadiene-methyl acetylene, Methyl acetylene-propadiene mixture (stabilized), Propyne-allene mixture, Propyne-propadiene mixture

**Exposure Limits:**
**NIOSH REL:** TWA 1000 ppm (1800 mg/m$^3$)
ST 1250 ppm (2250 mg/m$^3$)
**OSHA PEL†:** TWA 1000 ppm (1800 mg/m$^3$)

**Measurement Methods (see Table 1):**
**NIOSH** S85 (II-6)
**OSHA** 7

**Physical Description:** Colorless gas with a strong, characteristic, foul odor.
[**Note:** A fuel that is shipped as a liquefied compressed gas.]

| Chemical & Physical Properties: | Personal Protection/Sanitation (see Table 2): | Respirator Recommendations (see Tables 3 and 4): |
|---|---|---|
| **MW:** 40.1<br>**BP:** -36 to -4°F<br>**Sol:** Insoluble<br>**Fl.P:** NA (Gas)<br>**IP:** ?<br>**RGasD:** 1.48<br>**VP:** >1 atm<br>**FRZ:** -213°F<br>**UEL:** 10.8%<br>**LEL:** 3.4%<br>Flammable Gas | **Skin:** Frostbite<br>**Eyes:** Frostbite<br>**Wash skin:** N.R.<br>**Remove:** When wet (flamm)<br>**Change:** N.R.<br>**Provide:** Frostbite wash | **NIOSH/OSHA**<br>**3400 ppm:** Sa/ScbaF<br>**§:** ScbaF:Pd,Pp/SaF:Pd,Pp:AScba<br>**Escape:** GmFS/ScbaE |

**Incompatibilities and Reactivities:** Strong oxidizers, copper alloys
[**Note:** Forms explosive compounds at high pressure in contact with alloys containing more than 67% copper.]

| Exposure Routes, Symptoms, Target Organs (see Table 5): | First Aid (see Table 6): |
|---|---|
| **ER:** Inh, Con (liquid)<br>**SY:** Irrit resp sys; excitement, conf, anes; liquid: frostbite<br>**TO:** Resp sys, CNS | **Eye:** Frostbite<br>**Skin:** Frostbite<br>**Breath:** Resp support |

M

| Methyl acrylate | Formula: $CH_2{=}CHCOOCH_3$ | CAS#: 96-33-3 | RTECS#: AT2800000 | IDLH: 250 ppm |
|---|---|---|---|---|
| **Conversion:** 1 ppm = 3.52 mg/m$^3$ | **DOT:** 1919 129P (inhibited) | | | |

**Synonyms/Trade Names:** Methoxycarbonylethylene, Methyl ester of acrylic acid, Methyl propenoate

**Exposure Limits:**
**NIOSH REL:** TWA 10 ppm (35 mg/m$^3$) [skin]
**OSHA PEL:** TWA 10 ppm (35 mg/m$^3$) [skin]

**Measurement Methods (see Table 1):**
**NIOSH** 1459, 2552
**OSHA** 92

**Physical Description:** Colorless liquid with an acrid odor.

| Chemical & Physical Properties: | Personal Protection/Sanitation (see Table 2): | Respirator Recommendations (see Tables 3 and 4): |
|---|---|---|
| **MW:** 86.1<br>**BP:** 176°F<br>**Sol:** 6%<br>**Fl.P:** 27°F<br>**IP:** 9.90 eV<br>**Sp.Gr:** 0.96<br>**VP:** 65 mmHg<br>**FRZ:** -106°F<br>**UEL:** 25%<br>**LEL:** 2.8%<br>Class IB Flammable Liquid | **Skin:** Prevent skin contact<br>**Eyes:** Prevent eye contact<br>**Wash skin:** When contam<br>**Remove:** When wet (flamm)<br>**Change:** N.R.<br>**Provide:** Quick drench | **NIOSH/OSHA**<br>**100 ppm:** Sa*<br>**250 ppm:** Sa:Cf*/ScbaF/SaF<br>**§:** ScbaF:Pd,Pp/SaF:Pd,Pp:AScba<br>**Escape:** GmFOv/ScbaE |

**Incompatibilities and Reactivities:** Nitrates, oxidizers such as peroxides, strong alkalis
[**Note:** Polymerizes easily; usually contains an inhibitor such as hydroquinone.]

| Exposure Routes, Symptoms, Target Organs (see Table 5): | First Aid (see Table 6): |
|---|---|
| **ER:** Inh, Abs, Ing, Con<br>**SY:** Irrit eyes, skin, upper resp sys<br>**TO:** Eyes, skin, resp sys | **Eye:** Irr immed<br>**Skin:** Water flush immed<br>**Breath:** Resp support<br>**Swallow:** Medical attention immed |

| **Methylacrylonitrile** | **Formula:** $CH_2=C(CH_3)CN$ | **CAS#:** 126-98-7 | **RTECS#:** UD1400000 | **IDLH:** N.D. |
|---|---|---|---|---|
| **Conversion:** 1 ppm = 2.74 $mg/m^3$ | **DOT:** 3079 131P (inhibited) | | | |

**Synonyms/Trade Names:** 2-Cyanopropene-1, 2-Cyano-1-propene, Isoprene cyanide, Isopropenylnitrile, Methacrylonitrile, $\alpha$-Methylacrylonitrile, 2-Methylpropenenitrile

**Exposure Limits:**
**NIOSH REL:** TWA 1 ppm (3 $mg/m^3$) [skin]
**OSHA PEL†:** none

**Measurement Methods (see Table 1):**
None available

**Physical Description:** Colorless liquid with an odor like bitter almonds.

**Chemical & Physical Properties:**
**MW:** 67.1
**BP:** 195°F
**Sol:** 3%
**Fl.P:** 34°F
**IP:** ?
**Sp.Gr:** 0.80
**VP(77°F):** 71 mmHg
**FRZ:** -32°F
**UEL:** 6.8%
**LEL:** 2%
Class IB Flammable Liquid

**Personal Protection/Sanitation (see Table 2):**
**Skin:** Prevent skin contact
**Eyes:** Prevent eye contact
**Wash skin:** When contam
**Remove:** When wet (flamm)
**Change:** N.R.

**Respirator Recommendations (see Tables 3 and 4):**
Not available.

**Incompatibilities and Reactivities:** Strong acids, strong oxidizers, alkali, light
[**Note:** Polymerization may occur due to elevated temperature, visible light, or contact with a concentrated alkali.]

**Exposure Routes, Symptoms, Target Organs (see Table 5):**
**ER:** Inh, Abs, Ing, Con
**SY:** Irrit eyes, skin; lac; in animals: convuls, loss of motor control in hind limbs
**TO:** Eyes, skin, CNS

**First Aid (see Table 6):**
**Eye:** Irr immed
**Skin:** Soap wash immed
**Breath:** Resp support
**Swallow:** Medical attention immed



| **Methylal** | **Formula:** $CH_3OCH_2OCH_3$ | **CAS#:** 109-87-5 | **RTECS#:** PA8750000 | **IDLH:** 2200 ppm [10%LEL] |
|---|---|---|---|---|
| **Conversion:** 1 ppm = 3.11 $mg/m^3$ | **DOT:** 1234 127 | | | |

**Synonyms/Trade Names:** Dimethoxymethane, Formal, Formaldehyde dimethylacetal, Methoxymethyl methyl ether, Methylene dimethyl ether

**Exposure Limits:**
**NIOSH REL:** TWA 1000 ppm (3100 $mg/m^3$)
**OSHA PEL:** TWA 1000 ppm (3100 $mg/m^3$)

**Measurement Methods (see Table 1):**
**NIOSH** 1611

**Physical Description:** Colorless liquid with a chloroform-like odor.

**Chemical & Physical Properties:**
**MW:** 76.1
**BP:** 111°F
**Sol:** 33%
**Fl.P(oc):** -26°F
**IP:** 10.00 eV
**Sp.Gr:** 0.86
**VP:** 330 mmHg
**FRZ:** -157°F
**UEL:** 13.8%
**LEL:** 2.2%
Class IB Flammable Liquid

**Personal Protection/Sanitation (see Table 2):**
**Skin:** Prevent skin contact
**Eyes:** Prevent eye contact
**Wash skin:** When contam
**Remove:** When wet (flamm)
**Change:** N.R.

**Respirator Recommendations (see Tables 3 and 4):**
**NIOSH/OSHA**
**2200 ppm:** Sa/ScbaF
**§:** ScbaF:Pd,Pp/SaF:Pd,Pp:AScba
**Escape:** GmFOv/ScbaE

**Incompatibilities and Reactivities:** Strong oxidizers, acids

**Exposure Routes, Symptoms, Target Organs (see Table 5):**
**ER:** Inh, Ing, Con
**SY:** Irrit eyes, skin, upper resp sys; anes
**TO:** Eyes, skin, resp sys, CNS

**First Aid (see Table 6):**
**Eye:** Irr immed
**Skin:** Water flush prompt
**Breath:** Resp support
**Swallow:** Medical attention immed

## Methyl alcohol

**Formula:** $CH_3OH$ | **CAS#:** 67-56-1 | **RTECS#:** PC1400000 | **IDLH:** 6000 ppm

**Conversion:** 1 ppm = 1.31 mg/$m^3$ | **DOT:** 1230 131

**Synonyms/Trade Names:** Carbinol, Columbian spirits, Methanol, Pyroligneous spirit, Wood alcohol, Wood naphtha, Wood spirit

**Exposure Limits:**
**NIOSH REL:** TWA 200 ppm (260 mg/$m^3$)
ST 250 ppm (325 mg/$m^3$) [skin]
**OSHA PEL†:** TWA 200 ppm (260 mg/$m^3$)

**Measurement Methods (see Table 1):**
**NIOSH** 2000, 3800
**OSHA** 91

**Physical Description:** Colorless liquid with a characteristic pungent odor.

**Chemical & Physical Properties:**
**MW:** 32.1
**BP:** 147°F
**Sol:** Miscible
**Fl.P:** 52°F
**IP:** 10.84 eV
**Sp.Gr:** 0.79
**VP:** 96 mmHg
**FRZ:** -144°F
**UEL:** 36%
**LEL:** 6.0%
Class IB Flammable Liquid

**Personal Protection/Sanitation (see Table 2):**
**Skin:** Prevent skin contact
**Eyes:** Prevent eye contact
**Wash skin:** When contam
**Remove:** When wet (flamm)
**Change:** N.R.

**Respirator Recommendations (see Tables 3 and 4):**
**NIOSH/OSHA**
**2000 ppm:** Sa
**5000 ppm:** Sa:Cf
**6000 ppm:** SaT:Cf/ScbaF/SaF
**§:** ScbaF:Pd,Pp/SaF:Pd,Pp:AScba
**Escape:** ScbaE

**Incompatibilities and Reactivities:** Strong oxidizers



**Exposure Routes, Symptoms, Target Organs (see Table 5):**
**ER:** Inh, Abs, Ing, Con
**SY:** Irrit eyes, skin, upper resp sys; head, drow, dizz, nau, vomit; vis dist, optic nerve damage (blindness); derm
**TO:** Eyes, skin, resp sys, CNS, GI tract

**First Aid (see Table 6):**
**Eye:** Irr immed
**Skin:** Water flush prompt
**Breath:** Resp support
**Swallow:** Medical attention immed

## Methylamine

**Formula:** $CH_3NH_2$ | **CAS#:** 74-89-5 | **RTECS#:** PF6300000 | **IDLH:** 100 ppm

**Conversion:** 1 ppm = 1.27 mg/$m^3$ | **DOT:** 1061 118 (anhydrous); 1235 132 (aqueous)

**Synonyms/Trade Names:** Aminomethane, Methylamine (anhydrous), Methylamine (aqueous), Monomethylamine

**Exposure Limits:**
**NIOSH REL:** TWA 10 ppm (12 mg/$m^3$)
**OSHA PEL:** TWA 10 ppm (12 mg/$m^3$)

**Measurement Methods (see Table 1):**
**OSHA** 40

**Physical Description:** Colorless gas with a fish- or ammonia-like odor.
[**Note:** A liquid below 21°F. Shipped as a liquefied compressed gas.]

**Chemical & Physical Properties:**
**MW:** 31.1
**BP:** 21°F
**Sol:** Soluble
**Fl.P:** NA (Gas)
14°F (Liquid)
**IP:** 8.97 eV
**RGasD:** 1.08
**Sp.Gr:** 0.70 (Liquid at 13°F)
**VP:** 3.0 atm
**FRZ:** -136°F
**UEL:** 20.7%
**LEL:** 4.9%
Flammable Gas

**Personal Protection/Sanitation (see Table 2):**
**Skin:** Prevent skin contact (solution)
Frostbite
**Eyes:** Prevent eye contact (solution)
Frostbite
**Wash skin:** When contam (solution)
**Remove:** When wet (flamm)
**Change:** N.R.
**Provide:** Frostbite wash

**Respirator Recommendations (see Tables 3 and 4):**
**NIOSH/OSHA**
**100 ppm:** CcrFS/GmFS/PaprS£/
ScbaF/SaF
**§:** ScbaF:Pd,Pp/SaF:Pd,Pp:AScba
**Escape:** GmFS/ScbaE

**Incompatibilities and Reactivities:** Mercury, strong oxidizers, nitromethane
[**Note:** Corrosive to copper & zinc alloys, aluminum & galvanized surfaces.]

**Exposure Routes, Symptoms, Target Organs (see Table 5):**
**ER:** Inh, Abs (solution), Ing (solution), Con (solution/liquid)
**SY:** Irrit eyes, skin, resp sys; cough; skin, muc memb burns; derm; conj; liquid: frostbite
**TO:** Eyes, skin, resp sys

**First Aid (see Table 6):**
**Eye:** Irr immed (solution)/Frostbite
**Skin:** Water flush immed (solution)/Frostbite
**Breath:** Resp support
**Swallow:** Medical attention immed (solution)

## Methyl (n-amyl) ketone

| Formula: | CAS#: | RTECS#: | IDLH: |
|---|---|---|---|
| $CH_3CO[CH_2]_4CH_3$ | 110-43-0 | MJ5075000 | 800 ppm |

**Conversion:** 1 ppm = 4.67 mg/m$^3$ | **DOT:** 1110 127

**Synonyms/Trade Names:** Amyl methyl ketone, n-Amyl methyl ketone, 2-Heptanone

**Exposure Limits:**
**NIOSH REL:** TWA 100 ppm (465 mg/m$^3$)
**OSHA PEL:** TWA 100 ppm (465 mg/m$^3$)

**Measurement Methods (see Table 1):**
**NIOSH** 1301, 2553

**Physical Description:** Colorless to white liquid with a banana-like, fruity odor.

**Chemical & Physical Properties:**
**MW:** 114.2
**BP:** 305°F
**Sol:** 0.4%
**Fl.P:** 102°F
**IP:** 9.33 eV
**Sp.Gr:** 0.81
**VP:** 3 mmHg
**FRZ:** -32°F
**UEL(250°F):** 7.9%
**LEL(151°F):** 1.1%
Class II Combustible Liquid

**Personal Protection/Sanitation (see Table 2):**
**Skin:** Prevent skin contact
**Eyes:** Prevent eye contact
**Wash skin:** When contam
**Remove:** When wet or contam
**Change:** N.R.

**Respirator Recommendations (see Tables 3 and 4):**
**NIOSH/OSHA**
**800 ppm:** CcrOv*/PaprOv*/GmFOv/Sa*/ScbaF
**§:** ScbaF:Pd,Pp/SaF:Pd,Pp:AScba
**Escape:** GmFOv/ScbaE

**Incompatibilities and Reactivities:** Strong acids, alkalis & oxidizers [**Note:** Will attack some forms of plastic.]

**Exposure Routes, Symptoms, Target Organs (see Table 5):**
**ER:** Inh, Ing, Con
**SY:** Irrit eyes, skin, muc memb; head; narco, coma; derm
**TO:** Eyes, skin, resp sys, CNS, PNS

**First Aid (see Table 6):**
**Eye:** Irr immed
**Skin:** Soap wash
**Breath:** Fresh air
**Swallow:** Medical attention immed



## Methyl bromide

| Formula: | CAS#: | RTECS#: | IDLH: |
|---|---|---|---|
| $CH_3Br$ | 74-83-9 | PA4900000 | Ca [250 ppm] |

**Conversion:** 1 ppm = 3.89 mg/m$^3$ | **DOT:** 1062 123

**Synonyms/Trade Names:** Bromomethane, Monobromomethane

**Exposure Limits:**
**NIOSH REL:** Ca
See Appendix A
**OSHA PEL†:** C 20 ppm (80 mg/m$^3$) [skin]

**Measurement Methods (see Table 1):**
**NIOSH** 2520
**OSHA** PV2040

**Physical Description:** Colorless gas with a chloroform-like odor at high concentrations. [**Note:** A liquid below 38°F. Shipped as a liquefied compressed gas.]

**Chemical & Physical Properties:**
**MW:** 95.0
**BP:** 38°F
**Sol:** 2%
**Fl.P:** NA (Gas)
**IP:** 10.54 eV
**RGasD:** 3.36
**Sp.Gr:** 1.73 (Liquid at 32°F)
**VP:** 1.9 atm
**FRZ:** -137°F
**UEL:** 16.0%
**LEL:** 10%
Flammable Gas, but only in presence of a high energy ignition source.

**Personal Protection/Sanitation (see Table 2):**
**Skin:** Prevent skin contact (liquid)
**Eyes:** Prevent eye contact (liquid)
**Wash skin:** When contam (liquid)
**Remove:** When wet (flamm)
**Change:** N.R.
**Provide:** Quick drench (liquid)

**Respirator Recommendations (see Tables 3 and 4):**
**NIOSH**
**¥:** ScbaF:Pd,Pp/SaF:Pd,Pp:AScba
**Escape:** GmFOv/ScbaE

**Incompatibilities and Reactivities:** Aluminum, magnesium, strong oxidizers [**Note:** Attacks aluminum to form aluminum trimethyl, which is SPONTANEOUSLY flammable.]

**Exposure Routes, Symptoms, Target Organs (see Table 5):**
**ER:** Inh, Abs (liquid), Con (liquid)
**SY:** Irrit eyes, skin, resp sys; musc weak, inco, vis dist, dizz; nau, vomit, head; mal; hand tremor; convuls; dysp; skin vesic; liquid: frostbite; [carc]
**TO:** Eyes, skin, resp sys, CNS [in animals: lung, kidney & forestomach tumors]

**First Aid (see Table 6):**
**Eye:** Irr immed (liquid)
**Skin:** Water flush immed (liquid)
**Breath:** Resp support

| **Methyl Cellosolve®** | **Formula:** $CH_3OCH_2CH_2OH$ | **CAS#:** 109-86-4 | **RTECS#:** KL5775000 | **IDLH:** 200 ppm |
|---|---|---|---|---|
| **Conversion:** 1 ppm = 3.11 mg/$m^3$ | **DOT:** 1188 127 | | | |
| **Synonyms/Trade Names:** EGME, Ethylene glycol monomethyl ether, Glycol monomethyl ether, 2-Methoxyethanol | | | | |
| **Exposure Limits:**<br>**NIOSH REL:** TWA 0.1 ppm (0.3 mg/$m^3$) [skin]<br>**OSHA PEL:** TWA 25 ppm (80 mg/$m^3$) [skin] | | | **Measurement Methods (see Table 1):**<br>**NIOSH** 1403<br>**OSHA** 53, 79 | |
| **Physical Description:** Colorless liquid with a mild, ether-like odor. | | | | |
| **Chemical & Physical Properties:**<br>**MW:** 76.1<br>**BP:** 256°F<br>**Sol:** Miscible<br>**Fl.P:** 102°F<br>**IP:** 9.60 eV<br>**Sp.Gr:** 0.96<br>**VP:** 6 mmHg<br>**FRZ:** -121°F<br>**UEL:** 14%<br>**LEL:** 1.8%<br>Class II Combustible Liquid | **Personal Protection/Sanitation (see Table 2):**<br>**Skin:** Prevent skin contact<br>**Eyes:** Prevent eye contact<br>**Wash skin:** When contam<br>**Remove:** When wet or contam<br>**Change:** N.R.<br>**Provide:** Quick drench | | **Respirator Recommendations (see Tables 3 and 4):**<br>**NIOSH**<br>**1 ppm**: Sa*<br>**2.5 ppm:** Sa:Cf*<br>**5 ppm:** ScbaF/SaF<br>**100 ppm:** Sa:Pd,Pp*<br>**200 ppm:** SaF:Pd,Pp<br>**§:** ScbaF:Pd,Pp/SaF:Pd,Pp:AScba<br>**Escape:** GmFOv/ScbaE | |
| **Incompatibilities and Reactivities:** Strong oxidizers, caustics | | | | |
| **Exposure Routes, Symptoms, Target Organs (see Table 5):**<br>**ER:** Inh, Abs, Ing, Con<br>**SY:** Irrit eyes, nose, throat; head, drow, lass; ataxia, tremor; anemic pallor; in animals: repro, terato effects<br>**TO:** Eyes, resp sys, CNS, blood, kidneys, repro sys, hemato sys | | **First Aid (see Table 6):**<br>**Eye:** Irr immed<br>**Skin:** Water flush prompt<br>**Breath:** Resp support<br>**Swallow:** Medical attention immed | | |

M

| **Methyl Cellosolve® acetate** | **Formula:** $CH_3COOCH_2CH_2OCH_3$ | **CAS#:** 110-49-6 | **RTECS#:** KL5950000 | **IDLH:** 200 ppm |
|---|---|---|---|---|
| **Conversion:** 1 ppm = 4.83 mg/$m^3$ | **DOT:** 1189 129 | | | |
| **Synonyms/Trade Names:** EGMEA, Ethylene glycol monomethyl ether acetate, Glycol monomethyl ether acetate, 2-Methoxyethyl acetate | | | | |
| **Exposure Limits:**<br>**NIOSH REL:** TWA 0.1 ppm (0.5 mg/$m^3$ [skin]<br>**OSHA PEL:** TWA 25 ppm (120 mg/$m^3$) [skin] | | | **Measurement Methods (see Table 1):**<br>**NIOSH** 1451<br>**OSHA** 53, 79 | |
| **Physical Description:** Colorless liquid with a mild, ether-like odor. | | | | |
| **Chemical & Physical Properties:**<br>**MW:** 118.1<br>**BP:** 293°F<br>**Sol:** Miscible<br>**Fl.P:** 120°F<br>**IP:** ?<br>**Sp.Gr:** 1.01<br>**VP:** 2 mmHg<br>**FRZ:** -85°F<br>**UEL:** 8.2%<br>**LEL:** 1.7%<br>Class II Combustible Liquid | **Personal Protection/Sanitation (see Table 2):**<br>**Skin:** Prevent skin contact<br>**Eyes:** Prevent eye contact<br>**Wash skin:** When contam<br>**Remove:** When wet or contam<br>**Change:** N.R. | | **Respirator Recommendations (see Tables 3 and 4):**<br>**NIOSH**<br>**1 ppm**: Sa*<br>**2.5 ppm:** Sa:Cf*<br>**5 ppm:** ScbaF/SaF<br>**100 ppm:** Sa:Pd,Pp*<br>**200 ppm:** SaF:Pd,Pp<br>**§:** ScbaF:Pd,Pp/SaF:Pd,Pp:AScba<br>**Escape:** GmFOv/ScbaE | |
| **Incompatibilities and Reactivities:** Nitrates; strong oxidizers, alkalis & acids | | | | |
| **Exposure Routes, Symptoms, Target Organs (see Table 5):**<br>**ER:** Inh, Abs, Ing, Con<br>**SY:** Irrit eyes, nose, throat; kidney, brain damage; in animals: narco; repro, terato effects<br>**TO:** Eyes, resp sys, kidneys, brain, CNS, PNS, repro sys, hemato sys | | | **First Aid (see Table 6):**<br>**Eye:** Irr immed<br>**Skin:** Water flush prompt<br>**Breath:** Resp support<br>**Swallow:** Medical attention immed | |

## Methyl chloride

| Methyl chloride | Formula: $CH_3Cl$ | CAS#: 74-87-3 | RTECS#: PA6300000 | IDLH: Ca [2000 ppm] |
|---|---|---|---|---|
| **Conversion:** 1 ppm = 2.07 mg/m$^3$ | **DOT:** 1063 115 | | | |

**Synonyms/Trade Names:** Chloromethane, Monochloromethane

**Exposure Limits:**
**NIOSH REL:** Ca
See Appendix A
**OSHA PEL†:** TWA 100 ppm
C 200 ppm
300 ppm (5-minute maximum peak in any 3 hours)

**Measurement Methods (see Table 1):**
**NIOSH** 1001

**Physical Description:** Colorless gas with a faint, sweet odor which is not noticeable at dangerous concentrations. [**Note:** Shipped as a liquefied compressed gas.]

**Chemical & Physical Properties:**
**MW:** 50.5
**BP:** -12°F
**Sol:** 0.5%
**Fl.P:** NA (Gas)
**IP:** 11.28 eV
**RGasD:** 1.78
**VP:** 5.0 atm
**FRZ:** -144°F
**UEL:** 17.4%
**LEL:** 8.1%
Flammable Gas

**Personal Protection/Sanitation (see Table 2):**
**Skin:** Frostbite
**Eyes:** Frostbite
**Wash skin:** N.R.
**Remove:** When wet (flamm)
**Change:** N.R.
**Provide:** Frostbite wash

**Respirator Recommendations (see Tables 3 and 4):**
**NIOSH**
**¥:** ScbaF:Pd,Pp/SaF:Pd,Pp:AScba
**Escape:** ScbaE

**Incompatibilities and Reactivities:** Chemically-active metals such as potassium, powdered aluminum, zinc, and magnesium; water [**Note:** Reacts with water (hydrolyzes) to form hydrochloric acid.]

**Exposure Routes, Symptoms, Target Organs (see Table 5):**
**ER:** Inh, Con (liquid)
**SY:** Dizz, nau, vomit; vis dist, stagger, slurred speech, convuls, coma; liver, kidney damage; liquid: frostbite; repro, terato effects; [carc]
**TO:** CNS, liver, kidneys, repro sys [in animals: lung, kidney & forestomach tumors]

**First Aid (see Table 6):**
**Eye:** Frostbite
**Skin:** Frostbite
**Breath:** Resp support

M

## Methyl chloroform

| Methyl chloroform | Formula: $CH_3CCl_3$ | CAS#: 71-55-6 | RTECS#: KJ2975000 | IDLH: 700 ppm |
|---|---|---|---|---|
| **Conversion:** 1 ppm = 5.46 mg/m$^3$ | **DOT:** 2831 160 | | | |

**Synonyms/Trade Names:** Chlorothene; 1,1,1-Trichloroethane; 1,1,1-Trichloroethane (stabilized)

**Exposure Limits:**
**NIOSH REL:** C 350 ppm (1900 mg/m$^3$) [15-minute]
See Appendix C (Chloroethanes)
**OSHA PEL†:** TWA 350 ppm (1900 mg/m$^3$)

**Measurement Methods (see Table 1):**
**NIOSH** 1003

**Physical Description:** Colorless liquid with a mild, chloroform-like odor.

**Chemical & Physical Properties:**
**MW:** 133.4
**BP:** 165°F
**Sol:** 0.4%
**Fl.P:** ?
**IP:** 11.00 eV
**Sp.Gr:** 1.34
**VP:** 100 mmHg
**FRZ:** -23°F
**UEL:** 12.5%
**LEL:** 7.5%
Combustible Liquid, but burns with difficulty.

**Personal Protection/Sanitation (see Table 2):**
**Skin:** Prevent skin contact
**Eyes:** Prevent eye contact
**Wash skin:** When contam
**Remove:** When wet or contam
**Change:** N.R.

**Respirator Recommendations (see Tables 3 and 4):**
**NIOSH/OSHA**
**700 ppm:** Sa*/ScbaF
**§:** ScbaF:Pd,Pp/SaF:Pd,Pp:AScba
**Escape:** GmFOv/ScbaE

**Incompatibilities and Reactivities:** Strong caustics; strong oxidizers; chemically-active metals such as zinc, aluminum, magnesium powders, sodium & potassium; water
[**Note:** Reacts slowly with water to form hydrochloric acid.]

**Exposure Routes, Symptoms, Target Organs (see Table 5):**
**ER:** Inh, Ing, Con
**SY:** Irrit eyes, skin; head, lass, CNS depres, poor equi; derm; card arrhy; liver damage
**TO:** Eyes, skin, CNS, CVS, liver

**First Aid (see Table 6):**
**Eye:** Irr immed
**Skin:** Soap wash prompt
**Breath:** Resp support
**Swallow:** Medical attention immed

## Methyl-2-cyanoacrylate

| Methyl-2-cyanoacrylate | Formula: | CAS#: | RTECS#: | IDLH: |
|---|---|---|---|---|
| | $CH_2=C(CN)COOCH_3$ | 137-05-3 | AS7000000 | N.D. |

**Conversion:** 1 ppm = 4.54 mg/m$^3$ | **DOT:**

**Synonyms/Trade Names:** Mecrylate, Methyl cyanoacrylate, Methyl α-cyanoacrylate, Methyl ester of 2-cyanoacrylic acid

**Exposure Limits:**
**NIOSH REL:** TWA 2 ppm (8 mg/m$^3$)
ST 4 ppm (16 mg/m$^3$)
**OSHA PEL†:** none

**Measurement Methods (see Table 1):**
**OSHA** 55

**Physical Description:** Colorless liquid with a characteristic odor.

**Chemical & Physical Properties:**
**MW:** 111.1
**BP:** ?
**Sol:** 30%
**Fl.P:** 174°F
**IP:** ?
**Sp.Gr(81°F):** 1.10
**VP(77°F):** 0.2 mmHg
**FRZ:** ?
**UEL:** ?
**LEL:** ?
Class IIIA Combustible Liquid

**Personal Protection/Sanitation (see Table 2):**
**Skin:** Prevent skin contact
**Eyes:** Prevent eye contact
**Wash skin:** Daily
**Remove:** N.R.
**Change:** N.R.
**Provide:** Eyewash

**Respirator Recommendations (see Tables 3 and 4):**
Not available.

**Incompatibilities and Reactivities:** Moisture [**Note:** Contact with moisture causes rapid polymerization.]

**Exposure Routes, Symptoms, Target Organs (see Table 5):**
**ER:** Inh, Ing, Con
**SY:** Irrit eyes, skin, nose; blurred vision, lac; rhinitis
**TO:** Eyes, skin, resp sys

**First Aid (see Table 6):**
**Eye:** Irr immed
**Skin:** Water wash
**Breath:** Resp support
**Swallow:** Medical attention immed

M

## Methylcyclohexane

| Methylcyclohexane | Formula: | CAS#: | RTECS#: | IDLH: |
|---|---|---|---|---|
| | $CH_3C_6H_{11}$ | 108-87-2 | GV6125000 | 1200 ppm [LEL] |

**Conversion:** 1 ppm = 4.02 mg/m$^3$ | **DOT:** 2296 128

**Synonyms/Trade Names:** Cyclohexylmethane, Hexahydrotoluene

**Exposure Limits:**
**NIOSH REL:** TWA 400 ppm (1600 mg/m$^3$)
**OSHA PEL†:** TWA 500 ppm (2000 mg/m$^3$)

**Measurement Methods (see Table 1):**
**NIOSH** 1500
**OSHA** 7

**Physical Description:** Colorless liquid with a faint, benzene-like odor.

**Chemical & Physical Properties:**
**MW:** 98.2
**BP:** 214°F
**Sol:** Insoluble
**Fl.P:** 25°F
**IP:** 9.85 eV
**Sp.Gr:** 0.77
**VP:** 37 mmHg
**FRZ:** -196°F
**UEL:** 6.7%
**LEL:** 1.2%
Class IB Flammable Liquid

**Personal Protection/Sanitation (see Table 2):**
**Skin:** Prevent skin contact
**Eyes:** Prevent eye contact
**Wash skin:** When contam
**Remove:** When wet (flamm)
**Change:** N.R.

**Respirator Recommendations (see Tables 3 and 4):**
**NIOSH**
**1200 ppm:** Sa/ScbaF
**§:** ScbaF:Pd,Pp/SaF:Pd,Pp:AScba
**Escape:** GmFOv/ScbaE

**Incompatibilities and Reactivities:** Strong oxidizers

**Exposure Routes, Symptoms, Target Organs (see Table 5):**
**ER:** Inh, Ing, Con
**SY:** Irrit eyes, skin, nose, throat; dizz, drow; in animals: narco
**TO:** Eyes, skin, resp sys, CNS

**First Aid (see Table 6):**
**Eye:** Irr immed
**Skin:** Soap wash prompt
**Breath:** Resp support
**Swallow:** Medical attention immed

## Methylcyclohexanol

| Methylcyclohexanol | Formula: $CH_3C_6H_{10}OH$ | CAS#: 25639-42-3 | RTECS#: GW0175000 | IDLH: 500 ppm |
|---|---|---|---|---|
| **Conversion:** 1 ppm = 4.67 mg/m$^3$ | **DOT:** 2617 129 | | | |

**Synonyms/Trade Names:** Hexahydrocresol, Hexahydromethylphenol

**Exposure Limits:**
**NIOSH REL:** TWA 50 ppm (235 mg/m$^3$)
**OSHA PEL†:** TWA 100 ppm (470 mg/m$^3$)

**Measurement Methods (see Table 1):**
**NIOSH** 1404

**Physical Description:** Straw-colored liquid with a weak odor like coconut oil.

| Chemical & Physical Properties: | Personal Protection/Sanitation (see Table 2): | Respirator Recommendations (see Tables 3 and 4): |
|---|---|---|
| **MW:** 114.2<br>**BP:** 311-356°F<br>**Sol:** 4%<br>**Fl.P:** 149-158°F<br>**IP:** 9.80 eV<br>**Sp.Gr:** 0.92<br>**VP(86°F):** 2 mmHg<br>**FRZ:** -58°F<br>**UEL:** ?<br>**LEL:** ?<br>Class IIIA Combustible Liquid | **Skin:** Prevent skin contact<br>**Eyes:** Prevent eye contact<br>**Wash skin:** When contam<br>**Remove:** When wet or contam<br>**Change:** N.R. | **NIOSH**<br>**500 ppm:** Sa*/ScbaF<br>**§:** ScbaF:Pd,Pp/SaF:Pd,Pp:AScba<br>**Escape:** GmFOv/ScbaE |

**Incompatibilities and Reactivities:** Strong oxidizers

**Exposure Routes, Symptoms, Target Organs (see Table 5):**
**ER:** Inh, Abs, Ing, Con
**SY:** Irrit eyes, skin, upper resp sys; head; in animals: narco; liver, kidney damage
**TO:** Eyes, skin, resp sys, CNS, kidneys, liver

**First Aid (see Table 6):**
**Eye:** Irr immed
**Skin:** Soap wash prompt
**Breath:** Resp support
**Swallow:** Medical attention immed

M

## o-Methylcyclohexanone

| o-Methylcyclohexanone | Formula: $CH_3C_6H_9O$ | CAS#: 583-60-8 | RTECS#: GW1750000 | IDLH: 600 ppm |
|---|---|---|---|---|
| **Conversion:** 1 ppm = 4.59 mg/m$^3$ | **DOT:** 2297 128 | | | |

**Synonyms/Trade Names:** 2-Methylcyclohexanone

**Exposure Limits:**
**NIOSH REL:** TWA 50 ppm (230 mg/m$^3$) [skin]
ST 75 ppm (345 mg/m$^3$)
**OSHA PEL†:** TWA 100 ppm (460 mg/m$^3$) [skin]

**Measurement Methods (see Table 1):**
**NIOSH** 2521

**Physical Description:** Colorless liquid with a weak, peppermint-like odor.

| Chemical & Physical Properties: | Personal Protection/Sanitation (see Table 2): | Respirator Recommendations (see Tables 3 and 4): |
|---|---|---|
| **MW:** 112.2<br>**BP:** 325°F<br>**Sol:** Insoluble<br>**Fl.P:** 118°F<br>**IP:** ?<br>**Sp.Gr:** 0.93<br>**VP:** 1 mmHg<br>**FRZ:** 7°F<br>**UEL:** ?<br>**LEL:** ?<br>Class II Combustible Liquid | **Skin:** Prevent skin contact<br>**Eyes:** Prevent eye contact<br>**Wash skin:** When contam<br>**Remove:** When wet or contam<br>**Change:** N.R. | **NIOSH**<br>**500 ppm:** Sa*<br>**600 ppm:** Sa:Cf*/ScbaF/SaF<br>**§:** ScbaF:Pd,Pp/SaF:Pd,Pp:AScba<br>**Escape:** GmFOv/ScbaE |

**Incompatibilities and Reactivities:** Strong oxidizers

**Exposure Routes, Symptoms, Target Organs (see Table 5):**
**ER:** Inh, Abs, Ing, Con
**SY:** In animals: irrit eyes, muc memb; narco; derm
**TO:** Skin, resp sys, liver, kidneys, CNS

**First Aid (see Table 6):**
**Eye:** Irr immed
**Skin:** Soap wash prompt
**Breath:** Resp support
**Swallow:** Medical attention immed

## Methyl cyclopentadienyl manganese tricarbonyl (as Mn)

**Formula:** $CH_3C_5H_4Mn(CO)_3$ | **CAS#:** 12108-13-3 | **RTECS#:** OP1450000 | **IDLH:** N.D.

**Conversion:** | **DOT:**

**Synonyms/Trade Names:** CI-2, Combustion Improver-2, Manganese tricarbonylmethylcyclopentadienyl, 2-Methylcyclopentadienyl manganese tricarbonyl, MMT

**Exposure Limits:**
**NIOSH REL:** TWA 0.2 mg/m$^3$ [skin]
**OSHA PEL†:** C 5 mg/m$^3$

**Measurement Methods (see Table 1):**
None available

**Physical Description:** Yellow to dark-orange liquid with a faint, pleasant odor.
[**Note:** A solid below 36°F.]

**Chemical & Physical Properties:**
**MW:** 218.1
**BP:** 449°F
**Sol:** Insoluble
**Fl.P:** 230°F
**IP:** ?
**Sp.Gr:** 1.39
**VP(212°F):** 7 mmHg
**FRZ:** 36°F
**UEL:** ?
**LEL:** ?
Class IIIB Combustible Liquid

**Personal Protection/Sanitation (see Table 2):**
**Skin:** Prevent skin contact
**Eyes:** Prevent eye contact
**Wash skin:** When contam
**Remove:** When wet or contam
**Change:** N.R.

**Respirator Recommendations (see Tables 3 and 4):**
Not available.

**Incompatibilities and Reactivities:** Light (decomposes)

M

**Exposure Routes, Symptoms, Target Organs (see Table 5):**
**ER:** Inh, Abs, Ing, Con
**SY:** Irrit eyes; dizz, nau, head; in animals: tremor, severe clonic spasms, lass, slow respiration; liver, kidney inj
**TO:** Eyes, CNS, liver, kidneys

**First Aid (see Table 6):**
**Eye:** Irr immed
**Skin:** Soap wash immed
**Breath:** Resp support
**Swallow:** Medical attention immed

## Methyl demeton

**Formula:** $C_6H_{15}O_3PS_2$ | **CAS#:** 8022-00-2 | **RTECS#:** TG1760000 | **IDLH:** N.D.

**Conversion:** | **DOT:**

**Synonyms/Trade Names:** Demeton methyl; O,O-Dimethyl 2-ethylmercaptoethyl thiophosphate; Metasystox®; Methyl mercaptophos; Methyl systox®

**Exposure Limits:**
**NIOSH REL:** TWA 0.5 mg/m$^3$ [skin]
**OSHA PEL†:** none

**Measurement Methods (see Table 1):**
None available

**Physical Description:** Oily, colorless to pale-yellow liquid with an unpleasant odor.
[insecticide] [**Note:** Technical grade consists of 2 isomers: thiono & thiolo.]

**Chemical & Physical Properties:**
**MW:** 230.3
**BP:** Decomposes
**Sol:** 0.03-0.3%
**Fl.P:** ?
**IP:** ?
**Sp.Gr:** 1.20
**VP:** 0.0004 mmHg
**FRZ:** ?
**UEL:** ?
**LEL:** ?
Combustible Liquid

**Personal Protection/Sanitation (see Table 2):**
**Skin:** Prevent skin contact
**Eyes:** Prevent eye contact
**Wash skin:** When contam
**Remove:** When wet or contam
**Change:** Daily
**Provide:** Eyewash
Quick drench

**Respirator Recommendations (see Tables 3 and 4):**
Not available.

**Incompatibilities and Reactivities:** Strong oxidizers, alkalis, water

**Exposure Routes, Symptoms, Target Organs (see Table 5):**
**ER:** Inh, Abs, Ing, Con
**SY:** Irrit eyes, skin; ache eyes, rhin; nau, head, dizz, vomit
**TO:** Eyes, skin, resp sys, CNS, CVS, blood chol

**First Aid (see Table 6):**
**Eye:** Irr immed
**Skin:** Soap wash immed
**Breath:** Resp support
**Swallow:** Medical attention immed

## 4,4'-Methylenebis(2-chloroaniline)

**Formula:** $CH_2(C_6H_4ClNH_2)_2$
**CAS#:** 101-14-4
**RTECS#:** CY1050000
**IDLH:** Ca [N.D.]

**Conversion:**
**DOT:**

**Synonyms/Trade Names:** DACPM; 3,3'-Dichloro-4,4'-diaminodiphenylmethane; MBOCA; 4,4'-Methylenebis(o-chloro aniline); 4,4'-Methylenebis(2-chlorobenzenamine); MOCA

**Exposure Limits:**
**NIOSH REL:** Ca
TWA 0.003 mg/m³ [skin]
See Appendix A
**OSHA PEL†:** none

**Measurement Methods (see Table 1):**
**OSHA** 24, 71

**Physical Description:** Tan-colored pellets or flakes with a faint, amine-like odor.

**Chemical & Physical Properties:**
**MW:** 267.2
**BP:** ?
**Sol:** Slight
**Fl.P:** ?
**IP:** ?
**Sp.Gr:** 1.44
**VP(77°F):** 0.00001 mmHg
**MLT:** 230°F
**UEL:** ?
**LEL:** ?

**Personal Protection/Sanitation (see Table 2):**
**Skin:** Prevent skin contact
**Eyes:** Prevent eye contact
**Wash skin:** When contam/Daily
**Remove:** When wet or contam
**Change:** Daily
**Provide:** Eyewash
Quick drench

**Respirator Recommendations (see Tables 3 and 4):**
**NIOSH**
**¥:** ScbaF:Pd,Pp/SaF:Pd,Pp:AScba
**Escape:** GmFOv100/ScbaE

**Incompatibilities and Reactivities:** Chemically-active metals (e.g., potassium, sodium, magnesium, zinc)

**Exposure Routes, Symptoms, Target Organs (see Table 5):**
**ER:** Inh, Abs, Ing, Con
**SY:** Hema, cyan, nau, methemo, kidney irrit; [carc]
**TO:** Liver, blood, kidneys [in animals: liver, lung & bladder tumors]

**First Aid (see Table 6):**
**Eye:** Irr immed
**Skin:** Soap flush immed
**Breath:** Resp support
**Swallow:** Medical attention immed

M

## Methylene bis(4-cyclohexylisocyanate)

**Formula:** $CH_2[(C_6H_{10})NCO]_2$
**CAS#:** 5124-30-1
**RTECS#:** NQ9250000
**IDLH:** N.D.

**Conversion:** 1 ppm = 10.73 mg/m³
**DOT:**

**Synonyms/Trade Names:** Dicyclohexylmethane 4,4'-diisocyanate; DMDI; bis(4-Isocyanatocyclohexyl)methane; HMDI; Hydrogenated MDI; Reduced MDI; Saturated MDI

**Exposure Limits:**
**NIOSH REL:** C 0.01 ppm (0.11 mg/m³)
**OSHA PEL†:** none

**Measurement Methods (see Table 1):**
**NIOSH** 5525
**OSHA** PV2092

**Physical Description:** Clear, colorless to light-yellow liquid.

**Chemical & Physical Properties:**
**MW:** 262.4
**BP:** ?
**Sol:** Reacts
**Fl.P:** >395°F
**IP:** ?
**Sp.Gr(77°F):** 1.07
**VP(77°F):** 0.001 mmHg
**FRZ:** <14°F
**UEL:** ?
**LEL:** ?
Class IIIB Combustible Liquid

**Personal Protection/Sanitation (see Table 2):**
**Skin:** Prevent skin contact
**Eyes:** Prevent eye contact
**Wash skin:** When contam
**Remove:** When wet or contam
**Change:** N.R.
**Provide:** Quick drench

**Respirator Recommendations (see Tables 3 and 4):**
**NIOSH**
**0.1 ppm:** Sa*
**0.25 ppm:** Sa:Cf*
**0.5 ppm:** ScbaF/SaF
**1 ppm:** SaF:Pd,Pp
**§:** ScbaF:Pd,Pp/SaF:Pd,Pp:AScba
**Escape:** GmFOv/ScbaE

**Incompatibilities and Reactivities:** Water, ethanol, alcohols, amines, bases, acids, organotin catalysts [**Note:** May slowly polymerize if heated above 122°F.]

**Exposure Routes, Symptoms, Target Organs (see Table 5):**
**ER:** Inh, Ing, Con
**SY:** Irrit eyes, skin, resp sys; skin, resp sens; chest tight, dysp, cough, dry throat, wheez, pulm edema; skin blisters
**TO:** Eyes, skin, resp sys

**First Aid (see Table 6):**
**Eye:** Irr immed
**Skin:** Water flush immed
**Breath:** Resp support
**Swallow:** Medical attention immed

## Methylene bisphenyl isocyanate

| Formula: | CAS#: | RTECS#: | IDLH: |
|---|---|---|---|
| $CH_2(C_6H_4NCO)_2$ | 101-68-8 | NQ9350000 | 75 mg/$m^3$ |

**Conversion:** 1 ppm = 10.24 mg/$m^3$ | **DOT:**

**Synonyms/Trade Names:** 4,4'-Diphenylmethane diisocyanate; MDI; Methylene bis(4-phenyl isocyanate); Methylene di-p-phenylene ester of isocyanic acid

**Exposure Limits:**
**NIOSH REL:** TWA 0.05 mg/$m^3$ (0.005 ppm)
C 0.2 mg/$m^3$ (0.020 ppm) [10-minute]
**OSHA PEL:** C 0.2 mg/$m^3$ (0.02 ppm)

**Measurement Methods (see Table 1):**
**NIOSH** 5521, 5522, 5525
**OSHA** 18

**Physical Description:** White to light-yellow, odorless flakes. [**Note:** A liquid above 99°F.]

**Chemical & Physical Properties:**
**MW:** 250.3
**BP:** 597°F
**Sol:** 0.2%
**Fl.P:** 390°F
**IP:** ?
**Sp.Gr:** 1.23 (Solid at 77°F)
1.19 (Liquid at 122°F)
**VP(77°F):** 0.000005 mmHg
**MLT:** 99°F
**UEL:** ?
**LEL:** ?
Combustible Solid

**Personal Protection/Sanitation (see Table 2):**
**Skin:** Prevent skin contact
**Eyes:** Prevent eye contact
**Wash skin:** When contam
**Remove:** When wet or contam
**Change:** Daily

**Respirator Recommendations (see Tables 3 and 4):**
**NIOSH**
**0.5 mg/$m^3$:** Sa*
**1.25 mg/$m^3$:** Sa:Cf*
**2.5 mg/$m^3$:** ScbaF/SaF
**75 mg/$m^3$:** SaF:Pd,Pp
**§:** ScbaF:Pd,Pp/SaF:Pd,Pp:AScba
**Escape:** GmFOv100/ScbaE

**Incompatibilities and Reactivities:** Strong alkalis, acids, alcohol [**Note:** Polymerizes at 450°F.]



**Exposure Routes, Symptoms, Target Organs (see Table 5):**
**ER:** Inh, Ing, Con
**SY:** Irrit eyes, nose, throat; resp sens; cough, pulm secretions, chest pain, dysp; asthma
**TO:** Eyes, resp sys

**First Aid (see Table 6):**
**Eye:** Irr immed
**Skin:** Soap wash immed
**Breath:** Resp support
**Swallow:** Medical attention immed

## Methylene chloride

| Formula: | CAS#: | RTECS#: | IDLH: |
|---|---|---|---|
| $CH_2Cl_2$ | 75-09-2 | PA8050000 | Ca [2300 ppm] |

**Conversion:** 1 ppm = 3.47 mg/$m^3$ | **DOT:** 1593 160

**Synonyms/Trade Names:** Dichloromethane, Methylene dichloride

**Exposure Limits:**
**NIOSH REL:** Ca
See Appendix A
**OSHA PEL:** [1910.1052] TWA 25 ppm
ST 125 ppm

**Measurement Methods (see Table 1):**
**NIOSH** 1005, 3800
**OSHA** 59, 80

**Physical Description:** Colorless liquid with a chloroform-like odor. [**Note:** A gas above 104°F.]

**Chemical & Physical Properties:**
**MW:** 84.9
**BP:** 104°F
**Sol:** 2%
**Fl.P:** ?
**IP:** 11.32 eV
**Sp.Gr:** 1.33
**VP:** 350 mmHg
**FRZ:** -139°F
**UEL:** 23%
**LEL:** 13%
Combustible Liquid

**Personal Protection/Sanitation (see Table 2):**
**Skin:** Prevent skin contact
**Eyes:** Prevent eye contact
**Wash skin:** When contam
**Remove:** When wet or contam
**Change:** N.R.
**Provide:** Eyewash
Quick drench

**Respirator Recommendations (see Tables 3 and 4):**
**NIOSH**
**¥:** ScbaF:Pd,Pp/SaF:Pd,Pp:AScba
**Escape:** GmFOv/ScbaE

**See Appendix E** (page 351)

**Incompatibilities and Reactivities:** Strong oxidizers; caustics; chemically-active metals such as aluminum, magnesium powders, potassium & sodium; concentrated nitric acid

**Exposure Routes, Symptoms, Target Organs (see Table 5):**
**ER:** Inh, Abs, Ing, Con
**SY:** Irrit eyes, skin; lass, drow, dizz; numb, tingle limbs; nau; [carc]
**TO:** Eyes, skin, CVS, CNS [in animals: lung, liver, salivary & mammary gland tumors]

**First Aid (see Table 6):**
**Eye:** Irr immed
**Skin:** Soap wash prompt
**Breath:** Resp support
**Swallow:** Medical attention immed

## 4,4'-Methylenedianiline

| | |
|---|---|
| **Formula:** | $CH_2(C_6H_4NH_2)_2$ |
| **CAS#:** | 101-77-9 |
| **RTECS#:** | BY5425000 |
| **IDLH:** | Ca [N.D.] |
| **Conversion:** | |
| **DOT:** | |

**Synonyms/Trade Names:** 4,4'-Diaminodiphenylmethane; para, para'-Diaminodiphenyl-methane; Dianilinomethane; 4,4'-Diphenylmethanediamine; MDA

**Exposure Limits:**
**NIOSH REL:** Ca
See Appendix A
**OSHA PEL:** [1910.1050] TWA 0.010 ppm
ST 0.100 ppm

**Measurement Methods (see Table 1):**
**NIOSH** 5029

**Physical Description:** Pale-brown, crystalline solid with a faint, amine-like odor.

**Chemical & Physical Properties:**
**MW:** 198.3
**BP:** 748°F
**Sol:** 0.1%
**Fl.P:** 374°F
**IP:** 10.70 eV
**Sp.Gr:** 1.06 (Liquid at 212°F)
**VP(77°F):** 0.0000002 mmHg
**MLT:** 198°F
**UEL:** ?
**LEL:** ?
Combustible Solid

**Personal Protection/Sanitation (see Table 2):**
**Skin:** Prevent skin contact
**Eyes:** Prevent eye contact
**Wash skin:** When contam/Daily
**Remove:** When wet or contam
**Change:** Daily
**Provide:** Eyewash
Quick drench

**Respirator Recommendations (see Tables 3 and 4):**
**NIOSH**
**¥:** ScbaF:Pd,Pp/SaF:Pd,Pp:AScba
**Escape:** GmFOv100/ScbaE

**See Appendix E** (page 351)

**Incompatibilities and Reactivities:** Strong oxidizers

**Exposure Routes, Symptoms, Target Organs (see Table 5):**
**ER:** Inh, Abs, Ing, Con
**SY:** Irrit eyes; jaun, hepatitis; myocardial damage; in animals: heart, liver, spleen damage; [carc]
**TO:** Eyes, liver, CVS, spleen [in animals: bladder cancer]

**First Aid (see Table 6):**
**Eye:** Irr immed
**Skin:** Soap wash immed
**Breath:** Resp support
**Swallow:** Medical attention immed



## Methyl ethyl ketone peroxide

| | |
|---|---|
| **Formula:** | $C_8H_{16}O_4$ |
| **CAS#:** | 1338-23-4 |
| **RTECS#:** | EL9450000 |
| **IDLH:** | N.D. |
| **Conversion:** | 1 ppm = 7.21 mg/$m^3$ |
| **DOT:** | |

**Synonyms/Trade Names:** 2-Butanone peroxide, Ethyl methyl ketone peroxide, MEKP, MEK peroxide, Methyl ethyl ketone hydroperoxide

**Exposure Limits:**
**NIOSH REL:** C 0.2 ppm (1.5 mg/$m^3$)
**OSHA PEL†:** none

**Measurement Methods (see Table 1):**
**NIOSH** 3508
**OSHA** 77

**Physical Description:** Colorless liquid with a characteristic odor.
[**Note:** Explosive decomposition occurs at 230°F.]

**Chemical & Physical Properties:**
**MW:** 176.2
**BP:** 244°F (Decomposes)
**Sol:** Soluble
**Fl.P(oc):** 125-200°F (60% MEKP)
**IP:** ?
**Sp.Gr(59°F):** 1.12
**VP:** ?
**FRZ:** ?
**UEL:** ?
**LEL:** ?
Combustible Liquid

**Personal Protection/Sanitation (see Table 2):**
**Skin:** Prevent skin contact
**Eyes:** Prevent eye contact
**Wash skin:** When contam
**Remove:** When wet or contam
**Change:** N.R.
**Provide:** Eyewash
Quick drench

**Respirator Recommendations (see Tables 3 and 4):**
Not available.

**Incompatibilities and Reactivities:** Organic materials, heat, flames, sunlight, trace contaminants
[**Note:** A strong oxidizing agent. Pure MEKP is shock sensitive. Commercial product is diluted with 40% dimethyl phthalate, cyclohexane peroxide, or diallyl phthalate to reduce sensitivity to shock.]

**Exposure Routes, Symptoms, Target Organs (see Table 5):**
**ER:** Inh, Ing, Con
**SY:** Irrit eyes, skin, nose, throat; cough, dysp, pulm edema; blurred vision; blisters, scars skin; abdom pain, vomit, diarr; derm; in animals: liver, kidney damage
**TO:** Eyes, skin, resp sys, liver, kidneys

**First Aid (see Table 6):**
**Eye:** Irr immed
**Skin:** Water wash immed
**Breath:** Resp support
**Swallow:** Medical attention immed

M

| Methyl formate | Formula: $HCOOCH_3$ | CAS#: 107-31-3 | RTECS#: LQ8925000 | IDLH: 4500 ppm |
|---|---|---|---|---|
| Conversion: 1 ppm = 2.46 mg/m$^3$ | DOT: 1243 129 | | | |

**Synonyms/Trade Names:** Methyl ester of formic acid, Methyl methanoate

**Exposure Limits:**
**NIOSH REL:** TWA 100 ppm (250 mg/m$^3$)
ST 150 ppm (375 mg/m$^3$)
**OSHA PEL†:** TWA 100 ppm (250 mg/m$^3$)

**Measurement Methods (see Table 1):**
**NIOSH** S291 (II-5)
**OSHA** PV2041

**Physical Description:** Colorless liquid with a pleasant odor.
[**Note:** A gas above 89°F.]

| Chemical & Physical Properties: | Personal Protection/Sanitation (see Table 2): | Respirator Recommendations (see Tables 3 and 4): |
|---|---|---|
| **MW:** 60.1<br>**BP:** 89°F<br>**Sol:** 30%<br>**Fl.P:** -2°F<br>**IP:** 10.82 eV<br>**Sp.Gr:** 0.98<br>**VP:** 476 mmHg<br>**FRZ:** -148°F<br>**UEL:** 23%<br>**LEL:** 4.5%<br>Class IA Flammable Liquid | **Skin:** Prevent skin contact<br>**Eyes:** Prevent eye contact<br>**Wash skin:** When contam<br>**Remove:** When wet (flamm)<br>**Change:** N.R. | **NIOSH/OSHA**<br>**1000 ppm:** Sa*<br>**2500 ppm:** Sa:Cf*<br>**4500 ppm:** ScbaF/SaF<br>**§:** ScbaF:Pd,Pp/SaF:Pd,Pp:AScba<br>**Escape:** GmFOv/ScbaE |

**Incompatibilities and Reactivities:** Strong oxidizers
[**Note:** Reacts slowly with water to form methanol & formic acid.]

| Exposure Routes, Symptoms, Target Organs (see Table 5): | First Aid (see Table 6): |
|---|---|
| **ER:** Inh, Abs, Ing, Con<br>**SY:** Irrit eyes, nose; chest tight, dysp; vis dist; CNS depres; in animals: pulm edema; narco<br>**TO:** Eyes, resp sys, CNS | **Eye:** Irr immed<br>**Skin:** Soap wash immed<br>**Breath:** Resp support<br>**Swallow:** Medical attention immed |

| 5-Methyl-3-heptanone | Formula: $C_2H_5COCH_2CH(CH_3)CH_2CH_3$ | CAS#: 541-85-5 | RTECS#: MJ7350000 | IDLH: 100 ppm |
|---|---|---|---|---|
| Conversion: 1 ppm = 5.24 mg/m$^3$ | DOT: 2271 127 | | | |

**Synonyms/Trade Names:** Amyl ethyl ketone, Ethyl amyl ketone, 3-Methyl-5-heptanone

**Exposure Limits:**
**NIOSH REL:** TWA 25 ppm (130 mg/m$^3$)
**OSHA PEL:** TWA 25 ppm (130 mg/m$^3$)

**Measurement Methods (see Table 1):**
**NIOSH** 1301, 2553

**Physical Description:** Colorless liquid with a pungent odor.

| Chemical & Physical Properties: | Personal Protection/Sanitation (see Table 2): | Respirator Recommendations (see Tables 3 and 4): |
|---|---|---|
| **MW:** 128.2<br>**BP:** 315°F<br>**Sol:** Insoluble<br>**Fl.P:** 138°F<br>**IP:** ?<br>**Sp.Gr:** 0.82<br>**VP:** 2 mmHg<br>**FRZ:** -70°F<br>**UEL:** ?<br>**LEL:** ?<br>Class II Combustible Liquid | **Skin:** Prevent skin contact<br>**Eyes:** Prevent eye contact<br>**Wash skin:** When contam<br>**Remove:** When wet or contam<br>**Change:** N.R. | **NIOSH/OSHA**<br>**100 ppm:** CcrOv*/PaprOv*/GmFOv/ Sa*/ScbaF<br>**§:** ScbaF:Pd,Pp/SaF:Pd,Pp:AScba<br>**Escape:** GmFOv/ScbaE |

**Incompatibilities and Reactivities:** Strong oxidizers

| Exposure Routes, Symptoms, Target Organs (see Table 5): | First Aid (see Table 6): |
|---|---|
| **ER:** Inh, Ing, Con<br>**SY:** Irrit eyes, skin, muc memb; head; narco, coma; derm<br>**TO:** Eyes, skin, resp sys, CNS | **Eye:** Irr immed<br>**Skin:** Water flush<br>**Breath:** Resp support<br>**Swallow:** Medical attention immed |

## Methyl hydrazine

**Formula:** $CH_3NHNH_2$ | **CAS#:** 60-34-4 | **RTECS#:** MV5600000 | **IDLH:** Ca [20 ppm]

**Conversion:** 1 ppm = 1.89 mg/m$^3$ | **DOT:** 1244 131

**Synonyms/Trade Names:** MMH, Monomethylhydrazine

**Exposure Limits:**
**NIOSH REL:** Ca
C 0.04 ppm (0.08 mg/m$^3$) [2-hr]
See Appendix A
**OSHA PEL:** C 0.2 ppm (0.35 mg/m$^3$) [skin]

**Measurement Methods (see Table 1):**
**NIOSH** 3510

**Physical Description:** Fuming, colorless liquid with an ammonia-like odor.

**Chemical & Physical Properties:**
**MW:** 46.1
**BP:** 190°F
**Sol:** Miscible
**Fl.P:** 17°F
**IP:** 8.00 eV
**Sp.Gr(77°F):** 0.87
**VP:** 38 mmHg
**FRZ:** -62°F
**UEL:** 92%
**LEL:** 2.5%
Class IB Flammable Liquid

**Personal Protection/Sanitation (see Table 2):**
**Skin:** Prevent skin contact
**Eyes:** Prevent eye contact
**Wash skin:** When contam
**Remove:** When wet (flamm)
**Change:** N.R.
**Provide:** Eyewash
Quick drench

**Respirator Recommendations (see Tables 3 and 4):**
**NIOSH**
**¥:** ScbaF:Pd,Pp/SaF:Pd,Pp:AScba
**Escape:** ScbaE

**Incompatibilities and Reactivities:** Oxides of iron; copper; manganese; lead; copper alloys; porous materials such as earth, asbestos, wood & cloth; strong oxidizers such as fluorine & chlorine; nitric acid; hydrogen peroxide

**Exposure Routes, Symptoms, Target Organs (see Table 5):**
**ER:** Inh, Abs, Ing, Con
**SY:** Irrit eyes, skin, resp sys; vomit, diarr, tremor, ataxia; anoxia, cyan; convuls; [carc]
**TO:** Eyes, skin, resp sys, CNS, liver, blood, CVS [in animals: lung, liver, blood vessel & intestine tumors]

**First Aid (see Table 6):**
**Eye:** Irr immed
**Skin:** Water flush immed
**Breath:** Resp support
**Swallow:** Medical attention immed



## Methyl iodide

**Formula:** $CH_3I$ | **CAS#:** 74-88-4 | **RTECS#:** PA9450000 | **IDLH:** Ca [100 ppm]

**Conversion:** 1 ppm = 5.80 mg/m$^3$ | **DOT:** 2644 151

**Synonyms/Trade Names:** Iodomethane, Monoiodomethane

**Exposure Limits:**
**NIOSH REL:** Ca
TWA 2 ppm (10 mg/m$^3$) [skin]
See Appendix A
**OSHA PEL:** TWA 5 ppm (28 mg/m$^3$) [skin]

**Measurement Methods (see Table 1):**
**NIOSH** 1014

**Physical Description:** Colorless liquid with a pungent, ether-like odor.
[**Note:** Turns yellow, red, or brown on exposure to light & moisture.]

**Chemical & Physical Properties:**
**MW:** 141.9
**BP:** 109°F
**Sol:** 1%
**Fl.P:** NA
**IP:** 9.54 eV
**Sp.Gr:** 2.28
**VP:** 400 mmHg
**FRZ:** -88°F
**UEL:** NA
**LEL:** NA
Noncombustible Liquid

**Personal Protection/Sanitation (see Table 2):**
**Skin:** Prevent skin contact
**Eyes:** Prevent eye contact
**Wash skin:** When contam
**Remove:** When wet or contam
**Change:** N.R.
**Provide:** Eyewash
Quick drench

**Respirator Recommendations (see Tables 3 and 4):**
**NIOSH**
**¥:** ScbaF:Pd,Pp/SaF:Pd,Pp:AScba
**Escape:** GmFOv/ScbaE

**Incompatibilities and Reactivities:** Strong oxidizers [**Note:** Decomposes at 518°F.]

**Exposure Routes, Symptoms, Target Organs (see Table 5):**
**ER:** Inh, Abs, Ing, Con
**SY:** Irrit eyes, skin, resp sys; nau, vomit; dizz, ataxia; slurred speech, drow; derm; [carc]
**TO:** Eyes, skin, resp sys, CNS [in animals: lung, kidney & forestomach tumors]

**First Aid (see Table 6):**
**Eye:** Irr immed
**Skin:** Soap flush immed
**Breath:** Resp support
**Swallow:** Medical attention immed

## Methyl isoamyl ketone

**Formula:** $CH_3COCH_2CH_2CH(CH_3)_2$ | **CAS#:** 110-12-3 | **RTECS#:** MP3850000 | **IDLH:** N.D.

**Conversion:** 1 ppm = 4.67 mg/m$^3$ | **DOT:** 2302 127

**Synonyms/Trade Names:** Isoamyl methyl ketone, Isopentyl methyl ketone, 2-Methyl-5-hexanone, 5-Methyl-2-hexanone, MIAK

**Exposure Limits:**
**NIOSH REL:** TWA 50 ppm (240 mg/m$^3$)
**OSHA PEL†:** TWA 100 ppm (475 mg/m$^3$)

**Measurement Methods (see Table 1):**
**OSHA** PV2042

**Physical Description:** Colorless, clear liquid with a pleasant, fruity odor.

**Chemical & Physical Properties:**
**MW:** 114.2
**BP:** 291°F
**Sol:** 0.5%
**Fl.P:** 97°F
**IP:** 9.284 eV
**Sp.Gr:** 0.81
**VP:** 5 mmHg
**FRZ:** -101°F
**UEL(200°F):** 8.2%
**LEL(200°F):** 1.0%
Class IC Flammable Liquid

**Personal Protection/Sanitation (see Table 2):**
**Skin:** Prevent skin contact
**Eyes:** Prevent eye contact
**Wash skin:** When contam
**Remove:** When wet (flamm)
**Change:** N.R.

**Respirator Recommendations (see Tables 3 and 4):**
**NIOSH**
**500 ppm:** CcrOv*/Sa*
**1250 ppm:** Sa:Cf*/PaprOv*
**2500 ppm:** CcrFOv/GmFOv/PaprTOv*/SaT:Cf*/ScbaF/SaF
**5000 ppm:** SaF:Pd,Pp
**§:** ScbaF:Pd,Pp/SaF:Pd,Pp:AScba
**Escape:** GmFOv/ScbaE

**Incompatibilities and Reactivities:** Oxidizers

M

**Exposure Routes, Symptoms, Target Organs (see Table 5):**
**ER:** Inh, Ing, Con
**SY:** Irrit eyes, skin, muc memb; head, narco, coma; derm; in animals: liver, kidney damage
**TO:** Eyes, skin, resp sys, CNS, liver, kidneys

**First Aid (see Table 6):**
**Eye:** Irr immed
**Skin:** Soap flush prompt
**Breath:** Resp support
**Swallow:** Medical attention immed

## Methyl isobutyl carbinol

**Formula:** $(CH_3)_2CHCH_2CH(OH)CH_3$ | **CAS#:** 108-11-2 | **RTECS#:** SA7350000 | **IDLH:** 400 ppm

**Conversion:** 1 ppm = 4.18 mg/m$^3$ | **DOT:** 2053 129

**Synonyms/Trade Names:** Isobutylmethylcarbinol, Methyl amyl alcohol, 4-Methyl-2-pentanol, MIBC

**Exposure Limits:**
**NIOSH REL:** TWA 25 ppm (100 mg/m$^3$)
ST 40 ppm (165 mg/m$^3$) [skin]
**OSHA PEL†:** TWA 25 ppm (100 mg/m$^3$) [skin]

**Measurement Methods (see Table 1):**
**NIOSH** 1402, 1405
**OSHA** 7

**Physical Description:** Colorless liquid with a mild odor.

**Chemical & Physical Properties:**
**MW:** 102.2
**BP:** 271°F
**Sol:** 2%
**Fl.P:** 106°F
**IP:** ?
**Sp.Gr:** 0.81
**VP:** 3 mmHg
**FRZ:** -130°F
**UEL:** 5.5%
**LEL:** 1.0%
Class II Combustible Liquid

**Personal Protection/Sanitation (see Table 2):**
**Skin:** Prevent skin contact
**Eyes:** Prevent eye contact
**Wash skin:** When contam
**Remove:** When wet or contam
**Change:** N.R.

**Respirator Recommendations (see Tables 3 and 4):**
**NIOSH/OSHA**
**250 ppm:** Sa*
**400 ppm:** Sa:Cf*/ScbaF/SaF
**§:** ScbaF:Pd,Pp/SaF:Pd,Pp:AScba
**Escape:** GmFOv/ScbaE

**Incompatibilities and Reactivities:** Strong oxidizers

**Exposure Routes, Symptoms, Target Organs (see Table 5):**
**ER:** Inh, Abs, Ing, Con
**SY:** Irrit eyes, skin; head, drow; derm; in animals: narco
**TO:** Eyes, skin, CNS

**First Aid (see Table 6):**
**Eye:** Irr immed
**Skin:** Water flush prompt
**Breath:** Resp support
**Swallow:** Medical attention immed

## Methyl isocyanate

| Methyl isocyanate | Formula: $CH_3NCO$ | CAS#: 624-83-9 | RTECS#: NQ9450000 | IDLH: 3 ppm |
|---|---|---|---|---|
| **Conversion:** 1 ppm = 2.34 mg/m$^3$ | **DOT:** 2480 155 | | | |

**Synonyms/Trade Names:** Methyl ester of isocyanic acid, MIC

**Exposure Limits:**
**NIOSH REL:** TWA 0.02 ppm (0.05 mg/m$^3$) [skin]
**OSHA PEL:** TWA 0.02 ppm (0.05 mg/m$^3$) [skin]

**Measurement Methods (see Table 1):**
OSHA 54

**Physical Description:** Colorless liquid with a sharp, pungent odor.

**Chemical & Physical Properties:**
**MW:** 57.1
**BP:** 102-104°F
**Sol(59°F):** 10%
**Fl.P:** 19°F
**IP:** 10.67 eV
**Sp.Gr:** 0.96
**VP:** 348 mmHg
**FRZ:** -49°F
**UEL:** 26%
**LEL:** 5.3%
Class IB Flammable Liquid

**Personal Protection/Sanitation (see Table 2):**
**Skin:** Prevent skin contact
**Eyes:** Prevent eye contact
**Wash skin:** When contam
**Remove:** When wet (flamm)
**Change:** N.R.
**Provide:** Eyewash
Quick drench

**Respirator Recommendations (see Tables 3 and 4):**
**NIOSH/OSHA**
**0.2 ppm:** Sa*
**0.5 ppm:** Sa:Cf*
**1 ppm:** ScbaF/SaF
**3 ppm:** SaF:Pd,Pp
**§:** ScbaF:Pd,Pp/SaF:Pd,Pp:AScba
**Escape:** GmFOv/ScbaE

**Incompatibilities and Reactivities:** Water, oxidizers, acids, alkalis, amines, iron, tin, copper
[**Note:** Usually contains inhibitors to prevent polymerization.]

**Exposure Routes, Symptoms, Target Organs (see Table 5):**
**ER:** Inh, Abs, Ing, Con
**SY:** Irrit eyes, skin, nose, throat; resp sens, cough, pulm secretions, chest pain, dysp; asthma; eye, skin damage; in animals: pulm edema
**TO:** Eyes, skin, resp sys

**First Aid (see Table 6):**
**Eye:** Irr immed
**Skin:** Water flush immed
**Breath:** Resp support
**Swallow:** Medical attention immed

M

## Methyl isopropyl ketone

| Methyl isopropyl ketone | Formula: $CH_3COCH(CH_3)_2$ | CAS#: 563-80-4 | RTECS#: EL9100000 | IDLH: N.D. |
|---|---|---|---|---|
| **Conversion:** 1 ppm = 3.53 mg/m$^3$ | **DOT:** 2397 127 | | | |

**Synonyms/Trade Names:** 2-Acetyl propane, Isopropyl methyl ketone, 3-Methyl-2-butanone, 3-Methyl butan-2-one, MIPK

**Exposure Limits:**
**NIOSH REL:** TWA 200 ppm (705 mg/m$^3$)
**OSHA PEL†:** none

**Measurement Methods (see Table 1):**
None available

**Physical Description:** Colorless liquid with an acetone-like odor.

**Chemical & Physical Properties:**
**MW:** 86.2
**BP:** 199°F
**Sol:** Very slight
**Fl.P:** ?
**IP:** 9.32 eV
**Sp.Gr:** 0.81
**VP:** 42 mmHg
**FRZ:** -134°F
**UEL:** ?
**LEL:** ?
Combustible Liquid

**Personal Protection/Sanitation (see Table 2):**
**Skin:** Prevent skin contact
**Eyes:** Prevent eye contact
**Wash skin:** When contam
**Remove:** When wet or contam
**Change:** N.R.

**Respirator Recommendations (see Tables 3 and 4):**
Not available.

**Incompatibilities and Reactivities:** Oxidizers

**Exposure Routes, Symptoms, Target Organs (see Table 5):**
**ER:** Inh, Ing, Con
**SY:** Irrit eyes, skin, muc memb, resp sys; cough
**TO:** Eyes, skin, resp sys

**First Aid (see Table 6):**
**Eye:** Irr immed
**Skin:** Soap wash immed
**Breath:** Resp support
**Swallow:** Medical attention immed

## Methyl mercaptan

| Formula: | CAS#: | RTECS#: | IDLH: |
|---|---|---|---|
| $CH_3SH$ | 74-93-1 | PB4375000 | 150 ppm |

**Conversion:** 1 ppm = 1.97 mg/m$^3$ | **DOT:** 1064 117

**Synonyms/Trade Names:** Mercaptomethane, Methanethiol, Methyl sulfhydrate

**Exposure Limits:**
**NIOSH REL:** C 0.5 ppm (1 mg/m$^3$) [15-minute]
**OSHA PEL†:** C 10 ppm (20 mg/m$^3$)

**Measurement Methods (see Table 1):**
**NIOSH** 2542
**OSHA** 26

**Physical Description:** Colorless gas with a disagreeable odor like garlic or rotten cabbage. [**Note:** A liquid below 43°F. Shipped as a liquefied compressed gas.]

**Chemical & Physical Properties:**
**MW:** 48.1
**BP:** 43°F
**Sol:** 2%
**Fl.P:** NA (Gas)
(oc) 0°F (Liquid)
**IP:** 9.44 eV
**RGasD:** 1.66
**Sp.Gr:** 0.90 (Liquid at 32°F)
**VP:** 1.7 atm
**FRZ:** -186°F
**UEL:** 21.8%
**LEL:** 3.9%
Flammable Gas

**Personal Protection/Sanitation (see Table 2):**
**Skin:** Prevent skin contact (liquid)
Frostbite
**Eyes:** Prevent eye contact (liquid)
Frostbite
**Wash skin:** N.R.
**Remove:** When wet (flamm)
**Change:** N.R.
**Provide:** Eyewash (liquid)
Quick drench (liquid)
Frostbite wash

**Respirator Recommendations (see Tables 3 and 4):**
**NIOSH**
**5 ppm:** CcrOv/Sa
**12.5 ppm:** Sa:Cf/PaprOv
**25 ppm:** CcrFOv/GmFOv/PaprTOv/
SaT:Cf/ScbaF/SaF
**150 ppm:** Sa:Pd,Pp
**§:** ScbaF:Pd,Pp/SaF:Pd,Pp:AScba
**Escape:** GmFOv/ScbaE

M

**Incompatibilities and Reactivities:** Strong oxidizers, bleaches, copper, aluminum, nickel-copper alloys

**Exposure Routes, Symptoms, Target Organs (see Table 5):**
**ER:** Inh, Con (liquid)
**SY:** Irrit eyes, skin, resp sys; narco; cyan; convuls; liquid: frostbite
**TO:** Eyes, skin, resp sys, CNS, blood

**First Aid (see Table 6):**
**Eye:** Irr immed (liquid)/Frostbite
**Skin:** Water flush immed (liquid)/Frostbite
**Breath:** Resp support

## Methyl methacrylate

| Formula: | CAS#: | RTECS#: | IDLH: |
|---|---|---|---|
| $CH_2{=}C(CH_3)COOCH_3$ | 80-62-6 | OZ5075000 | 1000 ppm |

**Conversion:** 1 ppm = 4.09 mg/m$^3$ | **DOT:** 1247 129P (inhibited)

**Synonyms/Trade Names:** Methacrylate monomer, Methyl ester of methacrylic acid, Methyl-2-methyl-2-propenoate

**Exposure Limits:**
**NIOSH REL:** TWA 100 ppm (410 mg/m$^3$)
**OSHA PEL:** TWA 100 ppm (410 mg/m$^3$)

**Measurement Methods (see Table 1):**
**NIOSH** 2537
**OSHA** 94

**Physical Description:** Colorless liquid with an acrid, fruity odor.

**Chemical & Physical Properties:**
**MW:** 100.1
**BP:** 214°F
**Sol:** 1.5%
**Fl.P(oc):** 50°F
**IP:** 9.70 eV
**Sp.Gr:** 0.94
**VP:** 29 mmHg
**FRZ:** -54°F
**UEL:** 8.2%
**LEL:** 1.7%
Class IB Flammable Liquid

**Personal Protection/Sanitation (see Table 2):**
**Skin:** Prevent skin contact
**Eyes:** Prevent eye contact
**Wash skin:** When contam
**Remove:** When wet (flamm)
**Change:** N.R.

**Respirator Recommendations (see Tables 3 and 4):**
**NIOSH/OSHA**
**1000 ppm:** Sa:Cf£/CcrFOv/GmFOv/
PaprOv£/ScbaF/SaF
**§:** ScbaF:Pd,Pp/SaF:Pd,Pp:AScba
**Escape:** GmFOv/ScbaE

**Incompatibilities and Reactivities:** Nitrates, oxidizers, peroxides, strong alkalis, moisture [**Note:** May polymerize if subjected to heat, oxidizers, or ultraviolet light. Usually contains an inhibitor such as hydroquinone.]

**Exposure Routes, Symptoms, Target Organs (see Table 5):**
**ER:** Inh, Ing, Con
**SY:** Irrit eyes, skin, nose, throat; derm
**TO:** Eyes, skin, resp sys

**First Aid (see Table 6):**
**Eye:** Irr immed
**Skin:** Water flush prompt
**Breath:** Resp support
**Swallow:** Medical attention immed

## Methyl parathion

| | | | | |
|---|---|---|---|---|
| **Methyl parathion** | **Formula:** $(CH_3O)_2P(S)OC_6H_4NO_2$ | **CAS#:** 298-00-0 | **RTECS#:** TG0175000 | **IDLH:** N.D. |
| **Conversion:** | **DOT:** 2783 152 (solid); 3018 152 (liquid) | | | |

**Synonyms/Trade Names:** Azophos®; O,O-Dimethyl-O-p-nitrophenylphosphorothioate; Parathion methyl

**Exposure Limits:**
**NIOSH REL:** TWA 0.2 mg/m$^3$ [skin]
**OSHA PEL†:** none

**Measurement Methods (see Table 1):**
**NIOSH** 5600
**OSHA** PV2112

**Physical Description:** White to tan, crystalline solid or powder with a pungent, garlic-like odor. [pesticide] [**Note:** The commercial product in xylene is a tan liquid.]

| Chemical & Physical Properties: | Personal Protection/Sanitation (see Table 2): | Respirator Recommendations (see Tables 3 and 4): |
|---|---|---|
| **MW:** 263.2<br>**BP:** 289°F<br>**Sol(77°F):** 0.006%<br>**Fl.P:** ?<br>**IP:** ?<br>**Sp.Gr:** 1.36<br>**VP:** 0.00001 mmHg<br>**MLT:** 99°F<br>**UEL:** ?<br>**LEL:** ?<br>Combustible Solid | **Skin:** Prevent skin contact<br>**Eyes:** Prevent eye contact<br>**Wash skin:** When contam/Daily<br>**Remove:** When wet or contam<br>**Change:** Daily<br>**Provide:** Eyewash<br>Quick drench | **NIOSH**<br>**2 mg/m$^3$:** CcrOv95/Sa<br>**5 mg/m$^3$:** Sa:Cf/PaprOvHie<br>**10 mg/m$^3$:** CcrFOv100/GmFOv100/PaprTOvHie/SaT:Cf/ScbaF/SaF<br>**200 mg/m$^3$:** SaF:Pd,Pp<br>**§:** ScbaF:Pd,Pp/SaF:Pd,Pp:AScba<br>**Escape:** GmFOv100/ScbaE |

**Incompatibilities and Reactivities:** Strong oxidizers, water [**Note:** Explosive risk when heated above 122°F.]

**Exposure Routes, Symptoms, Target Organs (see Table 5):**
**ER:** Inh, Abs, Ing, Con
**SY:** Irrit eyes, skin; nau, vomit, abdom cramps, diarr, salv; head, dizz, lass; rhin, chest tight; blurred vision, miosis; card irreg; musc fasc; dysp
**TO:** Eyes, skin, resp sys, CNS, CVS, blood chol

**First Aid (see Table 6):**
**Eye:** Irr immed
**Skin:** Soap wash immed
**Breath:** Resp support
**Swallow:** Medical attention immed

M

## Methyl silicate

| | | | | |
|---|---|---|---|---|
| **Methyl silicate** | **Formula:** $(CH_3O)_4Si$ | **CAS#:** 681-84-5 | **RTECS#:** VV9800000 | **IDLH:** N.D. |
| **Conversion:** 1 ppm = 6.23 mg/m$^3$ | **DOT:** 2606 155 | | | |

**Synonyms/Trade Names:** Methyl orthosilicate, Tetramethoxysilane, Tetramethyl ester of silicic acid, Tetramethyl silicate

**Exposure Limits:**
**NIOSH REL:** TWA 1 ppm (6 mg/m$^3$)
**OSHA PEL†:** none

**Measurement Methods (see Table 1):**
None available

**Physical Description:** Clear, colorless liquid. [**Note:** A solid below 28°F.]

| Chemical & Physical Properties: | Personal Protection/Sanitation (see Table 2): | Respirator Recommendations (see Tables 3 and 4): |
|---|---|---|
| **MW:** 152.3<br>**BP:** 250°F<br>**Sol:** Soluble<br>**Fl.P:** 205°F<br>**IP:** ?<br>**Sp.Gr:** 1.02<br>**VP(77°F):** 12 mmHg<br>**FRZ:** 28°F<br>**UEL:** ?<br>**LEL:** ?<br>Class IIIB Combustible Liquid | **Skin:** Prevent skin contact<br>**Eyes:** Prevent eye contact<br>**Wash skin:** Daily<br>**Remove:** When wet or contam<br>**Change:** N.R.<br>**Provide:** Eyewash | Not available. |

**Incompatibilities and Reactivities:** Oxidizers; hexafluorides of rhenium, molybdenum & tungsten

**Exposure Routes, Symptoms, Target Organs (see Table 5):**
**ER:** Inh, Ing, Con
**SY:** Irrit eyes, corn damage (following even short-term exposure to the vapor); lung, kidney inj; pulm edema
**TO:** Eyes, resp sys, kidneys

**First Aid (see Table 6):**
**Eye:** Irr immed
**Skin:** Soap wash
**Breath:** Resp support
**Swallow:** Medical attention immed

## α-Methyl styrene

| α-Methyl styrene | Formula: $C_6H_5C(CH_3){=}CH_2$ | CAS#: 98-83-9 | RTECS#: WL5075300 | IDLH: 700 ppm |
|---|---|---|---|---|
| **Conversion:** 1 ppm = 4.83 mg/m$^3$ | **DOT:** | | | |

**Synonyms/Trade Names:** AMS, Isopropenyl benzene, 1-Methyl-1-phenylethylene, 2-Phenyl propylene

**Exposure Limits:**
**NIOSH REL:** TWA 50 ppm (240 mg/m$^3$)
ST 100 ppm (485 mg/m$^3$)
**OSHA PEL†:** C 100 ppm (480 mg/m$^3$)

**Measurement Methods (see Table 1):**
**NIOSH** 1501
**OSHA** 7

**Physical Description:** Colorless liquid with a characteristic odor.

**Chemical & Physical Properties:**
**MW:** 118.2
**BP:** 330°F
**Sol:** Insoluble
**Fl.P:** 129°F
**IP:** 8.35 eV
**Sp.Gr:** 0.91
**VP:** 2 mmHg
**FRZ:** -10°F
**UEL:** 6.1%
**LEL:** 1.9%
Class II Combustible Liquid

**Personal Protection/Sanitation (see Table 2):**
**Skin:** Prevent skin contact
**Eyes:** Prevent eye contact
**Wash skin:** When contam
**Remove:** When wet or contam
**Change:** N.R.

**Respirator Recommendations (see Tables 3 and 4):**
**NIOSH**
**500 ppm:** CcrOv*/Sa*
**700 ppm:** Sa:Cf*/CcrFOv/GmFOv/PaprOv*/ScbaF/SaF
**§:** ScbaF:Pd,Pp/SaF:Pd,Pp:AScba
**Escape:** GmFOv/ScbaE

**Incompatibilities and Reactivities:** Oxidizers, peroxides, halogens, catalysts for vinyl or ionic polymers; aluminum, iron chloride, copper [**Note:** Usually contains an inhibitor such as tert-butyl catechol.]

M

**Exposure Routes, Symptoms, Target Organs (see Table 5):**
**ER:** Inh, Ing, Con
**SY:** Irrit eyes, skin, nose, throat; drow; derm
**TO:** Eyes, skin, resp sys, CNS

**First Aid (see Table 6):**
**Eye:** Irr immed
**Skin:** Water flush prompt
**Breath:** Resp support
**Swallow:** Medical attention immed

## Metribuzin

| Metribuzin | Formula: $C_8H_{14}N_4OS$ | CAS#: 21087-64-9 | RTECS#: XZ2990000 | IDLH: N.D. |
|---|---|---|---|---|
| **Conversion:** | **DOT:** | | | |

**Synonyms/Trade Names:** 4-Amino-6-(1,1-dimethylethyl)-3-(methylthio)-1,2,4-triazin-5(4H)-one

**Exposure Limits:**
**NIOSH REL:** TWA 5 mg/m$^3$
**OSHA PEL†:** none

**Measurement Methods (see Table 1):**
**OSHA** PV2044

**Physical Description:** Colorless, crystalline solid. [herbicide]

**Chemical & Physical Properties:**
**MW:** 214.3
**BP:** ?
**Sol:** 0.1%
**Fl.P:** NA
**IP:** ?
**Sp.Gr:** 1.31
**VP:** 0.0000004 mmHg
**MLT:** 257°F
**UEL:** NA
**LEL:** NA
Noncombustible Solid

**Personal Protection/Sanitation (see Table 2):**
**Skin:** Prevent skin contact
**Eyes:** Prevent eye contact
**Wash skin:** When contam/Daily
**Remove:** When wet or contam
**Change:** Daily

**Respirator Recommendations (see Tables 3 and 4):**
Not available.

**Incompatibilities and Reactivities:** None reported

**Exposure Routes, Symptoms, Target Organs (see Table 5):**
**ER:** Inh, Ing, Con
**SY:** In animals: CNS depres; thyroid, liver enzyme changes
**TO:** CNS, thyroid, liver

**First Aid (see Table 6):**
**Eye:** Irr immed
**Skin:** Soap wash
**Breath:** Fresh air
**Swallow:** Medical attention immed

## Mica (containing less than 1% quartz)

| Mica (containing less than 1% quartz) | Formula: | CAS#: 12001-26-2 | RTECS#: VV8760000 | IDLH: 1500 mg/m³ |
|---|---|---|---|---|
| **Conversion:** | **DOT:** | | | |

**Synonyms/Trade Names:** Biotite, Lepidolite, Margarite, Muscovite, Phlogopite, Roscoelite, Zimmwaldite

**Exposure Limits:**
**NIOSH REL:** TWA 3 mg/$m^3$ (resp)
**OSHA PEL†:** TWA 20 mppcf

**Measurement Methods (see Table 1):**
**NIOSH** 0600

**Physical Description:** Colorless, odorless flakes or sheets of hydrous silicates.

| Chemical & Physical Properties: | Personal Protection/Sanitation (see Table 2): | Respirator Recommendations (see Tables 3 and 4): |
|---|---|---|
| **MW:** 797 (approx)<br>**BP:** ?<br>**Sol:** Insoluble<br>**Fl.P:** NA<br>**IP:** NA<br>**Sp.Gr:** 2.6-3.2<br>**VP:** 0 mmHg (approx)<br>**MLT:** ?<br>**UEL:** NA<br>**LEL:** NA<br>Noncombustible Solid | **Skin:** N.R.<br>**Eyes:** N.R.<br>**Wash skin:** N.R.<br>**Remove:** N.R.<br>**Change:** N.R. | **NIOSH**<br>**15 mg/m³:** Qm<br>**30 mg/m³:** 95XQ/Sa<br>**75 mg/m³:** Sa:Cf/PaprHie<br>**150 mg/m³:** 100F/SaT:Cf/PaprTHie/ScbaF/SaF<br>**1500 mg/m³:** Sa:Pd,Pp<br>**§:** ScbaF:Pd,Pp/SaF:Pd,Pp:AScba<br>**Escape:** 100F/ScbaE |

**Incompatibilities and Reactivities:** None reported

| Exposure Routes, Symptoms, Target Organs (see Table 5): | First Aid (see Table 6): |
|---|---|
| **ER:** Inh, Con<br>**SY:** Irrit eyes; pneumoconiosis, cough, dysp; lass; low-wgt<br>**TO:** Resp sys | **Eye:** Irr immed<br>**Breath:** Fresh air |

M

## Mineral wool fiber

| Mineral wool fiber | Formula: | CAS#: | RTECS#: PY8070000 | IDLH: N.D. |
|---|---|---|---|---|
| **Conversion:** | **DOT:** | | | |

**Synonyms/Trade Names:** Manmade mineral fibers, Rock wool, Slag wool, Synthetic vitreous fibers
[**Note:** Produced by blowing steam or air through molten rock (rock wool) or various furnace slags that are by-products of metal smelting or refining processes (slag wool).]

**Exposure Limits:**
**NIOSH REL:** TWA 3 fibers/$cm^3$ (fibers ≤ 3.5 µm diameter & ≥ 10 µm in length)
TWA 5 mg/$m^3$ (total)
**OSHA PEL:** TWA 15 mg/$m^3$ (total)
TWA 5 mg/$m^3$ (resp)

**Measurement Methods (see Table 1):**
**NIOSH** 0500, 7400

**Physical Description:** Typically, a mineral "wool" with diameters >0.5 µm & >1.5 µm in length.

| Chemical & Physical Properties: | Personal Protection/Sanitation (see Table 2): | Respirator Recommendations (see Tables 3 and 4): |
|---|---|---|
| **MW:** varies<br>**BP:** NA<br>**Sol:** Insoluble<br>**Fl.P:** NA<br>**IP:** NA<br>**Sp.Gr:** ?<br>**VP:** 0 mmHg (approx)<br>**MLT:** ?<br>**UEL:** NA<br>**LEL:** NA<br>Noncombustible Fibers | **Skin:** Prevent skin contact<br>**Eyes:** Prevent eye contact<br>**Wash skin:** Daily<br>**Remove:** N.R.<br>**Change:** Daily | **NIOSH**<br>**5X REL:** Qm<br>**10X REL:** 95XQ/Sa<br>**25X REL:** Sa:Cf/PaprHie<br>**50X REL:** 100F/PaprTHie/ScbaF/SaF<br>**1000X REL:** SaF:Pd,Pp<br>**§:** ScbaF:Pd,Pp/SaF:Pd,Pp:AScba<br>**Escape:** 100F/ScbaE |

**Incompatibilities and Reactivities:** None reported

| Exposure Routes, Symptoms, Target Organs (see Table 5): | First Aid (see Table 6): |
|---|---|
| **ER:** Inh, Con<br>**SY:** Irrit eyes, skin, resp sys; dysp<br>**TO:** Eyes, skin, resp sys | **Eye:** Irr immed<br>**Breath:** Fresh air |

## Molybdenum

| Formula: | CAS#: | RTECS#: | IDLH: |
|---|---|---|---|
| Mo | 7439-98-7 | QA4680000 | 5000 mg/m³ (as Mo) |

**Conversion:**

**DOT:**

**Synonyms/Trade Names:** Molybdenum metal

**Exposure Limits:**
**NIOSH REL*:** See Appendix D
**OSHA PEL*†:** TWA 15 mg/m³
[***Note:** The REL and PEL also apply to other insoluble molybdenum compounds (as Mo).]

**Measurement Methods (see Table 1):**
**NIOSH** 7300, 7301, 7303, 9102
**OSHA** ID121, ID125G

**Physical Description:** Dark gray or black powder with a metallic luster.

**Chemical & Physical Properties:**
**MW:** 95.9
**BP:** 8717°F
**Sol:** Insoluble
**Fl.P:** NA
**IP:** NA
**Sp.Gr:** 10.28
**VP:** 0 mmHg (approx)
**MLT:** 4752°F
**UEL:** NA
**LEL:** NA
Combustible Solid in form of dust or powder.

**Personal Protection/Sanitation (see Table 2):**
**Skin:** N.R.
**Eyes:** N.R.
**Wash skin:** N.R.
**Remove:** N.R.
**Change:** N.R.

**Respirator Recommendations (see Tables 3 and 4):**
**OSHA**
**75 mg/m³:** Qm
**150 mg/m³:** 95XQ/Sa
**375 mg/m³:** Sa:Cf/PaprHie
**750 mg/m³:** 100F/SaT:Cf/PaprTHie/ScbaF/SaF
**5000 mg/m³:** Sa:Pd,Pp
**§:** ScbaF:Pd,Pp/SaF:Pd,Pp:AScba
**Escape:** 100F/ScbaE

**Incompatibilities and Reactivities:** Strong oxidizers

M

**Exposure Routes, Symptoms, Target Organs (see Table 5):**
**ER:** Inh, Ing, Con
**SY:** In animals: irrit eyes, nose, throat; anor, diarr, low-wgt; listlessness; liver, kidney damage
**TO:** Eyes, resp sys, liver, kidneys

**First Aid (see Table 6):**
**Eye:** Irr immed
**Breath:** Resp support
**Swallow:** Medical attention immed

## Molybdenum (soluble compounds, as Mo)

| Formula: | CAS#: | RTECS#: | IDLH: |
|---|---|---|---|
| | | | 1000 mg/m³ (as Mo) |

**Conversion:**

**DOT:**

**Synonyms/Trade Names:** Synonyms vary depending upon the specific soluble molybdenum compound.

**Exposure Limits:**
**NIOSH REL:** See Appendix D
**OSHA PEL:** TWA 5 mg/m³

**Measurement Methods (see Table 1):**
**NIOSH** 7300, 7301, 7303, 9102
**OSHA** ID121, ID125G

**Physical Description:** Appearance and odor vary depending upon the specific soluble molybdenum compound.

**Chemical & Physical Properties:**
Properties vary depending upon the specific soluble molybdenum compound.

**Personal Protection/Sanitation (see Table 2):**
**Skin:** Prevent skin contact
**Eyes:** Prevent eye contact
**Wash skin:** When contam
**Remove:** When wet or contam
**Change:** N.R.

**Respirator Recommendations (see Tables 3 and 4):**
**OSHA**
**25 mg/m³:** Qm*
**50 mg/m³:** 95XQ*/Sa*
**125 mg/m³:** Sa:Cf*/PaprHie*
**250 mg/m³:** 100F/SaT:Cf*/PaprTHie*/ScbaF/SaF
**1000 mg/m³:** SaF:Pd,Pp
**§:** ScbaF:Pd,Pp/SaF:Pd,Pp:AScba
**Escape:** 100F/ScbaE

**Incompatibilities and Reactivities:** Varies

**Exposure Routes, Symptoms, Target Organs (see Table 5):**
**ER:** Inh, Ing, Con
**SY:** In animals: irrit eyes, nose, throat; anor; inco; dysp; anemia
**TO:** Eyes, resp sys, kidneys, blood

**First Aid (see Table 6):**
**Eye:** Irr immed
**Skin:** Water flush
**Breath:** Resp support
**Swallow:** Medical attention immed

## Monocrotophos

| Monocrotophos | Formula: $C_7H_{14}NO_5P$ | CAS#: 6923-22-4 | RTECS#: TC4375000 | IDLH: N.D. |
|---|---|---|---|---|
| **Conversion:** | **DOT:** 2783 152 (organophosphorus pesticide, solid) | | | |

**Synonyms/Trade Names:** Azodrin®, 3-Hydroxy-N-methylcrotonamide dimethylphosphate, Monocron

**Exposure Limits:**
**NIOSH REL:** TWA 0.25 mg/m$^3$
**OSHA PEL†:** none

**Measurement Methods (see Table 1):**
**NIOSH** 5600
**OSHA** PV2045

**Physical Description:** Colorless to reddish-brown solid with a mild, ester odor. [insecticide]

| Chemical & Physical Properties: | Personal Protection/Sanitation (see Table 2): | Respirator Recommendations (see Tables 3 and 4): |
|---|---|---|
| **MW:** 223.2<br>**BP:** 257°F<br>**Sol:** Miscible<br>**Fl.P:** >200°F<br>**IP:** ?<br>**Sp.Gr:** ?<br>**VP:** 0.000007 mmHg<br>**MLT:** 129°F<br>**UEL:** ?<br>**LEL:** ?<br>Combustible Solid | **Skin:** Prevent skin contact<br>**Eyes:** Prevent eye contact<br>**Wash skin:** When contam<br>**Remove:** When wet or contam<br>**Change:** Daily | Not available. |

**Incompatibilities and Reactivities:** Metals, low molecular weight alcohols & glycols
[**Note:** Corrosive to black iron, drum steel, stainless steel 304 & brass. Should be stored at 70-80°F.]

**Exposure Routes, Symptoms, Target Organs (see Table 5):**
**ER:** Inh, Abs, Ing, Con
**SY:** Irrit eyes, miosis, blurred vision; dizz, convuls; dysp; salv, abdom cramps, nau, diarr, vomit; in animals: possible terato effects
**TO:** Eyes, resp sys, CNS, CVS, blood chol, repro sys

**First Aid (see Table 6):**
**Eye:** Irr immed
**Skin:** Water flush immed
**Breath:** Resp support
**Swallow:** Medical attention immed

M

## Monomethyl aniline

| Monomethyl aniline | Formula: $C_6H_5NHCH_3$ | CAS#: 100-61-8 | RTECS#: BY4550000 | IDLH: 100 ppm |
|---|---|---|---|---|
| **Conversion:** 1 ppm = 4.38 mg/m$^3$ | **DOT:** 2294 153 | | | |

**Synonyms/Trade Names:** MA, (Methylamino)benzene, N-Methyl aniline, Methylphenylamine, N-Phenylmethylamine

**Exposure Limits:**
**NIOSH REL:** TWA 0.5 ppm (2 mg/m$^3$) [skin]
**OSHA PEL†:** TWA 2 ppm (9 mg/m$^3$) [skin]

**Measurement Methods (see Table 1):**
**NIOSH** 3511

**Physical Description:** Yellow to light-brown liquid with a weak, ammonia-like odor.

| Chemical & Physical Properties: | Personal Protection/Sanitation (see Table 2): | Respirator Recommendations (see Tables 3 and 4): |
|---|---|---|
| **MW:** 107.2<br>**BP:** 384°F<br>**Sol:** Insoluble<br>**Fl.P:** 175°F<br>**IP:** 7.32 eV<br>**Sp.Gr:** 0.99<br>**VP:** 0.3 mmHg<br>**FRZ:** -71°F<br>**UEL:** ?<br>**LEL:** ?<br>Class IIIA Combustible Liquid | **Skin:** Prevent skin contact<br>**Eyes:** Prevent eye contact<br>**Wash skin:** When contam<br>**Remove:** When wet or contam<br>**Change:** N.R. | **NIOSH**<br>**5 ppm:** Sa<br>**12.5 ppm:** Sa:Cf<br>**25 ppm:** SaT:Cf/ScbaF/SaF<br>**100 ppm:** SaF:Pd,Pp<br>**§:** ScbaF:Pd,Pp/SaF:Pd,Pp:AScba<br>**Escape:** GmFS/ScbaE |

**Incompatibilities and Reactivities:** Strong acids, strong oxidizers

**Exposure Routes, Symptoms, Target Organs (see Table 5):**
**ER:** Inh, Abs, Ing, Con
**SY:** Lass, dizz, head; dysp, cyan; methemo; pulm edema; liver, kidney damage
**TO:** Resp sys, liver, kidneys, blood, CNS

**First Aid (see Table 6):**
**Eye:** Irr immed
**Skin:** Soap wash immed
**Breath:** Resp support
**Swallow:** Medical attention immed

## Morpholine

| Morpholine | Formula: $C_4H_9ON$ | CAS#: 110-91-8 | RTECS#: QD6475000 | IDLH: 1400 ppm [10%LEL] |
|---|---|---|---|---|
| **Conversion:** 1 ppm = 3.56 mg/m$^3$ | **DOT:** 2054 132 | | | |

**Synonyms/Trade Names:** Diethylene imidoxide; Diethylene oximide; Tetrahydro-1,4-oxazine; Tetrahydro-p-oxazine

**Exposure Limits:**
**NIOSH REL:** TWA 20 ppm (70 mg/m$^3$) [skin]
ST 30 ppm (105 mg/m$^3$)
**OSHA PEL†:** TWA 20 ppm (70 mg/m$^3$) [skin]

**Measurement Methods (see Table 1):**
**NIOSH** S150 (II-3)

**Physical Description:** Colorless liquid with a weak, ammonia- or fish-like odor.
[**Note:** A solid below 23°F.]

**Chemical & Physical Properties:**
**MW:** 87.1
**BP:** 264°F
**Sol:** Miscible
**Fl.P(oc):** 98°F
**IP:** 8.88 eV
**Sp.Gr:** 1.007
**VP:** 6 mmHg
**FRZ:** 23°F
**UEL:** 11.2%
**LEL:** 1.4%
Class IC Flammable Liquid

**Personal Protection/Sanitation (see Table 2):**
**Skin:** Prevent skin contact
**Eyes:** Prevent eye contact
**Wash skin:** When contam
**Remove:** When wet (flamm)
**Change:** N.R.
**Provide:** Eyewash (>15%)
Quick drench (>25%)

**Respirator Recommendations (see Tables 3 and 4):**
**NIOSH/OSHA**
**500 ppm:** Sa:Cf£/PaprOv£
**1000 ppm:** CcrFOv/GmFOv/PaprTOv£/ScbaF/SaF
**1400 ppm:** SaF:Pd,Pp
**§:** ScbaF:Pd,Pp/SaF:Pd,Pp:AScba
**Escape:** GmFOv/ScbaE

**Incompatibilities and Reactivities:** Strong acids, strong oxidizers, metals, nitro compounds
[**Note:** Corrosive to metals.]

**Exposure Routes, Symptoms, Target Organs (see Table 5):**
**ER:** Inh, Abs, Ing, Con
**SY:** Irrit eyes, skin, nose, resp sys; vis dist; cough; in animals: liver, kidney damage
**TO:** Eyes, skin, resp sys, liver, kidneys

**First Aid (see Table 6):**
**Eye:** Irr immed
**Skin:** Water flush immed
**Breath:** Resp support
**Swallow:** Medical attention immed

N

## Naphtha (coal tar)

| Naphtha (coal tar) | Formula: | CAS#: 8030-30-6 | RTECS#: DE3030000 | IDLH: 1000 ppm [10%LEL] |
|---|---|---|---|---|
| **Conversion:** 1 ppm = 4.50 mg/m$^3$ (approx) | **DOT:** | | | |

**Synonyms/Trade Names:** Crude solvent coal tar naphtha, High solvent naphtha, Naphtha

**Exposure Limits:**
**NIOSH REL:** TWA 100 ppm (400 mg/m$^3$)
**OSHA PEL:** TWA 100 ppm (400 mg/m$^3$)

**Measurement Methods (see Table 1):**
**NIOSH** 1550

**Physical Description:** Reddish-brown, mobile liquid with an aromatic odor.

**Chemical & Physical Properties:**
**MW:** 110 (approx)
**BP:** 320-428°F
**Sol:** Insoluble
**Fl.P:** 100-109°F
**IP:** ?
**Sp.Gr:** 0.89-0.97
**VP:** <5 mmHg
**FRZ:** ?
**UEL:** ?
**LEL:** ?
Class II Combustible Liquid

**Personal Protection/Sanitation (see Table 2):**
**Skin:** Prevent skin contact
**Eyes:** Prevent eye contact
**Wash skin:** When contam
**Remove:** When wet or contam
**Change:** N.R.

**Respirator Recommendations (see Tables 3 and 4):**
**NIOSH/OSHA**
**1000 ppm:** Sa:Cf£/CcrFOv/GmFOv/PaprOv£/ScbaF/SaF
**§:** ScbaF:Pd,Pp/SaF:Pd,Pp:AScba
**Escape:** GmFOv/ScbaE

**Incompatibilities and Reactivities:** Strong oxidizers

**Exposure Routes, Symptoms, Target Organs (see Table 5):**
**ER:** Inh, Ing, Con
**SY:** Irrit eyes, skin, nose; dizz, drow; derm; in animals: liver, kidney damage
**TO:** Eyes, skin, resp sys, CNS, liver, kidneys

**First Aid (see Table 6):**
**Eye:** Irr immed
**Skin:** Soap wash prompt
**Breath:** Resp support
**Swallow:** Medical attention immed

| Naphthalene | Formula: $C_{10}H_8$ | CAS#: 91-20-3 | RTECS#: QJ0525000 | IDLH: 250 ppm |
|---|---|---|---|---|

**Conversion:** 1 ppm = 5.24 mg/m$^3$ | **DOT:** 1334 133 (crude or refined); 2304 133 (molten)

**Synonyms/Trade Names:** Naphthalin, Tar camphor, White tar

**Exposure Limits:**
**NIOSH REL:** TWA 10 ppm (50 mg/m$^3$)
ST 15 ppm (75 mg/m$^3$)
**OSHA PEL†:** TWA 10 ppm (50 mg/m$^3$)

**Measurement Methods (see Table 1):**
**NIOSH** 1501
**OSHA** 35

**Physical Description:** Colorless to brown solid with an odor of mothballs.
[**Note:** Shipped as a molten solid.]

**Chemical & Physical Properties:**
**MW:** 128.2
**BP:** 424°F
**Sol:** 0.003%
**Fl.P:** 174°F
**IP:** 8.12 eV
**Sp.Gr:** 1.15
**VP:** 0.08 mmHg
**MLT:** 176°F
**UEL:** 5.9%
**LEL:** 0.9%
Combustible Solid, but will take some effort to ignite.

**Personal Protection/Sanitation (see Table 2):**
**Skin:** Prevent skin contact
**Eyes:** Prevent eye contact
**Wash skin:** When contam
**Remove:** When wet or contam
**Change:** Daily

**Respirator Recommendations (see Tables 3 and 4):**
**NIOSH/OSHA**
**100 ppm:** CcrOv95*/Sa*
**250 ppm:** Sa:Cf*/CcrFOv100/PaprOvHie*/ScbaF/SaF
**§:** ScbaF:Pd,Pp/SaF:Pd,Pp:AScba
**Escape:** GmFOv100/ScbaE

**Incompatibilities and Reactivities:** Strong oxidizers, chromic anhydride

**Exposure Routes, Symptoms, Target Organs (see Table 5):**
**ER:** Inh, Abs, Ing, Con
**SY:** Irrit eyes; head, conf, excitement, mal; nau, vomit, abdom pain; irrit bladder; profuse sweat; jaun; hema, renal shutdown; derm, optical neuritis, corn damage
**TO:** Eyes, skin, blood, liver, kidneys, CNS

**First Aid (see Table 6):**
**Eye:** Irr immed
**Skin:** Molten flush immed/sol-liq soap wash prompt
**Breath:** Resp support
**Swallow:** Medical attention immed

N

| Naphthalene diisocyanate | Formula: $C_{10}H_6(NCO)_2$ | CAS#: 3173-72-6 | RTECS#: NQ9600000 | IDLH: N.D. |
|---|---|---|---|---|

**Conversion:** 1 ppm = 8.60 mg/m$^3$ | **DOT:**

**Synonyms/Trade Names:** 1,5-Diisocyanatonaphthalene; 1,5-Naphthalene diisocyanate; 1,5-Naphthalene ester of isocyanic acid; NDI

**Exposure Limits:**
**NIOSH REL:** TWA 0.040 mg/m$^3$ (0.005 ppm)
C 0.170 mg/m$^3$ (0.020 ppm) [10-minute]
**OSHA PEL:** none

**Measurement Methods (see Table 1):**
**NIOSH** 5525
**OSHA** PV2046

**Physical Description:** White to light-yellow, crystalline flakes.

**Chemical & Physical Properties:**
**MW:** 210.2
**BP:** 505°F
**Sol:** ?
**Fl.P(oc):** 311°F
**IP:** ?
**Sp.Gr:** ?
**VP(75°F):** 0.003 mmHg
**MLT:** 261°F
**UEL:** ?
**LEL:** ?
Combustible Solid

**Personal Protection/Sanitation (see Table 2):**
**Skin:** Prevent skin contact
**Eyes:** Prevent eye contact
**Wash skin:** When contam
**Remove:** When wet or contam
**Change:** Daily

**Respirator Recommendations (see Tables 3 and 4):**
**NIOSH**
**0.05 ppm:** Sa*
**0.125 ppm:** Sa:Cf*
**0.25 ppm:** ScbaF/SaF
**1 ppm:** SaF:Pd,Pp
**§:** ScbaF:Pd,Pp/SaF:Pd,Pp:AScba
**Escape:** GmFOv/ScbaE

**Incompatibilities and Reactivities:** None reported

**Exposure Routes, Symptoms, Target Organs (see Table 5):**
**ER:** Inh, Ing, Con
**SY:** Irrit eyes, nose, throat; resp sens, cough, pulm secretions, chest pain, dysp; asthma
**TO:** Eyes, resp sys

**First Aid (see Table 6):**
**Eye:** Irr immed
**Skin:** Soap wash immed
**Breath:** Resp support
**Swallow:** Medical attention immed

N

## α-Naphthylamine

| α-Naphthylamine | Formula: $C_{10}H_7NH_2$ | CAS#: 134-32-7 | RTECS#: QM1400000 | IDLH: Ca [N.D.] |
|---|---|---|---|---|
| **Conversion:** | **DOT:** 2077 153 | | | |

**Synonyms/Trade Names:** 1-Aminonaphthalene, 1-Naphthylamine

**Exposure Limits:**
**NIOSH REL:** Ca
See Appendix A
**OSHA PEL:** [1910.1004] See Appendix B

**Measurement Methods (see Table 1):**
**NIOSH** 5518
**OSHA** 93

**Physical Description:** Colorless crystals with an ammonia-like odor.
[**Note:** Darkens in air to a reddish-purple color.]

**Chemical & Physical Properties:**
**MW:** 143.2
**BP:** 573°F
**Sol:** 0.002%
**Fl.P:** 315°F
**IP:** 7.30 eV
**Sp.Gr:** 1.12
**VP(220°F):** 1 mmHg
**MLT:** 122°F
**UEL:** ?
**LEL:** ?
Combustible Solid

**Personal Protection/Sanitation (see Table 2):**
**Skin:** Prevent skin contact
**Eyes:** Prevent eye contact
**Wash skin:** When contam/Daily
**Remove:** When wet or contam
**Change:** Daily
**Provide:** Eyewash
Quick drench

**Respirator Recommendations (see Tables 3 and 4):**
**NIOSH**
**¥:** ScbaF:Pd,Pp/SaF:Pd,Pp:AScba
**Escape:** 100F/ScbaE

**See Appendix E** (page 351)

**Incompatibilities and Reactivities:** Oxidizes in air

**Exposure Routes, Symptoms, Target Organs (see Table 5):**
**ER:** Inh, Abs, Ing, Con
**SY:** Derm; hemorrhagic cystitis; dysp, ataxia, methemo; hema; dysuria; [carc]
**TO:** Bladder, skin [bladder cancer]

**First Aid (see Table 6):**
**Eye:** Irr immed
**Skin:** Soap wash immed
**Breath:** Resp support
**Swallow:** Medical attention immed

## β-Naphthylamine

| β-Naphthylamine | Formula: $C_{10}H_7NH_2$ | CAS#: 91-59-8 | RTECS#: QM2100000 | IDLH: Ca [N.D.] |
|---|---|---|---|---|
| **Conversion:** | **DOT:** 1650 153 | | | |

**Synonyms/Trade Names:** 2-Aminonaphthalene, 2-Naphthylamine

**Exposure Limits:**
**NIOSH REL:** Ca
See Appendix A
**OSHA PEL:** [1910.1009] See Appendix B

**Measurement Methods (see Table 1):**
**NIOSH** 5518
**OSHA** 93

**Physical Description:** Odorless, white to red crystals with a faint, aromatic odor.
[**Note:** Darkens in air to a reddish-purple color.]

**Chemical & Physical Properties:**
**MW:** 143.2
**BP:** 583°F
**Sol:** Miscible in hot water
**Fl.P:** 315°F
**IP:** 9.71 eV
**Sp.Gr(208°F):** 1.06
**VP(226°F):** 1 mmHg
**MLT:** 232°F
**UEL:** ?
**LEL:** ?
Combustible Solid

**Personal Protection/Sanitation (see Table 2):**
**Skin:** Prevent skin contact
**Eyes:** Prevent eye contact
**Wash skin:** When contam/Daily
**Remove:** When wet or contam
**Change:** Daily
**Provide:** Eyewash
Quick drench

**Respirator Recommendations (see Tables 3 and 4):**
**NIOSH**
**¥:** ScbaF:Pd,Pp/SaF:Pd,Pp:AScba
**Escape:** 100F/ScbaE

**See Appendix E** (page 351)

**Incompatibilities and Reactivities:** None reported

**Exposure Routes, Symptoms, Target Organs (see Table 5):**
**ER:** Inh, Abs, Ing, Con
**SY:** Derm; hemorrhagic cystitis; dysp; ataxia; methemo, hema; dysuria; [carc]
**TO:** Bladder, skin [bladder cancer]

**First Aid (see Table 6):**
**Eye:** Irr immed
**Skin:** Soap wash immed
**Breath:** Resp support
**Swallow:** Medical attention immed

| Niax® Catalyst ESN | Formula: | CAS#: 62765-93-9 | RTECS#: QR3900000 | IDLH: N.D. |
|---|---|---|---|---|
| **Conversion:** | **DOT:** | | | |

**Synonyms/Trade Names:** None
[**Note:** A mixture of 95% dimethylaminopropionitrile & 5% bis(2-dimethylamino)ethyl ether.]

**Exposure Limits:**
**NIOSH REL:** See Appendix C
**OSHA PEL:** See Appendix C

**Measurement Methods (see Table 1):**
None available

**Physical Description:** A liquid mixture.
[**Note:** Used in the past as a catalyst in the manufacture of flexible polyurethane foams.]

| Chemical & Physical Properties: | Personal Protection/Sanitation (see Table 2): | Respirator Recommendations (see Tables 3 and 4): |
|---|---|---|
| **MW:** mixture<br>**BP:** ?<br>**Sol:** ?<br>**Fl.P:** ?<br>**IP:** ?<br>**Sp.Gr:** ?<br>**VP:** ?<br>**FRZ:** ?<br>**UEL:** ?<br>**LEL:** ? | **Skin:** Prevent skin contact<br>**Eyes:** Prevent eye contact<br>**Wash skin:** When contam<br>**Remove:** When wet or contam<br>**Change:** Daily<br>**Provide:** Eyewash<br>Quick drench | **NIOSH**<br>**¥:** ScbaF:Pd,Pp/SaF:Pd,Pp:AScba<br>**Escape:** GmFOv/ScbaE |

**Incompatibilities and Reactivities:** Oxidizers

**Exposure Routes, Symptoms, Target Organs (see Table 5):**
**ER:** Inh, Abs, Ing, Con
**SY:** Irrit eyes, skin; urinary dist; neurological disorders; pins & needles in hands & feet; musc weak, lass, nau, vomit; decr nerve conduction in lower legs
**TO:** Eyes, skin, urinary tract, PNS

**First Aid (see Table 6):**
**Eye:** Irr immed
**Skin:** Soap flush immed
**Breath:** Resp support
**Swallow:** Medical attention immed

N

| Nickel carbonyl | Formula: $Ni(CO)_4$ | CAS#: 13463-39-3 | RTECS#: QR6300000 | IDLH: Ca [2 ppm] |
|---|---|---|---|---|
| **Conversion:** 1 ppm = 6.98 mg/m$^3$ | **DOT:** 1259 131 | | | |

**Synonyms/Trade Names:** Nickel tetracarbonyl, Tetracarbonyl nickel

**Exposure Limits:**
**NIOSH REL:** Ca
TWA 0.001 ppm (0.007 mg/m$^3$)
See Appendix A
**OSHA PEL:** TWA 0.001 ppm (0.007 mg/m$^3$)

**Measurement Methods (see Table 1):**
**NIOSH** 6007

**Physical Description:** Colorless to yellow liquid with a musty odor. [**Note:** A gas above 110°F.]

| Chemical & Physical Properties: | Personal Protection/Sanitation (see Table 2): | Respirator Recommendations (see Tables 3 and 4): |
|---|---|---|
| **MW:** 170.7<br>**BP:** 110°F<br>**Sol:** 0.05%<br>**Fl.P:** <-4°F<br>**IP:** 8.28 eV<br>**Sp.Gr(63°F):** 1.32<br>**VP:** 315 mmHg<br>**FRZ:** -13°F<br>**UEL:** ?<br>**LEL:** 2%<br>Class IB Flammable Liquid | **Skin:** Prevent skin contact<br>**Eyes:** Prevent eye contact<br>**Wash skin:** When contam<br>**Remove:** When wet (flamm)<br>**Change:** N.R.<br>**Provide:** Eyewash<br>Quick drench | **NIOSH**<br>**¥:** ScbaF:Pd,Pp/SaF:Pd,Pp:AScba<br>**Escape:** GmFS/ScbaE |

**Incompatibilities and Reactivities:** Nitric acid, bromine, chlorine & other oxidizers; flammable materials

**Exposure Routes, Symptoms, Target Organs (see Table 5):**
**ER:** Inh, Ing, Abs, Con
**SY:** Head, dizz; nau, vomit, epigastric pain; substernal pain; cough, hyperpnea; cyan; lass; leucyt, pneu; delirium, convuls; [carc]; in animals: repro, terato effects
**TO:** Lungs, paranasal sinus, CNS, repro sys [lung & nasal cancer]

**First Aid (see Table 6):**
**Eye:** Irr immed
**Skin:** Soap wash immed
**Breath:** Resp support
**Swallow:** Medical attention immed

N

## Nickel metal and other compounds (as Ni)

| | | | | |
|---|---|---|---|---|
| **Nickel metal and other compounds (as Ni)** | **Formula:** Ni (metal) | **CAS#:** 7440-02-0 (metal) | **RTECS#:** QR5950000 (metal) | **IDLH:** Ca [10 mg/m$^3$ (as Ni)] |
| **Conversion:** | **DOT:** | | | |

**Synonyms/Trade Names: Nickel metal:** Elemental nickel, Nickel catalyst
Synonyms of other nickel compounds vary depending upon the specific compound.

**Exposure Limits:**
**NIOSH REL*:** Ca
TWA 0.015 mg/m$^3$
See Appendix A
**OSHA PEL*†:** TWA 1 mg/m$^3$
[***Note**: The REL and PEL do not apply to Nickel carbonyl.]

**Measurement Methods (see Table 1):**
**NIOSH** 7300, 7301, 7303, 9102
**OSHA** ID121, ID125G

**Physical Description:** Metal: Lustrous, silvery, odorless solid.

**Chemical & Physical Properties:**
**MW:** 58.7
**BP:** 5139°F
**Sol:** Insoluble
**Fl.P:** NA
**IP:** NA
**Sp.Gr:** 8.90 (Metal)
**VP:** 0 mmHg (approx)
**MLT:** 2831°F
**UEL:** NA
**LEL:** NA
Metal: Combustible Solid; nickel sponge catalyst may ignite SPONTANEOUSLY in air.

**Personal Protection/Sanitation (see Table 2):**
**Skin:** Prevent skin contact
**Eyes:** N.R.
**Wash skin:** When contam/Daily
**Remove:** When wet or contam
**Change:** Daily

**Respirator Recommendations (see Tables 3 and 4):**
**NIOSH**
**¥:** ScbaF:Pd,Pp/SaF:Pd,Pp:AScba
**Escape:** 100F/ScbaE

**Incompatibilities and Reactivities:** Strong acids, sulfur, selenium, wood & other combustibles, nickel nitrate

**Exposure Routes, Symptoms, Target Organs (see Table 5):**
**ER:** Inh, Ing, Con
**SY:** Sens derm, allergic asthma, pneu; [carc]
**TO:** Nasal cavities, lungs, skin [lung and nasal cancer]

**First Aid (see Table 6):**
**Skin:** Water flush immed
**Breath:** Resp support
**Swallow:** Medical attention immed

## Nicotine

| | | | | |
|---|---|---|---|---|
| **Nicotine** | **Formula:** $C_5H_4NC_4H_7NCH_3$ | **CAS#:** 54-11-5 | **RTECS#:** QS5250000 | **IDLH:** 5 mg/m$^3$ |
| **Conversion:** | **DOT:** 1654 151 | | | |

**Synonyms/Trade Names:** 3-(1-Methyl-2-pyrrolidyl)pyridine

**Exposure Limits:**
**NIOSH REL:** TWA 0.5 mg/m$^3$ [skin]
**OSHA PEL:** TWA 0.5 mg/m$^3$ [skin]

**Measurement Methods (see Table 1):**
**NIOSH** 2544, 2551

**Physical Description:** Pale-yellow to dark-brown liquid with a fish-like odor when warm. [insecticide]

**Chemical & Physical Properties:**
**MW:** 162.2
**BP:** 482°F (Decomposes)
**Sol:** Miscible
**Fl.P:** 203°F
**IP:** 8.01 eV
**Sp.Gr:** 1.01
**VP:** 0.08 mmHg
**FRZ:** -110°F
**UEL:** 4.0%
**LEL:** 0.7%
Class IIIB Combustible Liquid

**Personal Protection/Sanitation (see Table 2):**
**Skin:** Prevent skin contact
**Eyes:** Prevent eye contact
**Wash skin:** When contam
**Remove:** When wet or contam
**Change:** N.R.
**Provide:** Eyewash
Quick drench

**Respirator Recommendations (see Tables 3 and 4):**
**NIOSH/OSHA**
**5 mg/m$^3$:** Sa/ScbaF
**§:** ScbaF:Pd,Pp/SaF:Pd,Pp:AScba
**Escape:** GmFOv/ScbaE

**Incompatibilities and Reactivities:** Strong oxidizers, strong acids

**Exposure Routes, Symptoms, Target Organs (see Table 5):**
**ER:** Inh, Abs, Ing, Con
**SY:** Nau, salv, abdom pain, vomit, diarr; head, dizz, hearing, vis dist; conf, lass, inco; card arrhy; convuls, dysp; in animals: terato effects
**TO:** CNS, CVS, lungs, GI tract, repro sys

**First Aid (see Table 6):**
**Eye:** Irr immed
**Skin:** Water flush immed
**Breath:** Resp support
**Swallow:** Medical attention immed

## Nitric acid

| Nitric acid | Formula: $HNO_3$ | CAS#: 7697-37-2 | RTECS#: QU5775000 | IDLH: 25 ppm |
|---|---|---|---|---|

**Conversion:** 1 ppm = 2.58 mg/m$^3$ | **DOT:** 2032 157 (fuming); 2031 157 (other than red fuming)

**Synonyms/Trade Names:** Aqua fortis, Engravers acid, Hydrogen nitrate, Red fuming nitric acid (RFNA), White fuming nitric acid (WFNA)

**Exposure Limits:**
**NIOSH REL:** TWA 2 ppm (5 mg/m$^3$)
ST 4 ppm (10 mg/m$^3$)
**OSHA PEL†:** TWA 2 ppm (5 mg/m$^3$)

**Measurement Methods (see Table 1):**
**NIOSH** 7903
**OSHA** ID165SG

**Physical Description:** Colorless, yellow, or red, fuming liquid with an acrid, suffocating odor.
[**Note:** Often used in an aqueous solution. Fuming nitric acid is concentrated nitric acid that contains dissolved nitrogen dioxide.]

**Chemical & Physical Properties:**
**MW:** 63.0
**BP:** 181°F
**Sol:** Miscible
**Fl.P:** NA
**IP:** 11.95 eV
**Sp.Gr(77°F):** 1.50
**VP:** 48 mmHg
**FRZ:** -44°F
**UEL:** NA
**LEL:** NA
Noncombustible Liquid, but increases the flammability of combustible materials.

**Personal Protection/Sanitation (see Table 2):**
**Skin:** Prevent skin contact
**Eyes:** Prevent eye contact
**Wash skin:** When contam
**Remove:** When wet or contam
**Change:** N.R.
**Provide:** Eyewash (pH<2.5)
Quick drench (pH<2.5)

**Respirator Recommendations (see Tables 3 and 4):**
**NIOSH/OSHA**
**25 ppm:** Sa:Cf*/CcrFS¿/GmFS¿/ScbaF/SaF
**§:** ScbaF:Pd,Pp/SaF:Pd,Pp:AScba
**Escape:** GmFS¿/ScbaE

**Incompatibilities and Reactivities:** Combustible materials, metallic powders, hydrogen sulfide, carbides, alcohols
[**Note:** Reacts with water to produce heat. Corrosive to metals.]

**Exposure Routes, Symptoms, Target Organs (see Table 5):**
**ER:** Inh, Ing, Con
**SY:** Irrit eyes, skin, muc memb; delayed pulm edema, pneu, bron; dental erosion
**TO:** Eyes, skin, resp sys, teeth

**First Aid (see Table 6):**
**Eye:** Irr immed
**Skin:** Water flush immed
**Breath:** Resp support
**Swallow:** Medical attention immed

N

## Nitric oxide

| Nitric oxide | Formula: NO | CAS#: 10102-43-9 | RTECS#: QX0525000 | IDLH: 100 ppm |
|---|---|---|---|---|

**Conversion:** 1 ppm = 1.23 mg/m$^3$ | **DOT:** 1660 124

**Synonyms/Trade Names:** Mononitrogen monoxide, Nitrogen monoxide

**Exposure Limits:**
**NIOSH REL:** TWA 25 ppm (30 mg/m$^3$)
**OSHA PEL:** TWA 25 ppm (30 mg/m$^3$)

**Measurement Methods (see Table 1):**
**NIOSH** 6014
**OSHA** ID190

**Physical Description:** Colorless gas.
[**Note:** Shipped as a nonliquefied compressed gas.]

**Chemical & Physical Properties:**
**MW:** 30.0
**BP:** -241°F
**Sol:** 5%
**Fl.P:** NA
**IP:** 9.27 eV
**RGasD:** 1.04
**VP:** 34.2 atm
**FRZ:** -263°F
**UEL:** NA
**LEL:** NA
Nonflammable Gas, but will accelerate the burning of combustible materials.

**Personal Protection/Sanitation (see Table 2):**
**Skin:** N.R.
**Eyes:** N.R.
**Wash skin:** N.R.
**Remove:** N.R.
**Change:** N.R.

**Respirator Recommendations (see Tables 3 and 4):**
**NIOSH/OSHA**
**100 ppm:** Sa:Cf*/CcrFS¿/PaprS*¿/GmFS¿/Sa*/ScbaF
**§:** ScbaF:Pd,Pp/SaF:Pd,Pp:AScba
**Escape:** GmFS¿/ScbaE

**Incompatibilities and Reactivities:** Fluorine, combustible materials, ozone, $NH_3$, chlorinated hydrocarbons, metals, carbon disulfide [**Note:** Reacts with water to form nitric acid. Rapidly converted in air to nitrogen dioxide.]

**Exposure Routes, Symptoms, Target Organs (see Table 5):**
**ER:** Inh
**SY:** Irrit eyes, wet skin, nose, throat; drow, uncon; methemo
**TO:** Eyes, skin, resp sys, blood, CNS

**First Aid (see Table 6):**
**Breath:** Resp support

N

## p-Nitroaniline

| p-Nitroaniline | Formula: $NO_2C_6H_4NH_2$ | CAS#: 100-01-6 | RTECS#: BY7000000 | IDLH: 300 mg/m³ |
|---|---|---|---|---|
| Conversion: | DOT: 1661 153 | | | |

**Synonyms/Trade Names:** para-Aminonitrobenzene, 4-Nitroaniline, 4-Nitrobenzenamine, p-Nitrophenylamine, PNA

**Exposure Limits:**
**NIOSH REL:** TWA 3 $mg/m^3$ [skin]
**OSHA PEL†:** TWA 6 $mg/m^3$ (1 ppm) [skin]

**Measurement Methods (see Table 1):**
**NIOSH** 5033

**Physical Description:** Bright yellow, crystalline powder with a slight ammonia-like odor.

**Chemical & Physical Properties:**
**MW:** 138.1
**BP:** 630°F
**Sol:** 0.08%
**Fl.P:** 390°F
**IP:** 8.85 eV
**Sp.Gr:** 1.42
**VP:** 0.00002 mmHg
**MLT:** 295°F
**UEL:** ?
**LEL:** ?
Combustible Solid

**Personal Protection/Sanitation (see Table 2):**
**Skin:** Prevent skin contact
**Eyes:** Prevent eye contact
**Wash skin:** When contam/Daily
**Remove:** When wet or contam
**Change:** Daily
**Provide:** Quick drench

**Respirator Recommendations (see Tables 3 and 4):**
**NIOSH**
**30 $mg/m^3$:** Sa*
**75 $mg/m^3$:** Sa:Cf*
**150 $mg/m^3$:** ScbaF/SaF
**300 $mg/m^3$:** SaF:Pd,Pp
**§:** ScbaF:Pd,Pp/SaF:Pd,Pp:AScba
**Escape:** GmFOv100/ScbaE

**Incompatibilities and Reactivities:** Strong oxidizers, strong reducers
[**Note:** May result in spontaneous heating of organic materials in the presence of moisture.]

**Exposure Routes, Symptoms, Target Organs (see Table 5):**
**ER:** Inh, Abs, Ing, Con
**SY:** Irrit nose, throat; cyan, ataxia; tacar, tachypnea; dysp; irrity; vomit, diarr; convuls; resp arrest; anemia; methemo; jaundice
**TO:** Resp sys, blood, heart, liver

**First Aid (see Table 6):**
**Eye:** Irr immed
**Skin:** Water flush immed
**Breath:** Resp support
**Swallow:** Medical attention immed

## Nitrobenzene

| Nitrobenzene | Formula: $C_6H_5NO_2$ | CAS#: 98-95-3 | RTECS#: DA6475000 | IDLH: 200 ppm |
|---|---|---|---|---|
| Conversion: 1 ppm = 5.04 mg/m³ | DOT: 1662 152 | | | |

**Synonyms/Trade Names:** Essence of mirbane, Nitrobenzol, Oil of mirbane

**Exposure Limits:**
**NIOSH REL:** TWA 1 ppm (5 $mg/m^3$) [skin]
**OSHA PEL:** TWA 1 ppm (5 $mg/m^3$) [skin]

**Measurement Methods (see Table 1):**
**NIOSH** 2005, 2017

**Physical Description:** Yellow, oily liquid with a pungent odor like paste shoe polish.
[**Note:** A solid below 42°F.]

**Chemical & Physical Properties:**
**MW:** 123.1
**BP:** 411°F
**Sol:** 0.2%
**Fl.P:** 190°F
**IP:** 9.92 eV
**Sp.Gr:** 1.20
**VP(77°F):** 0.3 mmHg
**FRZ:** 42°F
**UEL:** ?
**LEL(200°F):** 1.8%
Class IIIA Combustible Liquid

**Personal Protection/Sanitation (see Table 2):**
**Skin:** Prevent skin contact
**Eyes:** Prevent eye contact
**Wash skin:** N.R.
**Remove:** When wet or contam
**Change:** Daily
**Provide:** Quick drench

**Respirator Recommendations (see Tables 3 and 4):**
**NIOSH/OSHA**
**10 ppm:** CcrOv*/Sa*
**25 ppm:** Sa:Cf*/PaprOv*
**50 ppm:** CcrFOv/GmFOv/PaprTOv*/ScbaF/SaF
**200 ppm:** SaF:Pd,Pp
**§:** ScbaF:Pd,Pp/SaF:Pd,Pp:AScba
**Escape:** GmFOv/ScbaE

**Incompatibilities and Reactivities:** Concentrated nitric acid, nitrogen tetroxide, caustics, phosphorus pentachloride, chemically-active metals such as tin or zinc

**Exposure Routes, Symptoms, Target Organs (see Table 5):**
**ER:** Inh, Abs, Ing, Con
**SY:** Irrit eyes, skin; anoxia; derm; anemia; methemo; in animals: liver, kidney damage; testicular effects
**TO:** Eyes, skin, blood, liver, kidneys, CVS, repro sys

**First Aid (see Table 6):**
**Eye:** Irr immed
**Skin:** Soap wash immed
**Breath:** Resp support
**Swallow:** Medical attention immed

| 4-Nitrobiphenyl | Formula: $C_6H_5C_6H_4NO_2$ | CAS#: 92-93-3 | RTECS#: DV5600000 | IDLH: Ca [N.D.] |
|---|---|---|---|---|
| **Conversion:** | **DOT:** | | | |

**Synonyms/Trade Names:** p-Nitrobiphenyl, p-Nitrodiphenyl, 4-Nitrodiphenyl, p-Phenylnitrobenzene, 4-Phenylnitrobenzene, PNB

**Exposure Limits:**
**NIOSH REL:** Ca
See Appendix A
**OSHA PEL:** [1910.1003] See Appendix B

**Measurement Methods (see Table 1):**
**NIOSH** P&CAM273 (II-4)
**OSHA** PV2082

**Physical Description:** White to yellow, needle-like, crystalline solid with a sweetish odor.

| Chemical & Physical Properties: | Personal Protection/Sanitation (see Table 2): | Respirator Recommendations (see Tables 3 and 4): |
|---|---|---|
| **MW:** 199.2<br>**BP:** 644°F<br>**Sol:** Insoluble<br>**Fl.P:** 290°F<br>**IP:** ?<br>**Sp.Gr:** ?<br>**VP:** ?<br>**MLT:** 237°F<br>**UEL:** ?<br>**LEL:** ?<br>Combustible Solid | **Skin:** Prevent skin contact<br>**Eyes:** Prevent eye contact<br>**Wash skin:** When contam/Daily<br>**Remove:** When wet or contam<br>**Change:** Daily<br>**Provide:** Eyewash<br>Quick drench | **NIOSH**<br>**¥:** ScbaF:Pd,Pp/SaF:Pd,Pp:AScba<br>**Escape:** 100F/ScbaE<br><br>**See Appendix E** (page 351) |

**Incompatibilities and Reactivities:** Strong reducers

| Exposure Routes, Symptoms, Target Organs (see Table 5): | First Aid (see Table 6): |
|---|---|
| **ER:** Inh, Abs, Ing, Con<br>**SY:** Head, drow, dizz; dysp; ataxia, lass; methemo; urinary burning; acute hemorrhagic cystitis; [carc]<br>**TO:** Bladder, blood [in animals: bladder tumors] | **Eye:** Irr immed<br>**Skin:** Soap wash immed<br>**Breath:** Resp support<br>**Swallow:** Medical attention immed |



| p-Nitrochlorobenzene | Formula: $ClC_6H_4NO_2$ | CAS#: 100-00-5 | RTECS#: CZ1050000 | IDLH: Ca [100 $mg/m^3$] |
|---|---|---|---|---|
| **Conversion:** | **DOT:** 1578 152 | | | |

**Synonyms/Trade Names:** p-Chloronitrobenzene, 4-Chloronitrobenzene, 1-Chloro-4-nitrobenzene, 4-Nitrochlorobenzene, PCNB, PNCB

**Exposure Limits:**
**NIOSH REL:** Ca
See Appendix A [skin]
**OSHA PEL:** TWA 1 $mg/m^3$ [skin]

**Measurement Methods (see Table 1):**
**NIOSH** 2005

**Physical Description:** Yellow, crystalline solid with a sweet odor.

| Chemical & Physical Properties: | Personal Protection/Sanitation (see Table 2): | Respirator Recommendations (see Tables 3 and 4): |
|---|---|---|
| **MW:** 157.6<br>**BP:** 468°F<br>**Sol:** Slight<br>**Fl.P:** 261°F<br>**IP:** 9.96 eV<br>**Sp.Gr:** 1.52<br>**VP(86°F):** 0.2 mmHg<br>**MLT:** 182°F<br>**UEL:** ?<br>**LEL:** ?<br>Solid that does not burn, or burns with difficulty. | **Skin:** Prevent skin contact<br>**Eyes:** Prevent eye contact<br>**Wash skin:** When contam/Daily<br>**Remove:** When wet or contam<br>**Change:** Daily<br>**Provide:** Eyewash<br>Quick drench | **NIOSH**<br>**¥:** ScbaF:Pd,Pp/SaF:Pd,Pp:AScba<br>**Escape:** 100F/ScbaE |

**Incompatibilities and Reactivities:** Strong oxidizers, alkalis

| Exposure Routes, Symptoms, Target Organs (see Table 5): | First Aid (see Table 6): |
|---|---|
| **ER:** Inh, Abs, Ing, Con<br>**SY:** Anoxia; unpleasant taste; anemia; methemo; in animals: hema; spleen, kidney, bone marrow changes; repro effects; [carc]<br>**TO:** Blood, liver, kidneys, CVS, spleen, bone marrow, repro sys [in animals: vascular & liver tumors] | **Eye:** Irr immed<br>**Skin:** Soap wash immed<br>**Breath:** Resp support<br>**Swallow:** Medical attention immed |

## Nitroethane

| Formula: | CAS#: | RTECS#: | IDLH: |
|---|---|---|---|
| $CH_3CH_2NO_2$ | 79-24-3 | KI5600000 | 1000 ppm |

**Conversion:** 1 ppm = 3.07 mg/m$^3$ **DOT:** 2842 129

**Synonyms/Trade Names:** Nitroetan

**Exposure Limits:**
**NIOSH REL:** TWA 100 ppm (310 mg/m$^3$)
**OSHA PEL:** TWA 100 ppm (310 mg/m$^3$)

**Measurement Methods (see Table 1):**
**NIOSH** 2526

**Physical Description:** Colorless, oily liquid with a mild, fruity odor.

**Chemical & Physical Properties:**
**MW:** 75.1
**BP:** 237°F
**Sol:** 5%
**Fl.P:** 82°F
**IP:** 10.88 eV
**Sp.Gr:** 1.05
**VP(77°F):** 21 mmHg
**FRZ:** -130°F
**UEL:** ?
**LEL:** 3.4%
Class IC Flammable Liquid

**Personal Protection/Sanitation (see Table 2):**
**Skin:** Prevent skin contact
**Eyes:** Prevent eye contact
**Wash skin:** When contam
**Remove:** When wet (flamm)
**Change:** N.R.

**Respirator Recommendations (see Tables 3 and 4):**
**NIOSH/OSHA**
**1000 ppm:** ScbaF/SaF
**§:** ScbaF:Pd,Pp/SaF:Pd,Pp:AScba
**Escape:** ScbaE

**Incompatibilities and Reactivities:** Amines; strong acids, alkalis & oxidizers; hydrocarbons; combustibles; metal oxides

**Exposure Routes, Symptoms, Target Organs (see Table 5):**
**ER:** Inh, Ing, Con
**SY:** Derm; in animals: lac; dysp, pulm rales, edema; liver, kidney inj; narco
**TO:** Skin, resp sys, CNS, kidneys, liver

**First Aid (see Table 6):**
**Eye:** Irr immed
**Skin:** Soap wash prompt
**Breath:** Resp support
**Swallow:** Medical attention immed



## Nitrogen dioxide

| Formula: | CAS#: | RTECS#: | IDLH: |
|---|---|---|---|
| $NO_2$ | 10102-44-0 | QW9800000 | 20 ppm |

**Conversion:** 1 ppm = 1.88 mg/m$^3$ **DOT:** 1067 124

**Synonyms/Trade Names:** Dinitrogen tetroxide ($N_2O_4$), Nitrogen peroxide

**Exposure Limits:**
**NIOSH REL:** ST 1 ppm (1.8 mg/m$^3$)
**OSHA PEL†:** C 5 ppm (9 mg/m$^3$)

**Measurement Methods (see Table 1):**
**NIOSH** 6014
**OSHA** ID182

**Physical Description:** Yellowish-brown liquid or reddish-brown gas (above 70°F) with a pungent, acrid odor. [**Note:** In solid form (below 15°F) it is found structurally as $N_2O_4$.]

**Chemical & Physical Properties:**
**MW:** 46.0
**BP:** 70°F
**Sol:** Reacts
**Fl.P:** NA
**IP:** 9.75 eV
**RGasD:** 2.62
**Sp.Gr:** 1.44 (Liquid at 68°F)
**VP:** 720 mmHg
**FRZ:** 15°F
**UEL:** NA
**LEL:** NA
Noncombustible Liquid/Gas, but will accelerate the burning of combustible materials.

**Personal Protection/Sanitation (see Table 2):**
**Skin:** Prevent skin contact
**Eyes:** Prevent eye contact
**Wash skin:** When contam
**Remove:** When wet or contam
**Change:** N.R.
**Provide:** Eyewash
Quick drench

**Respirator Recommendations (see Tables 3 and 4):**
**NIOSH**
**20 ppm:** Sa:Cf£/ScbaF/SaF
**§:** ScbaF:Pd,Pp/SaF:Pd,Pp:AScba
**Escape:** GmFS¿/ScbaE

**Incompatibilities and Reactivities:** Combustible material, water, chlorinated hydrocarbons, carbon disulfide, ammonia [**Note:** Reacts with water to form nitric acid.]

**Exposure Routes, Symptoms, Target Organs (see Table 5):**
**ER:** Inh, Ing, Con
**SY:** Irrit eyes, nose, throat; cough, mucoid frothy sputum, decr pulm func, chronic bron, dysp; chest pain; pulm edema, cyan, tachypnea, tacar
**TO:** Eyes, resp sys, CVS

**First Aid (see Table 6):**
**Eye:** Irr immed
**Skin:** Water flush immed
**Breath:** Resp support
**Swallow:** Medical attention immed

## Nitrogen trifluoride

| Nitrogen trifluoride | Formula: $NF_3$ | CAS#: 7783-54-2 | RTECS#: QX1925000 | IDLH: 1000 ppm |
|---|---|---|---|---|
| **Conversion:** 1 ppm = 2.90 mg/$m^3$ | **DOT:** 2451 122 | | | |

**Synonyms/Trade Names:** Nitrogen fluoride, Trifluoramine, Trifluorammonia

**Exposure Limits:**
**NIOSH REL:** TWA 10 ppm (29 mg/$m^3$)
**OSHA PEL:** TWA 10 ppm (29 mg/$m^3$)

**Measurement Methods (see Table 1):**
None available

**Physical Description:** Colorless gas with a moldy odor.
[**Note:** Shipped as a nonliquefied compressed gas.]

| Chemical & Physical Properties: | Personal Protection/Sanitation (see Table 2): | Respirator Recommendations (see Tables 3 and 4): |
|---|---|---|
| **MW:** 71.0<br>**BP:** -200°F<br>**Sol:** Slight<br>**Fl.P:** NA<br>**IP:** 12.97 eV<br>**RGasD:** 2.46<br>**VP:** >1 atm<br>**FRZ:** -340°F<br>**UEL:** NA<br>**LEL:** NA<br>Nonflammable Gas | **Skin:** N.R.<br>**Eyes:** N.R.<br>**Wash skin:** N.R.<br>**Remove:** N.R.<br>**Change:** N.R. | **NIOSH/OSHA**<br>**100 ppm:** CcrS/Sa<br>**250 ppm:** Sa:Cf/PaprS<br>**500 ppm:** CcrFS/GmFS/PaprTS*/SaT:Cf*/ScbaF/SaF<br>**1000 ppm:** SaF:Pd,Pp<br>**§:** ScbaF:Pd,Pp/SaF:Pd,Pp:AScba<br>**Escape:** GmFS/ScbaE |

**Incompatibilities and Reactivities:** Water, oil, grease, oxidizable materials, ammonia, carbon monoxide, methane, hydrogen, hydrogen sulfide, activated charcoal, diborane

**Exposure Routes, Symptoms, Target Organs (see Table 5):**
**ER:** Inh
**SY:** In animals: anoxia, cyan; methemo; lass, dizz, head; liver, kidney inj
**TO:** Blood, liver, kidneys

**First Aid (see Table 6):**
**Breath:** Resp support

N

## Nitroglycerine

| Nitroglycerine | Formula: $CH_2NO_3CHNO_3CH_2NO_3$ | CAS#: 55-63-0 | RTECS#: QX2100000 | IDLH: 75 mg/$m^3$ |
|---|---|---|---|---|
| **Conversion:** 1 ppm = 9.29 mg/$m^3$ | **DOT:** 1204 127 (≤ 1% solution in alcohol); 3064 127 (1-5% solution in alcohol) | | | |

**Synonyms/Trade Names:** Glyceryl trinitrate; NG; 1,2,3-Propanetriol trinitrate; Trinitroglycerine

**Exposure Limits:**
**NIOSH REL:** ST 0.1 mg/$m^3$ [skin]
**OSHA PEL†:** C 0.2 ppm (2 mg/$m^3$) [skin]

**Measurement Methods (see Table 1):**
**NIOSH** 2507
**OSHA** 43

**Physical Description:** Colorless to pale-yellow, viscous liquid or solid (below 56°F).
[**Note:** An explosive ingredient in dynamite (20-40%) with ethylene glycol dinitrate (80-60%).]

| Chemical & Physical Properties: | Personal Protection/Sanitation (see Table 2): | Respirator Recommendations (see Tables 3 and 4): |
|---|---|---|
| **MW:** 227.1<br>**BP:** Begins to decompose at 122-140°F<br>**Sol:** 0.1%<br>**Fl.P:** Explodes<br>**IP:** ?<br>**Sp.Gr:** 1.60<br>**VP:** 0.0003 mmHg<br>**FRZ:** 56°F<br>**UEL:** ?<br>**LEL:** ?<br>Explosive Liquid | **Skin:** Prevent skin contact<br>**Eyes:** Prevent eye contact<br>**Wash skin:** When contam<br>**Remove:** When wet (flamm)<br>**Change:** Daily<br>**Provide:** Quick drench | **NIOSH**<br>**1 mg/$m^3$:** Sa*<br>**2.5 mg/$m^3$:** Sa:Cf*<br>**5 mg/$m^3$:** SaT:Cf*/ScbaF/SaF<br>**75 mg/$m^3$:** SaF:Pd,Pp<br>**§:** ScbaF:Pd,Pp/SaF:Pd,Pp:AScba<br>**Escape:** GmFOv100/ScbaE |

**Incompatibilities and Reactivities:** Heat, ozone, shock, acids [**Note:** An OSHA Class A Explosive (1910.109).]

**Exposure Routes, Symptoms, Target Organs (see Table 5):**
**ER:** Inh, Abs, Ing, Con
**SY:** Throb head; dizz; nau, vomit, abdom pain; hypotension; flush; palp; methemo; delirium, CNS depres; angina; skin irrit
**TO:** CVS, blood, skin, CNS

**First Aid (see Table 6):**
**Eye:** Irr immed
**Skin:** Soap wash immed
**Breath:** Resp support
**Swallow:** Medical attention immed

| Nitromethane | Formula: $CH_3NO_2$ | CAS#: 75-52-5 | RTECS#: PA9800000 | IDLH: 750 ppm |
|---|---|---|---|---|
| Conversion: 1 ppm = 2.50 mg/m$^3$ | DOT: 1261 129 | | | |

**Synonyms/Trade Names:** Nitrocarbol

**Exposure Limits:**
**NIOSH REL:** See Appendix D
**OSHA PEL:** TWA 100 ppm (250 mg/m$^3$)

**Measurement Methods (see Table 1):**
NIOSH 2527

**Physical Description:** Colorless, oily liquid with a disagreeable odor.

| Chemical & Physical Properties: | Personal Protection/Sanitation (see Table 2): | Respirator Recommendations (see Tables 3 and 4): |
|---|---|---|
| **MW:** 61.0<br>**BP:** 214°F<br>**Sol:** 10%<br>**Fl.P:** 95°F<br>**IP:** 11.08 eV<br>**Sp.Gr:** 1.14<br>**VP:** 28 mmHg<br>**FRZ:** -20°F<br>**UEL:** ?<br>**LEL:** 7.3%<br>Class IC Flammable Liquid | **Skin:** Prevent skin contact<br>**Eyes:** Prevent eye contact<br>**Wash skin:** When contam<br>**Remove:** When wet (flamm)<br>**Change:** N.R. | **OSHA**<br>**750 ppm:** Sa:Cf£/ScbaF/SaF<br>**§:** ScbaF:Pd,Pp/SaF:Pd,Pp:AScba<br>**Escape:** ScbaE |

**Incompatibilities and Reactivities:** Amines; strong acids, alkalis & oxidizers; hydrocarbons & other combustible materials; metallic oxides [**Note:** Slowly corrodes steel & copper when wet.]

| Exposure Routes, Symptoms, Target Organs (see Table 5): | First Aid (see Table 6): |
|---|---|
| **ER:** Inh, Ing, Con<br>**SY:** Derm; in animals: irrit eyes, resp sys; convuls, narco; liver damage<br>**TO:** Eyes, skin, CNS, liver | **Eye:** Irr immed<br>**Skin:** Soap wash prompt<br>**Breath:** Resp support<br>**Swallow:** Medical attention immed |



| 2-Nitronaphthalene | Formula: $C_{10}H_7NO_2$ | CAS#: 581-89-5 | RTECS#: QJ9760000 | IDLH: Ca [N.D.] |
|---|---|---|---|---|
| Conversion: | DOT: 2538 133 | | | |

**Synonyms/Trade Names:** β-Nitronaphthalene

**Exposure Limits:**
**NIOSH REL:** Ca*
See Appendix A
[***Note:** Since metabolized to β-Naphthylamine.]
**OSHA PEL:** none

**Measurement Methods (see Table 1):**
None available

**Physical Description:** Colorless solid.

| Chemical & Physical Properties: | Personal Protection/Sanitation (see Table 2): | Respirator Recommendations (see Tables 3 and 4): |
|---|---|---|
| **MW:** 178.2<br>**BP:** ?<br>**Sol:** Insoluble<br>**Fl.P:** ?<br>**IP:** 8.67 eV<br>**Sp.Gr:** ?<br>**VP:** ?<br>**MLT:** 174°F<br>**UEL:** ?<br>**LEL:** ?<br>Combustible Solid | **Skin:** Prevent skin contact<br>**Eyes:** Prevent eye contact<br>**Wash skin:** When contam/Daily<br>**Remove:** When wet or contam<br>**Change:** Daily<br>**Provide:** Eyewash<br>Quick drench | **NIOSH**<br>**¥:** ScbaF:Pd,Pp/SaF:Pd,Pp:AScba<br>**Escape:** GmFOv100/ScbaE |

**Incompatibilities and Reactivities:** For "Nitrates" in general: Aluminum, cyanides, esters, phosphorus, tin chlorides, thiocyanates, sodium hypophosphite

| Exposure Routes, Symptoms, Target Organs (see Table 5): | First Aid (see Table 6): |
|---|---|
| **ER:** Inh, Abs, Ing, Con<br>**SY:** Irrit skin, resp sys; derm; [carc]<br>**TO:** Skin, resp sys [bladder cancer] | **Eye:** Irr immed<br>**Skin:** Soap wash immed<br>**Breath:** Resp support<br>**Swallow:** Medical attention immed |

## 1-Nitropropane

| 1-Nitropropane | Formula: $CH_3CH_2CH_2NO_2$ | CAS#: 108-03-2 | RTECS#: TZ5075000 | IDLH: 1000 ppm |
|---|---|---|---|---|
| **Conversion:** 1 ppm = 3.64 mg/m$^3$ | **DOT:** 2608 129 | | | |

**Synonyms/Trade Names:** Nitropropane, 1-NP

**Exposure Limits:**
**NIOSH REL:** TWA 25 ppm (90 mg/m$^3$)
**OSHA PEL:** TWA 25 ppm (90 mg/m$^3$)

**Measurement Methods (see Table 1):**
**OSHA** 46

**Physical Description:** Colorless liquid with a somewhat disagreeable odor.

**Chemical & Physical Properties:**
**MW:** 89.1
**BP:** 269°F
**Sol:** 1%
**Fl.P:** 96°F
**IP:** 10.81 eV
**Sp.Gr:** 1.00
**VP:** 8 mmHg
**FRZ:** -162°F
**UEL:** ?
**LEL:** 2.2%
Class IC Flammable Liquid

**Personal Protection/Sanitation (see Table 2):**
**Skin:** N.R.
**Eyes:** Prevent eye contact
**Wash skin:** N.R.
**Remove:** When wet (flamm)
**Change:** N.R.

**Respirator Recommendations (see Tables 3 and 4):**
**NIOSH/OSHA**
**250 ppm:** Sa*
**625 ppm:** Sa:Cf*
**1000 ppm:** ScbaF/SaF
**§:** ScbaF:Pd,Pp/SaF:Pd,Pp:AScba
**Escape:** ScbaE

**Incompatibilities and Reactivities:** Amines; strong acids, alkalis & oxidizers; hydrocarbons & other combustible materials; metal oxides

**Exposure Routes, Symptoms, Target Organs (see Table 5):**
**ER:** Inh, Ing, Con
**SY:** Irrit eyes; head, nau, vomit, diarr; in animals: liver, kidney damage
**TO:** Eyes, CNS, liver, kidneys

**First Aid (see Table 6):**
**Eye:** Irr immed
**Skin:** Soap wash prompt
**Breath:** Resp support
**Swallow:** Medical attention immed



## 2-Nitropropane

| 2-Nitropropane | Formula: $(CH_3)_2CH(NO_2)$ | CAS#: 79-46-9 | RTECS#: TZ5250000 | IDLH: Ca [100 ppm] |
|---|---|---|---|---|
| **Conversion:** 1 ppm = 3.64 mg/m$^3$ | **DOT:** 2608 129 | | | |

**Synonyms/Trade Names:** Dimethylnitromethane, iso-Nitropropane, 2-NP

**Exposure Limits:**
**NIOSH REL:** Ca
See Appendix A
**OSHA PEL†:** TWA 25 ppm (90 mg/m$^3$)

**Measurement Methods (see Table 1):**
**NIOSH** 2528
**OSHA** 15, 46

**Physical Description:** Colorless liquid with a pleasant, fruity odor.

**Chemical & Physical Properties:**
**MW:** 89.1
**BP:** 249°F
**Sol:** 2%
**Fl.P:** 75°F
**IP:** 10.71 eV
**Sp.Gr:** 0.99
**VP:** 13 mmHg
**FRZ:** -135°F
**UEL:** 11.0%
**LEL:** 2.6%
Class IC Flammable Liquid

**Personal Protection/Sanitation (see Table 2):**
**Skin:** Prevent skin contact
**Eyes:** Prevent eye contact
**Wash skin:** When contam
**Remove:** When wet (flamm)
**Change:** N.R.

**Respirator Recommendations (see Tables 3 and 4):**
**NIOSH**
**¥:** ScbaF:Pd,Pp/SaF:Pd,Pp:AScba
**Escape:** ScbaE

**Incompatibilities and Reactivities:** Amines; strong acids, alkalis & oxidizers; metal oxides; combustible materials

**Exposure Routes, Symptoms, Target Organs (see Table 5):**
**ER:** Inh, Ing, Con
**SY:** Irrit eyes, skin, nose, resp sys; head, anor, nau, vomit, diarr; kidney, liver damage; [carc]
**TO:** Eyes, skin, resp sys, CNS, kidneys, liver [in animals: liver tumors]

**First Aid (see Table 6):**
**Eye:** Irr immed
**Skin:** Soap wash prompt
**Breath:** Resp support
**Swallow:** Medical attention immed

N

## N-Nitrosodimethylamine

| | | | | |
|---|---|---|---|---|
| **N-Nitrosodimethylamine** | **Formula:** $(CH_3)_2N_2O$ | **CAS#:** 62-75-9 | **RTECS#:** IQ0525000 | **IDLH:** Ca [N.D.] |
| **Conversion:** | **DOT:** | | | |

**Synonyms/Trade Names:** Dimethylnitrosamine; N,N-Dimethylnitrosamine; DMNA; N-Methyl-N-nitroso-methanamine; NDMA; N-Nitroso-N,N-dimethylamine

**Exposure Limits:**
**NIOSH REL:** Ca
See Appendix A
**OSHA PEL:** [1910.1016] See Appendix B

**Measurement Methods (see Table 1):**
**NIOSH** 2522
**OSHA** 38

**Physical Description:** Yellow, oily liquid with a faint, characteristic odor.

**Chemical & Physical Properties:**
**MW:** 74.1
**BP:** 306°F
**Sol:** Soluble
**Fl.P:** ?
**IP:** 8.69 eV
**Sp.Gr:** 1.005
**VP:** 3 mmHg
**FRZ:** ?
**UEL:** ?
**LEL:** ?
Combustible Liquid

**Personal Protection/Sanitation (see Table 2):**
**Skin:** Prevent skin contact
**Eyes:** Prevent eye contact
**Wash skin:** When contam/Daily
**Remove:** When wet or contam
**Change:** Daily
**Provide:** Eyewash
Quick drench

**Respirator Recommendations (see Tables 3 and 4):**
**NIOSH**
**¥:** ScbaF:Pd,Pp/SaF:Pd,Pp:AScba
**Escape:** 100F/ScbaE

**See Appendix E** (page 351)

**Incompatibilities and Reactivities:** Strong oxidizers [**Note:** Should be stored in dark bottles.]

**Exposure Routes, Symptoms, Target Organs (see Table 5):**
**ER:** Inh, Abs, Ing, Con
**SY:** Nau, vomit, diarr, abdom cramps; head; fever; enlarged liver, jaun; decr liver, kidney, pulm func; [carc]
**TO:** Liver, kidneys,lungs [in animals; lung, kidney, liver & nasal cavity tumors]

**First Aid (see Table 6):**
**Eye:** Irr immed
**Skin:** Soap wash immed
**Breath:** Resp support
**Swallow:** Medical attention immed

## m-Nitrotoluene

| | | | | |
|---|---|---|---|---|
| **m-Nitrotoluene** | **Formula:** $NO_2C_6H_4CH_3$ | **CAS#:** 99-08-1 | **RTECS#:** XT2975000 | **IDLH:** 200 ppm |
| **Conversion:** 1 ppm = 5.61 mg/m$^3$ | **DOT:** 1664 152 | | | |

**Synonyms/Trade Names:** m-Methylnitrobenzene, 3-Methylnitrobenzene, meta-Nitrotoluene, 3-Nitrotoluene

**Exposure Limits:**
**NIOSH REL:** TWA 2 ppm (11 mg/m$^3$) [skin]
**OSHA PEL†:** TWA 5 ppm (30 mg/m$^3$) [skin]

**Measurement Methods (see Table 1):**
**NIOSH** 2005

**Physical Description:** Yellow liquid with a weak, aromatic odor.
[**Note:** A solid below 59°F.]

**Chemical & Physical Properties:**
**MW:** 137.1
**BP:** 450°F
**Sol:** 0.05%
**Fl.P:** 223°F
**IP:** 9.48 eV
**Sp.Gr:** 1.16
**VP:** 0.1 mmHg
**FRZ:** 59°F
**UEL:** ?
**LEL:** 1.6%
Class IIIB Combustible Liquid

**Personal Protection/Sanitation (see Table 2):**
**Skin:** Prevent skin contact
**Eyes:** Prevent eye contact
**Wash skin:** When contam
**Remove:** When wet or contam
**Change:** N.R.

**Respirator Recommendations (see Tables 3 and 4):**
**NIOSH**
**20 ppm:** Sa*
**50 ppm:** Sa:Cf*
**100 ppm:** SaT:Cf*/ScbaF/SaF
**200 ppm:** SaF:Pd,Pp
**§:** ScbaF:Pd,Pp/SaF:Pd,Pp:AScba
**Escape:** GmFOv100/ScbaE

**Incompatibilities and Reactivities:** Strong oxidizers, sulfuric acid

**Exposure Routes, Symptoms, Target Organs (see Table 5):**
**ER:** Inh, Abs, Ing, Con
**SY:** Anoxia, cyan; head, lass, dizz; ataxia; dysp; tacar; nau, vomit
**TO:** Blood, CNS, CVS, skin, GI tract

**First Aid (see Table 6):**
**Eye:** Irr immed
**Skin:** Soap wash immed
**Breath:** Resp support
**Swallow:** Medical attention immed

## o-Nitrotoluene

**Formula:** $NO_2C_6H_4CH_3$ | **CAS#:** 88-72-2 | **RTECS#:** XT3150000 | **IDLH:** 200 ppm

**Conversion:** 1 ppm = 5.61 mg/m$^3$ | **DOT:** 1664 152

**Synonyms/Trade Names:** o-Methylnitrobenzene, 2-Methylnitrobenzene, ortho-Nitrotoluene, 2-Nitrotoluene

**Exposure Limits:**
**NIOSH REL:** TWA 2 ppm (11 mg/m$^3$) [skin]
**OSHA PEL†:** TWA 5 ppm (30 mg/m$^3$) [skin]

**Measurement Methods (see Table 1):**
**NIOSH** 2005

**Physical Description:** Yellow liquid with a weak, aromatic odor.
[**Note:** A solid below 25°F.]

**Chemical & Physical Properties:**
**MW:** 137.1
**BP:** 432°F
**Sol:** 0.07%
**Fl.P:** 223°F
**IP:** 9.43 eV
**Sp.Gr:** 1.16
**VP:** 0.1 mmHg
**FRZ:** 25°F
**UEL:** ?
**LEL:** 2.2%
Class IIIB Combustible Liquid

**Personal Protection/Sanitation (see Table 2):**
**Skin:** Prevent skin contact
**Eyes:** Prevent eye contact
**Wash skin:** When contam
**Remove:** When wet or contam
**Change:** N.R.

**Respirator Recommendations (see Tables 3 and 4):**
**NIOSH**
**20 ppm:** Sa*
**50 ppm:** Sa:Cf*
**100 ppm:** SaT:Cf*/ScbaF/SaF
**200 ppm:** SaF:Pd,Pp
**§:** ScbaF:Pd,Pp/SaF:Pd,Pp:AScba
**Escape:** GmFOv100/ScbaE

**Incompatibilities and Reactivities:** Strong oxidizers, sulfuric acid

**Exposure Routes, Symptoms, Target Organs (see Table 5):**
**ER:** Inh, Abs, Ing, Con
**SY:** Anoxia, cyan; head, lass, dizz; ataxia; dysp; tacar; nau, vomit
**TO:** Blood, CNS, CVS, skin, GI tract

**First Aid (see Table 6):**
**Eye:** Irr immed
**Skin:** Soap wash immed
**Breath:** Resp support
**Swallow:** Medical attention immed



## p-Nitrotoluene

**Formula:** $NO_2C_6H_4CH_3$ | **CAS#:** 99-99-0 | **RTECS#:** XT3325000 | **IDLH:** 200 ppm

**Conversion:** 1 ppm = 5.61 mg/m$^3$ | **DOT:** 1664 152

**Synonyms/Trade Names:** p-Methylnitrobenzene, 4-Methylnitrobenzene, para-Nitrotoluene, 4-Nitrotoluene

**Exposure Limits:**
**NIOSH REL:** TWA 2 ppm (11 mg/m$^3$) [skin]
**OSHA PEL†:** TWA 5 ppm (30 mg/m$^3$) [skin]

**Measurement Methods (see Table 1):**
**NIOSH** 2005

**Physical Description:** Crystalline solid with a weak, aromatic odor.

**Chemical & Physical Properties:**
**MW:** 137.1
**BP:** 460°F
**Sol:** 0.04%
**Fl.P:** 223°F
**IP:** 9.50 eV
**Sp.Gr:** 1.12
**VP:** 0.1 mmHg
**MLT:** 126°F
**UEL:** ?
**LEL:** 1.6%
Combustible Solid

**Personal Protection/Sanitation (see Table 2):**
**Skin:** Prevent skin contact
**Eyes:** Prevent eye contact
**Wash skin:** When contam
**Remove:** When wet or contam
**Change:** Daily

**Respirator Recommendations (see Tables 3 and 4):**
**NIOSH**
**20 ppm:** Sa*
**50 ppm:** Sa:Cf*
**100 ppm:** SaT:Cf*/ScbaF/SaF
**200 ppm:** SaF:Pd,Pp
**§:** ScbaF:Pd,Pp/SaF:Pd,Pp:AScba
**Escape:** GmFOv100/ScbaE

**Incompatibilities and Reactivities:** Strong oxidizers, sulfuric acid

**Exposure Routes, Symptoms, Target Organs (see Table 5):**
**ER:** Inh, Abs, Ing, Con
**SY:** Anoxia, cyan; head, lass, dizz; ataxia; dysp; tacar; nau, vomit
**TO:** Blood, CNS, CVS, skin, GI tract

**First Aid (see Table 6):**
**Eye:** Irr immed
**Skin:** Soap wash immed
**Breath:** Resp support
**Swallow:** Medical attention immed

N

## Nitrous oxide

| Formula: | CAS#: | RTECS#: | IDLH: |
|---|---|---|---|
| $N_2O$ | 10024-97-2 | QX1350000 | N.D. |

**Conversion:** 1 ppm = 1.80 mg/m$^3$
**DOT:** 1070 122; 2201 122 (refrigerated liquid)

**Synonyms/Trade Names:** Dinitrogen monoxide, Hyponitrous acid anhydride, Laughing gas

**Exposure Limits:**
**NIOSH REL*:** TWA 25 ppm (46 mg/m$^3$) (TWA over the time exposed)
[***Note:** REL for exposure to waste anesthetic gas.]
**OSHA PEL:** none

**Measurement Methods (see Table 1):**
**NIOSH** 3800, 6600
**OSHA** ID166

**Physical Description:** Colorless gas with a slightly sweet odor. [inhalation anesthetic]
[**Note:** Shipped as a liquefied compressed gas.]

**Chemical & Physical Properties:**
**MW:** 44.0
**BP:** -127°F
**Sol(77°F):** 0.1%
**Fl.P:** NA
**IP:** 12.89 eV
**RGasD:** 1.53
**VP:** 51.3 atm
**FRZ:** -132°F
**UEL:** NA
**LEL:** NA
Nonflammable Gas, but supports combustion at elevated temperatures.

**Personal Protection/Sanitation (see Table 2):**
**Skin:** Frostbite
**Eyes:** Frostbite
**Wash skin:** N.R.
**Remove:** N.R.
**Change:** N.R.
**Provide:** Frostbite wash

**Respirator Recommendations (see Tables 3 and 4):**
Not available.

**Incompatibilities and Reactivities:** Aluminum, boron, hydrazine, lithium hydride, phosphine, sodium

**Exposure Routes, Symptoms, Target Organs (see Table 5):**
**ER:** Inh, Con (liquid)
**SY:** Dysp; drow, head; asphy; repro effects; liquid: frostbite
**TO:** Resp sys, CNS, repro sys

**First Aid (see Table 6):**
**Eye:** Frostbite
**Skin:** Frostbite
**Breath:** Fresh air

## Nonane

| Formula: | CAS#: | RTECS#: | IDLH: |
|---|---|---|---|
| $CH_3(CH_2)_7CH_3$ | 111-84-2 | RA6115000 | N.D. |

**Conversion:** 1 ppm = 5.25 mg/m$^3$
**DOT:** 1920 128

**Synonyms/Trade Names:** n-Nonane, Nonyl hydride

**Exposure Limits:**
**NIOSH REL:** TWA 200 ppm (1050 mg/m$^3$)
**OSHA PEL†:** none

**Measurement Methods (see Table 1):**
None available

**Physical Description:** Colorless liquid with a gasoline-like odor.

**Chemical & Physical Properties:**
**MW:** 128.3
**BP:** 303°F
**Sol:** Insoluble
**Fl.P:** 88°F
**IP:** 10.21 eV
**Sp.Gr:** 0.72
**VP:** 3 mmHg
**FRZ:** -60°F
**UEL:** 2.9%
**LEL:** 0.8%
Class IC Flammable Liquid

**Personal Protection/Sanitation (see Table 2):**
**Skin:** N.R.
**Eyes:** Prevent eye contact
**Wash skin:** Daily
**Remove:** When wet (flamm)
**Change:** N.R.
**Provide:** Eyewash

**Respirator Recommendations (see Tables 3 and 4):**
Not available.

**Incompatibilities and Reactivities:** Strong oxidizers (e.g., peroxides, nitrates, perchlorates)

**Exposure Routes, Symptoms, Target Organs (see Table 5):**
**ER:** Inh, Ing, Con
**SY:** Irrit eyes, skin, nose, throat; head, drow, dizz, conf, nau, tremor, inco; chemical pneu (aspir liquid)
**TO:** Eyes, skin, resp sys, CNS

**First Aid (see Table 6):**
**Eye:** Irr immed
**Skin:** Soap wash immed
**Breath:** Resp support
**Swallow:** Medical attention immed

## 1-Nonanethiol

| 1-Nonanethiol | Formula: $CH_3(CH_2)_8SH$ | CAS#: 1455-21-6 | RTECS#: | IDLH: N.D. |
|---|---|---|---|---|
| Conversion: 1 ppm = 6.56 mg/m$^3$ | DOT: 1228 131 | | | |

**Synonyms/Trade Names:** 1-Mercaptononane, n-Nonyl mercaptan, Nonylthiol

**Exposure Limits:**
**NIOSH REL:** C 0.5 ppm (3.3 mg/m$^3$) [15-minute]
**OSHA PEL:** none

**Measurement Methods (see Table 1):**
None available

**Physical Description:** Liquid.

**Chemical & Physical Properties:**
**MW:** 160.3
**BP:** ?
**Sol:** Insoluble
**Fl.P:** ?
**IP:** ?
**Sp.Gr:** ?
**VP:** ?
**FRZ:** ?
**UEL:** ?
**LEL:** ?
Combustible Liquid

**Personal Protection/Sanitation (see Table 2):**
**Skin:** Prevent skin contact
**Eyes:** Prevent eye contact
**Wash skin:** When contam
**Remove:** When wet or contam
**Change:** N.R.

**Respirator Recommendations (see Tables 3 and 4):**
**NIOSH**
**5 ppm:** CcrOv/Sa
**12.5 ppm:** Sa:Cf/PaprOv
**25 ppm:** CcrFOv/GmFOv/PaprTOv/ScbaF/SaF
**§:** ScbaF:Pd,Pp/SaF:Pd,Pp:AScba
**Escape:** GmFOv/ScbaE

**Incompatibilities and Reactivities:** Oxidizers, reducing agents, strong acids & bases, alkali metals

**Exposure Routes, Symptoms, Target Organs (see Table 5):**
**ER:** Inh, Ing, Con
**SY:** Irrit eyes, skin, nose, throat; lass, cyan, incr respiration, nau, drow, head, vomit
**TO:** Eyes, skin, resp sys, blood, CNS

**First Aid (see Table 6):**
**Eye:** Irr immed
**Skin:** Soap wash
**Breath:** Resp support
**Swallow:** Medical attention immed

O

## Octachloronaphthalene

| Octachloronaphthalene | Formula: $C_{10}Cl_8$ | CAS#: 2234-13-1 | RTECS#: QK0250000 | IDLH: See Appendix F |
|---|---|---|---|---|
| Conversion: | DOT: | | | |

**Synonyms/Trade Names:** Halowax® 1051; 1,2,3,4,5,6,7,8-Octachloronaphthalene; Perchloronaphthalene

**Exposure Limits:**
**NIOSH REL:** TWA 0.1 mg/m$^3$
ST 0.3 mg/m$^3$ [skin]
**OSHA PEL†:** TWA 0.1 mg/m$^3$ [skin]

**Measurement Methods (see Table 1):**
**NIOSH** S97 (II-2)

**Physical Description:** Waxy, pale-yellow solid with an aromatic odor.

**Chemical & Physical Properties:**
**MW:** 403.7
**BP:** 770°F
**Sol:** Insoluble
**Fl.P:** NA
**IP:** ?
**Sp.Gr:** 2.00
**VP:** <1 mmHg
**MLT:** 365°F
**UEL:** NA
**LEL:** NA
Noncombustible Solid

**Personal Protection/Sanitation (see Table 2):**
**Skin:** Prevent skin contact
**Eyes:** Prevent eye contact
**Wash skin:** When contam/Daily
**Remove:** When wet or contam
**Change:** Daily

**Respirator Recommendations (see Tables 3 and 4):**
**NIOSH/OSHA**
**1 mg/m$^3$:** Sa/ScbaF
**§:** ScbaF:Pd,Pp/SaF:Pd,Pp:AScba
**Escape:** GmFOv100/ScbaE

**See Appendix F**

**Incompatibilities and Reactivities:** Strong oxidizers

**Exposure Routes, Symptoms, Target Organs (see Table 5):**
**ER:** Inh, Abs, Ing, Con
**SY:** Acne-form derm; liver damage, jaun
**TO:** Skin, liver

**First Aid (see Table 6):**
**Eye:** Irr immed
**Skin:** Water flush immed
**Breath:** Resp support
**Swallow:** Medical attention immed

| 1-Octadecanethiol | Formula: $CH_3(CH_2)_{17}SH$ | CAS#: 2885-00-9 | RTECS#: | IDLH: N.D. |
|---|---|---|---|---|
| **Conversion:** 1 ppm = 11.72 mg/m$^3$ | **DOT:** 1228 131 (liquid) | | | |

**Synonyms/Trade Names:** 1-Mercaptooctadecane, Octadecyl mercaptan, Stearyl mercaptan

**Exposure Limits:**
**NIOSH REL:** C 0.5 ppm (5.9 mg/m$^3$) [15-minute]
**OSHA PEL:** none

**Measurement Methods (see Table 1):**
None available

**Physical Description:** Solid or liquid (above 77°F).

| Chemical & Physical Properties: | Personal Protection/Sanitation (see Table 2): | Respirator Recommendations (see Tables 3 and 4): |
|---|---|---|
| **MW:** 286.6<br>**BP:** ?<br>**Sol:** Insoluble<br>**Fl.P:** ?<br>**IP:** ?<br>**Sp.Gr:** 0.85<br>**VP:** ?<br>**MLT:** 77°F<br>**UEL:** ?<br>**LEL:** ?<br>Combustible Solid<br>Combustible Liquid | **Skin:** Prevent skin contact<br>**Eyes:** Prevent eye contact<br>**Wash skin:** When contam<br>**Remove:** When wet or contam<br>**Change:** Daily | **NIOSH**<br>**5 ppm:** CcrOv/Sa<br>**12.5 ppm:** Sa:Cf/PaprOv<br>**25 ppm:** CcrFOv/GmFOv/PaprTOv/ ScbaF/SaF<br>**§:** ScbaF:Pd,Pp/SaF:Pd,Pp:AScba<br>**Escape:** GmFOv/ScbaE |

**Incompatibilities and Reactivities:** Oxidizers, reducing agents, strong acids & bases, alkali metals

**Exposure Routes, Symptoms, Target Organs (see Table 5):**
**ER:** Inh, Abs, Ing, Con
**SY:** Irrit eyes, skin, resp sys; head, dizz, lass, cyan, nau, convuls
**TO:** Eyes, skin, resp sys, CNS, blood

**First Aid (see Table 6):**
**Eye:** Irr immed
**Skin:** Soap wash immed
**Breath:** Resp support
**Swallow:** Medical attention immed



| Octane | Formula: $CH_3[CH_2]_6CH_3$ | CAS#: 111-65-9 | RTECS#: RG8400000 | IDLH: 1000 ppm [10%LEL] |
|---|---|---|---|---|
| **Conversion:** 1 ppm = 4.67 mg/m$^3$ | **DOT:** 1262 128 | | | |

**Synonyms/Trade Names:** n-Octane, normal-Octane

**Exposure Limits:**
**NIOSH REL:** TWA 75 ppm (350 mg/m$^3$)
C 385 ppm (1800 mg/m$^3$) [15-minute]
**OSHA PEL†:** TWA 500 ppm (2350 mg/m$^3$)

**Measurement Methods (see Table 1):**
**NIOSH** 1500
**OSHA** 7

**Physical Description:** Colorless liquid with a gasoline-like odor.

| Chemical & Physical Properties: | Personal Protection/Sanitation (see Table 2): | Respirator Recommendations (see Tables 3 and 4): |
|---|---|---|
| **MW:** 114.2<br>**BP:** 258°F<br>**Sol(77°F):** 0.00007%<br>**Fl.P:** 56°F<br>**IP:** 9.82 eV<br>**Sp.Gr:** 0.70<br>**VP:** 10 mmHg<br>**FRZ:** -70°F<br>**UEL:** 6.5%<br>**LEL:** 1.0%<br>Class IB Flammable Liquid | **Skin:** Prevent skin contact<br>**Eyes:** Prevent eye contact<br>**Wash skin:** When contam<br>**Remove:** When wet (flamm)<br>**Change:** N.R. | **NIOSH**<br>**750 ppm:** Sa*<br>**1000 ppm:** Sa:Cf*/ScbaF/SaF<br>**§:** ScbaF:Pd,Pp/SaF:Pd,Pp:AScba<br>**Escape:** GmFOv/ScbaE |

**Incompatibilities and Reactivities:** Strong oxidizers

**Exposure Routes, Symptoms, Target Organs (see Table 5):**
**ER:** Inh, Ing, Con
**SY:** Irrit eyes, nose; drow; derm; chemical pneu (aspir liquid); in animals: narco
**TO:** Eyes, skin, resp sys, CNS

**First Aid (see Table 6):**
**Eye:** Irr immed
**Skin:** Soap wash prompt
**Breath:** Resp support
**Swallow:** Medical attention immed

## 1-Octanethiol

| | | | | |
|---|---|---|---|---|
| **1-Octanethiol** | **Formula:** $CH_3(CH_2)_7SH$ | **CAS#:** 111-88-6 | **RTECS#:** | **IDLH:** N.D. |
| **Conversion:** 1 ppm = 5.98 mg/m$^3$ | **DOT:** 1228 131 | | | |

**Synonyms/Trade Names:** 1-Mercaptooctane, n-Octyl mercaptan, Octylthiol, 1-Octylthiol

**Exposure Limits:**
**NIOSH REL:** C 0.5 ppm (3.0 mg/m$^3$) [15-minute]
**OSHA PEL:** none

**Measurement Methods (see Table 1):**
**NIOSH** 2510

**Physical Description:** Water-white liquid with a mild odor.

| **Chemical & Physical Properties:** | **Personal Protection/Sanitation (see Table 2):** | **Respirator Recommendations (see Tables 3 and 4):** |
|---|---|---|
| **MW:** 146.3<br>**BP:** 390°F<br>**Sol:** Insoluble<br>**Fl.P(oc):** 115°F<br>**IP:** ?<br>**Sp.Gr:** 0.84<br>**VP(212°F):** 3 mmHg<br>**FRZ:** -57°F<br>**UEL:** ?<br>**LEL:** ?<br>Class II Combustible Liquid | **Skin:** Prevent skin contact<br>**Eyes:** Prevent eye contact<br>**Wash skin:** When contam<br>**Remove:** When wet or contam<br>**Change:** N.R. | **NIOSH**<br>**5 ppm:** CcrOv/Sa<br>**12.5 ppm:** Sa:Cf/PaprOv<br>**25 ppm:** CcrFOv/GmFOv/PaprTOv/ScbaF/SaF<br>**§:** ScbaF:Pd,Pp/SaF:Pd,Pp:AScba<br>**Escape:** GmFOv/ScbaE |

**Incompatibilities and Reactivities:** Oxidizers, reducing agents, strong acids & bases, alkali metals

| **Exposure Routes, Symptoms, Target Organs (see Table 5):** | **First Aid (see Table 6):** |
|---|---|
| **ER:** Inh, Ing, Con<br>**SY:** Irrit eyes, skin, nose, throat; lass, cyan, incr respiration, nau, drow, head, vomit<br>**TO:** Eyes, skin, resp sys, blood, CNS | **Eye:** Irr immed<br>**Skin:** Soap wash immed<br>**Breath:** Resp support<br>**Swallow:** Medical attention immed |

O

## Oil mist (mineral)

| | | | | |
|---|---|---|---|---|
| **Oil mist (mineral)** | **Formula:** | **CAS#:** 8012-95-1 | **RTECS#:** PY8030000 | **IDLH:** 2500 mg/m$^3$ |
| **Conversion:** | **DOT:** | | | |

**Synonyms/Trade Names:** Heavy mineral oil mist, Paraffin oil mist, White mineral oil mist

**Exposure Limits:**
**NIOSH REL:** TWA 5 mg/m$^3$
ST 10 mg/m$^3$
**OSHA PEL:** TWA 5 mg/m$^3$

**Measurement Methods (see Table 1):**
**NIOSH** 5026, 5524

**Physical Description:** Colorless, oily liquid aerosol dispersed in air.
[**Note:** Has an odor like burned lubricating oil.]

| **Chemical & Physical Properties:** | **Personal Protection/Sanitation (see Table 2):** | **Respirator Recommendations (see Tables 3 and 4):** |
|---|---|---|
| **MW:** Varies<br>**BP:** 680°F<br>**Sol:** Insoluble<br>**Fl.P(oc):** 380°F<br>**IP:** ?<br>**Sp.Gr:** 0.90<br>**VP:** <0.5 mmHg<br>**FRZ:** 0°F<br>**UEL:** ?<br>**LEL:** ?<br>Class IIIB Combustible Liquid | **Skin:** Prevent skin contact<br>**Eyes:** N.R.<br>**Wash skin:** When contam<br>**Remove:** When wet or contam<br>**Change:** Daily | **NIOSH/OSHA**<br>**50 mg/m$^3$:** 100XQ/Sa<br>**125 mg/m$^3$:** Sa:Cf/PaprHie<br>**250 mg/m$^3$:** 100F/SaT:Cf/PaprTHie/ScbaF/SaF<br>**2500 mg/m$^3$:** Sa:Pd,Pp<br>**§:** ScbaF:Pd,Pp/SaF:Pd,Pp:AScba<br>**Escape:** 100F/ScbaE |

**Incompatibilities and Reactivities:** None reported

| **Exposure Routes, Symptoms, Target Organs (see Table 5):** | **First Aid (see Table 6):** |
|---|---|
| **ER:** Inh, Con<br>**SY:** Irrit eyes, skin, resp sys<br>**TO:** Eyes, skin, resp sys | **Skin:** Soap wash<br>**Breath:** Fresh air |

O

| Osmium tetroxide | Formula: $OsO_4$ | CAS#: 20816-12-0 | RTECS#: RN1140000 | IDLH: 1 mg/m$^3$ |
|---|---|---|---|---|
| **Conversion:** 1 ppm = 10.40 mg/m$^3$ | **DOT:** 2471 154 | | | |

**Synonyms/Trade Names:** Osmic acid anhydride, Osmium oxide

**Exposure Limits:**
**NIOSH REL:** TWA 0.002 mg/m$^3$ (0.0002 ppm)
ST 0.006 mg/m$^3$ (0.0006 ppm)
**OSHA PEL†:** TWA 0.002 mg/m$^3$

**Measurement Methods (see Table 1):**
None available

**Physical Description:** Colorless, crystalline solid or pale-yellow mass with an unpleasant, acrid, chlorine-like odor. [**Note:** A liquid above 105°F.]

**Chemical & Physical Properties:**
**MW:** 254.2
**BP:** 266°F
**Sol(77°F):** 6%
**Fl.P:** NA
**IP:** 12.60 eV
**Sp.Gr:** 5.10
**VP:** 7 mmHg
**MLT:** 105°F
**UEL:** NA
**LEL:** NA
Noncombustible Solid

**Personal Protection/Sanitation (see Table 2):**
**Skin:** Prevent skin contact
**Eyes:** Prevent eye contact
**Wash skin:** When contam
**Remove:** When wet or contam
**Change:** Daily
**Provide:** Eyewash

**Respirator Recommendations (see Tables 3 and 4):**
**NIOSH/OSHA**
**0.1 mg/m$^3$:** CcrFS100/GmFS100/ScbaF/SaF
**1 mg/m$^3$:** SaF:Pd,Pp
**§:** ScbaF:Pd,Pp/SaF:Pd,Pp:AScba
**Escape:** GmFS100/ScbaE

**Incompatibilities and Reactivities:** Hydrochloric acid, easily oxidized organic materials
[**Note:** Begins to sublime below BP. Contact with other materials may cause fire.]

**Exposure Routes, Symptoms, Target Organs (see Table 5):**
**ER:** Inh, Ing, Con
**SY:** Irrit eyes, resp sys; lac, vis dist; conj; head; cough, dysp; derm
**TO:** Eyes, skin, resp sys

**First Aid (see Table 6):**
**Eye:** Irr immed
**Skin:** Soap wash immed
**Breath:** Resp support
**Swallow:** Medical attention immed

| Oxalic acid | Formula: HOOCCOOH×$2H_2O$ | CAS#: 144-62-7 | RTECS#: RO2450000 | IDLH: 500 mg/m$^3$ |
|---|---|---|---|---|
| **Conversion:** | **DOT:** | | | |

**Synonyms/Trade Names:** Ethanedioic acid, Oxalic acid (aqueous), Oxalic acid dihydrate

**Exposure Limits:**
**NIOSH REL:** TWA 1 mg/m$^3$
ST 2 mg/m$^3$
**OSHA PEL†:** TWA 1 mg/m$^3$

**Measurement Methods (see Table 1):**
None available

**Physical Description:** Colorless, odorless powder or granular solid.
[**Note:** The anhydrous form $(COOH)_2$ is an odorless, white solid.]

**Chemical & Physical Properties:**
**MW:** 126.1
**BP:** Sublimes
**Sol:** 14%
**Fl.P:** ?
**IP:** ?
**Sp.Gr:** 1.90
**VP:** <0.001 mmHg
**MLT:** 215°F (Sublimes)
**UEL:** ?
**LEL:** ?
Combustible Solid

**Personal Protection/Sanitation (see Table 2):**
**Skin:** Prevent skin contact
**Eyes:** Prevent eye contact
**Wash skin:** When contam
**Remove:** When wet or contam
**Change:** Daily
**Provide:** Eyewash

**Respirator Recommendations (see Tables 3 and 4):**
**NIOSH/OSHA**
**25 mg/m$^3$:** Sa:Cf£/PaprHie£
**50 mg/m$^3$:** 100F/ScbaF/SaF
**500 mg/m$^3$:** SaF:Pd,Pp
**§:** ScbaF:Pd,Pp/SaF:Pd,Pp:AScba
**Escape:** 100F/ScbaE

**Incompatibilities and Reactivities:** Strong oxidizers, silver compounds, strong alkalis, chlorites
[**Note:** Gives off water of crystallization at 215°F and begins to sublime.]

**Exposure Routes, Symptoms, Target Organs (see Table 5):**
**ER:** Inh, Ing, Con
**SY:** Irrit eyes, skin, muc memb; eye burns; local pain, cyan; shock, collapse, convuls; kidney damage
**TO:** Eyes, skin, resp sys, kidneys

**First Aid (see Table 6):**
**Eye:** Irr immed
**Skin:** Water flush prompt
**Breath:** Resp support
**Swallow:** Medical attention immed

| Oxygen difluoride | Formula: $OF_2$ | CAS#: 7783-41-7 | RTECS#: RS2100000 | IDLH: 0.5 ppm |
|---|---|---|---|---|
| **Conversion:** 1 ppm = 2.21 mg/$m^3$ | **DOT:** 2190 124 | | | |

**Synonyms/Trade Names:** Difluorine monoxide, Fluorine monoxide, Oxygen fluoride

**Exposure Limits:**
**NIOSH REL:** C 0.05 ppm (0.1 mg/$m^3$)
**OSHA PEL†:** TWA 0.05 ppm (0.1 mg/$m^3$)

**Measurement Methods (see Table 1):**
None available

**Physical Description:** Colorless gas with a peculiar, foul odor.
[**Note:** Shipped as a nonliquefied compressed gas.]

| Chemical & Physical Properties: | Personal Protection/Sanitation (see Table 2): | Respirator Recommendations (see Tables 3 and 4): |
|---|---|---|
| **MW:** 54.0<br>**BP:** -230°F<br>**Sol:** 0.02%<br>**Fl.P:** NA<br>**IP:** 13.11 eV<br>**RGasD:** 1.88<br>**VP:** >1 atm<br>**FRZ:** -371°F<br>**UEL:** NA<br>**LEL:** NA<br>Nonflammable Gas, but a strong oxidizer. | **Skin:** N.R.<br>**Eyes:** N.R.<br>**Wash skin:** N.R.<br>**Remove:** N.R.<br>**Change:** N.R. | **NIOSH/OSHA**<br>**0.5 ppm:** Sa/ScbaF<br>**§:** ScbaF:Pd,Pp/SaF:Pd,Pp:AScba<br>**Escape:** GmFS¿/ScbaE |

**Incompatibilities and Reactivities:** Combustible materials, chlorine, bromine, iodine, platinum, metal oxides, moist air, hydrogen sulfide, hydrocarbons, water [**Note:** Reacts very slowly with water to form hydrofluoric acid.]

| Exposure Routes, Symptoms, Target Organs (see Table 5): | First Aid (see Table 6): |
|---|---|
| **ER:** Inh, Con<br>**SY:** Irrit eyes, skin, resp sys; head; pulm edema; eye, skin burns (from contact with the gas under pressure)<br>**TO:** Eyes, skin, resp sys | **Eye:** Irr immed<br>**Skin:** Water flush immed<br>**Breath:** Resp support |

O

| Ozone | Formula: $O_3$ | CAS#: 10028-15-6 | RTECS#: RS8225000 | IDLH: 5 ppm |
|---|---|---|---|---|
| **Conversion:** 1 ppm = 1.96 mg/$m^3$ | **DOT:** | | | |

**Synonyms/Trade Names:** Triatomic oxygen

**Exposure Limits:**
**NIOSH REL:** C 0.1 ppm (0.2 mg/$m^3$)
**OSHA PEL†:** TWA 0.1 ppm (0.2 mg/$m^3$)

**Measurement Methods (see Table 1):**
**OSHA** ID214

**Physical Description:** Colorless to blue gas with a very pungent odor.

| Chemical & Physical Properties: | Personal Protection/Sanitation (see Table 2): | Respirator Recommendations (see Tables 3 and 4): |
|---|---|---|
| **MW:** 48.0<br>**BP:** -169°F<br>**Sol(32°F):** 0.001%<br>**Fl.P:** NA<br>**IP:** 12.52 eV<br>**RGasD:** 1.66<br>**VP:** >1 atm<br>**FRZ:** -315°F<br>**UEL:** NA<br>**LEL:** NA<br>Nonflammable Gas, but a powerful oxidizer. | **Skin:** N.R.<br>**Eyes:** N.R.<br>**Wash skin:** N.R.<br>**Remove:** N.R.<br>**Change:** N.R. | **NIOSH/OSHA**<br>**1 ppm:** CcrS¿/Sa<br>**2.5 ppm:** Sa:Cf/PaprS¿<br>**5 ppm:** CcrFS¿/GmFS¿/SaT:Cf/ScbaF/SaF<br>**§:** ScbaF:Pd,Pp/SaF:Pd,Pp:AScba<br>**Escape:** GmFS¿/ScbaE |

**Incompatibilities and Reactivities:** All oxidizable materials (both organic & inorganic)

| Exposure Routes, Symptoms, Target Organs (see Table 5): | First Aid (see Table 6): |
|---|---|
| **ER:** Inh, Con<br>**SY:** Irrit eyes, muc memb; pulm edema; chronic resp disease<br>**TO:** Eyes, resp sys | **Eye:** Medical attention<br>**Breath:** Fresh air; 100% $O_2$ |

## Paraffin wax fume

**Formula:** $C_nH_{2n+2}$ | **CAS#:** 8002-74-2 | **RTECS#:** RV0350000 | **IDLH:** N.D.

**Conversion:** | **DOT:**

**Synonyms/Trade Names:** Paraffin fume, Paraffin scale fume

**Exposure Limits:**
**NIOSH REL:** TWA 2 mg/m$^3$
**OSHA PEL†:** none

**Measurement Methods (see Table 1):**
**OSHA** PV2047

**Physical Description:** Paraffin wax is a white to slightly yellowish, odorless solid.
[**Note:** Consists of a mixture of high molecular weight hydrocarbons (e.g., $C_{36}H_{74}$).]

**Chemical & Physical Properties:**
**MW:** 350-420
**BP:** ?
**Sol:** Insoluble
**Fl.P:** 390°F
**IP:** ?
**Sp.Gr:** 0.88-0.92
**VP:** ?
**MLT:** 115-154°F
**UEL:** ?
**LEL:** ?
Combustible Solid

**Personal Protection/Sanitation (see Table 2):**
**Skin:** N.R.
**Eyes:** Prevent eye contact
**Wash skin:** N.R.
**Remove:** N.R.
**Change:** N.R.

**Respirator Recommendations (see Tables 3 and 4):**
Not available.

**Incompatibilities and Reactivities:** None reported

**Exposure Routes, Symptoms, Target Organs (see Table 5):**
**ER:** Inh, Con
**SY:** Irrit eyes, skin, resp sys; discomfort, nau
**TO:** Eyes, skin, resp sys

**First Aid (see Table 6):**
**Eye:** Irr immed
**Breath:** Resp support

P

## Paraquat (Paraquat dichloride)

**Formula:** $CH_3(C_5H_4N)_2CH_3 \times 2Cl$ | **CAS#:** 1910-42-5 | **RTECS#:** DW2275000 | **IDLH:** 1 mg/m$^3$

**Conversion:** | **DOT:**

**Synonyms/Trade Names:** 1,1'-Dimethyl-4,4'-bipyridinium dichloride; N,N'-Dimethyl-4,4'-bipyridinium dichloride; Paraquat chloride; Paraquat dichloride
[**Note:** Paraquat is a cation ($C_{12}H_{14}N_2^{++}$; 1,1-Dimethyl-4,4-bipyridinium ion); the commercial product is the dichloride salt of paraquat.]

**Exposure Limits:**
**NIOSH REL:** TWA 0.1 mg/m$^3$ (resp) [skin]
**OSHA PEL†:** TWA 0.5 mg/m$^3$ (resp) [skin]

**Measurement Methods (see Table 1):**
**NIOSH** 5003

**Physical Description:** Yellow solid with a faint, ammonia-like odor. [herbicide]
[**Note:** Paraquat may also be found commercially as a methyl sulfate salt $C_{12}H_{14}N_2 \times 2CH_3SO_4$.]

**Chemical & Physical Properties:**
**MW:** 257.2
**BP:** Decomposes
**Sol:** Miscible
**Fl.P:** NA
**IP:** ?
**Sp.Gr:** 1.24
**VP:** <0.0000001 mmHg
**MLT:** 572°F (Decomposes)
**UEL:** NA
**LEL:** NA
Noncombustible Solid

**Personal Protection/Sanitation (see Table 2):**
**Skin:** Prevent skin contact
**Eyes:** Prevent eye contact
**Wash skin:** When contam
**Remove:** When wet or contam
**Change:** N.R.
**Provide:** Quick drench

**Respirator Recommendations (see Tables 3 and 4):**
**NIOSH**
**1 mg/m$^3$:** CcrOv95*/PaprOvHie*/Sa*/ScbaF
**§:** ScbaF:Pd,Pp/SaF:Pd,Pp:AScba
**Escape:** GmFOv100/ScbaE

**Incompatibilities and Reactivities:** Strong oxidizers, alkylaryl-sulfonate wetting agents
[**Note:** Corrosive to metals. Decomposes in presence of ultraviolet light.]

**Exposure Routes, Symptoms, Target Organs (see Table 5):**
**ER:** Inh, Abs, Ing, Con
**SY:** Irrit eyes, skin, nose, throat, resp sys; epis; derm; fingernail damage; irrit GI tract; heart, liver, kidney damage
**TO:** Eyes, skin, resp sys, heart, liver, kidneys, GI tract

**First Aid (see Table 6):**
**Eye:** Irr immed
**Skin:** Water flush immed
**Breath:** Resp support
**Swallow:** Medical attention immed

## Parathion

**Formula:** $(C_2H_5O)_2P(S)OC_6H_4NO_2$ | **CAS#:** 56-38-2 | **RTECS#:** TF4550000 | **IDLH:** 10 mg/m$^3$

**Conversion:** | **DOT:** 2783 152

**Synonyms/Trade Names:** O,O-Diethyl-O(p-nitrophenyl) phosphorothioate; Diethyl parathion; Ethyl parathion; Parathion-ethyl

**Exposure Limits:**
**NIOSH REL:** TWA 0.05 mg/m$^3$ [skin]
**OSHA PEL:** TWA 0.1 mg/m$^3$ [skin]

**Measurement Methods (see Table 1):**
**NIOSH** 5600
**OSHA** 62

**Physical Description:** Pale-yellow to dark-brown liquid with a garlic-like odor.
[**Note:** A solid below 43°F. Pesticide that may be absorbed on a dry carrier.]

**Chemical & Physical Properties:**
**MW:** 291.3
**BP:** 707°F
**Sol:** 0.001%
**Fl.P(oc):** 392°F
**IP:** ?
**Sp.Gr:** 1.27
**VP:** 0.00004 mmHg
**FRZ:** 43°F
**UEL:** ?
**LEL:** ?
Class IIIB Combustible Liquid

**Personal Protection/Sanitation (see Table 2):**
**Skin:** Prevent skin contact
**Eyes:** Prevent eye contact
**Wash skin:** When contam
**Remove:** When wet or contam
**Change:** Daily
**Provide:** Eyewash
Quick drench

**Respirator Recommendations (see Tables 3 and 4):**
**NIOSH**
**0.5 mg/m$^3$:** CcrOv95/Sa
**1.25 mg/m$^3$:** Sa:Cf/PaprOvHie
**2.5 mg/m$^3$:** CcrFOv100/SaT:Cf/PaprTOvHie/ScbaF/SaF
**10 mg/m$^3$:** Sa:Pd,Pp
**§:** ScbaF:Pd,Pp/SaF:Pd,Pp:AScba
**Escape:** GmFOv100/ScbaE

**Incompatibilities and Reactivities:** Strong oxidizers, alkaline materials

**Exposure Routes, Symptoms, Target Organs (see Table 5):**
**ER:** Inh, Abs, Ing, Con
**SY:** Irrit eyes, skin, resp sys; miosis; rhin; head; chest tight, wheez, lar spasm, salv, cyan; anor, nau, vomit, abdom cramps, diarr; sweat; musc fasc, lass, para; dizz, conf, ataxia; convuls, coma; low BP; card irreg
**TO:** Eyes, skin, resp sys, CNS, CVS, blood chol

**First Aid (see Table 6):**
**Eye:** Irr immed
**Skin:** Soap wash immed
**Breath:** Resp support
**Swallow:** Medical attention immed

P

## Particulates not otherwise regulated

**Formula:** | **CAS#:** | **RTECS#:** | **IDLH:** N.D.

**Conversion:** | **DOT:**

**Synonyms/Trade Names:** "Inert" dusts, Nuisance dusts, PNOR
[**Note:** Includes all inert or nuisance dusts, whether mineral, inorganic, not listed specifically in 1910.1000.]

**Exposure Limits:**
**NIOSH REL:** See Appendix D
**OSHA PEL:** TWA 15 mg/m$^3$ (total)
TWA 5 mg/m$^3$ (resp)

**Measurement Methods (see Table 1):**
**NIOSH** 0500, 0600

**Physical Description:** Dusts from solid substances without specific occupational exposure standards.

**Chemical & Physical Properties:**
Properties vary depending upon the specific solid.

**Personal Protection/Sanitation (see Table 2):**
**Skin:** N.R.
**Eyes:** N.R.
**Wash skin:** N.R.
**Remove:** N.R.
**Change:** N.R.

**Respirator Recommendations (see Tables 3 and 4):**
Not available.

**Incompatibilities and Reactivities:** Varies

**Exposure Routes, Symptoms, Target Organs (see Table 5):**
**ER:** Inh, Con
**SY:** Irrit eyes, skin, throat, upper resp sys
**TO:** Eyes, skin, resp sys

**First Aid (see Table 6):**
**Eye:** Irr immed
**Breath:** Fresh air

P

## Pentaborane

**Formula:** $B_5H_9$ | **CAS#:** 19624-22-7 | **RTECS#:** RY8925000 | **IDLH:** 1 ppm

**Conversion:** 1 ppm = 2.58 mg/m$^3$ | **DOT:** 1380 135

**Synonyms/Trade Names:** Pentaboron nonahydride

**Exposure Limits:**
**NIOSH REL:** TWA 0.005 ppm (0.01 mg/m$^3$)
ST 0.015 ppm (0.03 mg/m$^3$)
**OSHA PEL†:** TWA 0.005 ppm (0.01 mg/m$^3$)

**Measurement Methods (see Table 1):**
None available

**Physical Description:** Colorless liquid with a pungent odor like sour milk.

**Chemical & Physical Properties:**
**MW:** 63.1
**BP:** 140°F
**Sol:** Reacts
**Fl.P:** 86°F
**IP:** 9.90 eV
**Sp.Gr:** 0.62
**VP:** 171 mmHg
**FRZ:** -52°F
**UEL:** ?
**LEL:** 0.42%
Class IC Flammable Liquid

**Personal Protection/Sanitation (see Table 2):**
**Skin:** Prevent skin contact
**Eyes:** Prevent eye contact
**Wash skin:** When contam
**Remove:** When wet (flamm)
**Change:** N.R.
**Provide:** Eyewash
Quick drench

**Respirator Recommendations (see Tables 3 and 4):**
**NIOSH/OSHA**
**0.05 ppm:** Sa
**0.125 ppm:** Sa:Cf
**0.25 ppm:** SaT:Cf/ScbaF/SaF
**1 ppm:** Sa:Pd,Pp
**§:** ScbaF:Pd,Pp/SaF:Pd,Pp:AScba
**Escape:** GmFS/ScbaE

**Incompatibilities and Reactivities:** Oxidizers, halogens, water, halogenated hydrocarbons
[**Note:** May ignite SPONTANEOUSLY in moist air. Corrosive to natural rubber. Hydrolyzes slowly with heat in water to form boric acid.]

**Exposure Routes, Symptoms, Target Organs (see Table 5):**
**ER:** Inh, Abs, Ing, Con
**SY:** Irrit eyes, skin; dizz, head, drow, inco, tremor, convuls, behavioral changes; tonic spasm face, neck, abdom, limbs
**TO:** Eyes, skin, CNS

**First Aid (see Table 6):**
**Eye:** Irr immed
**Skin:** Soap wash immed
**Breath:** Resp support
**Swallow:** Medical attention immed

## Pentachloroethane

**Formula:** $CHCl_2CCl_3$ | **CAS#:** 76-01-7 | **RTECS#:** KI6300000 | **IDLH:** N.D.

**Conversion:** | **DOT:** 1669 151

**Synonyms/Trade Names:** Ethane pentachloride, Pentalin

**Exposure Limits:**
**NIOSH REL:** Handle with care in the workplace.
See Appendix C (Chloroethanes)
**OSHA PEL:** none

**Measurement Methods (see Table 1):**
**NIOSH** 2517

**Physical Description:** Colorless liquid with a sweetish, chloroform-like odor.

**Chemical & Physical Properties:**
**MW:** 202.3
**BP:** 322°F
**Sol:** 0.05%
**Fl.P:** ?
**IP:** 11.28 eV
**Sp.Gr:** 1.68
**VP:** 3 mmHg
**FRZ:** -20°F
**UEL:** ?
**LEL:** ?
Combustible Liquid

**Personal Protection/Sanitation (see Table 2):**
**Skin:** Prevent skin contact
**Eyes:** Prevent eye contact
**Wash skin:** When contam
**Remove:** When wet or contam
**Change:** N.R.
**Provide:** Eyewash
Quick drench

**Respirator Recommendations (see Tables 3 and 4):**
Not available.

**Incompatibilities and Reactivities:** (Sodium-potassium alloy + bromoform), alkalis, metals, water
[**Note:** Hydrolysis produces dichloroacetic acid. Reaction with alkalis & metals produces spontaneously explosive chloroacetylenes.]

**Exposure Routes, Symptoms, Target Organs (see Table 5):**
**ER:** Inh, Ing, Con
**SY:** In animals: irrit eyes, skin; lass, restless, irreg respiration, musc inco; liver, kidney, lung changes
**TO:** Eyes, skin, resp sys, CNS, liver, kidneys

**First Aid (see Table 6):**
**Eye:** Irr immed
**Skin:** Soap wash
**Breath:** Resp support
**Swallow:** Medical attention immed

## Pentachloronaphthalene

| Pentachloronaphthalene | Formula: $C_{10}H_3Cl_5$ | CAS#: 1321-64-8 | RTECS#: QK0300000 | IDLH: See Appendix F |
|---|---|---|---|---|
| Conversion: | DOT: | | | |

**Synonyms/Trade Names:** Halowax® 1013; 1,2,3,4,5-Pentachloronaphthalene

**Exposure Limits:**
**NIOSH REL:** TWA 0.5 mg/m$^3$ [skin]
**OSHA PEL:** TWA 0.5 mg/m$^3$ [skin]

**Measurement Methods (see Table 1):**
**NIOSH** S96 (II-2)

**Physical Description:** Pale-yellow or white solid or powder with an aromatic odor.

**Chemical & Physical Properties:**
**MW:** 300.4
**BP:** 636°F
**Sol:** Insoluble
**Fl.P:** NA
**IP:** ?
**Sp.Gr:** 1.67
**VP:** <1 mmHg
**MLT:** 248°F
**UEL:** NA
**LEL:** NA
Noncombustible Solid

**Personal Protection/Sanitation (see Table 2):**
**Skin:** Prevent skin contact
**Eyes:** Prevent eye contact
**Wash skin:** When contam
**Remove:** When wet or contam
**Change:** Daily

**Respirator Recommendations (see Tables 3 and 4):**
**NIOSH/OSHA**
**5 mg/m$^3$:** Sa*/ScbaF
**§:** ScbaF:Pd,Pp/SaF:Pd,Pp:AScba
**Escape:** GmFOv100/ScbaE

**See Appendix F**

**Incompatibilities and Reactivities:** Strong oxidizers

**Exposure Routes, Symptoms, Target Organs (see Table 5):**
**ER:** Inh, Abs, Ing, Con
**SY:** Head, lass, dizz, anor; pruritus, acne-form skin eruptions; jaun, liver nec
**TO:** Skin, liver, CNS

**First Aid (see Table 6):**
**Eye:** Irr immed
**Skin:** Soap prompt/molten flush immed
**Breath:** Resp support
**Swallow:** Medical Attention immed

## Pentachlorophenol

| Pentachlorophenol | Formula: $C_6Cl_5OH$ | CAS#: 87-86-5 | RTECS#: SM6300000 | IDLH: 2.5 mg/m$^3$ |
|---|---|---|---|---|
| Conversion: | DOT: 3155 154 | | | |

P

**Synonyms/Trade Names:** PCP; Penta; 2,3,4,5,6-Pentachlorophenol

**Exposure Limits:**
**NIOSH REL:** TWA 0.5 mg/m$^3$ [skin]
**OSHA PEL:** TWA 0.5 mg/m$^3$ [skin]

**Measurement Methods (see Table 1):**
**NIOSH** 5512

**Physical Description:** Colorless to white, crystalline solid with a benzene-like odor. [fungicide]

**Chemical & Physical Properties:**
**MW:** 266.4
**BP:** 588°F (Decomposes)
**Sol:** 0.001%
**Fl.P:** NA
**IP:** NA
**Sp.Gr:** 1.98
**VP(77°F):** 0.0001 mmHg
**MLT:** 374°F
**UEL:** NA
**LEL:** NA
Noncombustible Solid

**Personal Protection/Sanitation (see Table 2):**
**Skin:** Prevent skin contact
**Eyes:** Prevent eye contact
**Wash skin:** When contam
**Remove:** When wet or contam
**Change:** Daily
**Provide:** Eyewash
Quick drench

**Respirator Recommendations (see Tables 3 and 4):**
**NIOSH/OSHA**
**2.5 mg/m$^3$:** CcrOv95*/PaprOvHie*/
Sa*/ScbaF
**§:** ScbaF:Pd,Pp/SaF:Pd,Pp:AScba
**Escape:** GmFOv100/ScbaE

**Incompatibilities and Reactivities:** Strong oxidizers, acids, alkalis

**Exposure Routes, Symptoms, Target Organs (see Table 5):**
**ER:** Inh, Abs, Ing, Con
**SY:** Irrit eyes, nose, throat; sneez, cough; lass, anor, low-wgt; sweat; head, dizz; nau, vomit; dysp, chest pain; high fever; derm
**TO:** Eyes, skin, resp sys, CVS, liver, kidneys, CNS

**First Aid (see Table 6):**
**Eye:** Irr immed
**Skin:** Soap wash immed
**Breath:** Resp support
**Swallow:** Medical attention immed

## Pentaerythritol

| Pentaerythritol | Formula: $C(CH_2OH)_4$ | CAS#: 115-77-5 | RTECS#: RZ2490000 | IDLH: N.D. |
|---|---|---|---|---|

**Conversion:**

**DOT:**

**Synonyms/Trade Names:** 2,2-bis(Hydroxymethyl)-1,3-propanediol; Methane tetramethylol; Monopentaerythritol; PE; Tetrahydroxymethylolmethane; Tetramethylolmethane

**Exposure Limits:**
**NIOSH REL:** TWA 10 mg/m$^3$ (total)
TWA 5 mg/m$^3$ (resp)
**OSHA PEL†:** TWA 15 mg/m$^3$ (total)
TWA 5 mg/m$^3$ (resp)

**Measurement Methods (see Table 1):**
**NIOSH** 0500, 0600

**Physical Description:** Colorless to white, crystalline, odorless powder.
[**Note:** Technical grade is 88% monopentaerythritol & 12% dipentaerythritol.]

**Chemical & Physical Properties:**
**MW:** 136.2
**BP:** Sublimes
**Sol(59°F):** 6%
**Fl.P:** ?
**IP:** ?
**Sp.Gr:** 1.38
**VP:** 0.00000008 mmHg
**MLT:** 500°F (Sublimes)
**UEL:** ?
**LEL:** ?
Combustible Solid

**Personal Protection/Sanitation (see Table 2):**
**Skin:** N.R.
**Eyes:** N.R.
**Wash skin:** N.R.
**Remove:** N.R.
**Change:** N.R.

**Respirator Recommendations (see Tables 3 and 4):**
Not available.

**Incompatibilities and Reactivities:** Organic acids, oxidizers
[**Note:** Explosive compound is formed when a mixture of PE & thiophosphoryl chloride is heated.]

**Exposure Routes, Symptoms, Target Organs (see Table 5):**
**ER:** Inh, Ing, Con
**SY:** Irrit eyes, resp sys
**TO:** Eyes, resp sys

**First Aid (see Table 6):**
**Eye:** Irr immed
**Skin:** Water wash
**Breath:** Fresh air
**Swallow:** Medical attention immed



## n-Pentane

| n-Pentane | Formula: $CH_3[CH_2]_3CH_3$ | CAS#: 109-66-0 | RTECS#: RZ9450000 | IDLH: 1500 ppm [10%LEL] |
|---|---|---|---|---|

**Conversion:** 1 ppm = 2.95 mg/m$^3$

**DOT:** 1265 128

**Synonyms/Trade Names:** Pentane, normal-Pentane

**Exposure Limits:**
**NIOSH REL:** TWA 120 ppm (350 mg/m$^3$)
C 610 ppm (1800 mg/m$^3$) [15-minute]
**OSHA PEL†:** TWA 1000 ppm (2950 mg/m$^3$)

**Measurement Methods (see Table 1):**
**NIOSH** 1500
**OSHA** 7

**Physical Description:** Colorless liquid with a gasoline-like odor.
[**Note:** A gas above 97°F. May be utilized as a fuel.]

**Chemical & Physical Properties:**
**MW:** 72.2
**BP:** 97°F
**Sol:** 0.04%
**Fl.P:** -57°F
**IP:** 10.34 eV
**Sp.Gr:** 0.63
**VP:** 420 mmHg
**FRZ:** -202°F
**UEL:** 7.8%
**LEL:** 1.5%
Class IA Flammable Liquid

**Personal Protection/Sanitation (see Table 2):**
**Skin:** Prevent skin contact
**Eyes:** Prevent eye contact
**Wash skin:** When contam
**Remove:** When wet (flamm)
**Change:** N.R.

**Respirator Recommendations (see Tables 3 and 4):**
**NIOSH**
**1200 ppm:** Sa
**1500 ppm:** Sa:Cf/ScbaF/SaF
**§:** ScbaF:Pd,Pp/SaF:Pd,Pp:AScba
**Escape:** GmFOv/ScbaE

**Incompatibilities and Reactivities:** Strong oxidizers

**Exposure Routes, Symptoms, Target Organs (see Table 5):**
**ER:** Inh, Ing, Con
**SY:** Irrit eyes, skin, nose; derm; chemical pneu (aspir liquid); drow; in animals: narco
**TO:** Eyes, skin, resp sys, CNS

**First Aid (see Table 6):**
**Eye:** Irr immed
**Skin:** Water wash prompt
**Breath:** Resp support
**Swallow:** Medical attention immed

## 1-Pentanethiol

| Formula: | CAS#: | RTECS#: | IDLH: |
|---|---|---|---|
| $CH_3(CH_2)_4SH$ | 110-66-7 | SA3150000 | N.D. |

**Conversion:** 1 ppm = 4.26 mg/m$^3$ | **DOT:** 1111 130

**Synonyms/Trade Names:** Amyl hydrosulfide, Amyl mercaptan, Amyl sulfhydrate, Pentyl mercaptan

**Exposure Limits:**
**NIOSH REL:** C 0.5 ppm (2.1 mg/m$^3$) [15-minute]
**OSHA PEL:** none

**Measurement Methods (see Table 1):**
None available

**Physical Description:** Water-white to yellowish liquid with a strong, garlic-like odor.

**Chemical & Physical Properties:**
**MW:** 104.2
**BP:** 260°F
**Sol:** Insoluble
**Fl.P(oc):** 65°F
**IP:** ?
**Sp.Gr:** 0.84
**VP(77°F):** 14 mmHg
**FRZ:** -104°F
**UEL:** ?
**LEL:** ?
Class IB Flammable Liquid

**Personal Protection/Sanitation (see Table 2):**
**Skin:** Prevent skin contact
**Eyes:** Prevent eye contact
**Wash skin:** When contam
**Remove:** When wet (flamm)
**Change:** N.R.

**Respirator Recommendations (see Tables 3 and 4):**
**NIOSH**
**5 ppm:** CcrOv/Sa
**12.5 ppm:** Sa:Cf/PaprOv
**25 ppm:** CcrFOv/GmFOv/PaprTOv/ScbaF/SaF
**§:** ScbaF:Pd,Pp/SaF:Pd,Pp:AScba
**Escape:** GmFOv/ScbaE

**Incompatibilities and Reactivities:** Oxidizers, reducing agents, alkali metals, calcium hypochlorite, concentrated nitric acid

**Exposure Routes, Symptoms, Target Organs (see Table 5):**
**ER:** Inh, Ing, Con
**SY:** Irrit eyes, skin, nose, throat, resp sys; head, nau, dizz; vomit, diarr; derm, skin sens
**TO:** Eyes, skin, resp sys, CNS

**First Aid (see Table 6):**
**Eye:** Irr immed
**Skin:** Soap wash immed
**Breath:** Resp support
**Swallow:** Medical attention immed

## 2-Pentanone

| Formula: | CAS#: | RTECS#: | IDLH: |
|---|---|---|---|
| $CH_3COCH_2CH_2CH_3$ | 107-87-9 | SA7875000 | 1500 ppm |

**Conversion:** 1 ppm = 3.52 mg/m$^3$ | **DOT:** 1249 127

**Synonyms/Trade Names:** Ethyl acetone, Methyl propyl ketone, MPK

**Exposure Limits:**
**NIOSH REL:** TWA 150 ppm (530 mg/m$^3$)
**OSHA PEL†:** TWA 200 ppm (700 mg/m$^3$)

**Measurement Methods (see Table 1):**
**NIOSH** 1300, 2555

**Physical Description:** Colorless to water-white liquid with a characteristic acetone-like odor.

**Chemical & Physical Properties:**
**MW:** 86.1
**BP:** 215°F
**Sol:** 6%
**Fl.P:** 45°F
**IP:** 9.39 eV
**Sp.Gr:** 0.81
**VP:** 27 mmHg
**FRZ:** -108°F
**UEL:** 8.2%
**LEL:** 1.5%
Class IB Flammable Liquid

**Personal Protection/Sanitation (see Table 2):**
**Skin:** Prevent skin contact
**Eyes:** Prevent eye contact
**Wash skin:** When contam
**Remove:** When wet (flamm)
**Change:** N.R.

**Respirator Recommendations (see Tables 3 and 4):**
**NIOSH**
**1500 ppm:** CcrOv*/PaprOv*/GmFOv/Sa*/ScbaF
**§:** ScbaF:Pd,Pp/SaF:Pd,Pp:AScba
**Escape:** GmFOv/ScbaE

**Incompatibilities and Reactivities:** Oxidizers, bromine trifluoride

**Exposure Routes, Symptoms, Target Organs (see Table 5):**
**ER:** Inh, Ing, Con
**SY:** Irrit eyes, skin, muc memb; head; derm; narco, coma
**TO:** Eyes, skin, resp sys, CNS

**First Aid (see Table 6):**
**Eye:** Irr immed
**Skin:** Water flush
**Breath:** Resp support
**Swallow:** Medical attention immed

## Perchloromethyl mercaptan

| Formula: | CAS#: | RTECS#: | IDLH: |
|---|---|---|---|
| $Cl_3CSCl$ | 594-42-3 | PB0370000 | 10 ppm |

**Conversion:** 1 ppm = 7.60 mg/m$^3$ **DOT:** 1670 157

**Synonyms/Trade Names:** PCM, PMM, Trichloromethane sulfenyl chloride, Trichloromethyl sulfur chloride

**Exposure Limits:**
**NIOSH REL:** TWA 0.1 ppm (0.8 mg/m$^3$)
**OSHA PEL:** TWA 0.1 ppm (0.8 mg/m$^3$)

**Measurement Methods (see Table 1):**
None available

**Physical Description:** Pale-yellow, oily liquid with an unbearable, acrid odor.

**Chemical & Physical Properties:**
**MW:** 185.9
**BP:** 297°F (Decomposes)
**Sol:** Insoluble
**Fl.P:** NA
**IP:** ?
**Sp.Gr:** 1.69
**VP:** 3 mmHg
**FRZ:** ?
**UEL:** NA
**LEL:** NA
Noncombustible Liquid, but will support combustion.

**Personal Protection/Sanitation (see Table 2):**
**Skin:** Prevent skin contact
**Eyes:** Prevent eye contact
**Wash skin:** When contam
**Remove:** When wet or contam
**Change:** N.R.

**Respirator Recommendations (see Tables 3 and 4):**
**NIOSH/OSHA**
**1 ppm:** CcrOv*/Sa*
**2.5 ppm:** Sa:Cf*/PaprOv*
**5 ppm:** CcrFOv/GmFOv/PaprTOv*/SaT:Cf*/ScbaF/SaF
**10 ppm:** SaF:Pd,Pp
**§:** ScbaF:Pd,Pp/SaF:Pd,Pp:AScba
**Escape:** GmFOv/ScbaE

**Incompatibilities and Reactivities:** Alkalis, amines, hot iron, water
[**Note:** Corrosive to most metals. Forms HCl, sulfur & $CO_2$ on contact with water.]

**Exposure Routes, Symptoms, Target Organs (see Table 5):**
**ER:** Inh, Abs, Ing, Con
**SY:** Irrit eyes, skin, nose, throat; lac; cough, dysp, deep breath pain, coarse rales; vomit; pallor, tacar; acidosis; anuria; liver, kidney damage
**TO:** Eyes, skin, resp sys, liver, kidneys

**First Aid (see Table 6):**
**Eye:** Irr immed
**Skin:** Soap wash immed
**Breath:** Resp support
**Swallow:** Medical attention immed

P

## Perchloryl fluoride

| Formula: | CAS#: | RTECS#: | IDLH: |
|---|---|---|---|
| $ClO_3F$ | 7616-94-6 | SD1925000 | 100 ppm |

**Conversion:** 1 ppm = 4.19 mg/m$^3$ **DOT:** 3083 124

**Synonyms/Trade Names:** Chlorine fluoride oxide, Chlorine oxyfluoride, Trioxychlorofluoride

**Exposure Limits:**
**NIOSH REL:** TWA 3 ppm (14 mg/m$^3$)
ST 6 ppm (28 mg/m$^3$)
**OSHA PEL†:** TWA 3 ppm (13.5 mg/m$^3$)

**Measurement Methods (see Table 1):**
None available

**Physical Description:** Colorless gas with a characteristic, sweet odor.
[**Note:** Shipped as a liquefied compressed gas.]

**Chemical & Physical Properties:**
**MW:** 102.5
**BP:** -52°F
**Sol:** 0.06%
**Fl.P:** NA
**IP:** 13.60 eV
**RGasD:** 3.64
**VP:** 10.5 atm
**FRZ:** -234°F
**UEL:** NA
**LEL:** NA
Nonflammable Gas, but will support combustion.

**Personal Protection/Sanitation (see Table 2):**
**Skin:** Frostbite
**Eyes:** Frostbite
**Wash skin:** N.R.
**Remove:** N.R.
**Change:** N.R.
**Provide:** Frostbite wash

**Respirator Recommendations (see Tables 3 and 4):**
**NIOSH/OSHA**
**30 ppm:** Sa
**75 ppm:** Sa:Cf*
**100 ppm:** ScbaF/SaF
**§:** ScbaF:Pd,Pp/SaF:Pd,Pp:AScba
**Escape:** GmFS¿/ScbaE

**Incompatibilities and Reactivities:** Combustibles, strong bases, amines, finely divided metals, reducing agents, alcohols

**Exposure Routes, Symptoms, Target Organs (see Table 5):**
**ER:** Inh, Con (liquid)
**SY:** Irrit resp sys; liquid: frostbite; in animals: methemo; cyan; lass, dizz, head; pulm edema; pneu; anoxia
**TO:** Skin, resp sys, blood

**First Aid (see Table 6):**
**Eye:** Frostbite
**Skin:** Frostbite
**Breath:** Resp support

## Perlite

| Formula: | CAS#: 93763-70-3 | RTECS#: SD5254000 | IDLH: N.D. |
|---|---|---|---|

**Conversion:**

**DOT:**

**Synonyms/Trade Names:** Expanded perlite
[**Note:** An amorphous material consisting of fused sodium potassium aluminum silicate.]

**Exposure Limits:**
**NIOSH REL:** TWA 10 mg/m$^3$ (total)
TWA 5 mg/m$^3$ (resp)
**OSHA PEL:** TWA 15 mg/m$^3$ (total)
TWA 5 mg/m$^3$ (resp)

**Measurement Methods (see Table 1):**
**NIOSH** 0500, 0600

**Physical Description:** Odorless, light-gray to glassy-black solid.
[**Note:** Expanded perlite is a fluffy, white particulate.]

**Chemical & Physical Properties:**
**MW:** varies
**BP:** ?
**Sol:** <1%
**Fl.P:** NA
**IP:** NA
**Sp.Gr:** 2.2 - 2.4 (crude)
0.05 - 0.3 (expanded)
**VP:** 0 mmHg (approx)
**MLT:** >2000°F
**UEL:** NA
**LEL:** NA
Noncombustible Solid

**Personal Protection/Sanitation (see Table 2):**
**Skin:** N.R.
**Eyes:** N.R.
**Wash skin:** N.R.
**Remove:** N.R.
**Change:** N.R.

**Respirator Recommendations (see Tables 3 and 4):**
Not available.

**Incompatibilities and Reactivities:** None reported

**Exposure Routes, Symptoms, Target Organs (see Table 5):**
**ER:** Inh, Con
**SY:** Irrit eyes, skin, throat, upper resp sys
**TO:** Eyes, skin, resp sys

**First Aid (see Table 6):**
**Eye:** Irr immed
**Breath:** Fresh air



## Petroleum distillates (naphtha)

| Formula: | CAS#: 8002-05-9 | RTECS#: SE7449000 | IDLH: 1100 ppm [10%LEL] |
|---|---|---|---|

**Conversion:** 1 ppm = 4.05 mg/m$^3$

**DOT:**

**Synonyms/Trade Names:** Aliphatic petroleum naphtha, Petroleum naphtha, Rubber solvent

**Exposure Limits:**
**NIOSH REL:** TWA 350 mg/m$^3$
C 1800 mg/m$^3$ [15-minute]
**OSHA PEL†:** TWA 500 ppm (2000 mg/m$^3$)

**Measurement Methods (see Table 1):**
**NIOSH** 1550

**Physical Description:** Colorless liquid with a gasoline- or kerosene-like odor.
[**Note:** A mixture of paraffins ($C_5$ to $C_{13}$) that may contain a small amount of aromatic hydrocarbons.]

**Chemical & Physical Properties:**
**MW:** 99 (approx)
**BP:** 86-460°F
**Sol:** Insoluble
**Fl.P:** -40 to -86°F
**IP:** ?
**Sp.Gr:** 0.63-0.66
**VP:** 40 mmHg (approx)
**FRZ:** -99°F
**UEL:** 5.9%
**LEL:** 1.1%
Flammable Liquid

**Personal Protection/Sanitation (see Table 2):**
**Skin:** Prevent skin contact
**Eyes:** Prevent eye contact
**Wash skin:** When contam
**Remove:** When wet (flamm)
**Change:** N.R.

**Respirator Recommendations (see Tables 3 and 4):**
**NIOSH**
**850 ppm:** Sa
**1100 ppm:** Sa:Cf*/ScbaF/SaF
**§:** ScbaF:Pd,Pp/SaF:Pd,Pp:AScba
**Escape:** GmFOv/ScbaE

**Incompatibilities and Reactivities:** Strong oxidizers

**Exposure Routes, Symptoms, Target Organs (see Table 5):**
**ER:** Inh, Ing, Con
**SY:** Irrit eyes, nose, throat; dizz, drow, head, nau; dry cracked skin; chemical pneu (aspir liquid)
**TO:** Eyes, skin, resp sys, CNS

**First Aid (see Table 6):**
**Eye:** Irr immed
**Skin:** Soap wash prompt
**Breath:** Resp support
**Swallow:** Medical attention immed

## Phenol

| Phenol | Formula: $C_6H_5OH$ | CAS#: 108-95-2 | RTECS#: SJ3325000 | IDLH: 250 ppm |
|---|---|---|---|---|

**Conversion:** 1 ppm = 3.85 mg/m$^3$

**DOT:** 1671 153 (solid); 2312 153 (molten); 2821 153 (solution)

**Synonyms/Trade Names:** Carbolic acid, Hydroxybenzene, Monohydroxybenzene, Phenyl alcohol, Phenyl hydroxide

**Exposure Limits:**
**NIOSH REL:** TWA 5 ppm (19 mg/m$^3$) [skin]
C 15.6 ppm (60 mg/m$^3$) [15-minute]
**OSHA PEL:** TWA 5 ppm (19 mg/m$^3$) [skin]

**Measurement Methods (see Table 1):**
**NIOSH** 2546
**OSHA** 32

**Physical Description:** Colorless to light-pink, crystalline solid with a sweet, acrid odor. [**Note:** Phenol liquefies by mixing with about 8% water.]

**Chemical & Physical Properties:**
**MW:** 94.1
**BP:** 359°F
**Sol(77°F):** 9%
**Fl.P:** 175°F
**IP:** 8.50 eV
**Sp.Gr:** 1.06
**VP:** 0.4 mmHg
**MLT:** 109°F
**UEL:** 8.6%
**LEL:** 1.8%
Combustible Solid

**Personal Protection/Sanitation (see Table 2):**
**Skin:** Prevent skin contact
**Eyes:** Prevent eye contact
**Wash skin:** When contam
**Remove:** When wet or contam
**Change:** Daily
**Provide:** Eyewash
Quick drench

**Respirator Recommendations (see Tables 3 and 4):**
**NIOSH/OSHA**
**50 ppm:** CcrOv95/Sa
**125 ppm:** Sa:Cf/PaprOvHie
**250 ppm:** CcrFOv100/GmFOv100/PaprTOvHie/ScbaF/SaF
**§:** ScbaF:Pd,Pp/SaF:Pd,Pp:AScba
**Escape:** GmFOv100/ScbaE

**Incompatibilities and Reactivities:** Strong oxidizers, calcium hypochlorite, aluminum chloride, acids

**Exposure Routes, Symptoms, Target Organs (see Table 5):**
**ER:** Inh, Abs, Ing, Con
**SY:** Irrit eyes, nose, throat; anor, low-wgt; lass, musc ache, pain; dark urine; cyan; liver, kidney damage; skin burns; derm; ochronosis; tremor, convuls, twitch
**TO:** Eyes, skin, resp sys, liver, kidneys

**First Aid (see Table 6):**
**Eye:** Irr immed
**Skin:** Soap wash immed
**Breath:** Resp support
**Swallow:** Medical attention immed



## Phenothiazine

| Phenothiazine | Formula: $S(C_6H_4)_2NH$ | CAS#: 92-84-2 | RTECS#: SN5075000 | IDLH: N.D. |
|---|---|---|---|---|

**Conversion:**

**DOT:**

**Synonyms/Trade Names:** Dibenzothiazine, Fenothiazine, Thiodiphenylamine

**Exposure Limits:**
**NIOSH REL:** TWA 5 mg/m$^3$ [skin]
**OSHA PEL†:** none

**Measurement Methods (see Table 1):**
**OSHA** PV2048

**Physical Description:** Grayish-green to greenish-yellow solid. [insecticide]

**Chemical & Physical Properties:**
**MW:** 199.3
**BP:** 700°F
**Sol:** Insoluble
**Fl.P:** ?
**IP:** ?
**Sp.Gr:** ?
**VP:** 0 mmHg (approx)
**MLT:** 365°F
**UEL:** ?
**LEL:** ?
Combustible Solid, but not a high fire risk.

**Personal Protection/Sanitation (see Table 2):**
**Skin:** Prevent skin contact
**Eyes:** N.R.
**Wash skin:** When contam/Daily
**Remove:** When wet or contam
**Change:** Daily

**Respirator Recommendations (see Tables 3 and 4):**
Not available.

**Incompatibilities and Reactivities:** None reported

**Exposure Routes, Symptoms, Target Organs (see Table 5):**
**ER:** Inh, Abs, Ing, Con
**SY:** Itching, irrit, reddening skin; hepatitis, hemolytic anemia, abdom cramps, tacar; kidney damage; skin photo sens
**TO:** Skin, CVS, liver, kidneys

**First Aid (see Table 6):**
**Eye:** Irr immed
**Skin:** Soap wash prompt
**Breath:** Resp support
**Swallow:** Medical attention immed

## p-Phenylene diamine

| p-Phenylene diamine | Formula: | CAS#: | RTECS#: | IDLH: |
|---|---|---|---|---|
| | $C_6H_4(NH_2)_2$ | 106-50-3 | SS8050000 | 25 mg/m$^3$ |
| Conversion: | DOT: 1673 153 | | | |

**Synonyms/Trade Names:** 4-Aminoaniline; 1,4-Benzenediamine; p-Diaminobenzene; 1,4-Diaminobenzene; 1,4-Phenylene diamine

**Exposure Limits:**
**NIOSH REL:** TWA 0.1 mg/m$^3$ [skin]
**OSHA PEL:** TWA 0.1 mg/m$^3$ [skin]

**Measurement Methods (see Table 1):**
**OSHA** 87

**Physical Description:** White to slightly red, crystalline solid.

**Chemical & Physical Properties:**
**MW:** 108.2
**BP:** 513°F
**Sol(75°F):** 4%
**Fl.P:** 312°F
**IP:** 6.89 eV
**Sp.Gr:** ?
**VP:** <1 mmHg
**MLT:** 295°F
**UEL:** ?
**LEL:** ?
Combustible Solid

**Personal Protection/Sanitation (see Table 2):**
**Skin:** Prevent skin contact
**Eyes:** Prevent eye contact
**Wash skin:** When contam/Daily
**Remove:** When wet or contam
**Change:** Daily

**Respirator Recommendations (see Tables 3 and 4):**
**NIOSH/OSHA**
**2.5 mg/m$^3$:** Sa:Cf£
**5 mg/m$^3$:** ScbaF/SaF
**25 mg/m$^3$:** SaF:Pd,Pp
**§:** ScbaF:Pd,Pp/SaF:Pd,Pp:AScba
**Escape:** GmFS100/ScbaE

**Incompatibilities and Reactivities:** Strong oxidizers

**Exposure Routes, Symptoms, Target Organs (see Table 5):**
**ER:** Inh, Abs, Ing, Con
**SY:** Irrit pharynx, larynx; bronchial asthma; sens derm
**TO:** Resp sys, skin

**First Aid (see Table 6):**
**Eye:** Irr immed
**Skin:** Soap wash prompt
**Breath:** Resp support
**Swallow:** Medical attention immed

## Phenyl ether (vapor)



| Phenyl ether (vapor) | Formula: | CAS#: | RTECS#: | IDLH: |
|---|---|---|---|---|
| | $C_6H_5OC_6H_5$ | 101-84-8 | KN8970000 | 100 ppm |
| Conversion: 1 ppm = 6.96 mg/m$^3$ | DOT: | | | |

**Synonyms/Trade Names:** Diphenyl ether, Diphenyl oxide, Phenoxy benzene, Phenyl oxide

**Exposure Limits:**
**NIOSH REL:** TWA 1 ppm (7 mg/m$^3$)
**OSHA PEL:** TWA 1 ppm (7 mg/m$^3$)

**Measurement Methods (see Table 1):**
**NIOSH** 1617
**OSHA** PV2022

**Physical Description:** Colorless, crystalline solid or liquid (above 82°F) with a geranium-like odor.

**Chemical & Physical Properties:**
**MW:** 170.2
**BP:** 498°F
**Sol:** Insoluble
**Fl.P:** 239°F
**IP:** 8.09 eV
**Sp.Gr:** 1.08
**VP(77°F):** 0.02 mmHg
**MLT:** 82°F
**UEL:** 6.0%
**LEL:** 0.7%
Combustible Solid
Class IIIB Combustible Liquid

**Personal Protection/Sanitation (see Table 2):**
**Skin:** Prevent skin contact
**Eyes:** Prevent eye contact
**Wash skin:** When contam
**Remove:** When wet or contam
**Change:** N.R.

**Respirator Recommendations (see Tables 3 and 4):**
**NIOSH/OSHA**
**25 ppm:** Sa:Cf£/PaprOvHie£
**50 ppm:** CcrFOv100/GmFOv100/ScbaF/SaF
**100 ppm:** SaF:Pd,Pp
**§:** ScbaF:Pd,Pp/SaF:Pd,Pp:AScba
**Escape:** GmFOv100/ScbaE

**Incompatibilities and Reactivities:** Strong oxidizers

**Exposure Routes, Symptoms, Target Organs (see Table 5):**
**ER:** Inh, Con
**SY:** Irrit eyes, nose, skin; nau
**TO:** Eyes, skin, resp sys

**First Aid (see Table 6):**
**Eye:** Irr immed
**Skin:** Soap wash prompt
**Breath:** Resp support

## Phenyl ether-biphenyl mixture (vapor)

| | | | |
|---|---|---|---|
| **Formula:** $C_6H_5OC_6H_5/C_6H_5C_6H_5$ | **CAS#:** 8004-13-5 | **RTECS#:** DV1500000 | **IDLH:** 10 ppm |

**Conversion:** 1 ppm = 6.79 mg/m$^3$ (approx) | **DOT:**

**Synonyms/Trade Names:** Diphenyl oxide-diphenyl mixture, Dowtherm® A

**Exposure Limits:**
**NIOSH REL:** TWA 1 ppm (7 mg/m$^3$)
**OSHA PEL:** TWA 1 ppm (7 mg/m$^3$)

**Measurement Methods (see Table 1):**
**NIOSH** 2013

**Physical Description:** Colorless to straw-colored liquid or solid (below 54°F) with a disagreeable, aromatic odor.
[**Note:** A mixture typically contains 75% phenyl ether & 25% biphenyl.]

**Chemical & Physical Properties:**
**MW:** 166 (approx)
**BP:** 495°F
**Sol:** Insoluble
**Fl.P:** 239°F
**IP:** ?
**Sp.Gr(77°F):** 1.06
**VP(77°F):** 0.08 mmHg
**FRZ:** 54°F
**UEL:** ?
**LEL:** ?
Class IIIB Combustible Liquid

**Personal Protection/Sanitation (see Table 2):**
**Skin:** Prevent skin contact
**Eyes:** Prevent eye contact
**Wash skin:** When contam
**Remove:** When wet or contam
**Change:** N.R.

**Respirator Recommendations (see Tables 3 and 4):**
**NIOSH/OSHA**
**10 ppm:** Sa:Cf£/CcrFOv100/GmFOv100/PaprOvHie£/ScbaF/SaF
**§:** ScbaF:Pd,Pp/SaF:Pd,Pp:AScba
**Escape:** GmFOv100/ScbaE

**Incompatibilities and Reactivities:** Strong oxidizers

**Exposure Routes, Symptoms, Target Organs (see Table 5):**
**ER:** Inh, Con
**SY:** Irrit eyes, nose, skin; nau
**TO:** Eyes, skin, resp sys

**First Aid (see Table 6):**
**Eye:** Irr immed
**Skin:** Soap wash prompt
**Breath:** Resp support

P

## Phenyl glycidyl ether

| | | | |
|---|---|---|---|
| **Formula:** $C_9H_{10}O_2$ | **CAS#:** 122-60-1 | **RTECS#:** TZ3675000 | **IDLH:** Ca [100 ppm] |

**Conversion:** 1 ppm = 6.14 mg/m$^3$ | **DOT:**

**Synonyms/Trade Names:** 1,2-Epoxy-3-phenoxy propane; Glycidyl phenyl ether; PGE; Phenyl 2,3-epoxypropyl ether

**Exposure Limits:**
**NIOSH REL:** Ca
C 1 ppm (6 mg/m$^3$) [15-minute]
See Appendix A
**OSHA PEL†:** TWA 10 ppm (60 mg/m$^3$)

**Measurement Methods (see Table 1):**
**NIOSH** 1619
**OSHA** 7

**Physical Description:** Colorless liquid. [**Note:** A solid below 38°F.]

**Chemical & Physical Properties:**
**MW:** 150.1
**BP:** 473°F
**Sol:** 0.2%
**Fl.P:** 248°F
**IP:** ?
**Sp.Gr:** 1.11
**VP:** 0.01 mmHg
**FRZ:** 38°F
**UEL:** ?
**LEL:** ?
Class IIIB Combustible Liquid

**Personal Protection/Sanitation (see Table 2):**
**Skin:** Prevent skin contact
**Eyes:** Prevent eye contact
**Wash skin:** When contam
**Remove:** When wet or contam
**Change:** N.R.
**Provide:** Eyewash
Quick drench

**Respirator Recommendations (see Tables 3 and 4):**
**NIOSH**
**¥:** ScbaF:Pd,Pp/SaF:Pd,Pp:AScba
**Escape:** GmFOv/ScbaE

**Incompatibilities and Reactivities:** Strong oxidizers, amines, strong acids, strong bases

**Exposure Routes, Symptoms, Target Organs (see Table 5):**
**ER:** Inh, Abs, Ing, Con
**SY:** Irrit eyes, skin; upper resp sys; skin sens; narco; possible hemato, repro effects; [carc]
**TO:** Eyes, skin, CNS, hemato sys, repro sys [in animals: nasal cancer]

**First Aid (see Table 6):**
**Eye:** Irr immed
**Skin:** Soap wash prompt
**Breath:** Resp support
**Swallow:** Medical attention immed

| Phenylhydrazine | Formula: $C_6H_5NHNH_2$ | CAS#: 100-63-0 | RTECS#: MV8925000 | IDLH: Ca [15 ppm] |
|---|---|---|---|---|
| **Conversion:** 1 ppm = 4.42 mg/m$^3$ | **DOT:** 2572 153 | | | |

**Synonyms/Trade Names:** Hydrazinobenzene, Monophenylhydrazine

**Exposure Limits:**
**NIOSH REL:** Ca
C 0.14 ppm (0.6 mg/m$^3$) [2-hr] [skin]
See Appendix A
**OSHA PEL†:** TWA 5 ppm (22 mg/m$^3$) [skin]

**Measurement Methods (see Table 1):**
**NIOSH** 3518

**Physical Description:** Colorless to pale-yellow liquid or solid (below 67°F) with a faint, aromatic odor.

**Chemical & Physical Properties:**
**MW:** 108.1
**BP:** 470°F (Decomposes)
**Sol:** Slight
**Fl.P:** 190°F
**IP:** 7.64 eV
**Sp.Gr:** 1.10
**VP(77°F):** 0.04 mmHg
**FRZ:** 67°F
**UEL:** ?
**LEL:** ?
Class IIIA Combustible Liquid
Combustible Solid

**Personal Protection/Sanitation (see Table 2):**
**Skin:** Prevent skin contact
**Eyes:** Prevent eye contact
**Wash skin:** When contam/Daily
**Remove:** When wet or contam
**Change:** Daily
**Provide:** Eyewash
Quick drench

**Respirator Recommendations (see Tables 3 and 4):**
**NIOSH**
**¥:** ScbaF:Pd,Pp/SaF:Pd,Pp:AScba
**Escape:** ScbaE

**Incompatibilities and Reactivities:** Strong oxidizers, lead dioxide

**Exposure Routes, Symptoms, Target Organs (see Table 5):**
**ER:** Inh, Abs, Ing, Con
**SY:** Skin sens, hemolytic anemia, dysp, cyan; jaun; kidney damage; vascular thrombosis; [carc]
**TO:** Blood, resp sys, liver, kidneys, skin [in animals: tumors of the lungs, liver, blood vessels & intestine]

**First Aid (see Table 6):**
**Eye:** Irr immed
**Skin:** Soap wash immed
**Breath:** Resp support
**Swallow:** Medical attention immed

P

| N-Phenyl-β-naphthylamine | Formula: $C_{10}H_7NHC_6H_5$ | CAS#: 135-88-6 | RTECS#: QM4550000 | IDLH: Ca [N.D.] |
|---|---|---|---|---|
| **Conversion:** | **DOT:** | | | |

**Synonyms/Trade Names:** 2-Anilinonaphthalene, β-Naphthylphenylamine, PBNA, 2-Phenylaminonaphthalene, Phenyl-β-naphthylamine

**Exposure Limits:**
**NIOSH REL:** Ca*
See Appendix A
[***Note:** Since metabolized to β-Naphthylamine.]
**OSHA PEL:** none

**Measurement Methods (see Table 1):**
**OSHA** 96

**Physical Description:** White to yellow crystals or gray to tan flakes or powder.
[**Note:** Commercial product may contain 20-30 ppm of β-Naphthylamine.]

**Chemical & Physical Properties:**
**MW:** 219.3
**BP:** 743°F
**Sol:** Insoluble
**Fl.P:** ?
**IP:** ?
**Sp.Gr:** 1.24
**VP:** ?
**MLT:** 226°F
**UEL:** ?
**LEL:** ?
Combustible Solid

**Personal Protection/Sanitation (see Table 2):**
**Skin:** Prevent skin contact
**Eyes:** Prevent eye contact
**Wash skin:** When contam/Daily
**Remove:** When wet or contam
**Change:** Daily
**Provide:** Eyewash
Quick drench

**Respirator Recommendations (see Tables 3 and 4):**
**NIOSH**
**¥:** ScbaF:Pd,Pp/SaF:Pd,Pp:AScba
**Escape:** GmFOv100/ScbaE

**Incompatibilities and Reactivities:** Oxidizers

**Exposure Routes, Symptoms, Target Organs (see Table 5):**
**ER:** Inh, Abs, Ing, Con
**SY:** Irritation; leucoplakia; acne, hypersensitivity to sunlight; [carc]
**TO:** Eyes, skin, bladder [bladder cancer]

**First Aid (see Table 6):**
**Eye:** Irr immed
**Skin:** Soap wash immed
**Breath:** Resp support
**Swallow:** Medical attention immed

| Phenylphosphine | Formula: $C_6H_5PH_2$ | CAS#: 638-21-1 | RTECS#: SZ2100000 | IDLH: N.D. |
|---|---|---|---|---|
| **Conversion:** 1 ppm = 4.50 mg/$m^3$ | **DOT:** | | | |

**Synonyms/Trade Names:** Fenylfosfin, PF, Phosphaniline

**Exposure Limits:**
**NIOSH REL:** C 0.05 ppm (0.25 mg/$m^3$)
**OSHA PEL†:** none

**Measurement Methods (see Table 1):** None available

**Physical Description:** Clear, colorless liquid with a foul odor.

| Chemical & Physical Properties: | Personal Protection/Sanitation (see Table 2): | Respirator Recommendations (see Tables 3 and 4): |
|---|---|---|
| **MW:** 110.1<br>**BP:** 320°F<br>**Sol:** Insoluble<br>**Fl.P:** ?<br>**IP:** ?<br>**Sp.Gr(59°F):** 1.001<br>**VP:** ?<br>**FRZ:** ?<br>**UEL:** ?<br>**LEL:** ?<br>Combustible Liquid | **Skin:** Prevent skin contact<br>**Eyes:** Prevent eye contact<br>**Wash skin:** Daily<br>**Remove:** When wet or contam<br>**Change:** N.R. | Not available. |

**Incompatibilities and Reactivities:** None reported [**Note:** Spontaneously combustible in high concentrations in air. Potential exposure to gaseous PF when polyphosphinates are heated above 392°F.]

| Exposure Routes, Symptoms, Target Organs (see Table 5): | First Aid (see Table 6): |
|---|---|
| **ER:** Inh, Ing, Con<br>**SY:** In animals: blood changes, anemia, testicular degeneration; loss of appetite, diarr, lac, hind leg tremor; derm<br>**TO:** Blood, CNS, skin, repro sys | **Eye:** Irr immed<br>**Skin:** Soap wash<br>**Breath:** Resp support<br>**Swallow:** Medical attention immed |

P

| Phorate | Formula: $(C_2H_5O)_2P(S)SCH_2SC_2H_5$ | CAS#: 298-02-2 | RTECS#: TD9450000 | IDLH: N.D. |
|---|---|---|---|---|
| **Conversion:** | **DOT:** 3018 152 (organophosphorus pesticide, liquid, toxic) | | | |

**Synonyms/Trade Names:** O,O-Diethyl S-(ethylthio)methylphosphorodithioate; O,O-Diethyl S-ethylthiomethylthiothionophosphate; Thimet; Timet

**Exposure Limits:**
**NIOSH REL:** TWA 0.05 mg/$m^3$
ST 0.2 mg/$m^3$ [skin]
**OSHA PEL†:** none

**Measurement Methods (see Table 1):** NIOSH 5600

**Physical Description:** Clear liquid with a skunk-like odor. [insecticide]

| Chemical & Physical Properties: | Personal Protection/Sanitation (see Table 2): | Respirator Recommendations (see Tables 3 and 4): |
|---|---|---|
| **MW:** 260.4<br>**BP:** ?<br>**Sol:** 0.005%<br>**Fl.P(oc):** 320°F<br>**IP:** ?<br>**Sp.Gr(77°F):** 1.16<br>**VP:** 0.0008 mmHg<br>**FRZ:** -45°F<br>**UEL:** ?<br>**LEL:** ?<br>Class IIIB Combustible Liquid, but does not readily ignite. | **Skin:** Prevent skin contact<br>**Eyes:** Prevent eye contact<br>**Wash skin:** When contam<br>**Remove:** When wet or contam<br>**Change:** N.R.<br>**Provide:** Eyewash<br>Quick drench | Not available. |

**Incompatibilities and Reactivities:** Water, alkalis [**Note:** Hydrolyzed in the presence of moisture and by alkalis.]

| Exposure Routes, Symptoms, Target Organs (see Table 5): | First Aid (see Table 6): |
|---|---|
| **ER:** Inh, Abs, Ing, Con<br>**SY:** Irrit eyes, skin, resp sys; miosis; rhin; head; chest tight, wheez, lar spasm, salv, cyan; anor, nau, vomit, abdom cramps, diarr; sweat; musc fasc, lass, para; dizz, conf, ataxia; convuls, coma; low BP; card irreg<br>**TO:** Eyes, skin, resp sys, CNS, CVS, blood chol | **Eye:** Irr immed<br>**Skin:** Soap flush immed<br>**Breath:** Resp support<br>**Swallow:** Medical attention immed |

## Phosdrin

| Phosdrin | Formula: $C_7H_{13}PO_6$ | CAS#: 7786-34-7 | RTECS#: GQ5250000 | IDLH: 4 ppm |
|---|---|---|---|---|
| **Conversion:** 1 ppm = 9.17 mg/m$^3$ | **DOT:** 2783 152 | | | |

**Synonyms/Trade Names:** 2-Carbomethoxy-1-methylvinyl dimethyl phosphate, Mevinphos
[**Note:** Commercial product is a mixture of the cis- & trans-isomers.]

**Exposure Limits:**
**NIOSH REL:** TWA 0.01 ppm (0.1 mg/m$^3$) [skin]
ST 0.03 ppm (0.3 mg/m$^3$)
**OSHA PEL†:** TWA 0.1 mg/m$^3$ [skin]

**Measurement Methods (see Table 1):**
**NIOSH** 5600

**Physical Description:** Pale-yellow to orange liquid with a weak odor.
[**Note:** Insecticide that may be absorbed on a dry carrier.]

**Chemical & Physical Properties:**
**MW:** 224.2
**BP:** Decomposes
**Sol:** Miscible
**Fl.P(oc):** 347°F
**IP:** ?
**Sp.Gr:** 1.25
**VP:** 0.003 mmHg
**FRZ:** 44°F (trans-)
70°F (cis-)
**UEL:** ?
**LEL:** ?
Class IIIB Combustible Liquid

**Personal Protection/Sanitation (see Table 2):**
**Skin:** Prevent skin contact
**Eyes:** Prevent eye contact
**Wash skin:** When contam
**Remove:** When wet or contam
**Change:** N.R.
**Provide:** Eyewash
Quick drench

**Respirator Recommendations (see Tables 3 and 4):**
**NIOSH/OSHA**
**0.1 ppm:** Sa
**0.25 ppm:** Sa:Cf
**0.5 ppm:** SaT:Cf/ScbaF/SaF
**4 ppm:** Sa:Pd,Pp
**§:** ScbaF:Pd,Pp/SaF:Pd,Pp:AScba
**Escape:** GmFOv100/ScbaE

**Incompatibilities and Reactivities:** Strong oxidizers [**Note:** Corrosive to cast iron, some stainless steels & brass.]

**Exposure Routes, Symptoms, Target Organs (see Table 5):**
**ER:** Inh, Abs, Ing, Con
**SY:** Irrit eyes, skin, resp sys; miosis; rhin; head; chest tight, wheez, lar spasm, salv, cyan; anor, nau, vomit, abdom cramps, diarr; para; ataxia, convuls; low BP, card irreg
**TO:** Eyes, skin, resp sys, CNS, CVS, blood chol

**First Aid (see Table 6):**
**Eye:** Irr immed
**Skin:** Soap wash immed
**Breath:** Resp support
**Swallow:** Medical attention immed



## Phosgene

| Phosgene | Formula: $COCl_2$ | CAS#: 75-44-5 | RTECS#: SY5600000 | IDLH: 2 ppm |
|---|---|---|---|---|
| **Conversion:** 1 ppm = 4.05 mg/m$^3$ | **DOT:** 1076 125 | | | |

**Synonyms/Trade Names:** Carbon oxychloride, Carbonyl chloride, Carbonyl dichloride, Chloroformyl chloride

**Exposure Limits:**
**NIOSH REL:** TWA 0.1 ppm (0.4 mg/m$^3$)
C 0.2 ppm (0.8 mg/m$^3$) [15-minute]
**OSHA PEL:** TWA 0.1 ppm (0.4 mg/m$^3$)

**Measurement Methods (see Table 1):**
**OSHA** 61

**Physical Description:** Colorless gas with a suffocating odor like musty hay.
[**Note:** A fuming liquid below 47°F. Shipped as a liquefied compressed gas.]

**Chemical & Physical Properties:**
**MW:** 98.9
**BP:** 47°F
**Sol:** Slight
**Fl.P:** NA
**IP:** 11.55 eV
**RGasD:** 3.48
**Sp.Gr:** 1.43 (Liquid at 32°F)
**VP:** 1.6 atm
**FRZ:** -198°F
**UEL:** NA
**LEL:** NA
Nonflammable Gas

**Personal Protection/Sanitation (see Table 2):**
**Skin:** Prevent skin contact (liquid)
**Eyes:** Prevent eye contact (liquid)
**Wash skin:** When contam (liquid)
**Remove:** When wet or contam (liquid)
**Change:** N.R.
**Provide:** Quick drench (liquid)

**Respirator Recommendations (see Tables 3 and 4):**
**NIOSH/OSHA**
**1 ppm:** Sa*
**2 ppm:** ScbaF/SaF
**§:** ScbaF:Pd,Pp/SaF:Pd,Pp:AScba
**Escape:** GmFS/ScbaE

**Incompatibilities and Reactivities:** Moisture, alkalis, ammonia, alcohols, copper
[**Note:** Reacts slowly in water to form hydrochloric acid & carbon dioxide.]

**Exposure Routes, Symptoms, Target Organs (see Table 5):**
**ER:** Inh, Con (liquid)
**SY:** Irrit eyes; dry burning throat; vomit; cough, foamy sputum, dysp, chest pain, cyan; liquid: frostbite
**TO:** Eyes, skin, resp sys

**First Aid (see Table 6):**
**Eye:** Irr immed (liquid)
**Skin:** Water flush immed (liquid)
**Breath:** Resp support

## Phosphine

| Formula: | CAS#: | RTECS#: | IDLH: |
|---|---|---|---|
| $PH_3$ | 7803-51-2 | SY7525000 | 50 ppm |

**Conversion:** 1 ppm = 1.39 mg/m$^3$

**DOT:** 2199 119

**Synonyms/Trade Names:** Hydrogen phosphide, Phosphorated hydrogen, Phosphorus hydride, Phosphorus trihydride

**Exposure Limits:**
**NIOSH REL:** TWA 0.3 ppm (0.4 mg/m$^3$)
ST 1 ppm (1 mg/m$^3$)
**OSHA PEL†:** TWA 0.3 ppm (0.4 mg/m$^3$)

**Measurement Methods (see Table 1):**
**OSHA** 1003, ID180

**Physical Description:** Colorless gas with a fish- or garlic-like odor. [pesticide]
[**Note:** Shipped as a liquefied compressed gas. Pure compound is odorless.]

**Chemical & Physical Properties:**
**MW:** 34.0
**BP:** -126°F
**Sol:** Slight
**Fl.P:** NA (Gas)
**IP:** 9.96 eV
**RGasD:** 1.18
**VP:** 41.3 atm
**FRZ:** -209°F
**UEL:** ?
**LEL:** 1.79%
Flammable Gas

**Personal Protection/Sanitation (see Table 2):**
**Skin:** Frostbite
**Eyes:** Frostbite
**Wash skin:** N.R.
**Remove:** When wet (flamm)
**Change:** N.R.
**Provide:** Frostbite wash

**Respirator Recommendations (see Tables 3 and 4):**
**NIOSH/OSHA**
**3 ppm:** Sa
**7.5 ppm:** Sa:Cf
**15 ppm:** GmFS/ScbaF/SaF
**50 ppm:** Sa:Pd,Pp
**§:** ScbaF:Pd,Pp/SaF:Pd,Pp:AScba
**Escape:** GmFS/ScbaE

**Incompatibilities and Reactivities:** Air, oxidizers, chlorine, acids, moisture, halogenated hydrocarbons, copper
[**Note:** May ignite SPONTANEOUSLY on contact with air.]

**Exposure Routes, Symptoms, Target Organs (see Table 5):**
**ER:** Inh, Con (liquid)
**SY:** Nau, vomit, abdom pain, diarr; thirst; chest tight, dysp; musc pain, chills; stupor or syncope; pulm edema; liquid: frostbite
**TO:** Resp sys

**First Aid (see Table 6):**
**Eye:** Frostbite
**Skin:** Frostbite
**Breath:** Resp support



## Phosphoric acid

| Formula: | CAS#: | RTECS#: | IDLH: |
|---|---|---|---|
| $H_3PO_4$ | 7664-38-2 | TB6300000 | 1000 mg/m$^3$ |

**Conversion:**

**DOT:** 1805 154 (liquid or solution); 3453 154 (solid)

**Synonyms/Trade Names:** Orthophosphoric acid, Phosphoric acid (aqueous), White phosphoric acid

**Exposure Limits:**
**NIOSH REL:** TWA 1 mg/m$^3$
ST 3 mg/m$^3$
**OSHA PEL†:** TWA 1 mg/m$^3$

**Measurement Methods (see Table 1):**
**NIOSH** 7903
**OSHA** ID165SG

**Physical Description:** Thick, colorless, odorless, crystalline solid. [**Note:** Often used in an aqueous solution.]

**Chemical & Physical Properties:**
**MW:** 98.0
**BP:** 415°F
**Sol:** Miscible
**Fl.P:** NA
**IP:** ?
**Sp.Gr(77°F):** 1.87 (pure)
1.33 (50% solution)
**VP:** 0.03 mmHg
**MLT:** 108°F
**UEL:** NA
**LEL:** NA
Noncombustible Solid

**Personal Protection/Sanitation (see Table 2):**
**Skin:** Prevent skin contact
**Eyes:** Prevent eye contact
**Wash skin:** When contam
**Remove:** When wet or contam
**Change:** Daily
**Provide:** Eyewash (>1.6%)
Quick drench (>1.6%)

**Respirator Recommendations (see Tables 3 and 4):**
**NIOSH/OSHA**
**25 mg/m$^3$:** Sa:Cf*
**50 mg/m$^3$:** 100F/ScbaF/SaF
**1000 mg/m$^3$:** SaF:Pd,Pp
**§:** ScbaF:Pd,Pp/SaF:Pd,Pp:AScba
**Escape:** 100F/ScbaE

**Incompatibilities and Reactivities:** Strong caustics, most metals [**Note:** Readily reacts with metals to form flammable hydrogen gas. DO NOT MIX WITH SOLUTIONS CONTAINING BLEACH OR AMMONIA.]

**Exposure Routes, Symptoms, Target Organs (see Table 5):**
**ER:** Inh, Ing, Con
**SY:** Irrit eyes, skin, upper resp sys; eye, skin, burns; derm
**TO:** Eyes, skin, resp sys

**First Aid (see Table 6):**
**Eye:** Irr immed
**Skin:** Water flush Immed
**Breath:** Resp support
**Swallow:** Medical attention immed

## Phosphorus (yellow)

| Phosphorus (yellow) | Formula: $P_4$ | CAS#: 7723-14-0 | RTECS#: TH3500000 | IDLH: 5 mg/m$^3$ |
|---|---|---|---|---|
| Conversion: | DOT: 1381 136 | | | |

**Synonyms/Trade Names:** Elemental phosphorus, White phosphorus

**Exposure Limits:**
**NIOSH REL:** TWA 0.1 mg/m$^3$
**OSHA PEL:** TWA 0.1 mg/m$^3$

**Measurement Methods (see Table 1):**
**NIOSH** 7905

**Physical Description:** White to yellow, soft, waxy solid with acrid fumes in air.
[**Note:** Usually shipped or stored in water.]

**Chemical & Physical Properties:**
**MW:** 124.0
**BP:** 536°F
**Sol:** 0.0003%
**Fl.P:** ?
**IP:** ?
**Sp.Gr:** 1.82
**VP:** 0.03 mmHg
**MLT:** 111°F
**UEL:** ?
**LEL:** ?
Flammable Solid

**Personal Protection/Sanitation (see Table 2):**
**Skin:** Prevent skin contact*
[***Note:** Flame retardant personal protective equipment should be provided.]
**Eyes:** Prevent eye contact
**Wash skin:** When contam
**Remove:** When wet or contam
**Change:** Daily
**Provide:** Eyewash
Quick drench

**Respirator Recommendations (see Tables 3 and 4):**
**NIOSH/OSHA**
**1 mg/m$^3$:** Sa
**2.5 mg/m$^3$:** Sa:Cf£
**5 mg/m$^3$:** ScbaF/SaF
**§:** ScbaF:Pd,Pp/SaF:Pd,Pp:AScba
**Escape:** ScbaE

**Incompatibilities and Reactivities:** Air, oxidizers (including elemental sulfur & strong caustics), halogens
[**Note:** Ignites SPONTANEOUSLY in moist air.]

**Exposure Routes, Symptoms, Target Organs (see Table 5):**
**ER:** Inh, Ing, Con
**SY:** Irrit eyes, resp tract; eye, skin burns; abdom pain, nau, jaun; anemia; cachexia; dental pain, salv, jaw pain, swell
**TO:** Eyes, skin, resp sys, liver, kidneys, jaw, teeth, blood

**First Aid (see Table 6):**
**Eye:** Irr immed
**Skin:** Water flush immed
**Breath:** Resp support
**Swallow:** Medical attention immed

P

## Phosphorus oxychloride

| Phosphorus oxychloride | Formula: $POCl_3$ | CAS#: 10025-87-3 | RTECS#: TH4897000 | IDLH: N.D. |
|---|---|---|---|---|
| Conversion: 1 ppm = 6.27 mg/m$^3$ | DOT: 1810 137 | | | |

**Synonyms/Trade Names:** Phosphorus chloride, Phosphorus oxytrichloride, Phosphoryl chloride

**Exposure Limits:**
**NIOSH REL:** TWA 0.1 ppm (0.6 mg/m$^3$)
ST 0.5 ppm (3 mg/m$^3$)
**OSHA PEL†:** none

**Measurement Methods (see Table 1):**
None available

**Physical Description:** Clear, colorless to yellow, oily liquid with a pungent & musty odor.
[**Note:** A solid below 34°F.]

**Chemical & Physical Properties:**
**MW:** 153.3
**BP:** 222°F
**Sol:** Decomposes
**Fl.P:** NA
**IP:** ?
**Sp.Gr(77°F):** 1.65
**VP(81°F):** 40 mmHg
**FRZ:** 34°F
**UEL:** NA
**LEL:** NA
Noncombustible Liquid, but may set fire to combustible materials.

**Personal Protection/Sanitation (see Table 2):**
**Skin:** Prevent skin contact
**Eyes:** Prevent eye contact
**Wash skin:** When contam
**Remove:** When wet or contam
**Change:** N.R.
**Provide:** Eyewash
Quick drench

**Respirator Recommendations (see Tables 3 and 4):**
Not available.

**Incompatibilities and Reactivities:** Water, combustible materials, carbon disulfide, dimethyl-formamide, metals (except nickel & lead) [**Note:** Decomposes in water to hydrochloric & phosphoric acids.]

**Exposure Routes, Symptoms, Target Organs (see Table 5):**
**ER:** Inh, Ing, Con
**SY:** Irrit eyes, skin, resp sys; eye, skin burns; dysp, cough, pulm edema; dizz, head, lass; abdom pain, nau, vomit; neph
**TO:** Eyes, skin, resp sys, CNS, kidneys

**First Aid (see Table 6):**
**Eye:** Irr immed
**Skin:** Water flush immed
**Breath:** Resp support
**Swallow:** Medical attention immed

P

## Phosphorus pentachloride

| Phosphorus pentachloride | Formula: $PCl_5$ | CAS#: 10026-13-8 | RTECS#: TB6125000 | IDLH: 70 mg/m$^3$ |
|---|---|---|---|---|
| **Conversion:** | **DOT:** 1806 137 | | | |

**Synonyms/Trade Names:** Pentachlorophosphorus, Phosphoric chloride, Phosphorus perchloride

**Exposure Limits:**
**NIOSH REL:** TWA 1 mg/m$^3$
**OSHA PEL:** TWA 1 mg/m$^3$

**Measurement Methods (see Table 1):**
**NIOSH** S257 (II-5)

**Physical Description:** White to pale-yellow, crystalline solid with a pungent, unpleasant odor.

**Chemical & Physical Properties:**
**MW:** 208.3
**BP:** Sublimes
**Sol:** Reacts
**Fl.P:** NA
**IP:** ?
**Sp.Gr:** 3.60
**VP(132°F):** 1 mmHg
**MLT:** 324°F (Sublimes)
**UEL:** NA
**LEL:** NA
Noncombustible Solid

**Personal Protection/Sanitation (see Table 2):**
**Skin:** Prevent skin contact
**Eyes:** Prevent eye contact
**Wash skin:** When contam
**Remove:** When wet or contam
**Change:** Daily
**Provide:** Eyewash
Quick drench

**Respirator Recommendations (see Tables 3 and 4):**
**NIOSH/OSHA**
**10 mg/m$^3$:** Sa*
**25 mg/m$^3$:** Sa:Cf*
**50 mg/m$^3$:** ScbaF/SaF
**70 mg/m$^3$:** SaF:Pd,Pp
**§:** ScbaF:Pd,Pp/SaF:Pd,Pp:AScba
**Escape:** GmFOv100/ScbaE

**Incompatibilities and Reactivities:** Water, magnesium oxide, chemically-active metals such as sodium and potassium, alkalis, amines
[**Note:** Hydrolyzes in water (even in humid air) to form hydrochloric acid & phosphoric acid. Corrosive to metals.]

**Exposure Routes, Symptoms, Target Organs (see Table 5):**
**ER:** Inh, Ing, Con
**SY:** Irrit eyes, skin, resp sys; bron; derm
**TO:** Eyes, skin, resp sys

**First Aid (see Table 6):**
**Eye:** Irr immed
**Skin:** Water flush immed
**Breath:** Resp support
**Swallow:** Medical attention immed

## Phosphorus pentasulfide

| Phosphorus pentasulfide | Formula: $P_2S_5/P_4S_{10}$ | CAS#: 1314-80-3 | RTECS#: TH4375000 | IDLH: 250 mg/m$^3$ |
|---|---|---|---|---|
| **Conversion:** | **DOT:** 1340 139 | | | |

**Synonyms/Trade Names:** Phosphorus persulfide, Phosphorus sulfide, Sulfur phosphide

**Exposure Limits:**
**NIOSH REL:** TWA 1 mg/m$^3$
ST 3 mg/m$^3$
**OSHA PEL†:** TWA 1 mg/m$^3$

**Measurement Methods (see Table 1):**
None available

**Physical Description:** Greenish-gray to yellow, crystalline solid with an odor of rotten eggs.

**Chemical & Physical Properties:**
**MW:** 222.3 ($P_2S_5$)
444.6 ($P_4S_{10}$)
**BP:** 957°F
**Sol:** Reacts
**Fl.P:** ?
**IP:** ?
**Sp.Gr:** 2.09
**VP(572°F):** 1 mmHg
**MLT:** 550°F
**UEL:** ?
**LEL:** ?
Flammable Solid, which may SPONTANEOUSLY ignite in presence of moisture.

**Personal Protection/Sanitation (see Table 2):**
**Skin:** Prevent skin contact
**Eyes:** Prevent eye contact
**Wash skin:** When contam
**Remove:** When wet or contam
**Change:** Daily

**Respirator Recommendations (see Tables 3 and 4):**
**NIOSH/OSHA**
**10 mg/m$^3$:** Sa*
**25 mg/m$^3$:** Sa:Cf*
**50 mg/m$^3$:** ScbaF/SaF
**250 mg/m$^3$:** SaF:Pd,Pp
**§:** ScbaF:Pd,Pp/SaF:Pd,Pp:AScba
**Escape:** GmFS100/ScbaE

**Incompatibilities and Reactivities:** Water, alcohols, strong oxidizers, acids, alkalis [**Note:** Reacts with water to form hydrogen sulfide, sulfur dioxide, and phosphoric acid.]

**Exposure Routes, Symptoms, Target Organs (see Table 5):**
**ER:** Inh, Ing, Con
**SY:** Irrit eyes, skin, resp sys; apnea, coma, convuls; conj pain, lac, photo, kerato-conj, corn vesic; dizz; head; lass; irrity, insom; GI dist
**TO:** Eyes, skin, resp sys, CNS

**First Aid (see Table 6):**
**Eye:** Irr immed
**Skin:** Dust off solid; water flush
**Breath:** Resp support
**Swallow:** Medical attention immed

## Phosphorus trichloride

| Phosphorus trichloride | Formula: $PCl_3$ | CAS#: 7719-12-2 | RTECS#: TH3675000 | IDLH: 25 ppm |
|---|---|---|---|---|

**Conversion:** 1 ppm = 5.62 mg/m$^3$ | **DOT:** 1809 137

**Synonyms/Trade Names:** Phosphorus chloride

**Exposure Limits:**
**NIOSH REL:** TWA 0.2 ppm (1.5 mg/m$^3$)
ST 0.5 ppm (3 mg/m$^3$)
**OSHA PEL†:** TWA 0.5 ppm (3 mg/m$^3$)

**Measurement Methods (see Table 1):**
**NIOSH** 6402

**Physical Description:** Colorless to yellow, fuming liquid with an odor like hydrochloric acid.

**Chemical & Physical Properties:**
**MW:** 137.4
**BP:** 169°F
**Sol:** Reacts
**Fl.P:** NA
**IP:** 9.91 eV
**Sp.Gr:** 1.58
**VP:** 100 mmHg
**FRZ:** -170°F
**UEL:** NA
**LEL:** NA
Noncombustible Liquid; however, a strong oxidizer that may ignite combustibles upon contact.

**Personal Protection/Sanitation (see Table 2):**
**Skin:** Prevent skin contact
**Eyes:** Prevent eye contact
**Wash skin:** When contam
**Remove:** When wet or contam
**Change:** N.R.
**Provide:** Eyewash
Quick drench

**Respirator Recommendations (see Tables 3 and 4):**
**NIOSH**
**10 ppm:** ScbaF/SaF
**25 ppm:** SaF:Pd,Pp
**§:** ScbaF:Pd,Pp/SaF:Pd,Pp:AScba
**Escape:** GmFS¿/ScbaE

**Incompatibilities and Reactivities:** Water, chemically-active metals such as sodium & potassium, aluminum, strong nitric acid, acetic acid, organic matter [**Note:** Hydrolyzes in water to form hydrochloric acid and phosphoric acid.]

**Exposure Routes, Symptoms, Target Organs (see Table 5):**
**ER:** Inh, Ing, Con
**SY:** Irrit eyes, skin, nose, throat; pulm edema; eye, skin burns
**TO:** Eyes, skin, resp sys

**First Aid (see Table 6):**
**Eye:** Irr immed
**Skin:** Water flush immed
**Breath:** Resp support
**Swallow:** Medical attention immed

## Phthalic anhydride

| Phthalic anhydride | Formula: $C_6H_4(CO)_2O$ | CAS#: 85-44-9 | RTECS#: TI3150000 | IDLH: 60 mg/m$^3$ |
|---|---|---|---|---|

P

**Conversion:** 1 ppm = 6.06 mg/m$^3$ | **DOT:** 2214 156

**Synonyms/Trade Names:** 1,2-Benzenedicarboxylic anhydride; PAN; Phthalic acid anhydride

**Exposure Limits:**
**NIOSH REL:** TWA 6 mg/m$^3$ (1 ppm)
**OSHA PEL†:** TWA 12 mg/m$^3$ (2 ppm)

**Measurement Methods (see Table 1):**
**NIOSH** S179 (II-3)
**OSHA** 90

**Physical Description:** White solid (flake) or a clear, colorless, mobile liquid (molten) with a characteristic, acrid odor.

**Chemical & Physical Properties:**
**MW:** 148.1
**BP:** 563°F
**Sol:** 0.6%
**Fl.P:** 305°F
**IP:** 10.00 eV
**Sp.Gr:** 1.53 (Flake)
1.20 (Molten)
**VP:** 0.0015 mmHg
**MLT:** 267°F
**UEL:** 10.5%
**LEL:** 1.7%
Combustible Solid

**Personal Protection/Sanitation (see Table 2):**
**Skin:** Prevent skin contact
**Eyes:** Prevent eye contact
**Wash skin:** When contam
**Remove:** When wet or contam
**Change:** Daily

**Respirator Recommendations (see Tables 3 and 4):**
**NIOSH**
**30 mg/m$^3$:** Qm*
**60 mg/m$^3$:** 95XQ*/95F/PaprHie*/
Sa*/ScbaF
**§:** ScbaF:Pd,Pp/SaF:Pd,Pp:AScba
**Escape:** 100F/ScbaE

**Incompatibilities and Reactivities:** Strong oxidizers, water [**Note:** Converted to phthalic acid in hot water.]

**Exposure Routes, Symptoms, Target Organs (see Table 5):**
**ER:** Inh, Ing, Con
**SY:** Irrit eyes, skin, upper resp sys; conj; nasal ulcer bleeding; bron, bronchial asthma; derm; in animals: liver, kidney damage
**TO:** Eyes, skin, resp sys, liver, kidneys

**First Aid (see Table 6):**
**Eye:** Irr immed
**Skin:** Soap wash prompt
**Breath:** Resp support
**Swallow:** Medical attention immed

| m-Phthalodinitrile | Formula: $C_6H_4(CN)_2$ | CAS#: 626-17-5 | RTECS#: CZ1900000 | IDLH: N.D. |
|---|---|---|---|---|
| **Conversion:** | **DOT:** | | | |

**Synonyms/Trade Names:** 1,3-Benzenedicarbonitrile; m-Dicyanobenzene; 1,3-Dicyanobenzene; Isophthalodinitrile; m-PDN

**Exposure Limits:**
**NIOSH REL:** TWA 5 mg/m$^3$
**OSHA PEL†:** none

**Measurement Methods (see Table 1):**
None available

**Physical Description:** Needle-like, colorless to white, crystalline, flaky solid with an almond-like odor.

**Chemical & Physical Properties:**
**MW:** 128.1
**BP:** Sublimes
**Sol:** Slight
**Fl.P:** ?
**IP:** ?
**Sp.Gr:** 4.42
**VP:** 0.01 mmHg
**MLT:** 324°F (Sublimes)
**UEL:** ?
**LEL:** ?
Combustible Solid and a severe explosion hazard.

**Personal Protection/Sanitation (see Table 2):**
**Skin:** Prevent skin contact
**Eyes:** Prevent eye contact
**Wash skin:** Daily
**Remove:** When wet or contam
**Change:** Daily

**Respirator Recommendations (see Tables 3 and 4):**
Not available.

**Incompatibilities and Reactivities:** Strong oxidizers (e.g., chlorine, bromine, fluorine)

**Exposure Routes, Symptoms, Target Organs (see Table 5):**
**ER:** Inh, Abs, Ing, Con
**SY:** Head, nau, conf; in animals: irrit eyes, skin
**TO:** Eyes, skin, CNS

**First Aid (see Table 6):**
**Eye:** Irr immed
**Skin:** Soap wash immed
**Breath:** Resp support
**Swallow:** Medical attention immed

P

| Picloram | Formula: $C_6H_3Cl_3O_2N_2$ | CAS#: 1918-02-1 | RTECS#: TJ7525000 | IDLH: N.D. |
|---|---|---|---|---|
| **Conversion:** | **DOT:** | | | |

**Synonyms/Trade Names:** 4-Amino-3,5,6-trichloropicolinic acid; 4-Amino-3,5,6-trichloro-2-picolinic acid; ATCP; Tordon®

**Exposure Limits:**
**NIOSH REL:** See Appendix D
**OSHA PEL†:** TWA 15 mg/m$^3$ (total) TWA 5 mg/m$^3$ (resp)

**Measurement Methods (see Table 1):**
**NIOSH** 0500, 0600

**Physical Description:** Colorless to white crystals with a chlorine-like odor. [herbicide]

**Chemical & Physical Properties:**
**MW:** 241.5
**BP:** Decomposes
**Sol:** 0.04%
**Fl.P:** ?
**IP:** ?
**Sp.Gr:** ?
**VP(95°F):** 0.0000006 mmHg
**MLT:** 424°F (Decomposes)
**UEL:** ?
**LEL:** ?
Combustible Solid

**Personal Protection/Sanitation (see Table 2):**
**Skin:** Prevent skin contact
**Eyes:** Prevent eye contact
**Wash skin:** When contam
**Remove:** N.R.
**Change:** Daily

**Respirator Recommendations (see Tables 3 and 4):**
Not available.

**Incompatibilities and Reactivities:** Hot concentrated alkali (hydrolyzes)

**Exposure Routes, Symptoms, Target Organs (see Table 5):**
**ER:** Inh, Ing, Con
**SY:** Irrit eyes, skin, resp sys; nau; in animals: liver, kidney changes
**TO:** Eyes, skin, resp sys, liver, kidneys

**First Aid (see Table 6):**
**Eye:** Irr immed
**Skin:** Soap wash
**Breath:** Fresh air
**Swallow:** Medical attention immed

| Picric acid | Formula: $(NO_2)_3C_6H_2OH$ | CAS#: 88-89-1 | RTECS#: TJ7875000 | IDLH: 75 mg/m$^3$ |
|---|---|---|---|---|
| **Conversion:** 1 ppm = 9.37 mg/m$^3$ | **DOT:** 1344 113 (wet, ≥ 10% water); 3364 113 (wetted, ≥ 10% water) | | | |

**Synonyms/Trade Names:** Phenol trinitrate; 2,4,6-Trinitrophenol [**Note:** An OSHA Class A Explosive (1910.109).]

| Exposure Limits | Measurement Methods |
|---|---|
| **Exposure Limits:**<br>**NIOSH REL:** TWA 0.1 mg/m$^3$<br>ST 0.3 mg/m$^3$ [skin]<br>**OSHA PEL:** TWA 0.1 mg/m$^3$ [skin] | **Measurement Methods (see Table 1):**<br>**NIOSH** S228 (II-4) |
| **Physical Description:** Yellow, odorless solid. [**Note:** Usually used as an aqueous solution.] | |

| Chemical & Physical Properties | Personal Protection/Sanitation | Respirator Recommendations |
|---|---|---|
| **Chemical & Physical Properties:**<br>**MW:** 229.1<br>**BP:** Explodes above 572°F<br>**Sol:** 1%<br>**Fl.P:** 302°F<br>**IP:** ?<br>**Sp.Gr:** 1.76<br>**VP(383°F):** 1 mmHg<br>**MLT:** 252°F<br>**UEL:** ?<br>**LEL:** ?<br>Combustible Solid | **Personal Protection/Sanitation (see Table 2):**<br>**Skin:** Prevent skin contact<br>**Eyes:** Prevent eye contact<br>**Wash skin:** When contam/Daily<br>**Remove:** When wet or contam<br>**Change:** Daily | **Respirator Recommendations (see Tables 3 and 4):**<br>**NIOSH/OSHA**<br>**0.5 mg/m$^3$:** Qm<br>**1 mg/m$^3$:** 95XQ/Sa<br>**2.5 mg/m$^3$:** Sa:Cf/PaprHie<br>**5 mg/m$^3$:** 100F/SaT:Cf/PaprTHie/ ScbaF/SaF<br>**75 mg/m$^3$:** SaF:Pd,Pp<br>**§:** ScbaF:Pd,Pp/SaF:Pd,Pp:AScba<br>**Escape:** 100F/ScbaE |

**Incompatibilities and Reactivities:** Copper, lead, zinc & other metals; salts; plaster; concrete; ammonia [**Note:** Corrosive to metals. An explosive mixture results when the aqueous solution crystallizes.]

| Exposure Routes, Symptoms, Target Organs | First Aid |
|---|---|
| **Exposure Routes, Symptoms, Target Organs (see Table 5):**<br>**ER:** Inh, Abs, Ing, Con<br>**SY:** Irrit eyes, skin; sens derm; yellow-stained hair, skin; lass, myalgia, anuria, polyuria; bitter taste, GI dist; hepatitis, hema, album, neph<br>**TO:** Eyes, skin, kidneys, liver, blood | **First Aid (see Table 6):**<br>**Eye:** Irr immed<br>**Skin:** Soap wash prompt<br>**Breath:** Resp support<br>**Swallow:** Medical attention immed |



| Pindone | Formula: $C_9H_5O_2C(O)C(CH_3)_3$ | CAS#: 83-26-1 | RTECS#: NK6300000 | IDLH: 100 mg/m$^3$ |
|---|---|---|---|---|
| **Conversion:** | **DOT:** | | | |

**Synonyms/Trade Names:** tert-Butyl valone; 1,3-Dioxo-2-pivaloy-lindane; Pival®; Pivalyl; 2-Pivalyl-1,3-indandione

| Exposure Limits | Measurement Methods |
|---|---|
| **Exposure Limits:**<br>**NIOSH REL:** TWA 0.1 mg/m$^3$<br>**OSHA PEL:** TWA 0.1 mg/m$^3$ | **Measurement Methods (see Table 1):**<br>None available |
| **Physical Description:** Bright-yellow powder with almost no odor. [rodenticide] | |

| Chemical & Physical Properties | Personal Protection/Sanitation | Respirator Recommendations |
|---|---|---|
| **Chemical & Physical Properties:**<br>**MW:** 230.3<br>**BP:** Decomposes<br>**Sol(77°F):** 0.002%<br>**Fl.P:** ?<br>**IP:** ?<br>**Sp.Gr:** 1.06<br>**VP:** Very low<br>**MLT:** 230°F<br>**UEL:** ?<br>**LEL:** ? | **Personal Protection/Sanitation (see Table 2):**<br>**Skin:** N.R.<br>**Eyes:** N.R.<br>**Wash skin:** N.R.<br>**Remove:** N.R.<br>**Change:** Daily | **Respirator Recommendations (see Tables 3 and 4):**<br>**NIOSH/OSHA**<br>**0.5 mg/m$^3$:** Qm<br>**1 mg/m$^3$:** 95XQ/Sa<br>**2.5 mg/m$^3$:** Sa:Cf/PaprHie<br>**5 mg/m$^3$:** 100F/SaT:Cf/PaprTHie/ ScbaF/SaF<br>**100 mg/m$^3$:** SaF:Pd,Pp<br>**§:** ScbaF:Pd,Pp/SaF:Pd,Pp:AScba<br>**Escape:** 100F/ScbaE |

**Incompatibilities and Reactivities:** None reported

| Exposure Routes, Symptoms, Target Organs | First Aid |
|---|---|
| **Exposure Routes, Symptoms, Target Organs (see Table 5):**<br>**ER:** Inh, Ing<br>**SY:** Epis, excess bleeding from minor cuts, bruises; smoky urine, black tarry stools; abdom, back pain<br>**TO:** Blood prothrombin | **First Aid (see Table 6):**<br>**Eye:** Irr immed<br>**Breath:** Resp support<br>**Swallow:** Medical attention immed |

## Piperazine dihydrochloride

| Piperazine dihydrochloride | Formula: $C_4H_{10}N_2 \times 2HCl$ | CAS#: 142-64-3 | RTECS#: TL4025000 | IDLH: N.D. |
|---|---|---|---|---|
| Conversion: | DOT: | | | |

**Synonyms/Trade Names:** Piperazine hydrochloride
[**Note:** The monochloride, $C_4H_{10}N_2 \times HCl$ is also commercially available.]

**Exposure Limits:**
**NIOSH REL:** TWA 5 mg/m$^3$
**OSHA PEL†:** none

**Measurement Methods (see Table 1):**
None available

**Physical Description:** White to cream-colored needles or powder.

**Chemical & Physical Properties:**
**MW:** 159.1
**BP:** ?
**Sol:** 41%
**Fl.P:** ?
**IP:** ?
**Sp.Gr:** ?
**VP:** ?
**MLT:** 635°F
**UEL:** ?
**LEL:** ?
Combustible Solid, but does not ignite easily.

**Personal Protection/Sanitation (see Table 2):**
**Skin:** Prevent skin contact
**Eyes:** Prevent eye contact
**Wash skin:** When contam
**Remove:** When wet or contam
**Change:** Daily
**Provide:** Eyewash
Quick drench

**Respirator Recommendations (see Tables 3 and 4):**
Not available.

**Incompatibilities and Reactivities:** Water [**Note:** Slightly hygroscopic (i.e., absorbs moisture from the air).]

**Exposure Routes, Symptoms, Target Organs (see Table 5):**
**ER:** Inh, Abs, Ing, Con
**SY:** Irrit eyes, skin, resp sys; skin burns, sens; asthma; GI upset, head, nau, vomit, inco, musc weak
**TO:** Eyes, skin, resp sys, CNS

**First Aid (see Table 6):**
**Eye:** Irr immed
**Skin:** Water flush immed
**Breath:** Resp support
**Swallow:** Medical attention immed



## Plaster of Paris

| Plaster of Paris | Formula: $CaSO_4 \cdot 0.5H_2O$ | CAS#: 26499-65-0 | RTECS#: TP0700000 | IDLH: N.D. |
|---|---|---|---|---|
| Conversion: | DOT: | | | |

**Synonyms/Trade Names:** Calcium sulfate hemihydrate, Dried calcium sulfate, Gypsum hemihydrate, Hemihydrate gypsum
[**Note:** Plaster of Paris is the hemihydrate form of Calcium Sulfate & Gypsum is the dihydrate form.]

**Exposure Limits:**
**NIOSH REL:** TWA 10 mg/m$^3$ (total)
TWA 5 mg/m$^3$ (resp)
**OSHA PEL:** TWA 15 mg/m$^3$ (total)
TWA 5 mg/m$^3$ (resp)

**Measurement Methods (see Table 1):**
**NIOSH** 0500, 0600

**Physical Description:** White or yellowish, finely divided, odorless powder.

**Chemical & Physical Properties:**
**MW:** 145.2
**BP:** ?
**Sol(77°F):** 0.3%
**Fl.P:** NA
**IP:** NA
**Sp.Gr:** 2.5
**VP:** 0 mmHg (approx)
**MLT:** 325°F (Loses $H_2O$)
**UEL:** NA
**LEL:** NA
Noncombustible Solid

**Personal Protection/Sanitation (see Table 2):**
**Skin:** N.R.
**Eyes:** N.R.
**Wash skin:** N.R.
**Remove:** N.R.
**Change:** N.R.

**Respirator Recommendations (see Tables 3 and 4):**
Not available.

**Incompatibilities and Reactivities:** Moisture, water
[**Note:** Hygroscopic (i.e., absorbs moisture from the air). Reacts with water to form Gypsum.]

**Exposure Routes, Symptoms, Target Organs (see Table 5):**
**ER:** Inh, Ing, Con
**SY:** Irrit eyes, skin, muc memb, resp sys; cough
**TO:** Eyes, skin, resp sys

**First Aid (see Table 6):**
**Eye:** Irr immed
**Breath:** Resp support
**Swallow:** Medical attention immed

| Platinum | Formula: Pt | CAS#: 7440-06-4 | RTECS#: TP2160000 | IDLH: N.D. |
|---|---|---|---|---|
| Conversion: | DOT: | | | |

**Synonyms/Trade Names:** Platinum black, Platinum metal, Platinum sponge

**Exposure Limits:**
**NIOSH REL:** TWA 1 mg/m$^3$
**OSHA PEL†:** none

**Measurement Methods (see Table 1):**
**NIOSH** 7300, 7303
**OSHA** ID121, ID130SG

**Physical Description:** Silvery, whitish-gray, malleable, ductile metal.

| Chemical & Physical Properties: | Personal Protection/Sanitation (see Table 2): | Respirator Recommendations (see Tables 3 and 4): |
|---|---|---|
| **MW:** 195.1<br>**BP:** 6921°F<br>**Sol:** Insoluble<br>**Fl.P:** NA<br>**IP:** NA<br>**Sp.Gr:** 21.45<br>**VP:** 0 mmHg (approx)<br>**MLT:** 3222°F<br>**UEL:** NA<br>**LEL:** NA<br>Noncombustible Solid in bulk form, but finely divided powder can be dangerous to handle. | **Skin:** N.R.<br>**Eyes:** N.R.<br>**Wash skin:** N.R.<br>**Remove:** N.R.<br>**Change:** Daily | Not available. |

**Incompatibilities and Reactivities:** Aluminum, acetone, arsenic, ethane, hydrazine, hydrogen peroxide, lithium, phosphorus, selenium, tellurium, various fluorides

| Exposure Routes, Symptoms, Target Organs (see Table 5): | First Aid (see Table 6): |
|---|---|
| **ER:** Inh, Ing, Con<br>**SY:** Irrit skin, resp sys; derm<br>**TO:** Eyes, skin, resp sys | **Eye:** Irr immed<br>**Skin:** Soap wash<br>**Breath:** Resp support<br>**Swallow:** Medical attention immed |

P

| Platinum (soluble salts, as Pt) | Formula: | CAS#: | RTECS#: | IDLH: 4 mg/m$^3$ (as Pt) |
|---|---|---|---|---|
| Conversion: | DOT: | | | |

**Synonyms/Trade Names:** Synonyms vary depending upon the specific soluble platinum salt.

**Exposure Limits:**
**NIOSH REL:** TWA 0.002 mg/m$^3$
**OSHA PEL:** TWA 0.002 mg/m$^3$

**Measurement Methods (see Table 1):**
**NIOSH** 7300, 7303, S191 (II-7)

**Physical Description:** Appearance and odor vary depending upon the specific soluble platinum salt.

| Chemical & Physical Properties: | Personal Protection/Sanitation (see Table 2): | Respirator Recommendations (see Tables 3 and 4): |
|---|---|---|
| Properties vary depending upon the specific soluble platinum salt. | **Skin:** Prevent skin contact<br>**Eyes:** Prevent eye contact<br>**Wash skin:** When contam<br>**Remove:** When wet or contam<br>**Change:** Daily | **NIOSH/OSHA**<br>**0.05 mg/m$^3$:** Sa:Cf£<br>**0.1 mg/m$^3$:** 100F/ScbaF/SaF<br>**4 mg/m$^3$:** SaF:Pd,Pp<br>**§:** ScbaF:Pd,Pp/SaF:Pd,Pp:AScba<br>**Escape:** 100F/ScbaE |

**Incompatibilities and Reactivities:** Varies

| Exposure Routes, Symptoms, Target Organs (see Table 5): | First Aid (see Table 6): |
|---|---|
| **ER:** Inh, Ing, Con<br>**SY:** Irrit eyes, nose; cough, dysp, wheez, cyan; derm, sens skin; lymphocytosis<br>**TO:** Eyes, skin, resp sys | **Eye:** Irr immed<br>**Skin:** Water flush immed<br>**Breath:** Resp support<br>**Swallow:** Medical attention immed |

## Portland cement

| Formula: | CAS#: | RTECS#: | IDLH: |
|---|---|---|---|
| | 65997-15-1 | VV8770000 | 5000 mg/m$^3$ |

**Conversion:**

**DOT:**

**Synonyms/Trade Names:** Cement, Hydraulic cement, Portland cement silicate [**Note:** A class of hydraulic cements containing tri- and dicalcium silicate in addition to alumina, tricalcium aluminate, and iron oxide.]

**Exposure Limits:**
**NIOSH REL:** TWA 10 mg/m$^3$ (total)
TWA 5 mg/m$^3$ (resp)
**OSHA PEL†:** TWA 50 mppcf

**Measurement Methods (see Table 1):**
**NIOSH** 0500
**OSHA** ID207

**Physical Description:** Gray, odorless powder.

**Chemical & Physical Properties:**
**MW:** ?
**BP:** NA
**Sol:** Insoluble
**Fl.P:** NA
**IP:** NA
**Sp.Gr:** ?
**VP:** 0 mmHg (approx)
**MLT:** NA
**UEL:** NA
**LEL:** NA
Noncombustible Solid

**Personal Protection/Sanitation (see Table 2):**
**Skin:** Prevent skin contact
**Eyes:** Prevent eye contact
**Wash skin:** When contam
**Remove:** When wet or contam
**Change:** N.R.

**Respirator Recommendations (see Tables 3 and 4):**
**NIOSH**
**50 mg/m$^3$:** Qm
**100 mg/m$^3$:** 95XQ/Sa
**250 mg/m$^3$:** Sa:Cf/PaprHie
**500 mg/m$^3$:** 100F/SaT:Cf/PaprTHie/ScbaF/SaF
**5000 mg/m$^3$:** Sa:Pd,Pp
**§:** ScbaF:Pd,Pp/SaF:Pd,Pp:AScba
**Escape:** 100F/ScbaE

**Incompatibilities and Reactivities:** None reported

**Exposure Routes, Symptoms, Target Organs (see Table 5):**
**ER:** Inh, Ing, Con
**SY:** Irrit eyes, skin, nose; cough, expectoration; exertional dysp, wheez, chronic bron; derm
**TO:** Eyes, skin, resp sys

**First Aid (see Table 6):**
**Eye:** Irr immed
**Skin:** Soap wash prompt
**Breath:** Fresh air
**Swallow:** Medical attention immed



## Potassium cyanide (as CN)

| Formula: | CAS#: | RTECS#: | IDLH: |
|---|---|---|---|
| $KCN$ | 151-50-8 | TS8750000 | 25 mg/m$^3$ (as CN) |

**Conversion:**

**DOT:** 1680 157 (solid); 3413 157 (solution)

**Synonyms/Trade Names:** Potassium salt of hydrocyanic acid

**Exposure Limits:**
**NIOSH REL*:** C 5 mg/m$^3$ (4.7 ppm) [10-minute]
**OSHA PEL*:** TWA 5 mg/m$^3$
[***Note:** The REL and PEL also apply to other cyanides (as CN) except Hydrogen cyanide.]

**Measurement Methods (see Table 1):**
**NIOSH** 6010, 7904

**Physical Description:** White, granular or crystalline solid with a faint, almond-like odor.

**Chemical & Physical Properties:**
**MW:** 65.1
**BP:** 2957°F
**Sol(77°F):** 72%
**Fl.P:** NA
**IP:** NA
**Sp.Gr:** 1.55
**VP:** 0 mmHg (approx)
**MLT:** 1173°F
**UEL:** NA
**LEL:** NA
Noncombustible Solid, but contact with acids releases highly flammable hydrogen cyanide.

**Personal Protection/Sanitation (see Table 2):**
**Skin:** Prevent skin contact
**Eyes:** Prevent eye contact
**Wash skin:** When contam
**Remove:** When wet or contam
**Change:** Daily
**Provide:** Eyewash
Quick drench

**Respirator Recommendations (see Tables 3 and 4):**
**NIOSH/OSHA**
**25 mg/m$^3$:** Sa/ScbaF
**§:** ScbaF:Pd,Pp/SaF:Pd,Pp:AScba
**Escape:** GmFS100/ScbaE

**Incompatibilities and Reactivities:** Strong oxidizers (such as acids, acid salts, chlorates & nitrates)
[**Note:** Absorbs moisture from the air forming a syrup.]

**Exposure Routes, Symptoms, Target Organs (see Table 5):**
**ER:** Inh, Abs, Ing, Con
**SY:** Irrit eyes, skin, upper resp sys; asphy; lass, head, conf; nau, vomit; incr resp rate, slow gasping respiration; thyroid, blood changes
**TO:** Eyes, skin, resp sys, CVS, CNS, thyroid, blood

**First Aid (see Table 6):**
**Eye:** Irr immed
**Skin:** Soap wash immed
**Breath:** Resp support
**Swallow:** Medical attention immed

## Potassium hydroxide

| | | | | |
|---|---|---|---|---|
| **Potassium hydroxide** | **Formula:** KOH | **CAS#:** 1310-58-3 | **RTECS#:** TT2100000 | **IDLH:** N.D. |
| **Conversion:** | **DOT:** 1813 154 (dry, solid); 1814 154 (solution) | | | |

**Synonyms/Trade Names:** Caustic potash, Lye, Potassium hydrate

**Exposure Limits:**
**NIOSH REL:** C 2 mg/m$^3$
**OSHA PEL†:** none

**Measurement Methods (see Table 1):**
**NIOSH** 7401

**Physical Description:** Odorless, white or slightly yellow lumps, rods, flakes, sticks, or pellets. [**Note:** May be used as an aqueous solution.]

**Chemical & Physical Properties:**
**MW:** 56.1
**BP:** 2415°F
**Sol(59°F):** 107%
**Fl.P:** NA
**IP:** ?
**Sp.Gr:** 2.04
**VP(1317°F):** 1 mmHg
**MLT:** 716°F
**UEL:** NA
**LEL:** NA
Noncombustible Solid; however, may react with $H_2O$ & other substances and generate sufficient heat to ignite combustible materials.

**Personal Protection/Sanitation (see Table 2):**
**Skin:** Prevent skin contact
**Eyes:** Prevent eye contact
**Wash skin:** When contam
**Remove:** When wet or contam
**Change:** Daily
**Provide:** Eyewash
Quick drench

**Respirator Recommendations (see Tables 3 and 4):**
Not available.

**Incompatibilities and Reactivities:** Acids, water, metals (when wet), halogenated hydrocarbons, maleic anhydride [**Note:** Heat is generated if KOH comes in contact with $H_2O$ & $CO_2$ from the air.]

**Exposure Routes, Symptoms, Target Organs (see Table 5):**
**ER:** Inh, Ing, Con
**SY:** Irrit eyes, skin, resp sys; cough, sneez; eye, skin burns; vomit, diarr
**TO:** Eyes, skin, resp sys

**First Aid (see Table 6):**
**Eye:** Irr immed
**Skin:** Water flush immed
**Breath:** Resp support
**Swallow:** Medical attention immed

P

## Propane

| | | | | |
|---|---|---|---|---|
| **Propane** | **Formula:** $CH_3CH_2CH_3$ | **CAS#:** 74-98-6 | **RTECS#:** TX2275000 | **IDLH:** 2100 ppm [10%LEL] |
| **Conversion:** 1 ppm = 1.80 mg/m$^3$ | **DOT:** 1075 115; 1978 115 | | | |

**Synonyms/Trade Names:** Bottled gas, Dimethyl methane, n-Propane, Propyl hydride

**Exposure Limits:**
**NIOSH REL:** TWA 1000 ppm (1800 mg/m$^3$)
**OSHA PEL:** TWA 1000 ppm (1800 mg/m$^3$)

**Measurement Methods (see Table 1):**
**NIOSH** S87 (II-2)
**OSHA** PV2077

**Physical Description:** Colorless, odorless gas. [**Note:** A foul-smelling odorant is often added when used for fuel purposes. Shipped as a liquefied compressed gas.]

**Chemical & Physical Properties:**
**MW:** 44.1
**BP:** -44°F
**Sol:** 0.01%
**Fl.P:** NA (Gas)
**IP:** 11.07 eV
**RGasD:** 1.55
**VP(70°F):** 8.4 atm
**FRZ:** -306°F
**UEL:** 9.5%
**LEL:** 2.1%
Flammable Gas

**Personal Protection/Sanitation (see Table 2):**
**Skin:** Frostbite
**Eyes:** Frostbite
**Wash skin:** N.R.
**Remove:** When wet (flamm)
**Change:** N.R.
**Provide:** Frostbite wash

**Respirator Recommendations (see Tables 3 and 4):**
**NIOSH/OSHA**
**2100 ppm:** Sa/ScbaF
**§:** ScbaF:Pd,Pp/SaF:Pd,Pp:AScba
**Escape:** ScbaE

**Incompatibilities and Reactivities:** Strong oxidizers

**Exposure Routes, Symptoms, Target Organs (see Table 5):**
**ER:** Inh, Con (liquid)
**SY:** Dizz, conf, excitation, asphy; liquid: frostbite
**TO:** CNS

**First Aid (see Table 6):**
**Eye:** Frostbite
**Skin:** Frostbite
**Breath:** Resp support

## Propane sultone

| Propane sultone | Formula: $C_3H_6O_3S$ | CAS#: 1120-71-4 | RTECS#: RP5425000 | IDLH: Ca [N.D.] |
|---|---|---|---|---|

**Conversion:**

**DOT:**

**Synonyms/Trade Names:** 3-Hydroxy-1-propanesulphonic acid sultone; 1,3-Propane sultone

**Exposure Limits:**
**NIOSH REL:** Ca
See Appendix A
**OSHA PEL:** none

**Measurement Methods (see Table 1):**
None available

**Physical Description:** White, crystalline solid or a colorless liquid (above 86°F).
[**Note:** Releases a foul odor as it melts.]

**Chemical & Physical Properties:**
**MW:** 122.2
**BP:** ?
**Sol:** 10%
**Fl.P:** >235°F
**IP:** ?
**Sp.Gr:** 1.39
**VP:** ?
**MLT:** 86°F
**UEL:** ?
**LEL:** ?
Combustible Solid

**Personal Protection/Sanitation (see Table 2):**
**Skin:** Prevent skin contact
**Eyes:** Prevent eye contact
**Wash skin:** When contam/Daily
**Remove:** When wet or contam
**Change:** Daily
**Provide:** Eyewash
Quick drench

**Respirator Recommendations (see Tables 3 and 4):**
**NIOSH**
**¥:** ScbaF:Pd,Pp/SaF:Pd,Pp:AScba
**Escape:** GmFOv100/ScbaE

**Incompatibilities and Reactivities:** None reported

**Exposure Routes, Symptoms, Target Organs (see Table 5):**
**ER:** Inh, Abs, Ing, Con
**SY:** Irrit eyes, skin, resp sys; [carc]
**TO:** Eyes, skin, resp sys [in animals: skin tumors, leukemia, gliomas]

**First Aid (see Table 6):**
**Eye:** Irr immed
**Skin:** Water flush immed
**Breath:** Resp support
**Swallow:** Medical attention immed

P

## 1-Propanethiol

| 1-Propanethiol | Formula: $CH_3CH_2CH_2SH$ | CAS#: 107-03-9 | RTECS#: TZ7300000 | IDLH: N.D. |
|---|---|---|---|---|

**Conversion:** 1 ppm = 3.12 mg/$m^3$

**DOT:** 2402 130

**Synonyms/Trade Names:** 3-Mercaptopropane, Propane-1-thiol, Propyl mercaptan, n-Propyl mercaptan

**Exposure Limits:**
**NIOSH REL:** C 0.5 ppm (1.6 mg/$m^3$) [15-minute]
**OSHA PEL:** none

**Measurement Methods (see Table 1):**
None available

**Physical Description:** Colorless liquid with an offensive, cabbage-like odor.

**Chemical & Physical Properties:**
**MW:** 76.2
**BP:** 153°F
**Sol:** Slight
**Fl.P:** -5°F
**IP:** 9.195 eV
**Sp.Gr:** 0.84
**VP(77°F):** 155 mmHg
**FRZ:** -172°F
**UEL:** ?
**LEL:** ?
Class IB Flammable Liquid

**Personal Protection/Sanitation (see Table 2):**
**Skin:** N.R.
**Eyes:** Prevent eye contact
**Wash skin:** N.R.
**Remove:** When wet (flamm)
**Change:** N.R.
**Provide:** Eyewash

**Respirator Recommendations (see Tables 3 and 4):**
**NIOSH**
**5 ppm:** CcrOv/Sa
**12.5 ppm:** Sa:Cf/PaprOv
**25 ppm:** CcrFOv/GmFOv/PaprTOv/ScbaF/SaF
**§:** ScbaF:Pd,Pp/SaF:Pd,Pp:AScba
**Escape:** GmFOv/ScbaE

**Incompatibilities and Reactivities:** Oxidizers, reducing agents, strong acids & bases, alkali metals, calcium hypochlorite

**Exposure Routes, Symptoms, Target Organs (see Table 5):**
**ER:** Inh, Ing, Con
**SY:** Irrit eyes, skin, nose, throat, resp sys; head, nau, dizz, cyan; in animals: liver, kidney damage
**TO:** Eyes, skin, resp sys, CNS, blood, liver, kidneys

**First Aid (see Table 6):**
**Eye:** Irr immed
**Skin:** Soap wash
**Breath:** Resp support
**Swallow:** Medical attention immed

## Propargyl alcohol

| Formula: | CAS#: | RTECS#: | IDLH: |
|---|---|---|---|
| $C_3H_3OH$ | 107-19-7 | UK5075000 | N.D. |

**Conversion:** 1 ppm = 2.29 mg/m$^3$

**DOT:** 1986 131

**Synonyms/Trade Names:** 1-Propyn-3-ol; 2-Propyn-1-ol; 2-Propynyl alcohol

**Exposure Limits:**
**NIOSH REL:** TWA 1 ppm (2 mg/m$^3$) [skin]
**OSHA PEL†:** none

**Measurement Methods (see Table 1):**
**OSHA** 97

**Physical Description:** Colorless to straw-colored liquid with a mild, geranium odor.

**Chemical & Physical Properties:**
**MW:** 56.1
**BP:** 237°F
**Sol:** Miscible
**Fl.P(oc):** 97°F
**IP:** 10.51 eV
**Sp.Gr:** 0.97
**VP:** 12 mmHg
**FRZ:** -62°F
**UEL:** ?
**LEL:** ?
Class IC Flammable Liquid

**Personal Protection/Sanitation (see Table 2):**
**Skin:** Prevent skin contact
**Eyes:** Prevent eye contact
**Wash skin:** When contam
**Remove:** When wet or contam
**Change:** N.R.
**Provide:** Eyewash
Quick drench

**Respirator Recommendations (see Tables 3 and 4):**
Not available.

**Incompatibilities and Reactivities:** Phosphorus pentoxide, oxidizers

**Exposure Routes, Symptoms, Target Organs (see Table 5):**
**ER:** Inh, Abs, Ing, Con
**SY:** Irrit skin, muc memb; CNS depres; in animals: liver, kidney damage
**TO:** Skin, resp sys, CNS, liver, kidneys

**First Aid (see Table 6):**
**Eye:** Irr immed
**Skin:** Water flush prompt
**Breath:** Resp support
**Swallow:** Medical attention immed

## β-Propiolactone

| Formula: | CAS#: | RTECS#: | IDLH: |
|---|---|---|---|
| $C_3H_4O_2$ | 57-57-8 | RQ7350000 | Ca [N.D] |

**Conversion:**

**DOT:**

P

**Synonyms/Trade Names:** BPL; Hydroacrylic acid, β-lactone; 3-Hydroxy-β-lactone; 3-Hydroxy-propionic acid; β-Lactone; 2-Oxetanone; 3-Propiolactone

**Exposure Limits:**
**NIOSH REL:** Ca
See Appendix A
**OSHA PEL:** [1910.1013]
See Appendix B

**Measurement Methods (see Table 1):**
None available

**Physical Description:** Colorless liquid with a slightly sweet odor.

**Chemical & Physical Properties:**
**MW:** 72.1
**BP:** 323°F (Decomposes)
**Sol:** 37%
**Fl.P:** 165°F
**IP:** ?
**Sp.Gr:** 1.15
**VP(77°F):** 3 mmHg
**FRZ:** -28°F
**UEL:** ?
**LEL:** 2.9%
Class IIIA Combustible Liquid

**Personal Protection/Sanitation (see Table 2):**
**Skin:** Prevent skin contact
**Eyes:** Prevent eye contact
**Wash skin:** When contam/Daily
**Remove:** When wet or contam
**Change:** Daily
**Provide:** Eyewash
Quick drench

**Respirator Recommendations (see Tables 3 and 4):**
**NIOSH**
**¥:** ScbaF:Pd,Pp/SaF:Pd,Pp:AScba
**Escape:** GmFOv/ScbaE

**See Appendix E** (page 351)

**Incompatibilities and Reactivities:** Acetates, halogens, thiocyanates, thiosulfates
[**Note:** May polymerize upon storage.]

**Exposure Routes, Symptoms, Target Organs (see Table 5):**
**ER:** Inh, Abs, Ing, Con
**SY:** Skin irrit, blistering, burns; corn opac; frequent urination; dysuria; hema; [carc]
**TO:** Kidneys, skin, lungs, eyes [in animals: tumors of the liver, skin & stomach]

**First Aid (see Table 6):**
**Eye:** Irr immed
**Skin:** Soap wash immed
**Breath:** Resp support
**Swallow:** Medical attention immed

## Propionic acid

| Propionic acid | Formula: $CH_3CH_2COOH$ | CAS#: 79-09-4 | RTECS#: UE5950000 | IDLH: N.D. |
|---|---|---|---|---|
| Conversion: 1 ppm = 3.03 mg/m$^3$ | DOT: 1848 132 | | | |

**Synonyms/Trade Names:** Carboxyethane, Ethanecarboxylic acid, Ethylformic acid, Metacetonic acid, Methyl acetic acid, Propanoic acid

**Exposure Limits:**
**NIOSH REL:** TWA 10 ppm (30 mg/m$^3$)
ST 15 ppm (45 mg/m$^3$)
**OSHA PEL†:** none

**Measurement Methods (see Table 1):**
None available

**Physical Description:** Colorless, oily liquid with a pungent, disagreeable, rancid odor.
[**Note:** A solid below 5°F.]

**Chemical & Physical Properties:**
**MW:** 74.1
**BP:** 286°F
**Sol:** Miscible
**Fl.P:** 126°F
**IP:** 10.24 eV
**Sp.Gr:** 0.99
**VP:** 3 mmHg
**FRZ:** 5°F
**UEL:** 12.1%
**LEL:** 2.9%
Class II Combustible Liquid

**Personal Protection/Sanitation (see Table 2):**
**Skin:** Prevent skin contact
**Eyes:** Prevent eye contact
**Wash skin:** When contam
**Remove:** When wet or contam
**Change:** N.R.
**Provide:** Eyewash
Quick drench

**Respirator Recommendations (see Tables 3 and 4):**
Not available.

**Incompatibilities and Reactivities:** Alkalis, strong oxidizers (e.g., chromium trioxide) [**Note:** Corrosive to steel.]

**Exposure Routes, Symptoms, Target Organs (see Table 5):**
**ER:** Inh, Abs, Ing, Con
**SY:** Irrit eyes, skin, nose, throat; blurred vision, corn burns; skin burns; abdom pain, nau, vomit
**TO:** Eyes, skin, resp sys

**First Aid (see Table 6):**
**Eye:** Irr immed
**Skin:** Water flush immed
**Breath:** Resp support
**Swallow:** Medical attention immed

P

## Propionitrile

| Propionitrile | Formula: $CH_3CH_2CN$ | CAS#: 107-12-0 | RTECS#: UF9625000 | IDLH: N.D. |
|---|---|---|---|---|
| Conversion: 1 ppm = 2.25 mg/m$^3$ | DOT: 2404 131 | | | |

**Synonyms/Trade Names:** Cyanoethane, Ethyl cyanide, Propanenitrile, Propionic nitrile, Propiononitrile

**Exposure Limits:**
**NIOSH REL:** TWA 6 ppm (14 mg/m$^3$)
**OSHA PEL:** none

**Measurement Methods (see Table 1):**
**NIOSH** 1606 (adapt)

**Physical Description:** Colorless liquid with a pleasant, sweetish, ethereal odor.
[**Note:** Forms cyanide in the body.]

**Chemical & Physical Properties:**
**MW:** 55.1
**BP:** 207°F
**Sol:** 11.9%
**Fl.P:** 36°F
**IP:** 11.84 eV
**Sp.Gr:** 0.78
**VP:** 35 mmHg
**FRZ:** -133°F
**UEL:** ?
**LEL:** 3.1%
Class IB Flammable Liquid

**Personal Protection/Sanitation (see Table 2):**
**Skin:** Prevent skin contact
**Eyes:** Prevent eye contact
**Wash skin:** When contam
**Remove:** When wet or contam
**Change:** N.R.
**Provide:** Quick drench

**Respirator Recommendations (see Tables 3 and 4):**
**NIOSH**
**60 ppm:** CcrOv/Sa
**150 ppm:** Sa:Cf/PaprOv
**300 ppm:** CcrFOv/GmFOv/PaprTOv/ScbaF/SaF
**1000 ppm:** SaF:Pd,Pp
**§:** ScbaF:Pd,Pp/SaF:Pd,Pp:AScba
**Escape:** GmFOv/ScbaE

**Incompatibilities and Reactivities:** Strong oxidizers & reducing agents, strong acids & bases
[**Note:** Hydrogen cyanide is produced when propionitrile is heated to decomposition.]

**Exposure Routes, Symptoms, Target Organs (see Table 5):**
**ER:** Inh, Abs, Ing, Con
**SY:** Irrit eyes, skin, resp sys; nau, vomit; chest pain; lass; stupor, convuls; in animals: liver, kidney damage
**TO:** Eyes, skin, resp sys, CVS, CNS, liver, kidneys

**First Aid (see Table 6):**
**Eye:** Irr immed
**Skin:** Water flush immed
**Breath:** Resp support
**Swallow:** Medical attention immed

## Propoxur

**Formula:** $CH_3NHCOOC_6H_4OCH(CH_3)_2$ | **CAS#:** 114-26-1 | **RTECS#:** FC3150000 | **IDLH:** N.D.

**Conversion:**

**DOT:**

**Synonyms/Trade Names:** Aprocarb®, o-Isopropoxyphenyl-N-methylcarbamate, N-Methyl-2-isopropoxyphenyl-carbamate

**Exposure Limits:**
**NIOSH REL:** TWA 0.5 $mg/m^3$
**OSHA PEL†:** none

**Measurement Methods (see Table 1):**
**NIOSH** 5601
**OSHA** PV2007

**Physical Description:** White to tan, crystalline powder with a faint, characteristic odor. [insecticide]

**Chemical & Physical Properties:**
**MW:** 209.3
**BP:** Decomposes
**Sol:** 0.2%
**Fl.P:** >300°F
**IP:** ?
**Sp.Gr:** ?
**VP:** 0.000007 mmHg
**MLT:** 187-197°F
**UEL:** ?
**LEL:** ?
Class IIIB Combustible Liquid

**Personal Protection/Sanitation (see Table 2):**
**Skin:** Prevent skin contact
**Eyes:** Prevent eye contact
**Wash skin:** Daily
**Remove:** When wet or contam
**Change:** Daily

**Respirator Recommendations (see Tables 3 and 4):**
Not available.

**Incompatibilities and Reactivities:** Strong oxidizers, alkalis
[**Note:** Emits highly toxic methyl isocyanate fumes when heated to decomposition.]

**Exposure Routes, Symptoms, Target Organs (see Table 5):**
**ER:** Inh, Abs, Ing, Con
**SY:** Miosis, blurred vision; sweat, salv; abdom cramps, nau, diarr, vomit; head, lass, musc twitch
**TO:** CNS, liver, kidneys, GI tract, blood chol

**First Aid (see Table 6):**
**Eye:** Irr immed
**Skin:** Soap wash immed
**Breath:** Resp support
**Swallow:** Medical attention immed

P

## n-Propyl acetate

**Formula:** $CH_3COOCH_2CH_2CH_3$ | **CAS#:** 109-60-4 | **RTECS#:** AJ3675000 | **IDLH:** 1700 ppm

**Conversion:** 1 ppm = 4.18 $mg/m^3$

**DOT:** 1276 129

**Synonyms/Trade Names:** Propylacetate, n-Propyl ester of acetic acid

**Exposure Limits:**
**NIOSH REL:** TWA 200 ppm (840 $mg/m^3$)
ST 250 ppm (1050 $mg/m^3$)
**OSHA PEL†:** TWA 200 ppm (840 $mg/m^3$)

**Measurement Methods (see Table 1):**
**NIOSH** 1450
**OSHA** 7

**Physical Description:** Colorless liquid with a mild, fruity odor.

**Chemical & Physical Properties:**
**MW:** 102.2
**BP:** 215°F
**Sol:** 2%
**Fl.P:** 55°F
**IP:** 10.04 eV
**Sp.Gr:** 0.84
**VP:** 25 mmHg
**FRZ:** -134°F
**UEL:** 8%
**LEL(100°F):** 1.7%
Class IB Flammable Liquid

**Personal Protection/Sanitation (see Table 2):**
**Skin:** Prevent skin contact
**Eyes:** Prevent eye contact
**Wash skin:** When contam
**Remove:** When wet (flamm)
**Change:** N.R.

**Respirator Recommendations (see Tables 3 and 4):**
**NIOSH/OSHA**
**1700 ppm:** Sa:Cf£/CcrFOv/GmFOv/PaprOv£/ScbaF/SaF
**§:** ScbaF:Pd,Pp/SaF:Pd,Pp:AScba
**Escape:** GmFOv/ScbaE

**Incompatibilities and Reactivities:** Nitrates; strong oxidizers, alkalis & acids

**Exposure Routes, Symptoms, Target Organs (see Table 5):**
**ER:** Inh, Ing, Con
**SY:** In animals: irrit eyes, nose, throat; narco; derm
**TO:** Eyes, skin, resp sys, CNS

**First Aid (see Table 6):**
**Eye:** Irr immed
**Skin:** Water flush prompt
**Breath:** Resp support
**Swallow:** Medical attention immed

| n-Propyl alcohol | Formula: $CH_3CH_2CH_2OH$ | CAS#: 71-23-8 | RTECS#: UH8225000 | IDLH: 800 ppm |
|---|---|---|---|---|
| **Conversion:** 1 ppm = 2.46 mg/m$^3$ | **DOT:** 1274 129 | | | |

**Synonyms/Trade Names:** Ethyl carbinol, 1-Propanol, n-Propanol, Propyl alcohol

**Exposure Limits:**
**NIOSH REL:** TWA 200 ppm (500 mg/m$^3$) [skin]
ST 250 ppm (625 mg/m$^3$)
**OSHA PEL†:** TWA 200 ppm (500 mg/m$^3$)

**Measurement Methods (see Table 1):**
**NIOSH** 1401, 1405
**OSHA** 7

**Physical Description:** Colorless liquid with a mild, alcohol-like odor.

| Chemical & Physical Properties: | Personal Protection/Sanitation (see Table 2): | Respirator Recommendations (see Tables 3 and 4): |
|---|---|---|
| **MW:** 60.1<br>**BP:** 207°F<br>**Sol:** Miscible<br>**Fl.P:** 72°F<br>**IP:** 10.15 eV<br>**Sp.Gr:** 0.81<br>**VP:** 15 mmHg<br>**FRZ:** -196°F<br>**UEL:** 13.7%<br>**LEL:** 2.2%<br>Class IB Flammable Liquid | **Skin:** Prevent skin contact<br>**Eyes:** Prevent eye contact<br>**Wash skin:** When contam<br>**Remove:** When wet (flamm)<br>**Change:** N.R. | **NIOSH/OSHA**<br>**800 ppm:** CcrOv*/PaprOv*/GmFOv/Sa*/ScbaF<br>**§:** ScbaF:Pd,Pp/SaF:Pd,Pp:AScba<br>**Escape:** GmFOv/ScbaE |

**Incompatibilities and Reactivities:** Strong oxidizers

| Exposure Routes, Symptoms, Target Organs (see Table 5): | First Aid (see Table 6): |
|---|---|
| **ER:** Inh, Abs, Ing, Con<br>**SY:** Irrit eyes, nose, throat; dry cracking skin; drow, head; ataxia, GI pain; abdom cramps, nau, vomit, diarr; in animals: narco<br>**TO:** Eyes, skin, resp sys, GI tract, CNS | **Eye:** Irr immed<br>**Skin:** Water flush<br>**Breath:** Resp support<br>**Swallow:** Medical attention immed |



| Propylene dichloride | Formula: $CH_3CHClCH_2Cl$ | CAS#: 78-87-5 | RTECS#: TX9625000 | IDLH: Ca [400 ppm] |
|---|---|---|---|---|
| **Conversion:** 1 ppm = 4.62 mg/m$^3$ | **DOT:** 1279 130 | | | |

**Synonyms/Trade Names:** Dichloro-1,2-propane; 1,2-Dichloropropane

**Exposure Limits:**
**NIOSH REL:** Ca
See Appendix A
**OSHA PEL†:** TWA 75 ppm (350 mg/m$^3$)

**Measurement Methods (see Table 1):**
**NIOSH** 1013
**OSHA** 7

**Physical Description:** Colorless liquid with a chloroform-like odor. [pesticide]

| Chemical & Physical Properties: | Personal Protection/Sanitation (see Table 2): | Respirator Recommendations (see Tables 3 and 4): |
|---|---|---|
| **MW:** 113.0<br>**BP:** 206°F<br>**Sol:** 0.3%<br>**Fl.P:** 60°F<br>**IP:** 10.87 eV<br>**Sp.Gr:** 1.16<br>**VP:** 40 mmHg<br>**FRZ:** -149°F<br>**UEL:** 14.5%<br>**LEL:** 3.4%<br>Class IB Flammable Liquid | **Skin:** Prevent skin contact<br>**Eyes:** Prevent eye contact<br>**Wash skin:** When contam<br>**Remove:** When wet (flamm)<br>**Change:** N.R.<br>**Provide:** Eyewash<br>Quick drench | **NIOSH**<br>**¥:** ScbaF:Pd,Pp/SaF:Pd,Pp:AScba<br>**Escape:** GmFOv/ScbaE |

**Incompatibilities and Reactivities:** Strong oxidizers, strong acids, active metals

| Exposure Routes, Symptoms, Target Organs (see Table 5): | First Aid (see Table 6): |
|---|---|
| **ER:** Inh, Abs, Ing, Con<br>**SY:** Irrit eyes, skin, resp sys; drow, dizz; liver, kidney damage; in animals: CNS depres; [carc]<br>**TO:** Eyes, skin, resp sys, liver, kidneys, CNS [in animals: liver & mammary gland tumors] | **Eye:** Irr immed<br>**Skin:** Soap wash prompt<br>**Breath:** Resp support<br>**Swallow:** Medical attention immed |

## Propylene glycol dinitrate

| | | | | |
|---|---|---|---|---|
| **Propylene glycol dinitrate** | **Formula:** $CH_3CNO_2OHCHNO_2OH$ | **CAS#:** 6423-43-4 | **RTECS#:** TY6300000 | **IDLH:** N.D. |
| **Conversion:** 1 ppm = 6.79 mg/m$^3$ | **DOT:** | | | |

**Synonyms/Trade Names:** PGDN; Propylene glycol-1,2-dinitrate; 1,2-Propylene glycol dinitrate

**Exposure Limits:**
**NIOSH REL:** TWA 0.05 ppm (0.3 mg/m$^3$) [skin]
**OSHA PEL†:** none

**Measurement Methods (see Table 1):**
None available

**Physical Description:** Colorless liquid with a disagreeable odor.
[**Note:** A solid below 18°F.]

| Chemical & Physical Properties: | Personal Protection/Sanitation (see Table 2): | Respirator Recommendations (see Tables 3 and 4): |
|---|---|---|
| **MW:** 166.1<br>**BP:** ?<br>**Sol:** 0.1%<br>**Fl.P:** ?<br>**IP:** ?<br>**Sp.Gr(77°F):** 1.23<br>**VP(72°F):** 0.07 mmHg<br>**FRZ:** 18°F<br>**UEL:** ?<br>**LEL:** ?<br>Combustible Liquid | **Skin:** Prevent skin contact<br>**Eyes:** Prevent eye contact<br>**Wash skin:** N.R.<br>**Remove:** N.R.<br>**Change:** N.R. | Not available. |

**Incompatibilities and Reactivities:** Ammonia compounds, amines, oxidizers, reducing agents, combustible materials [**Note:** Similar to Ethylene glycol dinitrate in explosion potential.]

| Exposure Routes, Symptoms, Target Organs (see Table 5): | First Aid (see Table 6): |
|---|---|
| **ER:** Inh, Abs, Ing, Con<br>**SY:** Irrit eyes; conj; methemo; head, impaired balance, vis dist; in animals: liver, kidney damage<br>**TO:** Eyes, CNS, blood, liver, kidneys | **Eye:** Irr immed<br>**Skin:** Soap wash<br>**Breath:** Resp support<br>**Swallow:** Medical attention immed |

## Propylene glycol monomethyl ether

| | | | | |
|---|---|---|---|---|
| **Propylene glycol monomethyl ether** | **Formula:** $CH_3OCH_2CHOHCH_3$ | **CAS#:** 107-98-2 | **RTECS#:** UB7700000 | **IDLH:** N.D. |
| **Conversion:** 1 ppm = 3.69 mg/m$^3$ | **DOT:** | | | |

**Synonyms/Trade Names:** Dowtherm® 209, 1-Methoxy-2-hydroxypropane, 1-Methoxy-2-propanol, 2-Methoxy-1-methylethanol, Propylene glycol methyl ether

**Exposure Limits:**
**NIOSH REL:** TWA 100 ppm (360 mg/m$^3$)
ST 150 ppm (540 mg/m$^3$)
**OSHA PEL†:** none

**Measurement Methods (see Table 1):**
**NIOSH** 2554
**OSHA** 99

**Physical Description:** Clear, colorless liquid with a mild, ethereal odor.

| Chemical & Physical Properties: | Personal Protection/Sanitation (see Table 2): | Respirator Recommendations (see Tables 3 and 4): |
|---|---|---|
| **MW:** 90.1<br>**BP:** 248°F<br>**Sol:** Miscible<br>**Fl.P:** 97°F<br>**IP:** ?<br>**Sp.Gr:** 0.96<br>**VP(77°F):** 12 mmHg<br>**FRZ:** -139°F (Sets to glass)<br>**UEL(calc):** 13.8%<br>**LEL(calc.):** 1.6%<br>Class IC Flammable Liquid | **Skin:** N.R.<br>**Eyes:** Prevent eye contact<br>**Wash skin:** N.R.<br>**Remove:** When wet (flamm)<br>**Change:** N.R. | Not available. |

**Incompatibilities and Reactivities:** Oxidizers, strong acids [**Note:** Hygroscopic (i.e., absorbs moisture from air). May slowly form reactive peroxides during prolonged storage.]

| Exposure Routes, Symptoms, Target Organs (see Table 5): | First Aid (see Table 6): |
|---|---|
| **ER:** Inh, Ing, Con<br>**SY:** Irrit eyes, skin, nose, throat; head, nau, dizz, drow, inco; vomit, diarr<br>**TO:** Eyes, skin, resp sys, CNS | **Eye:** Irr immed<br>**Skin:** Water wash<br>**Breath:** Resp support<br>**Swallow:** Medical attention immed |

P

## Propylene imine

| Formula: | CAS#: | RTECS#: | IDLH: |
|---|---|---|---|
| $C_3H_7N$ | 75-55-8 | CM8050000 | Ca [100 ppm] |

**Conversion:** 1 ppm = 2.34 mg/m$^3$
**DOT:** 1921 131P (inhibited)

**Synonyms/Trade Names:** 2-Methylaziridine, 2-Methylethyleneimine, Propyleneimine, Propylene imine (inhibited), Propylenimine

**Exposure Limits:**
**NIOSH REL:** Ca
TWA 2 ppm (5 mg/m$^3$) [skin]
See Appendix A
**OSHA PEL:** TWA 2 ppm (5 mg/m$^3$) [skin]

**Measurement Methods (see Table 1):**
None available

**Physical Description:** Colorless, oily liquid with an ammonia-like odor.

**Chemical & Physical Properties:**
**MW:** 57.1
**BP:** 152°F
**Sol:** Miscible
**Fl.P:** 25°F
**IP:** 9.00 eV
**Sp.Gr:** 0.80
**VP:** 112 mmHg
**FRZ:** -85°F
**UEL:** ?
**LEL:** ?
Class IB Flammable Liquid

**Personal Protection/Sanitation (see Table 2):**
**Skin:** Prevent skin contact
**Eyes:** Prevent eye contact
**Wash skin:** When contam
**Remove:** When wet (flamm)
**Change:** N.R.
**Provide:** Eyewash
Quick drench

**Respirator Recommendations (see Tables 3 and 4):**
**NIOSH**
**¥:** ScbaF:Pd,Pp/SaF:Pd,Pp:AScba
**Escape:** GmFS/ScbaE

**Incompatibilities and Reactivities:** Acids, strong oxidizers, water, carbonyl compounds, quinones, sulfonyl halides
[**Note:** Subject to violent polymerization in contact with acids. Hydrolyzes in water to form methylethanolamine.]

**Exposure Routes, Symptoms, Target Organs (see Table 5):**
**ER:** Inh, Abs, Ing, Con
**SY:** Eye, skin burns; [carc]
**TO:** Eyes, skin [in animals: nasal tumors]

**First Aid (see Table 6):**
**Eye:** Irr immed
**Skin:** Water flush immed
**Breath:** Resp support
**Swallow:** Medical attention immed

## Propylene oxide

| Formula: | CAS#: | RTECS#: | IDLH: |
|---|---|---|---|
| $C_3H_6O$ | 75-56-9 | TZ2975000 | Ca [400 ppm] |

**Conversion:** 1 ppm = 2.38 mg/m$^3$
**DOT:** 1280 127P

**Synonyms/Trade Names:** 1,2-Epoxy propane; Methyl ethylene oxide; Methyloxirane; Propene oxide; 1,2-Propylene oxide

**Exposure Limits:**
**NIOSH REL:** Ca
See Appendix A
**OSHA PEL†:** TWA 100 ppm (240 mg/m$^3$)

**Measurement Methods (see Table 1):**
**NIOSH** 1612
**OSHA** 88

**Physical Description:** Colorless liquid with a benzene-like odor. [**Note:** A gas above 94°F.]

**Chemical & Physical Properties:**
**MW:** 58.1
**BP:** 94°F
**Sol:** 41%
**Fl.P:** -35°F
**IP:** 9.81 eV
**Sp.Gr:** 0.83
**VP:** 445 mmHg
**FRZ:** -170°F
**UEL:** 36%
**LEL:** 2.3%
Class IA Flammable Liquid

**Personal Protection/Sanitation (see Table 2):**
**Skin:** Prevent skin contact
**Eyes:** Prevent eye contact
**Wash skin:** When contam
**Remove:** When wet (flamm)
**Change:** N.R.
**Provide:** Quick drench

**Respirator Recommendations (see Tables 3 and 4):**
**NIOSH**
**¥:** ScbaF:Pd,Pp/SaF:Pd,Pp:AScba
**Escape:** GmFS/ScbaE

**Incompatibilities and Reactivities:** Anhydrous chlorides of iron, tin, and aluminum; peroxides of iron and aluminum; alkali metal hydroxides; iron; strong acids, caustics & peroxides [**Note:** Polymerization may occur due to high temperatures or contamination with alkalis, aqueous acids, amines & acidic alcohols.]

**Exposure Routes, Symptoms, Target Organs (see Table 5):**
**ER:** Inh, Ing, Con
**SY:** Irrit eyes, skin, resp sys; skin blisters, burns; [carc]
**TO:** Eyes, skin, resp sys [in animals: nasal tumors]

**First Aid (see Table 6):**
**Eye:** Irr immed
**Skin:** Water flush immed
**Breath:** Resp support
**Swallow:** Medical attention immed

| n-Propyl nitrate | Formula: $CH_3CH_2CH_2ONO_2$ | CAS#: 627-13-4 | RTECS#: UK0350000 | IDLH: 500 ppm |
|---|---|---|---|---|
| **Conversion:** 1 ppm = 4.30 mg/m$^3$ | **DOT:** 1865 131 | | | |

**Synonyms/Trade Names:** Propyl ester of nitric acid

**Exposure Limits:**
**NIOSH REL:** TWA 25 ppm (105 mg/m$^3$)
ST 40 ppm (170 mg/m$^3$)
**OSHA PEL†:** TWA 25 ppm (110 mg/m$^3$)

**Measurement Methods (see Table 1):**
**NIOSH** S227 (II-3)
**OSHA** 7

**Physical Description:** Colorless to straw-colored liquid with an ether-like odor.

| Chemical & Physical Properties: | Personal Protection/Sanitation (see Table 2): | Respirator Recommendations (see Tables 3 and 4): |
|---|---|---|
| **MW:** 105.1<br>**BP:** 231°F<br>**Sol:** Slight<br>**Fl.P:** 68°F<br>**IP:** 11.07 eV<br>**Sp.Gr:** 1.07<br>**VP:** 18 mmHg<br>**FRZ:** -148°F<br>**UEL:** 100%<br>**LEL:** 2%<br>Class IB Flammable Liquid | **Skin:** Prevent skin contact<br>**Eyes:** Prevent eye contact<br>**Wash skin:** When contam<br>**Remove:** When wet (flamm)<br>**Change:** N.R. | **NIOSH/OSHA**<br>**250 ppm:** Sa<br>**500 ppm:** Sa:Cf/ScbaF/SaF<br>**§:** ScbaF:Pd,Pp/SaF:Pd,Pp:AScba<br>**Escape:** GmFS¿/ScbaE |

**Incompatibilities and Reactivities:** Strong oxidizers, combustible materials
[**Note:** Forms explosive mixtures with combustible materials.]

| Exposure Routes, Symptoms, Target Organs (see Table 5): | First Aid (see Table 6): |
|---|---|
| **ER:** Inh, Ing, Con<br>**SY:** In animals: irrit eyes, skin; methemo, anoxia, cyan; dysp, lass, dizz, head<br>**TO:** Eyes, skin, blood | **Eye:** Irr immed<br>**Skin:** Soap wash prompt<br>**Breath:** Resp support<br>**Swallow:** Medical attention immed |

| Pyrethrum | Formula: $C_{20}H_{28}O_3$/$C_{21}H_{28}O_5$/$C_{21}H_{30}O_3$/ $C_{22}H_{30}O_5$/$C_{21}H_{28}O_3$/$C_{22}H_{28}O_5$ | CAS#: 8003-34-7 | RTECS#: UR4200000 | IDLH: 5000 mg/m$^3$ |
|---|---|---|---|---|
| **Conversion:** | **DOT:** | | | |

P

**Synonyms/Trade Names:** Cinerin I or II, Jasmolin I or II, Pyrethrin I or II, Pyrethrum I or II
[**Note:** Pyrethrum is a variable mixture of Cinerin, Jasmolin, and Pyrethrin.]

**Exposure Limits:**
**NIOSH REL:** TWA 5 mg/m$^3$
**OSHA PEL:** TWA 5 mg/m$^3$

**Measurement Methods (see Table 1):**
**NIOSH** 5008
**OSHA** 70

**Physical Description:** Brown, viscous oil or solid. [insecticide]

| Chemical & Physical Properties: | Personal Protection/Sanitation (see Table 2): | Respirator Recommendations (see Tables 3 and 4): |
|---|---|---|
| **MW:** 316-374<br>**BP:** ?<br>**Sol:** Insoluble<br>**Fl.P:** 180-190°F<br>**IP:** ?<br>**Sp.Gr:** 1 (approx)<br>**VP:** Low<br>**MLT:** ?<br>**UEL:** ?<br>**LEL:** ?<br>Class IIIA Combustible Liquid | **Skin:** Prevent skin contact<br>**Eyes:** Prevent eye contact<br>**Wash skin:** When contam<br>**Remove:** When wet or contam<br>**Change:** Daily | **NIOSH/OSHA**<br>**50 mg/m$^3$:** CcrOv95*/Sa*<br>**125 mg/m$^3$:** Sa:Cf*/PaprOvHie*<br>**250 mg/m$^3$:** CcrFOv100/PaprTOvHie*/ ScbaF/SaF<br>**5000 mg/m$^3$:** SaF:Pd,Pp<br>**§:** ScbaF:Pd,Pp/SaF:Pd,Pp:AScba<br>**Escape:** GmFOv100/ScbaE |

**Incompatibilities and Reactivities:** Strong oxidizers

| Exposure Routes, Symptoms, Target Organs (see Table 5): | First Aid (see Table 6): |
|---|---|
| **ER:** Inh, Ing, Con<br>**SY:** Erythema, derm, papules, pruritus, rhin; sneez; asthma<br>**TO:** Resp sys, skin, CNS | **Eye:** Irr immed<br>**Skin:** Soap wash immed<br>**Breath:** Resp support<br>**Swallow:** Medical attention immed |

| Pyridine | Formula: $C_5H_5N$ | CAS#: 110-86-1 | RTECS#: UR8400000 | IDLH: 1000 ppm |
|---|---|---|---|---|
| **Conversion:** 1 ppm = 3.24 mg/m$^3$ | **DOT:** 1282 129 | | | |

**Synonyms/Trade Names:** Azabenzene, Azine

**Exposure Limits:**
**NIOSH REL:** TWA 5 ppm (15 mg/m$^3$)
**OSHA PEL:** TWA 5 ppm (15 mg/m$^3$)

**Measurement Methods (see Table 1):**
**NIOSH** 1613
**OSHA** 7

**Physical Description:** Colorless to yellow liquid with a nauseating, fish-like odor.

**Chemical & Physical Properties:**
**MW:** 79.1
**BP:** 240°F
**Sol:** Miscible
**Fl.P:** 68°F
**IP:** 9.27 eV
**Sp.Gr:** 0.98
**VP:** 16 mmHg
**FRZ:** -44°F
**UEL:** 12.4%
**LEL:** 1.8%
Class IB Flammable Liquid

**Personal Protection/Sanitation (see Table 2):**
**Skin:** Prevent skin contact
**Eyes:** Prevent eye contact
**Wash skin:** When contam
**Remove:** When wet (flamm)
**Change:** N.R.
**Provide:** Eyewash
Quick drench

**Respirator Recommendations (see Tables 3 and 4):**
**NIOSH/OSHA**
**125 ppm:** Sa:Cf£/PaprOv£
**50 ppm:** CcrFOv/GmFOv/PaprTOv£/ScbaF/SaF
**1000 ppm:** SaF:Pd,Pp
**§:** ScbaF:Pd,Pp/SaF:Pd,Pp:AScba
**Escape:** GmFOv/ScbaE

**Incompatibilities and Reactivities:** Strong oxidizers, strong acids

**Exposure Routes, Symptoms, Target Organs (see Table 5):**
**ER:** Inh, Abs, Ing, Con
**SY:** Irrit eyes; head, anxi, dizz, insom; nau, anor; derm; liver, kidney damage
**TO:** Eyes, skin, CNS, liver, kidneys, GI tract,

**First Aid (see Table 6):**
**Eye:** Irr immed
**Skin:** Water flush immed
**Breath:** Resp support
**Swallow:** Medical attention immed

Q

| Quinone | Formula: $OC_6H_4O$ | CAS#: 106-51-4 | RTECS#: DK2625000 | IDLH: 100 mg/m$^3$ |
|---|---|---|---|---|
| **Conversion:** 1 ppm = 4.42 mg/m$^3$ | **DOT:** 2587 153 | | | |

**Synonyms/Trade Names:** 1,4-Benzoquinone; p-Benzoquinone; 1,4-Cyclohexadiene dioxide; p-Quinone

**Exposure Limits:**
**NIOSH REL:** TWA 0.4 mg/m$^3$ (0.1 ppm)
**OSHA PEL:** TWA 0.4 mg/m$^3$ (0.1 ppm)

**Measurement Methods (see Table 1):**
**NIOSH** S181 (II-4)

**Physical Description:** Pale-yellow solid with an acrid, chlorine-like odor.

**Chemical & Physical Properties:**
**MW:** 108.1
**BP:** Sublimes
**Sol:** Slight
**Fl.P:** 100-200°F
**IP:** 9.68 eV
**Sp.Gr:** 1.32
**VP(77°F):** 0.1 mmHg
**MLT:** 240°F
**UEL:** ?
**LEL:** ?
Combustible Solid

**Personal Protection/Sanitation (see Table 2):**
**Skin:** Prevent skin contact
**Eyes:** Prevent eye contact
**Wash skin:** When contam
**Remove:** When wet or contam
**Change:** Daily
**Provide:** Eyewash
Quick drench

**Respirator Recommendations (see Tables 3 and 4):**
**NIOSH/OSHA**
**10 mg/m$^3$:** Sa:Cf£
**20 mg/m$^3$:** ScbaF/SaF
**100 mg/m$^3$:** SaF:Pd,Pp
**§:** ScbaF:Pd,Pp/SaF:Pd,Pp:AScba
**Escape:** GmFOv100/ScbaE

**Incompatibilities and Reactivities:** Strong oxidizers

**Exposure Routes, Symptoms, Target Organs (see Table 5):**
**ER:** Inh, Ing, Con
**SY:** Eye irrit, conj; kera; skin irrit
**TO:** Eyes, skin

**First Aid (see Table 6):**
**Eye:** Irr immed
**Skin:** Soap wash immed
**Breath:** Resp support
**Swallow:** Medical attention immed

## Resorcinol

| Resorcinol | Formula: $C_6H_4(OH)_2$ | CAS#: 108-46-3 | RTECS#: VG9625000 | IDLH: N.D. |
|---|---|---|---|---|
| **Conversion:** 1 ppm = 4.50 mg/m$^3$ | **DOT:** 2876 153 | | | |

**Synonyms/Trade Names:** 1,3-Benzenediol; m-Benzenediol; 1,3-Dihydroxybenzene; m-Dihydroxybenzene; 3-Hydroxyphenol; m-Hydroxyphenol

**Exposure Limits:**
**NIOSH REL:** TWA 10 ppm (45 mg/m$^3$)
ST 20 ppm (90 mg/m$^3$)
**OSHA PEL†:** none

**Measurement Methods (see Table 1):**
**NIOSH** 5701
**OSHA** PV2053

**Physical Description:** White needles, plates, crystals, flakes, or powder with a faint odor.
[**Note:** Turns pink on exposure to air or light, or contact with iron.]

**Chemical & Physical Properties:**
**MW:** 110.1
**BP:** 531°F
**Sol:** 110%
**Fl.P:** 261°F
**IP:** 8.63 eV
**Sp.Gr:** 1.27
**VP(77°F):** 0.0002 mmHg
**MLT:** 228°F
**UEL:** ?
**LEL(392°F):** 1.4%
Class IIIB Combustible Liquid, but may be difficult to ignite.

**Personal Protection/Sanitation (see Table 2):**
**Skin:** Prevent skin contact
**Eyes:** Prevent eye contact
**Wash skin:** When contam
**Remove:** When wet or contam
**Change:** Daily
**Provide:** Eyewash

**Respirator Recommendations (see Tables 3 and 4):**
Not available.

**Incompatibilities and Reactivities:** Acetanilide, albumin, alkalis, antipyrine, camphor, ferric salts, menthol, spirit nitrous ether, strong oxidizers & bases
[**Note:** Hygroscopic (i.e., absorbs moisture from the air).]

**Exposure Routes, Symptoms, Target Organs (see Table 5):**
**ER:** Inh, Ing, Con
**SY:** Irrit eyes, skin, nose, throat, upper resp sys; methemo; cyan, convuls; restless, bluish skin, incr heart rate, dysp; dizz, drow, hypothermia, hema; spleen, kidney, liver changes; derm
**TO:** Eyes, skin, resp sys, CVS, CNS, blood, spleen, liver, kidneys

**First Aid (see Table 6):**
**Eye:** Irr immed
**Skin:** Water wash immed
**Breath:** Resp support
**Swallow:** Medical attention immed

## Rhodium (metal fume and insoluble compounds, as Rh)

| Rhodium (metal fume and insoluble compounds, as Rh) | Formula: Rh (metal) | CAS#: 7440-16-6 (metal) | RTECS#: VI9069000 | IDLH: 100 mg/m$^3$ (as Rh) |
|---|---|---|---|---|
| **Conversion:** | **DOT:** | | | |

**Synonyms/Trade Names: Rhodium metal:** Elemental rhodium
Synonyms of other insoluble rhodium compounds vary depending upon the specific compound.

**Exposure Limits:**
**NIOSH REL:** TWA 0.1 mg/m$^3$
**OSHA PEL:** TWA 0.1 mg/m$^3$

**Measurement Methods (see Table 1):**
**NIOSH** S188 (II-3)

**Physical Description:** Metal: White, hard, ductile, malleable solid with a bluish-gray luster.

**Chemical & Physical Properties:**
**MW:** 102.9
**BP:** 6741°F
**Sol:** Insoluble
**Fl.P:** NA
**IP:** NA
**Sp.Gr:** 12.41 (metal)
**VP:** 0 mmHg (approx)
**MLT:** 3571°F
**UEL:** NA
**LEL:** NA
Metal: Noncombustible Solid in bulk form, but flammable as dust or powder.

**Personal Protection/Sanitation (see Table 2):**
**Skin:** N.R.
**Eyes:** N.R.
**Wash skin:** N.R.
**Remove:** N.R.
**Change:** N.R.

**Respirator Recommendations (see Tables 3 and 4):**
**NIOSH/OSHA**
**0.5 mg/m$^3$:** Qm
**1 mg/m$^3$:** 95XQ/Sa
**2.5 mg/m$^3$:** Sa:Cf/PaprHie
**5 mg/m$^3$:** 100F/SaT:Cf/PaprTHie/ScbaF/SaF
**100 mg/m$^3$:** Sa:Pd,Pp
**§:** ScbaF:Pd,Pp/SaF:Pd,Pp:AScba
**Escape:** 100F/ScbaE

**Incompatibilities and Reactivities:** Chlorine trifluoride, oxygen difluoride

**Exposure Routes, Symptoms, Target Organs (see Table 5):**
**ER:** Inh
**SY:** Possible resp sens
**TO:** Resp sys

**First Aid (see Table 6):**
**Breath:** Resp support
**Swallow:** Medical attention immed

## Rhodium (soluble compounds, as Rh)

| Formula: | CAS#: | RTECS#: | IDLH: |
|---|---|---|---|
| | | | 2 mg/m³ (as Rh) |

**Conversion:**

**DOT:**

**Synonyms/Trade Names:** Synonyms vary depending upon the specific soluble rhodium compound.

**Exposure Limits:**
**NIOSH REL:** TWA 0.001 mg/m³
**OSHA PEL:** TWA 0.001 mg/m³

**Measurement Methods (see Table 1):**
**NIOSH** S189 (II-3)

**Physical Description:** Appearance and odor vary depending upon the specific soluble rhodium compound.

**Chemical & Physical Properties:**
Properties vary depending upon the specific soluble rhodium compound.

**Personal Protection/Sanitation (see Table 2):**
**Skin:** Prevent skin contact
**Eyes:** Prevent eye contact
**Wash skin:** When contam
**Remove:** When wet or contam
**Change:** N.R.

**Respirator Recommendations (see Tables 3 and 4):**
**NIOSH/OSHA**
**0.01 mg/m³:** 100XQ*/Sa*
**0.025 mg/m³:** Sa:Cf*/PaprHie*
**0.05 mg/m³:** 100F/PaprTHie*/ScbaF/SaF
**2 mg/m³:** SaF:Pd,Pp
**§:** ScbaF:Pd,Pp/SaF:Pd,Pp:AScba
**Escape:** 100F/ScbaE

**Incompatibilities and Reactivities:** Varies

**Exposure Routes, Symptoms, Target Organs (see Table 5):**
**ER:** Inh, Ing, Con
**SY:** In animals: irrit eyes; CNS damage
**TO:** Eyes, CNS

**First Aid (see Table 6):**
**Eye:** Irr immed
**Skin:** Water flush
**Breath:** Resp support
**Swallow:** Medical attention immed

## Ronnel

| Formula: | CAS#: | RTECS#: | IDLH: |
|---|---|---|---|
| $(CH_3O)_2P(S)OC_6H_2Cl_3$ | 299-84-3 | TG0525000 | 300 mg/m³ |

**Conversion:**

**DOT:**

**Synonyms/Trade Names:** O,O-Dimethyl O-(2,4,5-trichlorophenyl) phosphorothioate; Fenchlorophos

**Exposure Limits:**
**NIOSH REL:** TWA 10 mg/m³
**OSHA PEL†:** TWA 15 mg/m³

**Measurement Methods (see Table 1):**
**NIOSH** 5600
**OSHA** PV2054

**Physical Description:** White to light-tan, crystalline solid. [insecticide]
[Note: A liquid above 106°F.]

R

**Chemical & Physical Properties:**
**MW:** 321.6
**BP:** Decomposes
**Sol(77°F):** 0.004%
**Fl.P:** NA
**IP:** ?
**Sp.Gr(77°F):** 1.49
**VP(77°F):** 0.0008 mmHg
**MLT:** 106°F
**UEL:** NA
**LEL:** NA
Noncombustible Solid

**Personal Protection/Sanitation (see Table 2):**
**Skin:** Prevent skin contact
**Eyes:** Prevent eye contact
**Wash skin:** When contam
**Remove:** When wet or contam
**Change:** Daily

**Respirator Recommendations (see Tables 3 and 4):**
**NIOSH**
**100 mg/m³:** CcrOv95/Sa
**250 mg/m³:** Sa:Cf/PaprOvHie
**300 mg/m³:** CcrFOv100/GmFOv100/PaprTOvHie*/ScbaF/SaF
**§:** ScbaF:Pd,Pp/SaF:Pd,Pp:AScba
**Escape:** GmFOv100/ScbaE

**Incompatibilities and Reactivities:** Strong oxidizers

**Exposure Routes, Symptoms, Target Organs (see Table 5):**
**ER:** Inh, Ing, Con
**SY:** In animals: irrit eyes; chol inhibition; liver, kidney damage
**TO:** Eyes, liver, kidneys, blood plasma

**First Aid (see Table 6):**
**Eye:** Irr immed
**Skin:** Soap wash prompt
**Breath:** Resp support
**Swallow:** Medical attention immed

## Rosin core solder, pyrolysis products (as formaldehyde)

| Formula: | CAS#: | RTECS#: | IDLH: |
|---|---|---|---|
| | | | N.D. |

**Conversion:**
**DOT:**

**Synonyms/Trade Names:** Rosin flux pyrolysis products, Rosin core soldering flux pyrolysis products

**Exposure Limits:**
**NIOSH REL*:** TWA 0.1 mg/m$^3$
[***Note:** "Ca" in the presence of formaldehyde, acetaldehyde, or malonaldehyde. See Appendices A & C (Aldehydes).]
**OSHA PEL†:** none

**Measurement Methods (see Table 1):**
**NIOSH** 2541, 3500

**Physical Description:** Pyrolysis products of rosin core solder include acetone, aliphatic aldehydes, methyl alcohol, methane, ethane, various abietic acids (the major components of rosin), CO & $CO_2$.

**Chemical & Physical Properties:**
Properties vary depending upon the specific rosin core solder being used.

**Personal Protection/Sanitation (see Table 2):**
**Skin:** N.R.
**Eyes:** N.R.
**Wash skin:** N.R.
**Remove:** N.R.
**Change:** N.R.

**Respirator Recommendations (see Tables 3 and 4):**
Not available.

**In the presence of Formaldeyde, Acetaldehyde, or Malonaldehyde:**
**NIOSH**
**¥:** ScbaF:Pd,Pp/SaF:Pd,Pp:AScba
**Escape:** GmFOv100/ScbaE

**Incompatibilities and Reactivities:** Varies

**Exposure Routes, Symptoms, Target Organs (see Table 5):**
**ER:** Inh
**SY:** Irrit eyes, nose, throat, upper resp sys [carc (in the presence of Formaldehyde, Acetaldehyde, or Malonaldehyde)]
**TO:** Eyes, resp sys [nasal cancer; thyroid gland tumors in animals (in the presence of Formaldehyde, Acetaldehyde, or Malonaldehyde)]

**First Aid (see Table 6):**
**Eye:** Irr immed
**Breath:** Resp support

## Rotenone

| Formula: | CAS#: | RTECS#: | IDLH: |
|---|---|---|---|
| $C_{23}H_{22}O_6$ | 83-79-4 | DJ2800000 | 2500 mg/m$^3$ |

**Conversion:**
**DOT:**

**Synonyms/Trade Names:**
1,2,12,12a-Tetrahydro-8,9-dimethoxy-2-(1-methylethenyl)-[1]benzopyrano[3,4-b]furo[2,3-h][1]benzopyran-6(6aH)-one

**Exposure Limits:**
**NIOSH REL:** TWA 5 mg/m$^3$
**OSHA PEL:** TWA 5 mg/m$^3$

**Measurement Methods (see Table 1):**
**NIOSH** 5007

**Physical Description:** Colorless to red, odorless, crystalline solid. [insecticide]

**Chemical & Physical Properties:**
**MW:** 394.4
**BP:** Decomposes
**Sol:** Insoluble
**Fl.P:** ?
**IP:** ?
**Sp.Gr:** 1.27
**VP:** <0.00004 mmHg
**MLT:** 330°F
**UEL:** ?
**LEL:** ?
Combustible Solid

**Personal Protection/Sanitation (see Table 2):**
**Skin:** Prevent skin contact
**Eyes:** Prevent eye contact
**Wash skin:** When contam
**Remove:** When wet or contam
**Change:** Daily

**Respirator Recommendations (see Tables 3 and 4):**
**NIOSH/OSHA**
**5 mg/m$^3$:** CcrOv95/Sa
**125 mg/m$^3$:** Sa:Cf/PaprOvHie
**250 mg/m$^3$:** CcrFOv100/GmFOv100/PaprTOvHie/SaT:Cf/ScbaF/SaF
**2500 mg/m$^3$:** Sa:Pd,Pp
**§:** ScbaF:Pd,Pp/SaF:Pd,Pp:AScba
**Escape:** GmFOv100/ScbaE

**Incompatibilities and Reactivities:** Strong oxidizers, alkalis

**Exposure Routes, Symptoms, Target Organs (see Table 5):**
**ER:** Inh, Ing, Con
**SY:** Irrit eyes, skin, resp sys; numb muc memb; nau, vomit, abdom pain; musc tremor, inco, clonic convuls, stupor
**TO:** Eyes, skin, resp sys, CNS

**First Aid (see Table 6):**
**Eye:** Irr immed
**Skin:** Soap wash prompt
**Breath:** Resp support
**Swallow:** Medical attention immed

## Rouge

| Rouge | Formula: $Fe_2O_3$ | CAS#: 1309-37-1 | RTECS#: NO7400000 | IDLH: N.D. |
|---|---|---|---|---|

**Conversion:**

**DOT:**

**Synonyms/Trade Names:** Iron(III)oxide, Iron oxide red, Red iron oxide, Red oxide

**Exposure Limits:**
**NIOSH REL:** See Appendix D
**OSHA PEL†:** TWA 15 mg/m$^3$ (total)
TWA 5 mg/m$^3$ (resp)

**Measurement Methods (see Table 1):**
**NIOSH** 0500, 0600

**Physical Description:** A fine, red powder of ferric oxide.
[**Note:** Usually used in cake form or impregnated in paper or cloth.]

**Chemical & Physical Properties:**
**MW:** 159.7
**BP:** ?
**Sol:** Insoluble
**Fl.P:** NA
**IP:** NA
**Sp.Gr:** 5.24
**VP:** 0 mmHg (approx)
**MLT:** 2849°F
**UEL:** NA
**LEL:** NA
Noncombustible Solid

**Personal Protection/Sanitation (see Table 2):**
**Skin:** N.R.
**Eyes:** N.R.
**Wash skin:** N.R.
**Remove:** N.R.
**Change:** N.R.

**Respirator Recommendations (see Tables 3 and 4):**
Not available.

**Incompatibilities and Reactivities:** Calcium hypochlorite, carbon monoxide, hydrogen peroxide

**Exposure Routes, Symptoms, Target Organs (see Table 5):**
**ER:** Inh, Con
**SY:** Irrit eyes, skin, resp sys
**TO:** Eyes, skin, resp sys

**First Aid (see Table 6):**
**Eye:** Irr immed
**Breath:** Fresh air

## Selenium

| Selenium | Formula: Se | CAS#: 7782-49-2 | RTECS#: VS7700000 | IDLH: 1 mg/m$^3$ (as Se) |
|---|---|---|---|---|

**Conversion:**

**DOT:** 2658 152 (powder)

**Synonyms/Trade Names:** Elemental selenium, Selenium alloy

**Exposure Limits:**
**NIOSH REL*:** TWA 0.2 mg/m$^3$
**OSHA PEL*:** TWA 0.2 mg/m$^3$
[***Note:** The REL and PEL also apply to other selenium compounds (as Se) except Selenium hexafluoride.]

S

**Measurement Methods (see Table 1):**
**NIOSH** 7300, 7301, 7303, 9102, S190 (II-7)
**OSHA** ID121

**Physical Description:** Amorphous or crystalline, red to gray solid.
[**Note:** Occurs as an impurity in most sulfide ores.]

**Chemical & Physical Properties:**
**MW:** 79.0
**BP:** 1265°F
**Sol:** Insoluble
**Fl.P:** NA
**IP:** NA
**Sp.Gr:** 4.28
**VP:** 0 mmHg (approx)
**MLT:** 392°F
**UEL:** NA
**LEL:** NA
Combustible Solid

**Personal Protection/Sanitation (see Table 2):**
**Skin:** Prevent skin contact
**Eyes:** N.R.
**Wash skin:** When contam
**Remove:** When wet or contam
**Change:** N.R.
**Provide:** Quick drench

**Respirator Recommendations (see Tables 3 and 4):**
**NIOSH/OSHA**
**1 mg/m$^3$:** Qm*/95XQ*/100F/PaprHie*/PaprHie*/Sa*/ScbaF
**§:** ScbaF:Pd,Pp/SaF:Pd,Pp:AScba
**Escape:** 100F/ScbaE

**Incompatibilities and Reactivities:** Acids, strong oxidizers, chromium trioxide, potassium bromate, cadmium

**Exposure Routes, Symptoms, Target Organs (see Table 5):**
**ER:** Inh, Ing, Con
**SY:** Irrit eyes, skin, nose, throat; vis dist; head; chills, fever; dysp, bron; metallic taste, garlic breath, GI dist; derm; eye, skin burns; in animals: anemia; liver nec, cirr; kidney, spleen damage
**TO:** Eyes, skin, resp sys, liver, kidneys, blood, spleen

**First Aid (see Table 6):**
**Eye:** Irr immed
**Skin:** Soap wash immed
**Breath:** Resp support
**Swallow:** Medical attention immed

| **Selenium hexafluoride** | **Formula:** $SeF_6$ | **CAS#:** 7783-79-1 | **RTECS#:** VS9450000 | **IDLH:** 2 ppm |
|---|---|---|---|---|
| **Conversion:** 1 ppm = 7.89 mg/m$^3$ | **DOT:** 2194 125 | | | |

**Synonyms/Trade Names:** Selenium fluoride

**Exposure Limits:**
**NIOSH REL:** TWA 0.05 ppm
**OSHA PEL:** TWA 0.05 ppm (0.4 mg/m$^3$)

**Measurement Methods (see Table 1):**
None available

**Physical Description:** Colorless gas.

| Chemical & Physical Properties | Personal Protection/Sanitation (see Table 2) | Respirator Recommendations (see Tables 3 and 4) |
|---|---|---|
| **MW:** 193.0<br>**BP:** -30°F<br>**Sol:** Insoluble<br>**Fl.P:** NA<br>**IP:** ?<br>**RGasD:** 6.66<br>**VP:** >1 atm<br>**FRZ:** -59°F<br>**UEL:** NA<br>**LEL:** NA<br>Nonflammable Gas | **Skin:** N.R.<br>**Eyes:** N.R.<br>**Wash skin:** N.R.<br>**Remove:** N.R.<br>**Change:** N.R. | **NIOSH/OSHA**<br>**0.5 ppm:** Sa<br>**1.25 ppm:** Sa:Cf<br>**2 ppm:** SaT:Cf/ScbaF/SaF<br>**§:** ScbaF:Pd,Pp/SaF:Pd,Pp:AScba<br>**Escape:** GmFS/ScbaE |

**Incompatibilities and Reactivities:** Water [**Note:** Hydrolyzes very slowly in cold water.]

| Exposure Routes, Symptoms, Target Organs (see Table 5) | First Aid (see Table 6) |
|---|---|
| **ER:** Inh<br>**SY:** In animals: pulm irrit, edema<br>**TO:** Resp sys | **Breath:** Resp support |

| **Silica, amorphous** | **Formula:** $SiO_2$ | **CAS#:** 7631-86-9 | **RTECS#:** VV7310000 | **IDLH:** 3000 mg/m$^3$ |
|---|---|---|---|---|
| **Conversion:** | **DOT:** | | | |

**Synonyms/Trade Names:** Diatomaceous earth, Diatomaceous silica, Diatomite, Precipitated amorphous silica, Silica gel, Silicon dioxide (amorphous)

**Exposure Limits:**
**NIOSH REL:** TWA 6 mg/m$^3$
**OSHA PEL†:** TWA 20 mppcf [(80 mg/m$^3$)/%$SiO_2$]

**Measurement Methods (see Table 1):**
**NIOSH** 7501

**Physical Description:** Transparent to gray, odorless powder.
[**Note:** Amorphous silica is the non-crystalline form of $SiO_2$.]

| Chemical & Physical Properties | Personal Protection/Sanitation (see Table 2) | Respirator Recommendations (see Tables 3 and 4) |
|---|---|---|
| **MW:** 60.1<br>**BP:** 4046°F<br>**Sol:** Insoluble<br>**Fl.P:** NA<br>**IP:** NA<br>**Sp.Gr:** 2.20<br>**VP:** 0 mmHg (approx)<br>**MLT:** 3110°F<br>**UEL:** NA<br>**LEL:** NA<br>Noncombustible Solid | **Skin:** N.R.<br>**Eyes:** N.R.<br>**Wash skin:** N.R.<br>**Remove:** N.R.<br>**Change:** N.R. | **NIOSH**<br>**30 mg/m$^3$:** Qm<br>**60 mg/m$^3$:** 95XQ/Sa<br>**150 mg/m$^3$:** Sa:Cf/PaprHie<br>**300 mg/m$^3$:** 100F/SaT:Cf/PaprTHie/ScbaF/SaF<br>**3000 mg/m$^3$:** Sa:Pd,Pp<br>**§:** ScbaF:Pd,Pp/SaF:Pd,Pp:AScba<br>**Escape:** 100F/ScbaE |

**Incompatibilities and Reactivities:** Fluorine, oxygen difluoride, chlorine trifluoride

| Exposure Routes, Symptoms, Target Organs (see Table 5) | First Aid (see Table 6) |
|---|---|
| **ER:** Inh, Con<br>**SY:** Irrit eyes, pneumoconiosis<br>**TO:** Eyes, resp sys | **Eye:** Irr immed<br>**Breath:** Fresh air |

| Silica, crystalline (as respirable dust) | Formula: $SiO_2$ | CAS#: 14808-60-7 | RTECS#: VV7330000 | IDLH: Ca [25 mg/m$^3$ (cristobalite, tridymite); 50 mg/m$^3$ (quartz, tripoli)] |
|---|---|---|---|---|
| **Conversion:** | **DOT:** | | | |

**Synonyms/Trade Names:** Cristobalite, Quartz, Tridymite, Tripoli

**Exposure Limits:**
**NIOSH REL:** Ca
TWA 0.05 mg/m$^3$
See Appendix A
**OSHA PEL†:** See Appendix C (Mineral Dusts)

**Measurement Methods (see Table 1):**
**NIOSH** 7500, 7601, 7602
**OSHA** ID142

**Physical Description:** Colorless, odorless solid. [**Note:** A component of many mineral dusts.]

**Chemical & Physical Properties:**
**MW:** 60.1
**BP:** 4046°F
**Sol:** Insoluble
**Fl.P:** NA
**IP:** NA
**Sp.Gr:** 2.66
**VP:** 0 mmHg (approx)
**MLT:** 3110°F
**UEL:** NA
**LEL:** NA
Noncombustible Solid

**Personal Protection/Sanitation (see Table 2):**
**Skin:** N.R.
**Eyes:** N.R.
**Wash skin:** N.R.
**Remove:** N.R.
**Change:** N.R.

**Respirator Recommendations (see Tables 3 and 4):**
**NIOSH**
**0.5 mg/m$^3$:** 95XQ
**1.25 mg/m$^3$:** PaprHie/Sa:Cf
**2.5 mg/m$^3$:** 100F/PaprTHie
**25 mg/m$^3$:** Sa:Pd,Pp
**§:** ScbaF:Pd,Pp/SaF:Pd,Pp:AScba
**Escape:** 100F/ScbaE

**Incompatibilities and Reactivities:** Powerful oxidizers: fluorine, chlorine trifluoride, manganese trioxide, oxygen difluoride, hydrogen peroxide, etc.; acetylene; ammonia

**Exposure Routes, Symptoms, Target Organs (see Table 5):**
**ER:** Inh, Con
**SY:** Cough, dysp, wheez; decr pulm func, progressive resp symptoms (silicosis); irrit eyes; [carc]
**TO:** Eyes, resp sys [in animals: lung cancer]

**First Aid (see Table 6):**
**Eye:** Irr immed
**Breath:** Fresh air

| Silicon | Formula: Si | CAS#: 7440-21-3 | RTECS#: VW0400000 | IDLH: N.D. |
|---|---|---|---|---|
| **Conversion:** | **DOT:** 1346 170 (amorphous powder) | | | |

**Synonyms/Trade Names:** Elemental silicon
[**Note:** Does not occur free in nature, but is found in silicon dioxide (silica) & in various silicates.]

S

**Exposure Limits:**
**NIOSH REL:** TWA 10 mg/m$^3$ (total)
TWA 5 mg/m$^3$ (resp)
**OSHA PEL†:** TWA 15 mg/m$^3$ (total)
TWA 5 mg/m$^3$ (resp)

**Measurement Methods (see Table 1):**
**NIOSH** 0500, 0600

**Physical Description:** Black to gray, lustrous, needle-like crystals.
[**Note:** The amorphous form is a dark-brown powder.]

**Chemical & Physical Properties:**
**MW:** 28.1
**BP:** 4271°F
**Sol:** Insoluble
**Fl.P:** NA
**IP:** NA
**Sp.Gr(77°F):** 2.33
**VP:** 0 mmHg (approx)
**MLT:** 2570°F
**UEL:** NA
**LEL:** NA
**MEC:** 160 g/m$^3$
Combustible Solid in powder form.

**Personal Protection/Sanitation (see Table 2):**
**Skin:** N.R.
**Eyes:** Prevent eye contact
**Wash skin:** N.R.
**Remove:** N.R.
**Change:** N.R.

**Respirator Recommendations (see Tables 3 and 4):**
Not available.

**Incompatibilities and Reactivities:** Chlorine, fluorine, oxidizers, calcium, cesium carbide, alkaline carbonates

**Exposure Routes, Symptoms, Target Organs (see Table 5):**
**ER:** Inh, Ing, Con
**SY:** Irrit eyes, skin, upper resp sys; cough
**TO:** Eyes, skin, resp sys

**First Aid (see Table 6):**
**Eye:** Irr immed
**Breath:** Fresh air
**Swallow:** Medical attention immed

## Silicon carbide

| Formula: | CAS#: | RTECS#: | IDLH: |
|---|---|---|---|
| SiC | 409-21-2 | VW0450000 | N.D. |

**Conversion:**

**DOT:**

**Synonyms/Trade Names:** Carbon silicide, Carborundum®, Silicon monocarbide

**Exposure Limits:**
**NIOSH REL:** TWA 10 mg/m$^3$ (total)
TWA 5 mg/m$^3$ (resp)
**OSHA PEL†:** TWA 15 mg/m$^3$ (total)
TWA 5 mg/m$^3$ (resp)

**Measurement Methods (see Table 1):**
**NIOSH** 0500, 0600

**Physical Description:** Yellow to green to bluish-black, iridescent crystals.

**Chemical & Physical Properties:**
**MW:** 40.1
**BP:** Sublimes
**Sol:** Insoluble
**Fl.P:** NA
**IP:** 9.30 eV
**Sp.Gr:** 3.23
**VP:** 0 mmHg (approx)
**MLT:** 4892°F (Sublimes)
**UEL:** NA
**LEL:** NA
Noncombustible Solid

**Personal Protection/Sanitation (see Table 2):**
**Skin:** N.R.
**Eyes:** N.R.
**Wash skin:** N.R.
**Remove:** N.R.
**Change:** N.R.

**Respirator Recommendations (see Tables 3 and 4):**
Not available.

**Incompatibilities and Reactivities:** None reported [**Note:** Sublimes with decomposition at 4892°F.]

**Exposure Routes, Symptoms, Target Organs (see Table 5):**
**ER:** Inh, Ing, Con
**SY:** Irrit eyes, skin, upper resp sys; cough
**TO:** Eyes, skin, resp sys

**First Aid (see Table 6):**
**Eye:** Irr immed
**Breath:** Fresh air
**Swallow:** Medical attention immed

## Silicon tetrahydride

| Formula: | CAS#: | RTECS#: | IDLH: |
|---|---|---|---|
| $SiH_4$ | 7803-62-5 | VV1400000 | N.D. |

**Conversion:** 1 ppm = 1.31 mg/m$^3$

**DOT:** 2203 116

**Synonyms/Trade Names:** Monosilane, Silane, Silicane

**Exposure Limits:**
**NIOSH REL:** TWA 5 ppm (7 mg/m$^3$)
**OSHA PEL†:** none

**Measurement Methods (see Table 1):**
None available

**Physical Description:** Colorless gas with a repulsive odor.

**Chemical & Physical Properties:**
**MW:** 32.1
**BP:** -169°F
**Sol:** Decomposes
**Fl.P:** NA (Gas)
**IP:** ?
**RGasD:** 1.11
**VP:** >1 atm
**FRZ:** -301°F
**UEL:** ?
**LEL:** ?
Flammable Gas (may ignite SPONTANEOUSLY in air).

**Personal Protection/Sanitation (see Table 2):**
**Skin:** N.R.
**Eyes:** N.R.
**Wash skin:** N.R.
**Remove:** N.R.
**Change:** N.R.

**Respirator Recommendations (see Tables 3 and 4):**
Not available.

**Incompatibilities and Reactivities:** Halogens (bromine, chlorine, carbonyl chloride, antimony pentachloride, tin(IV) chloride), water

**Exposure Routes, Symptoms, Target Organs (see Table 5):**
**ER:** Inh
**SY:** Irrit eyes, skin, muc memb; nau, head
**TO:** Eyes, skin, resp sys, CNS

**First Aid (see Table 6):**
**Breath:** Resp support

## Silver (metal dust and soluble compounds, as Ag)

| Formula: | CAS#: | RTECS#: | IDLH: |
|---|---|---|---|
| Ag (metal) | 7440-22-4 (metal) | VW3500000 (metal) | 10 mg/m³ (as Ag) |

**Conversion:**

**DOT:**

**Synonyms/Trade Names: Silver metal:** Argentum
Synonyms of soluble silver compounds such as Silver nitrate ($AgNO_3$) vary depending upon the specific compound.

**Exposure Limits:**
**NIOSH REL:** TWA 0.01 mg/m³
**OSHA PEL:** TWA 0.01 mg/m³

**Measurement Methods (see Table 1):**
**NIOSH** 7300, 7301, 9102
**OSHA** ID121

**Physical Description:** Metal: White, lustrous solid.

**Chemical & Physical Properties:**
**MW:** 107.9
**BP:** 3632°F
**Sol:** Insoluble
**Fl.P:** NA
**IP:** NA
**Sp.Gr:** 10.49 (metal)
**VP:** 0 mmHg (approx)
**MLT:** 1761°F
**UEL:** NA
**LEL:** NA
Metal: Noncombustible Solid, but flammable in form of dust or powder.

**Personal Protection/Sanitation (see Table 2):**
**Skin:** Prevent skin contact
**Eyes:** Prevent eye contact
**Wash skin:** When contam
**Remove:** When wet or contam ($AgNO_3$)
**Change:** Daily
**Provide:** Eyewash

**Respirator Recommendations (see Tables 3 and 4):**
**NIOSH/OSHA**
**0.25 mg/m³:** Sa:Cf£/PaprHie£
**0.5 mg/m³:** 100F/ScbaF/SaF
**10 mg/m³:** SaF:Pd,Pp
**§:** ScbaF:Pd,Pp/SaF:Pd,Pp:AScba
**Escape:** 100F/ScbaE

**Incompatibilities and Reactivities:** Acetylene, ammonia, hydrogen peroxide, bromoazide, chlorine trifluoride, ethyleneimine, oxalic acid, tartaric acid

**Exposure Routes, Symptoms, Target Organs (see Table 5):**
**ER:** Inh, Ing, Con
**SY:** Blue-gray eyes, nasal septum, throat, skin; irrit, ulceration skin; GI dist
**TO:** Nasal septum, skin, eyes

**First Aid (see Table 6):**
**Eye:** Irr immed
**Skin:** Water flush
**Breath:** Resp support
**Swallow:** Medical attention immed

## Soapstone (containing less than 1% quartz)

| Formula: | CAS#: | RTECS#: | IDLH: |
|---|---|---|---|
| $3MgO\text{-}4SiO_2\text{-}H_2O$ | | VV8780000 | 3000 mg/m³ |

**Conversion:**

**DOT:**

**Synonyms/Trade Names:** Massive talc, Soapstone silicate, Steatite



**Exposure Limits:**
**NIOSH REL:** TWA 6 mg/m³ (total)
TWA 3 mg/m³ (resp)
**OSHA PEL†:** TWA 20 mppcf

**Measurement Methods (see Table 1):**
**NIOSH** 0500

**Physical Description:** Odorless, white-gray powder.

**Chemical & Physical Properties:**
**MW:** 379.3
**BP:** ?
**Sol:** Insoluble
**Fl.P:** NA
**IP:** NA
**Sp.Gr:** 2.7-2.8
**VP:** 0 mmHg (approx)
**MLT:** ?
**UEL:** NA
**LEL:** NA
Noncombustible Solid

**Personal Protection/Sanitation (see Table 2):**
**Skin:** N.R.
**Eyes:** N.R.
**Wash skin:** N.R.
**Remove:** N.R.
**Change:** N.R.

**Respirator Recommendations (see Tables 3 and 4):**
**NIOSH**
**30 mg/m³:** Qm
**60 mg/m³:** 95XQ/Sa
**150 mg/m³:** PaprHie
**300 mg/m³:** 100F/SaT:Cf*/PaprTHie*/ScbaF/SaF
**3000 mg/m³:** SaF:Pd,Pp
**§:** ScbaF:Pd,Pp/SaF:Pd,Pp:AScba
**Escape:** 100F/ScbaE

**Incompatibilities and Reactivities:** None reported

**Exposure Routes, Symptoms, Target Organs (see Table 5):**
**ER:** Inh, Con
**SY:** Pneumoconiosis: cough, dysp; digital clubbing; cyan; basal crackles, cor pulmonale
**TO:** Resp sys, CVS

**First Aid (see Table 6):**
**Eye:** Irr immed
**Breath:** Resp support

| Sodium aluminum fluoride (as F) | Formula: $Na_3AlF_6$ | CAS#: 15096-52-3 | RTECS#: WA9625000 | IDLH: 250 mg/m$^3$ (as F) |
|---|---|---|---|---|
| **Conversion:** | **DOT:** | | | |

**Synonyms/Trade Names:** Cryocide, Cryodust, Cryolite, Sodium hexafluoroaluminate

| Exposure Limits | Measurement Methods |
|---|---|
| **NIOSH REL*:** TWA 2.5 mg/m$^3$<br>**OSHA PEL*:** TWA 2.5 mg/m$^3$<br>[***Note:** The REL and PEL also apply to other inorganic, solid fluorides (as F).] | **(see Table 1):**<br>**NIOSH** 7902<br>**OSHA** ID110 |

**Physical Description:** Colorless to dark odorless solid. [pesticide]
[**Note:** Loses color on heating.]

| Chemical & Physical Properties: | Personal Protection/Sanitation (see Table 2): | Respirator Recommendations (see Tables 3 and 4): |
|---|---|---|
| **MW:** 209.9<br>**BP:** Decomposes<br>**Sol:** 0.04%<br>**Fl.P:** NA<br>**IP:** NA<br>**Sp.Gr:** 2.90<br>**VP:** 0 mmHg (approx)<br>**MLT:** 1832°F<br>**UEL:** NA<br>**LEL:** NA<br>Noncombustible Solid | **Skin:** Prevent skin contact<br>**Eyes:** Prevent eye contact<br>**Wash skin:** When contam<br>**Remove:** When wet or contam<br>**Change:** Daily | **NIOSH/OSHA**<br>**12.5 mg/m$^3$:** Qm<br>**25 mg/m$^3$:** 95XQ*/Sa*<br>**62.5 mg/m$^3$:** Sa:Cf*/PaprHie*+<br>**125 mg/m$^3$:** 100F+/ScbaF/SaF<br>**250 mg/m$^3$:** SaF:Pd,Pp<br>**§:** ScbaF:Pd,Pp/SaF:Pd,Pp:AScba<br>**Escape:** 100F+/ScbaE<br><br>**+Note:** May need acid gas sorbent |

**Incompatibilities and Reactivities:** Strong oxidizers

| Exposure Routes, Symptoms, Target Organs (see Table 5): | First Aid (see Table 6): |
|---|---|
| **ER:** Inh, Ing, Con<br>**SY:** Irrit eyes, resp sys; nau, abdom pain, diarr; salv, thirst, sweat; stiff spine; derm; calcification of ligaments of ribs, pelvis<br>**TO:** Eyes, skin, resp sys, CNS, skeleton, kidneys | **Eye:** Irr immed<br>**Skin:** Soap wash prompt<br>**Breath:** Fresh air<br>**Swallow:** Medical attention immed |

| Sodium azide | Formula: $NaN_3$ | CAS#: 26628-22-8 | RTECS#: VY8050000 | IDLH: N.D. |
|---|---|---|---|---|
| **Conversion:** | **DOT:** 1687 153 | | | |

**Synonyms/Trade Names:** Azide, Azium, Sodium salt of hydrazoic acid

| Exposure Limits | Measurement Methods |
|---|---|
| **NIOSH REL:** C 0.1 ppm (as $HN_3$) [skin]<br>C 0.3 mg/m$^3$ (as $NaN_3$) [skin]<br>**OSHA PEL†:** none | **(see Table 1):**<br>**OSHA** ID121, ID211 |

**Physical Description:** Colorless to white, odorless, crystalline solid. [pesticide]
[**Note:** Forms hydrazoic acid ($HN_3$) in water.]

| Chemical & Physical Properties: | Personal Protection/Sanitation (see Table 2): | Respirator Recommendations (see Tables 3 and 4): |
|---|---|---|
| **MW:** 65.0<br>**BP:** Decomposes<br>**Sol(63°F):** 42%<br>**Fl.P:** ?<br>**IP:** 11.70 eV<br>**Sp.Gr:** 1.85<br>**VP:** ?<br>**MLT:** 527°F (Decomposes)<br>**UEL:** ?<br>**LEL:** ?<br>Combustible Solid (if heated above 572°F). | **Skin:** Prevent skin contact<br>**Eyes:** Prevent eye contact<br>**Wash skin:** When contam<br>**Remove:** When wet or contam<br>**Change:** Daily<br>**Provide:** Eyewash<br>Quick drench | Not available. |

**Incompatibilities and Reactivities:** Acids, metals, water
[**Note:** Over a period of time, sodium azide may react with copper, lead, brass, or solder in plumbing systems to form an accumulation of the HIGHLY EXPLOSIVE compounds of lead azide & copper azide.]

| Exposure Routes, Symptoms, Target Organs (see Table 5): | First Aid (see Table 6): |
|---|---|
| **ER:** Inh, Abs, Ing, Con<br>**SY:** Irrit eyes, skin; head, dizz, lass, blurred vision; low BP, bradycardia; kidney changes<br>**TO:** Eyes, skin, CNS, CVS, kidneys | **Eye:** Irr immed<br>**Skin:** Water flush immed<br>**Breath:** Resp support<br>**Swallow:** Medical attention immed |

## Sodium bisulfite

| Sodium bisulfite | Formula: $NaHSO_3$ | CAS#: 7631-90-5 | RTECS#: VZ2000000 | IDLH: N.D. |
|---|---|---|---|---|
| **Conversion:** | **DOT:** 2693 154 (solution) | | | |

**Synonyms/Trade Names:** Monosodium salt of sulfurous acid, Sodium acid bisulfite, Sodium bisulphite, Sodium hydrogen sulfite

**Exposure Limits:**
**NIOSH REL:** TWA 5 mg/m$^3$
**OSHA PEL†:** none

**Measurement Methods (see Table 1):**
**NIOSH** 0500

**Physical Description:** White crystals or powder with a slight odor of sulfur dioxide.

| Chemical & Physical Properties: | Personal Protection/Sanitation (see Table 2): | Respirator Recommendations (see Tables 3 and 4): |
|---|---|---|
| **MW:** 104.1<br>**BP:** Decomposes<br>**Sol:** 29%<br>**Fl.P:** NA<br>**IP:** NA<br>**Sp.Gr:** 1.48<br>**VP:** ?<br>**MLT:** Decomposes<br>**UEL:** NA<br>**LEL:** NA<br>Noncombustible Solid | **Skin:** N.R.<br>**Eyes:** N.R.<br>**Wash skin:** N.R.<br>**Remove:** N.R.<br>**Change:** N.R. | Not available. |

**Incompatibilities and Reactivities:** Heat (decomposes) [**Note:** Slowly oxidized to the sulfate on exposure to air.]

| Exposure Routes, Symptoms, Target Organs (see Table 5): | First Aid (see Table 6): |
|---|---|
| **ER:** Inh, Ing, Con<br>**SY:** Irrit eyes, skin, muc memb<br>**TO:** Eyes, skin, resp sys | **Eye:** Irr immed<br>**Breath:** Fresh air<br>**Swallow:** Medical attention immed |

## Sodium cyanide (as CN)

| Sodium cyanide (as CN) | Formula: NaCN | CAS#: 143-33-9 | RTECS#: VZ7525000 | IDLH: 25 mg/m$^3$ (as CN) |
|---|---|---|---|---|
| **Conversion:** | **DOT:** 1689 157 (solid); 3414 157 (solution) | | | |

**Synonyms/Trade Names:** Sodium salt of hydrocyanic acid

**Exposure Limits:**
**NIOSH REL*:** C 5 mg/m$^3$ (4.7 ppm) [10-minute]
**OSHA PEL*:** TWA 5 mg/m$^3$
[***Note:** The REL and PEL also apply to other cyanides (as CN) except Hydrogen cyanide.]

**Measurement Methods (see Table 1):**
**NIOSH** 6010, 7904

S

**Physical Description:** White, granular or crystalline solid with a faint, almond-like odor.

| Chemical & Physical Properties: | Personal Protection/Sanitation (see Table 2): | Respirator Recommendations (see Tables 3 and 4): |
|---|---|---|
| **MW:** 49.0<br>**BP:** 2725°F<br>**Sol(77°F):** 58%<br>**Fl.P:** NA<br>**IP:** NA<br>**Sp.Gr:** 1.60<br>**VP:** 0 mmHg (approx)<br>**MLT:** 1047°F<br>**UEL:** NA<br>**LEL:** NA<br>Noncombustible Solid, but contact with acids releases highly flammable hydrogen cyanide. | **Skin:** Prevent skin contact<br>**Eyes:** Prevent eye contact<br>**Wash skin:** When contam<br>**Remove:** When wet or contam<br>**Change:** Daily<br>**Provide:** Eyewash<br>Quick drench | **NIOSH/OSHA**<br>**25 mg/m$^3$:** Sa/ScbaF<br>**§:** ScbaF:Pd,Pp/SaF:Pd,Pp:AScba<br>**Escape:** GmFS100/ScbaE |

**Incompatibilities and Reactivities:** Strong oxidizers (such as acids, acid salts, chlorates & nitrates)
[**Note:** Absorbs moisture from the air forming a syrup.]

| Exposure Routes, Symptoms, Target Organs (see Table 5): | First Aid (see Table 6): |
|---|---|
| **ER:** Inh, Abs, Ing, Con<br>**SY:** Irrit eyes, skin; asphy; lass, head, conf; nau, vomit; incr resp rate; slow gasping respiration; thyroid, blood changes<br>**TO:** Eyes, skin, CVS, CNS, thyroid, blood | **Eye:** Irr immed<br>**Skin:** Soap wash immed<br>**Breath:** Resp support<br>**Swallow:** Medical attention immed |

## Sodium fluoride (as F)

| Sodium fluoride (as F) | Formula: NaF | CAS#: 7681-49-4 | RTECS#: WB0350000 | IDLH: 250 mg/m$^3$ (as F) |
|---|---|---|---|---|
| **Conversion:** | **DOT:** 1690 154 | | | |

**Synonyms/Trade Names:** Floridine, Sodium monofluoride

**Exposure Limits:**
**NIOSH REL*:** TWA 2.5 mg/m$^3$
**OSHA PEL*:** TWA 2.5 mg/m$^3$
[***Note:** The REL and PEL also apply to other inorganic, solid fluorides (as F).]

**Measurement Methods (see Table 1):**
**NIOSH** 7902, 7906
**OSHA** ID110

**Physical Description:** Odorless, white powder or colorless crystals.
[**Note:** Pesticide grade is often dyed blue.]

**Chemical & Physical Properties:**
**MW:** 42.0
**BP:** 3099°F
**Sol:** 4%
**Fl.P:** NA
**IP:** NA
**Sp.Gr:** 2.78
**VP:** 0 mmHg (approx)
**MLT:** 1819°F
**UEL:** NA
**LEL:** NA
Noncombustible Solid

**Personal Protection/Sanitation (see Table 2):**
**Skin:** Prevent skin contact
**Eyes:** Prevent eye contact
**Wash skin:** When contam
**Remove:** When wet or contam
**Change:** Daily

**Respirator Recommendations (see Tables 3 and 4):**
**NIOSH/OSHA**
**12.5 mg/m$^3$:** Qm
**25 mg/m$^3$:** 95XQ*/Sa*
**62.5 mg/m$^3$:** Sa:Cf*/PaprHie*+
**125 mg/m$^3$:** 100F+/ScbaF/SaF
**250 mg/m$^3$:** SaF:Pd,Pp
**§:** ScbaF:Pd,Pp/SaF:Pd,Pp:AScba
**Escape:** 100F+/ScbaE

**+Note:** May need acid gas sorbent

**Incompatibilities and Reactivities:** Strong oxidizers

**Exposure Routes, Symptoms, Target Organs (see Table 5):**
**ER:** Inh, Ing, Con
**SY:** Irrit eyes, resp sys; nau, abdom pain, diarr; salv, thirst, sweat; stiff spine; derm; calcification of ligaments of ribs, pelvis
**TO:** Eyes, skin, resp sys, CNS, skeleton, kidneys

**First Aid (see Table 6):**
**Eye:** Irr immed
**Skin:** Soap wash prompt
**Breath:** Fresh air
**Swallow:** Medical attention immed

## Sodium fluoroacetate

| Sodium fluoroacetate | Formula: $FCH_2COONa$ | CAS#: 62-74-8 | RTECS#: AH9100000 | IDLH: 2.5 mg/m$^3$ |
|---|---|---|---|---|
| **Conversion:** | **DOT:** 2629 151 | | | |

**Synonyms/Trade Names:** SFA, Sodium monofluoroacetate

**Exposure Limits:**
**NIOSH REL:** TWA 0.05 mg/m$^3$
ST 0.15 mg/m$^3$ [skin]
**OSHA PEL†:** TWA 0.05 mg/m$^3$ [skin]

**Measurement Methods (see Table 1):**
**NIOSH** S301 (II-5)

**Physical Description:** Fluffy, colorless to white (sometimes dyed black), odorless powder. [**Note:** A liquid above 95°F.] [rodenticide]

**Chemical & Physical Properties:**
**MW:** 100.0
**BP:** Decomposes
**Sol:** Miscible
**Fl.P:** NA
**IP:** ?
**Sp.Gr:** ?
**VP:** Low
**MLT:** 392°F
**UEL:** NA
**LEL:** NA
Noncombustible Solid

**Personal Protection/Sanitation (see Table 2):**
**Skin:** Prevent skin contact
**Eyes:** Prevent eye contact
**Wash skin:** When contam
**Remove:** When wet or contam
**Change:** Daily
**Provide:** Quick drench

**Respirator Recommendations (see Tables 3 and 4):**
**NIOSH/OSHA**
**0.25 mg/m$^3$:** Qm
**0.5 mg/m$^3$:** 95XQ/Sa
**1.25 mg/m$^3$:** Sa:Cf/PaprHie
**2.5 mg/m$^3$:** 100F/SaT:Cf/PaprTHie/ScbaF/SaF
**§:** ScbaF:Pd,Pp/SaF:Pd,Pp:AScba
**Escape:** 100F/ScbaE

**Incompatibilities and Reactivities:** None reported

**Exposure Routes, Symptoms, Target Organs (see Table 5):**
**ER:** Inh, Abs, Ing, Con
**SY:** Vomit; anxi, auditory halu; facial pares; twitch face musc; pulsus altenans, ectopic heartbeat, tacar, card arrhy; pulm edema; nystagmus; convuls; liver, kidney damage
**TO:** Resp sys, CVS, liver, kidneys, CNS

**First Aid (see Table 6):**
**Eye:** Irr immed
**Skin:** Water flush immed
**Breath:** Resp support
**Swallow:** Medical attention immed

## Sodium hydroxide

| Formula: | CAS#: | RTECS#: | IDLH: |
|---|---|---|---|
| NaOH | 1310-73-2 | WB4900000 | 10 mg/m$^3$ |

**Conversion:**

**DOT:** 1823 154 (dry, solid); 1824 154 (solution)

**Synonyms/Trade Names:** Caustic soda, Lye, Soda lye, Sodium hydrate

**Exposure Limits:**
**NIOSH REL:** C 2 mg/m$^3$
**OSHA PEL†:** TWA 2 mg/m$^3$

**Measurement Methods (see Table 1):**
**NIOSH** 7401

**Physical Description:** Colorless to white, odorless solid (flakes, beads, granular form).

**Chemical & Physical Properties:**
**MW:** 40.0
**BP:** 2534°F
**Sol:** 111%
**Fl.P:** NA
**IP:** NA
**Sp.Gr:** 2.13
**VP:** 0 mmHg (approx)
**MLT:** 605°F
**UEL:** NA
**LEL:** NA
Noncombustible Solid, but when in contact with water may generate sufficient heat to ignite combustible materials.

**Personal Protection/Sanitation (see Table 2):**
**Skin:** Prevent skin contact
**Eyes:** Prevent eye contact
**Wash skin:** When contam
**Remove:** When wet or contam
**Change:** Daily
**Provide:** Eyewash
Quick drench

**Respirator Recommendations (see Tables 3 and 4):**
**NIOSH/OSHA**
**10 mg/m$^3$:** Sa:Cf£/100F/PaprHie£/ScbaF/SaF
**§:** ScbaF:Pd,Pp/SaF:Pd,Pp:AScba
**Escape:** 100F/ScbaE

**Incompatibilities and Reactivities:** Water; acids; flammable liquids; organic halogens; metals such as aluminum, tin & zinc; nitromethane [**Note:** Corrosive to metals.]

**Exposure Routes, Symptoms, Target Organs (see Table 5):**
**ER:** Inh, Ing, Con
**SY:** Irrit eyes, skin, muc memb; pneu; eye, skin burns; temporary loss of hair
**TO:** Eyes, skin, resp sys

**First Aid (see Table 6):**
**Eye:** Irr immed
**Skin:** Water flush immed
**Breath:** Resp support
**Swallow:** Medical attention immed

## Sodium metabisulfite

| Formula: | CAS#: | RTECS#: | IDLH: |
|---|---|---|---|
| $Na_2S_2O_5$ | 7681-57-4 | UX8225000 | N.D. |

**Conversion:**

**DOT:**



**Synonyms/Trade Names:** Disodium pyrosulfite, Sodium metabisulphite, Sodium pyrosulfite

**Exposure Limits:**
**NIOSH REL:** TWA 5 mg/m$^3$
**OSHA PEL†:** none

**Measurement Methods (see Table 1):**
**NIOSH** 0500

**Physical Description:** White to yellowish crystals or powder with an odor of sulfur dioxide.

**Chemical & Physical Properties:**
**MW:** 190.1
**BP:** Decomposes
**Sol:** 54%
**Fl.P:** NA
**IP:** NA
**Sp.Gr:** 1.4
**VP:** ?
**MLT:** >302°F (Decomposes)
**UEL:** NA
**LEL:** NA
Noncombustible Solid

**Personal Protection/Sanitation (see Table 2):**
**Skin:** N.R.
**Eyes:** N.R.
**Wash skin:** N.R.
**Remove:** N.R.
**Change:** N.R.

**Respirator Recommendations (see Tables 3 and 4):**
Not available.

**Incompatibilities and Reactivities:** Heat (decomposes)
[**Note:** Slowly oxidized to the sulfate on exposure to air & moisture.]

**Exposure Routes, Symptoms, Target Organs (see Table 5):**
**ER:** Inh, Ing, Con
**SY:** Irrit eyes, skin, muc memb
**TO:** Eyes, skin, resp sys

**First Aid (see Table 6):**
**Eye:** Irr immed
**Breath:** Fresh air
**Swallow:** Medical attention immed

## Starch

| Starch | Formula: $(C_6H_{10}O_5)n$ | CAS#: 9005-25-8 | RTECS#: GM5090000 | IDLH: N.D. |
|---|---|---|---|---|
| **Conversion:** | **DOT:** | | | |

**Synonyms/Trade Names:** Corn starch, Rice starch, Sorghum gum, α-Starch, Starch gum, Tapioca starch

**Exposure Limits:**
**NIOSH REL:** TWA 10 mg/m$^3$ (total)
TWA 5 mg/m$^3$ (resp)
**OSHA PEL:** TWA 15 mg/m$^3$ (total)
TWA 5 mg/m$^3$ (resp)

**Measurement Methods (see Table 1):**
**NIOSH** 0500, 0600

**Physical Description:** Fine, white, odorless powder.
[Note: A carbohydrate polymer composed of 25% amylose & 75% amylpectin.]

| Chemical & Physical Properties: | Personal Protection/Sanitation (see Table 2): | Respirator Recommendations (see Tables 3 and 4): |
|---|---|---|
| **MW:** varies<br>**BP:** Decomposes<br>**Sol:** Insoluble<br>**Fl.P:** NA<br>**IP:** NA<br>**Sp.Gr:** 1.45<br>**VP:** 0 mmHg (approx)<br>**MLT:** Decomposes<br>**UEL:** NA<br>**LEL:** NA<br>**MEC:** 50 g/m$^3$<br>Noncombustible Solid, but may form explosive mixture with air. | **Skin:** Prevent skin contact<br>**Eyes:** Prevent eye contact<br>**Wash skin:** Daily<br>**Remove:** When wet or contam<br>**Change:** Daily | Not available. |

**Incompatibilities and Reactivities:** Oxidizers, acids, iodine, alkalis

| Exposure Routes, Symptoms, Target Organs (see Table 5): | First Aid (see Table 6): |
|---|---|
| **ER:** Inh, Ing, Con<br>**SY:** Irrit eyes, skin, muc memb; cough, chest pain; derm; rhin<br>**TO:** Eyes, skin, resp sys | **Eye:** Irr immed<br>**Skin:** Soap wash<br>**Breath:** Fresh air<br>**Swallow:** Medical attention immed |

## Stibine

| Stibine | Formula: $SbH_3$ | CAS#: 7803-52-3 | RTECS#: WJ0700000 | IDLH: 5 ppm |
|---|---|---|---|---|
| **Conversion:** 1 ppm = 5.10 mg/m$^3$ | **DOT:** 2676 119 | | | |

**Synonyms/Trade Names:** Antimony hydride, Antimony trihydride, Hydrogen antimonide

**Exposure Limits:**
**NIOSH REL:** TWA 0.1 ppm (0.5 mg/m$^3$)
**OSHA PEL:** TWA 0.1 ppm (0.5 mg/m$^3$)

**Measurement Methods (see Table 1):**
**NIOSH** 6008

**Physical Description:** Colorless gas with a disagreeable odor like hydrogen sulfide.

| Chemical & Physical Properties: | Personal Protection/Sanitation (see Table 2): | Respirator Recommendations (see Tables 3 and 4): |
|---|---|---|
| **MW:** 124.8<br>**BP:** -1°F<br>**Sol:** Slight<br>**Fl.P:** NA (Gas)<br>**IP:** 9.51 eV<br>**RGasD:** 4.31<br>**VP:** >1 atm<br>**FRZ:** -126°F<br>**UEL:** ?<br>**LEL:** ?<br>Flammable Gas | **Skin:** N.R.<br>**Eyes:** N.R.<br>**Wash skin:** N.R.<br>**Remove:** N.R.<br>**Change:** N.R. | **NIOSH/OSHA**<br>**1 ppm:** Sa<br>**2.5 ppm:** Sa:Cf<br>**5 ppm:** SaT:Cf/ScbaF/SaF<br>**§:** ScbaF:Pd,Pp/SaF:Pd,Pp:AScba<br>**Escape:** GmFS/ScbaE |

**Incompatibilities and Reactivities:** Acids, halogenated hydrocarbons, oxidizers, moisture, chlorine, ozone, ammonia

| Exposure Routes, Symptoms, Target Organs (see Table 5): | First Aid (see Table 6): |
|---|---|
| **ER:** Inh<br>**SY:** Head, lass; nau, abdom pain; lumbar pain, hema, hemolytic anemia; jaun; pulm irrit<br>**TO:** Blood, liver, kidneys, resp sys | **Breath:** Resp support |

S

## Stoddard solvent

| Formula: | CAS#: | RTECS#: | IDLH: |
|---|---|---|---|
| | 8052-41-3 | WJ8925000 | 20,000 mg/m$^3$ |

**Conversion:**

**DOT:** 1268 128 (petroleum distillates, n.o.s.)

**Synonyms/Trade Names:** Dry cleaning safety solvent, Mineral spirits, Petroleum solvent, Spotting naphtha [**Note:** A refined petroleum solvent with a flash point of 102-110°F, boiling point of 309-396°F, and containing >65% $C_{10}$ or higher hydrocarbons.]

**Exposure Limits:**
**NIOSH REL:** TWA 350 mg/m$^3$
C 1800 mg/m$^3$ [15-minute]
**OSHA PEL†:** TWA 500 ppm (2900 mg/m$^3$)

**Measurement Methods (see Table 1):**
**NIOSH** 1550

**Physical Description:** Colorless liquid with a kerosene-like odor.

**Chemical & Physical Properties:**
**MW:** Varies
**BP:** 309-396°F
**Sol:** Insoluble
**Fl.P:** 102-110°F
**IP:** ?
**Sp.Gr:** 0.78
**VP:** ?
**FRZ:** ?
**UEL:** ?
**LEL:** ?
Class II Combustible Liquid

**Personal Protection/Sanitation (see Table 2):**
**Skin:** Prevent skin contact
**Eyes:** Prevent eye contact
**Wash skin:** When contam
**Remove:** When wet or contam
**Change:** N.R.

**Respirator Recommendations (see Tables 3 and 4):**
**NIOSH**
**3500 mg/m$^3$:** CcrOv*/Sa*
**8750 mg/m$^3$:** Sa:Cf*/PaprOv*
**17,500 mg/m$^3$:** CcrFOv/GmFOv/PaprTOv*/ScbaF/SaF
**20,000 mg/m$^3$:** SaF:Pd,Pp
**§:** ScbaF:Pd,Pp/SaF:Pd,Pp:AScba
**Escape:** GmFOv/ScbaE

**Incompatibilities and Reactivities:** Strong oxidizers

**Exposure Routes, Symptoms, Target Organs (see Table 5):**
**ER:** Inh, Ing, Con
**SY:** Irrit eyes, nose, throat; dizz; derm; chemical pneu (aspir liquid); in animals: kidney damage
**TO:** Eyes, skin, resp sys, CNS, kidneys

**First Aid (see Table 6):**
**Eye:** Irr immed
**Skin:** Soap wash prompt
**Breath:** Resp support
**Swallow:** Medical attention immed

## Strychnine

| Formula: | CAS#: | RTECS#: | IDLH: |
|---|---|---|---|
| $C_{21}H_{22}N_2O_2$ | 57-24-9 | WL2275000 | 3 mg/m$^3$ |

**Conversion:**

**DOT:** 1692 151

**Synonyms/Trade Names:** Nux vomica, Strynchnos



**Exposure Limits:**
**NIOSH REL:** TWA 0.15 mg/m$^3$
**OSHA PEL:** TWA 0.15 mg/m$^3$

**Measurement Methods (see Table 1):**
**NIOSH** 5016

**Physical Description:** Colorless to white, odorless, crystalline solid. [pesticide]

**Chemical & Physical Properties:**
**MW:** 334.4
**BP:** Decomposes
**Sol:** 0.02%
**Fl.P:** ?
**IP:** ?
**Sp.Gr:** 1.36
**VP:** Low
**MLT:** 514°F
**UEL:** ?
**LEL:** ?
Combustible Solid, but difficult to ignite.

**Personal Protection/Sanitation (see Table 2):**
**Skin:** Prevent skin contact
**Eyes:** N.R.
**Wash skin:** When contam
**Remove:** N.R.
**Change:** Daily

**Respirator Recommendations (see Tables 3 and 4):**
**NIOSH/OSHA**
**0.75 mg/m$^3$:** Qm
**1.5 mg/m$^3$:** 95XQ/Sa
**3 mg/m$^3$:** Sa:Cf/PaprHie/100F/ScbaF/SaF
**§:** ScbaF:Pd,Pp/SaF:Pd,Pp:AScba
**Escape:** 100F/ScbaE

**Incompatibilities and Reactivities:** Strong oxidizers

**Exposure Routes, Symptoms, Target Organs (see Table 5):**
**ER:** Inh, Ing, Con
**SY:** Stiff neck, facial musc; restless, anxi, incr acuity of perception; incr reflex excitability; cyan; tetanic convuls with opisthotonos
**TO:** CNS

**First Aid (see Table 6):**
**Eye:** Irr immed
**Skin:** Soap wash prompt
**Breath:** Resp support
**Swallow:** Medical attention immed

## Styrene

**Formula:** $C_6H_5CH=CH_2$ | **CAS#:** 100-42-5 | **RTECS#:** WL3675000 | **IDLH:** 700 ppm

**Conversion:** 1 ppm = 4.26 mg/m$^3$ | **DOT:** 2055 128P (inhibited)

**Synonyms/Trade Names:** Ethenyl benzene, Phenylethylene, Styrene monomer, Styrol, Vinyl benzene

**Exposure Limits:**
**NIOSH REL:** TWA 50 ppm (215 mg/m$^3$)
ST 100 ppm (425 mg/m$^3$)
**OSHA PEL†:** TWA 100 ppm
C 200 ppm
600 ppm (5-minute maximum peak in any 3 hours)

**Measurement Methods (see Table 1):**
**NIOSH** 1501, 3800
**OSHA** 9, 89

**Physical Description:** Colorless to yellow, oily liquid with a sweet, floral odor.

**Chemical & Physical Properties:**
**MW:** 104.2
**BP:** 293°F
**Sol:** 0.03%
**Fl.P:** 88°F
**IP:** 8.40 eV
**Sp.Gr:** 0.91
**VP:** 5 mmHg
**FRZ:** -23°F
**UEL:** 6.8%
**LEL:** 0.9%
Class IC Flammable Liquid

**Personal Protection/Sanitation (see Table 2):**
**Skin:** Prevent skin contact
**Eyes:** Prevent eye contact
**Wash skin:** When contam
**Remove:** When wet (flamm)
**Change:** N.R.

**Respirator Recommendations (see Tables 3 and 4):**
**NIOSH**
**500 ppm:** CcrOv*/Sa*
**700 ppm:** Sa:Cf*/CcrFOv/GmFOv/PaprOv*/ScbaF/SaF
**§:** ScbaF:Pd,Pp/SaF:Pd,Pp:AScba
**Escape:** GmFOv/ScbaE

**Incompatibilities and Reactivities:** Oxidizers, catalysts for vinyl polymers, peroxides, strong acids, aluminum chloride [**Note:** May polymerize if contaminated or subjected to heat. Usually contains an inhibitor such as tert-butylcatechol.]

**Exposure Routes, Symptoms, Target Organs (see Table 5):**
**ER:** Inh, Abs, Ing, Con
**SY:** Irrit eyes, nose, resp sys; head, lass, dizz, conf, mal, drow, unsteady gait; narco; defatting derm; possible liver inj; repro effects
**TO:** Eyes, skin, resp sys, CNS, liver, repro sys

**First Aid (see Table 6):**
**Eye:** Irr immed
**Skin:** Water flush
**Breath:** Resp support
**Swallow:** Medical attention immed

## Subtilisins

**Formula:** | **CAS#:** 1395-21-7 (BPN) 9014-01-1 (Carlsburg) | **RTECS#:** CO9450000 (BPN) CO9550000 (Carlsburg) | **IDLH:** N.D.

**Conversion:** | **DOT:**

**Synonyms/Trade Names:** Bacillus subtilis, Bacillus subtilis BPN, Bacillus subtilis Carlsburg, Proteolytic enzymes, Subtilisin BPN, Subtilisin Carlsburg [**Note:** Commercial proteolytic enzymes are used in laundry detergents.]

**Exposure Limits:**
**NIOSH REL:** ST 0.00006 mg/m$^3$ [60-minute]
**OSHA PEL†:** none

**Measurement Methods (see Table 1):**
None available

**Physical Description:** Light-colored, free-flowing powders.
[**Note:** A protein containing numerous amino acids.]

**Chemical & Physical Properties:**
**MW:** 28,000 (approx)
**BP:** ?
**Sol:** ?
**Fl.P:** NA
**IP:** NA
**Sp.Gr:** ?
**VP:** 0 mmHg (approx)
**MLT:** ?
**UEL:** NA
**LEL:** NA

**Personal Protection/Sanitation (see Table 2):**
**Skin:** Prevent skin contact
**Eyes:** Prevent eye contact
**Wash skin:** When contam
**Remove:** When wet or contam
**Change:** Daily

**Respirator Recommendations (see Tables 3 and 4):**
Not available.

**Incompatibilities and Reactivities:** None reported

**Exposure Routes, Symptoms, Target Organs (see Table 5):**
**ER:** Inh, Ing, Con
**SY:** Irrit eyes, skin, resp sys; resp sens (enzyme asthma): sweat, head, chest pain, flu-like symptoms, cough, breathlessness, wheez
**TO:** Eyes, skin, resp sys

**First Aid (see Table 6):**
**Eye:** Irr immed
**Skin:** Soap wash
**Breath:** Resp support
**Swallow:** Medical attention immed

## Succinonitrile

| Formula: | CAS#: | RTECS#: | IDLH: |
|---|---|---|---|
| $NCCH_2CH_2CN$ | 110-61-2 | WN3850000 | N.D. |

**Conversion:** 1 ppm = 3.28 mg/m$^3$

**DOT:**

**Synonyms/Trade Names:** Butanedinitrile; 1,2-Dicyanoethane; Dinile; Ethylene cyanide; Ethylene dicyanide; Succinic dinitrile

**Exposure Limits:**
**NIOSH REL:** TWA 6 ppm (20 mg/m$^3$)
**OSHA PEL:** none

**Measurement Methods (see Table 1):**
**NIOSH** Nitriles Criteria Document

**Physical Description:** Colorless, odorless, waxy solid.
[**Note:** Forms cyanide in the body.]

**Chemical & Physical Properties:**
**MW:** 80.1
**BP:** 509°F
**Sol:** 13%
**Fl.P:** 270°F
**IP:** ?
**Sp.Gr:** 0.99
**VP(212°F):** 2 mmHg
**MLT:** 134°F
**UEL:** ?
**LEL:** ?
Combustible Solid

**Personal Protection/Sanitation (see Table 2):**
**Skin:** Prevent skin contact
**Eyes:** Prevent eye contact
**Wash skin:** When contam
**Remove:** When wet or contam
**Change:** Daily
**Provide:** Eyewash

**Respirator Recommendations (see Tables 3 and 4):**
**NIOSH**
**60 ppm:** Sa
**150 ppm:** Sa:Cf
**250 ppm:** ScbaF/SaF
**§:** ScbaF:Pd,Pp/SaF:Pd,Pp:AScba
**Escape:** GmFOv/ScbaE

**Incompatibilities and Reactivities:** Oxidizers

**Exposure Routes, Symptoms, Target Organs (see Table 5):**
**ER:** Inh, Abs, Ing, Con
**SY:** Irrit eyes, skin, resp sys; head, dizz, lass, conf, convuls; blurred vision; dysp; abdom pain, nau, vomit
**TO:** Eyes, skin, resp sys, CNS, CVS

**First Aid (see Table 6):**
**Eye:** Irr immed
**Skin:** Water wash immed
**Breath:** Resp support
**Swallow:** Medical attention immed

## Sucrose

| Formula: | CAS#: | RTECS#: | IDLH: |
|---|---|---|---|
| $C_{12}H_{22}O_{11}$ | 57-50-1 | WN6500000 | N.D. |

**Conversion:**

**DOT:**

**Synonyms/Trade Names:** Beet sugar, Cane sugar, Confectioner's sugar, Granulated sugar, Rock candy, Saccarose, Sugar, Table sugar

S

**Exposure Limits:**
**NIOSH REL:** TWA 10 mg/m$^3$ (total)
TWA 5 mg/m$^3$ (resp)
**OSHA PEL:** TWA 15 mg/m$^3$ (total)
TWA 5 mg/m$^3$ (resp)

**Measurement Methods (see Table 1):**
**NIOSH** 0500, 0600

**Physical Description:** Hard, white, odorless crystals, lumps, or powder.
[**Note:** May have a characteristic, caramel odor when heated.]

**Chemical & Physical Properties:**
**MW:** 342.3
**BP:** Decomposes
**Sol:** 200%
**Fl.P:** NA
**IP:** NA
**Sp.Gr:** 1.59
**VP:** 0 mmHg (approx)
**MLT:** 320-367°F (Decomposes)
**UEL:** NA
**LEL:** NA
**MEC:** 45 g/m$^3$
Noncombustible Solid, but fine airborne dust may explode.

**Personal Protection/Sanitation (see Table 2):**
**Skin:** N.R.
**Eyes:** N.R.
**Wash skin:** N.R.
**Remove:** N.R.
**Change:** N.R.

**Respirator Recommendations (see Tables 3 and 4):**
Not available.

**Incompatibilities and Reactivities:** Oxidizers, sulfuric acid, nitric acid

**Exposure Routes, Symptoms, Target Organs (see Table 5):**
**ER:** Inh, Con
**SY:** Irrit eyes, skin, upper resp sys; cough
**TO:** Eyes, resp sys

**First Aid (see Table 6):**
**Eye:** Irr immed
**Breath:** Fresh air

## Sulfur dioxide

| | | | | |
|---|---|---|---|---|
| **Sulfur dioxide** | **Formula:** $SO_2$ | **CAS#:** 7446-09-5 | **RTECS#:** WS4550000 | **IDLH:** 100 ppm |
| **Conversion:** 1 ppm = 2.62 mg/$m^3$ | **DOT:** 1079 125 | | | |

**Synonyms/Trade Names:** Sulfurous acid anhydride, Sulfurous oxide, Sulfur oxide

**Exposure Limits:**
**NIOSH REL:** TWA 2 ppm (5 mg/$m^3$)
ST 5 ppm (13 mg/$m^3$)
**OSHA PEL†:** TWA 5 ppm (13 mg/$m^3$)

**Measurement Methods (see Table 1):**
**NIOSH** 3800, 6004
**OSHA** ID104, ID200

**Physical Description:** Colorless gas with a characteristic, irritating, pungent odor. [**Note:** A liquid below 14°F. Shipped as a liquefied compressed gas.]

**Chemical & Physical Properties:**
**MW:** 64.1
**BP:** 14°F
**Sol:** 10%
**Fl.P:** NA
**IP:** 12.30 eV
**RGasD:** 2.26
**VP:** 3.2 atm
**FRZ:** -104°F
**UEL:** NA
**LEL:** NA
Nonflammable Gas

**Personal Protection/Sanitation (see Table 2):**
**Skin:** Frostbite
**Eyes:** Frostbite
**Wash skin:** N.R.
**Remove:** When wet or contam (liquid)
**Change:** N.R.
**Provide:** Frostbite wash

**Respirator Recommendations (see Tables 3 and 4):**
**NIOSH**
**20 ppm:** CcrS*/Sa*
**50 ppm:** Sa:Cf*/PaprS*
**100 ppm:** CcrFS/GmFS/PaprTS*/SaT:Cf*/ScbaF/SaF
**§:** ScbaF:Pd,Pp/SaF:Pd,Pp:AScba
**Escape:** GmFS/ScbaE

**Incompatibilities and Reactivities:** Powdered alkali metals (such as sodium & potassium), water, ammonia, zinc, aluminum, brass, copper [**Note:** Reacts with water to form sulfurous acid ($H_2SO_3$).]

**Exposure Routes, Symptoms, Target Organs (see Table 5):**
**ER:** Inh, Con
**SY:** Irrit eyes, nose, throat; rhin; choking, cough; reflex bronchoconstriction; liquid: frostbite
**TO:** Eyes, skin, resp sys

**First Aid (see Table 6):**
**Eye:** Frostbite
**Skin:** Frostbite
**Breath:** Resp support

## Sulfur hexafluoride

| | | | | |
|---|---|---|---|---|
| **Sulfur hexafluoride** | **Formula:** $SF_6$ | **CAS#:** 2551-62-4 | **RTECS#:** WS4900000 | **IDLH:** N.D. |
| **Conversion:** 1 ppm = 5.98 mg/$m^3$ | **DOT:** 1080 126 | | | |

**Synonyms/Trade Names:** Sulfur fluoride [**Note:** May contain highly toxic sulfur pentafluoride as an impurity.]

**Exposure Limits:**
**NIOSH REL:** TWA 1000 ppm (6000 mg/$m^3$)
**OSHA PEL:** TWA 1000 ppm (6000 mg/$m^3$)

**Measurement Methods (see Table 1):**
**NIOSH** 6602

**Physical Description:** Colorless, odorless gas. [**Note:** Shipped as a liquefied compressed gas. Condenses directly to a solid upon cooling.]

**Chemical & Physical Properties:**
**MW:** 146.1
**BP:** Sublimes
**Sol(77°F):** 0.003%
**Fl.P:** NA
**IP:** 19.30 eV
**RGasD:** 5.11
**VP:** 21.5 atm
**FRZ:** -83°F (Sublimes)
**UEL:** NA
**LEL:** NA
Nonflammable Gas

**Personal Protection/Sanitation (see Table 2):**
**Skin:** Frostbite
**Eyes:** Frostbite
**Wash skin:** N.R.
**Remove:** N.R.
**Change:** N.R.
**Provide:** Frostbite wash

**Respirator Recommendations (see Tables 3 and 4):**
Not available.

**Incompatibilities and Reactivities:** Disilane

**Exposure Routes, Symptoms, Target Organs (see Table 5):**
**ER:** Inh
**SY:** Asphy: incr breath rate, pulse rate; slight musc inco, emotional upset; lass, nau, vomit, convuls; liquid: frostbite
**TO:** Resp sys

**First Aid (see Table 6):**
**Eye:** Frostbite
**Skin:** Frostbite
**Breath:** Resp support

## Sulfuric acid

| Sulfuric acid | Formula: $H_2SO_4$ | CAS#: 7664-93-9 | RTECS#: WS5600000 | IDLH: 15 mg/m$^3$ |
|---|---|---|---|---|

**Conversion:**

**DOT:** 1830 137; 1831 137 (fuming); 1832 137 (spent)

**Synonyms/Trade Names:** Battery acid, Hydrogen sulfate, Oil of vitriol, Sulfuric acid (aqueous)

**Exposure Limits:**
**NIOSH REL:** TWA 1 mg/m$^3$
**OSHA PEL:** TWA 1 mg/m$^3$

**Measurement Methods (see Table 1):**
**NIOSH** 7903
**OSHA** ID113, ID165SG

**Physical Description:** Colorless to dark-brown, oily, odorless liquid.
[**Note:** Pure compound is a solid below 51°F. Often used in an aqueous solution.]

**Chemical & Physical Properties:**
**MW:** 98.1
**BP:** 554°F
**Sol:** Miscible
**Fl.P:** NA
**IP:** ?
**Sp.Gr:** 1.84 (96-98% acid)
**VP:** 0.001 mmHg
**FRZ:** 51°F
**UEL:** NA
**LEL:** NA
Noncombustible Liquid, but capable of igniting finely divided combustible materials.

**Personal Protection/Sanitation (see Table 2):**
**Skin:** Prevent skin contact
**Eyes:** Prevent eye contact
**Wash skin:** When contam
**Remove:** When wet or contam
**Change:** N.R.
**Provide:** Eyewash (>1%)
Quick drench (>1%)

**Respirator Recommendations (see Tables 3 and 4):**
**NIOSH/OSHA**
**15 mg/m$^3$:** Sa:Cf£/PaprAgHie£/CcrFAg100/GmFAg100/ScbaF/SaF
**§:** ScbaF:Pd,Pp/SaF:Pd,Pp:AScba
**Escape:** GmFAg100/ScbaE

**Incompatibilities and Reactivities:** Organic materials, chlorates, carbides, fulminates, water, powdered metals
[**Note:** Reacts violently with water with evolution of heat. Corrosive to metals.]

**Exposure Routes, Symptoms, Target Organs (see Table 5):**
**ER:** Inh, Ing, Con
**SY:** Irrit eyes, skin, nose, throat; pulm edema, bron; emphy; conj; stomatis; dental erosion; eye, skin burns; derm
**TO:** Eyes, skin, resp sys, teeth

**First Aid (see Table 6):**
**Eye:** Irr immed
**Skin:** Water flush immed
**Breath:** Resp support
**Swallow:** Medical attention immed

## Sulfur monochloride

| Sulfur monochloride | Formula: $S_2Cl_2$ | CAS#: 10025-67-9 | RTECS#: WS4300000 | IDLH: 5 ppm |
|---|---|---|---|---|

**Conversion:** 1 ppm = 5.52 mg/m$^3$

**DOT:** 1828 137

**Synonyms/Trade Names:** Sulfur chloride, Sulfur subchloride, Thiosulfurous dichloride

**Exposure Limits:**
**NIOSH REL:** C 1 ppm (6 mg/m$^3$)
**OSHA PEL†:** TWA 1 ppm (6 mg/m$^3$)

**Measurement Methods (see Table 1):**
None available

**Physical Description:** Light-amber to yellow-red, oily liquid with a pungent, nauseating, irritating odor.

**Chemical & Physical Properties:**
**MW:** 135.0
**BP:** 280°F
**Sol:** Decomposes
**Fl.P:** 245°F
**IP:** 9.40 eV
**Sp.Gr:** 1.68
**VP:** 7 mmHg
**FRZ:** -107°F
**UEL:** ?
**LEL:** ?
Class IIIB Combustible Liquid

**Personal Protection/Sanitation (see Table 2):**
**Skin:** Prevent skin contact
**Eyes:** Prevent eye contact
**Wash skin:** When contam
**Remove:** When wet or contam
**Change:** N.R.
**Provide:** Eyewash
Quick drench

**Respirator Recommendations (see Tables 3 and 4):**
**NIOSH/OSHA**
**5 ppm:** CcrFS/GmFS/PaprS£/ScbaF/SaF
**§:** ScbaF:Pd,Pp/SaF:Pd,Pp:AScba
**Escape:** GmFS/ScbaE

**Incompatibilities and Reactivities:** Peroxides, oxides of phosphorous, organics, water
[**Note:** Decomposes violently in water to form hydrochloric acid, sulfur dioxide, sulfur, sulfite, thiosulfate, and hydrogen sulfide. Corrosive to metals.]

**Exposure Routes, Symptoms, Target Organs (see Table 5):**
**ER:** Inh, Ing, Con
**SY:** Irrit eyes, skin, muc memb; lac; cough; eye, skin burns; pulm edema
**TO:** Eyes, skin, resp sys

**First Aid (see Table 6):**
**Eye:** Irr immed
**Skin:** Water flush immed
**Breath:** Resp support
**Swallow:** Medical attention immed

## Sulfur pentafluoride

**Formula:** $S_2F_{10}$ | **CAS#:** 5714-22-7 | **RTECS#:** WS4480000 | **IDLH:** 1 ppm

**Conversion:** 1 ppm = 10.39 mg/m$^3$ | **DOT:**

**Synonyms/Trade Names:** Disulfur decafluoride, Sulfur decafluoride

**Exposure Limits:**
**NIOSH REL:** C 0.01 ppm (0.1 mg/m$^3$)
**OSHA PEL†:** TWA 0.025 ppm (0.25 mg/m$^3$)

**Measurement Methods (see Table 1):**
None available

**Physical Description:** Colorless liquid or gas (above 84°F) with an odor like sulfur dioxide.

**Chemical & Physical Properties:**
**MW:** 254.1
**BP:** 84°F
**Sol:** Insoluble
**Fl.P:** NA
**IP:** ?
**RGasD:** 8.77
**Sp.Gr(32°F):** 2.08
**VP:** 561 mmHg
**FRZ:** -134°F
**UEL:** NA
**LEL:** NA
Noncombustible Liquid
Nonflammable Gas

**Personal Protection/Sanitation (see Table 2):**
**Skin:** Prevent skin contact
**Eyes:** Prevent eye contact
**Wash skin:** N.R.
**Remove:** When wet or contam
**Change:** N.R.
**Provide:** Eyewash
Quick drench

**Respirator Recommendations (see Tables 3 and 4):**
**NIOSH**
**0.1 ppm:** Sa
**0.25 ppm:** Sa:Cf
**0.5 ppm:** SaT:Cf/ScbaF/SaF
**1 ppm:** Sa:Pd,Pp
**§:** ScbaF:Pd,Pp/SaF:Pd,Pp:AScba
**Escape:** GmFAg/ScbaE

**Incompatibilities and Reactivities:** None reported

**Exposure Routes, Symptoms, Target Organs (see Table 5):**
**ER:** Inh, Ing, Con
**SY:** Irrit eyes, skin, resp sys; in animals: pulm edema, hemorr
**TO:** Eyes, skin, resp sys, CNS

**First Aid (see Table 6):**
**Eye:** Irr immed
**Skin:** Soap wash immed
**Breath:** Resp support
**Swallow:** Medical attention immed

## Sulfur tetrafluoride

**Formula:** $SF_4$ | **CAS#:** 7783-60-0 | **RTECS#:** WT4800000 | **IDLH:** N.D.

**Conversion:** 1 ppm = 4.42 mg/m$^3$ | **DOT:** 2418 125

**Synonyms/Trade Names:** Tetrafluorosulfurane

**Exposure Limits:**
**NIOSH REL:** C 0.1 ppm (0.4 mg/m$^3$)
**OSHA PEL†:** none

**Measurement Methods (see Table 1):**
**OSHA** ID110

**Physical Description:** Colorless gas with an odor like sulfur dioxide.
[**Note:** Shipped as a liquefied compressed gas.]

**Chemical & Physical Properties:**
**MW:** 108.1
**BP:** -41°F
**Sol:** Reacts
**Fl.P:** NA
**IP:** 12.63 eV
**RGasD:** 3.78
**VP(70°F):** 10.5 atm
**FRZ:** -185°F
**UEL:** NA
**LEL:** NA
Nonflammable Gas

**Personal Protection/Sanitation (see Table 2):**
**Skin:** Frostbite
**Eyes:** Frostbite
**Wash skin:** N.R.
**Remove:** N.R.
**Change:** N.R.
**Provide:** Frostbite wash

**Respirator Recommendations (see Tables 3 and 4):**
Not available.

**Incompatibilities and Reactivities:** Moisture, concentrated sulfuric acid, dioxygen difluoride
[**Note:** Readily hydrolyzed by moisture, forming hydrofluoric acid & thionyl fluoride.]

**Exposure Routes, Symptoms, Target Organs (see Table 5):**
**ER:** Inh, Con
**SY:** Irrit eyes, skin, muc memb; eye, skin burns (from $SF_4$ releasing hydrofluoric acid on exposure to moisture); liquid: frostbite; in animals: dysp, lass, rhin
**TO:** Eyes, skin, resp sys

**First Aid (see Table 6):**
**Eye:** Frostbite
**Skin:** Frostbite
**Breath:** Resp support

## Sulfuryl fluoride

| Sulfuryl fluoride | Formula: $SO_2F_2$ | CAS#: 2699-79-8 | RTECS#: WT5075000 | IDLH: 200 ppm |
|---|---|---|---|---|
| **Conversion:** 1 ppm = 4.18 mg/m$^3$ | **DOT:** 2191 123 | | | |

**Synonyms/Trade Names:** Sulfur difluoride dioxide, Vikane®

**Exposure Limits:**
**NIOSH REL:** TWA 5 ppm (20 mg/m$^3$)
ST 10 ppm (40 mg/m$^3$)
**OSHA PEL†:** TWA 5 ppm (20 mg/m$^3$)

**Measurement Methods (see Table 1):**
**NIOSH** 6012

**Physical Description:** Colorless, odorless gas. [insecticide/fumigant]
[**Note:** Shipped as a liquefied compressed gas.]

| Chemical & Physical Properties: | Personal Protection/Sanitation (see Table 2): | Respirator Recommendations (see Tables 3 and 4): |
|---|---|---|
| **MW:** 102.1<br>**BP:** -68°F<br>**Sol(32°F):** 0.2%<br>**Fl.P:** NA<br>**IP:** 13.04 eV<br>**RGasD:** 3.72<br>**VP(70°F):** 15.8 atm<br>**FRZ:** -212°F<br>**UEL:** NA<br>**LEL:** NA<br>Nonflammable Gas | **Skin:** Frostbite<br>**Eyes:** Frostbite<br>**Wash skin:** N.R.<br>**Remove:** N.R.<br>**Change:** N.R.<br>**Provide:** Frostbite wash | **NIOSH/OSHA**<br>**50 ppm:** Sa*<br>**125 ppm:** Sa:Cf*<br>**200 ppm:** ScbaF/SaF<br>**§:** ScbaF:Pd,Pp/SaF:Pd,Pp:AScba<br>**Escape:** GmFS/ScbaE |

**Incompatibilities and Reactivities:** None reported

**Exposure Routes, Symptoms, Target Organs (see Table 5):**
**ER:** Inh, Con (liquid)
**SY:** Conj, rhinitis, pharyngitis, pares; liquid: frostbite: in animals: narco, tremor, convuls; pulm edema; kidney inj
**TO:** Eyes, skin, resp sys, CNS, kidneys

**First Aid (see Table 6):**
**Eye:** Frostbite
**Skin:** Frostbite
**Breath:** Resp support

## Sulprofos

| Sulprofos | Formula: $C_{12}H_{19}O_2PS_3$ | CAS#: 35400-43-2 | RTECS#: TE4165000 | IDLH: N.D. |
|---|---|---|---|---|
| **Conversion:** 1 ppm = 13.19 mg/m$^3$ | **DOT:** | | | |

**Synonyms/Trade Names:** Bolstar®, O-Ethyl O-(4-methylthio)phenyl S-propylphosphorodithioate

**Exposure Limits:**
**NIOSH REL:** TWA 1 mg/m$^3$
**OSHA PEL†:** none

**Measurement Methods (see Table 1):**
**NIOSH** 5600
**OSHA** PV2037

**Physical Description:** Tan-colored liquid with a sulfide-like odor.

| Chemical & Physical Properties: | Personal Protection/Sanitation (see Table 2): | Respirator Recommendations (see Tables 3 and 4): |
|---|---|---|
| **MW:** 322.5<br>**BP:** ?<br>**Sol:** Low<br>**Fl.P:** ?<br>**IP:** ?<br>**Sp.Gr:** 1.20<br>**VP:** <8 mmHg<br>**FRZ:** ?<br>**UEL:** ?<br>**LEL:** ? | **Skin:** Prevent skin contact<br>**Eyes:** N.R.<br>**Wash skin:** When contam<br>**Remove:** When wet or contam<br>**Change:** N.R. | Not available. |

**Incompatibilities and Reactivities:** None reported

**Exposure Routes, Symptoms, Target Organs (see Table 5):**
**ER:** Inh, Ing
**SY:** Nau, vomit, abdom cramps, diarr, salv; head, dizz, lass; rhin, chest tight; blurred vision, miosis; card irreg; musc fasc; dysp
**TO:** Resp sys, CNS, CVS, blood chol

**First Aid (see Table 6):**
**Eye:** Irr immed
**Skin:** Soap wash immed
**Breath:** Resp support
**Swallow:** Medical attention immed

| **2,4,5-T** | **Formula:** $Cl_3C_6H_2OCH_2COOH$ | **CAS#:** 93-76-5 | **RTECS#:** AJ8400000 | **IDLH:** 250 mg/m$^3$ |
|---|---|---|---|---|
| **Conversion:** | **DOT:** 2765 152 | | | |

**Synonyms/Trade Names:** 2,4,5-Trichlorophenoxyacetic acid

| | |
|---|---|
| **Exposure Limits:**<br>**NIOSH REL:** TWA 10 mg/m$^3$<br>**OSHA PEL:** TWA 10 mg/m$^3$ | **Measurement Methods (see Table 1):**<br>**NIOSH** 5001 |
| **Physical Description:** Colorless to tan, odorless, crystalline solid. [herbicide] | |

| **Chemical & Physical Properties:** | **Personal Protection/Sanitation (see Table 2):** | **Respirator Recommendations (see Tables 3 and 4):** |
|---|---|---|
| **MW:** 255.5<br>**BP:** Decomposes<br>**Sol(77°F):** 0.03%<br>**Fl.P:** ?<br>**IP:** ?<br>**Sp.Gr:** 1.80<br>**VP:** 1 x 10$^{-7}$ mmHg<br>**MLT:** 307°F<br>**UEL:** ?<br>**LEL:** ?<br>Combustible Solid, but burns with difficulty. | **Skin:** N.R.<br>**Eyes:** N.R.<br>**Wash skin:** N.R.<br>**Remove:** N.R.<br>**Change:** N.R. | **NIOSH/OSHA**<br>**50 mg/m$^3$:** Qm<br>**100 mg/m$^3$:** 95XQ/Sa<br>**250 mg/m$^3$:** Sa:Cf/100F/PaprHie/ ScbaF/SaF<br>**§:** ScbaF:Pd,Pp/SaF:Pd,Pp:AScba<br>**Escape:** 100F/ScbaE |

**Incompatibilities and Reactivities:** None reported

| **Exposure Routes, Symptoms, Target Organs (see Table 5):** | **First Aid (see Table 6):** |
|---|---|
| **ER:** Inh, Ing, Con<br>**SY:** In animals: ataxia; skin irrit, acne-like rash; liver damage<br>**TO:** Skin, liver, GI tract | **Eye:** Irr immed<br>**Skin:** Soap wash<br>**Breath:** Resp support<br>**Swallow:** Medical attention immed |

| **Talc (containing no asbestos and less than 1% quartz)** | **Formula:** $Mg_3Si_4O_{10}(OH)_2$ | **CAS#:** 14807-96-6 | **RTECS#:** WW2710000 | **IDLH:** 1000 mg/m$^3$ |
|---|---|---|---|---|
| **Conversion:** | **DOT:** | | | |

**Synonyms/Trade Names:** Hydrous magnesium silicate, Steatite talc

| | |
|---|---|
| **Exposure Limits:**<br>**NIOSH REL:** TWA 2 mg/m$^3$ (resp)<br>**OSHA PEL†:** TWA 20 mppcf | **Measurement Methods (see Table 1):**<br>**NIOSH** P&CAM355 (III) |
| **Physical Description:** Odorless, white powder. | |

| **Chemical & Physical Properties:** | **Personal Protection/Sanitation (see Table 2):** | **Respirator Recommendations (see Tables 3 and 4):** |
|---|---|---|
| **MW:** Varies<br>**BP:** ?<br>**Sol:** Insoluble<br>**Fl.P:** NA<br>**IP:** NA<br>**Sp.Gr:** 2.70-2.80<br>**VP:** 0 mmHg (approx)<br>**MLT:** 1652°F to 1832°F<br>**UEL:** NA<br>**LEL:** NA<br>Noncombustible Solid | **Skin:** N.R.<br>**Eyes:** N.R.<br>**Wash skin:** N.R.<br>**Remove:** N.R.<br>**Change:** N.R. | **NIOSH**<br>**10 mg/m$^3$:** Qm<br>**20 mg/m$^3$:** 95XQ/Sa<br>**50 mg/m$^3$:** PaprHie/Sa:Cf<br>**100 mg/m$^3$:** 100F/SaT:Cf/PaprTHie/ ScbaF/SaF<br>**1000 mg/m$^3$:** Sa:Pd,Pp<br>**§:** ScbaF:Pd,Pp/SaF:Pd,Pp:AScba<br>**Escape:** 100F/ScbaE |

**Incompatibilities and Reactivities:** None reported

| **Exposure Routes, Symptoms, Target Organs (see Table 5):** | **First Aid (see Table 6):** |
|---|---|
| **ER:** Inh, Con<br>**SY:** Fibrotic pneumoconiosis, irrit eyes<br>**TO:** Eyes, resp sys, CVS | **Eye:** Irr immed<br>**Breath:** Fresh air |

## Tantalum (metal and oxide dust, as Ta)

**Formula:** Ta (metal)
**CAS#:** 7440-25-7 (metal)
**RTECS#:** WW5505000 (metal)
**IDLH:** 2500 $mg/m^3$ (as Ta)

**Conversion:**
**DOT:**

**Synonyms/Trade Names: Tantalum metal:** Tantalum-181
Synonyms of other tantalum dusts (including oxide dusts) vary depending upon the specific compound.

**Exposure Limits:**
**NIOSH REL:** TWA 5 $mg/m^3$
ST 10 $mg/m^3$
**OSHA PEL:** TWA 5 $mg/m^3$

**Measurement Methods (see Table 1):**
NIOSH 0500

**Physical Description:** Metal: Steel-blue to gray solid or black, odorless powder.

**Chemical & Physical Properties:**
**MW:** 180.9
**BP:** 9797°F
**Sol:** Insoluble
**Fl.P:** NA
**IP:** NA
**Sp.Gr:** 16.65 (metal)
14.40 (powder)
**VP:** 0 mmHg (approx)
**MLT:** 5425°F
**UEL:** NA
**LEL:** NA
**MEC:** <200 $g/m^3$
Metal: Combustible Solid; powder ignites SPONTANEOUSLY in air.

**Personal Protection/Sanitation (see Table 2):**
**Skin:** N.R.
**Eyes:** N.R.
**Wash skin:** N.R.
**Remove:** N.R.
**Change:** N.R.

**Respirator Recommendations (see Tables 3 and 4):**
**NIOSH/OSHA**
**25 $mg/m^3$:** Qm
**50 $mg/m^3$:** 95XQ/Sa
**125 $mg/m^3$:** Sa:Cf/PaprHie
**250 $mg/m^3$:** 100F/SaT:Cf/PaprTHie/ScbaF/SaF
**2500 $mg/m^3$:** Sa:Pd,Pp
**§:** ScbaF:Pd,Pp/SaF:Pd,Pp:AScba
**Escape:** HieF/ScbaE

**Incompatibilities and Reactivities:** Strong oxidizers, bromine trifluoride, fluorine

**Exposure Routes, Symptoms, Target Organs (see Table 5):**
**ER:** Inh, Con
**SY:** Irrit eyes, skin; in animals: pulm irrit
**TO:** Eyes, skin, resp sys

**First Aid (see Table 6):**
**Eye:** Irr immed
**Breath:** Resp support

## TEDP

**Formula:** $[(CH_3CH_2O)_2PS]_2O$
**CAS#:** 3689-24-5
**RTECS#:** XN4375000
**IDLH:** 10 $mg/m^3$

**Conversion:** 1 ppm = 13.18 $mg/m^3$
**DOT:** 1704 153

**Synonyms/Trade Names:** Bladafum®, Dithion®, Sulfotep, Tetraethyl dithionopyrophosphate, Tetraethyl dithiopyrophosphate, Thiotepp®

**Exposure Limits:**
**NIOSH REL:** TWA 0.2 $mg/m^3$ [skin]
**OSHA PEL:** TWA 0.2 $mg/m^3$ [skin]

**Measurement Methods (see Table 1):**
None available



**Physical Description:** Pale-yellow liquid with a garlic-like odor. [**Note:** A pesticide that may be absorbed on a solid carrier or mixed in a more flammable liquid.]

**Chemical & Physical Properties:**
**MW:** 322.3
**BP:** Decomposes
**Sol:** 0.0007%
**Fl.P:** ?
**IP:** ?
**Sp.Gr(77°F):** 1.20
**VP:** 0.0002 mmHg
**FRZ:** ?
**UEL:** ?
**LEL:** ?
Combustible Liquid

**Personal Protection/Sanitation (see Table 2):**
**Skin:** Prevent skin contact
**Eyes:** Prevent eye contact
**Wash skin:** When contam
**Remove:** When wet or contam
**Change:** N.R.
**Provide:** Eyewash
Quick drench

**Respirator Recommendations (see Tables 3 and 4):**
**NIOSH/OSHA**
**2 $mg/m^3$:** Sa
**5 $mg/m^3$:** Sa:Cf
**10 $mg/m^3$:** ScbaF/SaF
**§:** ScbaF:Pd,Pp/SaF:Pd,Pp:AScba
**Escape:** GmFOv100/ScbaE

**Incompatibilities and Reactivities:** Strong oxidizers, iron [**Note:** Corrosive to iron.]

**Exposure Routes, Symptoms, Target Organs (see Table 5):**
**ER:** Inh, Abs, Ing, Con
**SY:** Irrit eyes, skin; eye pain, blurred vision, lac; rhin; head; cyan; anor, nau, vomit, diarr; local sweat, lass, twitch, para, Cheyne-Stokes respiration, convuls, low BP, card irreg
**TO:** Eyes, skin, resp sys, CNS, CVS, blood chol

**First Aid (see Table 6):**
**Eye:** Irr immed
**Skin:** Soap wash immed
**Breath:** Resp support
**Swallow:** Medical attention immed

## Tellurium

| Tellurium | Formula: Te | CAS#: 13494-80-9 | RTECS#: WY2625000 | IDLH: 25 mg/m³ (as Te) |
|---|---|---|---|---|
| **Conversion:** | **DOT:** | | | |

**Synonyms/Trade Names:** Aurum paradoxum, Metallum problematum

**Exposure Limits:**
**NIOSH REL*:** TWA 0.1 mg/m³
**OSHA PEL*:** TWA 0.1 mg/m³
[*Note: The REL and PEL also apply to other tellurium compounds (as Te) except Tellurium hexafluoride and Bismuth telluride.]

**Measurement Methods (see Table 1):**
**NIOSH** 7300, 7301, 7303, 9102
**OSHA** ID121

**Physical Description:** Odorless, dark-gray to brown, amorphous powder or grayish-white, brittle solid.

| Chemical & Physical Properties: | Personal Protection/Sanitation (see Table 2): | Respirator Recommendations (see Tables 3 and 4): |
|---|---|---|
| **MW:** 127.6<br>**BP:** 1814°F<br>**Sol:** Insoluble<br>**Fl.P:** NA<br>**IP:** NA<br>**Sp.Gr:** 6.24<br>**VP:** 0 mmHg (approx)<br>**MLT:** 842°F<br>**UEL:** NA<br>**LEL:** NA<br>Combustible Solid | **Skin:** N.R.<br>**Eyes:** N.R.<br>**Wash skin:** N.R.<br>**Remove:** N.R.<br>**Change:** N.R. | **NIOSH/OSHA**<br>**0.5 mg/m³:** Qm<br>**1 mg/m³:** 95XQ/Sa<br>**2.5 mg/m³:** Sa:Cf/PaprHie<br>**5 mg/m³:** 100F/SaT:Cf/PaprTHie/ScbaF/SaF<br>**25 mg/m³:** Sa:Pd,Pp<br>**§:** ScbaF:Pd,Pp/SaF:Pd,Pp:AScba<br>**Escape:** 100F/ScbaE |

**Incompatibilities and Reactivities:** Oxidizers, chlorine, cadmium

| Exposure Routes, Symptoms, Target Organs (see Table 5): | First Aid (see Table 6): |
|---|---|
| **ER:** Inh, Ing, Con<br>**SY:** Garlic breath, sweat; dry mouth, metallic taste; drow; anor, nau, no sweat; derm; in animals: CNS, red blood cell changes<br>**TO:** Skin, CNS, blood | **Eye:** Irr immed<br>**Skin:** Soap wash prompt<br>**Breath:** Resp support<br>**Swallow:** Medical attention immed |

## Tellurium hexafluoride

| Tellurium hexafluoride | Formula: $TeF_6$ | CAS#: 7783-80-4 | RTECS#: WY2800000 | IDLH: 1 ppm |
|---|---|---|---|---|
| **Conversion:** 1 ppm = 9.88 mg/m³ | **DOT:** 2195 125 | | | |

**Synonyms/Trade Names:** Tellurium fluoride

**Exposure Limits:**
**NIOSH REL:** TWA 0.02 ppm (0.2 mg/m³)
**OSHA PEL:** TWA 0.02 ppm (0.2 mg/m³)

**Measurement Methods (see Table 1):**
**NIOSH** S187 (II-3)

**Physical Description:** Colorless gas with a repulsive odor.

| Chemical & Physical Properties: | Personal Protection/Sanitation (see Table 2): | Respirator Recommendations (see Tables 3 and 4): |
|---|---|---|
| **MW:** 241.6<br>**BP:** Sublimes<br>**Sol:** Decomposes<br>**Fl.P:** NA<br>**IP:** ?<br>**RGasD:** 8.34<br>**VP:** >1 atm<br>**FRZ:** -36°F (Sublimes)<br>**UEL:** NA<br>**LEL:** NA<br>Nonflammable Gas | **Skin:** N.R.<br>**Eyes:** N.R.<br>**Wash skin:** N.R.<br>**Remove:** N.R.<br>**Change:** N.R. | **NIOSH/OSHA**<br>**0.2 ppm:** Sa<br>**0.5 ppm:** Sa:Cf<br>**1 ppm:** SaT:Cf/ScbaF/SaF<br>**§:** ScbaF:Pd,Pp/SaF:Pd,Pp:AScba<br>**Escape:** GmFS/ScbaE |

**Incompatibilities and Reactivities:** Water [Note: Hydrolyzes slowly in water to telluric acid.]

| Exposure Routes, Symptoms, Target Organs (see Table 5): | First Aid (see Table 6): |
|---|---|
| **ER:** Inh<br>**SY:** Head; dysp; garlic breath; in animals: pulm edema<br>**TO:** Resp sys | **Breath:** Resp support |

## Temephos

| Temephos | Formula: $S[C_6H_4OP(S)(OCH_3)_2]_2$ | CAS#: 3383-96-8 | RTECS#: TF6890000 | IDLH: N.D. |
|---|---|---|---|---|
| **Conversion:** | **DOT:** | | | |

**Synonyms/Trade Names:** Abate®; Temefos; O,O,O'O'-Tetramethyl O,O'-thiodi-p-phenylene phosphorothioate

**Exposure Limits:**
**NIOSH REL:** TWA 10 mg/m$^3$ (total)
TWA 5 mg/m$^3$ (resp)
**OSHA PEL†:** TWA 15 mg/m$^3$ (total)
TWA 5 mg/m$^3$ (resp)

**Measurement Methods (see Table 1):**
**NIOSH** 0500, 0600
**OSHA** PV2056

**Physical Description:** White, crystalline solid or liquid (above 87°F). [insecticide]
[**Note:** Technical grade is a viscous, brown liquid.]

| Chemical & Physical Properties: | Personal Protection/Sanitation (see Table 2): | Respirator Recommendations (see Tables 3 and 4): |
|---|---|---|
| **MW:** 466.5<br>**BP:** 248-257°F (Decomposes)<br>**Sol:** Insoluble<br>**Fl.P:** ?<br>**IP:** ?<br>**Sp.Gr:** 1.32<br>**VP(77°F):** 0.00000007 mmHg<br>**MLT:** 87°F<br>**UEL:** ?<br>**LEL:** ?<br>Combustible Solid | **Skin:** Prevent skin contact<br>**Eyes:** Prevent eye contact<br>**Wash skin:** When contam<br>**Remove:** When wet or contam<br>**Change:** Daily | Not available. |

**Incompatibilities and Reactivities:** None reported

**Exposure Routes, Symptoms, Target Organs (see Table 5):**
**ER:** Inh, Abs, Ing, Con
**SY:** Irrit eyes, blurred vision; dizz; dysp; salv; abdom cramps, nau, diarr, vomit
**TO:** Eyes, resp sys, CNS, CVS, blood chol

**First Aid (see Table 6):**
**Eye:** Irr immed
**Skin:** Soap wash immed
**Breath:** Resp support
**Swallow:** Medical attention immed

## TEPP

| TEPP | Formula: $[(CH_3CH_2O)_2PO]_2O$ | CAS#: 107-49-3 | RTECS#: UX6825000 | IDLH: 5 mg/m$^3$ |
|---|---|---|---|---|
| **Conversion:** 1 ppm = 11.87 mg/m$^3$ | **DOT:** 2783 152 (solid); 3018 152 (liquid) | | | |

**Synonyms/Trade Names:** Ethyl pyrophosphate, Tetraethyl pyrophosphate, Tetron®

**Exposure Limits:**
**NIOSH REL:** TWA 0.05 mg/m$^3$ [skin]
**OSHA PEL:** TWA 0.05 mg/m$^3$ [skin]

**Measurement Methods (see Table 1):**
**NIOSH** 2504

**Physical Description:** Colorless to amber liquid with a faint, fruity odor. [insecticide]
[**Note:** A solid below 32°F.]

T

| Chemical & Physical Properties: | Personal Protection/Sanitation (see Table 2): | Respirator Recommendations (see Tables 3 and 4): |
|---|---|---|
| **MW:** 290.2<br>**BP:** Decomposes<br>**Sol:** Miscible<br>**Fl.P:** NA<br>**IP:** ?<br>**Sp.Gr:** 1.19<br>**VP:** 0.0002 mmHg<br>**FRZ:** 32°F<br>**UEL:** NA<br>**LEL:** NA<br>Noncombustible Liquid | **Skin:** Prevent skin contact<br>**Eyes:** Prevent eye contact<br>**Wash skin:** When contam<br>**Remove:** When wet or contam<br>**Change:** N.R.<br>**Provide:** Eyewash<br>Quick drench | **NIOSH/OSHA**<br>**0.5 mg/m$^3$:** Sa<br>**1.25 mg/m$^3$:** Sa:Cf<br>**2.5 mg/m$^3$:** SaT:Cf/ScbaF/SaF<br>**5 mg/m$^3$:** Sa:Pd,Pp<br>**§:** ScbaF:Pd,Pp/SaF:Pd,Pp:AScba<br>**Escape:** GmFOv100/ScbaE |

**Incompatibilities and Reactivities:** Strong oxidizers, alkalis, water
[**Note:** Hydrolyzes quickly in water to form pyrophosphoric acid.]

**Exposure Routes, Symptoms, Target Organs (see Table 5):**
**ER:** Inh, Abs, Ing, Con
**SY:** Eye pain, blurred vision, lac; rhin; head, chest tight, cyan; anor, nau, vomit, diarr; lass, twitch, para, Cheyne-Stokes respiration, convuls; low BP, card irreg; sweat
**TO:** Eyes, resp sys, CNS, CVS, GI tract, blood chol

**First Aid (see Table 6):**
**Eye:** Irr immed
**Skin:** Water flush immed
**Breath:** Resp support
**Swallow:** Medical attention immed

## m-Terphenyl

| m-Terphenyl | Formula: $C_6H_5C_6H_4C_6H_5$ | CAS#: 92-06-8 | RTECS#: WZ6470000 | IDLH: 500 mg/m$^3$ |
|---|---|---|---|---|
| Conversion: 1 ppm = 9.57 mg/m$^3$ | DOT: | | | |

**Synonyms/Trade Names:** m-Diphenylbenzene; 1,3-Diphenylbenzene; Isodiphenylbenzene; 3-Phenylbiphenyl; 1,3-Terphenyl; meta-Terphenyl; m-Triphenyl

**Exposure Limits:**
**NIOSH REL:** C 5 mg/m$^3$ (0.5 ppm)
**OSHA PEL†:** C 9 mg/m$^3$ (1 ppm)

**Measurement Methods (see Table 1):**
**NIOSH** 5021

**Physical Description:** Yellow solid (needles).

| Chemical & Physical Properties: | Personal Protection/Sanitation (see Table 2): | Respirator Recommendations (see Tables 3 and 4): |
|---|---|---|
| **MW:** 230.3<br>**BP:** 689°F<br>**Sol:** Insoluble<br>**Fl.P(oc):** 375°F<br>**IP:** 8.01<br>**Sp.Gr:** 1.23<br>**VP(200°F):** 0.01 mmHg<br>**MLT:** 192°F<br>**UEL:** ?<br>**LEL:** ?<br>Combustible Solid | **Skin:** Prevent skin contact<br>**Eyes:** Prevent eye contact<br>**Wash skin:** When contam<br>**Remove:** When wet or contam<br>**Change:** Daily<br>**Provide:** Eyewash<br>Quick drench | **NIOSH**<br>**25 mg/m$^3$:** Qm£<br>**50 mg/m$^3$:** 95XQ£/Sa£<br>**125 mg/m$^3$:** Sa:Cf£/PaprHie£<br>**250 mg/m$^3$:** 100F/ScbaF/SaF<br>**500 mg/m$^3$:** SaF:Pd,Pp<br>**§:** ScbaF:Pd,Pp/SaF:Pd,Pp:AScba<br>**Escape:** 100F/ScbaE |

**Incompatibilities and Reactivities:** None reported

| Exposure Routes, Symptoms, Target Organs (see Table 5): | First Aid (see Table 6): |
|---|---|
| **ER:** Inh, Ing, Con<br>**SY:** Irrit eyes, skin, muc memb; thermal skin burns; head; sore throat; in animals: liver, kidney damage<br>**TO:** Eyes, skin, resp sys, liver, kidneys | **Eye:** Irr immed<br>**Skin:** Water flush immed<br>**Breath:** Resp support<br>**Swallow:** Medical attention immed |

## o-Terphenyl

| o-Terphenyl | Formula: $C_6H_5C_6H_4C_6H_5$ | CAS#: 84-15-1 | RTECS#: WZ6472000 | IDLH: 500 mg/m$^3$ |
|---|---|---|---|---|
| Conversion: 1 ppm = 9.42 mg/m$^3$ | DOT: | | | |

**Synonyms/Trade Names:** o-Diphenylbenzene; 1,2-Diphenylbenzene; 2-Phenylbiphenyl; 1,2-Terphenyl; ortho-Terphenyl; o-Triphenyl

**Exposure Limits:**
**NIOSH REL:** C 5 mg/m$^3$ (0.5 ppm)
**OSHA PEL†:** C 9 mg/m$^3$ (1 ppm)

**Measurement Methods (see Table 1):**
**NIOSH** 5021

**Physical Description:** Colorless or light-yellow solid.

| Chemical & Physical Properties: | Personal Protection/Sanitation (see Table 2): | Respirator Recommendations (see Tables 3 and 4): |
|---|---|---|
| **MW:** 230.3<br>**BP:** 630°F<br>**Sol:** Insoluble<br>**Fl.P(oc):** 325°F<br>**IP:** 7.99 eV<br>**Sp.Gr:** 1.1<br>**VP(200°F):** 0.09 mmHg<br>**MLT:** 136°F<br>**UEL:** ?<br>**LEL:** ?<br>Combustible Solid | **Skin:** Prevent skin contact<br>**Eyes:** Prevent eye contact<br>**Wash skin:** When contam<br>**Remove:** When wet or contam<br>**Change:** Daily<br>**Provide:** Eyewash<br>Quick drench | **NIOSH**<br>**25 mg/m$^3$:** Qm£<br>**50 mg/m$^3$:** 95XQ£/Sa£<br>**125 mg/m$^3$:** Sa:Cf£/PaprHie£<br>**250 mg/m$^3$:** 100F/ScbaF/SaF<br>**500 mg/m$^3$:** SaF:Pd,Pp<br>**§:** ScbaF:Pd,Pp/SaF:Pd,Pp:AScba<br>**Escape:** 100F/ScbaE |

**Incompatibilities and Reactivities:** None reported

| Exposure Routes, Symptoms, Target Organs (see Table 5): | First Aid (see Table 6): |
|---|---|
| **ER:** Inh, Ing, Con<br>**SY:** Irrit eyes, skin, muc memb; thermal skin burns; head; sore throat; in animals: liver, kidney damage<br>**TO:** Eyes, skin, resp sys, liver, kidneys | **Eye:** Irr immed<br>**Skin:** Water flush immed<br>**Breath:** Resp support<br>**Swallow:** Medical attention immed |

## p-Terphenyl

| p-Terphenyl | Formula: $C_6H_5C_6H_4C_6H_5$ | CAS#: 92-94-4 | RTECS#: WZ6475000 | IDLH: 500 $mg/m^3$ |
|---|---|---|---|---|
| **Conversion:** 1 ppm = 9.57 $mg/m^3$ | **DOT:** | | | |

**Synonyms/Trade Names:** p-Diphenylbenzene; 1,4-Diphenylbenzene; 4-Phenylbiphenyl; 1,4-Terphenyl; para-Terphenyl; p-Triphenyl

**Exposure Limits:**
**NIOSH REL:** C 5 $mg/m^3$ (0.5 ppm)
**OSHA PEL†:** C 9 $mg/m^3$ (1 ppm)

**Measurement Methods (see Table 1):**
**NIOSH** 5021

**Physical Description:** White or light-yellow solid.

**Chemical & Physical Properties:**
**MW:** 230.3
**BP:** 761°F
**Sol:** Insoluble
**Fl.P:** 405°F
**IP:** 7.78
**Sp.Gr:** 1.23
**VP:** Very low
**MLT:** 415°F
**UEL:** ?
**LEL:** ?
Combustible Solid

**Personal Protection/Sanitation (see Table 2):**
**Skin:** Prevent skin contact
**Eyes:** Prevent eye contact
**Wash skin:** When contam
**Remove:** When wet or contam
**Change:** Daily
**Provide:** Eyewash
Quick drench

**Respirator Recommendations (see Tables 3 and 4):**
**NIOSH**
**25 $mg/m^3$:** Qm£
**50 $mg/m^3$:** 95XQ£/Sa£
**125 $mg/m^3$:** Sa:Cf£/PaprHie£
**250 $mg/m^3$:** 100F/ScbaF/SaF
**500 $mg/m^3$:** SaF:Pd,Pp
**§:** ScbaF:Pd,Pp/SaF:Pd,Pp:AScba
**Escape:** 100F/ScbaE

**Incompatibilities and Reactivities:** None reported

**Exposure Routes, Symptoms, Target Organs (see Table 5):**
**ER:** Inh, Ing, Con
**SY:** Irrit eyes, skin, muc memb; thermal skin burns; head; sore throat; in animals: liver, kidney damage
**TO:** Eyes, skin, resp sys, liver, kidneys

**First Aid (see Table 6):**
**Eye:** Irr immed
**Skin:** Water flush immed
**Breath:** Resp support
**Swallow:** Medical attention immed

## 2,3,7,8-Tetrachloro-dibenzo-p-dioxin

| 2,3,7,8-Tetrachloro-dibenzo-p-dioxin | Formula: $C_{12}H_4Cl_4O_2$ | CAS#: 1746-01-6 | RTECS#: HP3500000 | IDLH: Ca [N.D.] |
|---|---|---|---|---|
| **Conversion:** | **DOT:** | | | |

**Synonyms/Trade Names:** Dioxin; Dioxine; TCDBD; TCDD; 2,3,7,8-TCDD
[**Note:** Formed during past production of 2,4,5-trichlorophenol, 2,4,5-T & 2(2,4,5-trichlorophenoxy)propionic acid.]

**Exposure Limits:**
**NIOSH REL:** Ca
See Appendix A
**OSHA PEL:** none

**Measurement Methods (see Table 1):**
None available

**Physical Description:** Colorless to white, crystalline solid.
[**Note:** Exposure may occur through contact at previously contaminated worksites.]

T

**Chemical & Physical Properties:**
**MW:** 322.0
**BP:** Decomposes
**Sol:** 0.00000002%
**Fl.P:** ?
**IP:** ?
**Sp.Gr:** ?
**VP(77°F):** 0.000002 mmHg
**MLT:** 581°F
**UEL:** ?
**LEL:** ?

**Personal Protection/Sanitation (see Table 2):**
**Skin:** Prevent skin contact
**Eyes:** Prevent eye contact
**Wash skin:** When contam/Daily
**Remove:** When wet or contam
**Change:** Daily
**Provide:** Eyewash
Quick drench

**Respirator Recommendations (see Tables 3 and 4):**
**NIOSH**
**¥:** ScbaF:Pd,Pp/SaF:Pd,Pp:AScba
**Escape:** GmFOv100/ScbaE

**Incompatibilities and Reactivities:** UV light (decomposes)

**Exposure Routes, Symptoms, Target Organs (see Table 5):**
**ER:** Inh, Abs, Ing, Con
**SY:** Irrit eyes; allergic derm, chloracne; porphyria; GI dist; possible repro, terato effects; in animals: liver, kidney damage; hemorr; [carc]
**TO:** Eyes, skin, liver, kidneys, repro sys [in animals: tumors at many sites]

**First Aid (see Table 6):**
**Eye:** Irr immed
**Skin:** Soap flush immed
**Breath:** Resp support
**Swallow:** Medical attention immed

| 1,1,1,2-Tetrachloro-2,2-difluoroethane | Formula: $CCl_3CClF_2$ | CAS#: 76-11-9 | RTECS#: KI1425000 | IDLH: 2000 ppm |
|---|---|---|---|---|
| **Conversion:** 1 ppm = 8.34 mg/$m^3$ | **DOT:** | | | |

**Synonyms/Trade Names:** 2,2-Difluoro-1,1,1,2-tetrachloroethane; Freon® 112a; Halocarbon 112a; Refrigerant 112a

**Exposure Limits:**
**NIOSH REL:** TWA 500 ppm (4170 mg/$m^3$)
**OSHA PEL:** TWA 500 ppm (4170 mg/$m^3$)

**Measurement Methods (see Table 1):**
**NIOSH** 1016
**OSHA** 7

**Physical Description:** Colorless solid with a slight, ether-like odor.
[**Note:** A liquid above 105°F.]

| Chemical & Physical Properties: | Personal Protection/Sanitation (see Table 2): | Respirator Recommendations (see Tables 3 and 4): |
|---|---|---|
| **MW:** 203.8<br>**BP:** 197°F<br>**Sol:** 0.01%<br>**Fl.P:** NA<br>**IP:** ?<br>**Sp.Gr:** 1.65<br>**VP:** 40 mmHg<br>**MLT:** 105°F<br>**UEL:** NA<br>**LEL:** NA<br>Noncombustible Solid | **Skin:** Prevent skin contact<br>**Eyes:** Prevent eye contact<br>**Wash skin:** When contam<br>**Remove:** When wet or contam<br>**Change:** N.R. | **NIOSH/OSHA**<br>**2000 ppm:** Sa/ScbaF<br>**§:** ScbaF:Pd,Pp/SaF:Pd,Pp:AScba<br>**Escape:** GmFOv/ScbaE |

**Incompatibilities and Reactivities:** Chemically-active metals such as potassium, beryllium, powdered aluminum, zinc, calcium, magnesium & sodium; acids

| Exposure Routes, Symptoms, Target Organs (see Table 5): | First Aid (see Table 6): |
|---|---|
| **ER:** Inh, Ing, Con<br>**SY:** Irrit eyes, skin; CNS depres; pulm edema; drow; dysp<br>**TO:** Eyes, skin, resp sys, CNS | **Eye:** Irr immed<br>**Skin:** Soap wash prompt<br>**Breath:** Resp support<br>**Swallow:** Medical attention immed |

| 1,1,2,2-Tetrachloro-1,2-difluoroethane | Formula: $CCl_2FCCl_2F$ | CAS#: 76-12-0 | RTECS#: KI1420000 | IDLH: 2000 ppm |
|---|---|---|---|---|
| **Conversion:** 1 ppm = 8.34 mg/$m^3$ | **DOT:** | | | |

**Synonyms/Trade Names:** 1,2-Difluoro-1,1,2,2-tetrachloroethane; Freon® 112; Halocarbon 112; Refrigerant 112

**Exposure Limits:**
**NIOSH REL:** TWA 500 ppm (4170 mg/$m^3$)
**OSHA PEL:** TWA 500 ppm (4170 mg/$m^3$)

**Measurement Methods (see Table 1):**
**NIOSH** 1016
**OSHA** 7

**Physical Description:** Colorless solid or liquid (above 77°F) with a slight, ether-like odor.

| Chemical & Physical Properties: | Personal Protection/Sanitation (see Table 2): | Respirator Recommendations (see Tables 3 and 4): |
|---|---|---|
| **MW:** 203.8<br>**BP:** 199°F<br>**Sol(77°F):** 0.01%<br>**Fl.P:** NA<br>**IP:** 11.30 eV<br>**Sp.Gr:** 1.65<br>**VP:** 40 mmHg<br>**MLT:** 77°F<br>**UEL:** NA<br>**LEL:** NA<br>Noncombustible Solid | **Skin:** Prevent skin contact<br>**Eyes:** Prevent eye contact<br>**Wash skin:** When contam<br>**Remove:** When wet or contam<br>**Change:** N.R. | **NIOSH/OSHA**<br>**2000 ppm:** Sa/ScbaF<br>**§:** ScbaF:Pd,Pp/SaF:Pd,Pp:AScba<br>**Escape:** GmFOv/ScbaE |

**Incompatibilities and Reactivities:** Chemically-active metals such as potassium, beryllium, powdered aluminum, zinc, magnesium, calcium & sodium; acids

| Exposure Routes, Symptoms, Target Organs (see Table 5): | First Aid (see Table 6): |
|---|---|
| **ER:** Inh, Ing, Con<br>**SY:** In animals: irrit eyes, skin; conj; pulm edema; narco<br>**TO:** Eyes, skin, resp sys, CNS | **Eye:** Irr immed<br>**Skin:** Soap wash prompt<br>**Breath:** Resp support<br>**Swallow:** Medical attention immed |

| 1,1,1,2-Tetrachloroethane | Formula: $CCl_3CH_2Cl$ | CAS#: 630-20-6 | RTECS#: KI8450000 | IDLH: N.D. |
|---|---|---|---|---|
| **Conversion:** | **DOT:** 1702 151 | | | |
| **Synonyms/Trade Names:** None | | | | |
| **Exposure Limits:**<br>**NIOSH REL:** Handle with caution in the workplace.<br>See Appendix C (Chloroethanes)<br>**OSHA PEL:** none | | | **Measurement Methods (see Table 1):**<br>None available | |
| **Physical Description:** Yellowish-red liquid. | | | | |
| **Chemical & Physical Properties:**<br>**MW:** 167.9<br>**BP:** 267°F<br>**Sol:** 0.1%<br>**Fl.P:** ?<br>**IP:** ?<br>**Sp.Gr:** 1.54<br>**VP(77°F):** 14 mmHg<br>**FRZ:** -94°F<br>**UEL:** ?<br>**LEL:** ? | **Personal Protection/Sanitation (see Table 2):**<br>**Skin:** Prevent skin contact<br>**Eyes:** Prevent eye contact<br>**Wash skin:** When contam<br>**Remove:** When wet or contam<br>**Change:** N.R.<br>**Provide:** Eyewash<br>Quick drench | | **Respirator Recommendations (see Tables 3 and 4):**<br>Not available. | |
| **Incompatibilities and Reactivities:** Potassium; sodium; dinitrogen tetraoxide; potassium hydroxide; nitrogen tetroxide; sodium potassium alloy; 2,4-dinitrophenyl disulfide | | | | |
| **Exposure Routes, Symptoms, Target Organs (see Table 5):**<br>**ER:** Inh, Ing, Con<br>**SY:** Irrit eyes, skin; lass, restless, irreg respiration, musc inco; in animals: liver changes<br>**TO:** Eyes, skin, CNS, liver | | **First Aid (see Table 6):**<br>**Eye:** Irr immed<br>**Skin:** Soap wash immed<br>**Breath:** Resp support<br>**Swallow:** Medical attention immed | | |

| 1,1,2,2-Tetrachloroethane | Formula: $CHCl_2CHCl_2$ | CAS#: 79-34-5 | RTECS#: KI8575000 | IDLH: Ca [100 ppm] |
|---|---|---|---|---|
| **Conversion:** 1 ppm = 6.87 $mg/m^3$ | **DOT:** 1702 151 | | | |
| **Synonyms/Trade Names:** Acetylene tetrachloride, Symmetrical tetrachloroethane | | | | |
| **Exposure Limits:**<br>**NIOSH REL:** Ca<br>TWA 1 ppm (7 $mg/m^3$) [skin]<br>See Appendix A<br>See Appendix C (Chloroethanes)<br>**OSHA PEL†:** TWA 5 ppm (35 $mg/m^3$) [skin] | | | **Measurement Methods (see Table 1):**<br>**NIOSH** 1019, 2562<br>**OSHA** 7 | |
| **Physical Description:** Colorless to pale-yellow liquid with a pungent, chloroform-like odor. | | | | |
| **Chemical & Physical Properties:**<br>**MW:** 167.9<br>**BP:** 296°F<br>**Sol:** 0.3%<br>**Fl.P:** NA<br>**IP:** 11.10 eV<br>**Sp.Gr(77°F):** 1.59<br>**VP:** 5 mmHg<br>**FRZ:** -33°F<br>**UEL:** NA<br>**LEL:** NA<br>Noncombustible Liquid | **Personal Protection/Sanitation (see Table 2):**<br>**Skin:** Prevent skin contact<br>**Eyes:** Prevent eye contact<br>**Wash skin:** When contam<br>**Remove:** When wet or contam<br>**Change:** N.R.<br>**Provide:** Eyewash<br>Quick drench | | **Respirator Recommendations (see Tables 3 and 4):**<br>**NIOSH**<br>**¥:** ScbaF:Pd,Pp/SaF:Pd,Pp:AScba<br>**Escape:** GmFOv/ScbaE | |
| **Incompatibilities and Reactivities:** Chemically-active metals, strong caustics, fuming sulfuric acid [**Note:** Degrades slowly when exposed to air.] | | | | |
| **Exposure Routes, Symptoms, Target Organs (see Table 5):**<br>**ER:** Inh, Abs, Ing, Con<br>**SY:** Nau, vomit, abdom pain; tremor fingers; jaun, hepatitis, liver tend; derm; leucyt; kidney damage; [carc]<br>**TO:** Skin, liver, kidneys, CNS, GI tract [in animals: liver tumors] | | **First Aid (see Table 6):**<br>**Eye:** Irr immed<br>**Skin:** Soap wash prompt<br>**Breath:** Resp support<br>**Swallow:** Medical attention immed | | |

## Tetrachloroethylene

| Tetrachloroethylene | Formula: $Cl_2C{=}CCl_2$ | CAS#: 127-18-4 | RTECS#: KX3850000 | IDLH: Ca [150 ppm] |
|---|---|---|---|---|
| **Conversion:** 1 ppm = 6.78 mg/m$^3$ | **DOT:** 1897 160 | | | |

**Synonyms/Trade Names:** Perchlorethylene, Perchloroethylene, Perk, Tetrachlorethylene

**Exposure Limits:**
**NIOSH REL:** Ca
Minimize workplace exposure concentrations.
See Appendix A
**OSHA PEL†:** TWA 100 ppm
C 200 ppm (for 5 mins. in any 3-hr. period), with a maximum peak of 300 ppm

**Measurement Methods (see Table 1):**
**NIOSH** 1003
**OSHA** 1001

**Physical Description:** Colorless liquid with a mild, chloroform-like odor.

**Chemical & Physical Properties:**
**MW:** 165.8
**BP:** 250°F
**Sol:** 0.02%
**Fl.P:** NA
**IP:** 9.32 eV
**Sp.Gr:** 1.62
**VP:** 14 mmHg
**FRZ:** -2°F
**UEL:** NA
**LEL:** NA
Noncombustible Liquid, but decomposes in a fire to hydrogen chloride and phosgene.

**Personal Protection/Sanitation (see Table 2):**
**Skin:** Prevent skin contact
**Eyes:** Prevent eye contact
**Wash skin:** When contam
**Remove:** When wet or contam
**Change:** N.R.
**Provide:** Eyewash
Quick drench

**Respirator Recommendations (see Tables 3 and 4):**
**NIOSH**
**¥:** ScbaF:Pd,Pp/SaF:Pd,Pp:AScba
**Escape:** GmFOv/ScbaE

**Incompatibilities and Reactivities:** Strong oxidizers; chemically-active metals such as lithium, beryllium & barium; caustic soda; sodium hydroxide; potash

**Exposure Routes, Symptoms, Target Organs (see Table 5):**
**ER:** Inh, Abs, Ing, Con
**SY:** Irrit eyes, skin, nose, throat, resp sys; nau; flush face, neck; dizz, inco; head, drow; skin eryt; liver damage; [carc]
**TO:** Eyes, skin, resp sys, liver, kidneys, CNS [in animals: liver tumors]

**First Aid (see Table 6):**
**Eye:** Irr immed
**Skin:** Soap wash prompt
**Breath:** Resp support
**Swallow:** Medical attention immed

## Tetrachloronaphthalene

| Tetrachloronaphthalene | Formula: $C_{10}H_4Cl_4$ | CAS#: 1335-88-2 | RTECS#: QK3700000 | IDLH: See Appendix F |
|---|---|---|---|---|
| **Conversion:** | **DOT:** | | | |

**Synonyms/Trade Names:** Halowax®, Nibren wax, Seekay wax

**Exposure Limits:**
**NIOSH REL:** TWA 2 mg/m$^3$ [skin]
**OSHA PEL:** TWA 2 mg/m$^3$ [skin]

**Measurement Methods (see Table 1):**
**NIOSH** S130 (II-2)

**Physical Description:** Colorless to pale-yellow solid with an aromatic odor.

**Chemical & Physical Properties:**
**MW:** 265.9
**BP:** 599-680°F
**Sol:** Insoluble
**Fl.P(oc):** 410°F
**IP:** ?
**Sp.Gr:** 1.59-1.65
**VP:** <1 mmHg
**MLT:** 360°F
**UEL:** ?
**LEL:** ?
Combustible Solid

**Personal Protection/Sanitation (see Table 2):**
**Skin:** Prevent skin contact
**Eyes:** Prevent eye contact
**Wash skin:** When contam
**Remove:** When wet or contam
**Change:** Daily

**Respirator Recommendations (see Tables 3 and 4):**
**NIOSH/OSHA**
**20 mg/m$^3$:** ScbaF/SaF
**§:** ScbaF:Pd,Pp/SaF:Pd,Pp:AScba
**Escape:** GmFOv100/ScbaE

**See Appendix F**

**Incompatibilities and Reactivities:** Strong oxidizers

**Exposure Routes, Symptoms, Target Organs (see Table 5):**
**ER:** Inh, Abs, Ing, Con
**SY:** Acne-form derm; head, lass, anor, dizz; jaun, liver inj
**TO:** Liver, skin, CNS

**First Aid (see Table 6):**
**Eye:** Irr immed
**Skin:** Soap wash immed
**Breath:** Resp support
**Swallow:** Medical attention immed

## Tetraethyl lead (as Pb)

| Formula: | CAS#: | RTECS#: | IDLH: |
|---|---|---|---|
| $Pb(C_2H_5)_4$ | 78-00-2 | TP4550000 | 40 mg/m$^3$ (as Pb) |

**Conversion:**

**DOT:** 1649 131

**Synonyms/Trade Names:** Lead tetraethyl, TEL, Tetraethylplumbane

**Exposure Limits:**
**NIOSH REL:** TWA 0.075 mg/m$^3$ [skin]
**OSHA PEL:** TWA 0.075 mg/m$^3$ [skin]

**Measurement Methods (see Table 1):**
**NIOSH** 2533

**Physical Description:** Colorless liquid (unless dyed red, orange, or blue) with a pleasant, sweet odor.
[**Note:** Main usage is in anti-knock additives for gasoline.]

**Chemical & Physical Properties:**
**MW:** 323.5
**BP:** 228°F (Decomposes)
**Sol:** 0.00002%
**Fl.P:** 200°F
**IP:** 11.10 eV
**Sp.Gr:** 1.65
**VP:** 0.2 mmHg
**FRZ:** -202°F
**UEL:** ?
**LEL:** 1.8%
Class IIIB Combustible Liquid

**Personal Protection/Sanitation (see Table 2):**
**Skin:** Prevent skin contact (>0.1%)
**Eyes:** Prevent eye contact
**Wash skin:** When contam (>0.1%)
**Remove:** When wet or contam (>0.1%)
**Change:** Daily
**Provide:** Quick drench (>0.1%)

**Respirator Recommendations (see Tables 3 and 4):**
**NIOSH/OSHA**
**0.75 mg/m$^3$:** Sa
**1.875 mg/m$^3$:** Sa:Cf
**3.75 mg/m$^3$:** SaT:Cf/ScbaF/SaF
**40 mg/m$^3$:** Sa:Pd,Pp
**§:** ScbaF:Pd,Pp/SaF:Pd,Pp:AScba
**Escape:** GmFOv/ScbaE

**Incompatibilities and Reactivities:** Strong oxidizers, sulfuryl chloride, rust, potassium permanganate
[**Note:** Decomposes slowly at room temperature and more rapidly at higher temperatures.]

**Exposure Routes, Symptoms, Target Organs (see Table 5):**
**ER:** Inh, Abs, Ing, Con
**SY:** Insom, lass, anxiety; tremor, hyper-reflexia, spasticity; bradycardia, hypotension, hypothermia, pallor, nau, anor, low-wgt; conf, halu, psychosis, mania, convuls, coma; eye irrit
**TO:** CNS, CVS, kidneys, eyes

**First Aid (see Table 6):**
**Eye:** Irr immed
**Skin:** Soap wash immed
**Breath:** Resp support
**Swallow:** Medical attention immed

## Tetrahydrofuran

| Formula: | CAS#: | RTECS#: | IDLH: |
|---|---|---|---|
| $C_4H_8O$ | 109-99-9 | LU5950000 | 2000 ppm [10%LEL] |

**Conversion:** 1 ppm = 2.95 mg/m$^3$

**DOT:** 2056 127

**Synonyms/Trade Names:** Diethylene oxide; 1,4-Epoxybutane; Tetramethylene oxide; THF

**Exposure Limits:**
**NIOSH REL:** TWA 200 ppm (590 mg/m$^3$)
ST 250 ppm (735 mg/m$^3$)
**OSHA PEL†:** TWA 200 ppm (590 mg/m$^3$)

**Measurement Methods (see Table 1):**
**NIOSH** 1609, 3800
**OSHA** 7



**Physical Description:** Colorless liquid with an ether-like odor.

**Chemical & Physical Properties:**
**MW:** 72.1
**BP:** 151°F
**Sol:** Miscible
**Fl.P:** 6°F
**IP:** 9.45 eV
**Sp.Gr:** 0.89
**VP:** 132 mmHg
**FRZ:** -163°F
**UEL:** 11.8%
**LEL:** 2%
Class IB Flammable Liquid

**Personal Protection/Sanitation (see Table 2):**
**Skin:** Prevent skin contact
**Eyes:** Prevent eye contact
**Wash skin:** When contam
**Remove:** When wet (flamm)
**Change:** N.R.

**Respirator Recommendations (see Tables 3 and 4):**
**NIOSH/OSHA**
**2000 ppm:** Sa:Cf£/CcrFOv/GmFOv/PaprOv£/ScbaF/SaF
**§:** ScbaF:Pd,Pp/SaF:Pd,Pp:AScba
**Escape:** GmFOv/ScbaE

**Incompatibilities and Reactivities:** Strong oxidizers, lithium-aluminum alloys
[**Note:** Peroxides may accumulate upon prolonged storage in presence of air.]

**Exposure Routes, Symptoms, Target Organs (see Table 5):**
**ER:** Inh, Con, Ing
**SY:** Irrit eyes, upper resp sys; nau, dizz, head, CNS depres
**TO:** Eyes, resp sys, CNS

**First Aid (see Table 6):**
**Eye:** Irr immed
**Skin:** Water flush prompt
**Breath:** Resp support
**Swallow:** Medical attention immed

## Tetramethyl lead (as Pb)

| Tetramethyl lead (as Pb) | Formula: $Pb(CH_3)_4$ | CAS#: 75-74-1 | RTECS#: TP4725000 | IDLH: 40 mg/m³ (as Pb) |
|---|---|---|---|---|
| **Conversion:** | **DOT:** | | | |

**Synonyms/Trade Names:** Lead tetramethyl, Tetramethylplumbane, TML

**Exposure Limits:**
**NIOSH REL:** TWA 0.075 mg/m³ [skin]
**OSHA PEL:** TWA 0.075 mg/m³ [skin]

**Measurement Methods (see Table 1):**
NIOSH 2534

**Physical Description:** Colorless liquid (unless dyed red, orange, or blue) with a fruity odor.
[**Note:** Main usage is in anti-knock additives for gasoline.]

| Chemical & Physical Properties: | Personal Protection/Sanitation (see Table 2): | Respirator Recommendations (see Tables 3 and 4): |
|---|---|---|
| **MW:** 267.3<br>**BP:** 212°F (Decomposes)<br>**Sol:** 0.002%<br>**Fl.P:** 100°F<br>**IP:** 8.50 eV<br>**Sp.Gr:** 2.00<br>**VP:** 23 mmHg<br>**FRZ:** -15°F<br>**UEL:** ?<br>**LEL:** ?<br>Class II Combustible Liquid | **Skin:** Prevent skin contact (>0.1%)<br>**Eyes:** Prevent eye contact<br>**Wash skin:** When contam (>0.1%)<br>**Remove:** When wet or contam (>0.1%)<br>**Change:** Daily<br>**Provide:** Quick drench (>0.1%) | **NIOSH/OSHA**<br>**0.75 mg/m³:** Sa<br>**1.875 mg/m³:** Sa:Cf<br>**3.75 mg/m³:** SaT:Cf/ScbaF/SaF<br>**40 mg/m³:** Sa:Pd,Pp<br>**§:** ScbaF:Pd,Pp/SaF:Pd,Pp:AScba<br>**Escape:** GmFOv/ScbaE |

**Incompatibilities and Reactivities:** Strong oxidizers such as sulfuryl chloride or potassium permanganate

**Exposure Routes, Symptoms, Target Organs (see Table 5):**
**ER:** Inh, Abs, Ing, Con
**SY:** Insom, bad dreams, restless, anxious; hypotension; nau, anor; delirium, mania, convuls; coma
**TO:** CNS, CVS, kidneys

**First Aid (see Table 6):**
**Eye:** Irr immed
**Skin:** Soap wash immed
**Breath:** Resp support
**Swallow:** Medical attention immed

## Tetramethyl succinonitrile

| Tetramethyl succinonitrile | Formula: $(CH_3)_2C(CN)C(CN)(CH_3)_2$ | CAS#: 3333-52-6 | RTECS#: WN4025000 | IDLH: 5 ppm |
|---|---|---|---|---|
| **Conversion:** 1 ppm = 5.57 mg/m³ | **DOT:** | | | |

**Synonyms/Trade Names:** Tetramethyl succinodinitrile, TMSN

**Exposure Limits:**
**NIOSH REL:** TWA 3 mg/m³ (0.5 ppm) [skin]
**OSHA PEL:** TWA 3 mg/m³ (0.5 ppm) [skin]

**Measurement Methods (see Table 1):**
NIOSH S155 (II-3)
OSHA 7

**Physical Description:** Colorless, odorless solid. [**Note:** Forms cyanide in the body.]

| Chemical & Physical Properties: | Personal Protection/Sanitation (see Table 2): | Respirator Recommendations (see Tables 3 and 4): |
|---|---|---|
| **MW:** 136.2<br>**BP:** Sublimes<br>**Sol:** Insoluble<br>**Fl.P:** ?<br>**IP:** ?<br>**Sp.Gr:** 1.07<br>**VP:** ?<br>**MLT:** 338°F (Sublimes)<br>**UEL:** ?<br>**LEL:** ?<br>Combustible Solid | **Skin:** Prevent skin contact<br>**Eyes:** Prevent eye contact<br>**Wash skin:** When contam<br>**Remove:** When wet or contam<br>**Change:** Daily | **NIOSH/OSHA**<br>**28 mg/m³:** Sa/ScbaF<br>**§:** ScbaF:Pd,Pp/SaF:Pd,Pp:AScba<br>**Escape:** GmFOv100/ScbaE |

**Incompatibilities and Reactivities:** Strong oxidizers

**Exposure Routes, Symptoms, Target Organs (see Table 5):**
**ER:** Inh, Abs, Ing, Con
**SY:** Head, nau; convuls, coma; liver, kidney, GI effects
**TO:** CNS, liver, kidneys, GI tract

**First Aid (see Table 6):**
**Eye:** Irr immed
**Skin:** Soap wash prompt
**Breath:** Resp support
**Swallow:** Medical attention immed

## Tetranitromethane

| Formula: | CAS#: | RTECS#: | IDLH: |
|---|---|---|---|
| $C(NO_2)_4$ | 509-14-8 | PB4025000 | 4 ppm |

**Conversion:** 1 ppm = 8.02 mg/m$^3$ | **DOT:** 1510 143

**Synonyms/Trade Names:** Tetan, TNM

**Exposure Limits:**
**NIOSH REL:** TWA 1 ppm (8 mg/m$^3$)
**OSHA PEL:** TWA 1 ppm (8 mg/m$^3$)

**Measurement Methods (see Table 1):**
**NIOSH** 3513

**Physical Description:** Colorless to pale-yellow liquid or solid (below 57°F) with a pungent odor.

**Chemical & Physical Properties:**
**MW:** 196.0
**BP:** 259°F
**Sol:** Insoluble
**Fl.P:** ?
**IP:** ?
**Sp.Gr:** 1.62
**VP:** 8 mmHg
**FRZ:** 57°F
**UEL:** ?
**LEL:** ?
Combustible Liquid, but difficult to ignite.

**Personal Protection/Sanitation (see Table 2):**
**Skin:** Prevent skin contact
**Eyes:** Prevent eye contact
**Wash skin:** When contam
**Remove:** When wet (flamm)
**Change:** Daily
**Provide:** Eyewash

**Respirator Recommendations (see Tables 3 and 4):**
**NIOSH/OSHA**
**4 ppm:** Sa:Cf£/CcrFS¿/GmFS¿/PaprS¿£/ScbaF/SaF
**§:** ScbaF:Pd,Pp/SaF:Pd,Pp:AScba
**Escape:** GmFS¿/ScbaE

**Incompatibilities and Reactivities:** Hydrocarbons, alkalis, metals, oxidizers, aluminum, toluene, cotton [**Note:** Combustible material wet with tetranitromethane may be highly explosive.]

**Exposure Routes, Symptoms, Target Organs (see Table 5):**
**ER:** Inh, Ing, Con
**SY:** Irrit eyes, skin, nose, throat; dizz, head; chest pain, dysp; methemo, cyan; skin burns
**TO:** Eyes, skin, resp sys, blood, CNS

**First Aid (see Table 6):**
**Eye:** Irr immed
**Skin:** Soap wash prompt
**Breath:** Resp support
**Swallow:** Medical attention immed

## Tetrasodium pyrophosphate

| Formula: | CAS#: | RTECS#: | IDLH: |
|---|---|---|---|
| $Na_4P_2O_7$ | 7722-88-5 | UX7350000 | N.D. |

**Conversion:** | **DOT:**

**Synonyms/Trade Names:** Pyrophosphate, Sodium pyrophosphate, Tetrasodium diphosphate, Tetrasodium pyrophosphate (anhydrous), TSPP

**Exposure Limits:**
**NIOSH REL:** TWA 5 mg/m$^3$
**OSHA PEL†:** none

**Measurement Methods (see Table 1):**
**NIOSH** 0500



**Physical Description:** Odorless, white powder or granules.
[**Note:** The decahydrate ($Na_4P_2O_7 \times 10H_2O$) is in the form of colorless, transparent crystals.]

**Chemical & Physical Properties:**
**MW:** 265.9
**BP:** Decomposes
**Sol(77°F):** 7%
**Fl.P:** NA
**IP:** NA
**Sp.Gr:** 2.45
**VP:** 0 mmHg (approx)
**MLT:** 1810°F
**UEL:** NA
**LEL:** NA
Noncombustible Solid

**Personal Protection/Sanitation (see Table 2):**
**Skin:** Prevent skin contact
**Eyes:** Prevent eye contact
**Wash skin:** When contam
**Remove:** When wet or contam
**Change:** Daily
**Provide:** Eyewash (solution)

**Respirator Recommendations (see Tables 3 and 4):**
Not available.

**Incompatibilities and Reactivities:** Strong acids

**Exposure Routes, Symptoms, Target Organs (see Table 5):**
**ER:** Inh, Ing, Con
**SY:** Irrit eyes, skin, nose, throat; derm
**TO:** Eyes, skin, resp sys

**First Aid (see Table 6):**
**Eye:** Irr immed
**Skin:** Water wash prompt
**Breath:** Resp support
**Swallow:** Medical attention immed

## Tetryl

| Tetryl | Formula: $(NO_2)_3C_6H_2N(NO_2)CH_3$ | CAS#: 479-45-8 | RTECS#: BY6300000 | IDLH: 750 mg/m$^3$ |
|---|---|---|---|---|
| Conversion: | DOT: | | | |

**Synonyms/Trade Names:** N-Methyl-N,2,4,6-tetranitroaniline; Nitramine; 2,4,6-Tetryl; 2,4,6-Trinitrophenyl-N-methylnitramine

**Exposure Limits:**
**NIOSH REL:** TWA 1.5 mg/m$^3$ [skin]
**OSHA PEL:** TWA 1.5 mg/m$^3$ [skin]

**Measurement Methods (see Table 1):**
**NIOSH** S225 (II-3)

**Physical Description:** Colorless to yellow, odorless, crystalline solid.

**Chemical & Physical Properties:**
**MW:** 287.2
**BP:** 356-374°F (Explodes)
**Sol:** 0.02%
**Fl.P:** Explodes
**IP:** ?
**Sp.Gr:** 1.57
**VP:** <1 mmHg
**MLT:** 268°F
**UEL:** ?
**LEL:** ?
Combustible Solid
(Class A Explosive)

**Personal Protection/Sanitation (see Table 2):**
**Skin:** Prevent skin contact
**Eyes:** Prevent eye contact
**Wash skin:** When contam/Daily
**Remove:** When wet or contam
**Change:** Daily

**Respirator Recommendations (see Tables 3 and 4):**
**NIOSH/OSHA**
**7.5 mg/m$^3$:** Qm
**15 mg/m$^3$:** 95XQ*/Sa*
**37.5 mg/m$^3$:** Sa:Cf*/PaprHie*
**75 mg/m$^3$:** 100F/ScbaF/SaF
**750 mg/m$^3$:** SaF:Pd,Pp
**§:** ScbaF:Pd,Pp/SaF:Pd,Pp:AScba
**Escape:** 100F/ScbaE

**Incompatibilities and Reactivities:** Oxidizable materials, hydrazine

**Exposure Routes, Symptoms, Target Organs (see Table 5):**
**ER:** Inh, Abs, Ing, Con
**SY:** Sens derm, itch, eryt; edema on nasal folds, cheeks, neck; kera; sneez; anemia; cough, coryza; irrity; mal, head, lass, insom; nau, vomit; liver, kidney damage
**TO:** Eyes, skin, resp sys, CNS, liver, kidneys

**First Aid (see Table 6):**
**Eye:** Irr immed
**Skin:** Soap wash prompt
**Breath:** Resp support
**Swallow:** Medical attention immed

## Thallium (soluble compounds, as Tl)

| Thallium (soluble compounds, as Tl) | Formula: | CAS#: | RTECS#: | IDLH: 15 mg/m$^3$ (as Tl) |
|---|---|---|---|---|
| Conversion: | DOT: 1707 151 (compounds, n.o.s.) | | | |

**Synonyms/Trade Names:** Synonyms vary depending upon the specific soluble thallium compound.

**Exposure Limits:**
**NIOSH REL:** TWA 0.1 mg/m$^3$ [skin]
**OSHA PEL:** TWA 0.1 mg/m$^3$ [skin]

**Measurement Methods (see Table 1):**
**NIOSH** 7300, 7301, 7303, 9102
**OSHA** ID121

**Physical Description:** Appearance and odor vary depending upon the specific soluble thallium compound.

**Chemical & Physical Properties:**
Properties vary depending upon the specific soluble thallium compound.

**Personal Protection/Sanitation (see Table 2):**
**Skin:** Prevent skin contact
**Eyes:** Prevent eye contact
**Wash skin:** When contam
**Remove:** When wet or contam
**Change:** Daily

**Respirator Recommendations (see Tables 3 and 4):**
**NIOSH/OSHA**
**0.5 mg/m$^3$:** Qm
**1 mg/m$^3$:** 95XQ/Sa
**2.5 mg/m$^3$:** Sa:Cf/PaprHie
**5 mg/m$^3$:** 100F/SaT:Cf/PaprTHie/ScbaF/SaF
**15 mg/m$^3$:** SaF:Pd,Pp
**§:** ScbaF:Pd,Pp/SaF:Pd,Pp:AScba
**Escape:** 100F/ScbaE

**Incompatibilities and Reactivities:** Varies

**Exposure Routes, Symptoms, Target Organs (see Table 5):**
**ER:** Inh, Abs, Ing, Con
**SY:** Nau, diarr, abdom pain, vomit; ptosis, strabismus; peri neuritis, tremor; retster tight, chest pain, pulm edema; convuls, chorea, psychosis; liver, kidney damage; alopecia; pares legs
**TO:** Eyes, resp sys, CNS, liver, kidneys, GI tract, body hair

**First Aid (see Table 6):**
**Eye:** Irr immed
**Skin:** Water flush prompt
**Breath:** Resp support
**Swallow:** Medical attention immed

## 4,4'-Thiobis(6-tert-butyl-m-cresol)

| Formula: | CAS#: | RTECS#: | IDLH: |
|---|---|---|---|
| $[CH_3(OH)C_6H_2C(CH_3)_3]_2S$ | 96-69-5 | GP3150000 | N.D. |

**Conversion:**

**DOT:**

**Synonyms/Trade Names:** 4,4'-Thiobis(3-methyl-6-tert-butylphenol); 1,1'-Thiobis(2-methyl-4-hydroxy-5-tert-butylbenzene)

**Exposure Limits:**
**NIOSH REL:** TWA 10 mg/m$^3$ (total)
TWA 5 mg/m$^3$ (resp)
**OSHA PEL†:** TWA 15 mg/m$^3$ (total)
TWA 5 mg/m$^3$ (resp)

**Measurement Methods (see Table 1):**
**NIOSH** 0500, 0600

**Physical Description:** Light-gray to tan powder with a slightly aromatic odor.

**Chemical & Physical Properties:**
**MW:** 358.6
**BP:** ?
**Sol:** 0.08%
**Fl.P:** 420°F
**IP:** ?
**Sp.Gr:** 1.10
**VP:** 0.0000006 mmHg
**MLT:** 302°F
**UEL:** NA
**LEL:** NA
Combustible Solid

**Personal Protection/Sanitation (see Table 2):**
**Skin:** N.R.
**Eyes:** N.R.
**Wash skin:** N.R.
**Remove:** N.R.
**Change:** N.R.

**Respirator Recommendations (see Tables 3 and 4):**
Not available.

**Incompatibilities and Reactivities:** None reported

**Exposure Routes, Symptoms, Target Organs (see Table 5):**
**ER:** Inh, Ing, Con
**SY:** Irrit eyes, skin, resp sys
**TO:** Eyes, skin, resp sys

**First Aid (see Table 6):**
**Eye:** Irr immed
**Breath:** Fresh air
**Swallow:** Medical attention immed

## Thioglycolic acid

| Formula: | CAS#: | RTECS#: | IDLH: |
|---|---|---|---|
| $HSCH_2COOH$ | 68-11-1 | AI5950000 | N.D. |

**Conversion:** 1 ppm = 3.77 mg/m$^3$

**DOT:** 1940 153

**Synonyms/Trade Names:** Acetyl mercaptan, Mercaptoacetate, Mercaptoacetic acid, 2-Mercaptoacetic acid, 2-Thioglycolic acid, Thiovanic acid

**Exposure Limits:**
**NIOSH REL:** TWA 1 ppm (4 mg/m$^3$) [skin]
**OSHA PEL†:** none

**Measurement Methods (see Table 1):**
None available



**Physical Description:** Colorless liquid with a strong, disagreeable odor characteristic of mercaptans. [**Note:** Olfactory fatigue may occur after short exposures.]

**Chemical & Physical Properties:**
**MW:** 92.1
**BP:** ?
**Sol:** Miscible
**Fl.P:** >230°F
**IP:** ?
**Sp.Gr:** 1.32
**VP(64°F):** 10 mmHg
**FRZ:** 2°F
**UEL:** ?
**LEL:** 5.9%
Class IIIB Combustible Liquid

**Personal Protection/Sanitation (see Table 2):**
**Skin:** Prevent skin contact
**Eyes:** Prevent eye contact
**Wash skin:** When contam
**Remove:** When wet or contam
**Change:** N.R.
**Provide:** Eyewash
Quick drench

**Respirator Recommendations (see Tables 3 and 4):**
Not available.

**Incompatibilities and Reactivities:** Air, strong oxidizers, bases, active metals (e.g., sodium potassium, magnesium, calcium) [**Note:** Readily oxidized by air.]

**Exposure Routes, Symptoms, Target Organs (see Table 5):**
**ER:** Inh, Abs, Ing, Con
**SY:** Irrit eyes, skin, nose, throat; lac, corn damage; skin burns, blisters; in animals: lass; gasping respirations; convuls
**TO:** Eyes, skin, resp sys

**First Aid (see Table 6):**
**Eye:** Irr immed
**Skin:** Water flush immed
**Breath:** Resp support
**Swallow:** Medical attention immed

## Thionyl chloride

**Formula:** $SOCl_2$ | **CAS#:** 7719-09-7 | **RTECS#:** XM5150000 | **IDLH:** N.D.

**Conversion:** 1 ppm = 4.87 mg/m$^3$ | **DOT:** 1836 137

**Synonyms/Trade Names:** Sulfinyl chloride, Sulfur chloride oxide, Sulfurous dichloride, Sulfurous oxychloride, Thionyl dichloride

**Exposure Limits:**
**NIOSH REL:** C 1 ppm (5 mg/m$^3$)
**OSHA PEL†:** none

**Measurement Methods (see Table 1):**
None available

**Physical Description:** Colorless to yellow to reddish liquid with a pungent odor like sulfur dioxide. [**Note:** Fumes form when exposed to moist air.]

**Chemical & Physical Properties:**
**MW:** 119.0
**BP:** 169°F
**Sol:** Reacts
**Fl.P:** NA
**IP:** ?
**Sp.Gr:** 1.64
**VP(70°F):** 100 mmHg
**FRZ:** -156°F
**UEL:** NA
**LEL:** NA
Noncombustible Liquid

**Personal Protection/Sanitation (see Table 2):**
**Skin:** Prevent skin contact
**Eyes:** Prevent eye contact
**Wash skin:** When contam
**Remove:** When wet or contam
**Change:** N.R.
**Provide:** Eyewash
Quick drench

**Respirator Recommendations (see Tables 3 and 4):**
Not available.

**Incompatibilities and Reactivities:** Water, acids, alkalis, ammonia, chloryl perchlorate
[**Note:** Reacts violently with water to form sulfur dioxide & hydrogen chloride.]

**Exposure Routes, Symptoms, Target Organs (see Table 5):**
**ER:** Inh, Ing, Con
**SY:** Irrit eyes, skin, muc memb; eye, skin burns
**TO:** Eyes, skin, resp sys

**First Aid (see Table 6):**
**Eye:** Irr immed
**Skin:** Water flush immed
**Breath:** Resp support
**Swallow:** Medical attention immed

## Thiram

**Formula:** $C_6H_{12}N_2S_4$ | **CAS#:** 137-26-8 | **RTECS#:** JO1400000 | **IDLH:** 100 mg/m$^3$

**Conversion:** | **DOT:** 2771 151

**Synonyms/Trade Names:** bis(Dimethylthiocarbamoyl) disulfide, Tetramethylthiuram disulfide

**Exposure Limits:**
**NIOSH REL:** TWA 5 mg/m$^3$
**OSHA PEL:** TWA 5 mg/m$^3$

**Measurement Methods (see Table 1):**
**NIOSH** 5005

**Physical Description:** Colorless to yellow, crystalline solid with a characteristic odor.
[**Note:** Commercial pesticide products may be dyed blue.]

**Chemical & Physical Properties:**
**MW:** 240.4
**BP:** Decomposes
**Sol:** 0.003%
**Fl.P:** ?
**IP:** ?
**Sp.Gr:** 1.29
**VP:** 0.000008 mmHg
**MLT:** 312°F
**UEL:** ?
**LEL:** ?
Combustible Solid

**Personal Protection/Sanitation (see Table 2):**
**Skin:** Prevent skin contact
**Eyes:** Prevent eye contact
**Wash skin:** When contam
**Remove:** When wet or contam
**Change:** Daily

**Respirator Recommendations (see Tables 3 and 4):**
**NIOSH/OSHA**
**50 mg/m$^3$:** CcrOv95*/Sa*
**100 mg/m$^3$:** Sa:Cf*/CcrFOv100/GmFOv100/PaprOvHie*/ScbaF/SaF
**§:** ScbaF:Pd,Pp/SaF:Pd,Pp:AScba
**Escape:** GmFOv100/ScbaE

**Incompatibilities and Reactivities:** Strong oxidizers, strong acids, oxidizable materials

**Exposure Routes, Symptoms, Target Organs (see Table 5):**
**ER:** Inh, Ing, Con
**SY:** Irrit eyes, skin, muc memb; derm; Antabuse-like effects
**TO:** Eyes, skin, resp sys, CNS

**First Aid (see Table 6):**
**Eye:** Irr immed
**Skin:** Soap wash prompt
**Breath:** Resp support
**Swallow:** Medical attention immed

## Tin

| Formula: | CAS#: | RTECS#: | IDLH: |
|---|---|---|---|
| Sn | 7440-31-5 | XP7320000 | 100 mg/m³ (as Sn) |

**Conversion:**

**DOT:**

**Synonyms/Trade Names:** Metallic tin, Tin flake, Tin metal, Tin powder

**Exposure Limits:**
**NIOSH REL*:** TWA 2 mg/m³
**OSHA PEL*:** TWA 2 mg/m³
[***Note:** The REL and PEL also apply to other inorganic tin compounds (as Sn) except tin oxides.]

**Measurement Methods (see Table 1):**
**NIOSH** 7300, 7301, 7303
**OSHA** ID121, ID206

**Physical Description:** Gray to almost silver-white, ductile, malleable, lustrous solid.

**Chemical & Physical Properties:**
**MW:** 118.7
**BP:** 4545°F
**Sol:** Insoluble
**Fl.P:** NA
**IP:** NA
**Sp.Gr:** 7.28
**VP:** 0 mmHg (approx)
**MLT:** 449°F
**UEL:** NA
**LEL:** NA
Noncombustible Solid, but powdered form may ignite.

**Personal Protection/Sanitation (see Table 2):**
**Skin:** N.R.
**Eyes:** N.R.
**Wash skin:** N.R.
**Remove:** N.R.
**Change:** N.R.

**Respirator Recommendations (see Tables 3 and 4):**
**NIOSH/OSHA**
**10 mg/m³:** Qm*
**20 mg/m³:** 95XQ*/Sa*
**50 mg/m³:** Sa:Cf*/PaprHie*
**100 mg/m³:** 100F/ScbaF/SaF
**§:** ScbaF:Pd,Pp/SaF:Pd,Pp:AScba
**Escape:** 100F/ScbaE

**Incompatibilities and Reactivities:** Chlorine, turpentine, acids, alkalis

**Exposure Routes, Symptoms, Target Organs (see Table 5):**
**ER:** Inh, Con
**SY:** Irrit eyes, skin, resp sys; in animals: vomit, diarr, para with musc twitch
**TO:** Eyes, skin, resp sys

**First Aid (see Table 6):**
**Eye:** Irr immed
**Skin:** Soap wash immed
**Breath:** Resp support
**Swallow:** Medical attention immed

## Tin (organic compounds, as Sn)

| Formula: | CAS#: | RTECS#: | IDLH: |
|---|---|---|---|
| | | | 25 mg/m³ (as Sn) |

**Conversion:**

**DOT:**

**Synonyms/Trade Names:** Synonyms vary depending upon the specific organic tin compound.
[**Note:** Also see specific listing for Cyhexatin.]

**Exposure Limits:**
**NIOSH REL*:** TWA 0.1 mg/m³ [skin]
[***Note:** The REL applies to all organic tin compounds except Cyhexatin.]
**OSHA PEL*:** TWA 0.1 mg/m³
[***Note:** The PEL applies to all organic tin compounds.]

**Measurement Methods (see Table 1):**
**NIOSH** 5504

**Physical Description:** Appearance and odor vary depending upon the specific organic tin compound.

**Chemical & Physical Properties:**
Properties vary depending upon the specific organic tin compound.

**Personal Protection/Sanitation (see Table 2):**
Recommendations regarding personal protective clothing vary depending upon the specific compound.

**Respirator Recommendations (see Tables 3 and 4):**
**NIOSH/OSHA**
**1 mg/m³:** CcrOv95/Sa
**2.5 mg/m³:** Sa:Cf/PaprOvHie
**5 mg/m³:** CcrFOv100/GmFOv100/PaprTOvHie/SaT:Cf/ScbaF/SaF
**25 mg/m³:** SaF:Pd,Pp
**§:** ScbaF:Pd,Pp/SaF:Pd,Pp:AScba
**Escape:** GmFOv100/ScbaE

**Incompatibilities and Reactivities:** Varies

**Exposure Routes, Symptoms, Target Organs (see Table 5):**
**ER:** Inh, Abs, Ing, Con
**SY:** Irrit eyes, skin, resp sys; head, dizz; psycho-neurologic dist; sore throat, cough; abdom pain, vomit; urine retention; paresis, focal anes; skin burns, pruritus; in animals: hemolysis; hepatic nec; kidney damage
**TO:** Eyes, skin, resp sys, CNS, liver, kidneys, urinary tract, blood

**First Aid (see Table 6):**
**Eye:** Irr immed
**Skin:** Water flush immed
**Breath:** Resp support
**Swallow:** Medical attention immed

| Tin(II) oxide (as Sn) | Formula: $SnO$ | CAS#: 21651-19-4 | RTECS#: XQ3700000 | IDLH: N.D. |
|---|---|---|---|---|
| **Conversion:** | **DOT:** | | | |

**Synonyms/Trade Names:** Stannous oxide, Tin protoxide [**Note:** Also see specific listing for Tin(IV) oxide (as Sn).]

**Exposure Limits:**
**NIOSH REL:** TWA 2 mg/m$^3$
**OSHA PEL†:** none

**Measurement Methods (see Table 1):**
**NIOSH** 7300, 7301, 7303

**Physical Description:** Brownish-black powder.

**Chemical & Physical Properties:**
**MW:** 134.7
**BP:** Decomposes
**Sol:** Insoluble
**Fl.P:** NA
**IP:** NA
**Sp.Gr:** 6.3
**VP:** 0 mmHg (approx)
**MLT(600 mmHg):** 1976°F (Decomposes)
**UEL:** NA
**LEL:** NA

**Personal Protection/Sanitation (see Table 2):**
**Skin:** N.R.
**Eyes:** N.R.
**Wash skin:** N.R.
**Remove:** N.R.
**Change:** N.R.

**Respirator Recommendations (see Tables 3 and 4):**
Not available.

**Incompatibilities and Reactivities:** None reported

**Exposure Routes, Symptoms, Target Organs (see Table 5):**
**ER:** Inh, Con
**SY:** Stannosis (benign pneumoconiosis): dysp, decr pulm func
**TO:** Resp sys

**First Aid (see Table 6):**
**Eye:** Irr immed
**Breath:** Fresh air

| Tin(IV) oxide (as Sn) | Formula: $SnO_2$ | CAS#: 18282-10-5 | RTECS#: XQ4000000 | IDLH: N.D. |
|---|---|---|---|---|
| **Conversion:** | **DOT:** | | | |

**Synonyms/Trade Names:** Stannic dioxide, Stannic oxide, White tin oxide
[**Note:** Also see specific listing for Tin(II) oxide (as Sn).]

**Exposure Limits:**
**NIOSH REL:** TWA 2 mg/m$^3$
**OSHA PEL†:** none

**Measurement Methods (see Table 1):**
**NIOSH** 7300, 7301, 7303

**Physical Description:** White or slightly gray powder.

**Chemical & Physical Properties:**
**MW:** 150.7
**BP:** Decomposes
**Sol:** Insoluble
**Fl.P:** NA
**IP:** NA
**Sp.Gr:** 6.95
**VP:** 0 mmHg (approx)
**MLT:** 2966°F (Decomposes)
**UEL:** NA
**LEL:** NA

**Personal Protection/Sanitation (see Table 2):**
**Skin:** N.R.
**Eyes:** N.R.
**Wash skin:** N.R.
**Remove:** N.R.
**Change:** N.R.

**Respirator Recommendations (see Tables 3 and 4):**
Not available.

**Incompatibilities and Reactivities:** Chlorine trifluoride

**Exposure Routes, Symptoms, Target Organs (see Table 5):**
**ER:** Inh, Con
**SY:** Stannosis (benign pneumoconiosis): dysp, decr pulm func
**TO:** Resp sys

**First Aid (see Table 6):**
**Eye:** Irr immed
**Breath:** Fresh air

## Titanium dioxide

**Formula:** $TiO_2$ | **CAS#:** 13463-67-7 | **RTECS#:** XR2275000 | **IDLH:** Ca [5000 mg/m$^3$]

**Conversion:** | **DOT:**

**Synonyms/Trade Names:** Rutile, Titanium oxide, Titanium peroxide

**Exposure Limits:**
**NIOSH REL:** Ca
See Appendix A
**OSHA PEL†:** TWA 15 mg/m$^3$

**Measurement Methods (see Table 1):**
**NIOSH** S385 (II-3)

**Physical Description:** White, odorless powder.

**Chemical & Physical Properties:**
**MW:** 79.9
**BP:** 4532-5432°F
**Sol:** Insoluble
**Fl.P:** NA
**IP:** NA
**Sp.Gr:** 4.26
**VP:** 0 mmHg (approx)
**MLT:** 3326-3362°F
**UEL:** NA
**LEL:** NA
Noncombustible Solid

**Personal Protection/Sanitation (see Table 2):**
**Skin:** N.R.
**Eyes:** N.R.
**Wash skin:** N.R.
**Remove:** N.R.
**Change:** Daily

**Respirator Recommendations (see Tables 3 and 4):**
**NIOSH**
**¥:** ScbaF:Pd,Pp/SaF:Pd,Pp:AScba
**Escape:** 100F/ScbaE

**Incompatibilities and Reactivities:** None reported

**Exposure Routes, Symptoms, Target Organs (see Table 5):**
**ER:** Inh
**SY:** Lung fib; [carc]
**TO:** Resp sys [in animals: lung tumors]

**First Aid (see Table 6):**
**Breath:** Resp support

## o-Tolidine

**Formula:** $C_{14}H_{16}N_2$ | **CAS#:** 119-93-7 | **RTECS#:** DD1225000 | **IDLH:** Ca [N.D.]

**Conversion:** | **DOT:**

**Synonyms/Trade Names:** 4,4'-Diamino-3,3'-dimethylbiphenyl; Diaminoditolyl; 3,3'-Dimethylbenzidine; 3,3'-Dimethyl-4,4'-diphenyldiamine; 3,3'-Tolidine

**Exposure Limits:**
**NIOSH REL:** Ca
C 0.02 mg/m$^3$ [60-minute] [skin]
See Appendix A
See Appendix C
**OSHA PEL:** See Appendix C

**Measurement Methods (see Table 1):**
**NIOSH** 5013
**OSHA** 71

T

**Physical Description:** White to reddish crystals or powder.
[**Note:** Darkens on exposure to air. Often used in paste or wet cake form. Used as a basis for many dyes.]

**Chemical & Physical Properties:**
**MW:** 212.3
**BP:** 572°F
**Sol:** 0.1%
**Fl.P:** ?
**IP:** ?
**Sp.Gr:** ?
**VP:** ?
**MLT:** 264°F
**UEL:** ?
**LEL:** ?
Combustible Solid

**Personal Protection/Sanitation (see Table 2):**
**Skin:** Prevent skin contact
**Eyes:** Prevent eye contact
**Wash skin:** When contam/Daily
**Remove:** When wet or contam
**Change:** Daily
**Provide:** Eyewash
Quick drench

**Respirator Recommendations (see Tables 3 and 4):**
**NIOSH**
**¥:** ScbaF:Pd,Pp/SaF:Pd,Pp:AScba
**Escape:** GmFOv100/ScbaE

**Incompatibilities and Reactivities:** Strong oxidizers

**Exposure Routes, Symptoms, Target Organs (see Table 5):**
**ER:** Inh, Abs, Ing, Con
**SY:** Irrit eyes, nose; in animals: liver, kidney damage; [carc]
**TO:** Eyes, resp sys, liver, kidneys [in animals: liver, bladder & mammary gland tumors]

**First Aid (see Table 6):**
**Eye:** Irr immed
**Skin:** Soap flush immed
**Breath:** Resp support
**Swallow:** Medical attention immed

## Toluene

| Toluene | Formula: $C_6H_5CH_3$ | CAS#: 108-88-3 | RTECS#: XS5250000 | IDLH: 500 ppm |
|---|---|---|---|---|
| Conversion: 1 ppm = 3.77 mg/m$^3$ | DOT: 1294 130 | | | |

**Synonyms/Trade Names:** Methyl benzene, Methyl benzol, Phenyl methane, Toluol

**Exposure Limits:**
**NIOSH REL:** TWA 100 ppm (375 mg/m$^3$)
ST 150 ppm (560 mg/m$^3$)
**OSHA PEL†:** TWA 200 ppm
C 300 ppm
500 ppm (10-minute maximum peak)

**Measurement Methods (see Table 1):**
**NIOSH** 1500, 1501, 3800, 4000
**OSHA** 111

**Physical Description:** Colorless liquid with a sweet, pungent, benzene-like odor.

**Chemical & Physical Properties:**
**MW:** 92.1
**BP:** 232°F
**Sol(74°F):** 0.07%
**Fl.P:** 40°F
**IP:** 8.82 eV
**Sp.Gr:** 0.87
**VP:** 21 mmHg
**FRZ:** -139°F
**UEL:** 7.1%
**LEL:** 1.1%
Class IB Flammable Liquid

**Personal Protection/Sanitation (see Table 2):**
**Skin:** Prevent skin contact
**Eyes:** Prevent eye contact
**Wash skin:** When contam
**Remove:** When wet (flamm)
**Change:** N.R.

**Respirator Recommendations (see Tables 3 and 4):**
**NIOSH**
**500 ppm:** CcrOv*/PaprOv*/GmFOv/Sa*/ScbaF
**§:** ScbaF:Pd,Pp/SaF:Pd,Pp:AScba
**Escape:** GmFOv/ScbaE

**Incompatibilities and Reactivities:** Strong oxidizers

**Exposure Routes, Symptoms, Target Organs (see Table 5):**
**ER:** Inh, Abs, Ing, Con
**SY:** Irrit eyes, nose; lass, conf, euph, dizz, head; dilated pupils, lac; anxi, musc ftg, insom; pares; derm; liver, kidney damage
**TO:** Eyes, skin, resp sys, CNS, liver, kidneys

**First Aid (see Table 6):**
**Eye:** Irr immed
**Skin:** Soap wash prompt
**Breath:** Resp support
**Swallow:** Medical attention immed

## Toluenediamine

| Toluenediamine | Formula: $CH_3C_6H_3(NH_2)_2$ | CAS#: 25376-45-8; 95-80-7 (2,4-TDA) | RTECS#: XS9445000; XS9625000 (2,4-TDA) | IDLH: Ca [N.D.] |
|---|---|---|---|---|
| Conversion: | DOT: 1709 151 (2,4-Toluenediamine) | | | |

**Synonyms/Trade Names:** Diaminotoluene, Methylphenylene diamine, TDA, Toluenediamine isomers, Tolylenediamine [**Note:** Various isomers of TDA exist.]

**Exposure Limits:**
**NIOSH REL:** Ca (all isomers)
See Appendix A
**OSHA PEL:** none

**Measurement Methods (see Table 1):**
**NIOSH** 5516
**OSHA** 65

**Physical Description:** Colorless to brown, needle-shaped crystals or powder.
[**Note:** Tends to darken on storage and exposure to air. Properties given are for 2,4-TDA.]

**Chemical & Physical Properties:**
**MW:** 122.2
**BP:** 558°F
**Sol:** Soluble
**Fl.P:** 300°F
**IP:** ?
**Sp.Gr:** 1.05 (Liquid at 212°F)
**VP(224°F):** 1 mmHg
**MLT:** 210°F
**UEL:** ?
**LEL:** ?
Combustible Solid

**Personal Protection/Sanitation (see Table 2):**
**Skin:** Prevent skin contact
**Eyes:** Prevent eye contact
**Wash skin:** When contam/Daily
**Remove:** When wet or contam
**Change:** Daily
**Provide:** Eyewash
Quick drench

**Respirator Recommendations (see Tables 3 and 4):**
**NIOSH**
**¥:** ScbaF:Pd,Pp/SaF:Pd,Pp:AScba
**Escape:** GmFOv/ScbaE

**Incompatibilities and Reactivities:** None reported

**Exposure Routes, Symptoms, Target Organs (see Table 5):**
**ER:** Inh, Abs, Ing, Con
**SY:** Irrit eyes, skin, nose, throat; derm; ataxia, tacar, nau, vomit, convuls, resp depres; methemo, cyan, head, lass, dizz, bluish skin; liver inj; [carc]
**TO:** Eyes, skin, resp sys, blood, CVS, liver [in animals: liver, skin & mammary gland tumors]

**First Aid (see Table 6):**
**Eye:** Irr immed
**Skin:** Water flush immed
**Breath:** Resp support
**Swallow:** Medical attention immed

## Toluene-2,4-diisocyanate

| Formula: | CAS#: | RTECS#: | IDLH: |
|---|---|---|---|
| $CH_3C_6H_3(NCO)_2$ | 584-84-9 | CZ6300000 | Ca [2.5 ppm] |

**Conversion:** 1 ppm = 7.13 mg/$m^3$ | **DOT:** 2078 156

**Synonyms/Trade Names:** TDI; 2,4-TDI; 2,4-Toluene diisocyanate

**Exposure Limits:**
**NIOSH REL:** Ca
See Appendix A
**OSHA PEL†:** C 0.02 ppm (0.14 mg/$m^3$)

**Measurement Methods (see Table 1):**
**NIOSH** 2535, 5521, 5522, 5525
**OSHA** 18, 33, 42

**Physical Description:** Colorless to pale-yellow solid or liquid (above 71°F) with a sharp, pungent odor.

**Chemical & Physical Properties:**
**MW:** 174.2
**BP:** 484°F
**Sol:** Insoluble
**Fl.P:** 260°F
**IP:** ?
**Sp.Gr:** 1.22
**VP(77°F):** 0.01 mmHg
**MLT:** 71°F
**UEL:** 9.5%
**LEL:** 0.9%
Class IIIB Combustible Liquid

**Personal Protection/Sanitation (see Table 2):**
**Skin:** Prevent skin contact
**Eyes:** Prevent eye contact
**Wash skin:** When contam/Daily
**Remove:** When wet or contam
**Change:** Daily
**Provide:** Eyewash
Quick drench

**Respirator Recommendations (see Tables 3 and 4):**
**NIOSH**
**¥:** ScbaF:Pd,Pp/SaF:Pd,Pp:AScba
**Escape:** GmFOv/ScbaE

**Incompatibilities and Reactivities:** Strong oxidizers, water, acids, bases & amines (may cause foam & spatter); alcohols [**Note:** Reacts slowly with water to form carbon dioxide and polyureas.]

**Exposure Routes, Symptoms, Target Organs (see Table 5):**
**ER:** Inh, Ing, Con
**SY:** Irrit eyes, skin, nose, throat; choke, paroxysmal cough; chest pain, retster soreness; nau, vomit, abdom pain; bron, bronchospasm, pulm edema; dysp, asthma; conj, lac; derm, skin sens; [carc]
**TO:** Eyes, skin, resp sys [in animals: pancreas, liver, mammary gland, circulatory sys & skin tumors]

**First Aid (see Table 6):**
**Eye:** Irr immed
**Skin:** Soap wash immed
**Breath:** Resp support
**Swallow:** Medical attention immed

## m-Toluidine

| Formula: | CAS#: | RTECS#: | IDLH: |
|---|---|---|---|
| $CH_3C_6H_4NH_2$ | 108-44-1 | XU2800000 | N.D. |

**Conversion:** | **DOT:** 1708 153

**Synonyms/Trade Names:** 3-Amino-1-methylbenzene, 1-Aminophenylmethane, m-Aminotoluene, 3-Methylaniline, 3-Methylbenzenamine, 3-Toluidine, meta-Toluidine, m-Tolylamine

**Exposure Limits:**
**NIOSH REL:** See Appendix D
**OSHA PEL†:** none

**Measurement Methods (see Table 1):**
**NIOSH** 2002
**OSHA** 73

T

**Physical Description:** Colorless to light-yellow liquid with an aromatic, amine-like odor. [**Note:** Used as a basis for many dyes.]

**Chemical & Physical Properties:**
**MW:** 107.2
**BP:** 397°F
**Sol:** 2%
**Fl.P:** 187°F
**IP:** 7.50 eV
**Sp.Gr:** 0.999
**VP(106°F):** 1 mmHg
**FRZ:** -23°F
**UEL:** ?
**LEL:** ?
Class IIIA Combustible Liquid

**Personal Protection/Sanitation (see Table 2):**
**Skin:** Prevent skin contact
**Eyes:** Prevent eye contact
**Wash skin:** When contam
**Remove:** When wet or contam
**Change:** N.R.

**Respirator Recommendations (see Tables 3 and 4):**
Not available.

**Incompatibilities and Reactivities:** Oxidizers, acids

**Exposure Routes, Symptoms, Target Organs (see Table 5):**
**ER:** Inh, Abs, Ing, Con
**SY:** Irrit eyes, skin; derm; hema, methemo; cyan, nau, vomit, low BP, convuls; anemia, lass
**TO:** Eyes, skin, blood, CVS

**First Aid (see Table 6):**
**Eye:** Irr immed
**Skin:** Soap wash immed
**Breath:** Resp support
**Swallow:** Medical attention immed

## o-Toluidine

| o-Toluidine | Formula: $CH_3C_6H_4NH_2$ | CAS#: 95-53-4 | RTECS#: XU2975000 | IDLH: Ca [50 ppm] |
|---|---|---|---|---|
| **Conversion:** 1 ppm = 4.38 mg/m$^3$ | **DOT:** 1708 153 | | | |

**Synonyms/Trade Names:** o-Aminotoluene, 2-Aminotoluene, 1-Methyl-2-aminobenzene, o-Methylaniline, 2-Methylaniline, ortho-Toluidine, o-Tolylamine

**Exposure Limits:**
**NIOSH REL:** Ca [skin]
See Appendix A
**OSHA PEL:** TWA 5 ppm (22 mg/m$^3$) [skin]

**Measurement Methods (see Table 1):**
**NIOSH** 2002, 2017, 8317
**OSHA** 73

**Physical Description:** Colorless to pale-yellow liquid with an aromatic, aniline-like odor.

**Chemical & Physical Properties:**
**MW:** 107.2
**BP:** 392°F
**Sol:** 2%
**Fl.P:** 185°F
**IP:** 7.44 eV
**Sp.Gr:** 1.01
**VP:** 0.3 mmHg
**FRZ:** 6°F
**UEL:** ?
**LEL:** ?
Class IIIA Combustible Liquid

**Personal Protection/Sanitation (see Table 2):**
**Skin:** Prevent skin contact
**Eyes:** Prevent eye contact
**Wash skin:** When contam
**Remove:** When wet or contam
**Change:** N.R.
**Provide:** Eyewash
Quick drench

**Respirator Recommendations (see Tables 3 and 4):**
**NIOSH**
**¥:** ScbaF:Pd,Pp/SaF:Pd,Pp:AScba
**Escape:** GmFOv/ScbaE

**Incompatibilities and Reactivities:** Strong oxidizers, nitric acid, bases

**Exposure Routes, Symptoms, Target Organs (see Table 5):**
**ER:** Inh, Abs, Ing, Con
**SY:** Irrit eyes; anoxia, head, cyan; lass, dizz, drow; micro hema; eye burns; derm; [carc]
**TO:** Eyes, skin, blood, kidneys, liver, CVS [bladder cancer]

**First Aid (see Table 6):**
**Eye:** Irr immed
**Skin:** Soap wash immed
**Breath:** Resp support
**Swallow:** Medical attention immed

## p-Toluidine

| p-Toluidine | Formula: $CH_3C_6H_4NH_2$ | CAS#: 106-49-0 | RTECS#: XU3150000 | IDLH: Ca [N.D.] |
|---|---|---|---|---|
| **Conversion:** | **DOT:** 1708 153 | | | |

**Synonyms/Trade Names:** 4-Aminotoluene, 4-Methylaniline, 4-Methylbenzenamine, 4-Toluidine, para-Toluidine, p-Tolylamine

**Exposure Limits:**
**NIOSH REL:** Ca
See Appendix A
**OSHA PEL†:** none

**Measurement Methods (see Table 1):**
**NIOSH** 2002
**OSHA** 73

**Physical Description:** White solid with an aromatic odor.
[**Note:** Used as a basis for many dyes.]

**Chemical & Physical Properties:**
**MW:** 107.2
**BP:** 393°F
**Sol:** 0.7%
**Fl.P:** 188°F
**IP:** 7.50 eV
**Sp.Gr:** 1.05
**VP(108°F):** 1 mmHg
**MLT:** 111°F
**UEL:** ?
**LEL:** ?
Combustible Solid

**Personal Protection/Sanitation (see Table 2):**
**Skin:** Prevent skin contact
**Eyes:** Prevent eye contact
**Wash skin:** When contam/Daily
**Remove:** When wet or contam
**Change:** Daily
**Provide:** Eyewash
Quick drench

**Respirator Recommendations (see Tables 3 and 4):**
**NIOSH**
**¥:** ScbaF:Pd,Pp/SaF:Pd,Pp:AScba
**Escape:** GmFOv100/ScbaE

**Incompatibilities and Reactivities:** Oxidizers, acids

**Exposure Routes, Symptoms, Target Organs (see Table 5):**
**ER:** Inh, Abs, Ing, Con
**SY:** Irrit eyes, skin; derm; hema, methemo; cyan, nau, vomit, low BP, convuls; anemia, lass; [carc]
**TO:** Eyes, skin, blood, CVS [in animals: liver tumors]

**First Aid (see Table 6):**
**Eye:** Irr immed
**Skin:** Soap wash immed
**Breath:** Resp support
**Swallow:** Medical attention immed

## Tributyl phosphate

| | | | |
|---|---|---|---|
| **Formula:** $(CH_3[CH_2]_3O)_3PO$ | **CAS#:** 126-73-8 | **RTECS#:** TC7700000 | **IDLH:** 30 ppm |

**Conversion:** 1 ppm = 10.89 mg/m$^3$ | **DOT:**

**Synonyms/Trade Names:** Butyl phosphate, TBP, Tributyl ester of phosphoric acid, Tri-n-butyl phosphate

**Exposure Limits:**
**NIOSH REL:** TWA 0.2 ppm (2.5 mg/m$^3$)
**OSHA PEL†:** TWA 5 mg/m$^3$

**Measurement Methods (see Table 1):**
**NIOSH** 5034

**Physical Description:** Colorless to pale-yellow, odorless liquid.

**Chemical & Physical Properties:**
**MW:** 266.3
**BP:** 552°F (Decomposes)
**Sol:** 0.6%
**Fl.P(oc):** 295°F
**IP:** ?
**Sp.Gr:** 0.98
**VP(77°F):** 0.004 mmHg
**FRZ:** -112°F
**UEL:** ?
**LEL:** ?
Class IIIB Combustible Liquid

**Personal Protection/Sanitation (see Table 2):**
**Skin:** Prevent skin contact
**Eyes:** Prevent eye contact
**Wash skin:** When contam
**Remove:** When wet or contam
**Change:** N.R.

**Respirator Recommendations (see Tables 3 and 4):**
**NIOSH**
**2 ppm:** Sa
**5 ppm:** Sa:Cf
**10 ppm:** ScbaF/SaF
**30 ppm:** SaF:Pd,Pp
**§:** ScbaF:Pd,Pp/SaF:Pd,Pp:AScba
**Escape:** GmFOv100/ScbaE

**Incompatibilities and Reactivities:** Alkalis, oxidizers, water, moist air

**Exposure Routes, Symptoms, Target Organs (see Table 5):**
**ER:** Inh, Ing, Con
**SY:** Irrit eyes, skin, resp sys, head; nau
**TO:** Eyes, skin, resp sys

**First Aid (see Table 6):**
**Eye:** Irr immed
**Skin:** Soap wash prompt
**Breath:** Resp support
**Swallow:** Medical attention immed

## Trichloroacetic acid

| | | | |
|---|---|---|---|
| **Formula:** $CCl_3COOH$ | **CAS#:** 76-03-9 | **RTECS#:** AJ7875000 | **IDLH:** N.D. |

**Conversion:** 1 ppm = 6.68 mg/m$^3$ | **DOT:** 1839 153 (solid); 2564 153 (solution)

**Synonyms/Trade Names:** TCA, Trichloroethanoic acid

**Exposure Limits:**
**NIOSH REL:** TWA 1 ppm (7 mg/m$^3$)
**OSHA PEL†:** none

**Measurement Methods (see Table 1):**
**OSHA** PV2017

**Physical Description:** Colorless to white, crystalline solid with a sharp, pungent odor.

**Chemical & Physical Properties:**
**MW:** 163.4
**BP:** 388°F
**Sol:** Miscible
**Fl.P:** NA
**IP:** ?
**Sp.Gr:** 1.62
**VP(124°F):** 1 mmHg
**MLT:** 136°F
**UEL:** NA
**LEL:** NA
Noncombustible Solid

**Personal Protection/Sanitation (see Table 2):**
**Skin:** Prevent skin contact
**Eyes:** Prevent eye contact
**Wash skin:** When contam
**Remove:** When wet or contam
**Change:** Daily
**Provide:** Eyewash
Quick drench

**Respirator Recommendations (see Tables 3 and 4):**
Not available.

**Incompatibilities and Reactivities:** Moisture, iron, zinc, aluminum, strong oxidizers
[**Note:** Decomposes on heating to form phosgene & hydrogen chloride. Corrosive to metals.]

**Exposure Routes, Symptoms, Target Organs (see Table 5):**
**ER:** Inh, Ing, Con
**SY:** Irrit eyes, skin, nose, throat, resp sys; cough, dysp, delayed pulm edema; eye, skin burns; derm; salv, vomit, diarr
**TO:** Eyes, skin, resp sys, GI tract

**First Aid (see Table 6):**
**Eye:** Irr immed
**Skin:** Water flush immed
**Breath:** Resp support
**Swallow:** Medical attention immed

## 1,2,4-Trichlorobenzene

| | | | | |
|---|---|---|---|---|
| **1,2,4-Trichlorobenzene** | **Formula:** $C_6H_3Cl_3$ | **CAS#:** 120-82-1 | **RTECS#:** DC2100000 | **IDLH:** N.D. |
| **Conversion:** 1 ppm = 7.42 mg/m$^3$ | **DOT:** 2321 153 (liquid) | | | |

**Synonyms/Trade Names:** unsym-Trichlorobenzene; 1,2,4-Trichlorobenzol

**Exposure Limits:**
**NIOSH REL:** C 5 ppm (40 mg/m$^3$)
**OSHA PEL†:** none

**Measurement Methods (see Table 1):**
**NIOSH** 5517

**Physical Description:** Colorless liquid or crystalline solid (below 63°F) with an aromatic odor.

**Chemical & Physical Properties:**
**MW:** 181.4
**BP:** 416°F
**Sol:** 0.003%
**Fl.P:** 222°F
**IP:** ?
**Sp.Gr:** 1.45
**VP:** 1 mmHg
**FRZ:** 63°F
**UEL(302°F):** 6.6%
**LEL(302°F):** 2.5%
Class IIIB Combustible Liquid
Combustible Solid

**Personal Protection/Sanitation (see Table 2):**
**Skin:** Prevent skin contact
**Eyes:** Prevent eye contact
**Wash skin:** When contam
**Remove:** When wet or contam
**Change:** N.R.

**Respirator Recommendations (see Tables 3 and 4):**
Not available.

**Incompatibilities and Reactivities:** Acids, acid fumes, oxidizers, steam

**Exposure Routes, Symptoms, Target Organs (see Table 5):**
**ER:** Inh, Abs, Ing, Con
**SY:** Irrit eyes, skin, muc memb; in animals: liver, kidney damage; possible terato effects
**TO:** Eyes, skin, resp sys, liver, repro sys

**First Aid (see Table 6):**
**Eye:** Irr immed
**Skin:** Soap wash
**Breath:** Resp support
**Swallow:** Medical attention immed

## 1,1,2-Trichloroethane

| | | | | |
|---|---|---|---|---|
| **1,1,2-Trichloroethane** | **Formula:** $CHCl_2CH_2Cl$ | **CAS#:** 79-00-5 | **RTECS#:** KJ3150000 | **IDLH:** Ca [100 ppm] |
| **Conversion:** 1 ppm = 5.46 mg/m$^3$ | **DOT:** | | | |

**Synonyms/Trade Names:** Ethane trichloride, β-Trichloroethane, Vinyl trichloride

**Exposure Limits:**
**NIOSH REL:** Ca
TWA 10 ppm (45 mg/m$^3$) [skin]
See Appendix A
See Appendix C (Chloroethanes)
**OSHA PEL:** TWA 10 ppm (45 mg/m$^3$) [skin]

**Measurement Methods (see Table 1):**
**NIOSH** 1003
**OSHA** 11

**Physical Description:** Colorless liquid with a sweet, chloroform-like odor.

**Chemical & Physical Properties:**
**MW:** 133.4
**BP:** 237°F
**Sol:** 0.4%
**Fl.P:** ?
**IP:** 11.00 eV
**Sp.Gr:** 1.44
**VP:** 19 mmHg
**FRZ:** -34°F
**UEL:** 15.5%
**LEL:** 6%
Combustible Liquid, forms dense soot.

**Personal Protection/Sanitation (see Table 2):**
**Skin:** Prevent skin contact
**Eyes:** Prevent eye contact
**Wash skin:** When contam
**Remove:** When wet or contam
**Change:** N.R.
**Provide:** Eyewash
Quick drench

**Respirator Recommendations (see Tables 3 and 4):**
**NIOSH**
**¥:** ScbaF:Pd,Pp/SaF:Pd,Pp:AScba
**Escape:** GmFOv/ScbaE

**Incompatibilities and Reactivities:** Strong oxidizers & caustics; chemically-active metals (such as aluminum, magnesium powders, sodium & potassium)

**Exposure Routes, Symptoms, Target Organs (see Table 5):**
**ER:** Inh, Abs, Ing, Con
**SY:** Irrit eyes, nose; CNS depres; liver, kidney damage; derm; [carc]
**TO:** Eyes, resp sys, CNS, liver, kidneys [in animals: liver cancer]

**First Aid (see Table 6):**
**Eye:** Irr immed
**Skin:** Soap wash prompt
**Breath:** Resp support
**Swallow:** Medical attention immed

## Trichloroethylene

| Trichloroethylene | Formula: ClCH=CCl$_2$ | CAS#: 79-01-6 | RTECS#: KX4550000 | IDLH: Ca [1000 ppm] |
|---|---|---|---|---|
| **Conversion:** 1 ppm = 5.37 mg/m$^3$ | **DOT:** 1710 160 | | | |

**Synonyms/Trade Names:** Ethylene trichloride, TCE, Trichloroethene, Trilene

**Exposure Limits:**
**NIOSH REL:** Ca
See Appendix A
See Appendix C
**OSHA PEL†:** TWA 100 ppm
C 200 ppm
300 ppm (5-minute maximum peak in any 2 hours)

**Measurement Methods (see Table 1):**
**NIOSH** 1022, 3800
**OSHA** 1001

**Physical Description:** Colorless liquid (unless dyed blue) with a chloroform-like odor.

**Chemical & Physical Properties:**
**MW:** 131.4
**BP:** 189°F
**Sol:** 0.1%
**Fl.P:** ?
**IP:** 9.45 eV
**Sp.Gr:** 1.46
**VP:** 58 mmHg
**FRZ:** -99°F
**UEL(77°F):** 10.5%
**LEL(77°F):** 8%
Combustible Liquid, but burns with difficulty.

**Personal Protection/Sanitation (see Table 2):**
**Skin:** Prevent skin contact
**Eyes:** Prevent eye contact
**Wash skin:** When contam
**Remove:** When wet or contam
**Change:** N.R.
**Provide:** Eyewash
Quick drench

**Respirator Recommendations (see Tables 3 and 4):**
**NIOSH**
**¥:** ScbaF:Pd,Pp/SaF:Pd,Pp:AScba
**Escape:** GmFOv/ScbaE

**Incompatibilities and Reactivities:** Strong caustics & alkalis; chemically-active metals (such as barium, lithium, sodium, magnesium, titanium & beryllium)

**Exposure Routes, Symptoms, Target Organs (see Table 5):**
**ER:** Inh, Abs, Ing, Con
**SY:** Irrit eyes, skin; head, vis dist, lass, dizz, tremor, drow, nau, vomit; derm; card arrhy, pares; liver inj; [carc]
**TO:** Eyes, skin, resp sys, heart, liver, kidneys, CNS [in animals: liver & kidney cancer]

**First Aid (see Table 6):**
**Eye:** Irr immed
**Skin:** Soap wash prompt
**Breath:** Resp support
**Swallow:** Medical attention immed

## Trichloronaphthalene

| Trichloronaphthalene | Formula: $C_{10}H_5Cl_3$ | CAS#: 1321-65-9 | RTECS#: QK4025000 | IDLH: See Appendix F |
|---|---|---|---|---|
| **Conversion:** | **DOT:** | | | |

T

**Synonyms/Trade Names:** Halowax®, Nibren wax, Seekay wax

**Exposure Limits:**
**NIOSH REL:** TWA 5 mg/m$^3$ [skin]
**OSHA PEL:** TWA 5 mg/m$^3$ [skin]

**Measurement Methods (see Table 1):**
**NIOSH** S128 (II-2)

**Physical Description:** Colorless to pale-yellow solid with an aromatic odor.

**Chemical & Physical Properties:**
**MW:** 231.5
**BP:** 579-669°F
**Sol:** Insoluble
**Fl.P(oc):** 392°F
**IP:** ?
**Sp.Gr:** 1.58
**VP:** <1 mmHg
**MLT:** 199°F
**UEL:** ?
**LEL:** ?
Combustible Solid

**Personal Protection/Sanitation (see Table 2):**
**Skin:** Prevent skin contact
**Eyes:** Prevent eye contact
**Wash skin:** When contam
**Remove:** When wet or contam
**Change:** Daily

**Respirator Recommendations (see Tables 3 and 4):**
**NIOSH/OSHA**
**50 mg/m$^3$:** ScbaF/SaF
**§:** ScbaF:Pd,Pp/SaF:Pd,Pp:AScba
**Escape:** GmFOv100/ScbaE

**See Appendix F**

**Incompatibilities and Reactivities:** Strong oxidizers

**Exposure Routes, Symptoms, Target Organs (see Table 5):**
**ER:** Inh, Abs, Ing, Con
**SY:** Anor, nau; dizz; jaun, liver inj
**TO:** Liver

**First Aid (see Table 6):**
**Eye:** Irr immed
**Skin:** Soap wash
**Breath:** Resp support
**Swallow:** Medical attention immed

## 1,2,3-Trichloropropane

| Formula: | CAS#: | RTECS#: | IDLH: |
|---|---|---|---|
| $CH_2ClCHClCH_2Cl$ | 96-18-4 | TZ9275000 | Ca [100 ppm] |

**Conversion:** 1 ppm = 6.03 mg/m$^3$

**DOT:**

**Synonyms/Trade Names:** Allyl trichloride, Glycerol trichlorohydrin, Glyceryl trichlorohydrin, Trichlorohydrin

**Exposure Limits:**
**NIOSH REL:** Ca
TWA 10 ppm (60 mg/m$^3$) [skin]
See Appendix A
**OSHA PEL†:** TWA 50 ppm (300 mg/m$^3$)

**Measurement Methods (see Table 1):**
**NIOSH** 1003
**OSHA** 7

**Physical Description:** Colorless liquid with a chloroform-like odor.

**Chemical & Physical Properties:**
**MW:** 147.4
**BP:** 314°F
**Sol:** 0.1%
**Fl.P:** 160°F
**IP:** ?
**Sp.Gr:** 1.39
**VP:** 3 mmHg
**FRZ:** 6°F
**UEL(302°F):** 12.6%
**LEL(248°F):** 3.2%
Class IIIA Combustible Liquid

**Personal Protection/Sanitation (see Table 2):**
**Skin:** Prevent skin contact
**Eyes:** Prevent eye contact
**Wash skin:** When contam
**Remove:** When wet or contam
**Change:** N.R.
**Provide:** Eyewash
Quick drench

**Respirator Recommendations (see Tables 3 and 4):**
**NIOSH**
**¥:** ScbaF:Pd,Pp/SaF:Pd,Pp:AScba
**Escape:** GmFOv/ScbaE

**Incompatibilities and Reactivities:** Chemically-active metals, strong caustics & oxidizers

**Exposure Routes, Symptoms, Target Organs (see Table 5):**
**ER:** Inh, Abs, Ing, Con
**SY:** Irrit eyes, nose, throat; CNS depres; in animals: liver, kidney inj; [carc]
**TO:** Eyes, skin, resp sys, CNS, liver, kidneys [in animals: forestomach, liver & mammary gland cancer]

**First Aid (see Table 6):**
**Eye:** Irr immed
**Skin:** Soap wash
**Breath:** Resp support
**Swallow:** Medical attention immed

## 1,1,2-Trichloro-1,2,2-trifluoroethane

| Formula: | CAS#: | RTECS#: | IDLH: |
|---|---|---|---|
| $CCl_2FCClF_2$ | 76-13-1 | KJ4000000 | 2000 ppm |

**Conversion:** 1 ppm = 7.67 mg/m$^3$

**DOT:**

**Synonyms/Trade Names:** Chlorofluorocarbon-113, CFC-113, Freon® 113, Genetron® 113, Halocarbon 113, Refrigerant 113, TTE

**Exposure Limits:**
**NIOSH REL:** TWA 1000 ppm (7600 mg/m$^3$)
ST 1250 ppm (9500 mg/m$^3$)
**OSHA PEL†:** TWA 1000 ppm (7600 mg/m$^3$)

**Measurement Methods (see Table 1):**
**NIOSH** 1020
**OSHA** 113

**Physical Description:** Colorless to water-white liquid with an odor like carbon tetrachloride at high concentrations. [**Note:** A gas above 118°F.]

**Chemical & Physical Properties:**
**MW:** 187.4
**BP:** 118°F
**Sol(77°F):** 0.02%
**Fl.P:** ?
**IP:** 11.99 eV
**Sp.Gr(77°F):** 1.56
**VP:** 285 mmHg
**FRZ:** -31°F
**UEL:** ?
**LEL:** ?
Noncombustible Liquid at ordinary temperatures, but the gas will ignite and burn weakly at 1256°F.

**Personal Protection/Sanitation (see Table 2):**
**Skin:** Prevent skin contact
**Eyes:** Prevent eye contact
**Wash skin:** When contam
**Remove:** When wet or contam
**Change:** N.R.

**Respirator Recommendations (see Tables 3 and 4):**
**NIOSH/OSHA**
**2000 ppm:** Sa/ScbaF
**§:** ScbaF:Pd,Pp/SaF:Pd,Pp:AScba
**Escape:** GmFOv/ScbaE

**Incompatibilities and Reactivities:** Chemically-active metals such as calcium, powdered aluminum, zinc, magnesium & beryllium
[**Note:** Decomposes if in contact with alloys containing >2% magnesium.]

**Exposure Routes, Symptoms, Target Organs (see Table 5):**
**ER:** Inh, Ing, Con
**SY:** Irrit skin, throat, drow, derm; CNS depres; in animals: card arrhy, narco
**TO:** Skin, heart, CNS, CVS

**First Aid (see Table 6):**
**Eye:** Irr immed
**Skin:** Soap wash prompt
**Breath:** Resp support
**Swallow:** Medical attention immed

## Triethylamine

**Formula:** $(C_2H_5)_3N$ | **CAS#:** 121-44-8 | **RTECS#:** YE0175000 | **IDLH:** 200 ppm

**Conversion:** 1 ppm = 4.14 mg/$m^3$ | **DOT:** 1296 132

**Synonyms/Trade Names:** TEA

**Exposure Limits:**
**NIOSH REL:** See Appendix D
**OSHA PEL†:** TWA 25 ppm (100 mg/$m^3$)

**Measurement Methods (see Table 1):**
**NIOSH** S152 (II-3)
**OSHA** PV2060

**Physical Description:** Colorless liquid with a strong, ammonia-like odor.

**Chemical & Physical Properties:**
**MW:** 101.2
**BP:** 193°F
**Sol:** 2%
**Fl.P:** 20°F
**IP:** 7.50 eV
**Sp.Gr:** 0.73
**VP:** 54 mmHg
**FRZ:** -175°F
**UEL:** 8.0%
**LEL:** 1.2%
Class IB Flammable Liquid

**Personal Protection/Sanitation (see Table 2):**
**Skin:** Prevent skin contact
**Eyes:** Prevent eye contact
**Wash skin:** When contam
**Remove:** When wet (flamm)
**Change:** N.R.
**Provide:** Eyewash (>1%)
Quick drench (>1%)

**Respirator Recommendations (see Tables 3 and 4):**
**OSHA**
**200 ppm:** Sa:Cf£/ScbaF/SaF
**§:** ScbaF:Pd,Pp/SaF:Pd,Pp:AScba
**Escape:** GmFS/ScbaE

**Incompatibilities and Reactivities:** Strong oxidizers, strong acids, chlorine, hypochlorite, halogenated compounds

**Exposure Routes, Symptoms, Target Organs (see Table 5):**
**ER:** Inh, Abs, Ing, Con
**SY:** Irrit eyes, skin, resp sys; in animals: myocardial, kidney, liver damage
**TO:** Eyes, skin, resp sys, CVS, liver, kidneys

**First Aid (see Table 6):**
**Eye:** Irr immed
**Skin:** Soap wash immed
**Breath:** Resp support
**Swallow:** Medical attention immed

## Trifluorobromomethane

**Formula:** $CBrF_3$ | **CAS#:** 75-63-8 | **RTECS#:** PA5425000 | **IDLH:** 40,000 ppm

**Conversion:** 1 ppm = 6.09 mg/$m^3$ | **DOT:** 1009 126

**Synonyms/Trade Names:** Bromotrifluoromethane, Fluorocarbon 1301, Freon® 13B1, Halocarbon 13B1, Halon® 1301, Monobromotrifluoromethane, Refrigerant 13B1, Trifluoromonobromomethane

**Exposure Limits:**
**NIOSH REL:** TWA 1000 ppm (6100 mg/$m^3$)
**OSHA PEL:** TWA 1000 ppm (6100 mg/$m^3$)

**Measurement Methods (see Table 1):**
**NIOSH** 1017

**Physical Description:** Colorless, odorless gas. [**Note:** Shipped as a liquefied compressed gas.]



**Chemical & Physical Properties:**
**MW:** 148.9
**BP:** -72°F
**Sol:** 0.03%
**Fl.P:** NA
**IP:** 11.78 eV
**RGasD:** 5.14
**VP:** >1 atm
**FRZ:** -267°F
**UEL:** NA
**LEL:** NA
Nonflammable Gas

**Personal Protection/Sanitation (see Table 2):**
**Skin:** Frostbite
**Eyes:** Frostbite
**Wash skin:** N.R.
**Remove:** N.R.
**Change:** N.R.
**Provide:** Frostbite wash

**Respirator Recommendations (see Tables 3 and 4):**
**NIOSH/OSHA**
**10,000 ppm:** Sa
**25,000 ppm:** Sa:Cf
**40,000 ppm:** SaT:Cf/ScbaF/SaF
**§:** ScbaF:Pd,Pp/SaF:Pd,Pp:AScba
**Escape:** GmFOv/ScbaE

**Incompatibilities and Reactivities:** Chemically-active metals (such as calcium, powdered aluminum, zinc, and magnesium)

**Exposure Routes, Symptoms, Target Organs (see Table 5):**
**ER:** Inh, Con (liquid)
**SY:** Dizz; card arrhy; liquid: frostbite
**TO:** CNS, heart

**First Aid (see Table 6):**
**Eye:** Frostbite
**Skin:** Frostbite
**Breath:** Resp support

## Trimellitic anhydride

**Formula:** $C_9H_4O_5$ | **CAS#:** 552-30-7 | **RTECS#:** DC2050000 | **IDLH:** N.D.

**Conversion:** 1 ppm = 7.86 mg/m$^3$ | **DOT:**

**Synonyms/Trade Names:** 1,2,4-Benzenetricarboxylic anhydride; 4-Carboxyphthalic anhydride; TMA; TMAN; Trimellic acid anhydride [**Note:** TMA is also a synonym for Trimethylamine.]

**Exposure Limits:**
**NIOSH REL:** TWA 0.005 ppm (0.04 mg/m$^3$)
Should be handled in the workplace as an extremely toxic substance.
**OSHA PEL†:** none

**Measurement Methods (see Table 1):**
**NIOSH** 5036
**OSHA** 98

**Physical Description:** Colorless solid.

**Chemical & Physical Properties:**
**MW:** 192.1
**BP:** ?
**Sol:** ?
**Fl.P:** NA
**IP:** ?
**Sp.Gr:** ?
**VP:** 0.000004 mmHg
**MLT:** 322°F
**UEL:** NA
**LEL:** NA
Combustible Solid

**Personal Protection/Sanitation (see Table 2):**
**Skin:** Prevent skin contact
**Eyes:** Prevent eye contact
**Wash skin:** When contam
**Remove:** When wet or contam
**Change:** Daily

**Respirator Recommendations (see Tables 3 and 4):**
Not available.

**Incompatibilities and Reactivities:** None reported

**Exposure Routes, Symptoms, Target Organs (see Table 5):**
**ER:** Inh, Ing, Con
**SY:** Irrit eyes, skin, nose, resp sys; pulm edema, resp sens; rhinitis, asthma, cough, wheez, dysp, mal, fever, musc aches, sneez
**TO:** Eyes, skin, resp sys

**First Aid (see Table 6):**
**Eye:** Irr immed
**Skin:** Soap wash
**Breath:** Resp support
**Swallow:** Medical attention immed

## Trimethylamine

**Formula:** $(CH_3)_3N$ | **CAS#:** 75-50-3 | **RTECS#:** PA0350000 | **IDLH:** N.D.

**Conversion:** 1 ppm = 2.42 mg/m$^3$ | **DOT:** 1083 118 (anhydrous); 1297 132 (aqueous solution)

**Synonyms/Trade Names:** N,N-Dimethylmethanamine; TMA
[**Note:** May be used in an aqueous solution (typically 25%, 30%, or 40% TMA.]

**Exposure Limits:**
**NIOSH REL:** TWA 10 ppm (24 mg/m$^3$)
ST 15 ppm (36 mg/m$^3$)
**OSHA PEL†:** none

**Measurement Methods (see Table 1):**
**OSHA** PV2060

**Physical Description:** Colorless gas with a fishy, amine odor.
[**Note:** A liquid below 37°F. Shipped as a liquefied compressed gas.]

**Chemical & Physical Properties:**
**MW:** 59.1
**BP:** 37°F
**Sol(86°F):** 48%
**Fl.P:** NA (Gas)
20°F (Liquid)
**IP:** 7.82 eV
**RGasD:** 2.09
**VP(70°F):** 1454 mmHg
**FRZ:** -179°F
**UEL:** 11.6%
**LEL:** 2.0%
Flammable Gas

**Personal Protection/Sanitation (see Table 2):**
**Skin:** Prevent skin contact (liquid/solution)
Frostbite
**Eyes:** Prevent eye contact (liquid/solution)
Frostbite
**Wash skin:** When contam (solution)
**Remove:** When wet (flamm)
**Change:** N.R.
**Provide:** Eyewash (liquid/solution)
Quick drench (liquid/solution)
Frostbite wash

**Respirator Recommendations (see Tables 3 and 4):**
Not available.

**Incompatibilities and Reactivities:** Strong oxidizers (including bromine), ethylene oxide, nitrosating agents (e.g., sodium nitrite), mercury, strong acids [**Note:** Corrosive to many metals (e.g., zinc, brass, aluminum, copper).]

**Exposure Routes, Symptoms, Target Organs (see Table 5):**
**ER:** Inh, Ing (solution), Con
**SY:** Irrit eyes, skin, nose, throat, resp sys; cough, dysp, delayed pulm edema; blurred vision, corn nec; skin burns; liquid: frostbite
**TO:** Eyes, skin, resp sys

**First Aid (see Table 6):**
**Eye:** Irr immed (liquid/solution)/Frostbite
**Skin:** Water flush immed (liquid/solution)/Frostbite
**Breath:** Resp support
**Swallow:** Medical attention immed (solution)

## 1,2,3-Trimethylbenzene

| Formula: | CAS#: | RTECS#: | IDLH: |
|---|---|---|---|
| $C_6H_3(CH_3)_3$ | 526-73-8 | DC3300000 | N.D. |

**Conversion:** 1 ppm = 4.92 mg/m$^3$ | **DOT:**

**Synonyms/Trade Names:** Hemellitol
[**Note:** Hemimellitene is a mixture of the 1,2,3-isomer with up to 10% of related aromatics such as the 1,2,4-isomer.]

**Exposure Limits:**
**NIOSH REL:** TWA 25 ppm (125 mg/m$^3$)
**OSHA PEL†:** none

**Measurement Methods (see Table 1):**
**OSHA** PV2091

**Physical Description:** Clear, colorless liquid with a distinctive, aromatic odor.

**Chemical & Physical Properties:**
**MW:** 120.2
**BP:** 349°F
**Sol:** Low
**Fl.P:** ?
**IP:** 8.48 eV
**Sp.Gr:** 0.89
**VP(62°F):** 1 mmHg
**FRZ:** -14°F
**UEL:** 6.6%
**LEL:** 0.8%
Flammable Liquid

**Personal Protection/Sanitation (see Table 2):**
**Skin:** Prevent skin contact
**Eyes:** Prevent eye contact
**Wash skin:** When contam
**Remove:** When wet or contam
**Change:** N.R.

**Respirator Recommendations (see Tables 3 and 4):**
Not available.

**Incompatibilities and Reactivities:** Oxidizers, nitric acid

**Exposure Routes, Symptoms, Target Organs (see Table 5):**
**ER:** Inh, Ing, Con
**SY:** Irrit eyes, skin, nose, throat, resp sys; bron; hypochromic anemia; head, drow, lass, dizz, nau, inco; vomit, conf; chemical pneu (aspir liquid)
**TO:** Eyes, skin, resp sys, CNS, blood

**First Aid (see Table 6):**
**Eye:** Irr immed
**Skin:** Soap wash
**Breath:** Resp support
**Swallow:** Medical attention immed

## 1,2,4-Trimethylbenzene

| Formula: | CAS#: | RTECS#: | IDLH: |
|---|---|---|---|
| $C_6H_3(CH_3)_3$ | 95-63-6 | DC3325000 | N.D. |

**Conversion:** 1 ppm = 4.92 mg/m$^3$ | **DOT:**

**Synonyms/Trade Names:** Asymetrical trimethylbenzene, psi-Cumene, Pseudocumene
[**Note:** Hemimellitene is a mixture of the 1,2,3-isomer with up to 10% of related aromatics such as the 1,2,4-isomer.]

**Exposure Limits:**
**NIOSH REL:** TWA 25 ppm (125 mg/m$^3$)
**OSHA PEL†:** none

**Measurement Methods (see Table 1):**
**OSHA** PV2091

**Physical Description:** Clear, colorless liquid with a distinctive, aromatic odor.



**Chemical & Physical Properties:**
**MW:** 120.2
**BP:** 337°F
**Sol:** 0.006%
**Fl.P:** 112°F
**IP:** 8.27 eV
**Sp.Gr:** 0.88
**VP(56°F):** 1 mmHg
**FRZ:** -77°F
**UEL:** 6.4%
**LEL:** 0.9%
Class II Flammable Liquid

**Personal Protection/Sanitation (see Table 2):**
**Skin:** Prevent skin contact
**Eyes:** Prevent eye contact
**Wash skin:** When contam
**Remove:** When wet or contam
**Change:** N.R.

**Respirator Recommendations (see Tables 3 and 4):**
Not available.

**Incompatibilities and Reactivities:** Oxidizers, nitric acid

**Exposure Routes, Symptoms, Target Organs (see Table 5):**
**ER:** Inh, Ing, Con
**SY:** Irrit eyes, skin, nose, throat, resp sys; bron; hypochromic anemia; head, drow, lass, dizz, nau, inco; vomit, conf; chemical pneu (aspir liquid)
**TO:** Eyes, skin, resp sys, CNS, blood

**First Aid (see Table 6):**
**Eye:** Irr immed
**Skin:** Soap wash
**Breath:** Resp support
**Swallow:** Medical attention immed

## 1,3,5-Trimethylbenzene

| 1,3,5-Trimethylbenzene | Formula: $C_6H_3(CH_3)_3$ | CAS#: 108-67-8 | RTECS#: OX6825000 | IDLH: N.D. |
|---|---|---|---|---|
| **Conversion:** 1 ppm = 4.92 mg/m³ | **DOT:** 2325 129 | | | |

**Synonyms/Trade Names:** Mesitylene, Symmetrical trimethylbenzene, sym-Trimethylbenzene

**Exposure Limits:**
**NIOSH REL:** TWA 25 ppm (125 mg/m³)
**OSHA PEL†:** none

**Measurement Methods (see Table 1):**
**OSHA** PV2091

**Physical Description:** Clear, colorless liquid with a distinctive, aromatic odor.

| Chemical & Physical Properties: | Personal Protection/Sanitation (see Table 2): | Respirator Recommendations (see Tables 3 and 4): |
|---|---|---|
| **MW:** 120.2<br>**BP:** 329°F<br>**Sol:** 0.002%<br>**Fl.P:** 122°F<br>**IP:** 8.39 eV<br>**Sp.Gr:** 0.86<br>**VP:** 2 mmHg<br>**FRZ:** -49°F<br>**UEL:** ?<br>**LEL:** ?<br>Class II Flammable Liquid | **Skin:** Prevent skin contact<br>**Eyes:** Prevent eye contact<br>**Wash skin:** When contam<br>**Remove:** When wet or contam<br>**Change:** N.R. | Not available. |

**Incompatibilities and Reactivities:** Oxidizers, nitric acid

**Exposure Routes, Symptoms, Target Organs (see Table 5):**
**ER:** Inh, Ing, Con
**SY:** Irrit eyes, skin, nose, throat, resp sys; bron; hypochromic anemia; head, drow, lass, dizz, nau, inco; vomit, conf; chemical pneu (aspir liquid)
**TO:** Eyes, skin, resp sys, CNS, blood

**First Aid (see Table 6):**
**Eye:** Irr immed
**Skin:** Soap wash
**Breath:** Resp support
**Swallow:** Medical attention immed

## Trimethyl phosphite

| Trimethyl phosphite | Formula: $(CH_3O)_3P$ | CAS#: 121-45-9 | RTECS#: TH1400000 | IDLH: N.D. |
|---|---|---|---|---|
| **Conversion:** 1 ppm = 5.08 mg/m³ | **DOT:** 2329 129 | | | |

**Synonyms/Trade Names:** Methyl phosphite, Trimethoxyphosphine, Trimethyl ester of phosphorous acid

**Exposure Limits:**
**NIOSH REL:** TWA 2 ppm (10 mg/m³)
**OSHA PEL†:** none

**Measurement Methods (see Table 1):**
None available

**Physical Description:** Colorless liquid with a distinctive, pungent odor.

| Chemical & Physical Properties: | Personal Protection/Sanitation (see Table 2): | Respirator Recommendations (see Tables 3 and 4): |
|---|---|---|
| **MW:** 124.1<br>**BP:** 232°F<br>**Sol:** Reacts<br>**Fl.P:** 82°F<br>**IP:** ?<br>**Sp.Gr:** 1.05<br>**VP(77°F):** 24 mmHg<br>**FRZ:** -108°F<br>**UEL:** ?<br>**LEL:** ?<br>Class IC Flammable Liquid | **Skin:** Prevent skin contact<br>**Eyes:** Prevent eye contact<br>**Wash skin:** When contam<br>**Remove:** When wet (flamm)<br>**Change:** N.R.<br>**Provide:** Quick drench | Not available. |

**Incompatibilities and Reactivities:** Magnesium perchlorate, water [**Note:** Reacts (hydrolyzes) with water.]

**Exposure Routes, Symptoms, Target Organs (see Table 5):**
**ER:** Inh, Ing, Con
**SY:** Irrit eyes, skin, upper resp sys; derm; in animals: terato effects
**TO:** Eyes, skin, resp sys, repro sys

**First Aid (see Table 6):**
**Eye:** Irr immed
**Skin:** Soap flush immed
**Breath:** Resp support
**Swallow:** Medical attention immed

## 2,4,6-Trinitrotoluene

| | | | |
|---|---|---|---|
| **Formula:** $CH_3C_6H_2(NO_2)_3$ | **CAS#:** 118-96-7 | **RTECS#:** XU0175000 | **IDLH:** 500 mg/m$^3$ |

**Conversion:**

**DOT:** 1356 113 (wet)

**Synonyms/Trade Names:** 1-Methyl-2,4,6-trinitrobenzene; TNT; Trinitrotoluene; sym-Trinitrotoluene; Trinitrotoluol

**Exposure Limits:**
**NIOSH REL:** TWA 0.5 mg/m$^3$ [skin]
**OSHA PEL†:** TWA 1.5 mg/m$^3$ [skin]

**Measurement Methods (see Table 1):**
OSHA 44

**Physical Description:** Colorless to pale-yellow, odorless solid or crushed flakes.

**Chemical & Physical Properties:**
**MW:** 227.1
**BP:** 464°F (Explodes)
**Sol(77°F):** 0.01%
**Fl.P:** ? (Explodes)
**IP:** 10.59 eV
**Sp.Gr:** 1.65
**VP:** 0.0002 mmHg
**MLT:** 176°F
**UEL:** ?
**LEL:** ?
Combustible Solid (Class A Explosive)

**Personal Protection/Sanitation (see Table 2):**
**Skin:** Prevent skin contact
**Eyes:** Prevent eye contact
**Wash skin:** When contam/Daily
**Remove:** When wet or contam
**Change:** Daily

**Respirator Recommendations (see Tables 3 and 4):**
**NIOSH**
**5 mg/m$^3$:** Sa*
**12.5 mg/m$^3$:** Sa:Cf*
**25 mg/m$^3$:** ScbaF/SaF
**500 mg/m$^3$:** SaF:Pd,Pp
**§:** ScbaF:Pd,Pp/SaF:Pd,Pp:AScba
**Escape:** GmFOv100/ScbaE

**Incompatibilities and Reactivities:** Strong oxidizers, ammonia, strong alkalis, combustible materials, heat [**Note:** Rapid heating will result in detonation.]

**Exposure Routes, Symptoms, Target Organs (see Table 5):**
**ER:** Inh, Abs, Ing, Con
**SY:** Irrit skin, muc memb; liver damage, jaun; cyan; sneez; cough, sore throat; peri neur, musc pain; kidney damage; cataract; sens derm; leucyt; anemia; card irreg
**TO:** Eyes, skin, resp sys, blood, liver, CVS, CNS, kidneys

**First Aid (see Table 6):**
**Eye:** Irr immed
**Skin:** Soap wash prompt
**Breath:** Resp support
**Swallow:** Medical attention immed

## Triorthocresyl phosphate

| | | | |
|---|---|---|---|
| **Formula:** $(CH_3C_6H_40)_3PO$ | **CAS#:** 78-30-8 | **RTECS#:** TD0350000 | **IDLH:** 40 mg/m$^3$ |

**Conversion:**

**DOT:** 2574 151

**Synonyms/Trade Names:** TCP, TOCP, Tri-o-cresyl ester of phosphoric acid, Tri-o-cresyl phosphate

**Exposure Limits:**
**NIOSH REL:** TWA 0.1 mg/m$^3$ [skin]
**OSHA PEL†:** TWA 0.1 mg/m$^3$

**Measurement Methods (see Table 1):**
NIOSH 5037

**Physical Description:** Colorless to pale-yellow, odorless liquid or solid (below 52°F).

T

**Chemical & Physical Properties:**
**MW:** 368.4
**BP:** 770°F (Decomposes)
**Sol:** Slight
**Fl.P:** 437°F
**IP:** ?
**Sp.Gr:** 1.20
**VP(77°F):** 0.00002 mmHg
**FRZ:** 52°F
**UEL:** ?
**LEL:** ?
Class IIIB Combustible Liquid

**Personal Protection/Sanitation (see Table 2):**
**Skin:** Prevent skin contact
**Eyes:** N.R.
**Wash skin:** When contam
**Remove:** When wet or contam
**Change:** N.R.

**Respirator Recommendations (see Tables 3 and 4):**
**NIOSH/OSHA**
**0.5 mg/m$^3$:** Qm
**1 mg/m$^3$:** 95XQ/Sa
**2.5 mg/m$^3$:** Sa:Cf/PaprHie
**5 mg/m$^3$:** 100F/SaT:Cf/PaprTHie/ScbaF/SaF
**40 mg/m$^3$:** Sa:Pd,Pp
**§:** ScbaF:Pd,Pp/SaF:Pd,Pp:AScba
**Escape:** 100F/ScbaE

**Incompatibilities and Reactivities:** Oxidizers

**Exposure Routes, Symptoms, Target Organs (see Table 5):**
**ER:** Inh, Abs, Ing, Con
**SY:** GI dist; peri neur; cramps in calves, pares in feet or hands; weak feet, wrist drop, para
**TO:** PNS, CNS

**First Aid (see Table 6):**
**Eye:** Irr immed
**Skin:** Soap wash immed
**Breath:** Resp support
**Swallow:** Medical attention immed

## Triphenylamine

| Formula: | CAS#: | RTECS#: | IDLH: |
|---|---|---|---|
| $(C_6H_5)_3N$ | 603-34-9 | YK2680000 | N.D. |

**Conversion:**

**DOT:**

**Synonyms/Trade Names:** N,N-Diphenylaniline; N,N-Diphenylbenzenamine

**Exposure Limits:**
**NIOSH REL:** TWA 5 mg/m$^3$
**OSHA PEL†:** none

**Measurement Methods (see Table 1):**
None available

**Physical Description:** Colorless solid.

**Chemical & Physical Properties:**
**MW:** 245.3
**BP:** 689°F
**Sol:** Insoluble
**Fl.P:** ?
**IP:** 7.60 eV
**Sp.Gr:** 0.77
**VP:** ?
**MLT:** 261°F
**UEL:** ?
**LEL:** ?

**Personal Protection/Sanitation (see Table 2):**
**Skin:** Prevent skin contact
**Eyes:** Prevent eye contact
**Wash skin:** Daily
**Remove:** N.R.
**Change:** Daily

**Respirator Recommendations (see Tables 3 and 4):**
Not available.

**Incompatibilities and Reactivities:** None reported

**Exposure Routes, Symptoms, Target Organs (see Table 5):**
**ER:** Inh, Ing, Con
**SY:** In animals: irrit skin
**TO:** Skin

**First Aid (see Table 6):**
**Eye:** Irr immed
**Skin:** Soap wash
**Breath:** Resp support
**Swallow:** Medical attention immed

## Triphenyl phosphate

| Formula: | CAS#: | RTECS#: | IDLH: |
|---|---|---|---|
| $(C_6H_5O)_3PO$ | 115-86-6 | TC8400000 | 1000 mg/m$^3$ |

**Conversion:**

**DOT:**

**Synonyms/Trade Names:** Phenyl phosphate, TPP, Triphenyl ester of phosphoric acid

**Exposure Limits:**
**NIOSH REL:** TWA 3 mg/m$^3$
**OSHA PEL:** TWA 3 mg/m$^3$

**Measurement Methods (see Table 1):**
**NIOSH** 5038

**Physical Description:** Colorless, crystalline powder with a phenol-like odor.

**Chemical & Physical Properties:**
**MW:** 326.3
**BP:** 776°F
**Sol(129°F):** 0.002%
**Fl.P:** 428°F
**IP:** ?
**Sp.Gr:** 1.29
**VP(380°F):** 1 mmHg
**MLT:** 120°F
**UEL:** ?
**LEL:** ?
Combustible Solid

**Personal Protection/Sanitation (see Table 2):**
**Skin:** N.R.
**Eyes:** N.R.
**Wash skin:** N.R.
**Remove:** N.R.
**Change:** N.R.

**Respirator Recommendations (see Tables 3 and 4):**
**NIOSH/OSHA**
**15 mg/m$^3$:** Qm
**30 mg/m$^3$:** 95XQ/Sa
**75 mg/m$^3$:** Sa:Cf/PaprHie
**150 mg/m$^3$:** 100F/SaT:Cf/PaprTHie/ScbaF/SaF
**1000 mg/m$^3$:** Sa:Pd,Pp
**§:** ScbaF:Pd,Pp/SaF:Pd,Pp:AScba
**Escape:** 100F/ScbaE

**Incompatibilities and Reactivities:** None reported

**Exposure Routes, Symptoms, Target Organs (see Table 5):**
**ER:** Inh, Ing
**SY:** Minor changes in blood enzymes; in animals: musc weak, para
**TO:** Blood, PNS

**First Aid (see Table 6):**
**Breath:** Resp support
**Swallow:** Medical attention immed

## Tungsten

| Tungsten | Formula: W | CAS#: 7440-33-7 | RTECS#: YO7175000 | IDLH: N.D. |
| --- | --- | --- | --- | --- |
| **Conversion:** | **DOT:** | | | |

**Synonyms/Trade Names:** Tungsten metal, Wolfram

**Exposure Limits:**
**NIOSH REL*:** TWA 5 mg/m$^3$
ST 10 mg/m$^3$
[***Note:** The REL also applies to other insoluble tungsten compounds (as W).]
**OSHA PEL†:** none

**Measurement Methods (see Table 1):**
**NIOSH** 7074, 7300, 7301
**OSHA** ID213

**Physical Description:** Hard, brittle, steel-gray to tin-white solid.

**Chemical & Physical Properties:**
**MW:** 183.9
**BP:** 10,701°F
**Sol:** Insoluble
**Fl.P:** NA
**IP:** NA
**Sp.Gr:** 19.3
**VP:** 0 mmHg (approx)
**MLT:** 6170°F
**UEL:** NA
**LEL:** NA
Combustible in the form of finely divided powder; may ignite spontaneously.

**Personal Protection/Sanitation (see Table 2):**
**Skin:** N.R.
**Eyes:** N.R.
**Wash skin:** N.R.
**Remove:** N.R.
**Change:** N.R.

**Respirator Recommendations (see Tables 3 and 4):**
**NIOSH**
**50 mg/m$^3$:** 100XQ/Sa/ScbaF
**§:** ScbaF:Pd,Pp/SaF:Pd,Pp:AScba
**Escape:** 100XQ/ScbaE

**Incompatibilities and Reactivities:** Bromine trifluoride, chlorine trifluoride, fluorine, iodine pentafluoride

**Exposure Routes, Symptoms, Target Organs (see Table 5):**
**ER:** Inh, Ing, Con
**SY:** Irrit eyes, skin, resp sys; diffuse pulm fib; loss of appetite, nau, cough; blood changes
**TO:** Eyes, skin, resp sys, blood

**First Aid (see Table 6):**
**Eye:** Irr immed
**Skin:** Soap wash
**Breath:** Fresh air
**Swallow:** Medical attention immed

## Tungsten (soluble compounds, as W)

| Tungsten (soluble compounds, as W) | Formula: | CAS#: | RTECS#: | IDLH: N.D. |
| --- | --- | --- | --- | --- |
| **Conversion:** | **DOT:** | | | |

**Synonyms/Trade Names:** Synonyms vary depending upon the specific soluble tungsten compound.

**Exposure Limits:**
**NIOSH REL:** TWA 1 mg/m$^3$
ST 3 mg/m$^3$
**OSHA PEL†:** none

**Measurement Methods (see Table 1):**
**NIOSH** 7074, 7300, 7301
**OSHA** ID213

T

**Physical Description:** Appearance and odor vary depending upon the specific soluble tungsten compound.

**Chemical & Physical Properties:**
Properties vary depending upon the specific soluble tungsten compound.

**Personal Protection/Sanitation (see Table 2):**
Recommendations regarding personal protective clothing vary depending upon the specific compound.

**Respirator Recommendations (see Tables 3 and 4):**
**NIOSH**
**10 mg/m$^3$:** 100XQ/Sa
**25 mg/m$^3$:** Sa:Cf
**50 mg/m$^3$:** 100F/ScbaF/SaF
**§:** ScbaF:Pd,Pp/SaF:Pd,Pp:AScba
**Escape:** 100F/ScbaE

**Incompatibilities and Reactivities:** Varies

**Exposure Routes, Symptoms, Target Organs (see Table 5):**
**ER:** Inh, Ing, Con
**SY:** Irrit eyes, skin, resp sys; in animals: CNS disturbances; diarr; resp failure; behavioral, body weight, blood changes
**TO:** Eyes, skin, resp sys, CNS, GI tract

**First Aid (see Table 6):**
**Eye:** Irr immed
**Skin:** Water wash
**Breath:** Resp support
**Swallow:** Medical attention immed

## Tungsten carbide (cemented)

**Formula:** WC/Co/Ni/Ti

**CAS#:** 1: 11107-01-0; 2: 12718-69-3; 3: 37329-49-0

**RTECS#:** 1: YO7350000; 2: YO7525000; 3: YO7700000

**IDLH:** N.D.

**Conversion:**

**DOT:**

**Synonyms/Trade Names:** Cemented tungsten carbide, Cemented WC, Hard metal
[**Note:** The tungsten carbide (WC) content is generally 85-95% & the cobalt content is generally 5-15%.]
[**1:** 85% WC, 15% Co; **2:** 92% WC, 8% Co; **3:** 78% WC, 14% Co, 8% Ti]

**Exposure Limits:**
**NIOSH REL:** See Appendix C  **OSHA PEL†:** See Appendix C

**Measurement Methods (see Table 1):**
None available

**Physical Description:** A mixture of tungsten carbide, cobalt, and sometimes other metals & metal oxides or carbides.

**Chemical & Physical Properties:**
Properties vary depending upon the specific mixture.

**Personal Protection/Sanitation (see Table 2):**
**Skin:** Prevent skin contact
**Eyes:** Prevent eye contact
**Wash skin:** When contam/Daily (Ni)
**Remove:** When wet or contam
**Change:** Daily

**Respirator Recommendations (see Tables 3 and 4):**
**NIOSH**
**0.25 mg Co/m$^3$:** Qm
**0.5 mg Co/m$^3$:** 95XQ*/Sa*
**1.25 mg Co/m$^3$:** Sa:Cf*/PaprHie*/PaprHie*
**2.5 mg Co/m$^3$:** 100F/ScbaF/SaF
**20 mg Co/m$^3$:** SaF:Pd,Pp
**§:** ScbaF:Pd,Pp/SaF:Pd,Pp:AScba
**Escape:** 100F/ScbaE

**Tungsten carbide (cemented) containing Nickel:**
**NIOSH**
**¥:**ScbaF:Pd,Pp/SaF:Pd,Pp:AScba
**Escape:** 100F/ScbaE

**Incompatibilities and Reactivities:** Tungsten carbide: Fluorine, chlorine trifluoride, oxides of nitrogen, lead dioxide

**Exposure Routes, Symptoms, Target Organs (see Table 5):**
**ER:** Inh, Ing, Con
**SY:** Irrit eyes, skin, resp sys; possible skin sens to cobalt, nickel; diffuse pulm fib; loss of appetite, nau, cough; blood changes
**TO:** Eyes, skin, resp sys, blood

**First Aid (see Table 6):**
**Eye:** Irr immed
**Skin:** Soap wash
**Breath:** Fresh air
**Swallow:** Medical attention immed

## Turpentine

**Formula:** $C_{10}H_{16}$ (approx)

**CAS#:** 8006-64-2

**RTECS#:** YO8400000

**IDLH:** 800 ppm

**Conversion:** 1 ppm = 5.56 mg/m$^3$ (approx)

**DOT:** 1299 128

**Synonyms/Trade Names:** Gumspirits, Gum turpentine, Spirits of turpentine, Steam distilled turpentine, Sulfate wood turpentine, Turps, Wood turpentine

**Exposure Limits:**
**NIOSH REL:** TWA 100 ppm (560 mg/m$^3$)
**OSHA PEL:** TWA 100 ppm (560 mg/m$^3$)

**Measurement Methods (see Table 1):**
**NIOSH** 1551

**Physical Description:** Colorless liquid with a characteristic odor.

**Chemical & Physical Properties:**
**MW:** 136 (approx)
**BP:** 309-338°F
**Sol:** Insoluble
**Fl.P:** 95°F
**IP:** ?
**Sp.Gr:** 0.86
**VP:** 4 mmHg
**FRZ:** -58 to -76°F
**UEL:** ?
**LEL:** 0.8%
Class IC Flammable Liquid

**Personal Protection/Sanitation (see Table 2):**
**Skin:** Prevent skin contact
**Eyes:** Prevent eye contact
**Wash skin:** When contam
**Remove:** When wet (flamm)
**Change:** N.R.

**Respirator Recommendations (see Tables 3 and 4):**
**NIOSH/OSHA**
**800 ppm:** Sa:Cf£/PaprOv£/CcrFOv/GmFOv/ScbaF/SaF
**§:** ScbaF:Pd,Pp/SaF:Pd,Pp:AScba
**Escape:** GmFOv/ScbaE

**Incompatibilities and Reactivities:** Strong oxidizers, chlorine, chromic anhydride, stannic chloride, chromyl chloride

**Exposure Routes, Symptoms, Target Organs (see Table 5):**
**ER:** Inh, Abs, Ing, Con
**SY:** Irrit eyes, skin, nose, throat; head, dizz, convuls; skin sens; hema, prot; kidney damage; abdom pain, nau, vomit, diarr; chemical pneu (aspir liquid)
**TO:** Eyes, skin, resp sys, CNS, kidneys

**First Aid (see Table 6):**
**Eye:** Irr immed
**Skin:** Soap wash prompt
**Breath:** Resp support
**Swallow:** Medical attention immed

| 1-Undecanethiol | Formula: $CH_3(CH_2)_{10}SH$ | CAS#: 5332-52-5 | RTECS#: | IDLH: N.D. |
|---|---|---|---|---|

**Conversion:** 1 ppm = 7.71 mg/m$^3$ | **DOT:** 1228 131

**Synonyms/Trade Names:** Undecyl mercaptan

**Exposure Limits:**
**NIOSH REL:** C 0.5 ppm (3.9 mg/m$^3$) [15-minute]
**OSHA PEL:** none

**Measurement Methods (see Table 1):**
None available

**Physical Description:** Liquid.

**Chemical & Physical Properties:**
**MW:** 188.4
**BP:** 495°F
**Sol:** Insoluble
**Fl.P:** ?
**IP:** ?
**Sp.Gr:** 0.84
**VP:** ?
**FRZ:** 27°F
**UEL:** ?
**LEL:** ?
Combustible Liquid

**Personal Protection/Sanitation (see Table 2):**
**Skin:** Prevent skin contact
**Eyes:** Prevent eye contact
**Wash skin:** When contam
**Remove:** When wet (flamm)
**Change:** N.R.

**Respirator Recommendations (see Tables 3 and 4):**
**NIOSH**
**5 ppm:** CcrOv/Sa
**12.5 ppm:** Sa:Cf/PaprOv
**25 ppm:** CcrFOv/GmFOv/PaprTOv/ScbaF/SaF
**§:** ScbaF:Pd,Pp/SaF:Pd,Pp:AScba
**Escape:** GmFOv/ScbaE

**Incompatibilities and Reactivities:** Oxidizers, reducing agents, strong acids & bases, alkali metals

**Exposure Routes, Symptoms, Target Organs (see Table 5):**
**ER:** Inh, Abs, Ing, Con
**SY:** Irrit eyes, skin, resp sys; conf, dizz, head, drow, nau, vomit, lass, convuls
**TO:** Eyes, skin, resp sys, CNS

**First Aid (see Table 6):**
**Eye:** Irr immed
**Skin:** Soap wash
**Breath:** Resp support
**Swallow:** Medical attention immed

| Uranium (insoluble compounds, as U) | Formula: U (metal) | CAS#: 7440-61-1 (metal) | RTECS#: YR3490000 (metal) | IDLH: Ca [10 mg/m$^3$ (as U)] |
|---|---|---|---|---|

**Conversion:** | **DOT:** 2979 162 (metal, pyrophoric)

**Synonyms/Trade Names: Uranium metal:** Uranium I
Synonyms of other insoluble uranium compounds vary depending upon the specific compound.

**Exposure Limits:**
**NIOSH REL:** Ca
TWA 0.2 mg/m$^3$
ST 0.6 mg/m$^3$
See Appendix A
**OSHA PEL†:** TWA 0.25 mg/m$^3$

**Measurement Methods (see Table 1):**
None available

**Physical Description:** Metal: Silver-white, malleable, ductile, lustrous solid. [**Note:** Weakly radioactive.]

U

**Chemical & Physical Properties:**
**MW:** 238.0
**BP:** 6895°F
**Sol:** Insoluble
**Fl.P:** NA
**IP:** NA
**Sp.Gr:** 19.05 (metal)
**VP:** 0 mmHg (approx)
**MLT:** 2097°F
**UEL:** NA
**LEL:** NA
**MEC:** 60 g/m$^3$
Metal: Combustible Solid, especially turnings and powder.

**Personal Protection/Sanitation (see Table 2):**
**Skin:** Prevent skin contact
**Eyes:** Prevent eye contact
**Wash skin:** When contam/Daily
**Remove:** When wet or contam
**Change:** Daily
**Provide:** Eyewash

**Respirator Recommendations (see Tables 3 and 4):**
**NIOSH**
**¥:** ScbaF:Pd,Pp/SaF:Pd,Pp:AScba
**Escape:** 100F/ScbaE

**Incompatibilities and Reactivities:** Carbon dioxide, carbon tetrachloride, nitric acid, fluorine [**Note:** Complete coverage of uranium metal scrap with oil is essential for prevention of fire.]

**Exposure Routes, Symptoms, Target Organs (see Table 5):**
**ER:** Inh, Ing, Con
**SY:** Derm; kidney damage; blood changes; [carc]; in animals: lung, lymph node damage; [carc] Potential for cancer is a result of alpha-emitting properties & radioactive decay products (e.g., radon).
**TO:** Skin, kidneys, bone marrow, lymphatic sys [lung cancer]

**First Aid (see Table 6):**
**Eye:** Irr immed
**Skin:** Soap wash prompt
**Breath:** Resp support
**Swallow:** Medical attention immed

## Uranium (soluble compounds, as U)

| Uranium (soluble compounds, as U) | Formula: | CAS#: | RTECS#: | IDLH: Ca [10 mg/m$^3$ (as U)] |
|---|---|---|---|---|
| **Conversion:** | **DOT:** | | | |

**Synonyms/Trade Names:** Synonyms vary depending upon the specific soluble uranium compound.

**Exposure Limits:**
**NIOSH REL:** Ca
TWA 0.05 mg/m$^3$
See Appendix A
**OSHA PEL:** TWA 0.05 mg/m$^3$

**Measurement Methods (see Table 1):**
None available

**Physical Description:** Appearance and odor vary depending upon the specific soluble uranium compound.

**Chemical & Physical Properties:**
Properties vary depending upon the specific soluble uranium compound.

**Personal Protection/Sanitation (see Table 2):**
**Skin:** Prevent skin contact
**Eyes:** Prevent eye contact
**Wash skin:** When contam/Daily
**Remove:** When wet or contam
**Change:** Daily
**Provide:** Eyewash ($UF_6$), Quick drench

**Respirator Recommendations (see Tables 3 and 4):**
**NIOSH**
**¥:** ScbaF:Pd,Pp/SaF:Pd,Pp:AScba
**Escape (Halides):** GmFAg100/ScbaE
**Escape (Non-halides):** 100F/ScbaE

**Incompatibilities and Reactivities: Uranyl nitrate:** combustibles; **Uranium hexafluoride:** water

**Exposure Routes, Symptoms, Target Organs (see Table 5):**
**ER:** Inh, Ing, Con
**SY:** Lac, conj; short breath, cough, chest rales; nau, vomit; skin burns; RBC, casts in urine; prot; high BUN; [carc] Potential for cancer is a result of alpha-emitting properties & radioactive decay products (e.g., radon).
**TO:** Resp sys, blood, liver, kidneys, lymphatic sys, skin, bone marrow [lung cancer]

**First Aid (see Table 6):**
**Eye:** Irr immed
**Skin:** Water flush immed
**Breath:** Resp support
**Swallow:** Medical attention immed

## n-Valeraldehyde

| n-Valeraldehyde | Formula: $CH_3(CH_2)_3CHO$ | CAS#: 110-62-3 | RTECS#: YV3600000 | IDLH: N.D. |
|---|---|---|---|---|
| **Conversion:** 1 ppm = 3.53 mg/m$^3$ | **DOT:** 2058 129 | | | |

**Synonyms/Trade Names:** Amyl aldehyde, Pentanal, Valeral, Valeraldehyde, Valeric aldehyde

**Exposure Limits:**
**NIOSH REL:** TWA 50 ppm (175 mg/m$^3$)
See Appendix C (Aldehydes)
**OSHA PEL†:** none

**Measurement Methods (see Table 1):**
**NIOSH** 2018, 2536
**OSHA** 85

**Physical Description:** Colorless liquid with a strong, acrid, pungent odor.

**Chemical & Physical Properties:**
**MW:** 86.2
**BP:** 217°F
**Sol:** Slight
**Fl.P:** 54°F
**IP:** 9.82 eV
**Sp.Gr:** 0.81
**VP:** 26 mmHg
**FRZ:** -133°F
**UEL:** ?
**LEL:** ?
Class IB Flammable Liquid

**Personal Protection/Sanitation (see Table 2):**
**Skin:** Prevent skin contact
**Eyes:** Prevent eye contact
**Wash skin:** When contam
**Remove:** When wet (flamm)
**Change:** N.R.
**Provide:** Eyewash
Quick drench

**Respirator Recommendations (see Tables 3 and 4):**
Not available.

**Incompatibilities and Reactivities:** None reported

**Exposure Routes, Symptoms, Target Organs (see Table 5):**
**ER:** Inh, Ing, Con
**SY:** Irrit eyes, skin, nose, throat
**TO:** Eyes, skin, resp sys

**First Aid (see Table 6):**
**Eye:** Irr immed
**Skin:** Soap flush immed
**Breath:** Resp support
**Swallow:** Medical attention immed

## Vanadium dust

| Vanadium dust | Formula: $V_2O_5$ | CAS#: 1314-62-1 | RTECS#: YW2450000 | IDLH: 35 mg/m³ (as V) |
|---|---|---|---|---|
| **Conversion:** | **DOT:** 2862 151 | | | |

**Synonyms/Trade Names:** Divanadium pentoxide dust, Vanadic anhydride dust, Vanadium oxide dust, Vanadium pentaoxide dust. Other synonyms vary depending upon the specific vanadium compound.

**Exposure Limits:**
**NIOSH REL*:** C 0.05 mg V/m³ [15-minute]
[***Note:** The REL applies to all vanadium compounds except Vanadium metal and Vanadium carbide (see Ferrovanadium dust).]
**OSHA PEL†:** C 0.5 mg $V_2O_5$/m³ (resp)

**Measurement Methods (see Table 1):**
**NIOSH** 7300, 7301, 7303, 7504, 9102
**OSHA** ID185

**Physical Description:** Yellow-orange powder or dark-gray, odorless flakes dispersed in air.

**Chemical & Physical Properties:**
**MW:** 181.9
**BP:** 3182°F (Decomposes)
**Sol:** 0.8%
**Fl.P:** NA
**IP:** NA
**Sp.Gr:** 3.36
**VP:** 0 mmHg (approx)
**MLT:** 1274°F
**UEL:** NA
**LEL:** NA
Noncombustible Solid, but may increase intensity of fire when in contact with combustible materials.

**Personal Protection/Sanitation (see Table 2):**
**Skin:** Prevent skin contact
**Eyes:** Prevent eye contact
**Wash skin:** When contam
**Remove:** When wet or contam
**Change:** N.R.

**Respirator Recommendations (see Tables 3 and 4):**
**NIOSH** (as V)
**0.5 mg/m³:** 100XQ*/Sa*
**1.25 mg/m³:** Sa:Cf*/PaprHie*
**2.5 mg/m³:** 100F/PaprTHie*/ScbaF/SaF
**35 mg/m³:** SaF:Pd,Pp
**§:** ScbaF:Pd,Pp/SaF:Pd,Pp:AScba
**Escape:** 100F/ScbaE

**Incompatibilities and Reactivities:** Lithium, chlorine trifluoride

**Exposure Routes, Symptoms, Target Organs (see Table 5):**
**ER:** Inh, Ing, Con
**SY:** Irrit eyes, skin, throat; green tongue, metallic taste, eczema; cough; fine rales, wheez, bron, dysp
**TO:** Eyes, skin, resp sys

**First Aid (see Table 6):**
**Eye:** Irr immed
**Skin:** Soap wash prompt
**Breath:** Resp support
**Swallow:** Medical attention immed

## Vanadium fume

| Vanadium fume | Formula: $V_2O_5$ | CAS#: 1314-62-1 | RTECS#: YW2460000 | IDLH: 35 mg/m³ (as V) |
|---|---|---|---|---|
| **Conversion:** | **DOT:** 2862 151 | | | |

**Synonyms/Trade Names:** Divanadium pentoxide fume, Vanadic anhydride fume, Vanadium oxide fume, Vanadium pentaoxide fume. Other synonyms vary depending upon the specific vanadium compound.

**Exposure Limits:**
**NIOSH REL:** C 0.05 mg V/m³ [15-minute]
**OSHA PEL†:** C 0.1 mg $V_2O_5$/m³

**Measurement Methods (see Table 1):**
**NIOSH** 7300, 7301, 7303, 7504
**OSHA** ID185

**Physical Description:** Finely divided particulate dispersed in air.

V

**Chemical & Physical Properties:**
**MW:** 181.9
**BP:** 3182°F (Decomposes)
**Sol:** 0.8%
**Fl.P:** NA
**IP:** NA
**Sp.Gr:** 3.36
**VP:** 0 mmHg (approx)
**MLT:** 1274°F
**UEL:** NA
**LEL:** NA
Noncombustible Solid

**Personal Protection/Sanitation (see Table 2):**
**Skin:** N.R.
**Eyes:** N.R.
**Wash skin:** N.R.
**Remove:** N.R.
**Change:** N.R.

**Respirator Recommendations (see Tables 3 and 4):**
**NIOSH** (as V)
**0.5 mg/m³:** 100XQ*/Sa*
**1.25 mg/m³:** Sa:Cf*/PaprHie*
**2.5 mg/m³:** 100F/PaprTHie*/ScbaF/SaF
**35 mg/m³:** SaF:Pd,Pp
**§:** ScbaF:Pd,Pp/SaF:Pd,Pp:AScba
**Escape:** 100F/ScbaE

**Incompatibilities and Reactivities:** Lithium, chlorine trifluoride

**Exposure Routes, Symptoms, Target Organs (see Table 5):**
**ER:** Inh, Con
**SY:** Irrit eyes, throat; green tongue, metallic taste; cough, fine rales, wheez, bron, dysp; eczema
**TO:** Eyes, skin, resp sys

**First Aid (see Table 6):**
**Breath:** Resp support

## Vegetable oil mist

| | | | | |
|---|---|---|---|---|
| **Vegetable oil mist** | **Formula:** | **CAS#:** 68956-68-3 | **RTECS#:** YX1850000 | **IDLH:** N.D. |
| **Conversion:** | **DOT:** | | | |

**Synonyms/Trade Names:** Vegetable mist

**Exposure Limits:**
**NIOSH REL:** TWA 10 mg/m$^3$ (total)
TWA 5 mg/m$^3$ (resp)
**OSHA PEL:** TWA 15 mg/m$^3$ (total)
TWA 5 mg/m$^3$ (resp)

**Measurement Methods (see Table 1):**
**NIOSH** 0500, 0600

**Physical Description:** An oil extracted from the seeds, fruit, or nuts of vegetables or other plant matter.

**Chemical & Physical Properties:**
**MW:** varies
**BP:** ?
**Sol:** Insoluble
**Fl.P:** 323-540°F
**IP:** ?
**Sp.Gr:** 0.91-0.95
**VP:** ?
**FRZ:** ?
**UEL:** ?
**LEL:** ?
Combustible Liquid

**Personal Protection/Sanitation (see Table 2):**
**Skin:** N.R.
**Eyes:** N.R.
**Wash skin:** N.R.
**Remove:** N.R.
**Change:** N.R.

**Respirator Recommendations (see Tables 3 and 4):**
Not available.

**Incompatibilities and Reactivities:** None reported

**Exposure Routes, Symptoms, Target Organs (see Table 5):**
**ER:** Inh, Con
**SY:** Irrit eyes, skin, resp sys; lac
**TO:** Eyes, skin, resp sys Determine based on working conditions

**First Aid (see Table 6):**
**Eye:** Irr immed
**Breath:** Fresh air

## Vinyl acetate

| | | | | |
|---|---|---|---|---|
| **Vinyl acetate** | **Formula:** $CH_2$=CHOOCC$H_3$ | **CAS#:** 108-05-4 | **RTECS#:** AK0875000 | **IDLH:** N.D. |
| **Conversion:** 1 ppm = 3.52 mg/m$^3$ | **DOT:** 1301 129P | | | |

**Synonyms/Trade Names:** 1-Acetoxyethylene, Ethenyl acetate, Ethenyl ethanoate, VAC, Vinyl acetate monomer, Vinyl ethanoate

**Exposure Limits:**
**NIOSH REL:** C 4 ppm (15 mg/m$^3$) [15-minute]
**OSHA PEL†:** none

**Measurement Methods (see Table 1):**
**NIOSH** 1453
**OSHA** 51

**Physical Description:** Colorless liquid with a pleasant, fruity odor.
[**Note:** Raw material for many polyvinyl resins.]

**Chemical & Physical Properties:**
**MW:** 86.1
**BP:** 162°F
**Sol:** 2%
**Fl.P:** 18°F
**IP:** 9.19 eV
**Sp.Gr:** 0.93
**VP:** 83 mmHg
**FRZ:** -136°F
**UEL:** 13.4%
**LEL:** 2.6%
Class IB Flammable Liquid

**Personal Protection/Sanitation (see Table 2):**
**Skin:** Prevent skin contact
**Eyes:** Prevent eye contact
**Wash skin:** When contam
**Remove:** When wet or contam
**Change:** N.R.
**Provide:** Eyewash
Quick drench

**Respirator Recommendations (see Tables 3 and 4):**
**NIOSH**
**40 ppm:** CcrOv*/Sa*
**100 ppm:** Sa:Cf*/PaprOv*
**200 ppm:** CcrFOv/GmFOv/PaprTOv*/ ScbaF/SaF
**4000 ppm:** Sa:Pd,Pp*
**§:** ScbaF:Pd,Pp/SaF:Pd,Pp:AScba
**Escape:** GmFOv/ScbaE

**Incompatibilities and Reactivities:** Acids, bases, silica gel, alumina, oxidizers, azo compounds, ozone
[**Note:** Usually contains a stabilizer (e.g., hydroquinone or diphenylamine) to prevent polymerization.]

**Exposure Routes, Symptoms, Target Organs (see Table 5):**
**ER:** Inh, Ing, Con
**SY:** Irrit eyes, skin, nose, throat; hoarseness, cough; loss of smell; eye burns, skin blisters
**TO:** Eyes, skin, resp sys

**First Aid (see Table 6):**
**Eye:** Irr immed
**Skin:** Soap flush immed
**Breath:** Resp support
**Swallow:** Medical attention immed

## Vinyl bromide

| Vinyl bromide | Formula: $CH_2=CHBr$ | CAS#: 593-60-2 | RTECS#: KU8400000 | IDLH: Ca [N.D.] |
|---|---|---|---|---|

**Conversion:** 1 ppm = 4.38 mg/$m^3$ | **DOT:** 1085 116P (inhibited)

**Synonyms/Trade Names:** Bromoethene, Bromoethylene, Monobromoethylene

**Exposure Limits:**
**NIOSH REL:** Ca
See Appendix A
**OSHA PEL†:** none

**Measurement Methods (see Table 1):**
**NIOSH** 1009
**OSHA** 8

**Physical Description:** Colorless gas or liquid (below 60°F) with a pleasant odor.
[**Note:** Shipped as a liquefied compressed gas with 0.1% phenol added to prevent polymerization.]

**Chemical & Physical Properties:**
**MW:** 107.0
**BP:** 60°F
**Sol:** Insoluble
**Fl.P:** NA (Gas)
**IP:** 9.80 eV
**RGasD:** 3.79
**Sp.Gr:** 1.49 (Liquid at 60°F)
**VP:** 1.4 atm
**FRZ:** -219°F
**UEL:** 15%
**LEL:** 9%
Flammable Gas

**Personal Protection/Sanitation (see Table 2):**
**Skin:** Prevent skin contact (liquid)
**Eyes:** Prevent eye contact (liquid)
**Wash skin:** When contam (liquid)
**Remove:** When wet (flamm)
**Change:** N.R.

**Respirator Recommendations (see Tables 3 and 4):**
**NIOSH**
**¥:** ScbaF:Pd,Pp/SaF:Pd,Pp:AScba
**Escape:** GmFOv/ScbaE

**Incompatibilities and Reactivities:** Strong oxidizers (e.g., perchlorates, peroxides, chlorates, permanganates & nitrates.)
[**Note:** May polymerize in sunlight.]

**Exposure Routes, Symptoms, Target Organs (see Table 5):**
**ER:** Inh, Ing (liquid), Con
**SY:** Irrit eyes, skin; dizz, conf, inco, narco, nau, vomit; liquid: frostbite; [carc]
**TO:** Eyes, skin, CNS, liver [in animals: liver & lymph node tumors]

**First Aid (see Table 6):**
**Eye:** Irr immed (liquid)
**Skin:** Water flush immed (liquid)
**Breath:** Resp support
**Swallow:** Medical attention immed (liquid)

## Vinyl chloride

| Vinyl chloride | Formula: $CH_2=CHCl$ | CAS#: 75-01-4 | RTECS#: KU9625000 | IDLH: Ca [N.D.] |
|---|---|---|---|---|

**Conversion:** 1 ppm = 2.56 mg/$m^3$ | **DOT:** 1086 116P (inhibited)

**Synonyms/Trade Names:** Chloroethene, Chloroethylene, Ethylene monochloride, Monochloroethene, Monochloroethylene, VC, Vinyl chloride monomer (VCM)

**Exposure Limits:**
**NIOSH REL:** Ca
See Appendix A
**OSHA PEL:** [1910.1017] TWA 1 ppm
C 5 ppm [15-minute]

**Measurement Methods (see Table 1):**
**NIOSH** 1007
**OSHA** 4, 75

**Physical Description:** Colorless gas or liquid (below 7°F) with a pleasant odor at high concentrations.
[**Note:** Shipped as a liquefied compressed gas.]

V

**Chemical & Physical Properties:**
**MW:** 62.5
**BP:** 7°F
**Sol(77°F):** 0.1%
**Fl.P:** NA (Gas)
**IP:** 9.99 eV
**RGasD:** 2.21
**VP:** 3.3 atm
**FRZ:** -256°F
**UEL:** 33.0%
**LEL:** 3.6%
Flammable Gas

**Personal Protection/Sanitation (see Table 2):**
**Skin:** Frostbite
**Eyes:** Frostbite
**Wash skin:** N.R.
**Remove:** When wet (flamm)
**Change:** N.R.
**Provide:** Frostbite wash

**Respirator Recommendations (see Tables 3 and 4):**
**NIOSH**
**¥:** ScbaF:Pd,Pp/SaF:Pd,Pp:AScba
**Escape:** GmFS/ScbaE

**See Appendix E** (page 351)

**Incompatibilities and Reactivities:** Copper, oxidizers, aluminum, peroxides, iron, steel [**Note:** Polymerizes in air, sunlight, or heat unless stabilized by inhibitors such as phenol. Attacks iron & steel in presence of moisture.]

**Exposure Routes, Symptoms, Target Organs (see Table 5):**
**ER:** Inh, Con (liquid)
**SY:** Lass; abdom pain, GI bleeding; enlarged liver; pallor or cyan of extremities; liquid: frostbite; [carc]
**TO:** Liver, CNS, blood, resp sys, lymphatic sys [liver cancer]

**First Aid (see Table 6):**
**Eye:** Frostbite
**Skin:** Frostbite
**Breath:** Resp support

## Vinyl cyclohexene dioxide

| Vinyl cyclohexene dioxide | Formula: $C_8H_{12}O_2$ | CAS#: 106-87-6 | RTECS#: RN8640000 | IDLH: Ca [N.D.] |
|---|---|---|---|---|
| **Conversion:** 1 ppm = 5.73 mg/m$^3$ | **DOT:** | | | |

**Synonyms/Trade Names:** 1-Epoxyethyl-3,4-epoxy-cyclohexane; 4-Vinylcyclohexene diepoxide; 4-Vinyl-1-cyclohexene dioxide

**Exposure Limits:**
**NIOSH REL:** Ca
TWA 10 ppm (60 mg/m$^3$) [skin]
See Appendix A
**OSHA PEL†:** none

**Measurement Methods (see Table 1):**
**OSHA** PV2083

**Physical Description:** Colorless liquid.

**Chemical & Physical Properties:**
**MW:** 140.2
**BP:** 441°F
**Sol:** High
**Fl.P(oc):** 230°F
**IP:** ?
**Sp.Gr:** 1.10
**VP:** 0.1 mmHg
**FRZ:** -164°F
**UEL:** ?
**LEL:** ?
Class IIIB Combustible Liquid

**Personal Protection/Sanitation (see Table 2):**
**Skin:** Prevent skin contact
**Eyes:** Prevent eye contact
**Wash skin:** When contam
**Remove:** When wet or contam
**Change:** N.R.
**Provide:** Eyewash
Quick drench

**Respirator Recommendations (see Tables 3 and 4):**
**NIOSH**
**¥:** ScbaF:Pd,Pp/SaF:Pd,Pp:AScba
**Escape:** GmFOv/ScbaE

**Incompatibilities and Reactivities:** Alcohols, amines, water [**Note:** Slowly hydrolyzes in water.]

**Exposure Routes, Symptoms, Target Organs (see Table 5):**
**ER:** Inh, Abs, Ing, Con
**SY:** In animals: irrit eyes, skin, resp sys; testicular atrophy; leupen, nec thymus; skin sens; [carc]
**TO:** Eyes, skin, resp sys, blood, thymus, repro sys [in animals: skin tumors]

**First Aid (see Table 6):**
**Eye:** Irr immed
**Skin:** Water wash immed
**Breath:** Resp support
**Swallow:** Medical attention immed

## Vinyl fluoride

| Vinyl fluoride | Formula: $CH_2$=CHF | CAS#: 75-02-5 | RTECS#: YZ3510000 | IDLH: N.D. |
|---|---|---|---|---|
| **Conversion:** 1 ppm = 1.89 mg/m$^3$ | **DOT:** 1860 116P (inhibited) | | | |

**Synonyms/Trade Names:** Fluoroethene, Fluoroethylene, Monofluoroethylene, Vinyl fluoride monomer

**Exposure Limits:**
**NIOSH REL:** TWA 1 ppm
C 5 ppm [use 1910.1017]
**OSHA PEL:** none

**Measurement Methods (see Table 1):**
None available

**Physical Description:** Colorless gas with a faint, ethereal odor.
[**Note:** Shipped as a liquefied compressed gas.]

**Chemical & Physical Properties:**
**MW:** 46.1
**BP:** -98°F
**Sol:** Insoluble
**Fl.P:** NA (Gas)
**IP:** 10.37 eV
**RGasD:** 1.60
**VP:** 25.2 atm
**FRZ:** -257°F
**UEL:** 21.7%
**LEL:** 2.6%
Flammable Gas

**Personal Protection/Sanitation (see Table 2):**
**Skin:** Frostbite
**Eyes:** Frostbite
**Wash skin:** N.R.
**Remove:** When wet (flamm)
**Change:** N.R.
**Provide:** Frostbite wash

**Respirator Recommendations (see Tables 3 and 4):**
**NIOSH**
**10 ppm:** CcrOv/Sa
**25 ppm:** Sa:Cf/PaprOv
**50 ppm:** CcrFOv/GmFOv/PaprTOv/ScbaF/SaF
**200 ppm:** SaF:Pd,Pp
**§:** ScbaF:Pd,Pp/SaF:Pd,Pp:AScba
**Escape:** GmFOv/ScbaE

**Incompatibilities and Reactivities:** None reported [**Note:** Inhibited with 0.2% terpenes to prevent polymerization.]

**Exposure Routes, Symptoms, Target Organs (see Table 5):**
**ER:** Inh, Con (liquid)
**SY:** Head, dizz, conf, inco, narco, nau, vomit; liquid: frostbite
**TO:** CNS

**First Aid (see Table 6):**
**Eye:** Frostbite
**Skin:** Frostbite
**Breath:** Resp support

| Vinylidene chloride | Formula: $CH_2=CCl_2$ | CAS#: 75-35-4 | RTECS#: KV9275000 | IDLH: Ca [N.D.] |
|---|---|---|---|---|
| **Conversion:** | **DOT:** 1303 130P (inhibited) | | | |

**Synonyms/Trade Names:** 1,1-DCE; 1,1-Dichloroethene; 1,1-Dichloroethylene; VDC; Vinylidene chloride monomer; Vinylidene dichloride

**Exposure Limits:**
**NIOSH REL:** Ca
See Appendix A
**OSHA PEL†:** none

**Measurement Methods (see Table 1):**
**NIOSH** 1015
**OSHA** 19

**Physical Description:** Colorless liquid or gas (above 89°F) with a mild, sweet, chloroform-like odor.

**Chemical & Physical Properties:**
**MW:** 96.9
**BP:** 89°F
**Sol:** 0.04%
**Fl.P:** -2°F
**IP:** 10.00 eV
**Sp.Gr:** 1.21
**VP:** 500 mmHg
**FRZ:** -189°F
**UEL:** 15.5%
**LEL:** 6.5%
Class IA Flammable Liquid

**Personal Protection/Sanitation (see Table 2):**
**Skin:** Prevent skin contact
**Eyes:** Prevent eye contact
**Wash skin:** When contam
**Remove:** When wet (flamm)
**Change:** N.R.
**Provide:** Eyewash
Quick drench

**Respirator Recommendations (see Tables 3 and 4):**
**NIOSH**
**¥:** ScbaF:Pd,Pp/SaF:Pd,Pp:AScba
**Escape:** GmFOv/ScbaE

**Incompatibilities and Reactivities:** Aluminum, sunlight, air, copper, heat
[**Note:** Polymerization may occur if exposed to oxidizers, chlorosulfonic acid, nitric acid, or oleum. Inhibitors such as the monomethyl ether of hydroquinone are added to prevent polymerization.]

**Exposure Routes, Symptoms, Target Organs (see Table 5):**
**ER:** Inh, Abs, Ing, Con
**SY:** Irrit eyes, skin, throat; dizz, head, nau, dysp; liver, kidney dist; pneu; [carc]
**TO:** Eyes, skin, resp sys, CNS, liver, kidneys [in animals: liver & kidney tumors]

**First Aid (see Table 6):**
**Eye:** Irr immed
**Skin:** Soap flush immed
**Breath:** Resp support
**Swallow:** Medical attention immed

| Vinylidene fluoride | Formula: $CH_2=CF_2$ | CAS#: 75-38-7 | RTECS#: KW0560000 | IDLH: N.D. |
|---|---|---|---|---|
| **Conversion:** 1 ppm = 2.62 mg/m$^3$ | **DOT:** 1959 116P | | | |

**Synonyms/Trade Names:** Difluoro-1,1-ethylene; 1,1-Difluoroethene; 1,1-Difluoroethylene; Halocarbon 1132A; VDF; Vinylidene difluoride

**Exposure Limits:**
**NIOSH REL:** TWA 1 ppm
C 5 ppm [use 1910.1017]
**OSHA PEL:** none

**Measurement Methods (see Table 1):**
**NIOSH** 3800



**Physical Description:** Colorless gas with a faint, ethereal odor. [**Note:** Shipped as a liquefied compressed gas.]

**Chemical & Physical Properties:**
**MW:** 64.0
**BP:** -122°F
**Sol:** Insoluble
**Fl.P:** NA (Gas)
**IP:** 10.29 eV
**RGasD:** 2.21
**VP:** 35.2 atm
**FRZ:** -227°F
**UEL:** 21.3%
**LEL:** 5.5%
Flammable Gas

**Personal Protection/Sanitation (see Table 2):**
**Skin:** Frostbite
**Eyes:** Frostbite
**Wash skin:** N.R.
**Remove:** When wet (flamm)
**Change:** N.R.
**Provide:** Frostbite wash

**Respirator Recommendations (see Tables 3 and 4):**
**NIOSH**
**10 ppm:** CcrOv/Sa
**25 ppm:** Sa:Cf/PaprOv
**50 ppm:** CcrFOv/GmFOv/PaprTOv/ScbaF/SaF
**200 ppm:** SaF:Pd,Pp
**§:** ScbaF:Pd,Pp/SaF:Pd,Pp:AScba
**Escape:** GmFOv/ScbaE

**Incompatibilities and Reactivities:** Oxidizers, aluminum chloride
[**Note:** Violent reaction with hydrogen chloride when heated under pressure.]

**Exposure Routes, Symptoms, Target Organs (see Table 5):**
**ER:** Inh, Con (liquid)
**SY:** Dizz, head, nau; liquid: frostbite
**TO:** CNS

**First Aid (see Table 6):**
**Eye:** Frostbite
**Skin:** Frostbite
**Breath:** Resp support

## Vinyl toluene

| | | | | |
|---|---|---|---|---|
| **Vinyl toluene** | **Formula:** $CH_2=CHC_6H_4CH_3$ | **CAS#:** 25013-15-4 (inhibited) | **RTECS#:** WL5075000 | **IDLH:** 400 ppm |
| **Conversion:** 1 ppm = 4.83 mg/m$^3$ | **DOT:** 2618 130P (inhibited) | | | |

**Synonyms/Trade Names:** Ethenylmethylbenzene, Methylstyrene, Tolyethylene

**Exposure Limits:**
**NIOSH REL:** TWA 100 ppm (480 mg/m$^3$)
**OSHA PEL:** TWA 100 ppm (480 mg/m$^3$)

**Measurement Methods (see Table 1):**
**NIOSH** 1501
**OSHA** 7

**Physical Description:** Colorless liquid with a strong, disagreeable odor.

**Chemical & Physical Properties:**
**MW:** 118.2
**BP:** 339°F
**Sol:** 0.009%
**Fl.P:** 127°F
**IP:** 8.20 eV
**Sp.Gr:** 0.89
**VP:** 1 mmHg
**FRZ:** -106°F
**UEL:** 11.0%
**LEL:** 0.8%
Class II Combustible Liquid

**Personal Protection/Sanitation (see Table 2):**
**Skin:** Prevent skin contact
**Eyes:** Prevent eye contact
**Wash skin:** When contam
**Remove:** When wet or contam
**Change:** N.R.

**Respirator Recommendations (see Tables 3 and 4):**
**NIOSH/OSHA**
**400 ppm:** CcrOv*/PaprOv*/GmFOv/Sa*/ScbaF
**§:** ScbaF:Pd,Pp/SaF:Pd,Pp:AScba
**Escape:** GmFOv/ScbaE

**Incompatibilities and Reactivities:** Oxidizers, peroxides, strong acids, iron or aluminum salts
[**Note:** Usually inhibited with tert-butyl catechol to prevent polymerization.]

**Exposure Routes, Symptoms, Target Organs (see Table 5):**
**ER:** Inh, Ing, Con
**SY:** Irrit eyes, skin, upper resp sys; drow; in animals: narco
**TO:** Eyes, skin, resp sys, CNS

**First Aid (see Table 6):**
**Eye:** Irr immed
**Skin:** Soap flush prompt
**Breath:** Resp support
**Swallow:** Medical attention immed

## VM & P Naphtha

| | | | | |
|---|---|---|---|---|
| **VM & P Naphtha** | **Formula:** | **CAS#:** 8032-32-4 | **RTECS#:** OI6180000 | **IDLH:** N.D. |
| **Conversion:** | **DOT:** 1268 128 (petroleum distillates, n.o.s.) | | | |

**Synonyms/Trade Names:** Ligroin, Painters naphtha, Petroleum ether, Petroleum spirit, Refined solvent naphtha, Varnish makers' & painters' naphtha

**Exposure Limits:**
**NIOSH REL:** TWA 350 mg/m$^3$
C 1800 mg/m$^3$ [15-minute]
**OSHA PEL†:** none

**Measurement Methods (see Table 1):**
**NIOSH** 1550
**OSHA** 48

**Physical Description:** Clear to yellowish liquid with a pleasant, aromatic odor.

**Chemical & Physical Properties:**
**MW:** 87-114 (approx)
**BP:** 203-320°F
**Sol:** Insoluble
**Fl.P:** 20-55°F
**IP:** ?
**Sp.Gr(60°F):** 0.73-0.76
**VP:** 2-20 mmHg
**FRZ:** ?
**UEL:** 6.0%
**LEL:** 1.2%
Class IB Flammable Liquid

**Personal Protection/Sanitation (see Table 2):**
**Skin:** Prevent skin contact
**Eyes:** Prevent eye contact
**Wash skin:** When contam
**Remove:** When wet (flamm)
**Change:** N.R.

**Respirator Recommendations (see Tables 3 and 4):**
**NIOSH**
**3500 mg/m$^3$:** CcrOv/Sa
**8750 mg/m$^3$:** Sa:Cf/PaprOv
**17,500 mg/m$^3$:** CcrFOv/GmFOv/PaprTOv/ScbaF/SaF
**§:** ScbaF:Pd,Pp/SaF:Pd,Pp:AScba
**Escape:** GmFOv/ScbaE

**Incompatibilities and Reactivities:** None reported
[**Note:** VM&P Naphtha is a refined petroleum solvent predominantly $C_7$-$C_{11}$ which is typically 55% paraffins, 30% monocycloparaffins, 2% dicycloparaffins & 12% alklybenzenes.]

**Exposure Routes, Symptoms, Target Organs (see Table 5):**
**ER:** Inh, Ing, Con
**SY:** Irrit eyes, upper resp sys; derm; CNS depres; chemical pneu (aspir liquid)
**TO:** Eyes, skin, resp sys, CNS

**First Aid (see Table 6):**
**Eye:** Irr immed
**Skin:** Soap wash prompt
**Breath:** Resp support
**Swallow:** Medical attention immed

## Warfarin

| Warfarin | Formula: $C_{19}H_{16}O_4$ | CAS#: 81-81-2 | RTECS#: GN4550000 | IDLH: 100 mg/m$^3$ |
|---|---|---|---|---|
| **Conversion:** | **DOT:** | | | |

**Synonyms/Trade Names:** 3-(α-Acetonyl)-benzyl-4-hydroxycoumarin; 4-Hydroxy-3-(3-oxo-1-phenyl butyl)-2H-1-benzopyran-2-one; WARF

**Exposure Limits:**
**NIOSH REL:** TWA 0.1 mg/m$^3$
**OSHA PEL:** TWA 0.1 mg/m$^3$

**Measurement Methods (see Table 1):**
**NIOSH** 5002

**Physical Description:** Colorless, odorless, crystalline powder. [rodenticide]

**Chemical & Physical Properties:**
**MW:** 308.3
**BP:** Decomposes
**Sol:** 0.002%
**Fl.P:** ?
**IP:** ?
**Sp.Gr:** ?
**VP(71°F):** 0.09 mmHg
**MLT:** 322°F
**UEL:** ?
**LEL:** ?
Combustible Solid

**Personal Protection/Sanitation (see Table 2):**
**Skin:** Prevent skin contact
**Eyes:** N.R.
**Wash skin:** When contam
**Remove:** When wet or contam
**Change:** Daily

**Respirator Recommendations (see Tables 3 and 4):**
**NIOSH/OSHA**
**0.5 mg/m$^3$:** Qm
**1 mg/m$^3$:** 95XQ/Sa
**2.5 mg/m$^3$:** Sa:Cf/PaprHie
**5 mg/m$^3$:** 100F/SaT:Cf/PaprTHie/ScbaF/SaF
**100 mg/m$^3$:** Sa:Pd,Pp
**§:** ScbaF:Pd,Pp/SaF:Pd,Pp:AScba
**Escape:** 100F/ScbaE

**Incompatibilities and Reactivities:** Strong oxidizers

**Exposure Routes, Symptoms, Target Organs (see Table 5):**
**ER:** Inh, Abs, Ing, Con
**SY:** Hema, back pain; hematoma arms, legs; epis, bleeding lips, muc memb hemorr; abdom pain, vomit, fecal blood; petechial rash; abnor hematologic indices
**TO:** Blood, CVS

**First Aid (see Table 6):**
**Eye:** Irr immed
**Skin:** Soap wash prompt
**Breath:** Resp support
**Swallow:** Medical attention immed

## Welding fumes

| Welding fumes | Formula: | CAS#: | RTECS#: ZC2550000 | IDLH: Ca [N.D.] |
|---|---|---|---|---|
| **Conversion:** | **DOT:** | | | |

**Synonyms/Trade Names:** Synonyms vary depending upon the specific component of the welding fumes.

**Exposure Limits:**
**NIOSH REL:** Ca
See Appendix A
**OSHA PEL†:** none

**Measurement Methods (see Table 1):**
**NIOSH** 7300, 7301, 7303

**Physical Description:** Fumes generated by the process of joining or cutting pieces of metal by heat, pressure, or both.

**Chemical & Physical Properties:**
Properties vary depending upon the specific component of the welding fumes.

**Personal Protection/Sanitation (see Table 2):**
**Skin:** N.R.
**Eyes:** N.R.
**Wash skin:** N.R.
**Remove:** N.R.
**Change:** N.R.

**Respirator Recommendations (see Tables 3 and 4):**
**NIOSH**
**¥:** ScbaF:Pd,Pp/SaF:Pd,Pp:AScba
**Escape:** GmFOv100/ScbaE

**Incompatibilities and Reactivities:** Varies

**Exposure Routes, Symptoms, Target Organs (see Table 5):**
**ER:** Inh, Con
**SY:** Symptoms vary depending upon the specific component of the welding fumes; metal fume fever: flu-like symptoms, dysp, cough, musc pain, fever, chills; interstitial pneu; [carc]
**TO:** Eyes, skin, resp sys, CNS [lung cancer]

**First Aid (see Table 6):**
**Eye:** Irr immed
**Skin:** Soap wash
**Breath:** Resp support

| **Wood dust** | **Formula:** | **CAS#:** | **RTECS#:** ZC9850000 | **IDLH:** Ca [N.D.] |
|---|---|---|---|---|
| **Conversion:** | **DOT:** | | | |

**Synonyms/Trade Names:** Hard wood dust, Soft wood dust, Western red cedar dust

**Exposure Limits:**
**NIOSH REL:** Ca
TWA 1 $mg/m^3$
See Appendix A
**OSHA PEL†:** TWA 15 $mg/m^3$ (total)
TWA 5 $mg/m^3$ (resp)

**Measurement Methods (see Table 1):**
**NIOSH** 0500

**Physical Description:** Dust from various types of wood.

| **Chemical & Physical Properties:** | **Personal Protection/Sanitation (see Table 2):** | **Respirator Recommendations (see Tables 3 and 4):** |
|---|---|---|
| **MW:** varies<br>**BP:** NA<br>**Sol:** ?<br>**Fl.P:** NA<br>**IP:** NA<br>**Sp.Gr:** ?<br>**VP:** 0 mmHg (approx)<br>**MLT:** NA<br>**UEL:** NA<br>**LEL:** NA<br>Combustible Solid | **Skin:** N.R.<br>**Eyes:** N.R.<br>**Wash skin:** N.R.<br>**Remove:** N.R.<br>**Change:** N.R. | **NIOSH**<br>**¥:** ScbaF:Pd,Pp/SaF:Pd,Pp:AScba<br>**Escape:** 100F/ScbaE |

**Incompatibilities and Reactivities:** None reported

| **Exposure Routes, Symptoms, Target Organs (see Table 5):** | **First Aid (see Table 6):** |
|---|---|
| **ER:** Inh, Con<br>**SY:** Irrit eyes; epis; derm; resp hypersensitivity; granulomatous pneu; asthma, cough, wheez, sinusitis; prolonged colds; [carc]<br>**TO:** Eyes, skin, resp sys [nasal cancer] | **Eye:** Irr immed<br>**Skin:** Soap wash<br>**Breath:** Fresh air |

| **m-Xylene** | **Formula:** $C_6H_4(CH_3)_2$ | **CAS#:** 108-38-3 | **RTECS#:** ZE2275000 | **IDLH:** 900 ppm |
|---|---|---|---|---|
| **Conversion:** 1 ppm = 4.34 $mg/m^3$ | **DOT:** 1307 130 | | | |

**Synonyms/Trade Names:** 1,3-Dimethylbenzene; meta-Xylene; m-Xylol

**Exposure Limits:**
**NIOSH REL:** TWA 100 ppm (435 $mg/m^3$)
ST 150 ppm (655 $mg/m^3$)
**OSHA PEL†:** TWA 100 ppm (435 $mg/m^3$)

**Measurement Methods (see Table 1):**
**NIOSH** 1501, 3800
**OSHA** 1002

**Physical Description:** Colorless liquid with an aromatic odor.

| **Chemical & Physical Properties:** | **Personal Protection/Sanitation (see Table 2):** | **Respirator Recommendations (see Tables 3 and 4):** |
|---|---|---|
| **MW:** 106.2<br>**BP:** 282°F<br>**Sol:** Slight<br>**Fl.P:** 82°F<br>**IP:** 8.56 eV<br>**Sp.Gr:** 0.86<br>**VP:** 9 mmHg<br>**FRZ:** -54°F<br>**UEL:** 7.0%<br>**LEL:** 1.1%<br>Class IC Flammable Liquid | **Skin:** Prevent skin contact<br>**Eyes:** Prevent eye contact<br>**Wash skin:** When contam<br>**Remove:** When wet (flamm)<br>**Change:** N.R. | **NIOSH/OSHA**<br>**900 ppm:** CcrOv*/PaprOv*/Sa*/ScbaF<br>**§:** ScbaF:Pd,Pp/SaF:Pd,Pp:AScba<br>**Escape:** GmFOv/ScbaE |

**Incompatibilities and Reactivities:** Strong oxidizers, strong acids

| **Exposure Routes, Symptoms, Target Organs (see Table 5):** | **First Aid (see Table 6):** |
|---|---|
| **ER:** Inh, Abs, Ing, Con<br>**SY:** Irrit eyes, skin, nose, throat; dizz, excitement, drow, inco, staggering gait; corn vacuolization; anor, nau, vomit, abdom pain; derm<br>**TO:** Eyes, skin, resp sys, CNS, GI tract, blood, liver, kidneys | **Eye:** Irr immed<br>**Skin:** Soap wash prompt<br>**Breath:** Resp support<br>**Swallow:** Medical attention immed |

| **o-Xylene** | **Formula:** $C_6H_4(CH_3)_2$ | **CAS#:** 95-47-6 | **RTECS#:** ZE2450000 | **IDLH:** 900 ppm |
|---|---|---|---|---|
| **Conversion:** 1 ppm = 4.34 mg/m$^3$ | **DOT:** 1307 130 | | | |

**Synonyms/Trade Names:** 1,2-Dimethylbenzene; ortho-Xylene; o-Xylol

| **Exposure Limits:** | **Measurement Methods (see Table 1):** |
|---|---|
| **NIOSH REL:** TWA 100 ppm (435 mg/m$^3$)<br>ST 150 ppm (655 mg/m$^3$)<br>**OSHA PEL†:** TWA 100 ppm (435 mg/m$^3$) | **NIOSH** 1501, 3800<br>**OSHA** 1002 |

**Physical Description:** Colorless liquid with an aromatic odor.

| **Chemical & Physical Properties:** | **Personal Protection/Sanitation (see Table 2):** | **Respirator Recommendations (see Tables 3 and 4):** |
|---|---|---|
| **MW:** 106.2<br>**BP:** 292°F<br>**Sol:** 0.02%<br>**Fl.P:** 90°F<br>**IP:** 8.56 eV<br>**Sp.Gr:** 0.88<br>**VP:** 7 mmHg<br>**FRZ:** -13°F<br>**UEL:** 6.7%<br>**LEL:** 0.9%<br>Class IC Flammable Liquid | **Skin:** Prevent skin contact<br>**Eyes:** Prevent eye contact<br>**Wash skin:** When contam<br>**Remove:** When wet (flamm)<br>**Change:** N.R. | **NIOSH/OSHA**<br>**900 ppm:** CcrOv*/PaprOv*/<br>Sa*/ScbaF<br>**§:** ScbaF:Pd,Pp/SaF:Pd,Pp:AScba<br>**Escape:** GmFOv/ScbaE |

**Incompatibilities and Reactivities:** Strong oxidizers, strong acids

| **Exposure Routes, Symptoms, Target Organs (see Table 5):** | **First Aid (see Table 6):** |
|---|---|
| **ER:** Inh, Abs, Ing, Con<br>**SY:** Irrit eyes, skin, nose, throat; dizz, excitement, drow, inco, staggering gait; corn vacuolization; anor, nau, vomit, abdom pain; derm<br>**TO:** Eyes, skin, resp sys, CNS, GI tract, blood, liver, kidneys | **Eye:** Irr immed<br>**Skin:** Soap wash prompt<br>**Breath:** Resp support<br>**Swallow:** Medical attention immed |

| **p-Xylene** | **Formula:** $C_6H_4(CH_3)_2$ | **CAS#:** 106-42-3 | **RTECS#:** ZE2625000 | **IDLH:** 900 ppm |
|---|---|---|---|---|
| **Conversion:** 1 ppm = 4.41 mg/m$^3$ | **DOT:** 1307 130 | | | |

**Synonyms/Trade Names:** 1,4-Dimethylbenzene; para-Xylene; p-Xylol

| **Exposure Limits:** | **Measurement Methods (see Table 1):** |
|---|---|
| **NIOSH REL:** TWA 100 ppm (435 mg/m$^3$)<br>ST 150 ppm (655 mg/m$^3$)<br>**OSHA PEL†:** TWA 100 ppm (435 mg/m$^3$) | **NIOSH** 1501, 3800<br>**OSHA** 1002 |

**Physical Description:** Colorless liquid with an aromatic odor. [**Note:** A solid below 56°F.]

| **Chemical & Physical Properties:** | **Personal Protection/Sanitation (see Table 2):** | **Respirator Recommendations (see Tables 3 and 4):** |
|---|---|---|
| **MW:** 106.2<br>**BP:** 281°F<br>**Sol:** 0.02%<br>**Fl.P:** 81°F<br>**IP:** 8.44 eV<br>**Sp.Gr:** 0.86<br>**VP:** 9 mmHg<br>**FRZ:** 56°F<br>**UEL:** 7.0%<br>**LEL:** 1.1%<br>Class IC Flammable Liquid | **Skin:** Prevent skin contact<br>**Eyes:** Prevent eye contact<br>**Wash skin:** When contam<br>**Remove:** When wet (flamm)<br>**Change:** N.R. | **NIOSH/OSHA**<br>**900 ppm:** CcrOv*/PaprOv*/<br>Sa*/ScbaF<br>**§:** ScbaF:Pd,Pp/SaF:Pd,Pp:AScba<br>**Escape:** GmFOv/ScbaE |

**Incompatibilities and Reactivities:** Strong oxidizers, strong acids

| **Exposure Routes, Symptoms, Target Organs (see Table 5):** | **First Aid (see Table 6):** |
|---|---|
| **ER:** Inh, Abs, Ing, Con<br>**SY:** Irrit eyes, skin, nose, throat; dizz, excitement, drow, inco, staggering gait; corn vacuolization; anor, nau, vomit, abdom pain; derm<br>**TO:** Eyes, skin, resp sys, CNS, GI tract, blood, liver, kidneys | **Eye:** Irr immed<br>**Skin:** Soap wash prompt<br>**Breath:** Resp support<br>**Swallow:** Medical attention immed |

| **m-Xylene α,α'-diamine** | **Formula:** $C_6H_4(CH_2NH_2)_2$ | **CAS#:** 1477-55-0 | **RTECS#:** PF8970000 | **IDLH:** N.D. |
|---|---|---|---|---|
| **Conversion:** | **DOT:** | | | |

**Synonyms/Trade Names:** 1,3-bis(Aminomethyl)benzene; 1,3-Benzenedimethanamine; MXDA; m-Phenylenebis(methylamine); m-Xylylenediamine

**Exposure Limits:**
**NIOSH REL:** C 0.1 mg/m$^3$ [skin]
**OSHA PEL†:** none

**Measurement Methods (see Table 1):**
**OSHA** 105

**Physical Description:** Colorless liquid.

| **Chemical & Physical Properties:** | **Personal Protection/Sanitation (see Table 2):** | **Respirator Recommendations (see Tables 3 and 4):** |
|---|---|---|
| **MW:** 136.2<br>**BP:** 477°F<br>**Sol:** Miscible<br>**Fl.P:** 243°F<br>**IP:** ?<br>**Sp.Gr:** 1.032<br>**VP(77°F):** 0.03 mmHg<br>**FRZ:** 58°F<br>**UEL:** ?<br>**LEL:** ?<br>Class IIIB Combustible Liquid | **Skin:** Prevent skin contact<br>**Eyes:** Prevent eye contact<br>**Wash skin:** When contam<br>**Remove:** When wet or contam<br>**Change:** N.R.<br>**Provide:** Eyewash<br>Quick drench | Not available. |

**Incompatibilities and Reactivities:** None reported

| **Exposure Routes, Symptoms, Target Organs (see Table 5):** | **First Aid (see Table 6):** |
|---|---|
| **ER:** Inh, Abs, Ing, Con<br>**SY:** In animals: irrit eyes, skin; liver, kidney, lung damage<br>**TO:** Eyes, skin, resp sys, liver, kidneys | **Eye:** Irr immed<br>**Skin:** Water flush immed<br>**Breath:** Resp support<br>**Swallow:** Medical attention immed |

| **Xylidine** | **Formula:** $(CH_3)_2C_6H_3NH_2$ | **CAS#:** 1300-73-8 | **RTECS#:** ZE8575000 | **IDLH:** 50 ppm |
|---|---|---|---|---|
| **Conversion:** 1 ppm = 4.96 mg/m$^3$ | **DOT:** 1711 153 | | | |

**Synonyms/Trade Names:** Aminodimethylbenzene, Aminoxylene, Dimethylaminobenzene, Dimethylaniline, Xylidine isomers (e.g., 2,4-Dimethylaniline)
[**Note:** Dimethylaniline is also used as a synonym for N,N-Dimethylaniline.]

**Exposure Limits:**
**NIOSH REL:** TWA 2 ppm (10 mg/m$^3$) [skin]
**OSHA PEL†:** TWA 5 ppm (25 mg/m$^3$) [skin]

**Measurement Methods (see Table 1):**
**NIOSH** 2002

**Physical Description:** Pale-yellow to brown liquid with a weak, aromatic, amine-like odor.

| **Chemical & Physical Properties:** | **Personal Protection/Sanitation (see Table 2):** | **Respirator Recommendations (see Tables 3 and 4):** |
|---|---|---|
| **MW:** 121.2<br>**BP:** 415-439°F<br>**Sol:** Slight<br>**Fl.P:** 206°F (2,3-)<br>**IP:** 7.65 eV (2,4-)<br>7.30 eV (2,6-)<br>**Sp.Gr:** 0.98<br>**VP:** <1 mmHg<br>**FRZ:** -33°F<br>**UEL:** ?<br>**LEL:** 1.0% (o-isomer)<br>Class IIIB Combustible Liquid (2,3-) | **Skin:** Prevent skin contact<br>**Eyes:** Prevent eye contact<br>**Wash skin:** When contam<br>**Remove:** When wet or contam<br>**Change:** N.R.<br>**Provide:** Eyewash<br>Quick drench | **NIOSH**<br>**20 ppm:** CcrOv/Sa<br>**50 ppm:** Sa:Cf/CcrFOv/GmFOv/<br>PaprOv/ScbaF/SaF<br>**§:** ScbaF:Pd,Pp/SaF:Pd,Pp:AScba<br>**Escape:** GmFOv/ScbaE |

**Incompatibilities and Reactivities:** Strong oxidizers, hypochlorite salts

| **Exposure Routes, Symptoms, Target Organs (see Table 5):** | **First Aid (see Table 6):** |
|---|---|
| **ER:** Inh, Abs, Ing, Con<br>**SY:** Anoxia, cyan, methemo; lung, liver, kidney damage<br>**TO:** Resp sys, blood, liver, kidneys, CVS | **Eye:** Irr immed<br>**Skin:** Soap wash immed<br>**Breath:** Resp support<br>**Swallow:** Medical attention immed |

| Yttrium | Formula: Y | CAS#: 7440-65-5 | RTECS#: ZG2980000 | IDLH: 500 mg/m$^3$ (as Y) |
|---|---|---|---|---|
| **Conversion:** | **DOT:** | | | |
| **Synonyms/Trade Names:** Yttrium metal | | | | |
| **Exposure Limits:**<br>**NIOSH REL*:** TWA 1 mg/m$^3$<br>**OSHA PEL*:** TWA 1 mg/m$^3$<br>[***Note:** The REL and PEL also apply to other yttrium compounds (as Y).] | | | **Measurement Methods (see Table 1):**<br>**NIOSH** 7300, 7301, 7303, 9102<br>**OSHA** ID121 | |
| **Physical Description:** Dark-gray to black, odorless solid. | | | | |
| **Chemical & Physical Properties:**<br>**MW:** 88.9<br>**BP:** 5301°F<br>**Sol:** Soluble in hot $H_2O$<br>**Fl.P:** NA<br>**IP:** NA<br>**Sp.Gr:** 4.47<br>**VP:** 0 mmHg (approx)<br>**MLT:** 2732°F<br>**UEL:** NA<br>**LEL:** NA<br>Noncombustible Solid in bulk form. | **Personal Protection/Sanitation (see Table 2):**<br>**Skin:** N.R.<br>**Eyes:** N.R.<br>**Wash skin:** N.R.<br>**Remove:** N.R.<br>**Change:** N.R. | **Respirator Recommendations (see Tables 3 and 4):**<br>**NIOSH/OSHA**<br>**5 mg/m$^3$:** Qm<br>**10 mg/m$^3$:** 95XQ/Sa<br>**25 mg/m$^3$:** Sa:Cf/PaprHie<br>**50 mg/m$^3$:** 100F/SaT:Cf/PaprTHie/ ScbaF/SaF<br>**500 mg/m$^3$:** Sa:Pd,Pp<br>**§:** ScbaF:Pd,Pp/SaF:Pd,Pp:AScba<br>**Escape:** 100F/ScbaE | | |
| **Incompatibilities and Reactivities:** Oxidizers | | | | |
| **Exposure Routes, Symptoms, Target Organs (see Table 5):**<br>**ER:** Inh, Ing, Con<br>**SY:** Irrit eyes; in animals: pulm irrit; eye inj; possible liver damage<br>**TO:** Eyes, resp sys, liver | | **First Aid (see Table 6):**<br>**Eye:** Irr immed<br>**Skin:** Soap wash prompt<br>**Breath:** Resp support<br>**Swallow:** Medical attention immed | | |

| Zinc chloride fume | Formula: $ZnCl_2$ | CAS#: 7646-85-7 | RTECS#: ZH1400000 | IDLH: 50 mg/m$^3$ |
|---|---|---|---|---|
| **Conversion:** | **DOT:** 2331 154 | | | |
| **Synonyms/Trade Names:** Zinc dichloride fume | | | | |
| **Exposure Limits:**<br>**NIOSH REL:** TWA 1 mg/m$^3$<br>ST 2 mg/m$^3$<br>**OSHA PEL†:** TWA 1 mg/m$^3$ | | | **Measurement Methods (see Table 1):**<br>**OSHA** ID121 | |
| **Physical Description:** White particulate dispersed in air. | | | | |
| **Chemical & Physical Properties:**<br>**MW:** 136.3<br>**BP:** 1350°F<br>**Sol(70°F):** 435%<br>**Fl.P:** NA<br>**IP:** NA<br>**Sp.Gr(77°F):** 2.91<br>**VP:** 0 mmHg (approx)<br>**MLT:** 554°F<br>**UEL:** NA<br>**LEL:** NA<br>Noncombustible Solid | **Personal Protection/Sanitation (see Table 2):**<br>**Skin:** N.R.<br>**Eyes:** N.R.<br>**Wash skin:** N.R.<br>**Remove:** N.R.<br>**Change:** N.R. | **Respirator Recommendations (see Tables 3 and 4):**<br>**NIOSH/OSHA**<br>**10 mg/m$^3$:** 95XQ*/Sa*<br>**25 mg/m$^3$:** Sa:Cf*/PaprHie*<br>**50 mg/m$^3$:** 100F/PaprTHie*/ScbaF/SaF<br>**§:** ScbaF:Pd,Pp/SaF:Pd,Pp:AScba<br>**Escape:** 100F/ScbaE | | |
| **Incompatibilities and Reactivities:** Potassium | | | | |
| **Exposure Routes, Symptoms, Target Organs (see Table 5):**<br>**ER:** Inh, Con<br>**SY:** Irrit eyes, skin, nose, throat; conj; cough, copious sputum; dysp, chest pain, pulm edema, pneu; pulm fib, cor pulmonale; fever; cyan; tachypnea; skin burns<br>**TO:** Eyes, skin, resp sys, CVS | | **First Aid (see Table 6):**<br>**Breath:** Resp support | | |

## Zinc oxide

**Formula:** ZnO | **CAS#:** 1314-13-2 | **RTECS#:** ZH4810000 | **IDLH:** 500 mg/m$^3$

**Conversion:** | **DOT:** 1516 143

**Synonyms/Trade Names:** Zinc peroxide

**Exposure Limits:**
**NIOSH REL:** Dust: TWA 5 mg/m$^3$
C 15 mg/m$^3$
Fume: TWA 5 mg/m$^3$
ST 10 mg/m$^3$
**OSHA PEL†:** TWA 5 mg/m$^3$ (fume)
TWA 15 mg/m$^3$ (total dust)
TWA 5 mg/m$^3$ (resp dust)

**Measurement Methods (see Table 1):**
**NIOSH** 7303, 7502
**OSHA** ID121, ID143

**Physical Description:** White, odorless solid.

**Chemical & Physical Properties:**
**MW:** 81.4
**BP:** ?
**Sol(64°F):** 0.0004%
**Fl.P:** NA
**IP:** NA
**Sp.Gr:** 5.61
**VP:** 0 mmHg (approx)
**MLT:** 3587°F
**UEL:** NA
**LEL:** NA
Noncombustible Solid

**Personal Protection/Sanitation (see Table 2):**
**Skin:** N.R.
**Eyes:** N.R.
**Wash skin:** N.R.
**Remove:** N.R.
**Change:** N.R.

**Respirator Recommendations (see Tables 3 and 4):**
**NIOSH/OSHA**
**50 mg/m$^3$:** 95XQ/Sa
**125 mg/m$^3$:** Sa:Cf/PaprHie
**250 mg/m$^3$:** 100F/SaT:Cf/PaprTHie/ScbaF/SaF
**500 mg/m$^3$:** Sa:Pd,Pp
**§:** ScbaF:Pd,Pp/SaF:Pd,Pp:AScba
**Escape:** 100F/ScbaE

**Incompatibilities and Reactivities:** Chlorinated rubber (at 419°F), water
[**Note:** Slowly decomposed by water.]

**Exposure Routes, Symptoms, Target Organs (see Table 5):**
**ER:** Inh
**SY:** Metal fume fever: chills, musc ache, nau, fever, dry throat, cough; lass; metallic taste; head; blurred vision; low back pain; vomit; mal; chest tight; dysp, rales, decr pulm func
**TO:** Resp sys

**First Aid (see Table 6):**
**Breath:** Resp support

## Zinc stearate

**Formula:** $Zn(C_{18}H_{35}O_2)_2$ | **CAS#:** 557-05-1 | **RTECS#:** ZH5200000 | **IDLH:** N.D.

**Conversion:** | **DOT:**

**Synonyms/Trade Names:** Dibasic zinc stearate, Zinc salt of stearic acid, Zinc distearate

**Exposure Limits:**
**NIOSH REL:** TWA 10 mg/m$^3$ (total)
TWA 5 mg/m$^3$ (resp)
**OSHA PEL†:** TWA 15 mg/m$^3$ (total)
TWA 5 mg/m$^3$ (resp)

**Measurement Methods (see Table 1):**
**NIOSH** 0500, 0600

**Physical Description:** Soft, white powder with a slight, characteristic odor.

**Chemical & Physical Properties:**
**MW:** 632.4
**BP:** ?
**Sol:** Insoluble
**Fl.P(oc):** 530°F
**IP:** NA
**Sp.Gr:** 1.10
**VP:** 0 mmHg (approx)
**MLT:** 266°F
**UEL:** ?
**LEL:** ?
**MEC:** 20 g/m$^3$
Combustible Solid

**Personal Protection/Sanitation (see Table 2):**
**Skin:** N.R.
**Eyes:** N.R.
**Wash skin:** N.R.
**Remove:** N.R.
**Change:** N.R.

**Respirator Recommendations (see Tables 3 and 4):**
Not available.

**Incompatibilities and Reactivities:** Oxidizers, dilute acids
[**Note:** Hydrophobic (i.e., repels water).]

**Exposure Routes, Symptoms, Target Organs (see Table 5):**
**ER:** Inh, Ing, Con
**SY:** Irrit eyes, skin, upper resp sys; cough
**TO:** Eyes, skin, resp sys

**First Aid (see Table 6):**
**Eye:** Irr immed
**Skin:** Soap wash
**Breath:** Fresh air
**Swallow:** Medical attention immed

| Zirconium compounds (as Zr) | Formula: Zr (metal) | CAS#: 7440-67-7 (metal) | RTECS#: ZH7070000 (metal) | IDLH: 50 mg/m$^3$ (as Zr) |
|---|---|---|---|---|
| **Conversion:** | **DOT:** 1358 170 (powder, wet); 1932 135 (scrap); 2008 135 (powder, dry) | | | |

**Synonyms/Trade Names: Zirconium metal:** Zirconium
Synonyms of other zirconium compounds vary depending upon the specific compound.

| Exposure Limits: | Measurement Methods (see Table 1): |
|---|---|
| **NIOSH REL*:** TWA 5 mg/m$^3$<br>ST 10 mg/m$^3$<br>[***Note:** The REL applies to all zirconium compounds (as Zr) except Zirconium tetrachloride.]<br>**OSHA PEL†:** TWA 5 mg/m$^3$ | **NIOSH** 7300, 7301, 9102<br>**OSHA** ID121 |

**Physical Description:** Metal: Soft, malleable, ductile, solid or gray to gold, amorphous powder.

| Chemical & Physical Properties: | Personal Protection/Sanitation (see Table 2): | Respirator Recommendations (see Tables 3 and 4): |
|---|---|---|
| **MW:** 91.2<br>**BP:** 6471°F<br>**Sol:** Insoluble<br>**Fl.P:** NA<br>**IP:** NA<br>**Sp.Gr:** 6.51 (Metal)<br>**VP:** 0 mmHg (approx)<br>**MLT:** 3375°F<br>**UEL:** NA<br>**LEL:** NA<br>Metal: Combustible, but solid form is difficult to ignite; however, powder form may ignite SPONTANEOUSLY and can continue burning under water. | Recommendations regarding personal protective clothing vary depending upon the specific compound. | **NIOSH/OSHA**<br>**25 mg/m$^3$:** Qm<br>**50 mg/m$^3$:** 95XQ/PaprHie/100F/Sa/ScbaF<br>**§:** ScbaF:Pd,Pp/SaF:Pd,Pp:AScba<br>**Escape:** 100F/ScbaE |

**Incompatibilities and Reactivities:** Potassium nitrate, oxidizers
[**Note:** Fine powder may be stored completely immersed in water.]

| Exposure Routes, Symptoms, Target Organs (see Table 5): | First Aid (see Table 6): |
|---|---|
| **ER:** Inh, Con<br>**SY:** Skin, lung granulomas; in animals: irrit skin, muc memb; X-ray evidence of retention in lungs<br>**TO:** Skin, resp sys | **Eye:** Irr immed<br>**Skin:** Soap wash<br>**Breath:** Resp support<br>**Swallow:** Medical attention immed |

Z

# APPENDICES

# Appendix A
# NIOSH POTENTIAL OCCUPATIONAL CARCINOGENS

## New Policy (Adopted September 1995)

For the past 20 plus years, NIOSH has subscribed to a carcinogen policy that was published in 1976 by Edward J. Fairchild, II, Associate Director for Cincinnati Operations, which called for "no detectable exposure levels for proven carcinogenic substances" (Annals of the New York Academy of Sciences, 271:200-207, 1976). This was in response to a generic OSHA rulemaking on carcinogens. Because of advances in science and in approaches to risk assessment and risk management, NIOSH has adopted a more inclusive policy. NIOSH recommended exposure limits (RELs) will be based on risk evaluations using human or animal health effects data, and on an assessment of what levels can be feasibly achieved by engineering controls and measured by analytical techniques. To the extent feasible, NIOSH will project not only a no effect exposure, but also exposure levels at which there may be residual risks. This policy applies to all workplace hazards, including carcinogens, and is responsive to Section 20(a)(3) of the Occupational Safety and Health Act of 1970, which charges NIOSH to ". . .describe exposure levels that are safe for various periods of employment, including but not limited to the exposure levels at which no employee will suffer impaired health or functional capacities or diminished life expectancy as a result of his work experience."

The effect of this new policy will be the development, whenever possible, of quantitative RELs that are based on human and/or animal data, as well as on the consideration of technological feasibility for controlling workplace exposures to the REL. Under the old policy, RELs for most carcinogens were non-quantitative values labeled "lowest feasible concentration (LFC)." [Note: There are a few exceptions to LFC RELs for carcinogens (e.g., RELs for asbestos, formaldehyde, benzene, and ethylene oxide are quantitative values based primarily on analytical limits of detection or technological feasibility). Also, in 1989, NIOSH adopted several quantitative RELs for carcinogens from OSHA's permissible exposure limit (PEL) update.]

Under the new policy, NIOSH will also recommend the complete range of respirators (as determined by the NIOSH Respirator Decision Logic) for carcinogens with quantitative RELs. In this way, respirators will be consistently recommended regardless of whether a substance is a carcinogen or a non-carcinogen.

# Appendix A (Continued)
# NIOSH POTENTIAL OCCUPATIONAL CARCINOGENS

## Old Policy

In the past, NIOSH identified numerous substances that should be treated as potential occupational carcinogens even though OSHA might not have identified them as such. In determining their carcinogenicity, NIOSH used the OSHA classification outlined in 29 CFR 1990.103, which states in part:

> Potential occupational carcinogen means any substance, or combination or mixture of substances, which causes an increased incidence of benign and/or malignant neoplasms, or a substantial decrease in the latency period between exposure and onset of neoplasms in humans or in one or more experimental mammalian species as the result of any oral, respiratory or dermal exposure, or any other exposure which results in the induction of tumors at a site other than the site of administration. This definition also includes any substance which is metabolized into one or more potential occupational carcinogens by mammals.

When thresholds for carcinogens that would protect 100% of the population had not been identified, NIOSH usually recommended that occupational exposures to carcinogens be limited to the lowest feasible concentration. To ensure maximum protection from carcinogens through the use of respiratory protection, NIOSH also recommended that only the most reliable and protective respirators be used. These respirators include (1) a self-contained breathing apparatus (SCBA) that has a full facepiece and is operated in a positive pressure mode, or (2) a supplied air respirator that has a full facepiece and is operated in a pressure demand or other positive pressure mode in combination with an auxiliary SCBA operated in a pressure demand or other positive pressure mode.

## Recommendations to be Revised

The RELs and respirator recommendations for carcinogens listed in this edition of the *Pocket Guide* still reflect the old policy. Changes in the RELs and respirator recommendations that reflect the new policy will be included in future editions.

APPENDIX B

# Appendix B
# THIRTEEN OSHA-REGULATED CARCINOGENS

Without establishing PELs, OSHA promulgated standards in 1974 to regulate the industrial use of the following 13 chemicals identified as potential occupational carcinogens:

- 2-Acetylaminofluorene
- 4-Aminodiphenyl
- Benzidine
- bis-Chloromethyl ether
- 3,3'-Dichlorobenzidine
- 4-Dimethylaminoazobenzene
- Ethyleneimine
- Methyl chloromethyl ether
- α-Naphthylamine
- β-Naphthylamine
- 4-Nitrobiphenyl
- N-Nitrosodimethylamine
- β-Propiolactone

Exposures of workers to these 13 chemicals are to be controlled through the required use of engineering controls, work practices, and personal protective equipment, including respirators. OSHA respirator requirements for these chemicals are provided in Appendix E (page 351). See 29 CFR 1910.1003 - 1910.1016 for more specific details of these requirements.

# Appendix C
# SUPPLEMENTARY EXPOSURE LIMITS

## Aldehydes (Low-Molecular-Weight)

Exposure to acetaldehyde has produced nasal tumors in rats and laryngeal tumors in hamsters, and exposure to malonaldehyde has produced thyroid gland and pancreatic islet cell tumors in rats. NIOSH therefore recommends that acetaldehyde and malonaldehyde be considered potential occupational carcinogens in conformance with the OSHA carcinogen policy. Testing has not been completed to determine the carcinogenicity of the following nine related low-molecular-weight aldehydes:



- Acrolein (CAS# 107-02-8)
- Butyraldehyde (CAS# 123-72-8)
- Crotonaldehyde (CAS# 4170-30-3)
- Glutaraldehyde (CAS# 111-30-8)
- Glyoxal (CAS# 107-22-2)
- Paraformaldehyde (CAS# 30525-89-4)
- Propiolaldehyde (CAS# 624-67-9)
- Propionaldehyde (CAS# 123-38-6)
- n-Valeraldehyde (CAS# 110-62-3)

However, the limited studies to date indicate that these substances have chemical reactivity and mutagenicity similar to acetaldehyde and malonaldehyde. Therefore, NIOSH recommends that careful consideration should be given to reducing exposures to these nine related aldehydes. Further information can be found in *NIOSH Current Intelligence Bulletin 55: Carcinogenicity of Acetaldehyde and Malonaldehyde, and Mutagenicity of Related Low-Molecular-Weight Aldehydes* [DHHS (NIOSH) Publication No. 91-112]. This document is available on the NIOSH Web site (http://www.cdc.gov/niosh/91112_55.html).

## Asbestos

NIOSH considers asbestos to be a potential occupational carcinogen and recommends that exposures be reduced to the lowest feasible concentration. For asbestos fibers >5 micrometers in length, NIOSH recommends a REL of 100,000 fibers per cubic meter of air (100,000 fibers/$m^3$), which is equal to 0.1 fiber per cubic centimeter of air (0.1 fiber/$cm^3$), as determined by a 400-liter air sample collected over 100 minutes in accordance with NIOSH Analytical Method #7400. Airborne asbestos fibers are defined as those particles having (1) an aspect ratio of 3 to 1 or greater and (2) the mineralogic characteristics (that is, the crystal structure and elemental composition) of the asbestos minerals and their nonasbestiform analogs. The asbestos minerals are defined as chrysotile, crocidolite, amosite (cummingtonite-grunerite), anthophyllite, tremolite, and actinolite. In addition, airborne cleavage fragments from the nonasbestiform habits of the serpentine minerals antigorite and lizardite, and the amphibole minerals contained in the series cummingtonite-grunerite, tremolite-ferroactinolite, and glaucophane-riebeckite should also be counted as fibers provided they meet the criteria for a fiber when viewed microscopically.

As found in 29 CFR 1910.1001, the OSHA PEL for asbestos fibers (i.e., actinolite asbestos, amosite, anthophyllite asbestos, chrysotile, crocidolite, and tremolite asbestos) is an 8-hour TWA airborne concentration of 0.1 fiber (longer than 5 micrometers and having a length to diameter ratio of at least 3 to 1) per cubic centimeter of air (0.1 fiber/$cm^3$), as determined by the membrane filter method at approximately 400X magnification with phase contrast illumination. No worker should be exposed in excess of 1 fiber/$cm^3$ (excursion limit) as averaged over a sampling period of 30 minutes.

## Appendix C (Continued)
## SUPPLEMENTARY EXPOSURE LIMITS

### Asphalt Fumes

The recommendations provided below are from *Health Effects of Occupational Exposure to Asphalt* [DHHS (NIOSH) Publication No. 2001-110] (http://www.cdc.gov/niosh/01-110pd.html).



Occupational exposure to asphalt fumes shall be controlled so that employees are not exposed to the airborne particulates at a concentration greater than 5 $mg/m^3$, determined during any 15-minute period.

Data regarding the potential carcinogenicity of paving asphalt fumes in humans are limited, and no animal studies have examined the carcinogenic potential of either field- or laboratory-generated samples of paving asphalt fume condensates. NIOSH concludes that the collective data currently available from studies on paving asphalt provide insufficient evidence for an association between lung cancer and exposure to asphalt during paving.

The results from epidemiologic studies indicate that roofers are at an increased risk of lung cancer, but it is uncertain whether this increase can be attributed to asphalt and/or to other exposures such as coal tar or asbestos. Data from experimental studies in animals and cultured mammalian cells indicate that laboratory-generated roofing asphalt fume condensates are genotoxic and cause skin tumors in mice when applied dermally. Furthermore, a known carcinogen (Benzo(a)pyrene) was detected in field-generated roofing fumes. The collective health and exposure data provide sufficient evidence for NIOSH to conclude that roofing asphalt fumes are a potential occupational carcinogen.

The available data indicate that although not all asphalt-based paint formulations may exert genotoxicity, some are genotoxic and carcinogenic in animals. No published data examine the carcinogenic potential of asphalt-based paints in humans, but NIOSH concludes that asphalt-based paints are potential occupational carcinogens.

### Benzidine-, o-Tolidine-, and o-Dianisidine-based Dyes

In December 1980, OSHA and NIOSH jointly published the *Health Hazard Alert: Benzidine-, o-Tolidine-, and o-Dianisidine-based Dyes* [DHHS (NIOSH) Publication No. 81-106] (http://www.cdc.gov/niosh/81-106.html). In this Alert, OSHA and NIOSH concluded that benzidine and benzidine-based dyes were potential occupational carcinogens and recommended that worker exposure be reduced to the lowest feasible level. OSHA and NIOSH further concluded that o-tolidine and o-dianisidine (and dyes based on them) may present a cancer risk to workers and should be handled with caution and exposure minimized.

### Carbon Black

NIOSH considers "Carbon Black" to be the material consisting of more than 80% elemental carbon, in the form of near-spherical colloidal particles and coalesced particle aggregates of colloidal size, that is obtained by the partial combustion or thermal decomposition of hydrocarbons. The NIOSH REL (10-hour TWA) for carbon black is 3.5 $mg/m^3$. Polycyclic aromatic hydrocarbons (PAHs), particulate polycyclic organic material (PPOM), and polynuclear aromatic hydrocarbons (PNAs) are terms frequently used to describe various petroleum-based substances that NIOSH considers to be potential occupational carcinogens. Since some of these aromatic hydrocarbons may be formed during the manufacture of carbon black (and become adsorbed on the carbon black), the NIOSH REL (10-hour TWA) for carbon black in the presence of PAHs is 0.1 mg PAHs/$m^3$ (measured as the cyclohexane-extractable fraction). The OSHA PEL (8-hour TWA) for carbon black is 3.5 mg/$m^3$.

### Chloroethanes

NIOSH considers the following four chemicals to be potential occupational carcinogens:

- Ethylene dichloride
- Hexachloroethane
- 1,1,2,2-Tetrachloroethane
- 1,1,2-Trichloroethane

Additionally, NIOSH recommends that the following five other chloroethane compounds be treated in the workplace with caution because of their structural similarity to the four chloroethanes shown to be carcinogenic in animals:

- 1,1-Dichloroethane
- Ethyl chloride
- Methyl chloroform
- Pentachloroethane
- 1,1,1,2-Tetrachloroethane



### Chromic Acid and Chromates (as $CrO_3$), Chromium(II) and Chromium(III) Compounds (as Cr), and Chromium Metal (as Cr)

The NIOSH REL (10-hour TWA) is 0.001 mg Cr(VI)/$m^3$ for all hexavalent chromium [Cr(VI)] compounds. NIOSH considers all Cr(VI) compounds (including chromic acid, tert-butyl chromate, zinc chromate, and chromyl chloride) to be potential occupational carcinogens. The NIOSH REL (8-hour TWA) is 0.5 mg Cr/$m^3$ for chromium metal and chromium(II) and chromium(III) compounds.

The OSHA PEL is 0.005 mg $CrO_3/m^3$ (8-hour TWA) for chromic acid and chromates (including tert-butyl chromate with a "skin" designation and zinc chromate); 0.5 mg Cr/$m^3$ (8-hour TWA) for chromium(II) and chromium(III) compounds; and 1 mg Cr/$m^3$ (8-hour TWA) for chromium metal and insoluble salts.

### Coal Dust and Coal Mine Dust

The NIOSH REL (10-hour TWA) for respirable coal mine dust is 1 mg/$m^3$, measured using a coal mine personal sampler unit (CPSU) as defined in 30 CFR 74.2. The REL is equivalent to 0.9 mg/$m^3$ measured according to the ISO/CEN/ACGIH (International Standards Organization/ Comité Européen de Normalisation/American Conference of Governmental Industrial Hygienists) definition of respirable dust. The REL applies to respirable coal mine dust and respirable coal dust in occupations other than mining. NIOSH recommends a separate REL for crystalline silica. See NIOSH publication 95-106 (*Criteria for a Recommended Standard - Occupational Exposure to Respirable Coal Mine Dust*) for more detailed information.

### Coal Tar Pitch Volatiles

NIOSH considers coal tar products (i.e., coal tar, coal tar pitch, or creosote) to be potential occupational carcinogens; the NIOSH REL (10-hour TWA) for coal tar products is 0.1 mg/$m^3$ (cyclohexane-extractable fraction).

The OSHA PEL (8-hour TWA) for coal tar pitch volatiles is 0.2 mg/$m^3$ (benzene-soluble fraction). OSHA defines "coal tar pitch volatiles" in 29 CFR 1910.1002 as the fused polycyclic hydrocarbons that volatilize from the distillation residues of coal, petroleum (excluding asphalt), wood, and other organic matter and includes substances such as anthracene, benzo(a)pyrene (BaP), phenanthrene, acridine, chrysene, pyrene, etc.

### Coke Oven Emissions



The production of coke by the carbonization of bituminous coal leads to the release of chemically-complex emissions from coke ovens that include both gases and particulate matter of varying chemical composition. The emissions include coal tar pitch volatiles (e.g., particulate polycyclic organic matter [PPOM], polycyclic aromatic hydrocarbons [PAHs], and polynuclear aromatic hydrocarbons [PNAs]), aromatic compounds (e.g., benzene and β-naphthylamine), trace metals (e.g., arsenic, beryllium, cadmium, chromium, lead, and nickel), and gases (e.g., nitric oxides and sulfur dioxide).

### Cotton Dust (raw)

NIOSH recommends reducing exposures to cotton dust to the lowest feasible concentration to reduce the prevalence and severity of byssinosis; the REL is <0.200 mg/m$^3$ (as lint free cotton dust).

As found in OSHA Table Z-1 (29 CFR 1910.1000), the PEL for cotton dust (raw) is 1 mg/m$^3$ for the cotton waste processing operations of waste recycling (sorting, blending, cleaning, and willowing) and garnetting. PELs for other sectors (as found in 29 CFR 1910.1043) are 0.200 mg/m$^3$ for yarn manufacturing and cotton washing operations, 0.500 mg/m$^3$ for textile mill waste house operations or for dust from "lower grade washed cotton" used during yarn manufacturing, and 0.750 mg/m$^3$ for textile slashing and weaving operations. The OSHA standard 29 CFR 1910.1043 does not apply to cotton harvesting, ginning, or the handling and processing of woven or knitted materials and washed cotton. All PELs for cotton dust are mean concentrations of lint-free, respirable cotton dust collected by a vertical elutriator or an equivalent method and averaged over an 8-hour period.

### Lead

NIOSH considers "Lead" to mean metallic lead, lead oxides, and lead salts (including organic salts such as lead soaps but excluding lead arsenate). The NIOSH REL for lead (8-hour TWA) is 0.050 mg/m$^3$; air concentrations should be maintained so that worker blood lead remains less than 0.060 mg Pb/100 g of whole blood.

OSHA considers "Lead" to mean metallic lead, all inorganic lead compounds (lead oxides and lead salts), and a class of organic compounds called soaps; all other lead compounds are excluded from this definition. The OSHA PEL (8-hour TWA) is 0.050 mg/m$^3$; other OSHA requirements can be found in 29 CFR 1910.1025. The OSHA PEL (8 hour-TWA) for lead in "non-ferrous foundries with less than 20 employees" is 0.075 mg/m$^3$.

### Mineral Dusts

The OSHA PELS for "mineral dusts" listed below are from Table Z-3 of 29 CFR 1910.1000. The OSHA PEL (8-hour TWA) for crystalline silica (as respirable quartz) is either 250 mppcf divided by the value "%$SiO_2$ + 5" or 10 mg/m$^3$ divided by the value

"%$SiO_2$ + 2." The OSHA PEL (8-hour TWA) for crystalline silica (as total quartz) is 30 mg/m$^3$ divided by the value "%$SiO_2$ + 2." The OSHA PELs (8-hour TWAs) for cristobalite and tridymite are ½ the values calculated above using the count or mass formulae for quartz.

The OSHA PEL (8-hour TWA) for amorphous silica (including diatomaceous earth) is either 80 mg/m$^3$ divided by the value "%$SiO_2$," or 20 mppcf.

The OSHA PELs (8-hour TWAs) for talc (not containing asbestos), mica, and soapstone are 20 mppcf. The OSHA PEL (8-hour TWA) for portland cement is 50 mppcf. The OSHA PEL (8-hour TWA) for graphite (natural) is 15 mppcf. The PELs for talc (not containing asbestos), mica, soapstone, and portland cement are applicable if the material contains less than 1% crystalline silica.



The OSHA PEL (8-hour TWA) for coal dust (as the respirable fraction) containing less than 5% $SiO_2$ is 2.4 mg/m$^3$ divided by the value "%$SiO_2$ + 2." The OSHA PEL (8-hour TWA) for coal dust (as the respirable fraction) containing greater than or equal to 5% $SiO_2$ is 10 mg/m$^3$ divided by the value "%$SiO_2$ + 2."

## NIAX® Catalyst ESN

In May 1978, OSHA and NIOSH jointly published *Current Intelligence Bulletin (CIB) 26: NIAX® Catalyst ESN.* In this CIB, OSHA and NIOSH recommended that occupational exposure to NIAX® Catalyst ESN, its components, dimethylaminopropionitrile and bis(2-(dimethylamino)ethyl)ether, as well as formulations containing either component, be minimized. Exposures should be limited to as few workers as possible, while minimizing workplace exposure concentrations with effective work practices and engineering controls. Exposed workers should be carefully monitored for potential disorders of the nervous and genitourinary system. Although substitution is a possible control measure, alternatives to NIAX® Catalyst ESN or its components should be carefully evaluated with regard to possible adverse health effects.

## Trichloroethylene

NIOSH considers trichloroethylene (TCE) to be a potential occupational carcinogen and recommends a REL of 2 ppm (as a 60-minute ceiling) during the use of TCE as an anesthetic agent, and 25 ppm (as a 10-hour TWA) during all other exposures.

## Tungsten Carbide (Cemented)

"Cemented tungsten carbide" or "hard metal" refers to a mixture of tungsten carbide, cobalt, and sometimes metal oxides or carbides and other metals (including nickel). When the cobalt (Co) content exceeds 2%, its contribution to the potential hazard is judged to exceed that of tungsten carbide. Therefore, the NIOSH REL (10-hour TWA) for cemented tungsten carbide containing >2% Co is 0.05 mg Co/m$^3$; the applicable OSHA PEL is 0.1 mg Co/m$^3$ (8-hour TWA). Nickel (Ni) may sometimes be used as a binder rather than cobalt. NIOSH considers cemented tungsten carbide containing nickel to be a potential occupational carcinogen and recommends a REL of 0.015 mg Ni/m$^3$ (10-hour TWA). The OSHA PEL for Insoluble Nickel (i.e., a 1 mg Ni/m$^3$ 8-hour TWA) applies to mixtures of tungsten carbide and nickel.

# Appendix D
## SUBSTANCES WITH NO ESTABLISHED RELs

After reviewing available published literature, NIOSH provided comments to OSHA on August 1, 1988, regarding the "Proposed Rule on Air Contaminants" (29 CFR 1910, Docket No. H-020). In these comments, NIOSH questioned whether the PELs proposed (and listed below) for the following substances included in the *Pocket Guide* were adequate to protect workers from recognized health hazards. The current PEL for each of these compounds is listed on the chemical page for each substance in the *Pocket Guide*. See pages *xi-xii* for a discussion of the vacated PELs.

APPENDIX D

- Acetylene tetrabromide [**TWA** 1 ppm]
- Chlorobenzene [**TWA** 75 ppm]
- Ethyl bromide [**TWA** 200 ppm, **STEL** 250 ppm]
- Ethylene glycol [**C** 50 ppm]
- Ethyl ether [**TWA** 400 ppm, **STEL** 500 ppm]
- Fenthion [**TWA** 0.2 $mg/m^3$ (skin)]
- Furfural [**TWA** 2 ppm (skin)]
- 2-Isopropoxyethanol [**TWA** 25 ppm]
- Isopropyl acetate [**TWA** 250 ppm, **STEL** 310 ppm]
- Isopropylamine [**TWA** 5 ppm, **STEL** 10 ppm]
- Manganese tetroxide (as Mn) [**TWA** 1 $mg/m^3$]
- Molybdenum (soluble compounds as Mo) [**TWA** 5 $mg/m^3$]
- Nitromethane [**TWA** 100 ppm]
- m-Toluidine [**TWA** 2 ppm (skin)]
- Triethylamine [**TWA** 10 ppm, **STEL** 15 ppm]

At that time, NIOSH also conducted a limited evaluation of the literature and concluded that the documentation cited by OSHA was inadequate to support the proposed PEL (as an 8-hour TWA) of 10 $mg/m^3$ for the compounds listed below. The current PEL for magnesium oxide fume is 15 $mg/m^3$ (8-hour TWA, total particulate), and the current PEL for molybdenum (insoluble compounds as Mo) is 15 $mg/m^3$ (8-hour TWA, total dust). For the other compounds listed below the current PEL is 15 $mg/m^3$ (8-hour TWA, total dust) and 5 $mg/m^3$ (8-hour TWA, respirable dust).

- α-Alumina
- Benomyl
- Emery
- Glycerine (mist)
- Graphite (synthetic)
- Magnesium oxide fume
- Molybdenum (insoluble compounds as Mo)
- Particulates not otherwise regulated
- Picloram
- Rouge

# Appendix E
# OSHA Respirator Requirements for Selected Chemicals

Revisions to the OSHA Respiratory Protection Standard (29 CFR 1910.134) became effective on April 8, 1998. Incorporated within the preamble of this ruling were changes to OSHA regulations for several chemicals or substances, which are listed as subheadings in blue text throughout this appendix. These subheadings, which are also the titles of the affected standards within 29 CFR 1910 and 29 CFR 1926, are followed by the standard number(s) in parentheses and the OSHA respirator requirements. Fit testing is required by OSHA for all tight-fitting air-purifying respirators. *Please consult 29 CFR 1910.134 for the full content of the changes that apply.* For all of the chemicals listed in this appendix, any respirators that are permitted at higher environmental concentrations can be used at lower concentrations.

## 13 Carcinogens (4-Nitrobiphenyl, etc.) (1910.1003)

Employees engaged in handling operations involving the carcinogens listed below must be provided with, and required to wear and use, a *half-mask* filter-type respirator for dusts, mists, and fumes. A respirator affording higher levels of protection than this respirator may be substituted.

- 2-Acetylaminofluorene
- 4-Aminodiphenyl
- Benzidine
- bis-Chloromethyl ether
- 3,3'-Dichlorobenzidine (and its salts)
- 4-Dimethylaminoazobenezene
- Ethyleneimine
- Methyl chloromethyl ether
- α-Naphthylamine
- β-Naphthylamine
- 4-Nitrobiphenyl
- N-Nitrosodimethylamine
- β-Propiolactone



## Acrylonitrile (1910.1045)

| Airborne Concentration or Condition of Use | Respirator Type |
|---|---|
| ≤ 20 ppm (parts per million) | **(1)** Chemical cartridge respirator with organic vapor cartridge(s) and half-mask facepiece; or **(2)** Supplied-air respirator with half-mask facepiece. |
| ≤ 100 ppm or maximum use concentration of cartridges or canisters, whichever is lower | **(1)** Full-facepiece respirator with *(A)* organic vapor cartridges, *(B)* organic vapor gas mask, chin-style, or *(C)* organic vapor gas mask canister, front- or back-mounted; **(2)** Supplied-air respirator with full facepiece; or **(3)** Self-contained breathing apparatus with full facepiece. |
| ≤ 4,000 ppm | Supplied-air respirator operated in positive-pressure mode with full facepiece, helmet, suit, or hood. |
| > 4,000 ppm or unknown concentration | **(1)** Supplied-air and auxiliary self-contained breathing apparatus with full facepiece in positive-pressure mode; or **(2)** Self-contained breathing apparatus with full facepiece in positive-pressure mode. |
| Firefighting | Self-contained breathing apparatus with full facepiece in positive-pressure mode. |
| Escape | **(1)** Any organic vapor respirator; or **(2)** Any self-contained breathing apparatus. |

## Appendix E (Continued)
## OSHA Respirator Requirements for Selected Chemicals

### Arsenic, inorganic (1910.1018)

**Requirements for Respiratory Protection for Inorganic Arsenic Particulate**
***Except for Those With Significant Vapor Pressure***

| Airborne Concentration (as As) or Condition of Use | Required Respirator |
|---|---|
| ≤ 100 μg/m³ (micrograms per cubic meter) | **(1)** Half-mask air-purifying respirator equipped with high-efficiency filter*; or **(2)** Any half-mask supplied air respirator. |
| ≤ 500 μg/m³ | **(1)** Full facepiece air-purifying respirator equipped with high-efficiency filter*; **(2)** Any full-facepiece supplied-air respirator; or **(3)** Any full-facepiece self-contained breathing apparatus. |
| ≤ 10,000 μg/m³ | **(1)** Powered air-purifying respirators in all inlet face coverings with high-efficiency filters*; or **(2)** Half-mask supplied-air respirators operated in positive-pressure mode. |
| ≤ 20,000 μg/m³ | Supplied-air respirator with full facepiece, hood, or helmet or suit, operated in positive-pressure mode. |
| > 20,000 μg/m³, unknown concentrations, or firefighting | Any full-facepiece self-contained breathing apparatus operated in positive-pressure mode. |

* A high-efficiency filter means a filter that is at least 99.97% efficient against mono-disperesed particles of 0.3 μm (micrometers) in diameter or higher.

APPENDIX E

**Requirements for Respiratory Protection for**
**Inorganic Arsenicals *With Significant Vapor Pressure***

| Airborne Concentration (as As) or Condition of Use | Required Respirator |
|---|---|
| ≤ 100 μg/m³ (micrograms per cubic meter) | **(1)** Half-mask* air-purifying respirator equipped with high-efficiency filter** and acid gas cartridge; or **(2)** Any half-mask* supplied-air respirator. |
| ≤ 500 μg/m³ | **(1)** Front- or back-mounted gas mask equipped with high-efficiency filter** and acid gas canister; **(2)** Any full-facepiece supplied-air respirator; or **(3)** Any full-facepiece self-contained breathing apparatus. |
| ≤ 10,000 μg/m³ | Half-mask* supplied-air respirator operated in positive-pressure mode. |
| ≤ 20,000 μg/m³ | Supplied-air respirator with full facepiece, hood, or helmet or suit, operated in positive-pressure mode. |
| > 20,000 μg/m³, unknown concentrations, or firefighting | Any full-facepiece self-contained breathing apparatus operated in positive-pressure mode. |

* Half-mask respirators shall not be used for protection against arsenic trichloride, as it is rapidly absorbed through the skin.
** A high-efficiency filter means a filter that is at least 99.97% efficient against mono-disperesed particles of 0.3 μm (micrometers) in diameter or higher.

## Appendix E (Continued)
## OSHA Respirator Requirements for Selected Chemicals

### Asbestos (1910.1001 & 1926.1101)

| Airborne Concentration or Condition of Use | Required Respirator |
|---|---|
| ≤ 1 f/cm$^3$ (fibers per cubic centimeter) (10 X PEL) | Half-mask air-purifying respirator other than a disposable respirator, equipped with high-efficiency filters*. |
| ≤ 5 f/cm$^3$ (50 X PEL) | Full-facepiece air-purifying respirator equipped with high-efficiency filters*. |
| ≤ 10 f/cm$^3$ (100 X PEL) | Any powered air-purifying respirator equipped with high-efficiency filters* or any supplied-air respirator operated in continuous-flow mode. |
| ≤ 100 f/cm$^3$ (1,000 X PEL) | Full-facepiece supplied air respirator operated in pressure-demand mode. |
| > 100 f/cm$^3$ (1,000 X PEL), or unknown concentrations | Full-facepiece supplied-air respirator operated in pressure-demand mode, equipped with an auxiliary positive-pressure self-contained breathing apparatus. |

* A high-efficiency filter means a filter that is at least 99.97% efficient against mono-dispersed particles of 0.3 μm (micrometers) in diameter or higher.

A P P E N D I X E

### Benzene (1910.1028)

| Airborne Concentration or Condition of Use | Required Respirator |
|---|---|
| ≤ 10 ppm (parts per million) | Half-mask air-purifying respirator with organic vapor cartridge. |
| ≤ 50 ppm | **(1)** Full-facepiece respirator with organic vapor cartridges; or **(2)** Full-facepiece gas mask with chin-style canisters*. |
| ≤ 100 ppm | Full-facepiece powered air-purifying respirator with organic vapor canister*. |
| ≤ 1,000 ppm | Supplied-air respirator with full facepiece in positive-pressure mode. |
| > 1,000 ppm or unknown concentration | **(1)** Self-contained breathing apparatus with full facepiece in positive-pressure mode; or **(2)** Full-facepiece positive-pressure supplied-air respirator with auxiliary self-contained air supply. |
| Escape | **(1)** Any organic vapor gas mask; or **(2)** Any self-contained breathing apparatus with full facepiece. |
| Firefighting | Full-facepiece self-contained breathing apparatus in positive-pressure mode. |

* Canisters must have a minimum service life of four (4) hours when tested at 150 ppm benzene, at a flow rate of 64 liters per minute (LPM), 25°C, and 85% relative humidity for non-powered air-purifying respirators. The flow rate shall be 115 LPM and 170 LPM, respectively, for tight-fitting and loose-fitting powered air-purifying respirators.

## Appendix E (Continued)
## OSHA Respirator Requirements for Selected Chemicals

**1,3-Butadiene (1910.1051)**

| Airborne Concentration or Condition of Use | Required Respirator |
|---|---|
| ≤ 5 ppm (parts per million) | Air-purifying half-mask or full-facepiece respirator equipped with approved butadiene or organic vapor cartridges or canisters. Cartridges or canisters shall be replaced every 4 hours. |
| ≤ 10 ppm | Air-purifying half-mask or full-facepiece respirator equipped with approved butadiene or organic vapor cartridges or canisters. Cartridges or canisters shall be replaced every 3 hours. |
| ≤ 25 ppm | **(1)** Air-purifying half-mask or full-facepiece respirator equipped with approved butadiene or organic vapor cartridges or canisters. Cartridges or canisters shall be replaced every 2 hours; **(2)** Any powered air-purifying respirator equipped with approved butadiene or organic vapor cartridges or canisters. Cartridges or canisters shall be replaced every [1] hour; or **(3)** Continuous-flow supplied-air respirator equipped with a hood or helmet. |
| ≤ 50 ppm | **(1)** Air-purifying full-facepiece respirator equipped with approved butadiene or organic vapor cartridges or canisters. Cartridges or canisters shall be replaced every [1] hour; or **(2)** Powered air-purifying respirator (PAPR) equipped with a tight-fitting facepiece and approved butadiene or organic vapor cartridges. PAPR cartridges shall be replaced every [1] hour. |
| ≤ 1,000 ppm | Supplied-air respirator equipped with a half-mask or full facepiece and operated in a pressure-demand or other positive-pressure mode. |
| > 1,000 ppm, unknown concentration, or firefighting | **(1)** Self-contained breathing apparatus equipped with a full facepiece and operated in a pressure-demand or other positive-pressure mode; or **(2)** Any supplied-air respirator equipped with a full facepiece and operated in a pressure-demand or other positive-pressure mode in combination with an auxiliary self-contained breathing apparatus operated in a pressure-demand or other positive-pressure mode. |
| Escape from IDLH conditions (IDLH is 2,000 ppm) | **(1)** Any positive-pressure self-contained breathing apparatus with an appropriate service life; or **(2)** Any air-purifying full-facepiece respirator equipped with a front- or back-mounted butadiene or organic vapor canister. |

# Appendix E (Continued)
# OSHA Respirator Requirements for Selected Chemicals

### Cadmium (1910.1027 & 1926.1127)

| Airborne Concentration or Condition of Use | Required Respirator |
|---|---|
| ≤ 50 μg/m$^3$ (micrograms per cubic meter) | Half-mask, air-purifying respirator equipped with a high-efficiency filter*. |
| ≤ 125 μg/m$^3$ | **(1)** Powered air-purifying respirator with a loose-fiting hood or helmet equipped with a high-efficiency filter*; or **(2)** Supplied-air respirator with a loose-fitting hood or helmet facepiece operated in continuous-flow mode. |
| ≤ 250 μg/m$^3$ | **(1)** Full-facepiece air-purifying respirator equipped with a high-efficiency filter*; **(2)** Powered air-purifying respirator with a tight-fitting half-mask equipped with a high-efficiency filter*; or **(3)** Supplied-air respirator with a tight-fitting half-mask operated in continuous-flow mode. |
| ≤ 1,250 μg/m$^3$ | **(1)** Powered air-purifying respirator with a tight-fitting full facepiece equipped with a high-efficiency filter*; or **(2)** Supplied-air respirator with a tight-fitting full facepiece operated in continuous-flow mode. |
| ≤ 5,000 μg/m$^3$ | Supplied-air respirator with half-mask or full facepiece operated in pressure-demand or other positive-pressure mode. |
| > 5,000 μg/m$^3$ or unknown concentration | **(1)** Self-contained breathing apparatus with a full facepiece operated in pressure-demand or other positive-pressure mode; or **(2)** Supplied-air respirator with a full facepiece operated in pressure-demand or other positive-pressure mode and equipped with an auxiliary escape-type self-contained breathing apparatus operated in pressure-demand mode. |
| Firefighting | Self-contained breathing apparatus with full facepiece operated in pressure-demand or other positive-pressure mode. |

**Note:** Quantitative fit testing is required for all tight-fitting air-purifying respirators where airborne concentration of cadmium exceeds 10 times the TWA PEL (10 X 5 μg/m$^3$ = 50 μg/m$^3$). A full-facepiece respirator is required when eye irritation is expected.

* A high-efficiency filter means a filter that is at least 99.97% efficient against mono-dispersed particles of 0.3 μm (micrometers) in diameter or higher.

### Coke oven emissions (1910.1029)

| Airborne Concentration | Required Respirator |
|---|---|
| ≤ 1500 μg/m$^3$ (micrograms per cubic meter) | **(1)** Any particulate filter respirator for dust and mist except single-use respirator; or **(2)** Any particulate filter respirator or combination chemical cartridge and particulate filter respirator for coke oven emissions. |
| Any concentrations | **(1)** Type C supplied-air respirator [see page 360] operated in pressure-demand or continuous-flow mode; **(2)** Powered air-purifying particulate filter respirator for dust and mist; or **(3)** Powered air-purifying particulate filter respirator or combination chemical cartridge and particulate filter respirator for coke oven emissions. |

## Appendix E (Continued)
## OSHA Respirator Requirements for Selected Chemicals

### Cotton dust (1910.1043)

| Airborne Concentration | Required Respirator |
|---|---|
| ≤ 5 X PEL | Disposable respirator* with a particulate filter. |
| ≤ 10 X PEL | Quarter- or half-mask respirator, other than a disposable respirator, equipped with particulate filters. |
| ≤ 100 X PEL | Full-facepiece respirator equipped with high-efficiency particulate filters**. |
| > 100 X PEL | Powered air-purifying respirator equipped with high-efficiency particulate filters. |

* A disposable respirator means the filter element is an inseparable part of the respirator.
** A high-efficiency filter means a filter that is at least 99.97% efficient against mono-dispersed particles of 0.3 μm (micrometers) in diameter or higher.

**Notes:**
Self-contained breathing apparatus are not required but are permitted respirators.

Supplied-air respirators are not required but are permitted under the following conditions:
Cotton dust concentration not greater than 10X the PEL: Any supplied air respirator; not greater than 100X the PEL: Any supplied-air respirator with full facepiece, helmet, or hood; greater than 100X the PEL: Supplied-air respirator operated in positive-pressure mode.

APPENDIX E

### 1,2-Dibromo-3-chloropropane (1910.1044)

| Airborne Concentration or Condition of Use | Required Respirator |
|---|---|
| ≤ 10 ppb (parts per billion) | **(1)** Any supplied-air respirator; or **(2)** any self-contained breathing apparatus. |
| ≤ 50 ppb | **(1)** Any supplied-air respirator with full facepiece, helmet, or hood; or **(2)** any self-contained breathing apparatus with full facepiece. |
| ≤ 1,000 ppb | Type C supplied-air respirator [see page 360] operated in pressure-demand or other positive-pressure or continuous-flow mode. |
| ≤ 2,000 ppb | Type C supplied-air respirator [see page 360] with full facepiece operated in pressure-demand or other positive-pressure mode, or with full facepiece, helmet, or hood operated in continuous-flow mode. |
| > 2,000 ppb or entry and escape from unknown concentrations | **(1)** A combination respirator which includes a Type C supplied-air respirator [see page 360] with full facepiece operated in pressure-demand or other positive pressure or continuous-flow mode and an auxiliary self-contained breathing apparatus operated in pressure-demand or positive-pressure mode; or **(2)** Self-contained breathing apparatus with full facepiece operated in pressure-demand or other positive-pressure mode. |
| Firefighting | Self-contained breathing apparatus with full facepiece operated in pressure-demand or other positive-pressure mode. |

# Appendix E (Continued)
## OSHA Respirator Requirements for Selected Chemicals

### Ethylene oxide (1910.1047)

| Airborne Concentration or Condition of Use | Required Respirator |
|---|---|
| ≤ 50 ppm (parts per million) | Full-facepiece respirator with ethylene oxide approved canister, front- or back-mounted. |
| ≤ 2,000 ppm | **(1)** Positive-pressure supplied-air respirator equipped with full facepiece, hood, or helmet; or **(2)** Continuous-flow supplied-air respirator (positive-pressure) equipped with hood, helmet, or suit. |
| > 2,000 ppm or unknown concentrations | **(1)** Positive-pressure self-contained breathing apparatus equipped with full facepiece; or **(2)** Positive-pressure full-facepiece supplied-air respirator equipped with an auxiliary positive-pressure self-contained breathing apparatus. |
| Firefighting | Positive-pressure self-contained breathing apparatus equipped with full facepiece. |
| Escape | Any respirator described above. |

### Formaldehyde (1910.1048)

| Airborne Concentration or Condition of Use | Required Respirator |
|---|---|
| ≤ 7.5 ppm (parts per million) (10 X PEL) | Full-facepiece respirator with cartridges or canisters specifically approved for protection against formaldehyde*. |
| ≤ 75 ppm (100 X PEL) | **(1)** Full-face mask respirator with chin style or chest- or back-mounted type with industrial size canister specifically approved for protection against formaldehyde; or **(2)** Type C supplied-air respirator [see page 360], demand type or continuous flow type, with full facepiece, hood, or helmet. |
| > 75 ppm (100 X PEL) or unknown concentrations (emergencies) | **(1)** Self-contained breathing apparatus with positive-pressure full-facepiece; or **(2)** Combination supplied-air, full-facepiece positive-pressure respirator with auxiliary self-contained air supply. |
| Firefighting | Self-contained breathing apparatus with positive-pressure in full facepiece. |
| Escape | **(1)** Self-contained breathing apparatus in demand or pressure-demand mode; or **(2)** Full-face mask respirator with chin-style or front- or back-mounted type industrial size canister specifically approved for protection against formaldehyde. |

* A half-mask respirator with cartridges specifically approved for protection against formaldehyde can be substituted for the full-facepiece respirator providing that effective gas-proof goggles are provided and used in combination with the half-mask respirator.

# Appendix E (Continued)
# OSHA Respirator Requirements for Selected Chemicals

## Lead (1910.1025 & 1926.62)

**Respirator Requirements of 1910.1025 (General Industry Lead Standard)**

| Airborne Concentration or Condition of Use | Required Respirator |
|---|---|
| ≤ 0.5 $mg/m^3$ (milligrams per cubic meter) (10 X PEL) | Half-mask* air-purifying respirator equipped with high-efficiency filters**. |
| ≤ 2.5 $mg/m^3$ (50 X PEL) | Full-facepiece air-purifying respirator with high-efficiency filters**. |
| ≤ 50 $mg/m^3$ (1000 X PEL) | **(1)** Any powered air-purifying respirator with high-efficiency filters**; or **(2)** Half-mask* supplied-air respirator operated in positive-pressure mode. |
| ≤ 100 $mg/m^3$ (2000 X PEL) | Supplied-air respirators with full facepiece, hood, helmet, or suit, operated in positive-pressure mode. |
| > 100 $mg/m^3$, unknown concentration, or firefighting | Full-facepiece, self-contained breathing apparatus operated in positive-pressure mode. |

* Full facepiece is required if the lead aerosols cause eye or skin irritation at the use concentrations.

** A high-efficiency filter means a filter that is at least 99.97% efficient against mono-dispersed particles of 0.3 µm (micrometers) in diameter or higher.

APPENDIX E

**Respirator Requirements of 1926.62 (Construction Lead Standard)**

| Airborne Concentration or Condition of Use | Required Respirator |
|---|---|
| ≤ 0.5 $mg/m^3$ (milligrams per cubic meter) | **(1)** Half-mask* air-purifying respirator with high-efficiency filters**; or **(2)** Half-mask* supplied-air respirator operated in demand (negative pressure) mode. |
| ≤ 1.25 $mg/m^3$ | **(1)** Loose-fitting hood or helmet powered air-purifying respirator with high-efficiency filters**; or **(2)** Hood or helmet supplied-air respirator operated in a continuous-flow mode (e.g., Type CE abrasive blasting respirators [see page 360] operated in a continuous-flow mode). |
| ≤ 2.5 $mg/m^3$ | **(1)** Full-facepiece air-purifying respirator with high-efficiency filters**; **(2)** Tight-fitting powered air-purifying respirator with high-efficiency filters**; **(3)** Full-facepiece supplied-air respirator operated in demand mode; **(4)** Half-mask* or full-facepiece supplied-air respirator operated in a continuous-flow mode; or **(5)** Full-facepiece self-contained breathing apparatus operated in demand mode. |
| ≤ 50 $mg/m^3$ | Half-mask* supplied-air respirator operated in pressure-demand or other positive-pressure mode. |
| ≤ 100 $mg/m^3$ | Full-facepiece supplied-air respirator operated in pressure-demand or other positive-pressure mode (e.g., Type CE abrasive blasting respirators [see page 360] operated in a continuous-flow mode). |
| > 100 $mg/m^3$, unknown concentration, or firefighting | Full-facepiece self-contained breathing apparatus in pressure-demand or other positive-pressure mode. |

* Full facepiece is required if the lead aerosols cause eye or skin irritation at the use concentrations.

** A high-efficiency filter means a filter that is at least 99.97% efficient against mono-dispersed particles of 0.3 µm (micrometers) in diameter or higher.

## Appendix E (Continued)
## OSHA Respirator Requirements for Selected Chemicals

### Methylene chloride (1910.1052)

| Airborne Concentration or Condition of Use | Required Respirator |
|---|---|
| ≤ 625 ppm (parts per million) (25 X PEL) | Continuous-flow supplied-air respirator, hood or helmet. |
| ≤ 1250 ppm (50 X PEL) | **(1)** Full-facepiece supplied-air respirator operated in negative-pressure (demand) mode; or **(2)** Full-facepiece self-contained breathing apparatus operated in negative-pressure (demand) mode. |
| ≤ 5,000 ppm (200 X PEL) | **(1)** Continuous-flow supplied-air respirator, full-facepiece; **(2)** Pressure-demand supplied-air respirator, full-facepiece; or **(3)** Positive-pressure full-facepiece self-contained breathing apparatus. |
| > 5,000 ppm or unknown concentration | **(1)** Positive-pressure full-facepiece self-contained breathing apparatus; or **(2)** Full-facepiece pressure-demand supplied-air respirator with an auxiliary self-contained air supply. |
| Firefighting | Positive-pressure full-facepiece self-contained breathing apparatus. |
| Emergency escape | **(1)** Any continuous-flow or pressure-demand self-contained breathing apparatus; or **(2)** Gas mask with organic vapor canister. |



### 4,4'-Methylenedianiline (1910.1050 & 1926.60)

| Airborne Concentration or Condition of Use | Required Respirator |
|---|---|
| ≤ 10 X PEL | Half-mask respirator with high-efficiency* cartridge**. |
| ≤ 50 X PEL | Full-facepiece respirator with high-efficiency* cartridge or canister**. |
| ≤ 1,000 X PEL | Full-facepiece powered air-purifying respirator with high-efficiency* cartridge**. |
| > 1,000 X PEL or unknown concentration | **(1)** Self-contained breathing apparatus with full facepiece in positive-pressure mode; or **(2)** Full-facepiece positive-pressure demand supplied-air respirator with auxiliary self-contained air supply. |
| Escape | **(1)** Any full-facepiece air-purifying respirator with high-efficiency* cartridges**; or **(2)** Any positive-pressure or continuous-flow self-contained breathing apparatus with full facepiece or hood. |
| Firefighting | Full-facepiece self-contained breathing apparatus in positive-pressure demand mode. |

* A high-efficiency filter means a filter that is at least 99.97% efficient against mono-dispersed particles of 0.3 µm (micrometers) in diameter or higher.
** Combination High-Efficiency/Organic Vapor Cartridges shall be used whenever Methylenedianiline is in liquid form or a process requiring heat is used.

## Appendix E (Continued)
## OSHA Respirator Requirements for Selected Chemicals

### Vinyl Chloride (1910.1017)

| Airborne Concentration or Condition of Use | Required Respirator |
|---|---|
| ≤ 10 ppm (parts per million) | **(1)** Combination Type C supplied-air respirator [see below], demand type, with half facepiece, and auxiliary self-contained air supply; **(2)** Type C supplied-air respirator [see below], demand type, with half facepiece; or **(3)** Any chemical cartridge respirator with an organic vapor cartridge which provides a service life of at least 1 hour for concentrations of vinyl chloride up to 10 ppm. |
| ≤ 25 ppm | **(1)** Powered air-purifying respirator with hood, helmet, full or half facepiece, and a canister which provides a service life of at least 4 hours for concentrations of vinyl chloride up to 25 ppm; or **(2)** Gas mask with front- or back-mounted canister which provides a service life of at least 4 hours for concentrations of vinyl chloride up to 25 ppm. |
| ≤ 100 ppm | **(1)** Combination Type C supplied-air respirator [see below], demand type, with full facepiece, and auxiliary self-contained air supply; or **(2)** Open-circuit self-contained breathing apparatus with full facepiece, in demand mode; or **(3)** Type C supplied-air respirator [see below], demand type, with full facepiece. |
| ≤ 1,000 ppm | Type C supplied-air respirator [see below], continuous-flow type, with full or half facepiece, helmet, or hood. |
| ≤ 3,600 ppm | **(1)** Combination Type C supplied-air respirator [see below], pressure demand type, with full or half facepiece, and auxiliary self-contained air supply; or **(2)** Combination type continuous-flow supplied-air respirator with full or half facepiece and auxiliary self-contained air supply. |
| > 3,600 ppm or unknown concentration | Open-circuit self-contained breathing apparatus, pressure-demand type, with full facepiece. |



### Definitions for Type C and Type CE Respirators

The definitions below were obtained from the NIOSH Certified Equipment List, which is available on the NIOSH Web site (http://www.cdc.gov/niosh/npptl/topics/respirators/cel).

**Type C Respirator:** An airline respirator, for entry into and escape from atmospheres not immediately dangerous to life or health, which consists of a source of respirable breathing air, a hose, a detachable coupling, a control valve, orifice, a demand valve or pressure demand valve, and arrangement for attaching the hose to the wearer and a facepiece, hood, or helmet.

**Type CE Respirator:** A Type C supplied-air respirator equipped with additional devices designed to protect the wearer's head and neck against impact and abrasion from rebounding abrasive material, and with shielding material such as plastic, glass, woven wire, sheet metal, or other suitable material to protect the window(s) of facepieces, hoods, and helmets which do not unduly interfere with the wearer's vision and permit easy access to the external surface of such window(s) for cleaning.

# Appendix F
# MISCELLANEOUS NOTES

**Benzene**

The final OSHA Benzene standard in 1910.1028 applies to all occupational exposures to benzene except some subsegments of industry where exposures are consistently under the action level (i.e., distribution and sales of fuels, sealed containers and pipelines, coke production, oil and gas drilling and production, natural gas processing, and the percentage exclusion for liquid mixtures); for the excepted subsegments, the benzene limits in Table Z-2 apply (i.e., an 8-hour TWA of 10 ppm, an acceptable ceiling of 25 ppm, and 50 ppm for a maximum duration of 10 minutes as an acceptable maximum peak above the acceptable ceiling).

**Octachloronaphthalene**
**Pentachloronaphthalene**
**Tetrachloronaphthalene**
**Trichloronaphthalene**

IDLH values for these four chloronaphthalene compounds are unknown. The *Documentation for Immediately Dangerous to Life or Health Concentrations* (NTIS Publication Number PB-94-195047) identified "Effective" IDLH values, based on analogy with other chloronaphthalenes and the then-effective *NIOSH Respirator Decision Logic* (DHHS [NIOSH] Publication No. 87-108; http://www.cdc.gov/niosh/docs/87-108). These values for respirator recommendations were determined by multiplying the NIOSH REL or OSHA PEL by an assigned protection factor of 10. This assigned protection factor was used during the Standards Completion Program for deciding when the "most protective" respirators should be used for these four chemicals. Listed below are the "Effective" IDLH values that were determined using 10 times the REL or PEL for each chemical. For more information please consult the *IDLH Documentation* on the NIOSH Web site (http://www.cdc.gov/niosh/idlh/idlh-1.html).

APPENDIX F

| Chemical | NIOSH REL/OSHA PEL | "Effective" IDLH (10 X REL/PEL) |
|---|---|---|
| Octachloronaphthalene | TWA 0.1 $mg/m^3$ * | 1 $mg/m^3$ |
| Pentachloronaphthalene | TWA 0.5 $mg/m^3$ | 5 $mg/m^3$ |
| Tetrachloronaphthalene | TWA 5 $mg/m^3$ | 50 $mg/m^3$ |
| Trichloronaphthalene | TWA 2 $mg/m^3$ | 20 $mg/m^3$ |

* NIOSH also recommends a STEL of 0.3 $mg/m^3$ for octachloronaphthalene; the TWA of 0.1 $mg/m^3$ was used to calculate the "Effective" IDLH of 1 $mg/m^3$.

# Appendix G
# VACATED 1989 OSHA PELs

(See pages xi and xii for an explanation of the vacated 1989 OSHA PELs.)

| Chemical | Vacated 1989 OSHA PEL |
|---|---|
| Acetaldehyde | **TWA** 100 ppm (180 mg/m$^3$), **ST** 150 ppm (270 mg/m$^3$) |
| Acetic anhydride | **C** 5 ppm (20 mg/m$^3$) |
| Acetone | **TWA** 750 ppm (1800 mg/m$^3$), **ST** 1000 ppm (2400 mg/m$^3$) |
| Acetonitrile | **TWA** 40 ppm (70 mg/m$^3$), **ST** 60 ppm (105 mg/m$^3$) |
| Acetylsalicyclic acid | **TWA** 5 mg/m$^3$ |
| Acrolein | **TWA** 0.1 ppm (0.25 mg/m$^3$), **ST** 0.3 ppm (0.8 mg/m$^3$) |
| Acrylamide | **TWA** 0.03 mg/m$^3$ [skin] |
| Acrylic acid | **TWA** 10 ppm (30 mg/m$^3$) [skin] |
| Allyl alcohol | **TWA** 2 ppm (5 mg/m$^3$), **ST** 4 ppm (10 mg/m$^3$) [skin] |
| Allyl chloride | **TWA** 1 ppm (3 mg/m$^3$), **ST** 2 ppm (6 mg/m$^3$) |
| Allyl glycidyl ether | **TWA** 5 ppm (22 mg/m$^3$), **ST** 10 ppm (44 mg/m$^3$) |
| Allyl propyl disulfide | **TWA** 2 ppm (12 mg/m$^3$), **ST** 3 ppm (18 mg/m$^3$) |
| $\alpha$-Alumina | **TWA** 10 mg/m$^3$ (total), **TWA** 5 mg/m$^3$ (resp) |
| Aluminum (pyro powders & welding fumes, as Al) | **TWA** 5 mg/m$^3$ |
| Aluminum (soluble salts & alkyls, as Al) | **TWA** 2 mg/m$^3$ |
| Amitrole | **TWA** 0.2 mg/m$^3$ |
| Ammonia | **ST** 35 ppm (27 mg/m$^3$) |
| Ammonium chloride fume | **TWA** 10 mg/m$^3$, **ST** 20 mg/m$^3$ |
| Ammonium sulfamate | **TWA** 10 mg/m$^3$ (total), **TWA** 5 mg/m$^3$ (resp) |
| Aniline (and homologs) | **TWA** 2 ppm (8 mg/m$^3$) [skin] |
| Atrazine | **TWA** 5 mg/m$^3$ |
| Barium sulfate | **TWA** 10 mg/m$^3$ (total), **TWA** 5 mg/m$^3$ (resp) |
| Benomyl | **TWA** 10 mg/m$^3$ (total), **TWA** 5 mg/m$^3$ (resp) |
| Benzenethiol | **TWA** 0.5 ppm (2 mg/m$^3$) |
| Bismuth telluride (doped with selenium sulfide, as $Bi_2Te_3$) | **TWA** 5 mg/m$^3$ |
| Borates, tetra, sodium salts (Anhydrous) | **TWA** 10 mg/m$^3$ |
| Borates, tetra, sodium salts (Decahydrate) | **TWA** 10 mg/m$^3$ |
| Borates, tetra, sodium salts (Pentahydrate) | **TWA** 10 mg/m$^3$ |
| Boron oxide | **TWA** 10 mg/m$^3$ |
| Boron tribromide | **C** 1 ppm (10 mg/m$^3$) |
| Bromacil | **TWA** 1 ppm (10 mg/m$^3$) |
| Bromine | **TWA** 0.1 ppm (0.7 mg/m$^3$), **ST** 0.3 ppm (2 mg/m$^3$) |

## Appendix G (Continued)
## VACATED 1989 OSHA PELs

(See pages xi and xii for an explanation of the vacated 1989 OSHA PELs.)

| Chemical | Vacated 1989 OSHA PEL |
|---|---|
| Bromine pentafluoride | **TWA** 0.1 ppm (0.7 $mg/m^3$) |
| n-Butane | **TWA** 800 ppm (1900 $mg/m^3$) |
| 2-Butanone | **TWA** 200 ppm (590 $mg/m^3$), **ST** 300 ppm (885 $mg/m^3$) |
| 2-Butoxyethanol | **TWA** 25 ppm (120 $mg/m^3$) [skin] |
| n-Butyl acetate | **TWA** 150 ppm (710 $mg/m^3$), **ST** 200 ppm (950 $mg/m^3$) |
| Butyl acrylate | **TWA** 10 ppm (55 $mg/m^3$) |
| n-Butyl alcohol | **C** 50 ppm (150 $mg/m^3$) [skin] |
| sec-Butyl alcohol | **TWA** 100 ppm (305 $mg/m^3$) |
| tert-Butyl alcohol | **TWA** 100 ppm (300 $mg/m^3$), **ST** 150 ppm (450 $mg/m^3$) |
| n-Butyl glycidyl ether | **TWA** 25 ppm (135 $mg/m^3$) |
| n-Butyl lactate | **TWA** 5 ppm (25 $mg/m^3$) |
| n-Butyl mercaptan | **TWA** 0.5 ppm (1.5 $mg/m^3$) |
| o-sec-Butylphenol | **TWA** 5 ppm (30 $mg/m^3$) [skin] |
| p-tert-Butyltoluene | **TWA** 10 ppm (60 $mg/m^3$), **ST** 20 ppm (120 $mg/m^3$) |
| Calcium cyanamide | **TWA** 0.5 $mg/m^3$ |
| Caprolactam | Dust: **TWA** 1 $mg/m^3$, **ST** 3 $mg/m^3$<br>Vapor: **TWA** 5 ppm (20 $mg/m^3$), **ST** 10 ppm (40 $mg/m^3$) |
| Captafol | **TWA** 0.1 $mg/m^3$ |
| Captan | **TWA** 5 $mg/m^3$ |
| Carbofuran | **TWA** 0.1 $mg/m^3$ |
| Carbon dioxide | **TWA** 10,000 ppm (18,000 $mg/m^3$)<br>**ST** 30,000 ppm (54,000 $mg/m^3$) |
| Carbon disulfide | **TWA** 4 ppm (12 $mg/m^3$), **ST** 12 ppm (36 $mg/m^3$) [skin] |
| Carbon monoxide | **TWA** 35 ppm (40 $mg/m^3$), C 200 ppm (229 $mg/m^3$) |
| Carbon tetrabromide | **TWA** 0.1 ppm (1.4 $mg/m^3$), **ST** 0.3 ppm (4 $mg/m^3$) |
| Carbon tetrachloride | **TWA** 2 ppm (12.6 $mg/m^3$) |
| Carbonyl fluoride | **TWA** 2 ppm (5 $mg/m^3$), **ST** 5 ppm (15 $mg/m^3$) |
| Catechol | **TWA** 5 ppm (20 $mg/m^3$) [skin] |
| Cesium hydroxide | **TWA** 2 $mg/m^3$ |
| Chlorinated camphene | **TWA** 0.5 $mg/m^3$, **ST** 1 $mg/m^3$ [skin] |
| Chlorine | **TWA** 0.5 ppm (1.5 $mg/m^3$), **ST** 1 ppm (3 $mg/m^3$) |
| Chlorine dioxide | **TWA** 0.1 ppm (0.3 $mg/m^3$), **ST** 0.3 ppm (0.9 $mg/m^3$) |
| Chloroacetyl chloride | **TWA** 0.05 ppm (0.2 $mg/m^3$) |
| o-Chlorobenzylidene malononitrile | **C** 0.05 ppm (0.4 $mg/m^3$) [skin] |
| Chlorodifluoromethane | **TWA** 1000 ppm (3500 $mg/m^3$) |
| Chloroform | **TWA** 2 ppm (9.78 $mg/m^3$) |

## Appendix G (Continued)
## VACATED 1989 OSHA PELs

(See pages xi and xii for an explanation of the vacated 1989 OSHA PELs.)

| Chemical | Vacated 1989 OSHA PEL |
|---|---|
| 1-Chloro-1-nitropropane | **TWA** 2 ppm (10 $mg/m^3$) |
| Chloropentafluoroethane | **TWA** 1000 ppm (6320 $mg/m^3$) |
| β-Chloroprene | **TWA** 10 ppm (35 $mg/m^3$) [skin] |
| o-Chlorostyrene | **TWA** 50 ppm (285 $mg/m^3$), **ST** 75 ppm (428 $mg/m^3$) |
| o-Chlorotoluene | **TWA** 50 ppm (250 $mg/m^3$) |
| Chlorpyrifos | **TWA** 0.2 $mg/m^3$ [skin] |
| Coal dust | **TWA** 2 $mg/m^3$ (<5% $SiO_2$) (resp dust)<br>**TWA** 0.1 $mg/m^3$ (≥ 5% $SiO_2$) (resp quartz) |
| Cobalt metal dust & fume (as Co) | **TWA** 0.05 $mg/m^3$ |
| Cobalt carbonyl (as Co) | **TWA** 0.1 $mg/m^3$ |
| Cobalt hydrocarbonyl (as Co) | **TWA** 0.1 $mg/m^3$ |
| Crag® herbicide | **TWA** 10 $mg/m^3$ (total), **TWA** 5 $mg/m^3$ (resp) |
| Crufomate | **TWA** 5 $mg/m^3$ |
| Cyanamide | **TWA** 2 $mg/m^3$ |
| Cyanogen | **TWA** 10 ppm (20 $mg/m^3$) |
| Cyanogen chloride | **C** 0.3 ppm (0.6 $mg/m^3$) |
| Cyclohexanol | **TWA** 50 ppm (200 $mg/m^3$) [skin] |
| Cyclohexanone | **TWA** 25 ppm (100 $mg/m^3$) [skin] |
| Cyclohexylamine | **TWA** 10 ppm (40 $mg/m^3$) |
| Cyclonite | **TWA** 1.5 $mg/m^3$ [skin] |
| Cyclopentane | **TWA** 600 ppm (1720 $mg/m^3$) |
| Cyhexatin | **TWA** 5 $mg/m^3$ |
| Decaborane | **TWA** 0.3 $mg/m^3$ (0.05 ppm), **ST** 0.9 $mg/m^3$ (0.15 ppm) [skin] |
| Diazinon® | **TWA** 0.1 $mg/m^3$ [skin] |
| 2-N-Dibutylaminoethanol | **TWA** 2 ppm (14 $mg/m^3$) |
| Dibutyl phosphate | **TWA** 1 ppm (5 $mg/m^3$), **ST** 2 ppm (10 $mg/m^3$) |
| Dichloroacetylene | **C** 0.1 ppm (0.4 $mg/m^3$) |
| p-Dichlorobenzene | **TWA** 75 ppm (450 $mg/m^3$), **ST** 110 ppm (675 $mg/m^3$) |
| 1,3-Dichloro-5,5-dimethylhydantoin | **TWA** 0.2 $mg/m^3$, **ST** 0.4 $mg/m^3$ |
| Dichloroethyl ether | **TWA** 5 ppm (30 $mg/m^3$), **ST** 10 ppm (60 $mg/m^3$) [skin] |
| Dichloromonofluoromethane | **TWA** 10 ppm (40 $mg/m^3$) |
| 1,1-Dichloro-1-nitroethane | **TWA** 2 ppm (10 $mg/m^3$) |
| 1,3-Dichloropropene | **TWA** 1 ppm (5 $mg/m^3$) [skin] |
| 2,2-Dichloropropionic acid | **TWA** 1 ppm (6 $mg/m^3$) |
| Dicrotophos | **TWA** 0.25 $mg/m^3$ [skin] |

## Appendix G (Continued)
## VACATED 1989 OSHA PELs

(See pages xi and xii for an explanation of the vacated 1989 OSHA PELs.)

| Chemical | Vacated 1989 OSHA PEL |
|---|---|
| Dicyclopentadiene | **TWA** 5 ppm (30 mg/m$^3$) |
| Dicyclopentadienyl iron | **TWA** 10 mg/m$^3$ (total), **TWA** 5 mg/m$^3$ (resp) |
| Diethanolamine | **TWA** 3 ppm (15 mg/m$^3$) |
| Diethylamine | **TWA** 10 ppm (30 mg/m$^3$), **ST** 25 ppm (75 mg/m$^3$) |
| Diethylenetriamine | **TWA** 1 ppm (4 mg/m$^3$) |
| Diethyl ketone | **TWA** 200 ppm (705 mg/m$^3$) |
| Diethyl phthalate | **TWA** 5 mg/m$^3$ |
| Diglycidyl ether | **TWA** 0.1 ppm (0.5 mg/m$^3$) |
| Diisobutyl ketone | **TWA** 25 ppm (150 mg/m$^3$) |
| N,N-Dimethylaniline | **TWA** 5 ppm (25 mg/m$^3$), **ST** 10 ppm (50 mg/m$^3$) [skin] |
| Dimethyl-1,2-dibromo-2,2-dichlorethyl phosphate | **TWA** 3 mg/m$^3$ [skin] |
| Dimethyl sulfate | **TWA** 0.1 ppm (0.5 mg/m$^3$) [skin] |
| Dinitolmide | **TWA** 5 mg/m$^3$ |
| Di-sec octyl phthalate | **TWA** 5 mg/m$^3$, **ST** 10 mg/m$^3$ |
| Dioxane | **TWA** 25 ppm (90 mg/m$^3$) [skin] |
| Dioxathion | **TWA** 0.2 mg/m$^3$ [skin] |
| Diphenylamine | **TWA** 10 mg/m$^3$ |
| Dipropylene glycol methyl ether | **TWA** 100 ppm (600 mg/m$^3$)<br>**ST** 150 ppm (900 mg/m$^3$) [skin] |
| Dipropyl ketone | **TWA** 50 ppm (235 mg/m$^3$) |
| Diquat (Diquat dibromide) | **TWA** 0.5 mg/m$^3$ |
| Disulfiram | **TWA** 2 mg/m$^3$ |
| Disulfoton | **TWA** 0.1 mg/m$^3$ [skin] |
| 2,6-Di-tert-butyl-p-cresol | **TWA** 10 mg/m$^3$ |
| Diuron | **TWA** 10 mg/m$^3$ |
| Divinyl benzene | **TWA** 10 ppm (50 mg/m$^3$) |
| Emery | **TWA** 10 mg/m$^3$ (total), **TWA** 5 mg/m$^3$ (resp) |
| Endosulfan | **TWA** 0.1 mg/m$^3$ [skin] |
| Epichlorohydrin | **TWA** 2 ppm (8 mg/m$^3$) [skin] |
| Ethanolamine | **TWA** 3 ppm (8 mg/m$^3$), **ST** 6 ppm (15 mg/m$^3$) |
| Ethion | **TWA** 0.4 mg/m$^3$ [skin] |
| Ethyl acrylate | **TWA** 5 ppm (20 mg/m$^3$), **ST** 25 ppm (100 mg/m$^3$) [skin] |
| Ethyl benzene | **TWA** 100 ppm (435 mg/m$^3$), **ST** 125 ppm (545 mg/m$^3$) |
| Ethyl bromide | **TWA** 200 ppm (890 mg/m$^3$), **ST** 250 ppm (1110 mg/m$^3$) |
| Ethylene chlorohydrin | **C** 1 ppm (3 mg/m$^3$) [skin] |
| Ethylene dichloride | **TWA** 1 ppm (4 mg/m$^3$), **ST** 2 ppm (8 mg/m$^3$) |

## Appendix G (Continued)
## VACATED 1989 OSHA PELs

(See pages xi and xii for an explanation of the vacated 1989 OSHA PELs.)

| Chemical | Vacated 1989 OSHA PEL |
|---|---|
| Ethylene glycol | **C** 50 ppm (125 mg/$m^3$) |
| Ethylene glycol dinitrate | **ST** 0.1 mg/$m^3$ [skin] |
| Ethyl ether | **TWA** 400 ppm (1200 mg/$m^3$), **ST** 500 ppm (1500 mg/$m^3$) |
| Ethylidene norbornene | **C** 5 ppm (25 mg/$m^3$) |
| Ethyl mercaptan | **TWA** 0.5 ppm (1 mg/$m^3$) |
| N-Ethylmorpholine | **TWA** 5 ppm (23 mg/$m^3$) [skin] |
| Ethyl silicate | **TWA** 10 ppm (85 mg/$m^3$) |
| Fenamiphos | **TWA** 0.1 mg/$m^3$ [skin] |
| Fensulfothion | **TWA** 0.1 mg/$m^3$ |
| Fenthion | **TWA** 0.2 mg/$m^3$ [skin] |
| Ferbam | **TWA** 10 mg/$m^3$ |
| Ferrovanadium dust | **TWA** 1 mg/$m^3$, **ST** 3 mg/$m^3$ |
| Fluorotrichloromethane | **C** 1000 ppm (5600 mg/$m^3$) |
| Fonofos | **TWA** 0.1 mg/$m^3$ [skin] |
| Formamide | **TWA** 20 ppm (30 mg/$m^3$), **ST** 30 ppm (45 mg/$m^3$) |
| Furfural | **TWA** 2 ppm (8 mg/$m^3$) [skin] |
| Furfuryl alcohol | **TWA** 10 ppm (40 mg/$m^3$), **ST** 15 ppm (60 mg/$m^3$) [skin] |
| Gasoline | **TWA** 300 ppm (900 mg/$m^3$), **ST** 500 ppm (1500 mg/$m^3$) |
| Germanium tetrahydride | **TWA** 0.2 ppm (0.6 mg/$m^3$) |
| Glutaraldehyde | **C** 0.2 ppm (0.8 mg/$m^3$) |
| Glycerin (mist) | **TWA** 10 mg/$m^3$ (total), **TWA** 5 mg/$m^3$ (resp) |
| Glycidol | **TWA** 25 ppm (75 mg/$m^3$) |
| Graphite (natural) | **TWA** 2.5 mg/$m^3$ (resp) |
| Graphite (synthetic) | **TWA** 10 mg/$m^3$ (total), **TWA** 5 mg/$m^3$ (resp) |
| n-Heptane | **TWA** 400 ppm (1600 mg/$m^3$), **ST** 500 ppm (2000 mg/$m^3$) |
| Hexachlorobutadiene | **TWA** 0.02 ppm (0.24 mg/$m^3$) |
| Hexachlorocyclopentadiene | **TWA** 0.01 ppm (0.1 mg/$m^3$) |
| Hexafluoroacetone | **TWA** 0.1 ppm (0.7 mg/$m^3$) [skin] |
| n-Hexane | **TWA** 50 ppm (180 mg/$m^3$) |
| Hexane isomers (except n-Hexane) | **TWA** 500 ppm (1800 mg/$m^3$), **ST** 1000 ppm (3600 mg/$m^3$) |
| 2-Hexanone | **TWA** 5 ppm (20 mg/$m^3$) |
| Hexone | **TWA** 50 ppm (205 mg/$m^3$), **ST** 75 ppm (300 mg/$m^3$) |
| Hexylene glycol | **C** 25 ppm (125 mg/$m^3$) |
| Hydrazine | **TWA** 0.1 ppm (0.1 mg/$m^3$) [skin] |
| Hydrogenated terphenyls | **TWA** 0.5 ppm (5 mg/$m^3$) |
| Hydrogen bromide | **C** 3 ppm (10 mg/$m^3$) |

## Appendix G (Continued)
## VACATED 1989 OSHA PELs

(See pages xi and xii for an explanation of the vacated 1989 OSHA PELs.)

| Chemical | Vacated 1989 OSHA PEL |
|---|---|
| Hydrogen cyanide | **ST** 4.7 ppm (5 mg/m$^3$) [skin] |
| Hydrogen fluoride (as F) | **TWA** 3 ppm, **ST** 6 ppm |
| Hydrogen sulfide | **TWA** 10 ppm (14 mg/m$^3$), **ST** 15 ppm (21 mg/m$^3$) |
| 2-Hydroxypropyl acrylate | **TWA** 0.5 ppm (3 mg/m$^3$) [skin] |
| Indene | **TWA** 10 ppm (45 mg/m$^3$) |
| Indium | **TWA** 0.1 mg/m$^3$ |
| Iodoform | **TWA** 0.6 ppm (10 mg/m$^3$) |
| Iron pentacarbonyl (as Fe) | **TWA** 0.1 ppm (0.8 mg/m$^3$), **ST** 0.2 ppm (1.6 mg/m$^3$) |
| Iron salts (soluble, as Fe) | **TWA** 1 mg/m$^3$ |
| Isoamyl alcohol (primary & secondary) | **TWA** 100 ppm (360 mg/m$^3$), **ST** 125 ppm (450 mg/m$^3$) |
| Isobutane | **TWA** 800 ppm (1900 mg/m$^3$) |
| Isobutyl alcohol | **TWA** 50 ppm (150 mg/m$^3$) |
| Isooctyl alcohol | **TWA** 50 ppm (270 mg/m$^3$) [skin] |
| Isophorone | **TWA** 4 ppm (23 mg/m$^3$) |
| Isophorone diisocyanate | **TWA** 0.005 ppm, **ST** 0.02 ppm [skin] |
| 2-Isopropoxyethanol | **TWA** 25 ppm (105 mg/m$^3$) |
| Isopropyl acetate | **TWA** 250 ppm (950 mg/m$^3$), **ST** 310 ppm (1185 mg/m$^3$) |
| Isopropyl alcohol | **TWA** 400 ppm (980 mg/m$^3$), **ST** 500 ppm (1225 mg/m$^3$) |
| Isopropylamine | **TWA** 5 ppm (12 mg/m$^3$), **ST** 10 ppm (24 mg/m$^3$) |
| N-Isopropylaniline | **TWA** 2 ppm (10 mg/m$^3$) [skin] |
| Isopropyl glycidyl ether | **TWA** 50 ppm (240 mg/m$^3$), **ST** 75 ppm (360 mg/m$^3$) |
| Kaolin | **TWA** 10 mg/m$^3$ (total), **TWA** 5 mg/m$^3$ (resp) |
| Ketene | **TWA** 0.5 ppm (0.9 mg/m$^3$), **ST** 1.5 ppm (3 mg/m$^3$) |
| Magnesium oxide fume | **TWA** 10 mg/m$^3$ |
| Malathion | **TWA** 10 mg/m$^3$ [skin] |
| Manganese compounds and fume (as Mn) | Compounds: **C** 5 mg/m$^3$<br>Fume: **TWA** 1 mg/m$^3$, **ST** 3 mg/m$^3$ |
| Manganese cyclopentadienyl tricarbonyl (as Mn) | **TWA** 0.1 mg/m$^3$ [skin] |
| Manganese tetroxide (as Mn) | **TWA** 1 mg/m$^3$ |
| Mercury compounds, as Hg [except (organo) alkyls] | Hg Vapor: **TWA** 0.05 mg/m$^3$ [skin]<br>Non-alkyl compounds: **C** 0.1 mg/m$^3$ [skin] |
| Mercury (organo) alkyl compounds (as Hg) | **TWA** 0.01 mg/m$^3$, **ST** 0.03 mg/m$^3$ [skin] |
| Mesityl oxide | **TWA** 15 ppm (60 mg/m$^3$), **ST** 25 ppm (100 mg/m$^3$) |
| Methacrylic acid | **TWA** 20 ppm (70 mg/m$^3$) [skin] |
| Methomyl | **TWA** 2.5 mg/m$^3$ |

## Appendix G (Continued)
## VACATED 1989 OSHA PELs

(See pages xi and xii for an explanation of the vacated 1989 OSHA PELs.)

| Chemical | Vacated 1989 OSHA PEL |
|---|---|
| Methoxychlor | **TWA** 10 mg/m$^3$ |
| 4-Methoxyphenol | **TWA** 5 mg/m$^3$ |
| Methyl acetate | **TWA** 200 ppm (610 mg/m$^3$), **ST** 250 ppm (760 mg/m$^3$) |
| Methyl acetylene-propadiene mixture | **TWA** 1000 ppm (1800 mg/m$^3$), **ST** 1250 ppm (2250 mg/m$^3$) |
| Methylacrylonitrile | **TWA** 1 ppm (3 mg/m$^3$) [skin] |
| Methyl alcohol | **TWA** 200 ppm (260 mg/m$^3$), **ST** 250 ppm (325 mg/m$^3$) [skin] |
| Methyl bromide | **TWA** 5 ppm (20 mg/m$^3$) [skin] |
| Methyl chloride | **TWA** 50 ppm (105 mg/m$^3$), **ST** 100 ppm (210 mg/m$^3$) |
| Methyl chloroform | **TWA** 350 ppm (1900 mg/m$^3$), **ST** 450 ppm (2450 mg/m$^3$) |
| Methyl-2-cyanoacrylate | **TWA** 2 ppm (8 mg/m$^3$), **ST** 4 ppm (16 mg/m$^3$) |
| Methylcyclohexane | **TWA** 400 ppm (1600 mg/m$^3$) |
| Methylcyclohexanol | **TWA** 50 ppm (235 mg/m$^3$) |
| o-Methylcyclohexanone | **TWA** 50 ppm (230 mg/m$^3$), **ST** 75 ppm (345 mg/m$^3$) [skin] |
| Methyl cyclopentadienyl manganese tricarbonyl (as Mn) | **TWA** 0.2 mg/m$^3$ [skin] |
| Methyl demeton | **TWA** 0.5 mg/m$^3$ [skin] |
| 4,4'-Methylenebis(2-chloro-aniline) | **TWA** 0.02 ppm (0.22 mg/m$^3$) [skin] |
| Methylene bis (4-cyclo-hexylisocyanate) | **C** 0.01 ppm (0.11 mg/m$^3$) [skin] |
| Methyl ethyl ketone peroxide | **C** 0.7 ppm (5 mg/m$^3$) |
| Methyl formate | **TWA** 100 ppm (250 mg/m$^3$), **ST** 150 ppm (375 mg/m$^3$) |
| Methyl iodide | **TWA** 2 ppm (10 mg/m$^3$) [skin] |
| Methyl isoamyl ketone | **TWA** 50 ppm (240 mg/m$^3$) |
| Methyl isobutyl carbinol | **TWA** 25 ppm (100 mg/m$^3$), **ST** 40 ppm (165 mg/m$^3$) [skin] |
| Methyl isopropyl ketone | **TWA** 200 ppm (705 mg/m$^3$) |
| Methyl mercaptan | **TWA** 0.5 ppm (1 mg/m$^3$) |
| Methyl parathion | **TWA** 0.2 mg/m$^3$ [skin] |
| Methyl silicate | **TWA** 1 ppm (6 mg/m$^3$) |
| α-Methyl styrene | **TWA** 50 ppm (240 mg/m$^3$), **ST** 100 ppm (485 mg/m$^3$) |
| Metribuzin | **TWA** 5 mg/m$^3$ |
| Mica | **TWA** 3 mg/m$^3$ (resp) |
| Molybdenum (insoluble compounds, as Mo) | **TWA** 10 mg/m$^3$ |
| Monocrotophos | **TWA** 0.25 mg/m$^3$ |
| Monomethyl aniline | **TWA** 0.5 ppm (2 mg/m$^3$) [skin] |
| Morpholine | **TWA** 20 ppm (70 mg/m$^3$), **ST** 30 ppm (105 mg/m$^3$) [skin] |

## Appendix G (Continued)
## VACATED 1989 OSHA PELs

(See pages xi and xii for an explanation of the vacated 1989 OSHA PELs.)

| Chemical | Vacated 1989 OSHA PEL |
|---|---|
| Naphthalene | **TWA** 10 ppm (50 mg/m$^3$), **ST** 15 ppm (75 mg/m$^3$) |
| Nickel metal & other compounds (as Ni) | Metal & insoluble compounds: **TWA** 1 mg/m$^3$<br>Soluble compounds: **TWA** 0.1 mg/m$^3$ |
| Nitric acid | **TWA** 2 ppm (5 mg/m$^3$), **ST** 4 ppm (10 mg/m$^3$) |
| p-Nitroaniline | **TWA** 3 mg/m$^3$ [skin] |
| Nitrogen dioxide | **ST** 1 ppm (1.8 mg/m$^3$) |
| Nitroglycerine | **ST** 0.1 mg/m$^3$ [skin] |
| 2-Nitropropane | **TWA** 10 ppm (35 mg/m$^3$) |
| Nitrotoluene (o-, m-, p-isomers) | **TWA** 2 ppm (11 mg/m$^3$) [skin] |
| Nonane | **TWA** 200 ppm (1050 mg/m$^3$) |
| Octachloronaphthalene | **TWA** 0.1 mg/m$^3$, **ST** 0.3 mg/m$^3$ [skin] |
| Octane | **TWA** 300 ppm (1450 mg/m$^3$), **ST** 375 ppm (1800 mg/m$^3$) |
| Osmium tetroxide (as Os) | **TWA** 0.002 mg/m$^3$ (0.0002 ppm), **ST** 0.006 mg/m$^3$ (0.0006 ppm) |
| Oxalic acid | **TWA** 1 mg/m$^3$, **ST** 2 mg/m$^3$ |
| Oxygen difluoride | **C** 0.05 ppm (0.1 mg/m$^3$) |
| Ozone | **TWA** 0.1 ppm (0.2 mg/m$^3$), **ST** 0.3 ppm (0.6 mg/m$^3$) |
| Paraffin wax fume | **TWA** 2 mg/m$^3$ |
| Paraquat | **TWA** 0.1 mg/m$^3$ (resp) [skin] |
| Pentaborane | **TWA** 0.005 ppm (0.01 mg/m$^3$), **ST** 0.015 ppm (0.03 mg/m$^3$) |
| Pentaerythritol | **TWA** 10 mg/m$^3$ (total), **TWA** 5 mg/m$^3$ (resp) |
| n-Pentane | **TWA** 600 ppm (1800 mg/m$^3$), **ST** 750 ppm (2250 mg/m$^3$) |
| 2-Pentanone | **TWA** 200 ppm (700 mg/m$^3$), **ST** 250 ppm (875 mg/m$^3$) |
| Perchloryl fluoride | **TWA** 3 ppm (14 mg/m$^3$), **ST** 6 ppm (28 mg/m$^3$) |
| Petroleum distillates (naphtha) | **TWA** 400 ppm (1600 mg/m$^3$) |
| Phenothiazine | **TWA** 5 mg/m$^3$ [skin] |
| Phenyl glycidyl ether | **TWA** 1 ppm (6 mg/m$^3$) |
| Phenylhydrazine | **TWA** 5 ppm (20 mg/m$^3$), **ST** 10 ppm (45 mg/m$^3$) [skin] |
| Phenylphosphine | **C** 0.05 ppm (0.25 mg/m$^3$) |
| Phorate | **TWA** 0.05 mg/m$^3$, **ST** 0.2 mg/m$^3$ [skin] |
| Phosdrin | **TWA** 0.01 ppm (0.1 mg/m$^3$), **ST** 0.03 ppm (0.3 mg/m$^3$) [skin] |
| Phosphine | **TWA** 0.3 ppm (0.4 mg/m$^3$), **ST** 1 ppm (1 mg/m$^3$) |
| Phosphoric acid | **TWA** 1 mg/m$^3$, **ST** 3 mg/m$^3$ |
| Phosphorus oxychloride | **TWA** 0.1 ppm (0.6 mg/m$^3$) |
| Phosphorus pentasulfide | **TWA** 1 mg/m$^3$, **ST** 3 mg/m$^3$ |
| Phosphorus trichloride | **TWA** 0.2 ppm (1.5 mg/m$^3$), **ST** 0.5 ppm (3 mg/m$^3$) |
| Phthalic anhydride | **TWA** 6 mg/m$^3$ (1 ppm) |
| m-Phthalodinitrile | **TWA** 5 mg/m$^3$ |

# Appendix G (Continued)
## VACATED 1989 OSHA PELs

(See pages xi and xii for an explanation of the vacated 1989 OSHA PELs.)

| Chemical | Vacated 1989 OSHA PEL |
|---|---|
| Picloram | **TWA** 10 $mg/m^3$ (total), **TWA** 5 $mg/m^3$ (resp) |
| Piperazine dihydrochloride | **TWA** 5 $mg/m^3$ |
| Platinum metal (as Pt) | **TWA** 1 $mg/m^3$ |
| Portland cement | **TWA** 10 $mg/m^3$ (total), **TWA** 5 $mg/m^3$ (resp) |
| Potassium hydroxide | **TWA** 2 $mg/m^3$ |
| Propargyl alcohol | **TWA** 1 ppm (2 $mg/m^3$) [skin] |
| Propionic acid | **TWA** 10 ppm (30 $mg/m^3$) |
| Propoxur | **TWA** 0.5 $mg/m^3$ |
| n-Propyl acetate | **TWA** 200 ppm (840 $mg/m^3$), **ST** 250 ppm (1050 $mg/m^3$) |
| n-Propyl alcohol | **TWA** 200 ppm (500 $mg/m^3$), **ST** 250 ppm (625 $mg/m^3$) |
| Propylene dichloride | **TWA** 75 ppm (350 $mg/m^3$), **ST** 110 ppm (510 $mg/m^3$) |
| Propylene glycol dinitrate | **TWA** 0.05 ppm (0.3 $mg/m^3$) |
| Propylene glycol monomethyl ether | **TWA** 100 ppm (360 $mg/m^3$), **ST** 150 ppm (540 $mg/m^3$) |
| Propylene oxide | **TWA** 20 ppm (50 $mg/m^3$) |
| n-Propyl nitrate | **TWA** 25 ppm (105 $mg/m^3$), **ST** 40 ppm (170 $mg/m^3$) |
| Resorcinol | **TWA** 10 ppm (45 $mg/m^3$), **ST** 20 ppm (90 $mg/m^3$) |
| Ronnel | **TWA** 10 $mg/m^3$ |
| Rosin core solder, pyrolysis products (as formaldehyde) | **TWA** 0.1 $mg/m^3$ |
| Rouge | **TWA** 10 $mg/m^3$ (total), **TWA** 5 $mg/m^3$ (resp) |
| Silica, amorphous | **TWA** 6 $mg/m^3$, **TWA** 0.1 $mg/m^3$ (fused) |
| Silica, crystalline (as respirable dust) | **TWA** 0.05 $mg/m^3$ (cristobalite), **TWA** 0.05 $mg/m^3$ (tridymite), **TWA** 0.1 $mg/m^3$ (quartz), **TWA** 0.1 $mg/m^3$ (tripoli) |
| Silicon | **TWA** 10 $mg/m^3$ (total), **TWA** 5 $mg/m^3$ (resp) |
| Silicon carbide | **TWA** 10 $mg/m^3$ (total), **TWA** 5 $mg/m^3$ (resp) |
| Silicon tetrahydride | **TWA** 5 ppm (7 $mg/m^3$) |
| Soapstone | **TWA** 6 $mg/m^3$ (total), **TWA** 3 $mg/m^3$ (resp) |
| Sodium azide | **C** 0.1 ppm (as $HN_3$) [skin], **C** 0.3 $mg/m^3$ (as $NaN_3$) [skin] |
| Sodium bisulfite | **TWA** 5 $mg/m^3$ |
| Sodium fluoroacetate | **TWA** 0.05 $mg/m^3$, **ST** 0.15 $mg/m^3$ [skin] |
| Sodium hydroxide | **C** 2 $mg/m^3$ |
| Sodium metabisulfite | **TWA** 5 $mg/m^3$ |
| Stoddard solvent | **TWA** 525 $mg/m^3$ (100 ppm) |
| Styrene | **TWA** 50 ppm (215 $mg/m^3$), **ST** 100 ppm (425 $mg/m^3$) |
| Subtilisins | **ST** 0.00006 $mg/m^3$ [60-minute] |
| Sulfur dioxide | **TWA** 2 ppm (5 $mg/m^3$), **ST** 5 ppm (13 $mg/m^3$) |

# Appendix G (Continued)
# VACATED 1989 OSHA PELs

(See pages xi and xii for an explanation of the vacated 1989 OSHA PELs.)

| Chemical | Vacated 1989 OSHA PEL |
|---|---|
| Sulfur monochloride | **C** 1 ppm (6 $mg/m^3$) |
| Sulfur tetrafluoride | **C** 0.1 ppm (0.4 $mg/m^3$) |
| Sulfuryl fluoride | **TWA** 5 ppm (20 $mg/m^3$), **ST** 10 ppm (40 $mg/m^3$) |
| Sulprofos | **TWA** 1 $mg/m^3$ |
| Talc | **TWA** 2 $mg/m^3$ (resp) |
| Temephos | **TWA** 10 $mg/m^3$ (total), **TWA** 5 $mg/m^3$ (resp) |
| Terphenyl (o-, m-, p-isomers) | **C** 5 $mg/m^3$ (0.5 ppm) |
| 1,1,2,2-Tetrachloroethane | **TWA** 1 ppm (7 $mg/m^3$) [skin] |
| Tetrachloroethylene | **TWA** 25 ppm (170 $mg/m^3$) |
| Tetrahydrofuran | **TWA** 200 ppm (590 $mg/m^3$), **ST** 250 ppm (735 $mg/m^3$) |
| Tetrasodium pyrophosphate | **TWA** 5 $mg/m^3$ |
| 4,4'-Thiobis(6-tert-butyl-m-cresol) | **TWA** 10 $mg/m^3$ (total), **TWA** 5 $mg/m^3$ (resp) |
| Thioglycolic acid | **TWA** 1 ppm (4 $mg/m^3$) [skin] |
| Thionyl chloride | **C** 1 ppm (5 $mg/m^3$) |
| Tin (organic compounds, as Sn) | **TWA** 0.1 $mg/m^3$ [skin] |
| Tin(II) oxide (as Sn) | **TWA** 2 $mg/m^3$ |
| Tin(IV) oxide (as Sn) | **TWA** 2 $mg/m^3$ |
| Titanium dioxide | **TWA** 10 $mg/m^3$ |
| Toluene | **TWA** 100 ppm (375 $mg/m^3$), **ST** 150 ppm (560 $mg/m^3$) |
| Toluene-2,4-diisocyanate | **TWA** 0.005 ppm (0.04 $mg/m^3$), **ST** 0.02 ppm (0.15 $mg/m^3$) |
| m-Toluidine | **TWA** 2 ppm (9 $mg/m^3$) [skin] |
| p-Toluidine | **TWA** 2 ppm (9 $mg/m^3$) [skin] |
| Tributyl phosphate | **TWA** 0.2 ppm (2.5 $mg/m^3$) |
| Trichloroacetic acid | **TWA** 1 ppm (7 $mg/m^3$) |
| 1,2,4-Trichlorobenzene | **C** 5 ppm (40 $mg/m^3$) |
| Trichloroethylene | **TWA** 50 ppm (270 $mg/m^3$), **ST** 200 ppm (1080 $mg/m^3$) |
| 1,2,3-Trichloropropane | **TWA** 10 ppm (60 $mg/m^3$) |
| 1,1,2-Trichloro-1,2,2-trifluoroethane | **TWA** 1000 ppm (7600 $mg/m^3$)<br>**ST** 1250 ppm (9500 $mg/m^3$) |
| Triethylamine | **TWA** 10 ppm (40 $mg/m^3$), **ST** 15 ppm (60 $mg/m^3$) |
| Trimellitic anhydride | **TWA** 0.005 ppm (0.04 $mg/m^3$) |
| Trimethylamine | **TWA** 10 ppm (24 $mg/m^3$), **ST** 15 ppm (36 $mg/m^3$) |
| 1,2,3-Trimethylbenzene | **TWA** 25 ppm (125 $mg/m^3$) |
| 1,2,4-Trimethylbenzene | **TWA** 25 ppm (125 $mg/m^3$) |
| 1,3,5-Trimethylbenzene | **TWA** 25 ppm (125 $mg/m^3$) |

## Appendix G (Continued)
## VACATED 1989 OSHA PELs

(See pages xi and xii for an explanation of the vacated 1989 OSHA PELs.)

| Chemical | Vacated 1989 OSHA PEL |
|---|---|
| Trimethyl phosphite | **TWA** 2 ppm (10 $mg/m^3$) |
| 2,4,6-Trinitrotoluene | **TWA** 0.5 $mg/m^3$ [skin] |
| Triorthocresyl phosphate | **TWA** 0.1 $mg/m^3$ [skin] |
| Triphenylamine | **TWA** 5 $mg/m^3$ |
| Tungsten (insoluble compounds, as W) | **TWA** 5 $mg/m^3$, **ST** 10 $mg/m^3$ |
| Tungsten (soluble compounds, as W) | **TWA** 1 $mg/m^3$, **ST** 3 $mg/m^3$ |
| Tungsten carbide (cemented) | **TWA** 5 $mg/m^3$ (as W), **ST** 10 $mg/m^3$ (as W), **TWA** 0.05 $mg/m^3$ (as Co), **TWA** 1 $mg/m^3$ (as Ni) |
| Uranium (insoluble compounds, as U) | **TWA** 0.2 $mg/m^3$, **ST** 0.6 $mg/m^3$ |
| n-Valeraldehyde | **TWA** 50 ppm (175 $mg/m^3$) |
| Vanadium dust | **TWA** 0.05 mg $V_2O_5/m^3$ (resp) |
| Vanadium fume | **C** 0.05 mg $V_2O_5/m^3$ |
| Vinyl acetate | **TWA** 10 ppm (30 $mg/m^3$), **ST** 20 ppm (60 $mg/m^3$) |
| Vinyl bromide | **TWA** 5 ppm (20 $mg/m^3$) |
| Vinyl cyclohexene dioxide | **TWA** 10 ppm (60 $mg/m^3$) [skin] |
| Vinylidene chloride | **TWA** 1 ppm (4 $mg/m^3$) |
| VM & P Naphtha | **TWA** 1350 $mg/m^3$ (300 ppm), **ST** 1800 $mg/m^3$ (400 ppm) |
| Welding fumes | **TWA** 5 $mg/m^3$ |
| Wood dust (all wood dusts except Western red cedar) | **TWA** 5 $mg/m^3$, **ST** 10 $mg/m^3$ |
| Wood dust (Western red cedar) | **TWA** 2.5 $mg/m^3$ |
| Xylene (o-, m-, p-isomers) | **TWA** 100 ppm (435 $mg/m^3$), **ST** 150 ppm (655 $mg/m^3$) |
| m-Xylene $\alpha,\alpha'$-diamine | **C** 0.1 $mg/m^3$ [skin] |
| Xylidine | **TWA** 2 ppm (10 $mg/m^3$) [skin] |
| Zinc chloride fume | **TWA** 1 $mg/m^3$, **ST** 2 $mg/m^3$ |
| Zinc oxide | **TWA** 5 $mg/m^3$ (fume), **ST** 10 $mg/m^3$ (fume), **TWA** 10 $mg/m^3$ (total dust), **TWA** 5 $mg/m^3$ (resp dust) |
| Zinc stearate | **TWA** 10 $mg/m^3$ (total), **TWA** 5 $mg/m^3$ (resp) |
| Zirconium compounds (as Zr) | **TWA** 5 $mg/m^3$, **ST** 10 $mg/m^3$ |

# INDICES

## CAS Number Index

## CAS Number Index (Continued)

CAS Number Index (Continued)

## CAS Number Index (Continued)

## CAS Number Index (Continued)

# DOT ID Number Index

# DOT ID Number Index (Continued)

## DOT ID Number Index (Continued)

## DOT ID Number Index (Continued)

# Chemical, Synonym, and Trade Name Index

**(Primary chemical names appear in blue text.)**

## Chemical, Synonym, and Trade Name Index (Continued)

(Primary chemical names appear in blue text.)

## Chemical, Synonym, and Trade Name Index (Continued)

(Primary chemical names appear in blue text.)

## Chemical, Synonym, and Trade Name Index (Continued)
(Primary chemical names appear in blue text.)



NAME INDEX

## Chemical, Synonym, and Trade Name Index (Continued)

**(Primary chemical names appear in blue text.)**



NAME INDEX

## Chemical, Synonym, and Trade Name Index (Continued)

**(Primary chemical names appear in blue text.)**



NAME INDEX

## Chemical, Synonym, and Trade Name Index (Continued)
**(Primary chemical names appear in blue text.)**

## Chemical, Synonym, and Trade Name Index (Continued)

(Primary chemical names appear in blue text.)

## Chemical, Synonym, and Trade Name Index (Continued)

**(Primary chemical names appear in blue text.)**

## Chemical, Synonym, and Trade Name Index (Continued)
**(Primary chemical names appear in blue text.)**



NAME INDEX

## Chemical, Synonym, and Trade Name Index (Continued)

**(Primary chemical names appear in blue text.)**



NAME INDEX

## Chemical, Synonym, and Trade Name Index (Continued)
**(Primary chemical names appear in blue text.)**



NAME INDEX

## Chemical, Synonym, and Trade Name Index (Continued)
**(Primary chemical names appear in blue text.)**

## Chemical, Synonym, and Trade Name Index (Continued)
**(Primary chemical names appear in blue text.)**



NAME INDEX

## Chemical, Synonym, and Trade Name Index (Continued)

**(Primary chemical names appear in blue text.)**

## Chemical, Synonym, and Trade Name Index (Continued)
(Primary chemical names appear in blue text.)

## Chemical, Synonym, and Trade Name Index (Continued)

(Primary chemical names appear in blue text.)



NAME INDEX

## Chemical, Synonym, and Trade Name Index (Continued)
**(Primary chemical names appear in blue text.)**

## Chemical, Synonym, and Trade Name Index (Continued)
(Primary chemical names appear in blue text.)

## Chemical, Synonym, and Trade Name Index (Continued)

**(Primary chemical names appear in blue text.)**

## Chemical, Synonym, and Trade Name Index (Continued)
**(Primary chemical names appear in blue text.)**



NAME INDEX

## Chemical, Synonym, and Trade Name Index (Continued)

**(Primary chemical names appear in blue text.)**



NAME INDEX

## Chemical, Synonym, and Trade Name Index (Continued)

**(Primary chemical names appear in blue text.)**

## Chemical, Synonym, and Trade Name Index (Continued)
(Primary chemical names appear in blue text.)



NAME INDEX

## **Chemical, Synonym, and Trade Name Index** (Continued)

**(Primary chemical names appear in blue text.)**



NAME INDEX

## Chemical, Synonym, and Trade Name Index (Continued)

**(Primary chemical names appear in blue text.)**

## Chemical, Synonym, and Trade Name Index (Continued)
(Primary chemical names appear in blue text.)

## Chemical, Synonym, and Trade Name Index (Continued)
**(Primary chemical names appear in blue text.)**

## Chemical, Synonym, and Trade Name Index (Continued)
(Primary chemical names appear in blue text.)



NAME INDEX

## Chemical, Synonym, and Trade Name Index (Continued)
**(Primary chemical names appear in blue text.)**



NAME INDEX

## Chemical, Synonym, and Trade Name Index (Continued)

(Primary chemical names appear in blue text.)

## Chemical, Synonym, and Trade Name Index (Continued)
**(Primary chemical names appear in blue text.)**



NAME INDEX

## Chemical, Synonym, and Trade Name Index (Continued)
**(Primary chemical names appear in blue text.)**



NAME INDEX

## Chemical, Synonym, and Trade Name Index (Continued)
**(Primary chemical names appear in blue text.)**

## Chemical, Synonym, and Trade Name Index (Continued)
(Primary chemical names appear in blue text.)

## Chemical, Synonym, and Trade Name Index (Continued)
### (Primary chemical names appear in blue text.)

## Chemical, Synonym, and Trade Name Index (Continued)

(Primary chemical names appear in blue text.)

## Chemical, Synonym, and Trade Name Index (Continued)
**(Primary chemical names appear in blue text.)**



NAME INDEX

## Chemical, Synonym, and Trade Name Index (Continued)
**(Primary chemical names appear in blue text.)**

www.ingramcontent.com/pod-product-compliance
Lightning Source LLC
Chambersburg PA
CBHW050453190326
41458CB00005B/1261